Carolyn G Brooks

United Kingdom Oil and Gas Fields
25 Years Commemorative Volume

Geological Society Memoirs
Series Editor J. Brooks

United Kingdom Oil and Gas Fields
25 Years Commemorative Volume

EDITED BY

I. L. ABBOTTS

Clyde Petroleum
Ledbury, UK

Memoir No. 14
1991

Published by
The Geological Society
London

THE GEOLOGICAL SOCIETY

The Society was founded in 1807 as the Geological Society of London and is the oldest geological society in the world. It received its Royal Charter in 1825 for the purpose of 'investigating the mineral structure of the Earth'. The Society is Britain's national learned society for geology with a Fellowship exceeding 6000. It has countrywide coverage and approximately one quarter of its membership resides overseas. The Society is responsible for promoting all aspects of the geological sciences and will also embrace professional matters on the completion of the reunification with the Institution of Geologists. The Society has its own publishing house to produce its international journals, books and maps, and is the European distributor for materials published by the American Association of Petroleum Geologists.

Fellowship is open to those holding a recognized honours degree in geology or cognate subject and who have at least two years relevant postgraduate experience, or have not less than six years relevant experience in geology or a cognate subject. A Fellow who has not less than five years relevant postgraduate experience in the practice of Geology may apply for validation and subject to approval will be able to use the designatory letters C. Geol (Chartered Geologists) from 1991.

Further information about the Society is available from the Membership Manager, Geological Society, Burlington House, London, United Kingdom W1V 0JU.

Published by the Geological Society from:
The Geological Society Publishing House
Unit 7
Brassmill Enterprise Centre
Brassmill Lane
Bath
Avon BA1 3JN
UK

(*Orders*: Tel. 0225 445046)

First published 1991

© The Geological Society 1991. All rights reserved.
No reproduction, copy or transmission of this publication may be made without written permission. No paragraph of this publication may be reproduced, copied or transmitted save with the written permission or in accordance with the provisions of the Copyright Act 1956 (as Amended) or under the terms of any licence permitting limited copying issued by the Copyright Licensing Agency, 90 Tottenham Court Road, London, W1P 9HE. Users registered with Copyright Clearance Center: this publication is registered with CCC, 27 Congress St., Salem, MA 01970, USA. 0435-4052/91 $03.50

British Library Cataloguing in Publication Data
United Kingdom oil and gas fields.
   1. Fuel resouces. Natural gas. Petroleum
   I. Abbotts, I. L.

ISBN 0-903317-62-1

Distributor USA:
AAPG Bookstore
PO Box 979
Tulsa
Oklahoma 74101-0979
USA
Tel. (918) 584-2555

Distributor Australia:
Australian Mineral Foundation
63 Conyngham Street
Glenside
South Australia 5065
Tel. (08) 379-0444

Cover illustration reproduced by kind permission of Oilfield Publications Limited, Ledbury, from *The North Sea Oil and Gas activity and concession map 1990*.

Contents

Foreword, vii

Part 1: Introduction

25 Years of UK North Sea Exploration *by* J. M. BOWEN, 1
Stratigraphy of the oil and gas reservoirs: UK continental shelf *by* S. BROWN, 9

Part 2: The Viking Graben

The Alwyn North Field, Blocks 3/9a, 3/4a, UK North Sea, *by* I. INGLIS & J. GERARD, 21
The Beryl Field, Block 9/13, UK North Sea *by* C. A. KNUTSON & I. C. MUNRO, 33
The North Brae Field, Block 16/7a, UK North Sea *by* M. A. STEPHENSON, 43
The Central Brae Field, Block 16/7a, UK North Sea *by* C. C. TURNER & P. J. ALLEN, 49
The South Brae Field, Block 16/7a, UK North Sea *by* M. J. ROBERTS, 55
The Brent Field, Block 211/29, UK North Sea *by* A. P. STRUIJK & R. T. GREEN, 63
The Comorant Field, Blocks 211/21a, 211/26a, UK North Sea *by* D. J. TAYLOR & J. P. A. DIETVORST, 73
The Deveron Field, Block 211/18a, UK North Sea *by* R. R. WILLIAMS, 83
The Don Field, Blocks 211/13a, 211/14, 211/18a, 211/19a, UK North Sea *by* D. MORRISON, G. G. BENNET & M. G. BAYAT, 89
The Dunlin Field, Blocks 211/23a, 211/24a, UK North Sea *by* A. BAUMANN & B. O'CATHAIN, 95
The Eider Field, Blocks 211/16a, 211/21a, UK North Sea *by* M. D. WENSRICH, K. M. EASTWOOD, C. D. VAN PANHUYS & J. M. SMART, 103
The Emerald Field, Blocks 2/10a, 2/15a, 3/11b, UK North Sea *by* D. M. STEWART & A. J. G. FAULKNER, 111
The Frigg Field, Block 10/1, UK North Sea and 25/1, Norwegian North Sea *by* J. BREWSTER, 117
The Heather Field, Block 2/5, UK North Sea *by* B. PENNY, 127
The Hutton Field, Blocks 211/28, 211/27, UK North Sea *by* D. B. HAIG, 135
The Northwest Hutton Field, Block 211/27, UK North Sea *by* L. H. JOHNES & M. B. GAUER, 145
The Magnus Field, Blocks 211/71, 211/12a, UK North Sea *by* M. SHEPHERD, 153
The Miller Field, Blocks 16/7b, 16/8b, UK North Sea *by* S. K. ROOKSBY, 159
The Murchison Field, Blocks 211/19a, UK North Sea *by* J. WARRENDER, 165
The Ninian Field, Blocks 3/3 & 3/8, UK North Sea *by* E. J. VAN VESSEM & T. L. GAN, 175
The Osprey Field, Blocks 211/18a, 211/23a, UK North Sea *by* J. W. ERICKSON & C. D. VAN PANHUYS, 183
The Tern Field, Block 210/25a, UK North Sea *by* M. VAN PANHUYS-SIGLER, A. BAUMANN & T. C. HOLLAND, 191
The Thistle Field, Blocks 211/18a and 211/19, UK North Sea *by* R. R. WILLIAMS & A. D. MILNE, 199

Part 3: The Central Graben and Moray Firth

The Arbroath and Montrose Field, Blocks 211/7a, 211/12a, UK North Sea *by* M. SHEPHERD, 211
The Argyll, Duncan and Innes Fields, Blocks 30/24, 30/25a, UK North Sea *by* D. ROBSON, 219
The Auk Field, Block 30/16, UK North Sea *by* N. H. TREWIN & M. G. BRAMWELL, 227
The Balmoral Field, Block 16/21, UK North Sea *by* P. C. TONKIN & A. R. FRASER, 237
The Beatrice Field, Block 11/30a, UK North Sea *by* V. STEVENS, 245
The Buchan Field, Blocks 20/5a, 21/1a, UK North Sea *by* C. W. EDWARDS, 253
The Chanter Field, Block 15/17, UK North Sea *by* H. R. H. SCHMITT, 261
The Claymore Field, Block 14/19, UK North Sea *by* S. D. HARKER, S. C. H. GREEN & R. S. ROMANI, 269
The Clyde Field, Block 30/17b, UK North Sea *by* D. A. STEVENS & R. J. WALLIS, 279
The Crawford Field, Block 9/28a, UK North Sea *by* A. YALIZ, 287
The Cyrus Field, Block 16/28, UK North Sea *by* D. G. MOUND, I. D. ROBERTSON & R. J. WALLIS, 295
The Forties Field, Blocks 21/10, 22/6a, UK North Sea *by* J. M. WILLS, 301
The Fulmar Field, Blocks 30/16, 30/11b, UK North Sea *by* C. P. STOCKBRIDGE & D. I. GRAY, 309
The Glamis Field, Block 16/21a, UK North Sea *by* A. R. FRASER & P. C. TONKIN, 317
The Highlander Field, Block 14/20b, UK North Sea *by* M. WHITEHEAD & S. J. PINNOCK, 323
The Ivanhoe and Rob Roy Fields, Block 15/21a-b, UK North Sea *by* R. H. PARKER, 331
The Kittiwake Field, Block 21/18, UK North Sea *by* K. W. GLENNIE & L. A. ARMSTRONG, 339
The Maureen Field, Block 16/29a, UK North Sea *by* P. L. CUTTS, 347
The Petronella Field, Block 14/20b, UK North Sea *by* P. WADDAMS & N. M. CLARK, 353
The Piper Field, Block 15/17, UK North Sea *by* H. R. H. SCHMITT & A. F. GORDON, 361
The Scapa Field, Block 14/19, UK North Sea *by* G. J. MCGANN, S. C. H. GREEN, S. D. HARKER & R. S. ROMANI, 369
The Tartan Field, Block 15/16, UK North Sea *by* R. N. COWARD, N. M. CLARK & S. J. PINNOCK, 377

Part 4: The Southern Gas Basin

The Amethyst Field, Blocks 47/8a, 47/9a, 47/13a, 47/14a, 47/15a, UK North Sea *by* C. R. GARLAND, 387
The Barque Field, Blocks 48/13a, 48/14, UK North Sea *by* R. T. FARMER & A. P. HILLIER, 395
The Camelot Fields, Blocks 53/1a, 53/2, UK North Sea *by* A. J. HOLMES, 401
The Cleeton Field, Block 42/29, UK North Sea *by* R. D. HEINRICH, 409
The Clipper Field, Blocks 48/19a, 48/19c, UK North Sea *by* R. T. FARMER & A. P. HILLIER, 417
The Esmond, Forbes and Gordon Fields, Blocks 43/8a, 48/13a, 48/15a, 48/20a, UK North Sea *by* F. J. KETTER, 425
The Hewett Field, Blocks 48/28–29–30, 52/4a–5a, UK North Sea *by* P. COOKE-YARBOROUGH, 433
The Indefatigable Field, Blocks 48/18, 48/19, 48/23, 48/24, UK North Sea *by* J. F. S. PEARSON, R. A. YOUNGS & A. SMITH, 443
The Leman Field, Blocks 49/26, 49/27, 49/28, 53/1, 53/2, UK North Sea *by* A. P. HILLIER & B. P. J. WILLIAMS, 451
The Ravenspurn North Field, Blocks 42/30, 43/26a, UK North Sea *by* F. J. KETTER, 387
The Ravenspurn South Field, Blocks 42/29, 42/30, 43/26, UK North Sea *by* R. D. HEINRICH, 469

The Rough Gas Storage Field, Blocks 47/3d, 47/8b, UK North Sea *by* I. A. Stuart, 477

The Sean North and Sean South Fields, Block 49/25a, UK North Sea *by* G. D. Hobson & A. P. Hillier, 485

The Thames, Yare and Bure Fields, Block 49/28, UK North Sea *by* O. C. Werngren, 491

The V-Fields, Blocks 49/16, 49/21, 48/20a, 48/25b, UK North Sea *by* M. J. Pritchard, 497

The Victor Field, Blocks 49/17, 49/22, UK North Sea *by* R. A. Lambert, 503

The Viking Complex Field, Blocks 49/12a, 49/16, 49/17, UK North Sea *by* C. P. Morgan, 509

The West Sole Field, Block 48/6, UK North Sea *by* D. A. Winter & B. King, 517

Part 5: The Morecambe Basin

The South Morecambe Field, Blocks 110/2a, 110/3a, 110/8a, UK East Irish Sea *by* I. A. Stuart & G. Cowan, 527

Part 6: Compilation Tables/Appendices

Data Summaries, 545

Abbreviations, 563

Index, 565

Foreword

This volume contains articles on 64 of the oil and gas fields discovered on the United Kingdom Continental Shelf. The articles have been written in a standardized layout with the aim of providing an easy to use databook for the Petroleum Geologist and Geophysicist. The idea for the compilation was conceived by the Petroleum Group of the Geological Society of London under the Chairmanship of R. F. Hardman.

The volume has been produced to commemorate the first 25 years of hydrocarbon exploration and production in the United Kingdom North Sea. The main aim of the volume is to provide a reference source of data relevant to the discovery and optimum development of the next generation of North Sea fields.

The standardized layout was largely taken from the recent compilation of Norwegian fields by the Norwegian Petroleum Society (Geology of the Norwegian Oil and Gas Fields, 1984, edited by A. M. SPENCER et al.). The format is designed to provide a succinct description, including relevant figures, of the petroleum geology of each field. The suggested layout was sent to 19 operating companies in November 1988 requesting articles on 67 fields which had, at that time, received Annex B approval. The 64 articles included here were received from July 1989 to March 1990. On receipt they were reviewed and edited by a special sub-committee (I. L. Abbotts, K. W. Glennie, R. F. Hardman, A. J. Martin, M. D. Thomas and B. A. Vining). In addition, two introductory articles, the first setting the fields in a historical perspective and the second placing them in a stratigraphic framework, are included, along with an appendix of abbreviations and various compilation tables.

It is hoped that the aims of the volume, the commemoration of past work and the assistance of future efforts, will be fulfilled. All who have helped the project to fruition are thanked, especially the authors, reviewers and the 19 operating companies:

Amerada Hess	Amoco	Arco	British Gas
BP	Chevron	Conoco	Elf
Hamilton	Mobil	Occidental	Phillips
Shell	Sovereign	Sun	Texaco
Total	Unocal.		

I. L. Abbotts, Editor
January 1991

25 years of UK North Sea exploration

J. M. BOWEN

Enterprise Oil plc, 5 Strand, London WC2N 5HU, UK

Any attempt to summarize 25 years of exploration for petroleum in the UK sector of the North Sea must be a daunting task. The outcome, in terms of the oil and gas fields discovered, is the subject of this volume. This introduction will attempt to outline, very briefly, some of the ups and downs of the exploration history which has led the industry to where it stands today, 25 years on (Fig. 1).

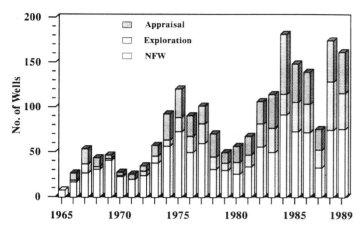

Wells drilled: by type

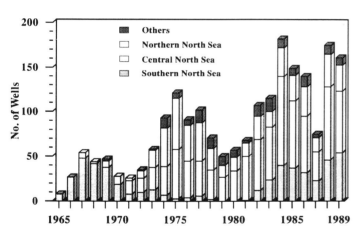

Wells drilled: by area

Fig. 1. UKCS drilling 25 year history.

The background

When the author was at university in the early 1950s the very idea that the United Kingdom would be likely to become a significant, let alone major world producer of petroleum would have been viewed as utterly ridiculous.

It is true that oil and gas indications had been encountered in wells and mines in such disparate areas as Sussex, the west Midlands and the Midland Valley of Scotland and as seepages in Dorset, Lancashire and West Lothian, but these had been thoroughly investigated without the discovery of any economically significant oil or gas fields. Indeed, the only economic production at that time came from BP's small east Midlands fields based on Eakring where the first discovery had been made in 1939.

The first serious attempt to explore for oil in the United Kingdom was initiated in 1918 for strategic reasons, when 11 relatively shallow wells were drilled on anticlinal features in various parts of the country. Of these only one, Hardstoft-1 in Derbyshire, discovered producible oil, but attempts to follow up the discovery were unsuccessful.

Exploration then effectively ceased until the mid-1930s when the passing of the Petroleum Production Act in 1934 made it possible for oil companies to obtain licences from the Government to explore for and produce hydrocarbons. This, and possibly the imminence of war, prompted a renewed effort by the D'Arcy company, the forerunner of BP, which resulted in success at Formby in Lancashire and at Eakring in Nottinghamshire, in 1939. Both were oil discoveries following significant, but non-commercial, gas discoveries at Cousland near Edinburgh and Eskdale in Yorkshire in 1938.

D'Arcy continued its exploration throughout the war and post-war years and made a number of further discoveries, albeit on a very small scale and all in the east Midlands where the Millstone Grit of the Carboniferous forms the principal reservoir. None of these finds had recoverable reserves of more than a few million barrels.

Fig. 2. Significant discoveries in Western Europe to 1958.

On the Continent there had been rather more success with the discovery during World War II of a number of oil and gas fields near the German/Netherlands border where the Schoonebeek field,

From Abbotts, I. L. (ed.), 1991, *United Kingdom Oil and Gas Fields,
25 Years Commemorative Volume,* Geological Society Memoir No. 14, pp. 1–7

discovered in 1943 had reserves of 385 MMBBL, and near Rotterdam and The Hague where Shell had first found oil indications in a shallow exhibition well drilled in 1936. In the Bordeaux area of France, the Lacq field discovered in 1949 had oil reserves of 29 MMBBL and 7.5 TCF gas and in 1954 the Parentis discovery found reserves of 210 MMBBL. The Dutch fields produced gas from the Zechstein and oil and gas from the Lower Cretaceous Wealden, while in France the production was from the Upper Cretaceous and Upper Jurassic. Thus, until the late 1950s oil and gas fields of major commercial significance were rare in western Europe and non-existent in the United Kingdom (Fig. 2). Such small UK production as there was came from the Carboniferous Millstone Grit, while gas had been found in the Permian Zechstein.

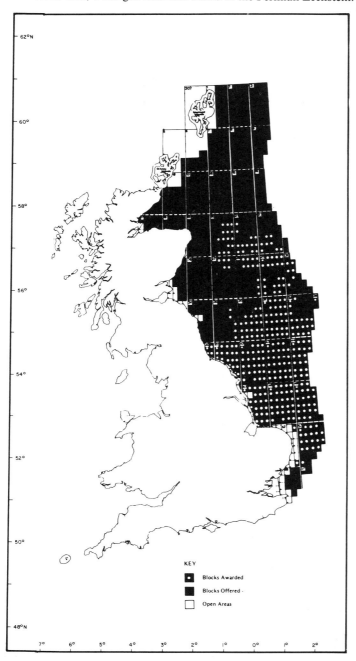

Fig. 3. UKCS first round licensing. Note: the map shows the designated area at 1964 with the exception of some small areas along the Norfolk, Lincolnshire and Yorkshire coastline.

The Zechstein was also an exploration objective in the northern Netherlands and it was here, at Slochteren-1 in 1959, that a major gas discovery was made near the city of Groningen, not in the marine Zechstein carbonates but in the underlying terrestrial Rotliegendes sands. It took some time before the magnitude of this discovery became apparent, even to the Shell geologists who were involved, but by the early 1960s following extensive appraisal the Groningen field was recognized as a world-ranking giant gas field. At last a field had been found that would clearly, even in those early days, have been profitable in an offshore environment such as the relatively shallow waters of the southern North Sea.

Exploitation of the continental shelves of the world outside the USA had had to await ratification of the Geneva Convention of 1958 which attempted to define national jurisdiction as far as the 200 m isobath. It required 22 countries to ratify the convention to give it the force of international law and it was not until 1964 that Britain, being the 22nd to ratify, actually set the ball rolling for the exploration of many continental shelves worldwide. In particular it triggered the start of exploration of the United Kingdom Continental Shelf (UKCS).

The division of the North Sea was agreed between governments, mainly on the basis of median lines, i.e. the principle of equal distance from their respective coasts. The United Kingdom, Norway, the Netherlands and Denmark reached agreement upon their common boundaries between March 1965 and March 1966; all accepted the median line formula.

Following the implementation of The Continental Shelf Act of 15 April 1964 and the Petroleum (Production) (Continental Shelf and Territorial Sea) Regulations 1964, the government put on offer virtually the entire area between the Dover Strait and the northern tip of the Shetlands at 61° North. Seismic surveys had been carried out in anticipation of this First Round and most of the blocks on offer in the southern area were allocated (up to 55°N), while a fair number were allocated up to 58°N; between 58° and 61°N no blocks were licensed. At the time of the First Round, the North Sea median lines had not been agreed (this came a year later) and hence blocks adjacent to the median lines were excluded from the Round (Fig. 3).

The southern gas boom

Little time was lost in pursuing the Groningen Rotliegendes play into the southern part of the UK North Sea. However, the exploration proved to be difficult owing to the presence of the variable thicknesses of mobile Zechstein salt, which also beneficially provided a very effective cap rock over most of the southern basin. This salt, however, also generally prevented the gas generated from the Carboniferous from reaching the excellent reservoirs and large structural closures at the Bunter Sandstone level. Disappointments were also caused by the absence of a salt seal in the south, the shaling out of the Rotliegendes reservoir to the north and areas of inversion with large attractive looking structures in which the expected favourable reservoir qualities had been destroyed by deep burial.

Nevertheless discoveries came thick and fast, starting with BP's West Sole and Conoco's South Viking in December 1965, and were quickly followed by Leman, Indefatigable (both Shell) and Hewett (Phillips) in 1966. Further discoveries followed in 1967 and 1968 and by the time this author arrived on the scene in early 1969, he was to be told 'Sorry, we have found all the gas in the south and we believe that there is little or no chance of finding hydrocarbons further north, (Fig. 4). Although only some 60% of the southern gas reserves, as currently known, had by then been defined, the remark reflected a negative attitude to further exploration then prevalent, not helped by the low gas prices then on offer from the monopoly buyer.

The Central Graben

1969, a peak year for exploration drilling in the south saw only desultory activity in what was then considered 'the north'. Most drilling was seen as fulfilment of licence obligations incurred in the first two rounds, nearing the end of their first terms, and no great optimism was displayed. However, many wells had oil or gas shows and to a few of us it seemed only a matter of time before a commercial discovery was made. In fact, as many as 14 wells had been drilled in the UK Central North Sea before the first modest UKCS discovery, the Montrose field, was made by Amoco in December 1969 (Fig. 4).

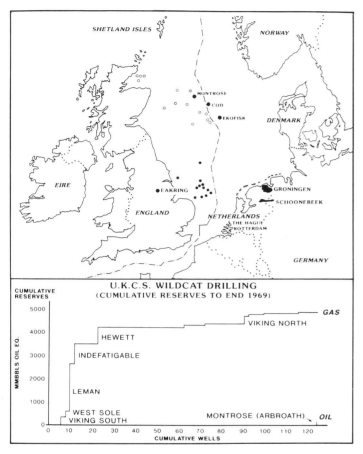

Fig. 4. UKCS new field discoveries to end of 1969.

The Montrose discovery was made in sandstones of late Palaeocene age and this seemed to be the most likely play to pursue at the time; the fact that these sandstones had been deposited in deep water was not generally recognized for several years to come. In December 1969, Phillips made the first major North Sea oil discovery in the Chalk, then considered a most improbable reservoir, at Ekofisk in Norwegian waters. Although this added a second play it turned out to be restricted to the deepest parts of the Tertiary basin and barely extended into the UK sector.

The Montrose and Ekofisk discoveries together with earlier finds in Norway, such as Phillip's Cod and unannounced discoveries at 25/11-1 (Balder), and 2/11-1 (Valhall), had the effect of significantly increasing the industry's interest in the northern areas. The announcements of the Montrose and Ekofisk discoveries coincided with the 3rd UKCS licensing Round which closed on the 5 January 1970. It was a considerable success with 106 of the 157 blocks being awarded. Interestingly more than one third of the 67 blocks awarded in the northern North Sea went to three groups, having associations with these discoveries.

The size of Ekofisk was not immediately apparent due to its geological complexity, but the discovery made by BP in December 1970 on block 21/10 in massive Palaeocene sands in a huge four way dip closed structure was clearly a giant field and immediately focused the attention of the industry on the central and northern North Sea. In the following year Auk and Argyll found by Shell and Hamilton added yet more diversity of objective in the form of the Zechstein and Rotliegendes, but these turned out to be small oil accumulations and the play has had little subsequent success.

The Third Round in 1969 had licensed several blocks north of 61°N and one of these, 211/29, became the site of what was then (1971) the world's most northerly offshore well, 300 km to the north of the nearest previous UKCS drilling. The well, drilled by Shell, discovered the Brent Field in Middle Jurassic deltaic sandstones, the second giant oil field after Forties in the UK sector; its discovery in June 1971 coincided with the announcement of the Fourth Round which, with 421 blocks, was the largest offering since the First Round. Coming on top of the major discoveries at Ekofisk and Forties (211/29-1 being a tight hole) the round created immense industry interest. In addition, for the first time, some blocks were available on a sealed cash bid basis. Less than 2 months were given for the industry to decide where to apply and how much to bid. The round was a resounding success, with 213 of the 421 discretionary blocks, and all 15 cash blocks awarded.

Applications were made by no less than 228 companies, approximately four times the number in any of the previous three rounds. Many newly formed independents were included. The government succeeded in awarding acreage to 213 of the 228 companies who applied, a clear reflection of the prospectivity of the blocks on offer and the diversity of play concepts perceived. The auction blocks brought in over £37m, ranging from £3200 for block 21/14 to the staggering amount of £21 050 001 for one block in the far north, 211/21, paid by Shell/Esso. It is interesting to note in hindsight that commercial discoveries so far exist on only three of the 15 auction blocks, namely: 211/21 (Cormorant), 9/13 (Beryl) and 48/15 (Audrey). By the conclusion of the Fourth Round, it may be said that the northern North Sea was established as a major new oil province.

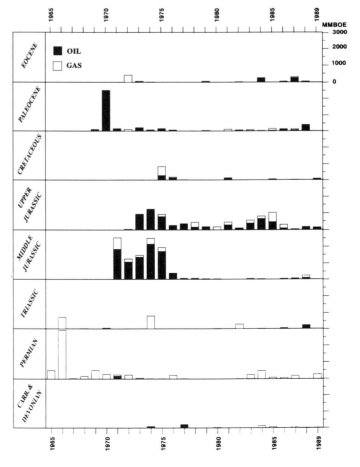

Fig. 5. UKCS geological diversity.

The Jurassic plays

Although the Middle Jurassic Brent reservoir was unknown to most Fourth Round applicants, the round opened the way to the discovery of numerous Middle and Upper Jurassic reservoirs which, in oil equivalent terms, now constitute approximately 30% and 20% respectively of total original UK reserves (Fig. 5). Following the Forties discovery most companies applied in the 4th Round for blocks with structural closure at base Tertiary level and many Jurassic discoveries, including such fields as Ninian and Magnus, were drilled with basal Tertiary sands as the main objective. In fact the Tertiary in the far north proved to be relatively unprospective, with the exception of the Frigg gas discovery in Eocene deep water sandstones which straddles the UK–Norwegian boundary.

The Fourth Round led to a fresh burst of exploration activity between 1972 and 1977, aided and abetted by OPEC which quadrupled crude oil prices in 1973 (Fig. 1). Suddenly, the possibility of economic development in water depths of 400–500 ft was realized, and this initiated what was to become one of the great technological achievements in oil exploration and production. After Brent, the Middle Jurassic deltaic sandstones also constituted reservoirs in such major northern fields as Ninian, Cormorant, Thistle, Dunlin, etc., while further south Beryl was also a major discovery. Unexpectedly, the Upper Jurassic also turned out to contain excellent clastic reservoirs mainly of deep water origin and gave rise to what is, perhaps, the most widespread UKCS play extending from Magnus in the north through Brae, Piper, Claymore and down to Fulmar in the south of the Central Graben. Most of these discoveries were made in the first 2–3 years of this period and by 1977 the rate at which reserves were discovered began to slacken. (Fig. 6)

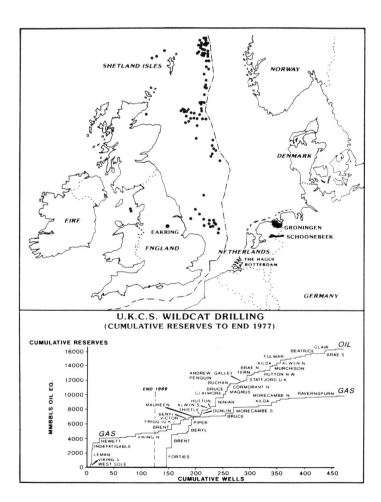

Fig. 6. UKCS new field discoveries to end of 1977.

Once again enthusiasm waned and the 'we have found everything worth finding' philosophy was widely supported. The introduction of Petroleum Revenue Tax (PRT) in 1975 significantly dampened exploration enthusiasm and led to the shelving of many developments. Not until the next oil price hike of 1977 and again in 1979 did previously marginal fields again become economic. The appearance of a national oil company at about this time with rights to 51% of all new licences, did little to lighten the gloom. There followed a period of low exploration activity, despite relatively high oil prices, until the early 1980s when activity gradually started to pick up again.

The 1980s revival

Following the oil price rise to $37 a barrel in 1980/81, activity slowly started to pick up again. This was spurred on in some cases by visions of $60–80 oil by 1990 following the Iranian revolution, but also by the imminent (or actual) decline of some of the major fields then on production and also by the presence of infrastructure, such as pipelines with available capacity. For these and the reasons that follow, by 1984 exploration activity had outstripped the highest levels of the mid-1970s and, with the exception of one year (1987) of very low activity (due to the 1986 oil price collapse) has remained uniformly high ever since, despite relatively low oil prices in recent years. The potential for selling gas at significantly higher prices also revived activity in the southern gas area.

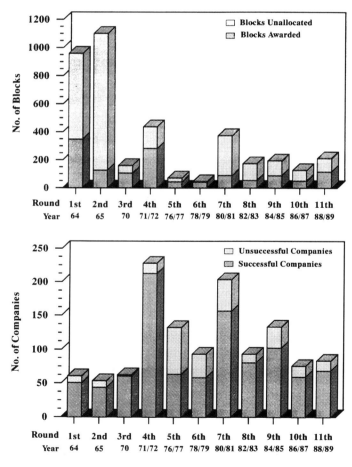

Fig. 7. UKCS licensing rounds.

Once again there had been a revival of confidence in the UK North Sea, brought about by the dawning realization that oil and particularly gas could still be found at very low cost. Also as fields became liable to PRT the ability to offset exploration costs against

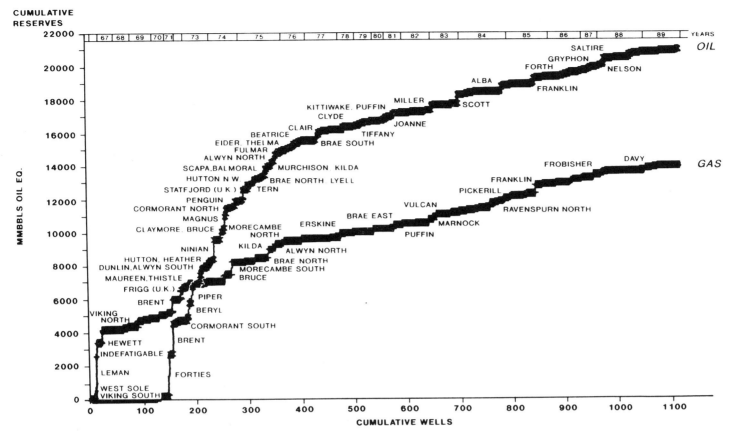

Fig. 8. UKCS wildcat drilling (cumulative reserves) 1965–1989.

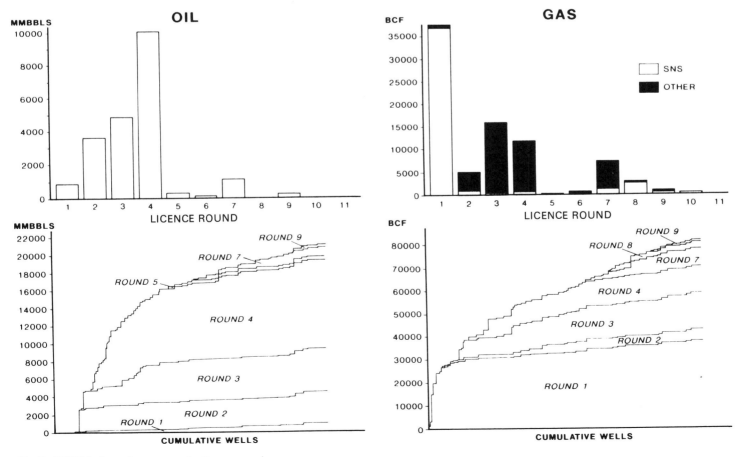

Fig. 9. UKCS hydrocarbon reserves by licence round.

PRT payments (made easier by 1983 tax changes) served to consolidate the revival. It was concluded that exploration and production activity would continue many decades into the next century. The gloomy prophecies of the mid-1970s that the UKCS would be 'finished' by the end of this century were recognized to be mistaken.

The Seventh Round in 1980/81, no longer involving mandatory state participation, demonstrated the renewed interest in the UK North Sea; 204 companies made applications (almost as high as in the 4th Round when 228 companies applied). This is to be compared with the low level of interest in the previous two rounds (applications from 64 and 59 companies respectively) (see Fig. 7). Subsequent rounds of licensing at two yearly intervals have been highly competitive, albeit with a smaller number of companies involved. The diminishing supply of attractive acreage has caused the intensified competition and the last round (the 11th) in 1989 achieved the highest ever drilling commitment, which averaged 2.38 wells per block.

The renewed burst of activity has resulted in a large number of discoveries, mostly small, but some of medium size (Fig. 8). It has also resulted in the development of new plays and the resuscitation of several of the older ones.

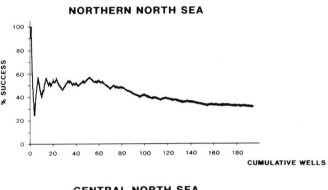
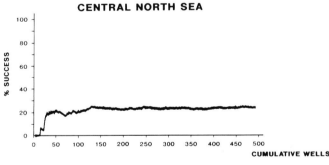
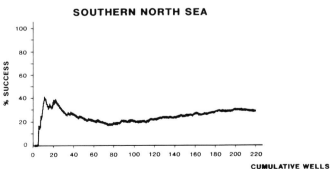

Fig. 10. UKCS historical success rate.

The accidental discovery of a major oil accumulation in Eocene sands at Alba, in 1984 while drilling to a deeper objective, revived on old play, initiated in 1971 with the Frigg gas discovery. This is still very much alive today, as exemplified by the subsequent discoveries at Forth and Gryphon. The Miller, Scott and Nelson discoveries might also be described as 'sleepers', i.e. known plays which had lain fallow for some years. Large volumes of gas and condensate have also been proven in the deeper, high-pressure and previously less accessible Jurassic and Triassic horizons in the Central Graben, examples being Marnock, Erskine and Franklin.

In the southern basin, as well as increasing success in locating Rotliegendes gas reservoirs, explorers turned to the Westphalian with some success in areas where the Permian sands are thin or absent. The recent discovery of such fields as North Ravenspurn, Murdoch/Caister, Frobisher, Camelot and Davy to name but a few, indicates that plenty of life remains in an area which had been described as 'fully mature' as long ago as 1969.

Analysis of the distribution of discovered reserves clearly shows the dominance of the early rounds; some 90% of all known reserves having been found on acreage licensed in rounds 1 to 4. Currently held acreage awarded in these early licence rounds will continue to play a significant role in terms of future discoveries, particularly if the level of drilling activity in these licences is increased (Fig. 9).

Geological diversity

Perhaps the most striking aspect of North Sea exploration has been the diversity of exploration objectives. What started in the south as a Permian gas play and, in the early years in the north as a Palaeocene oil play, has resulted in hydrocarbon discoveries in rocks ranging from Devonian to Eocene in age and from desert through deltaic to deep marine in environment. Carbonate reservoirs, apart from the unusual Chalk of the Ekofisk area and rare karstified Zechstein, are however lacking, the province being dominated by clastic rocks.

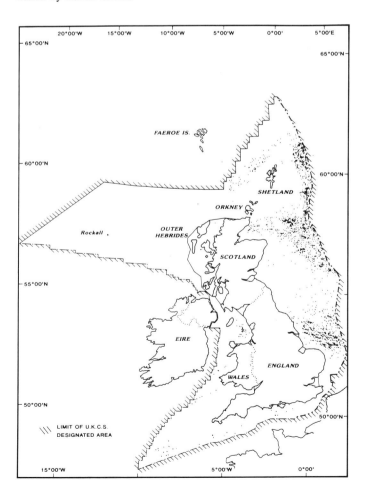

Fig. 11. 25 years of UKCS drilling.

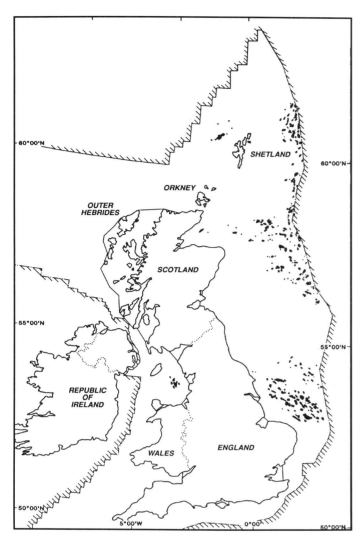

Fig. 12. UKCS oil and gas fields and discoveries 1965 to end of 1989.

In constrast to the reservoirs, there is little diversity of source rocks without which the whole North Sea phenomenon would not have happened. It is now generally accepted that the vast majority, if not all the hydrocarbons in the central and northern North Sea are derived from the widespread and richly organic Kimmeridge shales while in the southern areas Westphalian coals were the source for gas.

Given the diversity of reservoirs and abundance of source rock it is likely that further plays will still be developed whether by accident or design. As far as existing plays are concerned some, notably the Middle Jurassic of the Shetland Embayment, seem to have been quickly exhausted, while others such as the Palaeogene and Upper Jurassic in the north and the Permian and Carboniferous in the south continue to yield important discoveries (Fig. 5).

In general, however, success ratios have levelled off showing little sign of the decline which would result in the curtailment of the high level of exploration effort still evident today (Fig. 10).

The result of 25 years of exploration in the UKCS has produced a multitude of benefits for the UK, its government and industry but above all for geologists and geophysicists, for whom this volume has been produced, it has brought involvement in one of the world's most exciting and interesting petroleum provinces; excitement and interest which, I believe will continue for many decades to come (Figs 11 & 12).

The author wishes to express his thanks in particular to K. East who researched and prepared most of the statistical data presented in this paper and also to B. Evamy for his helpful comments and review of the text.

Stratigraphy of the oil and gas reservoirs: UK Continental Shelf

STEWART BROWN

The Petroleum Science and Technology Institute, Research Park, Riccarton, Edinburgh EH14 4AS, UK

The petroliferous sedimentary basins of the UK Continental Shelf are remarkable for the diversity of their reservoir strata. Reservoir rocks in fields currently in production range in age from Devonian to earliest Eocene, but significant hydrocarbon discoveries have also been made in rocks as young as the mid-Eocene. The reservoirs are predominantly siliciclastic rocks, with facies ranging from continental fluvial and aeolian, to marine gravity flow deposits from sub-wave base environments.

In this paper stratigraphic context of the producing horizons in the UK Continental Shelf (UKCS), principally the North Sea, is reviewed, and the sedimentation of the reservoir strata placed in an outline geological history. The main producing horizons are described in summary. Matters of stratigraphic terminology and correlation both between fields and between basins are discussed.

Lithostratigraphic terminology

A lithostratigraphy for the UK southern North Sea was established by Rhys (1974), and for the central and northern North Sea by Deegan & Scull (1977). Although these schemes have proved to be fairly robust, in the last 13 years the acquisition of new data plus a proliferation of new terms not fully documented in the public domain, argue strongly for a comprehensive revision and rationalization which is beyond the scope of this paper. Attempts in the public domain to standardize nomenclature across international boundaries in the North Sea, pursued by Deegan & Scull (1977) for the UK and Norwegian sectors, have lapsed for the most part in subsequent years.

A tectono-stratigraphic framework

Economic basement in the UK North Sea can be regarded at present as consisting of the pre-Devonian rocks, although fractured Precambrian basement contains oil in the undeveloped Clair Field, located to the west of the Shetland Islands (Ridd 1981). Evidence of pre-Devonian rocks in the North Sea area is provided by Frost *et al.* (1981) and pre-Devonian geological history is reviewed by Glennie (*in* Glennie 1990).

Old Red Sandstone (ORS) facies of Devonian age provide both source rocks in the western Moray Firth Basin (Fig. 1; Peters *et al.* 1989) and reservoir sandstones in the Buchan Field. Edwards (this volume) reports that the Buchan reservoir spans the late Devonian to Carboniferous, and notes biostratigraphic evidence for strata of Fammenian to latest Visean–earliest Namurian age based on palynology. In the Moray Firth area, these strata were deposited in a large lacustrine basin, the Orcadian Basin, which stretched eastwards to Norway (Trewin 1989; Richards, *in* Glennie 1990). The reservoir sandstones are of fluvial channel-fill and sheetflood origin.

In contrast to this, Middle Devonian marine carbonates occur in the Argyll Field, the deposits of a northern extension of the Devonian sea along what may have been an incipient structural low coincident with the site of the Mesozoic Central Graben. Late Devonian fluvial sandstone contributes to the pay zone in the Argyll Field (Robson, this volume).

Carboniferous sandstones in non-redbed facies are found locally in successor basins in the Moray Firth area and produce relatively minor amounts of oil in the Claymore Field (Harker *et al.* this volume). In contrast to the evidence of Carboniferous ages from ORS redbeds of the Buchan Field, Leeder & Boldy (1990) describe strata of the Forth Formation from the Moray Firth, of fluvial, lake/bay and marine origin, which are assigned a Visean age. These strata, developed in the area to the north of Buchan Field, are compared to strata in the Oil Shale Group of the Midland Valley of Scotland. Harker *et al.* (this volume) describe early Carboniferous coal-bearing strata in the Claymore Field.

However, Carboniferous strata are likely to be of more importance economically as reservoir rocks in the southern North Sea where, in the foreland area of the developing Variscan orogenic belt, Namurian and especially Westphalian strata are exploration targets for gas.

The stratigraphic succession is interrupted by a sub-Permian unconformity, the result of uplift and erosion associated with transpressive stress which promoted faulting and inversion of former Carboniferous basins following the Variscan (Hercynian) orogeny.

The beginning of a new tectonic regime in the North Sea can be traced back to the Permian, although there is some debate as to whether the rifting which came to dominate much of the Mesozoic history of the North Sea was initiated as early as the Permian or in the Triassic (see Glennie 1990). The post-Carboniferous strata can be sub-divided broadly into syn- and post-rift deposits, with the strata of pre-early Cretaceous age forming during a period of crustal extension in the North Sea, whereas the Cretaceous and younger strata formed during a time of post-rift thermal 'sag'. This description is however only a first approximation to the tectono-stratigraphic evolution of the basins. Evidence points to a number of rift phases during the Triassic and Jurassic, separated by times of little or no crustal extension (see Badley *et al.* 1988). Reservoir strata were deposited during both rift and post-rift intervals, for example during the Jurassic when the Brent Group sandstones (mid-Jurassic) were deposited during a phase of little or no crustal extension, whereas the various late Jurassic reservoirs found in fields such as Brae and Claymore were associated with syn-rift sedimentation. Furthermore, the influence of intra-basinal faults on sedimentation, so evident during the Triassic and especially the late Jurassic rifting phases, continued locally into the early Cretaceous, during the transition from rifting to thermal subsidence.

The southern gas basin and the southern part of the Central Graben suffered a phase of major inversion tectonics in the late Cretaceous which removed significant parts of the Mesozoic succession over the axes of inversion. In contrast, the central and northern North Sea basins suffered much less from inversion; major regional uplift, probably in the early Tertiary, was confined to areas such as the western Moray Firth (see McQuillin *et al.* 1982).

For much of the Tertiary, including the intervals of reservoir sand deposition during the Palaeocene and Eocene, the input of clastic sediment to the North Sea was strongly asymmetric. The bulk of the sands were derived from the west, from an area of uplift associated with the opening of the north Atlantic (see Rochow 1981; Lovell *in* Glennie 1990). In the less deformed strata of the Palaeocene and Eocene, and with high quality reflection seismic data, it has proved possible to erect very refined sequence stratigraphic schemes to sub-divide the section and to postulate a pronounced control on stratigraphic discontinuities by fluctuations in sea-level on at least a regional scale (see Rochow 1981; Stewart 1987).

Permian: Rotliegend

Permian strata, developed in continental aeolian and fluvial facies of the Rotliegend, provide the major gas reservoirs of the UK

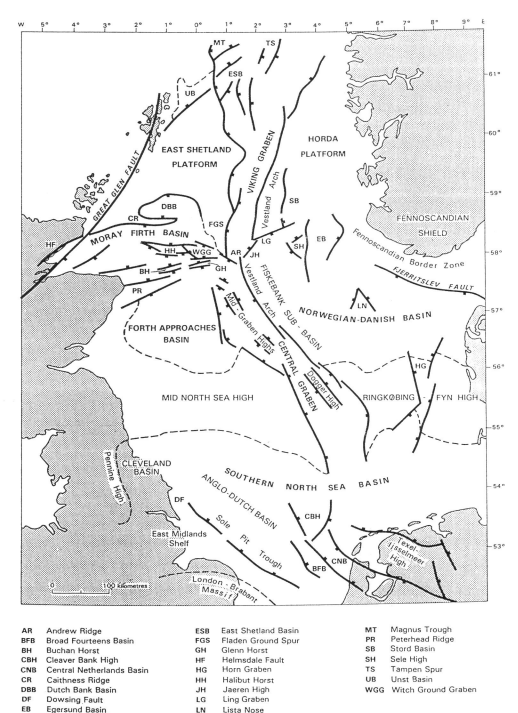

Fig. 1. Structural elements of the North Sea basins.

AR	Andrew Ridge	ESB	East Shetland Basin	MT	Magnus Trough
BFB	Broad Fourteens Basin	FGS	Fladen Ground Spur	PR	Peterhead Ridge
BH	Buchan Horst	GH	Glenn Horst	SB	Stord Basin
CBH	Cleaver Bank High	HF	Helmsdale Fault	SH	Sele High
CNB	Central Netherlands Basin	HG	Horn Graben	TS	Tampen Spur
CR	Caithness Ridge	HH	Halibut Horst	UB	Unst Basin
DBB	Dutch Bank Basin	JH	Jaeren High	WGG	Witch Ground Graben
DF	Dowsing Fault	LG	Ling Graben		
EB	Egersund Basin	LN	Lista Nose		

sector. Rotliegend sandstones also produce oil in relatively small amounts from the Auk, Argyll, and Innes fields in the central North Sea (see Trewin & Bramwell and Robson, this volume). Trewin & Bramwell (this volume) record a flora from the upper part of the Rotliegend sandstones in the Auk Field which give late Permian ages (see also Heward 1991). Pennington (1975) reported flora of late Permian age from the Argyll Field. The convention of placing the Rotliegend in the Lower Permian therefore is questioned.

Accounts of the southern North Sea gas fields in this volume almost without exception, place the Rotliegend reservoir sands in the Lower Permian. Winter & King (this volume) assign the Rotliegend reservoir in the West Sole Field variously to the early Permian and the early part of the late Permian. In the Clipper Field, Farmer and Hillier (this volume) assign the Leman Sandstone Formation reservoir to the Saxonian or the late part of the early Permian. Recent revision of the Permian time-scale places the Saxonian largely in the later Permian (see Table 1).

Deposition of the Rotliegend is now thought to extend from $c.300$ Ma to $c.258$ Ma (see Gast 1988; Glennie in Glennie 1990), that is from earliest Permian times to the late Permian, rather than within the latest Carboniferous to early Permian as previously postulated. Radiometric dates for Westphalian tuffs in Germany have resulted in the Carboniferous–Permian boundary being dated at $c.300$ Ma (Lippolt et al. 1984; Leeder 1988). The boundary between the early and late Permian is placed at 270 Ma. Deposition of the volcanic Lower Rotliegend strata, not represented in the UK area, occurred during the early Asselian. This is separated by the Saalian Unconformity from the Upper Rotliegend which, although the precise ages remain poorly constrained, was probably deposited during the mid-Artinskian to mid-Tartarian. Much of the Rotliegend reservoir sections may therefore be of late Permian age, with Zechstein deposition occurring only in the late Tartarian.

Jurassic

Stratigraphic terminology

In considering stratigraphic information on the Jurassic, a number of matters concerning chronostratigraphic terminology must be noted to avoid confusion and misinterpretation. First the basal stage of the Middle Jurassic is taken as the Bajocian by some authors whereas others sub-divide strata of equivalent age into an older Aalenian stage and a more restricted Bajocian stage (for example compare Bauman & O'Cathain, this volume, fig. 2, with Williams on Thistle, this volume, fig. 7). Also potentially confusing is the diversity of terms used to describe the chronostratigraphy of the uppermost Jurassic and the immediately overlying Cretaceous. The diversity reflects the biostratigrapher's recognition of provincialism of ammonite faunas at this time and the establishment of different zonal schemes for different parts of Europe, all of which have been applied by some authors to the UK North Sea. A comparison of the chronostratigraphic charts of Rooksby (fig. 3, this volume), Shepherd (fig. 3, this volume) and the chart of Stephenson (fig. 2, this volume) illustrates the variations in terminology.

Rooksby for the Miller Field uses the Portlandian as the uppermost Jurassic stage, a term adopted from a type area in southern

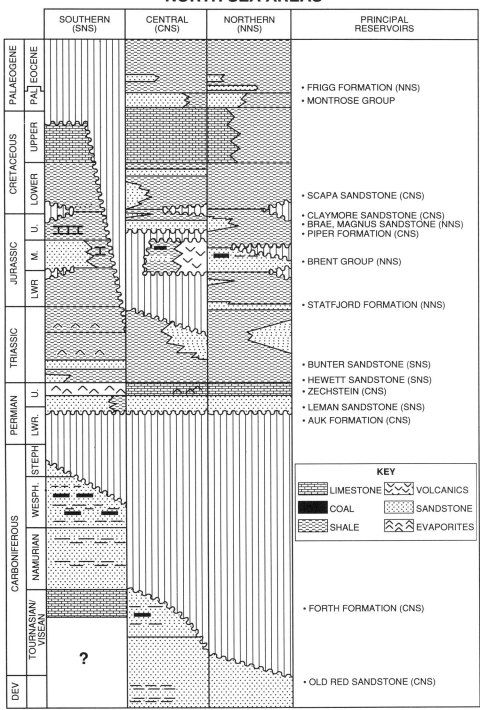

Fig. 2. Summary of North Sea stratigraphy.

Table 1. *Stratigraphic terminology for the Permian*

AGE Ma	International stage	Continental stage	Rock unit
	Triassic	Skythian	(Buntsandstein) Bacton GP
250	---	---	---
254		Thuringian	Zechstein
258		Tatarian	---

260	Late	Kazanian	
263		---Saxonian	U. Rotliegend
		Kungurian	
270	---		
		Artinskian	---?---
274		---	
280		Sakmarian	Saalian unconfomity
		Autunian	
287	Early	---	
290			
		Asselian	
			---?---
		Orenburgian	L. Rotliegend
300	---?---	---?---	
		Stephanian	Coal
306	Carboniferous	Silesian	---Measures
		Westphalian	

Table 2. *Approximate equivalence of stages around the Jurassic–Cretaceous boundary and an approximate correlation between the preferred North Sea stage nomenclature and dinoflagellate and ammonite zonal schemes*

'TETHYAN' SCHEME	ONSHORE ENGLAND SCHEME	PREFERRED 'BOREAL' SCHEME FOR NORTH SEA	DINOCYST ZONATION ZONES		SUB-ZONES	AMMONITE ZONES
Berriasian	Ryazanian	Ryazanian U	Dingodinium spinosum	V	V A	albidum
					V B	stenomphalus
						icenii
		L			V C	kochi
						runctoni
---?---					VI A	lamplughi
	Portlandian	U	Imbatodinium villosum	VI	VI B	preplicomphalus
						primitivus
---?---					VI C	oppressus
Tithonian		Volgian M	Muderongia sp A	VII	VII A	anguiformis
						kerberus
					VII B	okusensis
						glaucolithus
						albani
					VII C	fittoni
						rotunda
			Pareodinia mutabilis	VIII	VIII A	pallasioides
	Kimmeridgian U				VIII B	pectinatus
						hudlestoni
					IX A	wheatleyensis
		L	Gonyaulacysta longicornis	IX		scitulus
						elegans
---?---					IX B	autissiodorensis
Kimmeridgian	Kimmeridgian L		Gonyaulacysta cladophora	X		eudoxus
						mutabilis
						cymodoce
			Scriniodinium crystallinum (part)	XI	XI A	baylei

England. Shepherd for the Magnus Field uses the Tithonian as the top Jurassic stage, a term defined from Jurassic sections in the Tethyan realm, whereas Stephenson uses the term Volgian in the North Brae Field, defined from the Boreal realm. These stages are not equivalent in terms of age range and furthermore their use implies differences in the age range of the subjacent Kimmeridgian stage (see Table 2). Variations in usage seem to occur both between and within operating companies.

Mid-Jurassic: Brent Group

The Brent Group is the single most important oil reservoir in the UK sector of the North Sea. It is developed predominantly as a major regressive–transgressive clastic wedge or megasequence consisting of marine shoreface to delta plain deposits up to c. 300 m thick (Brown et al. 1987).

The Brent Group was defined formally by Deegan & Scull (1977) following closely the sub-divisions proposed by Bowen (1975) for the Brent Field succession. The five-fold division into formations established by Deegan & Scull (1977) has proved to be reasonably robust. A number of authors refer to Lower, Middle and Upper Brent Group strata; these units are equivalent to the Broom, Rannoch and Etive formations, to the Ness Formation, and to the Tarbert Formation respectively. The Broom Formation is considered to be genetically distinct from the rest of the Brent Group, part of a different depositional sequence (see Brown et al. 1987).

The age of the Brent Group in the northern North Sea is usually taken to be Aalenian to early Bajocian. Rather atypically, Struijk & Green (this volume) indicate that the Brent Group in Brent Field is Bajocian in age and rests on Dunlin Group strata which are in part Aalenian. Within the Brent Group the individual formations, especially the Rannoch to Tarbert units, are likely to be diachronous to some extent although precise documentation of this diachroneity has not been fully achieved in the literature to date (see Falt et al. 1989). A diachronous relationship between the Ness and Tarbert fomations in the Deveron Field however has been postulated by Williams (this volume).

(for sources see Brown in Glennie 1990).

The Brent Group usually rests on the marine shales of the Dunlin Group (early Jurassic) and is succeeded by marine shales of the Humber Group (Bathonian–Ryazanian). The lower boundary may be either at a rapid lithological transition or at a minor stratigraphic discontinuity whereas the upper boundary is very commonly an unconformity in wells drilled at or near the crests of tilted fault blocks subjected to erosion. In some places the erosion has been so deep that the Tarbert Formation has been completely removed. It is notable however that Williams (this volume) attributes thinning of the Upper Brent sequence over the crest of the Thistle Field to non-deposition and that in the Deveron Field Williams (this volume) reports onlap of the Ness Formation onto a fault-controlled high. There are exceptions to the erosional top to the Brent Group: a conformable boundary between the Tarbert Formation and the Heather Formation of the Humber Group is reported in the Alwyn North Field by Inglis & Gerard (this volume).

Reservoir stratigraphy and zonation. The Brent Group behaves as a strongly layered reservoir whose character is influenced by a number of stratigraphic features.

The basal unit, the Broom Formation, is a sandstone of variable reservoir quality (see Penhuys-Sigler *et al.* and Penny, both this volume). Thickest in the southwest where it can be up to *c.*30 m thick, it is thin or absent in the northeastern part of the East Shetland Basin.

The superjacent Rannoch Formation, deposited in offshore transition zone to shoreface environments which correspond largely to the pro-delta and delta front respectively of the prograding, wave-dominated Brent delta, consists of micaceous, commonly laminated sandstones with, in the north, a basal unit of marine shales and siltstones. The nature of the Broom–Rannoch succession varies across the basin therefore, from Rannoch sand on Broom in the south and west, to Rannoch shale on Broom further north, i.e. with a vertical transmissibility barrier between the sands, to Rannoch shale resting directly on Lower Jurassic Dunlin Group shale in the extreme north and northeast (e.g. in the Don Field: Morrison *et al.*, this volume).

In some sequences there is sufficient vertical change in the character of the Rannoch Formation sandstone to justify the subdivision into distinct reservoir zones. This happens in for example the Tern Field (Panhuys-Sigler *et al.*, this volume) where a lower zone of medium-coarse grained offshore bar sands are differentiated from an upper development of finer grained lower to middle shoreface sands.

The Etive Formation is usually of medium-grained sandstone and rests commonly at a sharp boundary on the Rannoch Formation. It has been variously interpreted as the deposits of a barrier bar complex (e.g. Vessem & Gan, this volume), as middle to upper shoreface deposits (Morrison *et al.*, this volume), and as a tidally-influenced distributary channel-fill deposit overlain by foreshore and backshore sandstones (Warrender, this volume). The Etive Formation is a polygenetic and commonly composite unit. In some places the Etive Formation has coarse-grained lag deposits at its base which form high permeability streaks. Overall, the stratigraphic juxtaposition of the generally higher permeability Etive Formation and the Rannoch Formation, with its low vertical permeability because of micaceous laminae, often leads to production problems within Rannoch during water flooding.

In most fields producing from the Brent Group the Etive Formation has a field-wide distribution. However in the Tern Field, Panhuys-Sigler *et al.* (this volume) report a lateral change from Etive sandstone to shales, coals and minor sands which behave as non-reservoir facies. A lateral change from relatively uniform, sandy Etive to sandstones with inclined shale laminae and shale rip-up clasts has been reported from the Northwest Hutton Field by Brown *et al.* (1987).

Haig (this volume) intoduces an additional formation/reservoir zone in the Hutton Field, encompassing the transition between the Rannoch Formation and the Etive Formation (the Etive/Rannoch Formation (sic) or the Massive/Mica Sand Unit). Usually the Rannoch–Etive boundary itself is a significant reservoir zone boundary in Brent Group reservoirs. This transitional zone in Hutton is described as consisting variously of ebb-tidal delta mouthbar and interbar sands, tidal-distributary channel sands and offshore barrier-bar sands. It is capped by micaceous siltstones which provide an important permeability barrier locally in the field.

The Ness Formation is the most heterolithic formation in the Brent Group, consisting of interbedded sandstones, siltstones, shales, and coals, deposited in back barrier lagoonal, distributary channel, and overbank environments on a deltaic/coastal plain. Determining the lateral continuity and vertical interconnectedness of reservoir sands is a major challenge for the reservoir geologist (see for example the study of Ness Formation stratigraphy, sedimentology and reservoir zonation in the Brent Field by Livera 1989). Over much of the Brent province, the Ness Formation has a correlatable shale horizon, the Mid Ness Shale, which can form a field-wide barrier to vertical fluid flow. This shale, noted by Budding & Inglin (1981) who considered it to be an isochronous horizon and interpreted it as the deposit of a widespread saline lagoon, is rather inappropriately named since it very often does not occur in the middle of the Ness Formation. It occurs progressively lower in the Ness section towards the north and in the Eider Field it rests directly on the Etive Formation (Wensrich *et al.*, this volume).

Towards the northern limits of Brent delta progradation, the heterolithic Ness Formation is absent and sandstones of the transgressive Tarbert Formation rest directly on massive sandstones. In their study of this northern area, Brown & Richards (1989) postulate that the lateral change from heterolithic Ness formation to this sand-dominated sequence occurred locally through the stacking of fluvial channel sands at the expense of overbank fines. In the Don Field, located in this northern area, Morrison *et al.* (this volume) adopt a contrasting hypothesis and describe a typically arenaceous Ness Formation consisting of stacked transgressive–regressive deposits.

The identification of the base of the Tarbert Formation is the principal problem in the application of the Deegan & Scull (1977) lithostratigraphic scheme. The Ness–Tarbert boundary, defined as the top of the last prominent shale bed in the Ness Formation, can be difficult to pick on geophysical well logs alone. It is known that locally the Tarbert Formation is heterolithic, with shales and coals as well as sandstones present, and that the base of Tarbert is commonly an erosional boundary. In well 211/18-A31 from the Thistle Field, Williams (fig. 2, this volume) chooses to place the Ness–Tarbert boundary at a relatively minor log peak, placing the basal boundary of the Reservoir Zone D lower in the sequence, close to the top of the uppermost significant shale in the Ness Formation. Williams (this volume) records an erosion surface at the base of the Tarbert Formation in core from the Thistle well to justify the formation boundary pick. Additional features such as the base of a thin transgressive lag deposit and/or the presence of marine bioturbated sandstone can also help identify the base of the Tarbert Formation in core (Brown *et al.* 1987: Graue *et al.* 1987).

The Tarbert Formation is of variable distribution and character. It is frequently thin or absent due to erosion at fault block crests, or locally to non-deposition (see account of the Thistle Field by Williams, this volume). It is found both as a rather uniform sandstone or, where more fully developed, as a stacked succession of transgressive–regressive cycles (see Rønning & Steel 1985; Graue *et al.* 1987; Inglis & Gerard, this volume). Deposited in part in a shoreface environment during overall transgression across the Brent delta, Tarbert has locally a micaceous sandstone facies similar to that encountered in the Rannoch Formation.

The stratigraphic resolution of seismic reflection data in the Brent Group interval has proved to be rather poor to date. The reservoir envelope can only be mapped directly from seismic data in a minority of fields and then perhaps only locally. In others either the top of the Brent Group or its base are mappable directly. In the Osprey Field (Erickson & Panhuys, this volume) the base of the Mid Ness Shale is picked on seismic.

Other mid-Jurassic reservoirs

Mid-Jurassic reservoirs are also important oil producers in the Beryl Embayment area of the Viking Graben, in the Beatrice Field in the western Moray Firth, and also in the Emerald Field located to the west of the Brent Province.

Emerald Sandstone. In the Emerald Field, the Bathonian–Callovian reservoir consists of a transgressive shallow marine sandstone, the Emerald sandstone (Stewart & Faulkner, this volume). Its presence to the west of the area of present-day Brent Group distribution, over a structurally higher fault terrace, suggests that its deposition was promoted by the continuation of the transgressive trend which had previously resulted in the deposition of the Tarbert Formation during the early Bathonian. Within the East Shetland Basin and the Viking Graben, Falt et al. (1989) describe a stacked and southward back-stepping series of transgressive–regressive sequences, ranging in age from Bathonian to Oxfordian, which form the Tarbert and Hugin formations (see Vollset & Dore 1984). It is appropriate to consider the Emerald sands as the product of the same transgressive trend, forming as the transgression spread westwards as well as southwards.

The Emerald sandstones are equivalent in age to part of the marine shale succession of the Heather Formation and also in part to the shallow marine sands of the Viking Group (Vollset & Dore 1984), developed over the Horda Platform on the eastern flank of the Viking Graben, which form the reservoir interval in the Norwegian Troll Field.

Mid-Jurassic reservoirs of the Beryl Embayment. Middle Jurassic reservoirs produce oil in the Beryl and Crawford fields to the south of the Brent Province, in the mid-Viking Graben (see Knutson & Munro and Yaliz, respectively, this volume). In the Beryl Field, the Middle Jurassic reservoir, deposited contemporaneously with intrabasinal faulting, spans the Aalenian to Callovian stages and is assigned by Knutson & Munro (this volume) to the informal Beryl Embayment Group. Two formations are recognized, with further reservoir sub-units established for reservoir management purposes. The basal Linnhe Formation is Aalenian to Bajocian in age and rests with minor unconformity on Toarcian marine shales of the Dunlin Group. The Linnhe Formation represents coastal/delta plain and mesotidal estuarine deposition.

Above the Linnhe Formation is the Beryl Formation, also consisting of estuarine to delta plain strata and deposited during northeastwards progradation. At the top of the Beryl Formation is a marine shelf sandstone unit (Unit 5 of Knutson & Munro, this volume), reflecting a late Bathonian to early Callovian regional rise in relative sea-level. Regionally a transgressive trend is evident from the early Bathonian with the deposition of the Tarbert Formation in the East Shetland Basin to the north: the late Bathonian to early Callovian transgressive pulse proposed by Knutson & Munro is broadly consistent with the regional findings of Falt et al. (1989) who postulate six transgressive–regressive depositional intervals between the Aalenian and Callovian. Unit 5 of the Beryl Formation is younger than the Tarbert Formation but appears to be in part at least contemporaneous with the Emerald sands.

The Linnhe Formation is equivalent in age to much of the Brent Group further north. This equivalence, together with the persistance of marine influence in the Beryl Embayment area during Brent delta progradation to the north, and its implication for mid-Jurassic palaeogeography, have been considered by Richards et al. (1989). These authors dispute the view of a dominantly northwards prograding Brent delta (Eynon 1981; Johnson & Stewart 1985) and postulate a greater degree of transverse drainage from the East Shetland Platform, with the preservation of a link between the marine environment lying to the north of the Brent delta during the Aalenian–Bajocian and the mesotidal estuarine environment reported by Knutson & Munro (this volume) in the Beryl Field. The late Bajocian palaeogeographic map of the Crawford area (Yaliz, this volume) indicates a marine embayment existing in the area to the northeast of the Crawford Field at this time. This is difficult to reconcile with palaeogeography proposed by Eynon (1981) and Johnson & Stewart (1985).

The Beryl Embayment Group is succeeded by the marine shales of the Heather Formation, with local intercalated sand developments, including the Callovian Angus Formation and Nevis Formation. These strata were formed during a time when structural control had a greater influence on the nature and distribution of lithofacies. The Angus Formation represents the products of fault block erosion whereas the Nevis Formation is described as a turbiditic unit (Knutson & Munro, this volume). During the deposition of these units in the Beryl Embayment, Heather Formation shales were being deposited in the Viking Graben to the east and transgressive shallow marine to coastal deposits of the Hugin Formation were being deposited in the Graben farther to the south (see Falt et al. 1989, fig. 7).

Correlation of the various Middle Jurassic lithostratigraphic units found in the Viking Graben is attempted by Yaliz (fig. 6, this volume). There appear to be some discrepancies in the age of the marine transgressive sands. Yaliz (this volume) assigns a marine sand in the Crawford Field to the 'Tarbert Formation equivalent' and gives a date of latest Bajocian to early Bathonian. The same age is proposed for equivalent strata in the Beryl Field, for similar strata in the South Viking Graben, and for the Tarbert Formation in its type area, the East Shetland Basin. According to this interpretation the transgressive sandstones which overlie the Ness/Sleipner paralic deposits in the northern North Sea form an essentially isochronous unit regionally, in contrast to the diachronous development implied by the work of Graue et al. (1987), Boote & Gustav (1987) and Falt et al. (1989). The presence of both Tarbert and Hugin formation transgressive sandstones in the South Viking Graben as proposed by Yaliz (this volume) is a radical departure from conventional wisdom (see Vollset & Dore 1984).

Beatrice Field reservoir. To complete the discussion of Middle Jurassic reservoirs in the UK North Sea it is appropriate to compare and contrast the stratigraphy of the Beatrice Field reservoir in the western Moray Firth Basin and the developments discussed above. The pay zone in the Beatrice Field spans the Hettangian to Callovian stages, with the main reservoir sandstones in the Callovian.

The Jurassic of the Beatrice Field can be correlated in part with the outcrops of Jurassic strata on the northern shores of the Moray Firth (see Linsley et al. 1980; Andrews & Brown 1987). However the solutions to Beatrice Field lithostratigraphy adopted by Stevens (this volume) differ from much that is in the literature where greater emphasis is placed on extending lithostratigraphic units defined at outcrop offshore. For example, the Pliensbachian marine shales referred by Stevens (this volume) to the Dronach Member are equivalent to the Lady's Walk Shale Member of the onshore succession. The overlying upward coarsening sandstones in the Beatrice succession (Beatrice Formation of Stevens, this volume) are not exposed/developed in onshore outcrop but the succeeding heterolithic unit (Pentland Formation of Stevens, this volume) is present. Andrews & Brown (1987) extend the onshore terminology offshore and refer the heterolithic unit to the Brora Coal Formation (see also MacLennan & Trewin 1988). The upward-coarsening sandstone is termed the Orrin Formation by Andrews & Brown (1987), a term first used by the Robertson Group.

Stevens (this volume) places the reservoir interval of Callovian age, consisting of marine sandstones and subordinate shales, in a Brora Formation, which is placed at the base of the Humber Group. Precise onshore–offshore lithostratigraphic correlation is not possible at Callovian level due to lateral facies changes. Strata of equivalent age onshore belong to the Brora Argillaceous Formation and Brora Arenaceous Formation (part). Andrews & Brown (1987) and MacLennan & Trewin (1988), in recognizing the importance of the Callovian sandstones to the reservoir of the Beatrice Field, adopted the term Beatrice Formation for these deposits. These sandstones are overlain by late Callovian to Oxford marine shales termed the Heather Formation by Stevens (this volume), but previously referred to the Uppat Formation by Andrews & Brown (1987) and MacLennan & Trewin (1988), applying a term first used in this context by the Robertson Group.

The above comparison of lithostratigraphic schemes exemplifies some of the terminological confusion found in the description of some parts of the UK North Sea succession.

Comparing the Beatrice succession with that in the northern North Sea, a number of differences emerge. Firstly the main transition from marine shales of the Lower Jurassic to superjacent coarser clastic deposits occurs around the Pliensbachian–Toarcian boundary in the Beatrice area but usually at the Toarcian–Aalenian boundary in the northern North Sea. Also, the onset of the principal mid-Jurassic transgression in the Beatrice area seems to be at the beginning of the Callovian. This is later than in the type area of the Brent Group but possibly more similar to that in places further south in the Viking Graben.

Callovian strata in the eastern Moray Firth have been regarded as absent by most authors (e.g. Boote & Gustav 1987; Andrews & Brown 1987; Schmitt & Gordon, this volume) but Ritchie *et al.* (1988) postulate that the Rattray Formation volcanics which occur around the confluence of the Moray Firth and Central Graben and which have been assigned a Bajocian–Bathonian age (Howitt *et al.* 1975), may be in part Callovian. O'Driscoll *et al.* (1990) postulate that coal-bearing Callovian strata, assigned to the lower part of the Sgiath Formation, are widespread in the eastern Moray Firth area. Coward *et al.* (this volume) suggest that the transgression which drowned the volcanic and related paralic strata of the mid-Jurassic Rattray and Pentland formations respectively in the eastern Moray Firth occurred as a brackish-water transgression during the Callovian–Oxfordian.

Late Jurassic reservoirs

Fields in the UK sector of the North Sea producing from late Jurassic reservoirs are located in the Viking Graben (including the East Shetland Basin), in the eastern part of the Moray Firth Basin, and on the western margins of the Central Graben. Late Jurassic reservoir sandstones include shallow marine to deltaic deposits which tend to be in rather widely distributed units, and marine gravity flow sediments forming submarine fans or aprons adjacent to contemporaneously active faults. A number of different play types are therefore associated with fields with late Jurassic reservoirs.

Late Jurassic of the Viking Graben. The principal phase of reservoir sand deposition from marine sediment gravity flows in the Viking Graben, occurred in the late Oxfordian to Volgian interval. This corresponds to a time of significant contemporaneous faulting especially along the western margin of the Viking Graben. It also corresponds in time for example with the deposition of the Helmsdale Boulder Beds and related deposits shed from the active Helmsdale Fault in the Inner Moray Firth (Brown *in* Glennie 1990).

In the East Shetland Basin, the Magnus Field (Shepherd, this volume) has an Oxfordian to Kimmeridgian reservoir sand unit deposited in a submarine fan setting. This unit is largely equivalent in age to the shallow marine-deltaic Piper Formation sands in the eastern Moray Firth Basin: the influx of fault-controlled gravity flow sands in the western Moray Firth commenced in the latest Oxfordian or earliest Kimmeridgian, with the deposition of the Helmsdale Boulder Beds and its correlative units but not until the late Kimmeridgian to Volgian in the Claymore Field in the eastern Moray Firth, with the deposition of the Claymore sands (see Harker *et al.*, this volume).

In the south Viking Graben, late Jurassic sandstone and conglomerate reservoirs in the Brae fields are broadly of late Oxfordian to Volgian age, but with some local differences in the age of the latest coarse clastic influx. In the South Brae Field Roberts, (this volume) dates the reservoir interval as late Oxfordian to mid-Volgian, referring the section to the 19 biostratigraphic zones which have been established locally for the Callovian to Ryazanian (see Riley *et al.* 1989). Roberts notes that the Callovian to Volgian section has a wedge shaped geometry in cross-section in contrast to the tabular package of subjacent Middle Jurassic strata. Stoker & Brown (1986) have suggested that the base of the wedging unit is late Oxfordian in age, closely corresponding to the initiation of the Brae submarine fan systems (Turner *et al.* 1987). Rooksby (this volume) reports that late Jurassic rifting began in the late Oxfordian in the nearby Miller Field.

In the South Brae Field, reservoir sands and conglomerates shed westwards from the Fladen Ground Spur persisted into the mid-Volgian. In the Central Brae Field the reservoir clastics are dated as no younger than early Volgian (Turner & Allen, this volume, fig. 2) with a minor sand unit enveloped in Kimmeridge Clay Formation shales of mid-Volgian to Ryazanian age. The deposition of coarse clastic deposits persisted for longest in the North Brae Field where Stephenson (this volume) reports reservoir deposits as young as late mid-Volgian. Therefore local controls on sediment supply along the edge of the Viking Graben resulted in the influx of coarse clastics being terminated earlier in some places than in others.

The Miller Field is situated to the east of the Brae structures and has a late Jurassic reservoir developed in more distal marine gravity flow facies. The late Jurassic succession in Miller is sub-divided into three members by Rooksby (this volume): the lower member of late Oxfordian–early Kimmeridgian age (i.e. approximately late Oxfordian to Kimmeridgian in Boreal terminology) deposited during a period of rapid subsidence; a middle member of early Kimmeridgian to early Portlandian age (i.e. Kimmeridgian to mid-Volgian); and an upper member of early to late Portlandian (i.e. mid–late Volgian). There are coarse clastics in each member but the main Miller reservoir is in the middle member. The youngest coarse clastic unit in Miller is equated by Rooksby to the North Brae reservoir.

Late Jurassic reservoirs: eastern Moray Firth. The succession in eight fields within the eastern Moray Firth is compared and contrasted, namely the Claymore, Chanter, Highlander, the closely associated Ivanhoe and Rob Roy, and the Petronella, Piper and Tartan fields. The lithostratigraphic terms employed in the accounts of these fields are in detail variable. The best documented stratigraphic scheme is that employed by Schmitt & Gordon (this volume) for the Piper Field (see also Deegan & Scull 1977; Maher 1980; Harker *et al.* 1987) and Schmitt (this volume) for the Chanter Field. A basal heterolithic, coal-bearing sequence, the Sgiath Formation is overlain by a widely distributed marine shale (The I-Shale) which forms the basal beds of the predominantly arenaceous Piper Formation. The Sgiath Formation forms a subsidiary reservoir in the Claymore Field where it is overlain by a predominantly shale and silt unit laterally equivalent to the Piper Formation sands (Harker *et al.*, this volume). In the Chanter Field, the Sgiath Formation is present but in non-reservoir facies (Schmitt, this volume). Regionally the Piper Formation is overlain, locally unconformably, by the Kimmeridge Clay Formation organic-rich marine shales of Kimmeridgian to Ryazanian age but locally by Kimmeridgian to Volgian sandstones of the Claymore Sand Member or its equivalents (Harker *et al.*, this volume). Schmitt & Gordon (this volume) interpret the Piper sands as the deposits of a wave-dominated delta whereas the Claymore sands are marine sediment gravity flow deposits.

The lithostratigraphy of the Ivanhoe and Rob Roy fields, although based on a different terminology (see Parker, this volume), can be compared with adjacent fields. A Basal Shale unit is equivalent to the Sgiath Formation plus the I-Shale, and is overlain by a sequence with two sand units, the Main Piper Sand Member and above, the Supra Piper Sand Member. O'Driscoll *et al.* (1990) postulate that the strata assigned to the Main Piper Sand Unit in the Ivanhoe and Rob Roy fields is equivalent in age to the upper part of the Sgiath Formation (Oxfordian to early Kimmeridgian), and further that the I-Shale and superjacent sands of the Piper Formation are wholly Kimmeridgian in age in contrast to the late Oxfordian to early Kimmeridgian age assigned by Boote & Gustav (1987) and Schmitt (this volume).

In the Tartan Field, the Sgiath Formation is probably represented by the Coal Marker Member to Lower Sand Member of Coward *et al.* (this volume), and the I-Shale by the Lower Shale Member. The superjacent Piper Formation in the Tartan Field is sub-divided into three principal sand units, namely the Main Sand at the base, the 15/16-6 Sand, and the Hot Sand. Whitehead & Pinnock (this volume) note that the Piper Formation in the Highlander Field is equivalent to the Main and 15/16-6 sand units only, based on biostratigraphy. No Sgiath Formation is noted in Highlander but it is regarded as equivalent to the Coal Marker and Coal Marker Shale at the base of the reservoir sequence in the Petronella Field (Waddams & Clark, this volume). This interpretation of the stratigraphy of the Highlander Field is at variance with the results of a regional stratigraphic study by Harker et al. (1987) which indicated that well 14/20-5 from the Highlander Field had Sgiath Formation above Triassic strata and had the Piper Formation of Whitehead & Pinnock (this volume) sub-divided in an alternative way into a lower section assigned to the Piper Formation and an upper section assigned to the Claymore Sandstone Member of the Kimmeridge Clay Formation (see Harker et al. 1987, fig. 7).

Shallow marine sands of Volgian age are developed over the Fladen Ground Spur in Quadrant 16 and form the reservoir in the Glamis Field (Fraser & Tonkin, this volume). The sands formed during a marine transgression which drowned the western flank of the Fladen Ground Spur. This occurred at the same time as Brae Formation turbidite sands were being deposited in the Viking Graben farther to the east and the Kimmeridge Clay Formation and localized submarine fan sands were being deposited in the Moray Firth Basin to the south and west.

Late Jurassic of the central North Sea. Jurassic reservoirs occur principally in three fields namely Clyde, Fulmar and Kittewake. The Clyde and Fulmar fields are located on fault terraces within the south-western part of the Central Graben, the Kittewake Field is located farther to the north, on the platform to the west of the Central Graben. The Jurassic reservoir sandstones rest unconformably on Triassic strata. Jurassic sandstones also form the reservoir in the Duncan Field where they have been assigned to the Duncan Sandstone Formation, of Oxfordian to Volgian age, by Robson (this volume) who considers them to be roughly equivalent to the Fulmar (Sandstone) Formation in the Clyde and Fulmar fields.

The Fulmar Formation sandstones are shallow marine deposits and pass laterally to deeper water, basinal shales to the north and east. They range in age from early Kimmeridgian to early Volgian in the Clyde Field (Stevens, this volume: Oxfordian to Volgian according to Smith 1987) and from late Oxfordian to early Kimmeridgian in the Fulmar Field (Stockbridge & Gray, this volume: Johnson et al. 1986). The Fulmar Formation in the Kittiwake Field is assigned an undifferentiated late Jurassic age by Glennis & Armstrong (this volume). Based on these variable age determinations, the Fulmar Formation appears to be similar in age to the Piper Formation sandstones in the Moray Firth Basin but with sand deposition persisting for longer in the Clyde Field. The Ula Formation shallow marine sandstones in the Norwegian sector of the Central Graben span the Oxfordian to Volgian stages (Vollset & Dore 1984). In the Fulmar Field, Stockbridge & Gray (this volume) and Johnson et al. (1986) describe a marine turbidite sand reservoir, the Ribble Formation of Volgian age, within the Kimmeridge Clay Formation shales which cap the Fulmar Formation reservoir. Turbidite sandstones of this age are found at a number of locations in the Moray Firth and Viking Graben (Brown 1990).

Zonation of the Fulmar Formation reservoir in the Clyde Field is into three main units on the basis of fluid communication between sand bodies (Stevens, this volume). Further subdivision of each of these units into up to three sub-units is established locally. In the Fulmar Field, the reservoir interval has been sub-divided by Stockbridge & Gray (this volume) into six members on lithostratigraphic criteria and also into fourteen units which reflect depositional cycles. This reservoir interval spans the Fulmar Formation and the overlying turbidite sands of the Ribble Member which lies within the Kimmeridge Clay Formation.

The lithostratigraphic classification of argillaceous strata in the Upper Jurassic to Ryazanian of the Central Graben area, not addressed by Deegan & Scull (1977) through lack of data, has been considered by Vollset & Dore (1984) in the Norwegian sector.

Vollset & Dore (1984) abandon the two-fold subdivision into Heather Formation and Kimmeridge Clay Formation which was established by Deegan & Scull (1977) in the Moray Firth and Viking Graben areas and extended into the UK Central Graben by some authors. They favour a three-fold division into Haugesund Formation (Callovian–early Kimmeridgian in age), Farsund Formation (Kimmeridgian–Volgian), and Mandal Formation (Volgian–Ryazanian).

Palaeogene reservoirs

The lithostratigraphy established for the Palaeogene by Deegan & Scull (1977) recognizes two sandstone-dominant groups of Palaeocene age, the Montrose Group, a sequence of overlapping submarine fans, overlain by the Moray Group, of shelf-deltaic sandstones, and a further sand-prone unit in the early Eocene, the Frigg Formation. Within the Palaeogene of the UK central and northern North Sea, hydrocarbons are currently produced from several horizons within the Palaeocene and from the early Eocene. More recent discoveries have encountered oil in younger Eocene sand bodies.

Submarine fan sandstones are producing horizons in the Maureen, Andrew and Forties formations of the Montrose Group (see Fig. 2). The Maureen Formation forms the reservoir in the Maureen Field (Cutts, this volume), the Andrew Formation forms the reservoir in the Balmoral Field (Fraser & Tonkin, this volume), the Cyrus Field (Mound et al., this volume) and one of the reservoir bodies in the Forties Field (Wills, this volume), and the Forties Formation is the reservoir in the Montrose and Arbroath structures (Crawford et al., this volume) and in the Forties Field. The Frigg Field reservoir is also a submarine fan sandstone, the Frigg Formation, of early Eocene age (Brewster, this volume).

A sequence stratigraphic analysis by Stewart (1987) recognizes ten depositional sequences, heterolithic in nature to varying degrees, within the Palaeocene to early Eocene (mid-Ypresian: pre-Frigg Formation) succession. Distinct sequences containing a Maureen Fan, an Andrew Fan and a Forties Fan are defined, bounded in part by stratigraphic discontinuities.

The oldest of the Palaeogene reservoir horizons, the Maureen Formation, consists of sandstones together with reworked carbonate clasts from the subjacent Chalk Group. Cutts (this volume) notes that the original definition by Deegan & Scull (1977) of the Maureen Formation was based on a section not typical of the Maureen Field. The Formation is dated as Danian to Thanetian by Cutts (fig. 3), falling into the age range of Stewart's (1987) Sequence 2.

The Andrew Formation is dated as early to mid-Thanetian in the Balmoral Field by Fraser & Tonkin (this volume) who assign it to the *Alisocysta margarita* dinocyst zone of Knox et al. 1981). Stewart (1987, figs 4 and 18) places the Andrew Fan in Sequence 3, but considers this to be older than the *Alisocysta margarita* Zone. The Andrew Formation together with the Forties Formation are placed within the Thanetian by Wills (this volume), with Stewart (1987) placing it in the late Thanetian. Wills subdivides the section from the top of the subjacent Chalk Group to the top of the Sele Formation, caprock to the Forties Field, into eleven units (A to M: see Fig. 2) based on seismic stratigraphic techniques. This provides a greater level of resolution locally than that attempted in the regional study by Stewart (1987) but comparable biostratigraphic data are not provided.

The Forties Formation is represented by laterally equivalent shale facies in the Cyrus Field (Mound et al., this volume) demonstrating the heterolithic character of the depositional sequences which contain the arenaceous formations.

Eocene strata in the Frigg Field have been assigned ages within a finely divided biostratigraphic zonation based on palynology and calcareous microfossils. The boundary between the Palaeocene and Eocene is taken at the base of locally derived palynozone Upper 2a and at the transition between standard calcareous nannoplankton zones NP10 on NP9 and planktonic foraminifera zones P6b on P6a (see Brewster, this volume, fig. 6). This is taken to correspond to the base of the Balder Tuff, a prominent stratigraphic marker in the central and northern North Sea. Note that Brewster (this volume) employs the term Landenian for the stage between the Danian and the Ypresian whereas Cutts (this volume), Fraser and Tonkin (this volume), and Wills (this volume) use the term Thanetian for this section. The terms imply similar stratigraphic intervals, the former coming from a type section in Belgium and the latter from the Thanet Sands of the Isle of Thanet, Kent in the UK.

Concluding remarks

Establishing a stratigraphy is a fundamental task for the geologist working in both exploration and production. The stratigraphic studies of reservoir horizons in the UK North Sea employ a range of techniques, including the conventional ones of lithostratigraphy and biostratigraphy, together with the use of valuable data on reservoir pressure depletion to determine the interconnectedness of sand bodies (see Roberts, this volume). Seismic (or sequence) stratigraphic techniques, increasingly employed as part of regional basin analysis and related exploration studies, are also used to assist in the determination of intra-field stratigraphy in some places (see Wills, this volume).

With the demands for greater stratigraphic resolution to promote better understanding of both basin and field scale geology, the integration of stratigraphic techniques, the refinement of existing biostratigraphic zonations, and the exploitation of additional techniques such as magnetostratigraphy and clay mineral stratigraphy, will be pursued. There is merit in reviews and rationalizations of diverse terminology in the public domain from time to time, to set standards, to establish high quality databases which help to promote relevant stratigraphic research in universities, and generally to aid clarity in communication. There is merit also in the establishment of better documented type or reference sections where the relationship between the various stratigraphic methods can be compared, contrasted, and used to calibrate other wells and seismic data.

The author is indebted to many former colleagues in the British Geological Survey for their contribution to his understanding of the stratigraphy of the UK Continental Shelf basins. Thanks are especially due to C. Deegan, M. Dean and P. Richards. K. Glennie is thanked for providing valuable information on the stratigraphy of the Rotliegend.

ANDREWS, I. J. & BROWN, S. 1987. Stratigraphic evolution of the Jurassic, Moray Firth. In: BROOKS J. & GLENNIE, K. W. (eds) q.v. 785–796.

BADLEY, M. E., PRICE, J. D. RAMBECH DAHL, C. & AGDSTEIN, T. 1988. The structural evolution of the north Viking Graben and its bearing on structural modes of basin formation. *Journal of the Geological Society, London*, **145**, 455–472.

BOOTE, D. R. D. & GUSTAV, S. H. 1987. Evolving depositional systems within an active rift, Witch Ground Graben, North Sea. In: BROOKS, J. & GLENNIE, K. W. (eds) q.v. 819–834.

BOWEN, J. M. 1975. The Brent oilfield. In: WOODLAND, A. W. (ed.) *Petroleum and the Continental Shelf of North-west Europe*, Vol 1 Applied Science Publishers, Barking, 356–362.

BROOKS, J. & GLENNIE, K. W. 1987. (eds) *Petroleum geology of north west Europe*. Graham & Trotman, London.

BROWN, S. 1990. Jurassic. *In* GLENNIE, K. W. (ed.) *Introduction to the Petroleum Geology of the North Sea*. Blackwell, 219–254.

—— & RICHARDS, P. C. 1989. Facies and development of the Middle Jurassic Brent Delta near its northern limit of progradation, UK North Sea. In: WHATLEY, M. K. G. & PICKERING, K. T. 1989 (eds) *Deltas: Sites and Traps for Fossil Fuels*. Geological Society, London, Special Publication, **41**, 253–267.

——, —— & THOMSON, A. R. 1987. Patterns in the deposition of the Brent Group (Middle Jurassic), UK North Sea. In: BROOKS, J. & GLENNIE, K. W. (eds) q.v. 899–914.

BUDDING, M. B. & INGLIN, H. F. 1981. A reservoir geological model of the Brent Sands in Southern Cormorant. In: ILLING, L. V. & HOBSON, G. D. (eds) q.v. 326–334.

DEEGAN, C. E. & SCULL, B. J. 1977. *A standard lithostratigraphic nomenclature for the Central and Northern North Sea*. Institute of Geological Sciences Report No. 77/25.

EYNON, G. 1981. Basin development and sedimentation in the Middle Jurassic of the northern North Sea. In: ILLING, L. V. & HOBSON, G. D. (eds) q.v. 196–204.

FALT, L. M. HELLEND, R. WIIK JACOBSON, V. & RENSHAW, D. 1989. Correlation of transgressive–regressive depositional sequences in the Middle Jurassic Brent/Viking Group megacycle, Viking Graben, Norwegian North Sea. In: COLLINSON, J. D. (ed.) *Correlation in Hydrocarbon Exploration*, Graham & Trotman, London, 191–201.

FROST, R. T. C., FITCH, F. J. & MILLER, J. A. 1981. The age and nature of the crystalline basement of the North Sea Basin. In: ILLING, L. V. & HOBSON, G. D. (ed.) q.v. 43–57.

GLENNIE, K. W. (ed.) 1990. *Introduction to the Petroleum Geology of the North Sea*. Blackwell, Oxford.

GAST, R. E. 1988. Rifting in Rotliegenden Niedersachsens. *Die Geowissenschaften*, **6**.

GRAUE, E., HELLEND-HANSEN, W., JOHNSON, L., LOMO, L., NOTTVEDT, A. RONNING, K., RYSETH, A. & STEEL, R. 1987. Advance and retreat of the Brent Delta system, Norwegian North Sea. In: BROOKS, J. & GLENNIE, K. W. (eds.) q.v. 915–938.

ILLING, L. V. & HOBSON, G. D. (eds) 1987. *Petroleum Geology of the Continental Shelf of North-west Europe*. Heyden & Son, London.

HARKER, S. D., GUSTAV, S. H. & RILEY, L. A. 1987. Triassic to Cenomanian stratigraphy of the Witch Ground Graben. In: BROOKS, J. & GLENNIE, K. W. (eds). q.v. 809–818.

HEWARD, A. P. 1991. Inside Auk—the anatomy of an aeolian oil reservoir. In: MIALL, D. & TYLER, N. (eds) *The three dimensional facies architecture of clastic sediments and its implications for hydrocarbon discovery and recovery*. Society of Economic Paleontologists and Mineralogists, in press.

HOWITT, F., ASTON, E. R. & JACQUE, M. 1975. The occurrence of Jurassic volcanics in the North Sea. In: WOODLAND, A. W. (ed.) *Petroleum and the Continental Shelf of North-west Europe*. Applied Science Publishers, Barking, 379–388.

JOHNSON, H. O. & STEWART, D. J. 1985. Role of clastic sedimentology in the exploration and production of oil and gas in the North Sea. In: BRENCHLEY, P. J. & WILLIAMS, B. P. J. (eds) *Sedimentology: Recent Developments and Applied Aspects*. Geological Society, London, Special Publication, **18**, 249–310.

——, MACKAY, T. A. & STEWART, D. J. 1986. The Fulmar oil-field (Central North Sea): geological aspects of its discovery, appraisal and development. *Marine and Petroleum Geology*, **3**, 99–125.

KNOX, R. W. O. B., MORTON, A. C. & HARLAND, R. 1981. Stratigraphical relationships of Palaeocene sands in the UK sector of the central North Sea. In: ILLING, L. V. & HOBSON, G. D. (eds) *Petroleum Geology of the Continental Shelf of North-west Europe*, Heyden, 267–281.

LEEDER, M. R. 1988. Recent developments in Carboniferous geology: a critical review with implications for the British Isles and North-West Europe. *Proceedings of the Geologists' Association*, **99**, 74–100.

—— & BOLDY, S. R. 1990. The Carboniferous of the Outer Moray Firth Basin, quadrants 14 and 15, Central North Sea. *Marine and Petroleum Geology*, **7**, 29–37.

LINSLEY, P. N., POTTER, H. C., MCNAB, G. & RACHER, D. 1980. The Beatrice Field, Inner Moray Firth, UK North Sea. In: HALBOUTY, M. T. (ed.) *Giant oil and gas fields of the decade 1968–78*. American Association of Petroleum Geologists, Memoir, **30**, Tulsa, 117–130.

LIPPOLT, H. J., HESS, J. C. & BURGER, K. 1984. Isotopiche Alter pyroclastischen Sanidinen aus Kaolin-kohlensteinene als Korrelationsmarker fur das mitteleuropaische Oberkarbon. *Fortschrifte in der Geologie von Rhineland und Westfalen*, **32**, 119–150.

LIVERA, S. E. 1989. Facies associations and sand-body geometries in the Ness Formation of the Brent Group, Brent Field. In: WHATELEY, M. K. G. & PICKERING, K. T. (eds) q.v. 269–288.

MACLENNAN, A. M. & TREWIN, N. H. 1988. Palaeoenvironments of the late Bathonian–mid Callovian in the Inner Moray Firth. In: BATTEN, D. J.

& KEEN, M. C. (ed.) *North-west European micropalaeontology and palynology*. British Micropalaeontology Society, 92–117.

MCQUILLIN, R. DONATO, J. A. & TULSTRUP, J. 1982. Dextral displacement of the Great Glen Fault as a factor in the development of the inner Moray Firth Basin. *Earth & Planetary Science Letters*, **60**, 127–139.

MAHER, C. E. 1980. The Piper oilfield. In: HALBOUTY, M. T. (ed.) *Giant oil and gas fields of the decade: 1968–78*. American Association of Petroleum Geologists, Memoir, **30**, 131–172.

O'DRISCOLL, D., HINDLE, A. D. & LONG, D. C. 1990. The structural controls on Upper Jurassic and Lower Cretaceous reservoir sandstones in the Witch Ground Graben, UK North Sea. In: HARDMAN, R. F. D. & BROOKS, J. (eds) *Tectonic Event Responsible for Britain's Oil and Gas Reserves*. Geological Society, London, Special Publication, **55**, 299–324.

PENNINGTON, J. J. 1975. The geology of the Argyll Field. In: WOODLAND, A. W. (ed.) *Petroleum and the Continental Shelf of North-west Europe*. Applied Science Publishers, Barking, 285–291.

PETERS, K. E., MOLDOWAN, J. M., DRISCOLE, A. R. and DEMAISON, G. J. 1989. Origin of Beatrice oil by co-sourcing from Devonian and Middle Jurassic source rocks, Inner Moray Firth, United Kingdom. *American Association of Petroleum Geologists' Bulletin*, **73.4**, 451–471.

PICKERING, K. T. 1984. The Upper Jurassic 'Boulder Beds' and related deposits: a fault controlled submarine slope, NE Scotland. *Journal of the Geological Society, London*, **141**, 357–374.

RHYS, G. H. 1974. *A proposed standard stratigraphic nomenclature for the Southern North Sea and an outline structural nomenclature for the whole of the (UK) North Sea*. Institute of Geological Sciences Report No **78/8**.

RICHARDS, P. C., BROWN, S., DEAN, J. M. & ANDERTON, R. 1988. A new palaeogeographic reconstruction for the Middle Jurassic of the northern North Sea. *Journal of the Geological Society, London*, **145**, 883–886.

RIDD, M. F. 1981. Petroleum geology west of the Shetlands. In: ILLING, L. V. & HOBSON, G. D. (ed.) q.v., 414–425.

RILEY, L. A., ROBERTS, M. J. and CONNELL, E. R. 1989. The application of palynology in the interpretation of Brae Formation stratigraphy and reservoir geology in the South Brae Field area, British North Sea. In: COLLINSON, J. D. (ed.) *Correlation in Hydrocarbon Exploration*. Graham & Trotman, London, 339–356.

RITCHIE, J. D., SWALLOW, J. L., MITCHELL, J. G. and MORTON, A. C. 1988. Jurassic ages from intrusives and extrusives within the Forties igneous province. *Scottish Journal of Geology*, **24**, 81–88.

ROCHOW, K. A. 1981. Seismic stratigraphy of the North Sea Palaeocene deposits. In: ILLING, L. V. & HOBSON, G. D. (ed.) q.v., 255–266.

RØNNING, K. & STEEL, R. J. 1987. Depositional sequences within a "transgressive" reservoir sandstone unit: the Middle Jurassic Tarbert Formation, Hild area, northern North Sea. In: *North Sea Oil and Gas Reservoirs*, Graham & Trotman, London, 169–176.

SMITH, R. L. 1987. The structural development of the Clyde Field. In: BROOKS, J. & GLENNIE, K. W. (eds) q.v., 523–531.

STEWART, I. J. 1987. A revised stratigraphic interpretation of the Early Palaeogene of the Central North Sea. In: BROOKS, J. & GLENNIE, K. W. (eds) q.v., 557–576.

STOKER, S. J. & BROWN, S. 1986. *Coarse clastic sediments of the Brae Field and adjacent areas, North Sea: a core workshop*. British Geological Survey Open File Report.

TREWIN, N. H. 1989. The petroleum potential of the Old Red Sandstone of northern Scotland. *Scottish Journal of Geology*, **25**, 201–225.

TURNER, C. C., COHEN, E. R. & COOPER, D. M. 1987. A depositional model for the South Brae oilfield. In: BROOKS, J. & GLENNIE, K. W. (eds) q.v., 853–864.

VOLLSET, J. & DORE, A. G. 1984. *A revised Jurassic and Triassic lithostratigraphic nomenclature for the Norwegian North Sea*. Bulletin of the Norwegian Petroleum Directorate **3**.

The Viking Graben

The Alwyn North Field, Blocks 3/9a, 3/4a, UK North Sea

I. INGLIS[1] & J. GERARD[2]

[1] Total Oil Marine, Altens, Aberdeen AB9 2AG, UK
[2] Total, Compagnie Francaise des Petroles, La Defense, Paris

Abstract: Situated in the southeastern part of the East Shetland Basin, the Alwyn North Field produces oil and gas from Brent Group reservoirs and gas and condensate from the Statfjord Formation. The structural style is of tilted and eroded fault blocks dipping to the west and aligned north–south conforming to the principal normal fault trend. NE–SW cross elements further separate the hydrocarbon accumulations. The hydrocarbon columns are restricted to the Tarbert and upper part of the Ness Formation of the Brent Group, in sediments associated with the retreat of the Brent delta. The Statfjord Formation was deposited in an alluvial, fan-delta setting with increasing marine influence towards the top of the formation.

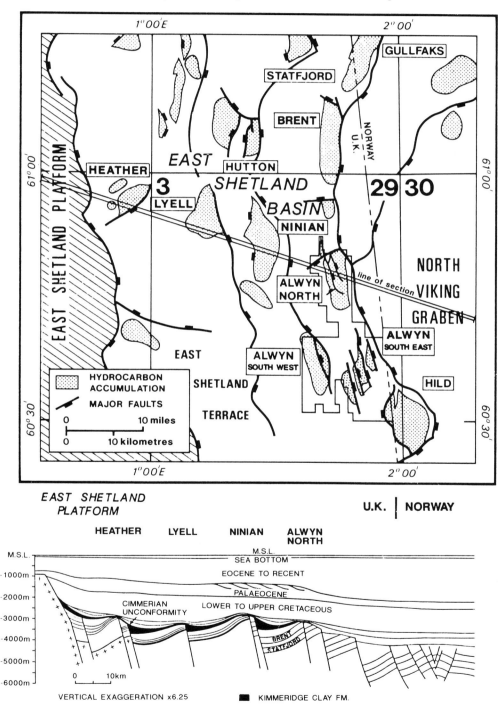

Fig. 1. Alwyn North Field location map and schematic structural cross-section showing principal hydrocarbon accumulations and major faults.

From Abbotts, I. L. (ed.), 1991, *United Kingdom Oil and Gas Fields, 25 Years Commemorative Volume*, Geological Society Memoir No. 14, pp. 21–32

Alwyn North Field is situated in blocks 3/9a and 3/4a in the southeastern part of the East Shetland Basin, some 25 km south of the Brent Field and 15 km southeast of the Ninian Field (Fig. 1). The UK–Norway median line is 9 km to the east. The field is 465 km from Aberdeen on the NE coast of Scotland, whilst the nearest landfall is the northern islands of the Shetlands, 140 km to the west.

Production is from tilted and faulted Brent Group and Statfjord Formation sandstone reservoirs with a cumulative total surface closure of 22 km². Initial recoverable reserves are estimated (1988 figures) to be $176 \times$ MMBBL of oil and condensate and 22×10^9 Sm³ of gas. This places Alwyn North Field above the midpoint of UKCS producing oil- and gas-fields in terms of size of recoverable oil and gas reserves (24th out of 60 producing UKCS oilfields and gasfields at the end of 1988). Alwyn is a name of Celtic origin, easily pronounceable in both English and French.

History

Block 3/9a belongs to the P.090 licence which was awarded to the Total/Elf Group in the second round of United Kingdom offshore licensing in 1965. At that time the consortium, known as the French Group, comprised chiefly the principal French oil companies of Total (operator), Elf, and Aquitane, each with nearly a one-third share. With the amalgamation of Aquitaine and Elf the interests have become: Total Oil Marine plc (operator) 33.33%; Elf UK plc 66.67%. Amongst the most northerly acreage so far awarded, it was not until 1971 that any wells were drilled north of the 60°N parallel. The first discoveries of hydrocarbons by Total were made in 1972 and 1973 on blocks 3/14 and 3/15; the area was called Alwyn. This area is situated in the southernmost part of the East Shetland Basin and is characterized by a complex pre-Cretaceous fault system splitting the hydrocarbon accumulations into many different com-

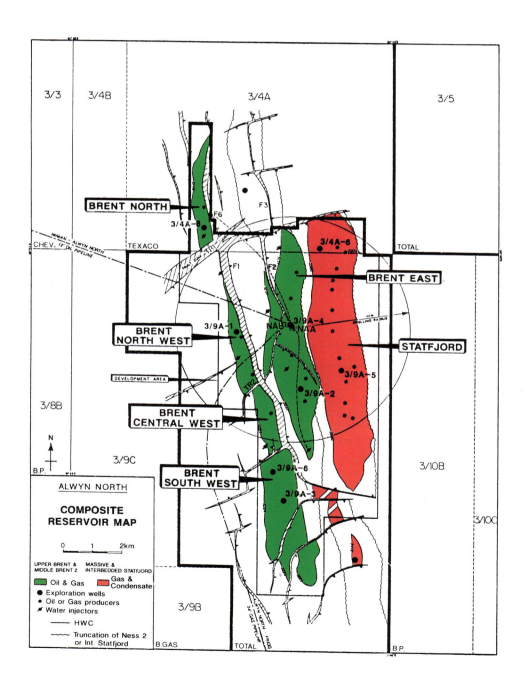

Fig. 2. Alwyn North Field composite reservoir map showing discovery and development wells.

partments. This did not permit a straightforward appraisal at that time. Further north, during the summer of 1975, two wells were drilled which effectively became the discovery wells for the Alwyn North Field. A joint-well, 3/4a-6, operated by Texaco encountered a thin, truncated oil-bearing Brent Group sequence before penetrating an 80 m gas-condensate column in the Statfjord Formation. While the 3/4a-6 well was underway, Total spudded the 3/9a-1 well and discovered a separate oil accumulation in the Brent Group (Fig. 2).

By the end of 1976, four wells had been drilled (3/4a-6, 3/9a-1, 2 and 3) which confirmed the presence of four different hydrocarbon accumulations in separate faulted compartments; three Brent oil reservoirs and the one gas-condensate accumulation in the Statfjord. Further seismic was shot before two appraisal wells were drilled in 1979–1980. These were the 3/9a-4 well, which confirmed the main Brent reservoir accumulation, called 'Brent East' and the 3/9a-5 well, the only appraisal well on the 10 km long structure of the Statfjord gas-condensate reservoir.

An extensive 3-D seismic survey was shot in the summer of 1980, and completed a year later. In June 1981 Texaco drilled the 3/4a-8 well which proved the presence of another separate oil accumulation in the Brent sequence, to be called 'Brent North'. In 1982 the Total/Elf Group purchased all hydrocarbon reserves within a defined area of the southern part of block 3/4. Government approval to develop the main part of the field was obtained on 29 October 1982.

The southernmost oil accumulation, 'Brent South-West', which had been identified by the 3/9a-3 well was delineated by the 3/9a-6 well during 1983 and proved the existence of complex hydrocarbon fluid close to critical conditions. This part of the Alwyn North Field is mostly outwith the drilling radius of the development platform. The additional reserves of 'Brent South-West' will therefore be produced by subsea wells back to the existing platform.

A deviated well was drilled in 1988 from the platform into what is now called 'Brent Central-West' and proved additional hydrocarbons which are more akin to the main Brent Group accumulations of the Alwyn North Field.

Both these additional reserves are to be exploited under an addendum to the current Alwyn North Field development approval, known as the Alwyn North Extension.

Alwyn North Field is being developed from two linked platforms, a 40-slot drilling and accommodation platform (NAA), and a separate processing and production platform (NAB). Gas export is via a 24 inch pipeline into the Frigg gas system and on to St Fergus on the Scottish mainland. Oil export is via a 12 inch oil pipeline to Ninian Field and thence to Sullom Voe. To date, two subsea well completions with 6 inch flowlines have been tied back to NAA.

The first of the two steel jackets, built at Methil in Fife, was positioned in 1985, the second a year later. First oil production was achieved in November 1987 just twelve months from the commencement of development drilling. First gas export was at the beginning of January 1988.

ALWYN NORTH FIELD LITHOSTRATIGRAPHY

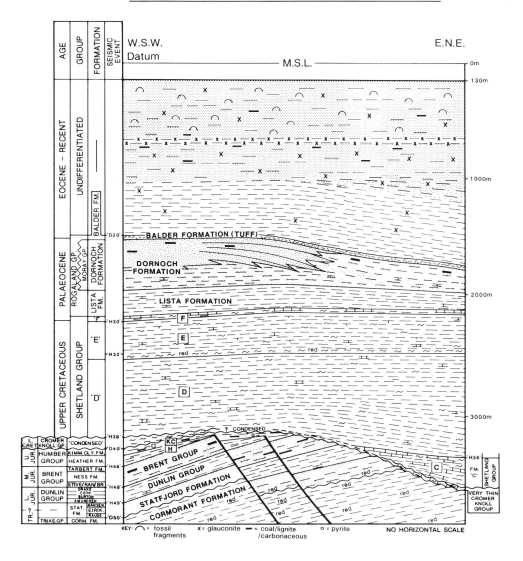

Fig. 3. Alwyn North Field lithostratigraphy.

Field stratigraphy

The deepest sediments penetrated in Alwyn North Field are the continental interbedded red shales, siltstones, and sandstones of the Cormorant Formation (Fig. 3). Age-diagnostic criteria are poor but yield Triassic and more specifically Rhaetian ages at the top of the sequence. Lithologically, the boundary between the Lower Jurassic/Latest Triassic Statfjord Formation and Triassic Cormorant Formation is not clear, particularly as variegated sand–shale sequences are also encountered in the Statfjord Formation. However, a sharp break in sonic log response and an associated strong seismic reflection can readily be seen where the shales become dominantly red brown (Vollset & Dore 1984).

The Raude, Eriksson and Nansen members of the Statfjord Formation (Deegan & Scull 1977) are all identifiable (Fig. 4). These are supplanted at Alwyn North by reservoir divisions A down to E.

Unit E forms a transitional 'overall coarsening-up' unit above the underlying Cormorant Formation. Overlying units D and C comprise interbedded sandstones, siltstones and shales deposited in an alluvial plain setting. The sandstones fluctuate between laterally extensive, strongly connected, braided channel sequences and more localized, thinner point bar sands distinguished by a more pronounced 'fining up' character. These are interspersed with flood plain deposits. This fluctuation from braided to meandering river styles is best explained by a complex interplay of sediment influx, subsidence rates and eustacy (Røe & Steel 1985). At the base of Unit B there is an abrupt change in depositional style and setting. This probably happened in response to localized tectonic activity associated with the Sinemurian transgression and resulted in the development of high-energy fan delta plain sediments consisting of coarse pebbly sandstones and only minor shale intercalations. As the transgression ensued, the fan delta plain ultimately became flooded. High energy waves eroded and reworked the underlying fluvial deposits into a series of coarsening-up retrogressive sandstone units (Unit A or Nansen Member). These are interpreted as prograding barrier sands laid down in a middle to upper shoreface environment.

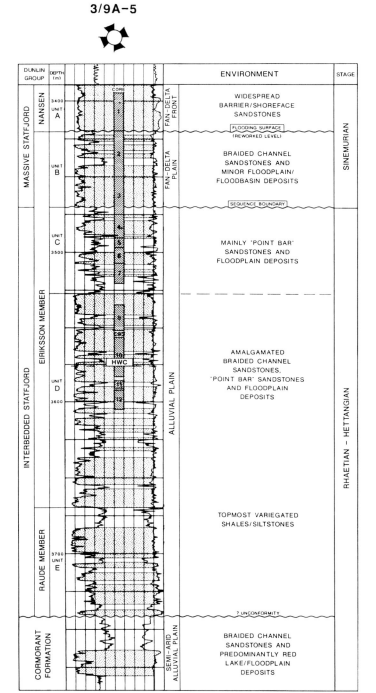

Fig. 4. 3/9a-5 Alwyn North Field, type Statfjord Formation well.

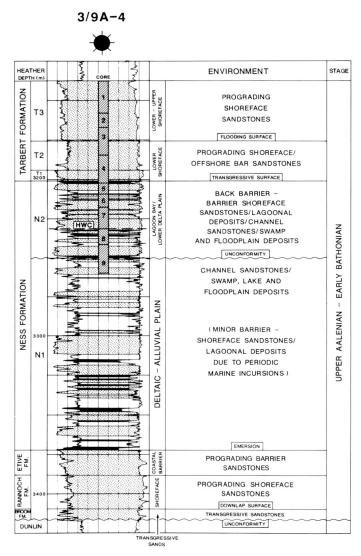

Fig. 5. 3/9a-4 Alwyn North Field, type Brent Group well.

The overlying Dunlin Group (Sinemurian–Toarcian) consists of marine shales which rest conformably on the Statfjord Formation and are a continuation of the marine deposition which commenced with the Nansen member (Unit A) of the Statfjord Formation. This Group, comprising four formations, Amundsen, Burton, Cooke and Drake (Deegan & Scull 1977), displays remarkable homogeneity in both facies and thickness. The formations are therefore easily distinguished by log character and are dominated by shales and siltstones with only minor sandstone horizons.

The Brent Group encompasses a series of sediments deposited in a wave/tide dominated delta system (Fig. 5). A fairly uniform total thickness of around 280 m is encountered whenever the full sequence has been drilled. The sand-dominated 'Lower Brent' is composed of three formations first described by Bowen (1975). At the base, the Broom Formation is interpreted as a brief transgressive lag deposit overlying the base Brent unconformity. The prograding sequence of the rest of the Brent Group began with the micaceous sands of the Rannoch Formation, which represent inner shelf to shoreface conditions (Brown et al. 1987). The continual advancement of a barrier protected coastline then led to widespread deposition of massive sands, the Etive Formation, prior to the emergence of the delta plain (Ness Formation). The Ness Formation consists of interbedded sandstones, siltstones, shales and coals typical of delta plain environments. However, several brief transgressive pulses can be identified throughout the succession, indicating periodic returns to lagoon, barrier or shoreface settings. Moreover, marine influences increase toward the top of the formation (intra-Bajocian times) and indicate the return of transgressive conditions and the retreat of the delta southwards. The Ness can therefore be split into two main units, the lower Ness which can be seen overall as a prograding sequence linked to the advance of the Brent Delta and the upper Ness which overall is a retrograding sequence linked to the retreat southwards. The two units are separated by an unconformity.

This is reflected in the reservoir layering, wherein the Ness Formation is divided into a lower Ness 1 unit followed by the overlying, more marine influenced Ness 2 possessing a higher sand/shale ratio.

The Tarbert Formation marks a full return to marine conditions. It is divided into three reservoir units, Tarbert 1, Tarbert 2, and Tarbert 3 echoing the tripartite division of the Lower Brent. The Ness/Tarbert boundary represents a major transgressive surface reflected by a thin conglomeratic lag at the base of the Tarbert 1 unit. The rest of this unit is representative of offshore bar/interbar sands. The Tarbert 2 unit mirrors the Rannoch Formation in so far as it represents a prograding sequence of laminated, micaceous sandstones, laid down under similar inner shelf to lower shoreface conditions. The remainder of the Tarbert Formation, the Tarbert 3 unit, is characterized by a series of retrogressive cycles, each following minor transgressions and usually indicated by basal pebbly lag deposits. This unit consists of essentially fine to medium grained sandstones deposited in a shoreface environment.

The marine transgression at the top of the Brent Group is continued by the conformable deposition of dark grey shales with limestone stringers. These form the Heather Formation (Bathonian–Oxfordian), the thickness of which is controlled by syntectonic sedimentation. It is seen to thin towards the crests of the tilted blocks, due in part to becoming condensed, but also to erosion associated with rotational movement. This gives a considerable variation to the thickness of the Heather Formation from a 200 m downflank close to the Ninian–Hutton fault to zero at the crests.

The Kimmeridge Clay Formation (late Oxfordian–late Ryaza-

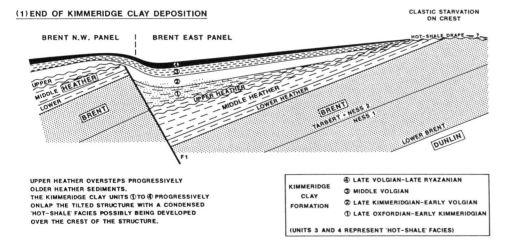

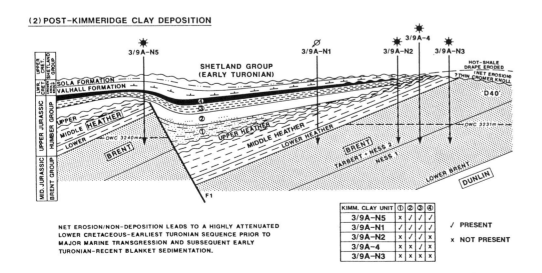

Fig. 6. Sedimentation model for the Heather and Kimmeridge Clay formations as evidenced from wells on Alwyn North Field.

nian) characterized by a high gamma-ray carbonaceous claystone facies, thins rapidly towards the crestal areas, progressively onlapping the underlying westerly dipping sediments (Fig. 6). On the more exposed blocks such as 'Brent East' and 'Brent South-West' where considerable erosion has taken place, the Kimmeridge Clay Formation is probably absent on the easterly erosion scarps and at least locally in crestal locations. The stratigraphic succession is continued by a thin blanket of marls and limestones of Lower Cretaceous to Early Turonian age draped across the Alwyn North Structure.

The remainder of the lithological succession is approximately 3000 m thick and is represented by the thick succession of Turonian to Recent claystones with siltstone and sandstone intercalations, particularly at the Palaeocene level, which accompanied the marine transgression and North Sea basin-fill that buries the pre-Cretaceous structure. Typical marker levels that are systematically logged on Alwyn North Field are: the Lower Campanian red marls, the calcareous claystones at the top of the Cretaceous, the tuffaceous claystones of the Balder Formation, and a strongly glauconitic marker bed in the Oligocene.

Geological history

The initiation of graben systems under a tensional regime in the Permo-Triassic period sets the subsequent tectonic control over the Alwyn North area. This tensional regime generated a complex fault pattern with two principal directions, a major N–S direction and a secondary ENE–WSW transverse direction. The ENE–WSW-trending faults are probably a re-activation of the much earlier Caledonian trend.

Such a fault system gives rise to the westward dipping rotated fault blocks which constitute the effective structural traps for hydrocarbons especially as their formation pre-dates any hydrocarbon migration.

For most of the Triassic period the N–S faults were particularly active and thick sedimentary sequences of mainly red-bed deposits, attributed to the Cormorant Formation, accumulated as the graben subsided. These sediments were deposited in a semi-arid continental regime subjected to periodic fluviatile/lacustrine influences. Towards the end of the Triassic, major tectonic activity of the early Cimmerian rifting resulted in increased basin subsidence and marine transgression leading to the end of continental deposition and a more humid climate. High surface run-off combined with steeper gradients created a moderately high-energy fluviatile system creating the alluvial fans, braided plains, and fan deltas of the Statfjord Formation. In the Alwyn North area these sediments were laid down on the eastern hanging wall of the active Ninian–Hutton Fault while erosion took place on the upthrown western side. As sea-level continued to rise, widespread coastal barrier sands (the Nansen member) transgressed the underlying sediments before being superseded by the shales and siltstones of the Dunlin Group.

At or about the Lower/Middle Jurassic transition, a second major episode of tectonic activity, the Mid-Cimmerian rift phase, was accompanied by widespread regression. This tectonic event appears to be synchronous with regional updoming and associated volcanic activity centred on the Outer Moray Firth. During the Middle Jurassic the Brent deltaic system prograded northwards away from the updomed area. A rapid eustatic rise in sea-level, initiated in the Bajocian stage, outstripped sediment supply, causing the delta to retreat southwards leading to the deposition of marine sands (The Tarbert Formation) in the Alwyn area.

Both lithofacies and biofacies criteria would suggest that there is a transitional boundary between the Tarbert and Heather Formations; however, during the deposition of the Heather Formation it is evident that tectonic activity increased as the fault blocks began to tilt significantly to the west. Partial erosion of the more exposed blocks as their crests were pushed above wave-base also commenced at this time. Collapse of poorly-consolidated fault scarps also took place creating locally complex sequences on the east flanks of the field. This tectonic phase may be coincident with collapse of the updomed area in the south which ensured fully marine conditions in the Alwyn North area hereafter. Another major pulse of structural tilting and subsidence at the end of the Heather Formation sedimentation initiated widespread Kimmeridge Clay Formation deposition. Late Cimmerian extensional tectonics resulted in a 'hot-shale' facies being developed towards the top of the formation, indicating (Rawson & Riley 1982), that the rate of subsidence exceeded both the rate of sea-level change and sedimentation, thus creating starved conditions and favouring the preservation of organic matter.

By the end of Kimmeridge Clay Formation deposition, major tectonic activity in the Alwyn North area had more or less ceased, although periods of structural rejuvenation are thought to have continued until the early part of the Upper Cretaceous. The early Turonian Shetland Group then blanketed the structure following major marine transgression. Subsequent Upper Cretaceous–Recent sedimentation was accompanied by gentle postrift subsidence linked to the general cooling of the upper mantle.

Geophysics

Early seismic data over Alwyn North Field was not of sufficient quality to permit clear interpretation of the structure underlying the Base Cretaceous Unconformity. Furthermore, no obvious closure at this level is visible in the Alwyn North area which explains the later exploration of Alwyn North area as compared to other parts of the Alwyn area further south (Blocks 3/14, 3/15). Instead the Base Cretaceous horizon rises progressively northwards through blocks 3/9, 3/4 and into the Brent and Statfjord Fields area. A second generation of 2-D seismic data in the mid-70s was of sufficient quality to define the structure and current fault pattern over the field area. Approximately 1000 km of seismic data were acquired on grid sizes of 1 km × 2 km and 0.5 km × 1 km. However, the size of the development programme, given the complexity of the area, and the need for better reserves estimation justified a full 3D seismic programme. Over 4000 km of 3-D seismic data were shot in two campaigns in 1980 and 1981, (bin size 25 m × 40 m) giving the necessary detail to improve definition of the structural configuration and later allowing optimization of the locations of up to 30 production wells (Fig. 7).

The several horizons relevant to the prospective Jurassic intervals can be picked over the development area with varying degrees of certainty and continuity (Fig. 8). The Base Cretaceous Unconformity is usually highly visible as a strong black peak. On the east flank of the structure, however, where the unconformity surface has been onlapped by the Turonian limestones and marls possessing high acoustic impedance contrasts, the event becomes locally confused by both constructive and destructive interferences. The strongest reflector below the Base Cretaceous Unconformity is the Top Dunlin Group, which can be mapped as a continuous event correlating with the centre of a strong black peak over most of the area. The Top Brent and Statfjord Formations are much less identifiable and are often resolved by isopaching a time interval up or down from the top Dunlin reflector.

Above the Base Cretaceous Unconformity, the varying lithologies cause corrections to be applied to the regional time/average velocity function when constructing an average velocity to the Base Cretaceous Unconformity, this being the most important component of the depth conversion. These variations result in part from the high interval velocity of the Turonian limestone sequence which onlaps the Base Cretaceous Unconformity and the prograding Palaeocene slope sequence giving an important sand-thickness variation over the northwestern part of the development area (Fig. 3).

Fig. 7. Isometric construction of the principal reservoir surfaces, Alwyn North Field.

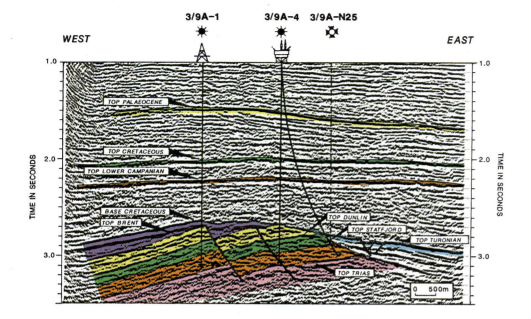

Fig. 8. West–east seismic section illustrating the principal reflectors on Alwyn North Field.

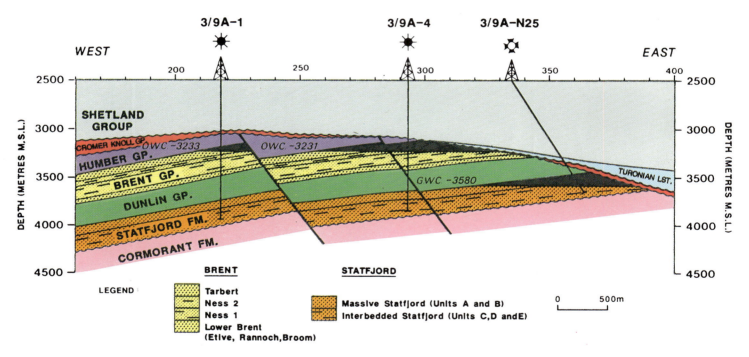

Fig. 9. West–east geological cross-section illustrating the pre-Cretaceous structure on alwyn North Field.

Trap

The predominant north–south system of normal faults downthrowing to the east, which creates the westerly dipping Middle Jurassic fault blocks, is the principal trap mechanism (Figs 8 & 9). The ENE–WSW system of cross faulting is partly responsible for splitting the area into smaller compartments. Each of the hydrocarbon accumulations, however, is limited by a combination of dip and faulting and/or truncation beneath the Base Cretaceous Unconformity.

Brent Group accumulations

Each of the accumulations is generally westerly dipping and is either truncated beneath Upper Jurassic shales or limited by the N–S faulting downthrowing to the east (Johnson & Eyssautier 1987).

To the north and south closure is either by dip or by erosion and faulting across one of the transverse ENE–WSW elements. The 'Brent East' accumulation is further split by a north–south fault downthrowing to the east with antithetic faults downthrowing to the southwest. The oil–water contact for 'Brent-East' is at 3231 m TVSS whilst that of 'Brent North-West' is at 3240 m TVSS.

The 'Brent North' accumulation is separated by a strong ENE–WSW fault from 'Brent North-West', evidenced by the stepping up of the oil–water contact by almost 100 m to 3144 m TVSS.

Statfjord Formation accumulation

This accumulation is structurally the most simple, being defined by dip to the west and erosional truncation to the east. Trapping at the northern and southern margins is effected mostly by the plunging

BRENT GROUP

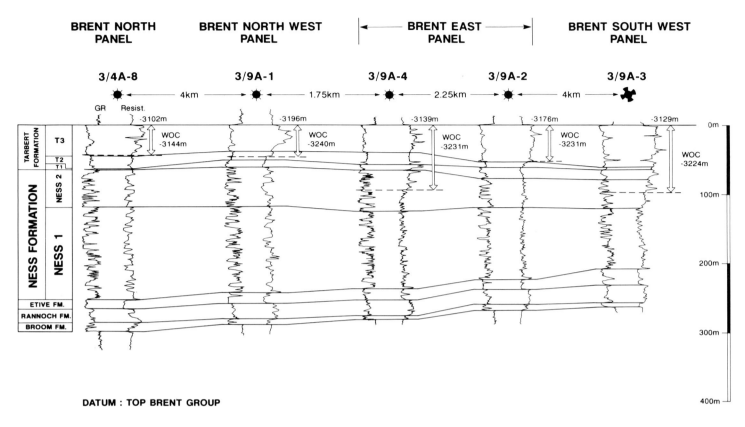

Fig. 10. Stratigraphic correlation of Brent Group wells, Alwyn North Field.

erosion surfaces, although a combination of faulting and dip again interplays. Low angle normal faulting, almost parallel to the Base Cretaceous erosion surface, has been indicated by some of the Statfjord accumulation wells' loss of reservoir section. The gas–water contact averages 3580 m TVSS. Interestingly, some wells have recorded a thin (few metres) oil rim.

Reservoir

Brent Group reservoirs

Reservoir information on the Brent Group accumulations was initially derived from the six exploration and delineation wells and has now been augmented by nearly twenty development wells. A characteristic of the Alwyn North Field is the considerably greater thickness (260–290 m) of the Brent Group Sandstones, as compared with the maximum hydrocarbon column (120 m). The hydrocarbons are thus confined almost entirely to the 'Upper Brent' (Tarbert Formation), itself further separated into Tarbert 1, 2 and 3 units and to the upper part of the 'Middle Brent' (Ness 2 Unit), (Fig. 10).

The 'Upper Brent' is a marine, shale free, shoreface complex of sandstones, laterally extensive, and displaying very good vertical communication as evidenced from pressure profiles. The basal transgressive sand, the Tarbert 1 unit is typically 6–12 m thick, and consists of fining upwards poorly-sorted medium to very coarse sandstone.

It is overlain by fine-grained well-sorted and well-laminated, micaceous sandstone, 10–22 m thick, of the Tarbert 2 unit. A massive unit, 35–60 m thick, at the top, forms the Tarbert 3 unit. It is a poorly sorted, fine to very coarse grained sandstone with fining upward cycles, containing abundant potassium feldspar and often with micaceous incursions, similar to the underlying Tarbert 2 unit. Good porosities and permeabilities exist in the Tarbert 1 and 3 units, typically 20% porosity and several hundreds of millidarcies, whilst the Tarbert 2 unit has around 16% porosity and 50–100 md permeability.

The underlying Ness Formation is a succession of delta-plain to littoral sediments: sandstones, siltstones, shales and coals. The base of each of these cyclical successions is marked by a flood surface where more marine influenced deposits typically overlie palaeosols of the deltaic deposits. Although of limited lateral extent and thus difficult to correlate, the constituent sand bodies, mainly channels, bars and splay deposits, display a high sand/shale ratio (0.6 in the Ness 2 unit contrasting with less than 0.4 in the underlying Ness 1 unit) and therefore allow a certain amount of lateral and vertical coalescence of the sand bodies. Porosities and in particular the permeabilities are variable, but on average, lower than the Tarbert units above. Average porosities in the Ness 2 range from 15% to 18%. The highest permeabilities (over 10 darcies) have been measured from the coarser lower portion of some of the Ness channel sands.

The petrography of the Brent Group Sandstones has been extensively studied from core material (Jourdan et al. 1987). The principal constituents are quite uniform, the differences being mainly due to subsequent diagenetic evolution. Apart from quartz grains, polycrystalline quartz as rock fragments and silica in the form of quartz overgrowths, the other minerals present are feldspars, micas, pyrite, organic matter and clays. The detrital clays are represented in even proportions by kaolinite and illite which are also the main diagenetic clays; illite becoming the more abundant toward the base of the hydrocarbon zone.

Three distinct phases in the diagenetic history of the Brent reservoir in the Alwyn North area have been recognized. A first

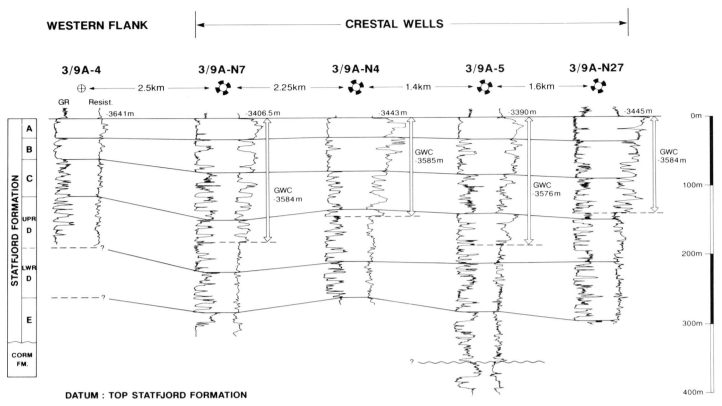

Fig. 11. Stratigraphic correlation of Statfjord Formation wells, Alwyn North Field.

kaolinitization and silicification phase induced by pore-water flow during Brent deposition and early compaction has limited influence on reservoir properties. The early silicification probably prevented porosity reduction by cementing the grain framework, which better resisted subsequent compaction. A second diagenetic stage is linked to the invasion of fresh meteoric waters into the Brent Group reservoir during the late Cimmerian tectonic phase. It has led to the development of kaolinite, both as a replacement of the leached feldspars and as pore infill. This stage is variably developed in the different units of the Brent Group, dependent on their original reservoir properties and the distance to the exposed crest of the tilted block. The Tarbert 3 unit, which had excellent original reservoir properties and was also close to the crest of the tilted block, experienced a stronger dissolution of feldspars than the underlying Tarbert 2 unit. Precipitation of kaolinite is, however, actually less developed in the Tarbert 3 unit indicating that the elements released by feldspar dissolution were drained downwards and away from this part of the reservoir. The third and most damaging stage in the diagenetic history is related to the circulation of hot acid waters expelled during the compaction of argillaceous series before and during hydrocarbon migration. The consequences of this acid water flow were the continuation of silicification and the precipitation of blocky kaolinite and fibrous illite, which by lining or bridging pore throats has a detrimental effect on permeability. It is likely that this diagenesis was inhibited by the movement of hydrocarbons into the reservoirs, but remained active in the aquifer.

Statfjord Formation reservoir

Units A and B at the top of Statfjord Formation were previously grouped together as the 'Massive Statfjord' and the underlying units as the 'Interbedded Statfjord'. Further division is possible from the twelve production wells that have now been drilled, mostly along the crest of the structure.

Unit A is a strongly correlatable 30 m thick unit, of shale-free marine sandstone possessing good reservoir chracteristics (Fig. 11). Its base is a marine transgression surface, as evidenced by a transgressive lag seen in the cored wells. Net to gross ratio is virtually unity, porosities averaging 12.5% and permeabilities averaging 80 md. Saturation exponent or 'n' values measured on core material over this interval confirm the individual nature of unit A, with a consistently low value of 1.48 for 'n' having been measured. Unit B is a 50 m thick distinct pebbly sandstone unit reflecting a braided fan-delta environment of deposition. Shale intercalations are not uncommon but a sequence boundary at the base again allows a smooth and isopachous correlation. This is further supported by the even pressure depletion of Units A and B as shown by repeat formation tester measurements. Correlation in the rapidly interbedded underlying C and D units is less certain. A higher sand to shale ratio exists in unit D as compared to unit C (0.65 versus 0.5), due to the presence of thicker amalgamated channel sands. Pressure depletion measurements are again helpful in refining the reservoir correlation. The lower half of the D unit for example is currently depleting at a much lower rate. It is only with the help of production wells drilled over the crest of the tilted block and penetrating the lower D unit in a structurally favourable position that such parts of the reservoir can be efficiently drained. Porosities in Units C and D are actually slightly higher, at around 14%, than in overlying units A and B. This is probably a result of the less efficient circulation of silica rich fluids, due to shale interbeds acting as barriers and therefore restricting the presence of authigenic silica. Permeabilities are conversely higher in units A and B than below, perhaps due to pressure solution in the latter which only minimally affects the pore volume, but adversely affects the pore throat size. The permeabilities overall are significantly lower than the Brent Group values, however, well deliverabilities are good with typical values in the order of 1.5×10^6 $Sm^3 d^{-1}$.

Source

The shales of the Kimmeridge Clay Formation are the principal hydrocarbon source rocks throughout the Viking Graben. Measured values in the Alwyn North area are 2–10% by weight for total organic carbon, a hydrogen index of between 100 and 800 mg g^{-1}, and a mean potential productivity of 8 kg per tonne. The coals of the middle Brent and the shales of the Dunlin Group also possess some reasonably good source rock qualities. In the area of 'Brent South-West' and further south on Alwyn Field it is likely that Brent and Statfjord oils may be a mixture from various source rocks. The Brent accumulations of Alwyn North Field are likely to have been sourced locally by oil moving updip from the west and south. A greater amount of diagenetic clay is found in 'Brent South-West', when compared to the main Alwyn North Field. This can be explained by a later phase of hydrocarbon migration, which allowed diagenetic processes in the present hydrocarbon leg of 'Brent South West' to continue over a longer time period. K-Ar dating of illites linked to the time of hydrocarbon migration, and fluid inclusion studies (Jourdan *et al.* 1987), enable definition of the palaeo-hydrodynamic regime in the Alwyn area. Hydrocarbon emplacement for Alwyn North is given at 60 Ma versus 45 Ma, for 'Brent South-West'. The Statfjord Formation reservoir is likely to have been charged with hydrocarbons sourced

Alwyn North Field data summary

	Brent East	Brent North–West	Brent North	Statfjord
Trap				
Type	Tilted fault block	Tilted fault block	Tilted fault block	Tilted fault block
Depth to crest	3110 mss	3115 mss	3010 mss	3340 mss
Lowest closing contour	3420 mss	3380 mss	3160 mss	3780 mss
Hydrocarbon contacts	3231 mss	3240 mss	3144 mss	3580 mss
Gas column	–	–	–	240 m
Oil column	121 m	125 m	134 m	–
Pay zone				
Formation or member	Tarbert and Ness	Tarbert and Ness	Tarbert and Ness	Nansen, Eiriksson and Raude
Age	Middle Jurassic	Middle Jurassic	Middle Jurassic	Lower Jurassic
Gross thickness	110 m	108 m	102 m	265 m
Net/gross ratio (average/range)	0.86/0.78–0.91	0.90/0.77–0.96	0.85/0.80–0.89	0.65/0.50–0.82
Cut off for n/g	$\varphi > 10\%$, Vsh $< 40\%$	$\varphi > 10\%$, Vsh $< 40\%$	$\varphi > 10\%$, Vsh $< 40\%$	$\varphi > 10\%$, Vsh $< 40\%$
Porosity (average/range)	17%/13–22%	17.7%/14–22%	17.8%/16–20%	13.5%/11–16%
Hydrocarbon saturation	81%	81%	77%	80%
Permeability (average/range)	500 md/0.06–10 000 md	500 md/0.1–6000 md	800 md/0.1–10 000 md	330 md/0.2–3000 md
Hydrocarbons				
Oil gravity	39° API	41.3–42.1° API	37.2–41.1° API	46–48° API
Oil type	Undersaturated	Undersaturated	Undersaturated	–
Gas gravity	0.68 g/cc	0.57 g/cc	0.62 g/cc	0.34 g/cc
Bubble point	255 bar at 3150 mss	330 bar at 3162 mss	300 bar at 3100 mss	–
Dew point	–	–	–	475 bar at 3500 mss
Gas/oil ratio	170 Sm^3/Sm^3	340 Sm^3/m^3	230 Sm^3/m^3	2000 Sm^3/m^3
Formation volume factor	1.55 m^3/m^3	2 m^3/m^3	1.85 m^3/m^3	1/275
Formation water				
Salinity	20 000 ppm	27 000 ppm	20 000 ppm	82 000–17 000 ppm
Resistivity	0.10 Ω m at 110°C	0.07 Ω m at 110°C	0.10 Ω m at 110°C	0.06 Ω m at 113°C to 0.20 Ω m at 121°C
Reservoir conditions				
Temperature	112°C	113°C	107°C	120°C
Initial pressure (absolute) at datum	453 bar (3231 mss)	452 bar (3240 mss)	444 bar (3144 mss)	496 bar (3580 mss)
Pressure gradient in reservoir	0.062 bar/m at 3150 mss	0.055 bar/m at 3162 mss	0.060 bar/m at 3100 mss	0.026 bar/m at 3500 mss
Field size				
Area	7.6 km^2	1.7 km^2	2 km^2	10.7 km^2
Gross rock volume	370.7 × 10^6 Sm^3	91.1 × 10^6 Sm^3	83.4 × 10^6 Sm^3	1195.8 × 10^6 Sm^3
Oil/condensate in place (surface conditions)	35.5 × 10^6 Sm^3	7.3 × 10^6 Sm^3	7.0 × 10^6 Sm^3	12.4 × 10^6 Sm^3
Gas in place (surface conditions)	6.0 × 10^9 Sm^3	2.5 × 10^9 Sm^3	1.6 × 10^9 Sm^3	24.6 × 10^9 Sm^3
Recovery factor oil/condensate	0.46	0.56	0.45	0.35
Recovery factor gas	0.45	0.56	0.45	0.70
Drive mechanism	Water injection	Water injection	Water injection	Natural depletion
Recoverable hydrocarbons				
Oil	16.3 × 10^6 Sm^3	4.1 × 10^6 Sm^3	3.2 × 10^6 Sm^3	–
NGL/condensate	–	–	–	4.3 × 10^6 Sm^3
Gas	2.8 × 10^9 Sm^3	1.4 × 10^9 Sm^3	0.7 × 10^9 Sm^3	17.2 × 10^9 Sm^3
Production				
Start up date	November 1987	November 1987	June 1988	December 1987
Development scheme	2 platforms	2 platforms	Subsea completion	2 platforms
Production rate	7950 Sm^3/d	2860 Sm^3/d	1270 Sm^3/d	5 × 10^6 Sm^3/d
Cumulative production to 31/03/90	6112 × 10^3 Sm^3	1309 × 10^3 Sm^3	842 × 10^3 Sm^3	4.4 × 10^9 Sm^3

from the trough of the Viking Graben to the east, where the Kimmeridge Clay formation has been rapidly and deeply buried.

Hydrocarbons

Brent Group accumulations

Slight variations occur in each of the Brent Group accumulations, probably due to different rates of hydrocarbon segregation occurring in each compartment. This is most marked when comparing the 'Brent South-West' accumulation with the other accumulations. A common fluid composition exists at the base of the hydrocarbon columns and is confirmed by geochemical fingerprinting of the oils. In 'Brent South-West' this evolves from an undersaturated oil (40° API, GOR 300 m^3/m^3) at the hydrocarbon contact, to an undersaturated gas condensate (48° API, GOR 1000 m^3/m^3) at the top of the reservoir. Since the fluid is undersaturated throughout no real gas–oil contact exists. Studies of this fluid at or close to critical conditions have been fully described, (Neveux & Sakthikumar 1986). Otherwise, the Brent reservoirs contain low sulphur, highly undersaturated oils, with high GORs: 150–350 m^3/m^3, for 39° to 42° API oil gravity. Oil viscosities are low, around 0.25 centipoise. A consistent pressure regime gives initial reservoir pressure of 450 bars at 3150 m TVSS with oil gradients ranging from 0.55 to 0.63 g cc^{-1}. Measured water salinities range from 15 000 to 28 000 ppm NaCl.

Statfjord Formation accumulation

The Statfjord Formation reservoir contains gas with condensate at an initial GOR of 2000 m^3/m^3 and a condensate gravity of 48° API. Initial reservoir pressure is 499 bars at 3580 m TVSS. Water salinities in the Statfjord Formation appear to vary from 15 000 to 80 000 ppm NaCl.

Reserves

Fortunately for both the Brent Group and Statfjord Formation accumulations, the better reservoir characteristics are located at the top of the limited hydrocarbon columns, thereby explaining the rather high expected recoveries.

In the case of Brent reservoirs, this situation, combined with a favourable mobility ratio, should give recoveries between 40% and 55% with the use of water injection. Due to the presure maintenance, no artificial lift is required. Natural depletion in the Statfjord reservoir is expected to achieve 70% recovery for the gas and 40% for the condensate.

Initial recoverable reserves (1988 figures) for the three Brent Group accumulations of Alwyn North Field and the Statfjord Formation accumulation amount to 176 MMBBL of oil including condensate, and 22 × 10^9 Sm^3 for gas. The amount of production as of end March 1989 is around 16% of the initial gas reserves and 19% of initial oil reserves. The main production period should yield a gas plateau of 6 years from the beginning of production at an average daily rate of 7.5 million cubic metres, and an oil plus condensate plateau of 4 years at 90 000 barrels of oil and condensate per day. Recoverable reserves for 'Brent Central' and 'Brent South-West' are expected to yield a further 34 × MMBBL of oil and 5.7 × 10^9 Sm^3 of gas, which will be produced through sub-sea completions to the existing process and transport facilities.

The authors wish to thank Total Oil Marine plc and Elf UK plc for permission to publish this report. Whilst the views and opinions expressed here are the responsibility of the authors they may not always necessarily reflect the standpoint of the operator, TOM or partner, ELF UK.

In particular, thanks are offered to all of those in the Petroleum Development Group within the Alwyn Division of Total Oil Marine and especially to reservoir geologists, S. Rae and A. Lepvraud for their contributions.

For a complex field such as Alwyn North Field the gestation period from initial discovery, through delineation, project, and on to development and operating phase is inevitably a long one. Many people therefore have been associated in bringing Alwyn North Field to fruition and are to be thanked. Special thanks, however, are required for the Total geological laboratory based in Bordeaux from where A. Jourdan has made his contributions on diagenetic studies. Similarly the contribution on all reservoir matters from G. Bousquet has been invaluable.

References

BOWEN, J. M. 1975. The Brent Oilfield, *In*: WOODLAND, A. W. (ed.) *Petroleum and the Continental Shelf of North-West Europe*. Blackwell Scientific Publications, Oxford, 353–361.

BROWN, S., RICHARDS, P. C. & THOMSON, A. R. 1987. Patterns in the deposition of the Brent Group (Middle Jurassic) U.K. North Sea, *In* BROOKS, J. & GLENNIE, K. (eds) *Petroleum Geology of North West Europe*, **2**, Graham & Trotman, 899–914.

DEEGAN, C. E. & SCULL, B. J. 1977. *A standard lithostratigraphic nomenclature for the Central and Northern North Sea*, Report of the Institute of Geological Sciences 77/25; Bulletin, Norwegian Petroleum Directorate 1.

JOHNSON, A. & EYSSAUTIER, M. 1987. Alwyn North Field and its regional geological context, *In* BROOKS, J. & GLENNIE, K. (eds) *Petroleum Geology of North West Europe*, **2**, Graham & Trotman, 963–970.

JOURDAN, A., THOMAS, M., BREVART, O., ROBSON, P., SOMMER, F. & SULLIVAN, M. 1987. Diagenesis as the control of Brent sandstone reservoir properties in the Greater Alwyn area (East Shetland Basin), *In* BROOKS, J. & GLENNIE, K. (eds) *Petroleum Geology of North West Europe*, **2**, Graham & Trotman, 951–961.

NEVEUX, A. R. & SAKTHIKUMAR, S. 1986. Delineation and evaluation of a North sea reservoir containing near critical fluids. *SPE European Petroleum Conference, London*, paper 15856.

RAWSON, P. F. and RILEY, L. A. 1982. Late Jurassic–Early Cretaceous Events and the 'Late Cimmerian Unconformity' in the North Sea Area, Bulletin American Association Petroleum Geologists 66, 2628–2648.

RØE, S. L. and STEEL, R. 1985. Sedimentation, sea-level rise and tectonics at the Triassic–Jurassic boundary (Statfjord Formation), Tampen Spur, Northern North Sea, Journal Petroleum Geology 8(2), 163–186.

VOLLSET, J. & DORÉ, A. G. 1984. *A revised Triassic and Jurassic lithostratigraphic nomenclature for the Norwegian North Sea*. Norwegian Petroleum Directorate, Bulletin No. 3.

The Beryl Field, Block 9/13, UK North Sea

C. A. KNUTSON & I. C. MUNRO

Mobil North Sea Ltd, 3 Clements Inn, London, UK

Abstract: The Beryl Field, the sixth largest oil field in the UK sector of the North Sea, is located within Block 9/13 in the west-central part of the Viking Graben. The block was awarded in 1971 to a Mobil operated partnership and the 9/13-1 discovery well was drilled in 1972. The Beryl A platform was emplaced in 1975 and the Beryl B platform in 1983. To date, ninety-five wells have been drilled in the field, and drilling activity is anticipated into the mid-1990s.

Commercial hydrocarbons occur in sandstone reservoirs ranging in age from Upper Triassic to Upper Jurassic. Structurally, the field consists of a NNE orientated horst in the Beryl A area and westward tilted fault blocks in the Beryl B area. The area is highly faulted and complicated by two major and four minor unconformities. The seal is provided by Upper Jurassic shales and Upper Cretaceous marls.

There are three prospective sedimentary sections in the Beryl Field ranked in importance as follows: the Middle Jurassic coastal deltaic sediments, the Upper Triassic to Lower Jurassic continental and marine sediments, and the Upper Jurassic turbidites. The total ultimate recovery of the field is about 800 MMBBL oil and 1.6 TCF gas. As of December 1989, the field has produced nearly 430 MMBBL oil (primarily from the Middle Jurassic Beryl Formation), or about 50% of the ultimate recovery. Gas sales are scheduled to begin in the early 1990s. Oil and gas production is forecast until licence expiration in 2018.

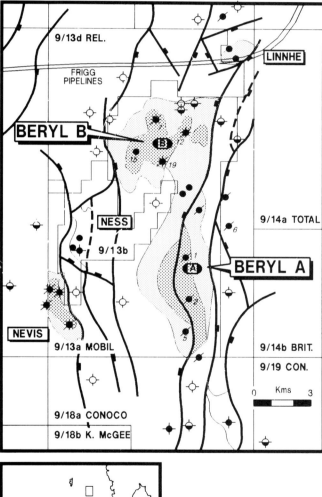

Fig. 1. Index map of the Beryl Field within Block 9/13.

The Beryl Field is located 215 miles northeast of Aberdeen, about 7 miles from the United Kingdom–Norwegian boundary. The field lies within Block 9/13 and covers an area of approximately 12 000 acres in water depths ranging from 350–400 ft. Block 9/13 contains several hydrocarbon-bearing structures, of which the Beryl Field is the largest (Fig. 1). The field is subdivided into two producing areas: the Beryl Alpha area which includes the initial discovery well, and the Beryl Bravo area located to the north. The estimate of oil originally in place is 1400 MMBBL for Beryl A and 700 MMBBL for Beryl B. The field has combined gas in place of 2.8 TCF, consisting primarily of solution gas. Hydrocarbon accumulations occur in six reservoir horizons ranging in age from Upper Triassic to Upper Jurassic. The Middle Jurassic (Bathonian to Callovian) age Beryl Formation is the main reservoir unit and contains 78% of the total ultimate recovery.

The field was named after Beryl Solomon, the wife of Charles Solomon, who was president of Mobil Europe in 1972 when the field was discovered. The satellite fields in Block 9/13 (Nevis, Ness and Linnhe) are named after Scottish lochs.

History

Block 9/13 was awarded in the United Kingdom Fourth Offshore Licensing Round as Licence P.139 in December 1971. Following mandatory relinquishments, the 9/13 Block is now sub-divided into four partial blocks designated 9/13a to 9/13d or Licences P.139 (9/13a), P.337 (9/13b and c) and P.629 (9/13d). Blocks 9/13a, b and c are held by the same Mobil operated partnership with current participating interests as follows:

Mobil North Sea Ltd (operator)	45%
Amerada Hess Ltd	20%
Enterprise Oil plc	20%
North Sea Ltd	10%
Oesterreichische Mineraloel Verwaltung	5%

In Block 9/13d the partnership is similar, but OMV's 5% interest is held by Mobil for a total operator's interest of 50%.

The block was acquired to test a large structural closure at the Palaeocene level and, secondarily, pre-Cretaceous sediments below a major unconformity. Well 9/13-1, spudded in September 1972, encountered abundant reservoir quality sandstones with minor traces of hydrocarbons within a 3000 ft Palaeocene section. Beneath the base Cretaceous unconformity, the well drilled to −10 300 ft TVSS and penetrated 350 ft of oil-bearing Jurassic sandstones; no oil–water contact was encountered. The Jurassic section flowed

3400 BOPD of 30° API oil through a 48/64 in choke. Well 9/13-2 was drilled 1.5 miles south of this location in April 1973 and encountered a 460 ft Triassic pay section with an oil–water contact at $-10\,770$ ft TVSS. The 9/13-2 tested oil similar in gravity to the discovery well at rates of 5500 BPOD. The Beryl Field was declared commercial and construction of a 40-slot 'Condeep' platform began in 1973. Two further delineation wells were drilled during 1973 to 1974. Well 9/13-5, situated 2 miles south of 9/13-2, and well 9/13-6, drilled 2 miles north of the 9/13-1, penetrated combined Jurassic and Triassic net oil sections of 359 and 344 ft, respectively. The Beryl A platform was emplaced in July 1975 and oil production commenced in July 1976.

A northern extension of the Beryl accumulation was discovered in April 1975 by well 9/13-7. The well penetrated Lower and Middle Jurassic sandstones with 125 net gas pay, 225 ft net oil pay and an oil–water contact of $-11\,304$ ft TVSS. Wells 9/13-12, 15 and 19 were drilled to further appraise this area and each encountered Middle Jurassic hydrocarbon sections. The 9/13-15 well was tied back to the Beryl A platform and began producing in January 1979 at a rate of 14 500 BOPD. Construction of the Beryl B platform, a 21-slot tubular steel jacket assembly, was initiated in March 1981. The platform was installed in June 1983 over a subsea drilling template from which six wells had been pre-drilled. First oil production was achieved in July 1984.

To date, 47 platform wells and 6 subsea producers have been drilled within the Beryl A area; and 25 platform wells, 3 subsea water injectors and one subsea producer in the Beryl B area. An additional 12 subsea wells have been drilled in the Beryl area as expendable appraisal wells. Cumulative oil production as of December 1989 was 330 MMBBL for Beryl A and 100 MMBBL for Beryl B.

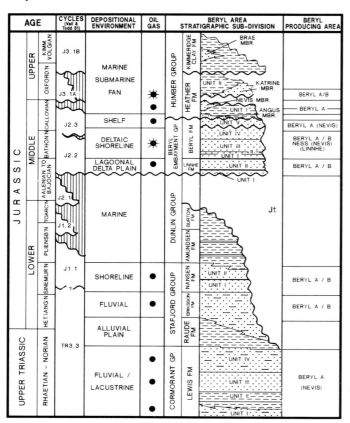

Fig. 2. Simplified lithostratigraphic column for the Beryl Field.

Field stratigraphy

The Beryl Field is situated in the west-central part of the Viking Graben within a westward re-entrant of the East Shetland Platform known as the Beryl Embayment. The strata penetrated in the embayment range from Carboniferous to Tertiary in age. A stratigraphic column for the Upper Triassic to Upper Jurassic sequence in the Beryl Field is given in Fig. 2. The lithostratigraphic scheme utilized by Mobil is modified from that proposed by Deegan & Scull (1977). Terminology introduced informally by Mobil for the Middle Jurassic section in the Beryl Embayment differs from standard North Sea nomenclature. In addition, the Statfjord Formation (as defined by Deegan & Scull 1977) has been elevated because of its regional continuity to Group status and the Raude, Eiriksson and Nansen Members to Formation status. Additionally the sediments of the Cormorant Group are known informally as the Lewis Formation.

The reservoir stratigraphy of the Beryl Field can be subdivided into three sedimentary packages bounded by unconformities. The oldest package consists of Upper Triassic to Lower Jurassic strata (Cormorant, Statfjord and Dunlin Groups) and records the transition from continental to marine conditions. Uplift in the early Middle Jurassic interrupted this sequence and resulted in a regional depositional hiatus. The second sedimentary package consists of Middle Jurassic coastal and deltaic sequences (Beryl Embayment Group) which onlap the older continental to marine sequence. The deltaic Middle Jurassic sedimentation was terminated in the late Bathonian to early Callovian by fault block rotation and marine transgression. Continued subsidence throughout the late Jurassic resulted in deposition of a thick marine sequence (Humber Group). Major Cimmerian tectonism in the late Jurassic terminated Jurassic sedimentation.

Geophysics

The primary seismic database for the Beryl Field comprises two 3-D seismic surveys acquired by GSI in 1979 and 1982. The 1979 Beryl B survey consists of 1430 recorded kilometres of data shot in an 11 × 9 km area with a 75 m line-spacing. The adjacent 1982 Beryl A seismic survey centred on the platform, covers an area of 11 × 10 km, totalling 1310 km, with a similar line-spacing. Based on the results of this later survey, a major reprocessing of the Beryl B data was initiated mid-1984.

The seismic data quality below the Base Cretaceous Unconformity is fair to poor because of a high density of faulting, lateral velocity variations and a lack of acoustic impedance contrasts. Acquisition factors contributing to the poor data quality include coarse (75 m) line-spacing, insufficient (2400 m) maximum receiver offset, field noise resulting from poor weather, and variable bin fold due to navigational inaccuracies. Seismic interpretation of the reservoir sections is complicated by lateral facies and thickness variations within specific horizons, and by the Jurassic unconformities. With the exception of the Base Cretaceous horizon, seismic interpretation throughout the stratigraphic section, particularly on the crestal parts of the field, is heavily dependent on well data.

A block wide 3-D survey is being acquired in 1990 to take advantage of recent innovations in acquisition and processing technology. The new survey will extend 4 km beyond the block boundaries and will be obtained using a dual source/dual streamer technique. This will provide 30 fold, 12.5 m CDP spacing × 25 m line-spacing, which is three times the line density of the older surveys. Preliminary results of a test program show marked improvement over existing data sets.

Trap

The Beryl Field has an overall length of 9 miles and a width which varies from about 3 miles in the Beryl A area to 5 miles in the Beryl B area. Hydrocarbons occur in six pre-Cretaceous reservoir horizons ranging in age from Upper Triassic to Upper Jurassic. A seal is provided by Upper Cretaceous marls and limestones above the

Base Cretaceous Unconformity surface. In areas where the Base Cretaceous Unconformity is less erosive, thick sections of Upper Jurassic shales contribute to the seal.

The present day structure of the reservoir horizons consists of westward tilted fault blocks in the Beryl B area and a NNE oriented horst in the Beryl A area (Figs 3, 4 and 5). Current interpretations suggest, however, that the present-day structures developed relatively late and that the structural configuration which controlled reservoir distribution was somewhat different. The early Jurassic structure of the area is thought to have been dominated by one or more large west-tilted fault blocks with the primary bounding fault near the Block 9/13 eastern boundary. Footwall uplift resulted in sedimentary thinning and crestal erosion especially east of the Beryl A platform. In the middle Jurassic, antithetic faulting, primarily in the Beryl A area, disrupted this simple west-tilted fault block scenario and initiated graben development west of the Beryl A platform where thick middle Jurassic deltaics were deposited. Sedimentary thinning to the east of the platform was probably still important. In the late Jurassic, structural collapse along the primary/bounding fault resulted in the westward migration of the crest toward its present location. Subsidence of the basin east of Beryl in the Cretaceous and Tertiary contributed to the inversion of the original crest and accentuated the expression of the present Jurassic–Triassic horst.

The structural elements of the Beryl Field result from the complex interplay of three fault trends: NE, NW and N. The structural interplay is thought to be related to three or more phases of extension. The first phase of extension is not clearly defined but is thought to have occurred in the Upper Triassic and probably rejuvenated NE and NW elements of the Caledonian and Tornquist

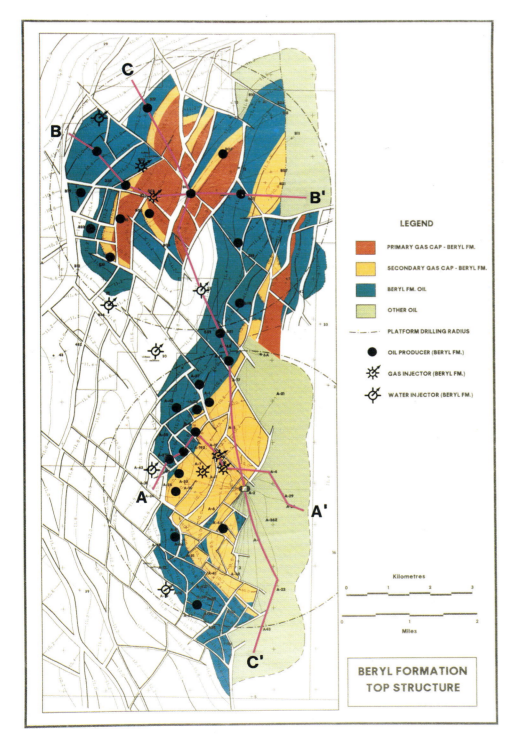

Fig. 3. Simplified top structure map of the Beryl Formation with lines of section marked.

trends. The second extensional phase, primarily in a northeastwards orientation, occurred during the Bajocian to Bathonian and resulted in well defined NW and NE structural trends. The third phase, primarily in an eastwards orientation, accentuated the NW and NE trends and introduced a compensating N structural trend. The third phase of extension occurred in the Callovian to Volgian and resulted in fault block rotation with footwall uplift and local crestal erosion.

The structural development of the Beryl area significantly influenced reservoir distribution. Most critically, the prolific Middle Jurassic reservoirs have been significantly eroded on and to the east of the present Beryl horst. Current production, therefore, is focused to the west of the structural high. The Middle Jurassic deltaic sediments contain three general hydrocarbon accumulations: one undersaturated and two with a primary gas cap (Figs 3 and 6). The variability of gas content is thought to reflect areally dissimilar hydrocarbon compositions due to diverse mixing during at least three episodes of hydrocarbon migration. The undersaturated accumulation extends along a 9 mile length from the southern Beryl A area to the eastern fault block of the Beryl B area. This area has an oil column of 1950 ft, ranging from −9600 ft TVSS on the crest of the Beryl horst to −11 550 ft TVSS at the oil–water contact. The structure spills to the south and appears to be filled to the mapped spill-point. The main saturated accumulation exists in the central and western areas of Beryl B. This area has a maximum hydrocarbon column of 1150 ft ranging from −10 400 ft TVSS at the crest to −11 550 ft TVSS at the deepest oil–water contact. Different oil–water contacts, ranging from −11 300 to 11 550 ft TVSS, are observed in this area and appear to be related to various structural spill-points. Different initial gas–oil contacts, ranging from −10 900 to −11 000 ft TVSS, are also observed. A second saturated accumulation exists along the east flank of the Beryl B area and is at present being appraised. This area is thought to have multiple oil–water contacts ranging from −11 425 ft TVSS to possibly deeper than −12 000 ft TVSS. The recently drilled reservoirs have virgin pressures and a primary gas–oil contact of −10 970 ft TVSS.

The Upper Triassic to Lower Jurassic continental and marine sediments are secondary reservoirs and are confined primarily to Beryl A and the eastern flank of Beryl B (Figs 4 and 5). The several accumulations have oil–water contacts varying from −10 500 to −12 250 ft TVSS. The contacts are thought to reflect different spill-points. Generally, the shallower contacts are associated with crestal accumulations, and the deeper with flank accumulations. All of these hydrocarbon accumulations are undersaturated. The thickest Lower Jurassic hydrocarbon column penetrated to date is 1350 ft, ranging from an estimated −10 700 ft TVSS at the crest to an oil–water contact of −11 550 ft TVSS.

The Upper Jurassic turbidites constitute a minor play and provide three separate hydrocarbon accumulations. Two accumulations occur in the far west of the Beryl B area where turbidites derived locally from the relatively high fault block to the west contain oil and gas. The first accumulation has an estimated hydrocrbon column of 400 ft, with the crest estimated at −10 400 ft TVSS and the gas–oil and oil–water contacts measured at −10 750 and −10 800 ft TVSS, respectively. The second accumulation has an oil column of at least 400 ft, ranging from an interpreted −10 400 ft TVSS crestal position to a lowest known oil of −10 800 ft TVSS. A third hydrocarbon accumulation occurs west of the Beryl A area (Fig. 4) and has an oil column of 1250 ft,

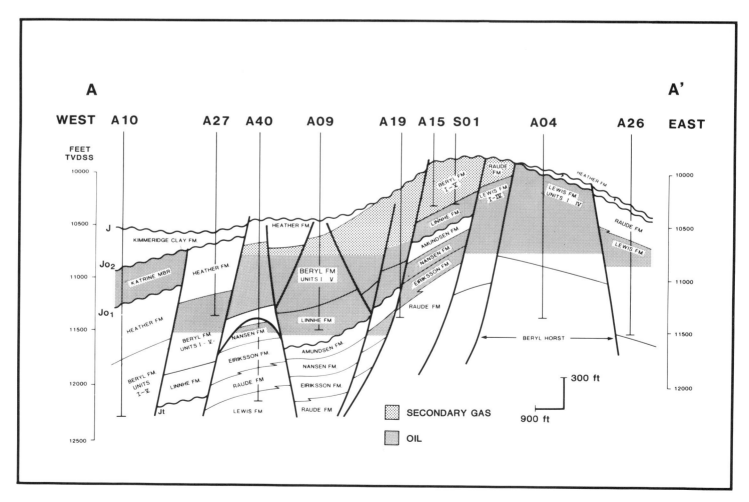

Fig. 4. Generalized east–west cross-section (A–A′) through the Beryl A area.

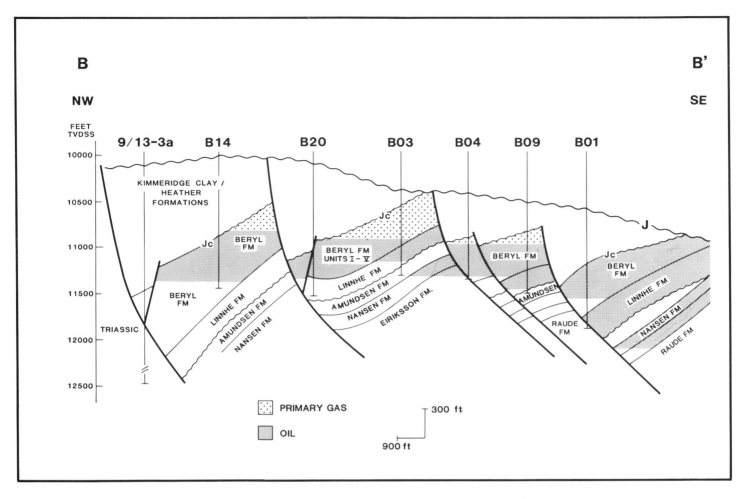

Fig. 5. Generalized east–west cross-section (B–B′) through the Beryl B area.

ranging from −9950 ft TVSS at the erosional limit on the crest to −11200 ft TVSS at the oil–water contact. The sandstones in this turbidite package are thought to have been sourced from the Beryl horst.

Reservoirs

The six reservoir formations in the Beryl Field include the Lewis Formation of the Cormorant Group, the Eiriksson and Nansen Formations of the Statfjord Group, the Linnhe and Beryl Formations of the Beryl Embayment Group, and the Katrine Member of the Heather Formation. The Middle Jurassic age Beryl Formation contains 78% of the total ultimate recovery of the field and is in a relatively mature stage of development. The Lower Jurassic age Statfjord Group and Triassic age Cormorant Group contain an additional 9% and 7% of ultimate recovery, respectively, and are at a less mature stage of development.

Cormorant Group

The informally defined Lewis Formation of the Cormorant Group has an average thickness of 1000 ft and is subdivided into four lithostratigraphic units (Units I–IV). The formation comprises a continental sequence of sandstones, siltstones, shales and caliche horizons. Facies analyses indicate the interaction of alluvial fans with a lacustrine environment in a semi-arid climate. Sheet flood sandstones, lacustrine deltaic sediments, fluvial sandstones and lacustrine shoreline facies are common.

Lewis Formation Unit I attains an average thickness of 400 ft and is interpreted as a series of small deltas prograding into a lake. This interpretation is inferred from the association of the overlying Lewis Formation Unit II, which is a 50 ft thick lacustrine shale. Evidence of calichification in the deltaic sediments suggests periods of subaerial exposure. The sandstones comprise coarsening upward sequences and are generally arkosic, grey to brown in colour. Lewis Formation Unit I has net/gross ratios ranging from 0.3 to 0.6, average porosities from 13–15% and permeabilities from 0.1–100 md. Sand discontinuity is thought to limit the productivity of this reservoir.

Lewis Formation Unit III is a relatively high quality Triassic reservoir. The unit (see Data Summary) is a laterally consistent, 100 ft to 150 ft thick, massive sandstone. The depositional environment is interpreted to be sheet-flood dominated fan deltas transgressing lacustrine mudstones. Sediment transport direction is varied but thought to be generally northwestwardly towards the main axis of the lake. The sandstones are heavily bioturbated, generally yellowish brown, fine- to medium-grained and well sorted. Intermittent calcrete horizons are relatively continuous and influence reservoir management. Lewis Formation Unit III demonstrates reservoir continuity between wells and good productivity.

Lewis Formation Unit IV is a reservoir of variable quality. The sequence attains an average thickness of 375 ft and represents a lacustrine delta dominated by stream channels with a progressively declining coarse sediment supply. Locally, the base of the unit contains some good quality massive channel sandstones up to 200 ft thick, but the sequence generally fines upwards and shales out to the northwest. The sandstones are grey and brown, locally greenish and typically medium- to fine-grained. Lewis Formation Unit IV has net/gross ratios ranging from 0.4 to 0.6, average porosities from 13–16% and permeabilities from 0.1–500 md. Sand continuity is less well developed than in Lewis Formation Unit III, but pressure

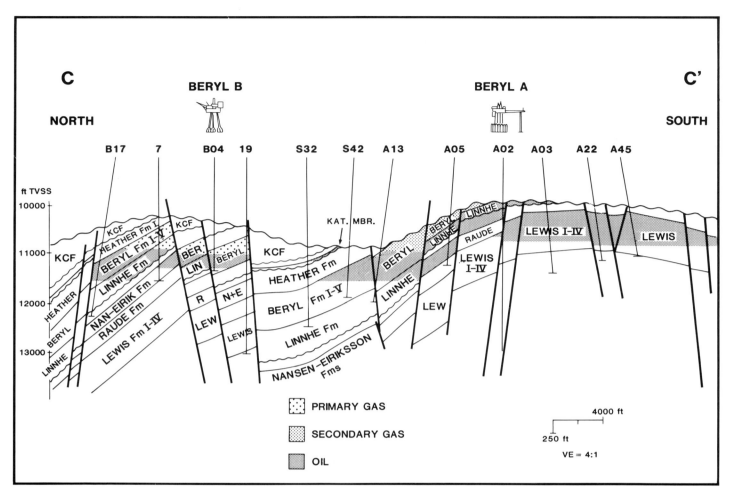

Fig. 6. Generalized north–south cross-section (C–C') through the Beryl Field.

information indicates reasonable reservoir continuity between wells. A water injection programme was initiated third quarter 1989 to maximize recovery from Lewis Formation Units III and IV.

Statfjord Group

The Statfjord Group is late Triassic to early Jurassic in age and consists of the Raude, Eiriksson and Nansen Formations. The sequence represents transgression over the Upper Triassic alluvial plain by a meandering fluvial system and, eventually, marine shoreline facies. The marine transgression was related to basinal subsidence coupled with eustatic sea level rise and a change in climate. The Rhaetian age Raude Formation is a non-reservoir unit containing interbedded siltstones (rare sandstones) and shales with a distinctive reddish colour. Facies analyses indicate that the deposits of the Raude Formation constitute a silty alluvial plain.

A major climatic change from semi-arid conditions to humid conditions appears to have occurred at the end of the Triassic (Manspeizer 1982). Coincident with the onset of more humid conditions was the deposition of the more sand-rich Eiriksson Formation. The Eiriksson Formation is Hettangian in age and varies in thickness from 70 ft to 200 ft. The formation comprises interbedded sandstones, siltstones and shales, the latter associated with localized carbonaceous horizons. Sedimentological studies indicate the presence of levee deposits, fluvial deposits and crevasse splays. The Eiriksson Formation is interpreted as a flood plain system with low-sinuosity fluvial channels. The sandstones are white or grey, generally medium grained and moderately sorted. The net/gross ratios range from 0.4 to 0.7, average porosities from 14–17% and permeabilities from 0.5–1000 md.

The Sinemurian age Nansen Formation is a uniform, good quality reservoir averaging 180 ft in thickness. The sandstone was deposited in a shallow marine environment and consists of trangressive marine shoreline deposits, shallow marine bar deposition and reworked subaqueous distributary channels. Typically the sandstones are white or grey to pale yellow in colour, medium-grained and moderately sorted. The Nansen Formation generally overlies the Eiriksson Formation with only a minor break in vertical sand continuity. In several areas, however, there is evidence that the formation boundary appears to be a permeability barrier and the horizons have different oil–water contacts. In other areas, water injection pathways appear to be unaffected by the formation boundary.

Dunlin Group

Conformably overlying the sediments of the Statfjord Group and indicating continued subsidence are the marine mudstones of the Dunlin Group. The Amundsen Formation at the base of the Dunlin Group consists of shales with some interbedded siltstones and localized sandstones. These sandstones represent periodic storm and bar apron deposits. One well has been successfully completed in the interbedded sandstones, but the reservoir quality is very low.

Beryl Embayment Group

Middle Cimmerian (Toarcian) events created a depositional hiatus (Jt Unconformity) which is associated with crestal erosion (50 to 200 ft). Overlying the Jt Unconformity and onlapping the Dunlin

Group are Middle Jurassic age sediments informally defined by Mobil as the Beryl Embayment Group. The Bajocian age Linnhe Formation is an interbedded sand, shale and coal sequence whereas the Bathonian to early Callovian age Beryl Formation is a sand-rich deltaic sequence. A late Bathonian to early Callovian regional sea-level rise (Jb2 event) ended delta progradation and resulted in marine transgression.

Facies analyses of the Linnhe Formation indicate the presence of distributary channels, mouth bar sandstones, levee deposits, crevasse splays, major coastal swamp deposits, lagoonal mudstones and shoreline sandstones. The Linnhe Formation represents coastal floodplain, mesotidal and delta plain deposits in a back barrier setting. Localized syndepositional fault movements result in a variation of formation thickness from 150 ft to 700 ft (see data summary). Sandstones are generally interbedded with shales but individual units achieve massive thicknesses of up to 300 ft. Typically, they are white or grey to brown in colour, fine- to course-grained and highly carbonaceous. To date, Linnhe Formation oil production has been primarily limited to the barrier bar system in the northeastern Beryl B area. Production performance and reservoir pressures from the two producing wells in this area indicate reasonable sand continuity and pressure communication.

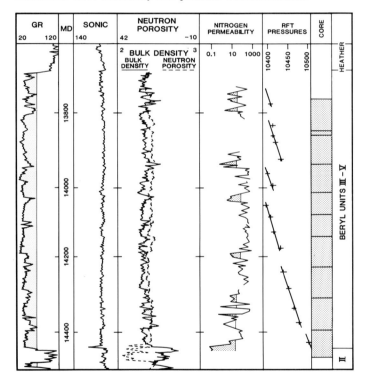

Fig. 7. Beryl Formation wireline log section with RFT pressure differentials associated with vertical permeability restrictions.

The Beryl Formation, the upper formation of the Beryl Embayment Group, is the primary reservoir in the Beryl Field (see data summary). The sedimentary facies reflect an estuarine deltaic system which prograded northeastwards and interacted with shallow marine wave and tidal processes. Lithostratigraphically the Beryl Formation is divided into five units (Units I through V) which, from a reservoir management viewpoint, may act independently. Beryl Formation Unit II is a 50 ft shale which forms an effective permeability restriction and results in Unit I being produced separately from Units III–V.

The Beryl Formation Unit I is a variable reservoir, 100 ft to 200 ft thick, with a net/gross ratio ranging from 0.15 to 0.95. Depositionally, the unit is similar to the Linnhe Formation and represents sedimentation primarily in a back barrier tidal flat system. Facies analyses indicate lagoonal mouthbars, lagoonal mudstones, distributary channels and shoreface sandstones. The primary barrier bar system is in the northeastern Beryl B area, from where most of the production has been to date. Sand continuity and pressure communication in this area is good, as indicated by the favourable response to water injection.

Beryl Formation Units III–V form a relatively continuous reservoir section and provide the majority of current production in the Beryl Field. The sequence varies in thickness from 150 ft on the crest of the Beryl horst to 800 ft on the flanks as a result of syndepositional faulting. Typically the sandstones are white or grey to buff in colour, fine- to medium-grained, moderately to well sorted and carbonaceous. Facies analyses of Beryl Formation Units III and IV indicate that braided fluvial and deltaic sequences predominate in the Beryl A area, while mixed deltaic and shallow marine sequences occur in the Beryl B area. Beryl Formation Unit V sediments represent ubiquitous marine sandstone deposition due to a regional sea-level rise and marine transgression denoted by the Jb2 event. Overall, the reservoir quality is excellent but production performance and pressure histories indicate some complications due to vertical permeability restrictions (C. Knutson et al. 1989). Thin interbedded coals, siltstones or shales can restrict vertical permeability and cause significant differential depletion. Core permeabilities and detailed RFT pressure logging indicate that even in apparently massive sandstone sequences, vertical heterogeneities support differential pressures (Fig. 7). In the Beryl B area, marine carbonate cements in Beryl Formation Unit III, interbedded shales in Beryl Formation Unit IV, and lateral variations from sandstone to siltstone in Beryl Formation Unit V, result in vertical heterogeneities and independent reservoir management for each unit.

Humber Group

The Beryl Formation sandstones were transgressed by the marine shales of the Upper Jurassic Humber Group beginning in Callovian time. Footwall uplift and fault block rotation caused local erosion and the deposition of turbidites (the Nevis and Katrine Members of the Heather Formation and the Brae Member of the Kimmeridge Clay Formation) within the thick marine shale section. Most of the turbidites are limited in extent and have poor reservoir quality. The Oxfordian age Katrine Member is a better quality reservoir and contains sandstone sequences up to 300 ft thick (see data summary). The thickest accumulations of the Oxfordian turbidites occur in the northwestern Beryl B area and west central Beryl A area. The Beryl B turbidites are sourced from the uplifted fault block to the west and consist of sequences up to 180 ft thick of white or grey to buff, generally fine-grained sandstone. The net/gross ratio is as high as 0.85 but the sandstones thin abruptly and generally pass to siltstones and shales within two miles. The Beryl A turbidites are thought to be sourced from the Beryl horst and are up to 300 ft thick with a maximum net/gross ratio of 0.70. Syndepositional faulting, sedimentary onlap and stratigraphic thinning result in rapid variations of reservoir thickness. Production performance indicates that the syndepositional faulting affects reservoir continuity.

Source

The geochemical properties of the oil from the Beryl Field suggest that the organic rich marine shales of the Kimmeridge Clay Formation are the primary source. Slight variations in the geochemical properties of the oils from the Beryl Field indicate that there are three basic oil types. The oils have a sulphur content ranging from 0.15–0.65%, a carbon isotopic ratio varying from −29.8 to −28.3 per mil, a pristane/phytane ratio ranging from 1.16 to 1.59 and a saturated/aromatic ratio between 1.0 and 2.0. The geochemical variations are thought to correspond to different organic facies and maturation histories within the Kimmeridge Clay Formation. One of these oil types appears to be derived from a mid-mature marine shale with a terrestrial organic component and is relatively rich in gas. The other two types are thought to be sourced from an anoxic marine shale; one derived from a mid-mature source, the other from a late mature source. The distribution of the three oil types indicates migration from two separate source areas during at least three separate episodes. Mixing of the oil types is common. Expulsion of mature hydrocarbons from the Kimmeridge Clay Formation is estimated to have begun in late Cretaceous or early Tertiary time. Probable migration pathways were along the Base Cretaceous Unconformity surface, within the complex fault system or through the various sandstones in the area.

Hydrocarbons

Oil gravity in the Beryl Field is generally 36–39° API, gas gravity is about 0.70. The initial reservoir conditions for the Jurassic reservoirs, as calculated from PVT data, were 207°F and 4900 psi (at −10 500 ft TVSS datum). Initial reservoir conditions for the Triassic reservoirs were 215°F and 5300 psi (at 10 500 ft TVSS datum). PVT analyses indicate that bubble point pressures vary by reservoir and within reservoirs (see data summary). The bubble point variations reflect compositional differences which in the Beryl Formation reservoirs appear to be a function of both area and depth, while in the Nansen and Eiriksson Formation reservoirs, seem to be simply a function of depth. Initial gas–oil ratios vary from about 750–950 SCF/STB for most of the reservoirs but is measured as high as 1500 SCF/STB in some wells in the Beryl reservoir (differential liberation data).

Water analyses for the various reservoirs are obtained primarily from produced water samples. The chemistry of the waters is quite variable (50 000–90 000 ppm TDS), but fractional analyses can usually differentiate the specific reservoir from which the water is derived. The chemical variation of water from an individual reservoir is generally less than 1500 ppm TDS and concentrations of Ca, Mg and Ba are especially definitive in identifying the source reservoir.

Reserves

The Beryl Field is estimated to have an original oil-in-place volume of about 2100 MMBBL (P1 + P2 + P3) with a risked ultimate recovery of about 800 MMBBL. The Beryl Formation is the primary reservoir in the field and has supplied 86% of the production to date. As of December 1989, cumulative production from the field has been 430 MMBBL oil with cumulative injection of 85 MMSCF gas and 250 MMBBL water. The average daily production in 1989 for the Beryl Field was 100 MBBL oil with 232 MMSCF of gas injection and 106 MBBL of water injection (Fig. 8). Reserves of about 365 MMBBL oil and 1600 MMSCF gas remain in the field.

The Beryl Field is produced from the Beryl A and Beryl B platforms in conjunction with a number of subsea wells. Three subsea water injectors and one subsea oil producer are tied by subsea flowlines to the Beryl B platform. Production from the Beryl B platform is then transferred to the Beryl A 'Condeep' platform along with production from three Beryl A subsea producers. Concrete cells at the base of the Beryl A platform capable of holding up to 864 MBBL store the crude until transfer into tankers through two single point mooring systems. The first oil was produced from the Beryl A platform in June 1976 and from the Beryl B platform in July 1984. Initial rates from early wells in the Beryl Formation were as high as 15 000 to 20 000 BOPD. Two early wells, the A09 (June 1977) and S15 (January 1979), are still producing at rates of 3000 BOPD and 13000 BOPD for cumulative totals of 35 MMBBL and 31 MMBBL oil, respectively.

The primary development strategy for the Beryl Formation has been displacement of oil by reinjection of associated gas into a secondary gas cap. Gas injection is augmented by water injection on the flanks, especially in the Beryl B area, as aquifer support is minor and limited to few areas. Because of the steep structural dip of the reservoir (10 to 25°), gravity drainage is also important. A gas injection scheme was chosen for the Beryl Formation because no gas export route was available and because early pseudo-compositional reservoir modelling estimated that an ultimate Beryl Formation recovery of 55% of the oil-in-place could be accomplished through gas injection (Steele & Adams 1984). Gas injection into the Beryl Formation began in November 1977; the first example of pressure maintenance through gas injection in the North Sea. Reservoir performance to date supports the prediction of a 55% recovery, even though vertical and horizontal permeability restrictions significantly complicate reservoir depletion (Knutson et al. 1989). Water injection into the Beryl Formation began in January 1979 and until recently was a minor component of pressure maintenance, limited primarily to the 9/13-A14 well in the southern Beryl A area. Between July 1986 and November 1987, five water injection wells in the Beryl B area and one additional injector in the

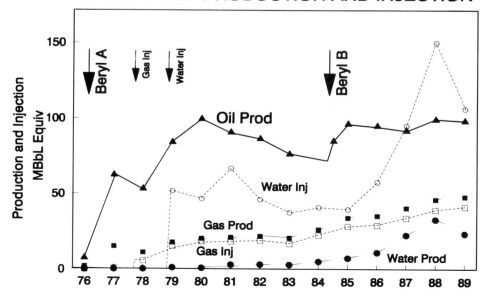

Fig. 8. Production and injection history of the Beryl Field.

Beryl Field data summary

	Lewis Fm Unit III Reservoir	Nansen Fm Reservoir	Linnhe Fm Reservoir	Beryl Fm Reservoir (Units I–V)	Katrine Reservoir
Trap					
Type	Structural/stratigraphic	Structural/stratigraphic	Structural/stratigraphic	Structural stratigraphic	Structural stratigraphic
Depth to crest	−10 200 ft TVSS	−9950 ft TVSS	−10 000 ft TVSS (Beryl A) −11 200 ft TVSS (Beryl B)	−9600 ft TVSS (Beryl A) −10 400 ft TVSS (Beryl B)	−9950 ft to −10 300 ft TVSS
Gas–oil contact	none	none	none	none (Beryl A) −10 900 ft to −11 000 ft TVSS (Beryl B)	none or −10 750 ft TVSS
Oil–water contact	−10 770 ft to −11 625 ft TVSS	−10 860 ft to −12 030 ft TVSS	−11 550 ft to est. 11 900 ft TVSS	−11 300 ft to −11 550 ft TVSS	−10 800 ft to −11 200 ft TVSS
Gas column	none	none	none	500 ft	350 ft max
Oil column	est. 570 ft	1150 ft max.	1550 ft (Beryl A) 600 ft (Beryl B)	1950 ft	1250 ft max
Pay zone					
Age	Upper Triassic (Norian)	Lower Jurassic (Sinemurian)	Middle Jurassic (Bajocian)	Middle Jurassic (Bajocian)	Upper Jurassic (Oxfordian)
Gross thickness	150 ft/100–200 ft	180 ft/120–230 ft	300 ft/100–700 ft	500 ft/200–800 ft	70 ft/0–300 ft
Net/gross ratio (average/range)	0.75/0.50–0.95	0.80/0.65–0.99	0.4/0.2–0.8	0.85/0.4–0.99	0.50/0.16–0.85
Porosity (average/range)	16%/10–24%	18%/13–26%	14%/10–20%	17%/10–23%	15%/9–20%
Hydrocarbon saturation	0.70	0.87	0.84	0.88	0.81
Permeability (average/range)	40 md/0.1–500 md	200 md/50–300 md	100 md/1.0–1000 md	350 md/30–4400 md	150 md/11–400 md
Productivity index	1–5 BOPD/psi	11 BOPD/psi		10–100 BOPD/psi	1 to 9 BOPD/psi
Hydrocarbons					
Oil gravity	36°API	27°API	38°API	37°API	36°API
Oil type	Low sulphur crude	Low sulphur crude	Low sulphur crude	Low sulphur crude	Low sulphur crude
Gas gravity	0.70	0.71	0.79	0.71	0.70
Bubble point	2900 psi	3200–3800 psi	3750 psi	3200–4900 psi	3100–4900 psi
Gas/oil ratio	750 SCF/STB	960 SCF/STB	960 SCF/STB	900–1500 SCF/STB	800 SCF/STB
Condensate yield				40	
Formation volume factor	1.41	1.5–1.6	1.6	1.5–1.7	1.46
Formation water					
Salinity	65 000–75 000 ppm TDS	80 000–90 000 ppm TDS	no samples, to date	65M–75M ppm TDS (Beryl A) 50M–65M ppm TDS (Beryl B)	75 000–90 000 ppm TDS
Reservoir conditions					
Temperature	215°F	207°F	207°F	207°F	207°F
Pressure	5300 psi (−10 500 ft TVSS Datum)	4900 psi (−10 500 ft TVSS Datum)	4900 psi	4900 psi (−10 500 ft TVSS Datum)	4900 psi (−10 500 ft TVSS Datum)
Pressure gradient in reservoir	—	0.29 psi/ft in oil	0.29 psi/ft	0.28 psi/ft	0.28 psi/ft
Field					
Original oil-in-place	324 MMBBL (Lewis Fm, Units I–IV)	290 MMBBL	131 MMBBL	1448 MMBBL	129 MMBBL
Recoverable oil	85 MMBBL (Lewis Fm, Units I–IV)	96 MMBBL	44 MMBBL	687 MMBBL	30 MMBBL
Drive mechanism	Solution gas drive	Solution gas drive with possible local water drive	Solution gas drive	Solution gas and gas cap drive	Solution gas and gas cap drive
Production					
Start-up date	November 1976	October 1976	September 1976	July 1976	October 1977
Current number of wells	4 producers, 1 water injector	5 producers, 3 water injectors	5 producers	27 producers, 5 gas injectors, 6 water injectors	1 producer
Development scheme	Primary depletion with water injection	Water injection on flanks with some crestal gas injection	Primary deletion with possible future water injection	Crestal injection of gas into primary or secondary gas cap, augmented by water injection on flanks	Primary depletion with possible future water injection
Cumulative production to 1.1.90	12.4 MMBBL (Lewis Fm, Units I–IV)	30.8 MMBBL (often comingled with Eiriksson Fm)	10.2 MMBBL	365.0 MMBBL	8.9 MMBBL

Beryl A area were completed to provide increased pressure support and accelerate oil production in anticipation of future gas sales. Presently water injection provides about 50% of the voidage replacement in the Beryl reservoir.

The development of the five other reservoirs in the Beryl Field is not as advanced as for the Beryl Formation reservoir. The Katrine Member, Linnhe Formation and Lewis Formation reservoirs have been historically produced by primary depletion and recovery factors in the range of 10–25% are expected. Water injection to enhance recovery from these reservoirs is planned and an injection programme for the Lewis Formation commenced late 1989. Production from the Nansen and Eiriksson Formation reservoirs has been supported by gas and water injection and recovery factors are estimated at 15–30%.

The authors wish to thank Mobil North Sea Ltd, Amerada Hess Ltd, Enterprise Oil plc, North Sea Ltd and Oesterreichische Mineraloel Verwaltung for permission to publish this paper. We also wish to acknowledge our many colleagues who have contributed to the Beryl Field interpretations summarized in this paper. In particular, we wish to thank R. E. Dunay and R. J. Hodgkinson of Mobil North Sea Ltd, who contributed significantly to the biostratigraphical and sedimentological understanding of the field. The statistics, interpretations and conclusions presented herein, however, should be considered the responsibility of the authors and do not necessarily reflect the views of all those knowledgeable of the Beryl Field.

References

DEEGAN, C. E. & SCULL, B. J. 1977. *A standard lithostratigraphic nomenclature for the Central and Northern North Sea.* Report of the Institute of Geological Science. 77/25; Bulletin Norwegian Petroleum Directorate No. 1.

KNUTSON, C. A., ERGA, R. & MCAULAY, F. M. 1989. Horizontal and Vertical Permeability Restrictions in the Beryl Reservoir, Beryl Field. *Proceedings from the Offshore Europe Conference, Aberdeen 1989*, SPE Paper No. 19299.

MANSPEIZER, W. 1982. Triassic–Liassic basin and climate of the Atlantic passive margins. *Geologische Rundschau*, **71**, 895–917.

STEELE, L. E. & ADAMS, G. W. 1984. A review of the northern North Sea's Beryl Field after seven years' production. *Proceedings from the European Petroleum conference, London, 1984*, SPE Paper No. 12960.

The North Brae Field, Block 16/7a, UK North Sea

MARK A. STEPHENSON

Marathon Oil UK, Ltd, Marathon House, Rubislaw Hill, Anderson Drive, Aberdeen AB2 4AZ, UK
Present address: Marathon Oil Company, PO Box 190168, Anchorage, Alaska 99519, USA

Abstract: North Brae is the first gas condensate field in the UK to be produced by gas recycling. The field lies at the western margin of the South Viking Graben in UK Block 16/7a. Estimated recoverable reserves are 178 MMBBL of condensate and 798 BCF of dry gas. First hydrocarbon production was in April 1988 from the Brae 'B' platform.

The reservoir is composed of coarse clastic sediments of the Upper Jurassic Brae Formation which were deposited by debris flows and turbidity currents in a submarine fan setting adjacent to an active fault scarp. The Brae Formation now abuts impermeable Devonian rocks of the Fladen Ground Spur to the west. The reservoir is capped by the Kimmeridge Clay Formation, which also provided the source of the hydrocarbons.

The North Brae Field is the most northern of three Upper Jurassic fields in UK Block 16/7a, at the western margin of the South Viking Graben, 169 miles northeast of Aberdeen (Fig. 1). The Brae 'B' platform stands in a water depth of 326 ft and is located just southwest of the discovery well 16/7-1. The reservoir extends over an area of approximately 4700 acres, and has a maximum hydrocarbon column of 645 ft. The reservoir fluid is a gas condensate which exhibits marked compositional change with depth. This is reflected in a gas–oil ratio variation of 6500–2500 SCF/STB with depth, and a corresponding range of gravity from 49–41° API. The reservoir is formed by the Brae Formation, and comprises thick accumulations of sand-matrix conglomerates deposited by debris flows and interbedded sandstone and mudstone brought into the basin by turbidity currents during the late Jurassic.

History

The general history of Licence P.108 is described by Roberts (this volume), and the history of North Brae is summarized as follows.

Over 500 miles of 2-D seismic data were acquired across the Brae area in the early 1970s, and the first exploration well, 16/7-1 (Fig. 1),

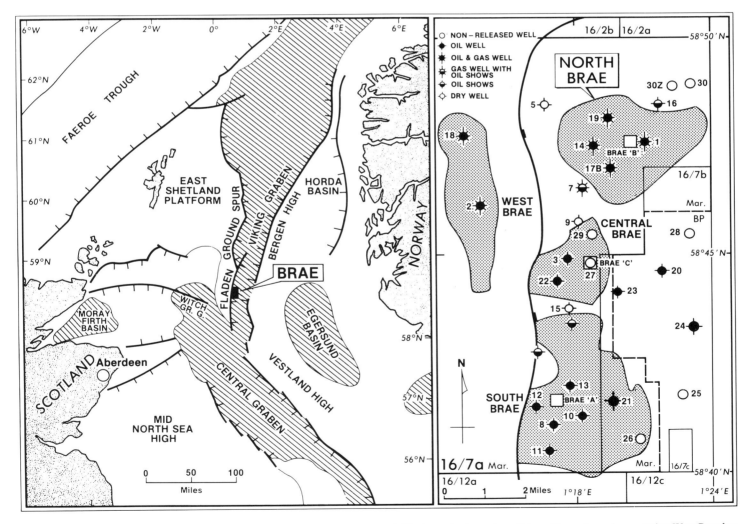

Fig. 1. Location of the North Brae Field in the South Viking Graben. The three fields shown in the graben all have Upper Jurassic reservoirs; West Brae has Palaeocene and Devonian reservoirs.

From Abbotts, I. L. (ed.), 1991, United Kingdom Oil and Gas Fields,
25 Years Commemorative Volume, Geological Society Memoir No. 14, pp. 43–48

was drilled in 1975 by Pan Ocean on the North Brae structure. This well discovered significant amounts of gas condensate, testing almost 100 MMSCFD of wet gas from eight intervals.

Pan Ocean was acquired by Marathon in 1976 and in the same year, well 16/7-5 was drilled to the west of the North Brae structure. This well encountered a sequence of tight Brae Formation conglomerates and breccias overlying Devonian basement and was abandoned as a dry well. In 1977, well 16/7a-7 was drilled to the southwest of 16/7-1 (Fig.1). Here, the Brae Formation is dominantly mudstone, however a 40 ft thick sandy interval flowed nearly 13 MMSCFD of wet gas.

It was not until 1980 that an appraisal well, 16/7a-14, was drilled on the same structure as 16/7-1 and confirmed the North Brae discovery. Two further successful appraisal wells (16/7a-17b and 19) and an unsuccessful (largely shaly) well, 16/7a-16, were drilled on the structure during 1981 and 1982. Following Annex B submission and approval, the Brae 'B' platform was installed in 1987 and first production began in April 1988. Brae 'B' is the first gas condensate recycling platform in the North Sea. The field is operated by Marathon Oil UK, Ltd on behalf of the Brae Group co-ventures; Britoil (BP), Bow Valley, Kerr McGee, Gas Council (Exploration), LL and E, Sovereign and Norsk Hydro. Up to May 1990, 16 development wells have been drilled from the platform.

Field stratigraphy

An overview of the geological history of the Brae area is described by Roberts (this volume) and the general stratigraphy of the North Brae Field is shown in Fig. 2 by reference to well 16/7a-14. In common with South Brae (Roberts, this volume) and Central Brae (Turner & Allen, this volume) the North Brae reservoir is a submarine fan sequence within the Brae Formation, which is overlain by, and regionally interdigitates with, the Kimmeridge Clay Formation.

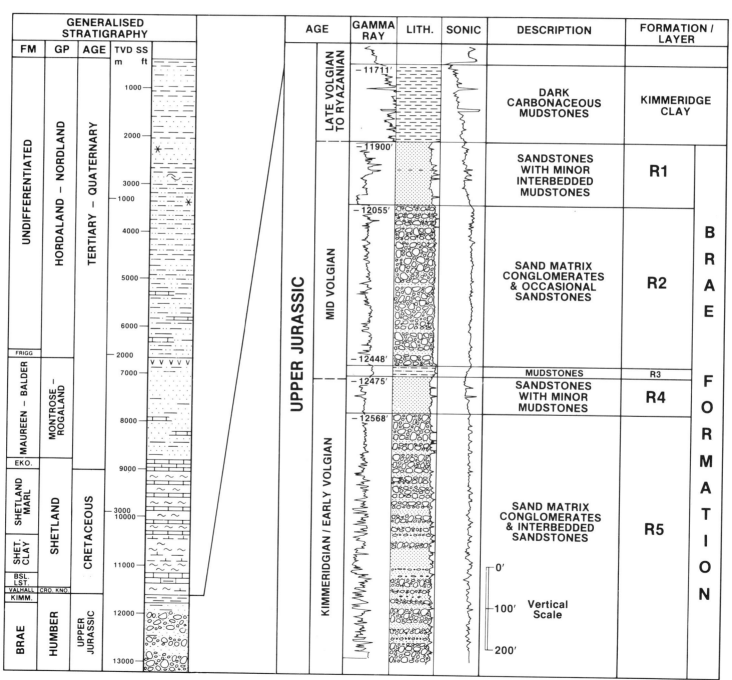

Fig. 2. Generalized North Brae stratigraphy and reservoir subdivision (16/7a-14).

The North Brae submarine fan has important differences in geometry and age compared with either the South Brae or Central Brae submarine fans. The North Brae fan is dominated by an east–west, elongate, conglomeratic channel-fill complex (the 'conglomerate core', see below), whereas South Brae contains a radiating pattern of channel-fill complexes, and Central Brae is an unchannelled cone of coarse clastic sediments.

The North Brae fan was initiated in the Kimmeridgian at about the same time as the South and Central Brae fans, but the North Brae fan remained active following cessation of deposition on the two fans to the south, and culminated in the late mid-Volgian (Turner & Allen, this volume, fig. 3).

The North Brae reservoir is subdivided into 5 layers (R1–R5) as identified in Fig. 2, using 16/7a-14 as the type well. The reservoir correlations are based on several criteria, including seismic evidence, palynological analysis (Riley et al. 1989), log character, lithology and pressure data. Individually, all of these data have limitations, but by using a combination of the data sets a relatively high degree of confidence in the correlations can be achieved.

Geophysics

Seismic surveys

Extensive seismic coverage exists in the North Brae area, much of which are 2-D data shot between 1970 and 1980. In 1984 and 1985 a further 1566 miles of 3-D data were acquired with a line spacing of 75 m (246 ft). These data were processed to ensure optimum spectral balancing and resolution for the Upper Jurassic.

Seismic interpretation

The North Brae 3-D seismic survey has proved an important tool in reservoir mapping and has successfully contributed to the understanding of the geology of the field.

Data quality of the 3-D seismic is excellent down to the Base Cretaceous reflector which, together with the shallower reflectors, can be mapped with confidence (Fig. 3). Additionally, the top reservoir has been mapped seismically, although with less confidence than the Base Cretaceous reflector, and mapping is particularly difficult in areas where the overlying Kimmeridge Clay Formation is thin (less than 100 ft). A top reservoir depth map was therefore constructed by adding a seismically and geologically derived Kimmeridge Clay Formation isochore to a seismically derived depth map of the Base Cretaceous.

Reflections within the reservoir sequence are not seen as continuous events across the field, a fact which can be attributed partially to abrupt lateral facies changes. Attempts have been made to map from the seismic attributes the distribution of the main reservoir lithofacies, including the margins of the 'conglomerate core'. This work has been of assistance in the construction of geological lithofacies models for North Brae. In some areas it is unclear whether disruptions in reflection continuity should be attributed to abrupt facies variations or faulting, or both.

One deeper seismic reflector at the top Middle Jurassic, has also been mapped. This reflector has a structural culmination near the 16/7a-16 well, and is offset to the northeast of the North Brae Field. This feature may have influenced the distribution of the late Jurassic submarine fan.

Trap

The trap is an east–west trending anticline within the graben-fill sequence which abuts the graben-margin fault zone (Fig. 4). The anticline was formed by a combination of faulting, drape, and differential compaction during the late Volgian to mid-Cretaceous. The reservoir is capped by the Kimmeridge Clay Formation.

The shape of the field is largely controlled by the presence of the 'conglomerate core' and by probable antithetic faulting beneath the syncline at the west of the field. Adjacent to the 'conglomerate core' finer-grained sediments have undergone considerable compaction compared with the conglomerates. The latest Jurassic and early Cretaceous sediments subsequently draped over the positive feature formed by the 'conglomerate core' (Fig. 3). These processes resulted in an elongate dome-shaped field structure (Fig. 4).

The field has dip closure on three sides, with a hydrocarbon–water contact at 12 475 ft TVSS. There is a stratigraphic pinch-out

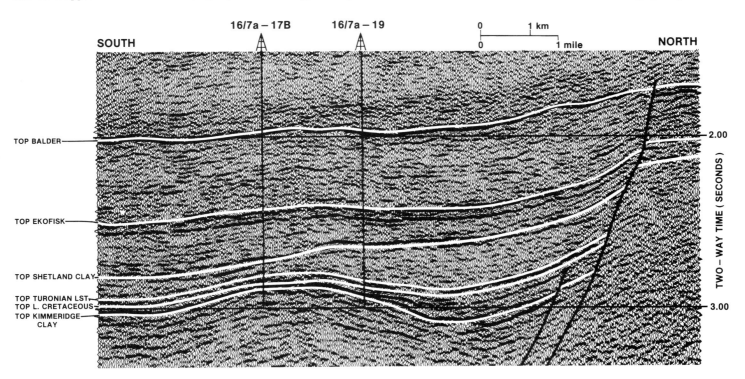

Fig. 3. North–south 3-D seismic section through wells 16/7a-17b and 16/7a-19, North Brae Field.

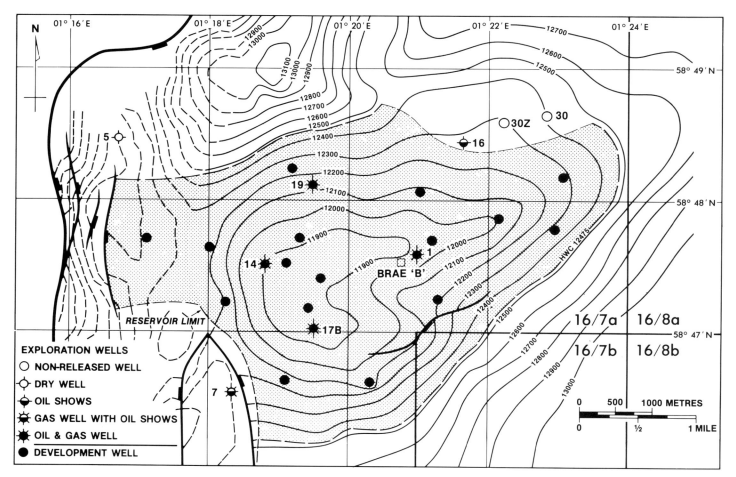

Fig. 4. North Brae top reservoir structure map and field outline. Depth contours in ft TVSS.

on the north-east side of the field close to well 16/7a-16. The western limit is poorly defined, and is either controlled by the abutment of reservoir rocks against the graben-margin fault zone (as is the case adjacent to the fault block on which well 16/7a-7 was drilled) or is caused by deteriorating reservoir quality in the very proximal part of the submarine fan sequence (as is suggested by the tight conglomerate and breccia drilled in well 16/7-5).

Reservoir

The North Brae reservoir was deposited by debris flows and turbidity currents in a marine environment. Sediments were derived from the Fladen Ground Spur to the west, which at the time of deposition existed as a land mass, and were transported down the marginal fault scarp of the Viking Graben and deposited as a

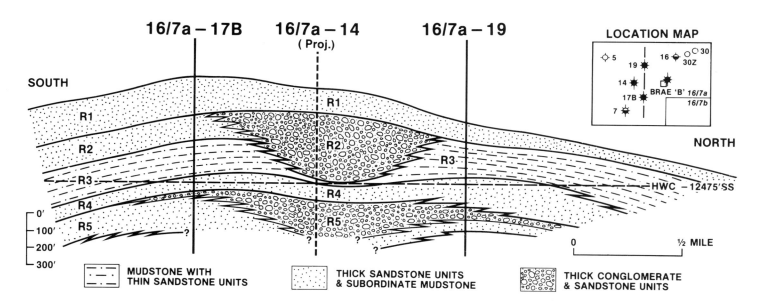

Fig. 5. North–south section illustrating the North Brae reservoir zonation scheme through wells 16/7a-17b and 16/7a-19.

submarine fan. Transportation was broadly eastwards, with coarser sediments being deposited proximal to the entry point and also distributed along the axis of the channel-fill complex through wells 16/7a-14 and 16/7-1.

The field can be split into two upward-fining major sedimentary cycles (Fig. 5); an upper unit consisting of R2 conglomerates and sandstones, overlain by R1 sandstones; and a lower unit comprising R5 conglomerates overlain by R4 sandstones. Separating the two cycles over much of the field is a mudstone interval, R3. The major hydrocarbon bearing layers are R1 and R2, with only a small proportion of R4 being elevated above the hydrocarbon water contact at 12 475 ft TVSS.

The characteristics of the North Brae reservoir lithologies are similar to those of South Brae (Turner et al. 1987; Roberts, this volume).

The topmost layer R1, is a complex sheet of predominantly fine- to medium-grained turbidite sandstones and thin (up to 30 ft thick) radiating channel-fill conglomerates. R1 blankets the underlying R2 layer and although it covers a greater areal extent than R2 its average vertical thickness (about 100 ft) is much less than the more complicated geometrical configuration of the lower unit. Porosity and permeability characteristics on the whole are good to excellent, with core porosities in the sandstones often ranging between 15% and 30%, and horizontal air permeabilities commonly between 200 md and 2000 md.

Layer R2 is dominated by an asymmetrical channel complex which is aligned west to east and is filled with a sequence which generally ranges between 200 and 400 ft thick, although the maximum thickness is 800 ft. Figure 5 shows the 'conglomerate core' of the R2 layer which passes into a partially age-equivalent mudstone sequence to the north and overbank sandstones to the south. Eastwards the sediments become more distal, with conglomerates passing into sandstones. Reservoir quality in the sandstones is good, but in the conglomeratic intervals it is the sandstone matrix component which contributes towards production. Where R2 conglomerate sequences dominate, a sharp contrast is seen in reservoir quality between the R2 and the overlying R1 layer. Core porosities in the sandstones are commonly 12% to 18% with horizontal air permeabilities ranging up to 600 md; for conglomerates the corresponding figures are 7% to 10% for porosity and 100 md for permeability.

Layer R3 constitutes non-reservoir rock, while R5 and R4 layers represent an upward-fining cycle similar to the upper cycle composed of layers R2 and R1. The vast majority of layer R4 and all of layer R5 is below the hydrocarbon–water contact.

Source

Comparison of mudstone extracts with condensate samples indicate that the Kimmeridge Clay Formation is the most likely source rock for the North Brae hydrocarbons. The organic components of these mudstones consist primarily of amorphous and algal kerogens with relatively minor amounts of inertinite, woody and herbaceous matter. Organic carbon content ranges up to 15.25%. Maturation indices for the Kimmeridge Clay Formation at North Brae are sufficiently high ($R_o = 0.6\%$) to indicate oil generation.

The pronounced difference in hydrocarbon composition between the gas condensate of North Brae and the black oil in both South Brae and Central Brae is not easy to explain. However, it is likely that the North Brae hydrocarbons had a longer migration pathway than the locally-sourced oils of South and Central Brae (Reitsema 1983). It is possible that gas generated from the Kimmeridge Clay in the deeper parts of the graben to the east could have migrated through the submarine fan sequence into the North Brae structure.

Organic-rich middle Oxfordian mudstones (of the Heather Formation) were penetrated by well 16/17a-16 below the North Brae reservoir. These mudstones exhibit levels of organic maturity well within the oil-generation window ($R_o \leqslant 0.72\%$) and may also have contributed to the North Brae hydrocarbon accumulation. Further research into these source aspects is ongoing.

North Brae Field data summary

Trap
Type	Structural/stratigraphic
Depth to crest	11 830 ft TVSS
Lowest closing contour	12 100 ft TVSS
Hydrocarbon–water contact	12 475 ft TVSS
Gas column	645 ft

Pay zone
Formation	Brae
Age	Late Jurassic (Kimmeridgian to Late Mid-Volgian)
Gross thickness (average; range)	120 ft; 0–645 ft
Net/gross ratio (average)	0.64
Cut off for net pay	7% porosity in all lithologies and an 85 API GR cut off in non-conglomerate lithologies.
Porosity (average)	14.5%
Original hydrocarbon saturation (average)	86%
Permeability (average; range)	360 md, 1–4000 md
Productivity index (Bl)	0.06 MMSCFD; psi

Hydrocarbons
Fluid type	Gas condensate
Oil gravity (range)	41–49° API
Gas gravity (relative to air)	0.78
Dew point	5500–6500 psia
Gas/oil ratio	2500–6500 SCF/STB
Condensate yield	230 SIB/MMSCF at 12 475 ft TVSS
Formation volume factor (average)	6.4×10^{-4} RB/SCF

Formation water
Salinity	77 000 ppm NaCl equivalent
Resistivity	0.12 ohm.m at 60°F

Reservoir conditions
Temperature	240°F at 12 475 ft TVSS
Pressure	6900 psia at 12 475 ft TVSS
Pressure gradient in reservoir	0.191 psi/ft

Field size
Area	4700 acres
Gross rock volume	1 571 160 ac.ft.
Hydrocarbon pore volume	102 980 ac.ft.
Recovery factor: condensate	0.66
Recovery factor: gas	0.72
Drive mechanism	Dry gas recycle
Recoverable hydrocarbons	
NGL/condensate	178 MMBBL
gas	798 BCF dry gas

Production
Start-up date	April 1988
Development scheme	Dry gas recycle followed by blowdown
Production rate (May 1990)	81 500 BCPD condensate
Cumulative production to end May 1990	48 MMBBL condensate
Number/type of wells	1 exploration
	4 appraisal
	16 development (to May 1990)
Secondary recovery method	Dry gas recycle

Hydrocarbons

Wells 16/7-1 and 16/7a-14 each flowed at aggregate rates of approximately 100 MMSCFD wet gas. Of the other appraisal wells, only 16/7a-17b and 19 tested significant quantities of gas condensate. The hydrocarbon fluid type is a retrograde gas condensate which exhibits a compositional variation with depth. The gas–oil ratio varies between 2500 and 6500 SCF/STB from the base to the top of the reservoir, and the dew point varies similarly between 6500–5500 psia. The original pressure was 6900 psia at the hydrocarbon–water contact. The formation water has a specific gravity of 1.06 at 60°F, a pH of 7.5, a salinity of 77 000 ppm NaCl equivalent and total dissolved solids content of 82 000 ppm.

Reserves

North Brae is the first gas condensate recycling field to be developed in the North Sea. It is being developed by injecting the produced dry gas, after condensate and natural gas liquids have been stripped out, in order to maintain reservoir pressure and to sweep the wet gas to the production wells. Additional volumes of gas are supplied from the South Brae Field to Brae 'B' for injection into the reservoir.

The estimated recoverable reserves are 178 MMBBL of condensate and 798 BCF of dry gas. Current condensate production rates (May 1990) are close to 81 500 BCPD. Current gas injection rates are approximately 440 MMSCFD.

This paper is published by permission of Marathon Oil UK, Ltd, the Operator of North Brae, and co-ventures; Britoil plc (BP), Bow Valley Exploration (UK) Limited, Kerr McGee Oil (UK) plc, Gas Council (Exploration) Limited, LL&E (UK) Inc, Sovereign Oil and Gas plc and Norsk Hydro Oil and Gas Limited. Although they cannot be acknowledged individually, many people have contributed to our sum knowledge of the North Brae Field over 19 years of exporation and production.

References

REITSEMA, R. H. 1983. Geochemistry of North and South Brae areas, North Sea. *In*: BROOKS, J. (ed.) *Petroleum Geochemistry and Exploration of Europe*, Geological Society, London, Special Publication, **12**, 203–212.

RILEY, L. A., ROBERTS, M. J. & CONNELL, E. R. 1989. The application of palynology in the interpretation of Brae Formation Stratigraphy and reservoir geology. Norwegian Petroleum Society, *In: Correlation in Hydrocarbon Exploration*, Graham & Trotman, London, 339–356.

ROBERTS, M. J. 1991. The South Brae Field, Block 16/7a, UK North Sea. *This volume*.

TURNER, C. C. & ALLEN, P. J. 1991. The Central Brae Field, Block 16/7a, UK North Sea. *This volume*.

—— COHEN, J. M., CONNELL, E. R. & COOPER, D. M. 1987. A depositional model for the South Brae Oilfield. *In*: BROOKS, J. & GLENNIE, K. W. (eds) *Petroleum Geology of North West Europe*, Graham and Trotman, London, 853–864.

The Central Brae Field, Block 16/7a, UK North Sea

COLIN C. TURNER & PHILIP J. ALLEN

Marathon Oil UK Ltd., 174 Marylebone Road, London NW1 5AT, UK

Abstract: Central Brae Oilfield is the smallest of three Upper Jurassic fields being developed in UK Block 16/7a. The field was discovered in 1976 and commenced production in September 1989 through a subsea template tied back to the Brae 'A' platform in the South Brae Oilfield. Recoverable reserves are estimated as 65 MMBBL of oil and 6 MMBBL of NGL. The Central Brae reservoir is a proximal submarine fan sequence, comprising dominantly sand-matrix conglomerate and sandstone with minor mudstone units. The sediments were shed eastwards off the Fladen Ground Spur and were deposited as a relatively small and steep sedimentary cone at the margin of the South Viking Graben. Mudstone facies border the submarine fan deposits to the north and south, forming stratigraphic seals. The structure is a faulted anticline developed during the latest Jurassic and early Cretaceous possibly in response to large-scale rotational slump movement within the easterly-dipping graben margin sequence. The western boundary of the field is formed by a sealing fault, whilst to the east, there is an oil-water contact at 13 426 ft TVSS. The overlying seal is the Kimmeridge Clay Formation, which also interdigitates with the coarser facies basinwards, and provides the source of the hydrocarbons.

Central Brae Field is located between the South Brae and North Brae Fields (Roberts, this volume; Stephenson, this volume) in UK Block 16/7a, some 165 miles northeast of Aberdeen (Fig. 1). Central Brae is an oilfield, with fluid properties which are similar to South Brae but distinctly different from the gas condensate of North Brae. The areal extent of the field (approximately 1800 acres) is considerably less than that of either of the other fields but there is high relief (maximum 1676 ft) which equals that of South Brae and greatly exceeds that of North Brae. Production commenced in September 1989 through a subsea template, set over well 16/7a-27 in 350 ft of water, tied back to the Brae 'A' Platform in the South Brae Field.

The Central Brae structure is a faulted anticline formed at the margin of the South Viking Graben. The reservoir comprises sandstone, conglomerate and minor mudstone deposited as a relatively small, discrete late Jurassic submarine fan, between the much larger submarine fans of South Brae and North Brae.

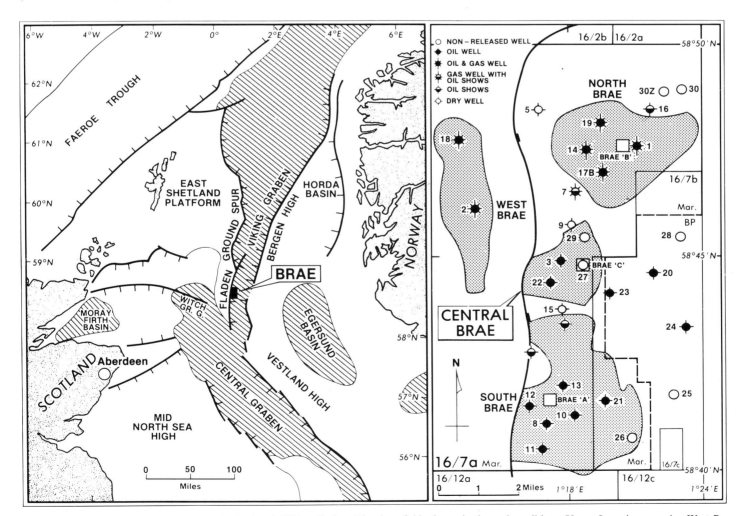

Fig. 1. Location of the Central Brae Field in the South Viking Graben. The three fields shown in the graben all have Upper Jurassic reservoirs; West Brae has Palaeocene and Devonian reservoirs.

From Abbotts, I. L. (ed.), 1991, United Kingdom Oil and Gas Fields,
25 Years Commemorative Volume, Geological Society Memoir No. 14, pp. 49–54

History

The history of Licence P.108, which covers Block 16/7a, is described by Roberts (this volume). Central Brae was discovered by well 16/7-3, which was drilled by Pan Ocean in 1975–6 as an appraisal well to the North Brae discovery, 16/7-1. Well 16/7-3 was sited on an anticline extending southwards from the North Brae domal feature, and tested oil at a cumulative rate of over 13 000 BOPD from five Upper Jurassic intervals. Following Marathon's acquisition of Pan Ocean in 1976, well 16/7a-9 was drilled in 1977 between Central Brae and North Brae, as an appraisal of the 16/7-3 discovery (Fig. 1). This well encountered an Upper Jurassic sequence dominated by mudstone and was plugged and abandoned without testing. In 1980, well 16/7a-15 was drilled to the south of 16/7-3, with similar results to well 16/7a-9.

Three successful Central Brae appraisal wells, 16/7a-22, 27 and 29 were then drilled between 1983 and 1985. Well 16/7a-29 was drilled as a deviated well from the 16/7a-27 location, and it was over these wellheads that the subsea template was placed in April 1989; both of these wells are now utilized for production. Two additional production wells have subsequently been drilled through the template.

In common with the other Brae fields, Central Brae is operated by Marathon Oil UK, Ltd on behalf of Britoil (BP), Bow Valley, Kerr-McGee, Gas Council (Exploration), LL & E, Sovereign and Norsk Hydro.

Field stratigraphy

The general geological history of Block 16/7a is described by Roberts (this volume) and the stratigraphy of the Central Brae

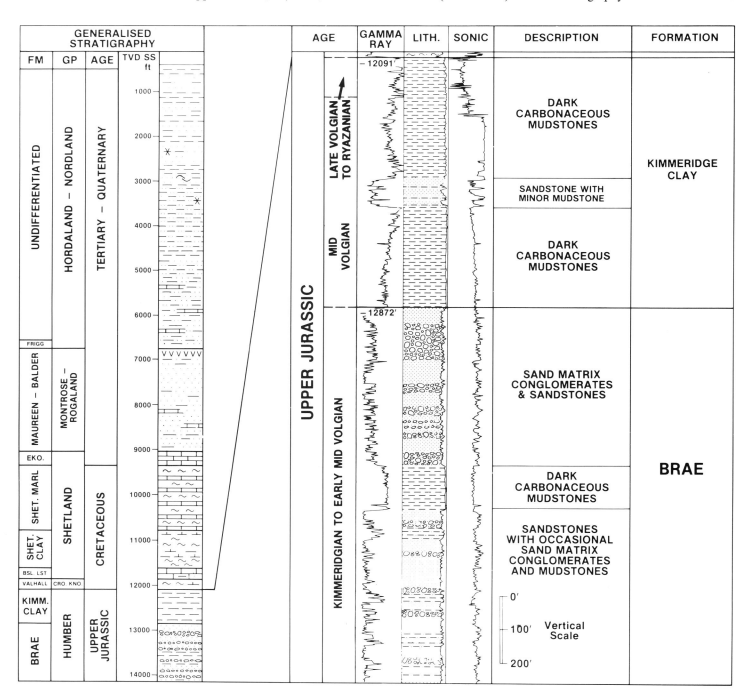

Fig. 2. Generalized Central Brae stratigraphic log and reservoir stratigraphy (16/7a-27).

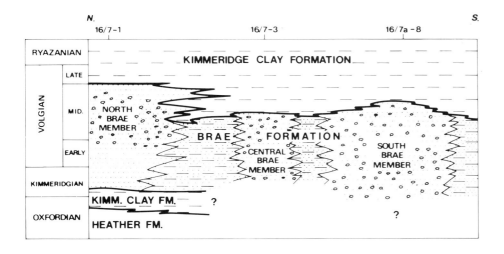

Fig. 3. Comparative Upper Jurassic stratigraphy in Central Brae, South Brae and North Brae.

Field is shown in Fig. 2 with reference to well 16/7a-27. The Central Brae Member of the Upper Jurassic Brae Formation forms the reservoir, which is overlain by, and in a regional sense interdigitates with, the Kimmeridge Clay Formation (Turner *et al.* 1987). The Central Brae Member is a submarine fan sequence, the bulk of which is Kimmeridgian to early Volgian in age. This fan system is separate from the South Brae and North Brae submarine fans, where deposition continued after the cessation of deposition on the Central Brae fan (Fig. 3).

Unlike the South Brae and North Brae reservoir sequences, the Central Brae Member is difficult to subdivide, due to the absence of distinct correlatable units across the entire field. The Central Brae Member in wells 16/7-3 and 16/7a-22 in the western and crestal part of the field is almost exclusively conglomerate and sandstone, whereas the more easterly, downdip wells, 16/7a-27 (Fig. 2) and 16/7a-29, also include mudstone units.

Geophysics

Seismic surveys

Several seismic surveys were acquired over the Central Brae area between 1970 and 1980. The most important of these early surveys was a 3-D survey shot with a line spacing of 100 m (328 ft) in 1977, which was reprocessed in 1983. This survey formed the major data set for field development planning. In 1985 a further 3-D seismic survey (the Brae–Miller survey; see Roberts, this volume) was acquired, which in part covered the Central Brae Field. This survey, shot with a dual source and a sail line spacing of 75 m (246 ft), has provided superior quality data to refine the reservoir interpretation.

Seismic interpretation

On the older vintages of seismic data, the Central Brae anticline at the Base Cretaceous level can be distinguished, which allowed the siting of the discovery well.

The 3-D migrated Brae–Miller data set has allowed more accurate definition of the Base Cretaceous, particularly in the structurally complex area towards the western part of the field, and, for the first time, the seismic mapping of the top of the reservoir (Fig. 4). However, data quality below the Base Cretaceous is still, at best, moderate, which together with the poorly layered nature of the reservoir, has allowed only limited seismic reservoir analysis. Several horizons within the relatively thick Kimmeridge Clay Formation on the field flanks however, can be correlated with reasonable confidence.

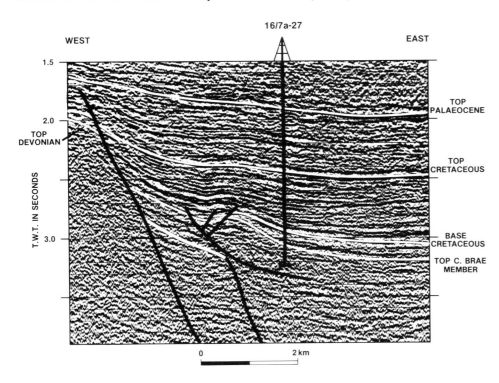

Fig. 4. West–east 3-D seismic section through well 16/7a-27, Central Brae Field.

Trap

Central Brae is a structural/stratigraphic trap formed by a combination of folding at the western margin of the South Viking Graben, lateral stratigraphic pinch-out of reservoir quality rocks and abutment against impermeable rocks of the Fladen Ground Spur. The western margin of the field is interpreted as an eastward-dipping listric fault (the western boundary fault) that overlies a possible fault terrace of Devonian rocks (Fig. 5). This fault, together with other faults in the field (Fig. 6) and the Central Brae anticline, are considered to have formed during the latest Jurassic to early Cretaceous by large-scale rotational slump movement within the easterly-dipping graben margin sequence, in response to continued subsidence and compaction within the basin.

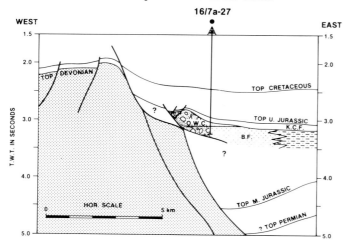

Fig. 5. Geoseismic cross section through well 16/7a-27, Central Brae Field.

The eastern margin of the field is defined by an oil-water contact at 13 426 ft TVSS (Fig. 6). It is possible that the structure is filled to a spill point at the northern apex of the field, where the Central Brae fan system may overlap with coarse clastic sediments of the North Brae fan system.

The northwestern and southwestern margins of the field are formed by the stratigraphic pinch-out of the reservoir sandstone and conglomerate into the siltstone and mudstone sequences encountered in wells 16/7a-9 and 15.

The reservoir is sealed by the overlying Kimmeridge Clay Formation and probably by the western boundary fault. However, there is a possibility of low quality (very conglomeratic) reservoir to the west of this fault (Fig. 5).

Reservoir

The Central Brae Member contains six lithofacies:

(1) sand-matrix conglomerate;
(2) mud-matrix breccia;
(3) medium to thick-bedded sandstone;
(4) alternating thin-bedded sandstone with interlaminated sandstone-mudstone;
(5) interlaminated mudstone-sandstone;
(6) laminated mudstone.

These facies are similar to those described in detail for the South Brae Field by Turner *et al.* (1987).

Sand-matrix conglomerate and medium to thick-bedded sandstone are the most common Central Brae reservoir facies. These were deposited in the proximal and central parts of a submarine fan system by high-density debris flows and high-density turbidity currents. Unlike the South Brae and North Brae submarine fans, no preferred channelways of coarse clastic sediments have been recognized in Central Brae.

The Central Brae submarine fan is thought to have been a relatively small and steep cone of sediment radiating basinwards from a sediment entry point on the faulted margin of the Fladen Ground Spur. The location of the sediment entry point may have been controlled by the intersection of east-west trending faults on the Fladen Ground Spur with the north-south trending graben-margin fault scarp.

In the marginal and distal parts of the fan, the finer grained, and more thinly bedded sediments become the dominant lithology,

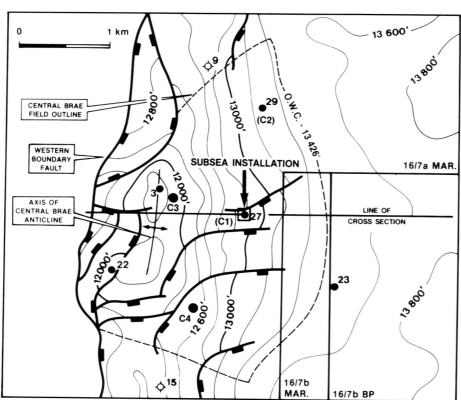

Fig. 6. Central Brae top reservoir structure map, showing Central Brae Field outline. Depth contours are in ft TVSS.

interspersed with units of mud-matrix breccia close to the graben-margin fault. The reservoir facies are completely replaced by the finer grained sediments in the inter-fan sequences drilled in wells 16/7a-9 and 15, and also eastwards in less than 4 miles from the graben-margin fault.

Correlatable units within the Central Brae reservoir are difficult to recognize and were not used for initial mapping purposes. The conformably overlying Kimmeridge Clay Formation, however, can be subdivided into numerous correlatable units. Despite a marked thickness change for this formation from the east flank of the field to the crest (e.g. 1030 ft in 16/7a-29, and 300 ft in well 16/7-3), correlations based on log character and detailed palynology (Riley et al. 1989) indicate that the lowest part of this formation is essentially the same age across almost the entire fan. Therefore the upper part of the underlying reservoir is also considered the same age (early Volgian, JB10 zone of Riley et al. 1989) across most of the field. Around well 16/7a-22 there is a localized development of younger sandstone (earliest middle Volgian, JB8 age) representing the final restricted phase of submarine fan deposition.

Core porosites range up to 16% in conglomerate and 18% in thick-bedded sandstone, and air horizontal permeabilities range up to 300 md and 1000 md respectively. Porosity generally decreases with depth, and in wells 16/7-3 and 16/7a-22 the lowest part of the Central Brae Member, comprising thick conglomerate with minor sandstone beds, is affected by pervasive calcite cementation, effectively forming a non-reservoir zone. The pervasive nature of the calcite cementation is not easy to explain, and is the subject of further research. However, early calcite cementation was probably caused by fluids being expelled from the basin at the graben margin. Secondary porosity development due to dissolution of early calcite cement occurred in the upper part of the reservoir, but is not evident lower in the submarine fan sequence.

Source

The Kimmeridge Clay Formation is the source rock for the oil accumulations in the Brae area (Cornford, 1984; MacKenzie et al. 1987). Compositional analyses of Central Brae hydrocarbons by Cornford (unpublished) indicate only a very short migration path, suggesting that the oil was generated on the flanks of the structure or in the immediately surrounding area. Mudstone at a depth of about 14 100 ft TVSS in well 16/7a-27 (Fig. 2) probably initially generated hydrocarbons during the early Eocene, and is presently at peak maturity ($R_0 = 0.8\%$). The Kimmeridge Clay overlying the reservoir in this well initially generated hydrocarbons during the late Eocene, and is presently in the mid-mature stage of oil generation ($R_0 = 0.7\%$).

Hydrocarbons

The maximum individual DST flow rate from the Central Brae wells was 5572 BOPD on a 1 inch choke from well 16/7-3. The oil composition is relatively uniform, and has an API gravity of approximately 33°. Associated gas has a CO_2 content of 25–30% and an H_2S content of about 25 ppm. Other hydrocarbon and water properties are shown on the Field Data Summary.

Reserves

Central Brae contains an estimated 64 MMBBL of recoverable oil, and a further 6 MMBBL of NGL. Production is currently (May 1990) from wells 16/7a-27 and 29 (re-named 16/7a-C1 and C2) and from two additional wells (16/7a-C3 and C4 on Fig. 6). Although the aquifer provides a certain amount of water drive, water injection will also be necessary. A further initial well is planned for this purpose, and it is also planned to use well 16/7a-29 subsequently as an injector.

This paper is published by permission of Marathon Oil UK Ltd, the Operator of Block 16/7a, and the Brae Group participants, Britoil plc (BP), Bow Valley Exploration (UK) Limited, Kerr-McGee Oil (UK) plc, Gas Council (Exploration) Limited, LL&E (UK) Inc, Sovereign Oil and Gas plc and Norsk Hydro Oil and Gas Limited. Thanks are also due to the following Marathon personnel whose work contributed significantly to the contents of this paper: N. S. Dorsey, B. H. McCarthy and T. Herman.

Central Brae Field data summary

Trap
Type — Structural/stratigraphic
Depth to crest — 11 750 ft TVSS
Lowest closing contour — 12 000 ft TVSS
Hydrocarbon–water contact — 13 426 ft TVSS
Oil column — 1676 ft

Pay zone
Formation — Brae
Age — Late Jurassic (Kimmeridgian–Early Volgian)
Gross thickness (average; range) — 800 ft; 0–1676 ft
Conglomerate + sandstone/shale ratio (average) — 0.7
Net/gross ratio (average) — 0.6
Cut off porosity for net pay — 8%
Porosity (average) — 11.5%
Hydrocarbon saturation — 80%
Permeability (average; range) — 100 md; 1–1000 md

Hydrocarbons
Fluid type — Black oil
Oil gravity — 33° API
Bubble point — 4112 psia
Gas/oil ratio — 1415 SCF/STB
Formation volume factor — 1.77 RB/STB

Formation water
Salinity — 79 000 ppm NaCl equivalent
Resistivity — 0.098 ohm m at 60°F

Reservoir conditions
Temperature — 246°F at 12 600 ft TVSS
Pressure — 7057 psia at 12 600 ft TVSS
Pressure gradient in reservoir — 0.327 psi/ft

Field size
Area — 1800 acres
Gross rock volume — 675 000 ac.ft
Hydrocarbon pore volume — 37 250 ac.ft
Recovery factor — 40%
Drive mechanism — water injection, aquifer
Recoverable hydrocarbons — Oil 64 MMBBL; NGL 6 MMBBL

Production
Start-up date — September 1989
Development scheme — subsea template and pipeline to Brae 'A'
Production rate (May 1990) — 15 500 BOPD
Cumulative production to May 1990 — approximately 3 MMBBL
Number/type of wells — 1 exploration; 3 appraisal; 4 development (to May 1990)
Secondary recovery method — water injection

References

CORNFORD, C. 1984. Source rocks and hydrocarbons of the North Sea. *In*: GLENNIE, K. W. (ed.) *Introduction to the Petroleum Geology of the North Sea*. Blackwell Scientific Publications, Oxford, 171–209.

MACKENZIE, A. S., PRICE, I., LEYTHAEUSER, D., MULLER, P., RADKE, M. & SCHAEFER, R. G. 1987. The expulsion of petroleum from Kimmeridge Clay source-rocks in the area of the Brae Oilfield, UK continental shelf. *In*: BROOKS, J. & GLENNIE, K. W. (eds) *Petroleum Geology of North West Europe*, Graham & Trotman, London, 865–877.

RILEY, L. A., ROBERTS, M. J. & CONNELL, E. R. 1989. The application of palynology in the interpretation of Brae Formation stratigraphy and reservoir geology in the South Brae Field area, British North Sea. *In*: Norwegian Petroleum Society, *Correlation in Hydrocarbon Exploration*, Graham & Trotman, London, 339–356.

ROBERTS, M. J. The South Brae Field, Block 16/7a UK North Sea. *This volume*.

STEPHENSON, M. A. The North Brae Field, Block 16/7a, UK North Sea. *This volume*.

TURNER, C. C., COHEN, J. M., CONNELL, E. R. & COOPER, D. M. 1987. A depositional model for the South Brae oilfield, *In*: BROOKS, J. & GLENNIE, K. W. (eds) *Petroleum Geology of North West Europe*. Graham & Trotman, London, 853–864.

The South Brae Field, Block 16/7a, UK North Sea

MARTIN J. ROBERTS

Marathon Oil UK Ltd., Marathon House, Rubislaw Hill, Anderson Drive, Aberdeen AB2 4AZ, UK

Present address: Shell Expro, 1 Altens Farm Road, Nigg, Aberdeen AB9 2HY, UK

Abstract: South Brae Oilfield lies at the western margin of the South Viking Graben, 161 miles northeast of Aberdeen. Oil production began in July 1983 from a single platform located in 368 ft of water. The field originally contained 312 MMBBL of recoverable reserves, and in May 1990, cumulative exports of oil and NGL reached 219 MMBBL. The reservoir lies at depths in excess of 11800 ft TVSS, has a maximum gross hydrocarbon column of 1670 ft, and covers an area of approximately 6000 acres.

The reservoir is the Upper Jurassic Brae Formation which is downfaulted against tight sealing rocks of probable Devonian age at the western margin of the field. The other field margins are constrained by a combination of structural dip and stratigraphic pinchout.

The reservoir is capped by the Kimmeridge Clay Formation, which is also the source of the oil.

The South Brae Oilfield is located in Block 16/7a adjacent to the western margin of the South Viking Graben, 161 miles northeast of Aberdeen, Scotland (Fig. 1). The field is being produced using a single, fixed, steel platform (Brae 'A') set in a water depth of 368 ft.

South Brae is one of the major oilfields of the UK North Sea, covering an area of approximately 6000 acres, having a maximum gross oil column of 1670 ft and originally containing 312 MMBBL of recoverable oil plus natural gas liquids (NGL). The reservoir sequence is of Upper Jurassic age and comprises thick units of sand-matrix conglomerate and sandstone, alternating with other thick units of mudstone and sandstone, which are commonly combined into large-scale fining upward sequences. The South Brae reservoir sequence is interpreted to have been deposited as the proximal part of a complex submarine fan system.

The name 'Brae' is derived from the old Scots word meaning hill.

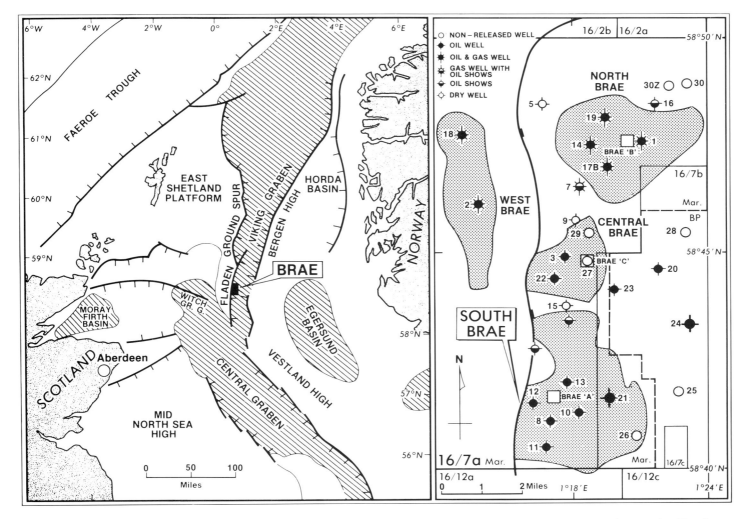

Fig. 1. Location of the South Brae Field in the South Viking Graben. The three fields shown in the graben all have Upper Jurassic reservoirs; West Brae has Palaeocene and Devonian reservoirs.

From Abbotts, I. L. (ed.), 1991, *United Kingdom Oil and Gas Fields,*
25 Years Commemorative Volume, Geological Society Memoir No. 14, pp. 55–62

History

Production Licence P.108, covering UK Blocks 16/7 and 16/3 was issued in 1970 as part of the third round of licence awards to Pan Ocean Oil (UK) Ltd, and Syracuse Oils (UK) Ltd. Subsequently, other participants joined in the venture and in 1976, Pan Ocean became a wholly owned subsidiary of Marathon Oil Company. In June 1976 a 50% relinquishment of Licence P.108 became effective with Block 16/7a and Block 16/3a constituting the remaining part of the Licence, which was then renewed for 40 years. Marathon Oil UK, Ltd operates the Licence on behalf of Brae Group co-venturers: Britoil (BP), Bow Valley, Kerr McGee, Gas Council (Exploration), LL and E, Sovereign and Norsk Hydro.

The South Brae discovery well, 16/7a-8, was drilled in 1977 at a location near the crest of an anticlinal structure mapped at Base Cretaceous level on 2-D seismic data. The well penetrated a gross hydrocarbon column of 1505 ft and black oil was flowed to surface from five drill-stem tests, with maximum flow rates of 7800 BOPD. Four additional appraisal wells were drilled in Block 16/7a during 1977 and 1978. An Annex B for field development was submitted in July 1979 and approved by the Department of Energy in January 1980. Two additional appraisal wells were drilled by BP in 1984 on the eastern flank of the field, in Block 16/7b. This portion of Block 16/7b subsequently came under Brae Group ownership in December 1988.

Following the installation of the Brae 'A' platform, first oil came on stream in July 1983. South Brae crude oil and NGL is exported from Brae 'A' via a 30 inch diameter pipeline 73 miles to BP's Forties 'C' platform, and then through the Forties pipeline system to Cruden Bay, northeast Scotland. A total of 33 development wells have been drilled into the South Brae reservoir (to May 1990) with field production now in decline from maximum rates in excess of 100 000 BOPD.

Field stratigraphy

A typical well illustrating the stratigraphy of the Upper Jurassic Brae Formation (Turner *et al.* 1987) and the overlying Upper Jurassic to Lower Cretaceous Kimmeridge Clay Formation is shown in Fig. 2. Palynological study of both formations has led to the establishment of a detailed biostratigraphic zonation scheme (Riley *et al.* 1989) which allows the major stratigraphic units to be dated (Fig. 2). A general stratigraphic log indicating the lithologies and formations of the post-Jurassic sequence of South Brae is shown in Fig. 3.

Geological history

In the South Brae area of the South Viking Graben the main period of fault movement along the graben-boundary fault can be broadly dated as Callovian to Volgian. Deep seismic data (6.0 s. TWT) beneath South Brae show some evidence of faulting in horizons that could be Triassic in age. These horizons are overlain by a Middle Jurassic, Bathonian–Bajocian, coal-bearing paralic sequence that can be correlated with nearby well control. This Middle Jurassic sequence is of relatively uniform thickness and was deposited under tectonically stable conditions. It is in the sequence immediately overlying the Middle Jurassic that a pronouced syntectonic wedge of sediments of Callovian to Volgian age can be seen on seismic to thin eastwards (Fig. 4). This wedge dates the main phase of fault movement along the boundary fault. In South Brae, sediments of the Brae Formation no older than late Oxfordian(?) have been penetrated by drilling, so the greater part of the syntectonic wedge of sediments has not been drilled close to the graben margin. The Brae Formation is composed of submarine fan sediments, deposited as late-stage graben fill, and is both overlain by, and is laterally equivalent to the organically rich hemipelagic mudstones of the Kimmeridge Clay Formation.

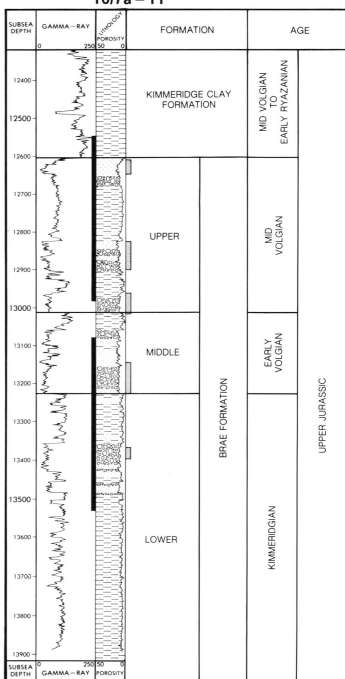

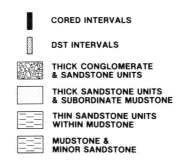

Fig. 2. Brae Formation stratigraphic terminology.

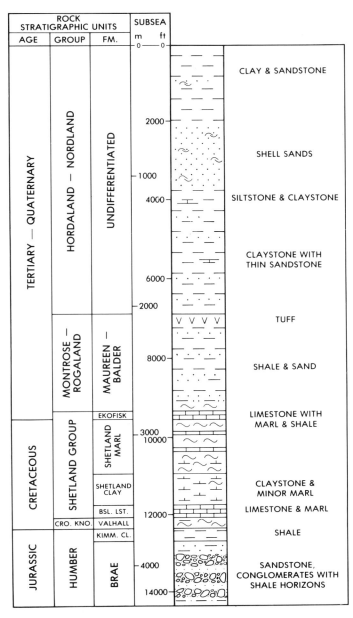

Fig. 3. Generalized South Brae stratigraphic log.

During the early Cretaceous, continued movement along the main boundary fault and associated 'antithetic' faulting, coupled with differential compaction over the coarse clastic sequence, led to the development of the South Brae Field structure. Thick late Cretaceous and Tertiary sequences (Fig. 3) were deposited under tectonically quieter conditions over a broad basin extending far beyond the South Viking Graben.

The last re-activation of the graben margin faults was during the late Palaeocene to early Eocene, which resulted in minor displacement of the lower Tertiary and older sequences.

Geophysics

Seismic surveys

Various vintages of 2-D seismic data were acquired between 1970 and 1976, and were used to map the structure prior to the drilling of the South Brae discovery well. During 1977 a 3-D seismic survey was shot over South Brae, and reprocessing of these data in 1983/84 provided the basis for initial development mapping.

During 1985, an extensive 3-D survey was acquired over Blocks 16/7a, 16/7b, 16/8a, 16/8b, and 16/8c in co-operation with the operators of these blocks (the Brae–Miller 3-D survey). Current interpretation of South Brae is based on this data set.

Seismic interpretation

The Top Kimmeridge Clay Formation (Base Cretaceous) reflector is clearly seen on all vintages of data, and can be unequivocally identified across the field. However, this reflector is a strong multiple generator, and on the older seismic data multiple energy largely masked or interfered with Upper Jurassic events. For the latest 3-D seismic data, careful selection of stacking velocities and deconvolution operators has attenuated multiples to a large degree (Fig. 5).

The 3-D data have allowed more accurate mapping of the Base Cretaceous, particularly in the structurally complex areas near the western margin of the South Viking Graben. Fault definition has been improved and, for the first time, seismic analysis of the detailed stratigraphy within the reservoir has been achieved.

Trap

The South Brae trap was formed by a combination of faulting, folding, and stratigraphic discontinuities. The maximum oil column is 1670 ft in height, of which structural closure on the Top Brae Fm accounts for only 200 ft. The main trapping element is located to the west, at the major fault zone marking the edge of the Fladen Ground Spur, where Brae Formation clastic rocks abut impermeable Devonian sandstones. South Brae is therefore a classic example of a hanging-wall fault closure trap.

South Brae comprises two main structural areas. The major part of the field is a N–S trending anticline with a gentle easterly dip and an associated syncline to the west which is more steeply dipping and faulted (Figs 4 and 6). To the north of the major structure, a low relief compactional saddle separates a smaller feature known as the 'northern lobe' (Fig. 6). Coarse clastic sediments within this 'northern lobe', which were delineated by well 16/7a-6 prior to development, are interpreted to be the product of a separate coarse clastic source to the west. The lobe shares the same original oil–water contact with the main part of the field, reservoir communication being transmitted through laterally overlapping coarse clastic layers.

To the south of the main field and north of the northern lobe, the South Brae reservoir passes laterally into fine grained, laminated mudstone sequences, while to the east, the field is in part bounded by an original oil–water contact at 13 488 ft TVSS and in part by reservoir pinchout. The overlying Kimmeridge Clay Formation provides an effective vertical seal.

Reservoir

The South Brae reservoir is the proximal part of a complex submarine fan sequence, which was the product of erosion of Devonian sandstones and possibly also younger sediments of the adjacent Fladen Ground Spur.

In order to analyse the varied lithologies making up the Brae Formation, over 24 000 ft of South Brae core have been described. From this extensive database, six major lithofacies are recognized, based on lithology and internal sedimentary structures, as follows:

(1) sand-matrix conglomerate;
(2) mud-matrix breccia;
(3) medium- to thick-bedded sandstone;
(4) alternating thin-bedded sandstone with interlaminated sandstone/mudstone;

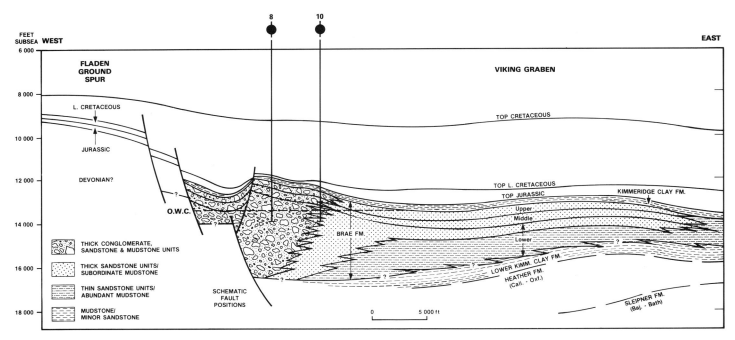

Fig. 4. South Brae structural cross-section.

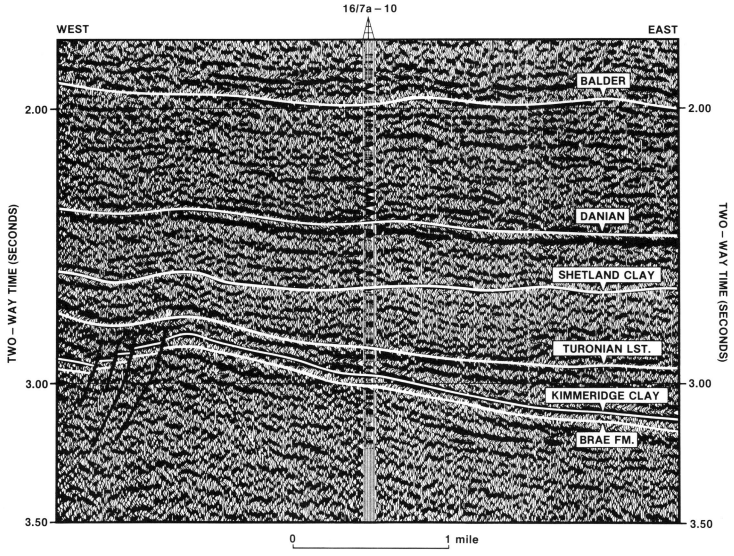

Fig. 5. West–east 3-D seismic section through well 16/7a-10, South Brae Field.

(5) interlaminated sandstone/mudstone;
(6) laminated mudstone.

These lithofacies and their depositional processes are described in detail by Turner *et al.* (1987). The conglomerate and sandstone facies are the product of various types of sediment gravity flows. The presence of variable but generally small components of comminuted marine fossil debris in all of these coarse clastic facies demonstrates that the sediment gravity flow processes were operating in an entirely submarine setting. Mudstones within the sequence are the product of hemipelagic settling of silt and clay-grade material.

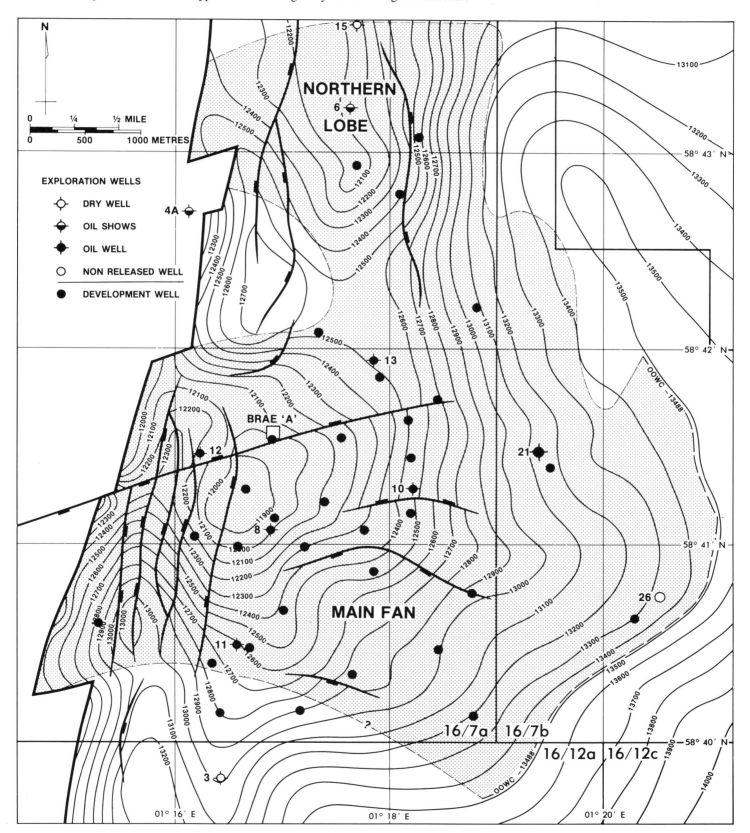

Fig. 6. Top Brae Formation structure map and South Brae Field outline. Depth contours are in ft TVSS.

There is a degree of vertical organization of facies within the South Brae Field. Large-scale upward-fining sequences can be recognized (Fig. 7), which are of the order of tens to hundreds of feet in thickness. The bases of these sequences are commonly erosional and sharply defined, while the overall upward fining results from interbedding and replacement of coarser facies by finer facies. The upward repetition of such sequences gives the reservoir a well-developed vertical succession of coarse grained packages (Facies 1 and 3) and finer-grained packages (Facies 4, 5 and 6), as shown in Fig. 8. In three dimensions, particularly in the upper Brae Formation, the coarse grained packages often occur as channel-like bodies which radiate basinwards from an apex to the west of the field, and which are separated by fine-grained interchannel areas. The channel-like bodies become less conglomeratic and more sandy basinwards. To the northwest and southwest of the fan system, and in close proximity to the graben margin fault zone, the dominant facies are mud-matrix breccia and mudstone.

Diagenetic processes have served both to reduce and subsequently enhance porosity within sandstones and conglomerates. Early carbonate cementation was probably initially extensive, although not pervasive, and in areas not affected by this process quartz overgrowths are common and sparse illitic rims occur on some quartz grains. Much of the early carbonate cement, together with some shell debris and significant amount of feldspar were dissolved during an important later stage of secondary porosity development, which preceded oil emplacement.

Porosity reduction due to diagenesis is more pronounced in the deeper parts of the reservoir, and in areas closest to the fault, where it may contribute to the seal of the reservoir against impermeable Devonian sandstones. Further research into diagenetic aspects of the reservoir is ongoing.

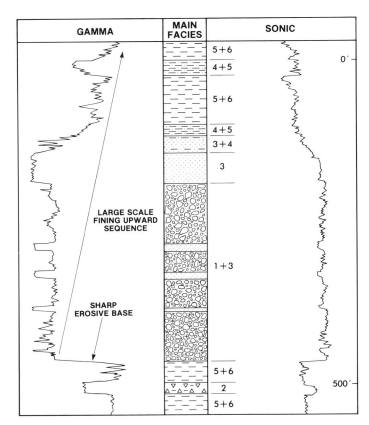

Fig. 7. Summary example of most common South Brae facies associations.

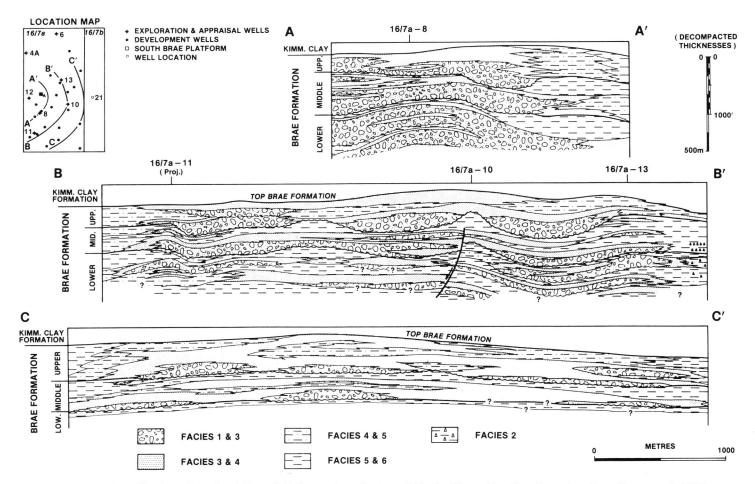

Fig. 8. North–south profiles through the South Brae Field, flattened at a horizon within the Kimmeridge Clay Formation. From Turner *et al.* (1987).

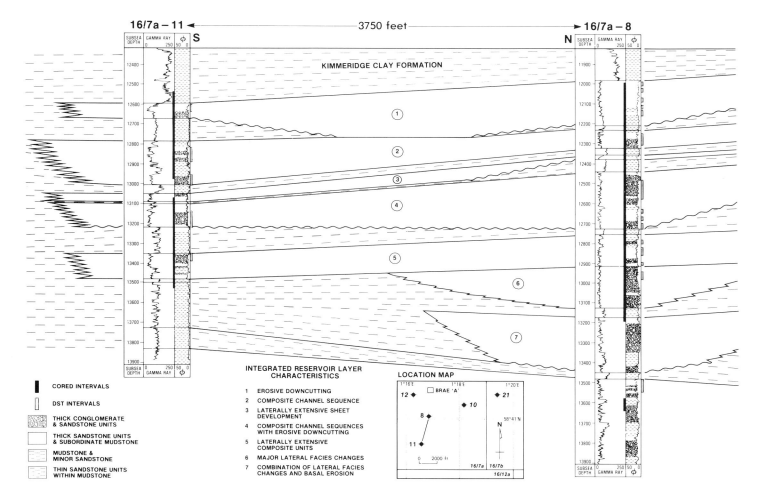

Fig. 9. Example of South Brae Field reservoir correlations.

Correlation within the reservoir

The techniques used for correlation within the South Brae reservoir can be grouped into three main categories, lithostratigraphy, biostratigraphy, and reservoir pressure techniques.

The large-scale upward-fining sequences aid correlations between adjacent wells. Electric-log responses clearly reflect these sequences and are used to facilitate correlation, particularly in the absence of core information.

In addition to vertical facies changes, lateral facies changes from conglomerate and sandstone to fine-grained facies in a direction perpendicular to sediment flow also occur. Where such lateral transitions can be identified, the fine-grained facies are grouped with the adjacent coarse clastic facies for reservoir mapping purposes.

For biostratigraphic purposes, palynomorphs have been extracted from a very large number of samples since exploration drilling began in South Brae. Preparations derived from conventional cores, sidewall cores, and ditch cuttings have consistently revealed palynomorphs from all lithofacies in varying degrees of abundance, but recovery is best from the mudstone facies.

Recently, the extensive palynological data set from the Brae area has been integrated into a biostratigraphic framework that appears to have regional application within Quadrant 16 (Riley *et al.* 1989). Nineteen biostratigraphic zones have been identified, which span the Callovian–Ryazanian interval of the Brae area. This biostratigraphic zonation scheme allows stratigraphic correlation to be made where significant lateral facies changes occur, both within the reservoir and adjacent to the margins of the submarine fan systems.

Since production began from the South Brae Field, a further tool for reservoir correlation has been available. Reservoir pressure depletion due to fluid production is routinely calculated in all wells from formation pressure tests. Prior to the start of production, a single oil pressure gradient and a single water pressure gradient were evident throughout the field. However, production and injection have affected reservoir layers in differing ways due to the interlayer mudstones forming local pressure barriers. Therefore, average pressure depletion of reservoir layers provides support for the correlation of reservoir units and the degree of connection within reservoir bodies.

Using the above techniques, a total of seven reservoir layers are correlated within the South Brae Field. An example of reservoir correlations is shown in Fig. 9. Porosity and permeability are very variable within each of the layers and are largely controlled by the facies distribution. The best reservoir rock is massive sandstone in the uppermost Brae Formation, where layer average porosities and air horizontal permeabilities range up to 21% and 2180 md respectively. Both of these parameters show a decline with depth.

Source rocks

The Brae Formation is both overlain by and regionally interdigitates with the organic-rich Kimmeridge Clay Formation which, together with mudstones within the Brae Formation, are considered to be the source rocks for the South Brae oil accumulation (Reitsema 1983). Relatively short migration pathways for the hydrocarbons into the South Brae structure are probable, and it is considered that hydrocarbon generation was dominantly from the Eocene onwards.

Hydrocarbons

The maximum DST flow rate on the South Brae Field discovery well 16/7a-8 was 7800 BOPD on a 1.25 inch choke. The crude oil gravity ranges from 33° to 37° API, and the gas–oil ratio at initial pressure is 1343 SCF/STB. Separator gases show specific gravities ranging from 1.02 to 1.05, CO_2 contents ranging from 32% to 35%, and low concentrations of H_2S (85 ppm). The oil formation volume factor ranges from 1.73 RB/STB at an initial reservoir pressure of 7128 psia, to 1.86 RB/STB at the bubble point pressure of 3702 psia. The crude oil therefore exhibits a relatively large shrinkage factor due to its high content of dissolved gas. This characteristic and good reservoir management has necessitated an extensive pressure maintenance programme during development.

Reserves

South Brae originally held 312 million barrels of recoverable oil plus NGL. The pressure maintenance scheme has been a combination of crestal injection of associated gas and peripheral water-flood. Gas export to North Brae commenced in April 1988 and when crestal gas injection ceases, maximum gas export to North Brae will occur. Thereafter South Brae will be produced from crestal wells utilizing water-flood from the downdip injectors for pressure support.

South Brae reached plateau oil production rates in excess of 100 000 BOPD in 1985 and held this level until late 1988 before decline began. As of May 1990, cumulative oil and NGL production was 219 MMBBL.

This paper is published with permission of Marathon Oil UK, Ltd, the Operator of South Brae, and Participants, Britoil plc (BP), Bow Valley Exploration (UK) Ltd, Kerr-McGee Oil (UK) plc, Gas Council (Exploration Ltd, LL&E (UK) Inc, Sovereign Oil and Gas plc and Norsk Hydro Oil and Gas Ltd. I should like to record my thanks to the numerous people who have all contributed to the knowledge of the South Brae Field throughout its exploration and production history.

References

REITSEMA, R. H. 1983. Geochemistry of North and South Brae areas, North Sea. *In*: BROOKS, J. (ed.) *Petroleum Geochemistry and Exploration of Europe*, Geological Society, London, Special Publication **12**, 203–212.

RILEY, L. A., ROBERTS, M. J. & CONNELL, E. R. 1989. The Application of palynology in the interpretation of Brae Formation stratigraphy and reservoir geology in the South Brae Field area, British North Sea. *In*: *Correlation in Hydrocarbon Exploration*. Norwegian Petroleum Society, Graham & Trotman, London, 339–356.

TURNER, C. C., COHEN, J. M., CONNELL, E. R. & COOPER, D. M. 1987. A depositional model for the South Brae Oilfield. *In*: BROOKS, J. & GLENNIE, K. W. (eds) *Petroleum Geology of North West Europe*. Graham & Trotman, London, 853–864.

South Brae Field data summary

Trap
Type	Combination structural/stratigraphic
Depth to crest	11821 ft TVSS
Lowest closing contour	12 100 ft TVSS
Hydrocarbon–water contact	13 488 ft TVSS
Oil column	1670 ft

Pay zone
Formation	Brae
Age	Upper Jurassic (Kimmeridgian to mid-Volgian)
Gross thickness (average; range)	800 ft; 0–1670 ft
Net/gross ratio (average)	0.75
Cut off for net pay	8% porosity in sandstones; 5% porosity in conglomerate
Porosity (average)	11.5%
Original hydrocarbon saturation (average)	80%
Permeability (average; range)	131 md; 2–500 md
Productivity index (range)	10–40 BOPD/psi

Hydrocarbons
Fluid type	Black oil
Oil gravity (range)	33° to 37° API
Gas gravity (relative to air)	1.02 to 1.05
Bubble point	3702 psia
Gas/oil ratio	1343 SCF/STB
Formation Volume Factor	1.73 RB/STB

Formation water
Salinity	75 000 ppm NaCl equivalent
Resistivity	0.120 ohm m at 60°F

Reservoir conditions
Temperature	253°F at 12 740 ft TVSS
Pressure	7128 psia at 12 740 ft TVSS
Pressure gradient in reservoir	0.295 psi/ft

Field size
Area	6000 acres
Gross rock volume	2 750 300 ac.ft
Recovery factor	33%
Drive mechanism	water injection, gas injection, aquifer
Recoverable hydrocarbons	Oil 274 MMBBL NGL 38 MMBBL

Production
Start-up date	July 1983
Development scheme	Production–processing–drilling platform
Production rate (May 1990)	34 400 BOPD (peak 115 000 BOPD, Nov. 1986)
Cumulative production to May 1990	219 MMBBL oil plus NGL
Number/type of wells	1 exploration 6 appraisal 33 development (to May 1990)
Secondary recovery method	water injection, gas injection

The Brent Field, Block 211/29, UK North Sea

A. P. STRUIJK & R. T. GREEN

Shell UK Exploration and Production, c/o 1 Altens Farm Road, Nigg, Aberdeen AB9 2HY, UK

Abstract: The Brent Field was the first discovery in the northern part of the North Sea, and is one of the largest hydrocarbon accumulations in the United Kingdom licence area. There are two separate major accumulations: one in the Middle Jurassic (Brent Group reservoir) and one in the Lower Jurassic/Triassic (Statfjord Formation reservoir). The field lies entirely within UK licence Block 211/29 at latitude 61°N and longitude 2°E. The water depth is 460 ft. The discovery well was drilled in 1971, and six further exploration and appraisal wells were drilled. Seismic data over the Brent Field has been acquired in three separate vintages. The latest acquisition is a 3-D grid recorded in 1986. Reprocessing of the entire 1986 3-D seismic data set was initiated in 1989.

The original oil/condensate-in-place, estimated on 1/1/89, is 3500 MMBBL, and the estimated original wet gas-in-place is 6700 TCF. Oil production is now in the decline phase. Average production in 1988 was 334,000 BOPD, with gas sales remaining at the plateau rate of 500 MMSCFD.

The field is being developed from four fixed platforms, each providing production, water injection and gas injection facilities for both Brent and Statfjord Formation reservoirs. Gas injection is distributed to achieve an intermediate oil rim development in some reservoir units. The platforms were installed between 1975 and 1978. Production commenced in 1976. The slump faulted crestal areas of both reservoirs have yet to be developed.

These crestal areas contain about 5% of the recoverable reserves. Appraisal drilling was carried out in the crest during 1988 and 1989.

The Brent Field is located approximately 100 miles north-east of the Shetland Islands and 300 miles NNE of Aberdeen (Fig. 1). The discovery well location is at latitude 61°05′53.87″ North longitude 1°41′30.11″ East. The water depth is 460 ft

The field comprises two distinct reservoirs, the Brent Group and the Statfjord Formation, which are of Middle Jurassic and Lower Jurassic/Triassic age respectively. The reservoirs occur in a westerly dipping tilted fault block in a fault controlled unconformity trap (Fig. 2).

The size of the hydrocarbon bearing area is approximately 10 miles from north to south and 3 miles from east to west (Fig. 3).

The reservoirs are in turn divided into seven separate reservoir units; four cycles in the Brent Group reservoir and three units in the Statfjord Formation reservoir. Laterally two major east–west orientated faults divide the field into three separate production areas. A fourth area is the north–south orientated crestal part of both reservoirs, which is faulted and has a series of down faulted slump blocks overlain by 'Reworked Sediment'. This area still has to be developed.

The Shell/Esso joint venture North Sea oilfields are named after water and waterside birds. The Brent Field is named after the Brent goose.

History (Table 1)

The Brent Field is located entirely in Block 211/29 of the United Kingdom licence area in the northern North Sea. The field was developed by Shell UK. Exploration and Production on behalf of the Shell/Esso joint venture under Production Licence P.117, granted in the 3rd round on 29 July 1970, for an initial term of 6 years, subsequently extended to 2018. The discovery well, 211/29-1, was drilled from May to July 1971 at SP 10 034 on seismic line U394.

The objective of this well was to test a monoclinally dipping sequence below a pronounced regional unconformity at a prognosed depth of 8850 ft TVSS in the Shetland Trough.

The well found the regional unconformity between clays of Lower Cretaceous Barremian age and Upper Jurassic Volgian–Kimmeridgian organic shale at 8462 ft TVSS. From 8848–9655 ft TVSS a section of interbedded sandstone, siltstone, shale and coal (later called the Brent reservoir) was encountered with 466 ft net

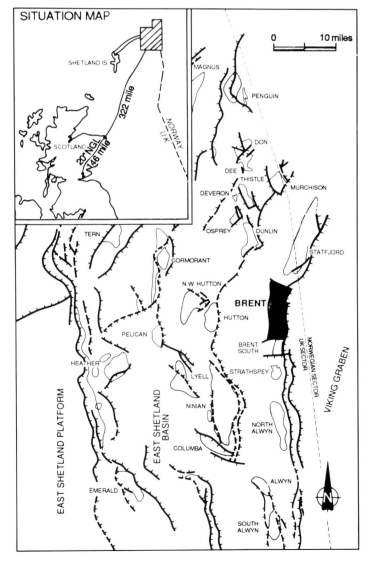

Fig. 1. Location map of the Brent Field.

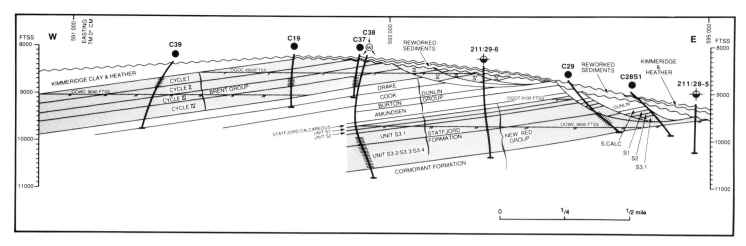

Fig. 2. Structural cross-section across the Brent Field.

Table 1. *Key dates in field development*

Discovery	:	1971
Declared commercial	:	1972
Production start-up	:	Nov. 1976 (Brent B, into storage)
First tanker loaded	:	December 1976
First oil to Sullom Voe	:	November 1979
First year peak production	:	1983 (oil), 1986 (NGL), 1984 (gas)
Last year peak production	:	1986 (oil)

		'A' Platform	'B' Platform
Contract placed	:	December 1972	August 1973
Installed	:	May 1976	August 1975
Production drilling started	:	December 1977	February 1976
First gas injection	:	April 1979	February 1980
First water injection	:	December 1979	October 1979
Production start-up	:	June 1978	November 1976
Start of gas export	:	March 1982	May 1982

		'C' Platform	'D' Platform
Contract placed	:	December 1973	May 1974
Installed	:	June 1978	July 1976
Production drilling started	:	Late 1979	January 1977
First gas injection	:	September 1981	October 1978
First water injection	:	September 1980	October 1979
Production start-up	:	June 1981	November 1977
Start of gas export	:	August 1982	September 1982

Spar			
Contract placed	:	March 1973	
Installed	:	June 1976	

sand including 140 ft of oil bearing sandstone. FITs confirmed light oil of 52.1 lbm/ft^3.

Well 211/29-1 penetrated the Brent Group reservoir in a downdip position. In June 1972 well 211/29-2 was drilled some 3 miles SSE of 211/29-1 and confirmed the whole Brent Group of about 870 ft to be hydrocarbon bearing including a 170 ft gas bearing interval. Production testing of the well yielded a flow of 6600 BOPD.

Four further appraisal wells were drilled in 1973 and 1974. Well 211/29-3 confirmed the northerly extent of the accumulation while 211/29-4 discovered hydrocarbons in the deeper 800 ft thick Statfjord Formation sands.

The exploration well 211/29-5 tested the deeper Triassic sequence, but found the sands water-bearing.

Appraisal well 211/29-6 confirmed the hydrocarbons in the Statfjord Formation sands proved by 211/29-4. 211/29-7 was drilled as a development well with an underwater completion.

The later well 211/29-8, drilled in 1982, tested the sedimentary sequence adjacent to the eastern boundary fault of the Brent structure. It found Upper Jurassic sandstones with residual oil saturations and water-bearing Brent Group sands.

The initial development plan for the Brent Field included development of both the oil zone and the gas cap of each of the reservoirs. The originally conceived basic strategy was to produce the Brent and Statfjord reservoirs using downdip water injection wells for pressure maintenance and updip for oil production. Considerable effort therefore went into replacing watered-out downdip oil completions with drier updip completions.

With gas production continuing to exceed sales commitments, some gas is reinjected into the upper reservoir units of both reservoirs. The two upper Brent cycles have been selected for gas injection because they both had sizeable initial gas caps. Oil recovery of these zones will be maximized by adopting an oil rim

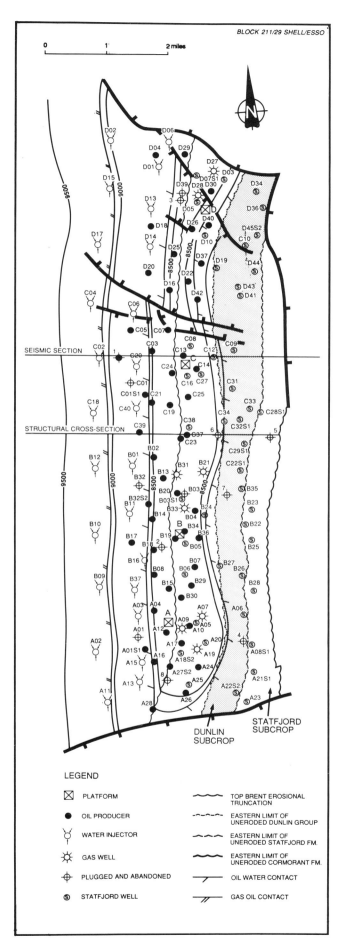

Fig. 3. Brent Field top reservoir map.

development strategy. For Statfjord Unit 1 there is the added advantage of a miscible displacement by gas (Jense et al. 1988).

Plateau production of some 400 000 BOPD was maintained from 1983 to 1986. Thereafter production declined from an average 372 000 BOPD in 1987 to 334 000 BOPD in 1988.

Produced gas is sold to British Gas at a rate of 500 MMSCFD.

Estimated ultimate recoveries are 1815 MMBBL of oil and condensate and 3.7 TCF of gas, with approximately two-thirds coming from the Brent Group reservoir and one-third from the Statfjord Formation reservoir.

The field is being developed from four fixed production platforms.

Oil was initially exported via tankers loading at the Brent Spar loading buoy. Since November 1979 oil has been mainly transported through the Brent System pipeline to the oil terminal at Sullom Voe in the Shetlands.

Brent Field gas has been exported via the FLAGS line to St Fergus in Scotland since May 1982.

Field stratigraphy

The stratigraphic setting of the Brent Field is illustrated in Fig. 2. Figure 4 shows a generalized stratigraphic column for the field.

Well 211/29-8, drilled to the east of the main field proved a thick sequence (greater than 1600 ft) of barren red siltstones and tight water bearing sandstones of the Triassic Cormorant Formation. Overlying these sediments is the older of the two reservoir horizons, the Rhaetian to Lower Sinemurian Statfjord Formation. The contact between the two formations is gradational and marked by an increase in net to gross sand ratio that reflects an increase in coarse clastic deposition in the basin. The Statfjord Formation has been formally subdivided from bottom to top into the Nansen, Eiriksson and Raude Members (Deegan & Scull 1977). In the Brent Field the Nansen Member (the youngest of the three) consists of a laterally homogeneous transgressive marine sand which is overlain by a carbonate cemented sand, informally termed the Statfjord Calcareous Unit.

The overlying mudstones and siltstones of the Dunlin Group separate the Brent Group and Statfjord Formation reservoir horizons. These fine-grained shallow marine sediments are very uniform in thickness in the fault terrace between North Alwyn Field to the south and Statfjord Field to the north, and are 850 ft thick in the Brent Field. Four formations are recognized within the Dunlin Group: the Amundsen, Burton, Cook and Drake, the youngest, the Amundsen, being Toarcian in age.

The second reservoir sequence, the Brent Group, is mostly of Bajocian age and is subdivided from bottom to top into five formations, the Broom, Rannoch, Etive, Ness and Tarbert Formations. The maximum thickness of the reservoir is 850 ft, and the sediments are a variety of coarse- to fine-grained sandstones, siltstones and mudstones together with significant thicknesses of coals (up to 30 ft in total) in the Ness Formation. The overlying Tarbert Formation consists of a varied assemblage of deposits that formed during the return to shallow marine sedimentation of the overlying Humber Group.

The Humber Group is subdivided into the Heather and the Kimmeridge Clay Formations, the latter being the main source rock of the region. The Heather Formation contains pale grey mudstones of mostly Bathonian age, the Callovian stage being largely missing due to the first of a number of unconformities and condensed sequences that characterize the late Jurassic and early Cretaceous in the Brent Field and elsewhere in the region. The dark organic rich shales of the Kimmeridge Clay Formation are of Oxfordian to Ryazanian age.

Lower Cretaceous sediments are generally thin over the Brent Field structure but include deposits of Albian and Aptian allochthonous chalky debris flows of uncertain source, some of which are characteristically red stained. Autochthonous carbonates and green and dark grey mudstones are also found in the Lower Cretaceous. The Upper Cretaceous and Tertiary sequence includes over 8200 ft of generally monotonous mudstones with thin chalky carbonate horizons near the base. Sandy Lower Tertiary foresets that represent the lateral equivalents of Palaeocene and Eocene turbidite submarine fans, and thin tuffaceous horizons are also present.

Tectonic history

The Brent Field is located at the western margin of the Viking Graben, the northern extension of a rift system 600 miles long which continues into the Central North Sea Graben to the south.

The Brent Field lies in a major fault terrace over 40 miles long that can be traced from the North Alwyn Field in the south to the Statfjord Field in the north. This terrace is bounded to the west by the Hutton–Dunlin–Murchison fault zone and to the east by faults that occur east of the crest of the Statfjord Formation in the Brent Field, making the terrace approximately 12 miles wide. Both these fault zones were active during the deposition of the reservoir sandstones.

The regional structural trends are dominantly orientated NE–SW. The Brent Field is characterized rather by north–south elements that define the eastern-most part of the fault block, bounding the Viking Graben. These elements formed by composite dip-slip normal faults which have a total throw of approximately 6000 ft and reflect extension within the Graben. Associated with these north–south elements are a series of east–west and NW–SE trending faults which display variable throw. These faults are mostly local to the Brent Field area and are accommodation features produced by differential tilting and uplift along the major fault terrace. Related faults mark the southern and northern boundaries of the field and define a 2–3 mile wide graben feature in the north of the field. The northern field boundary fault is an atypical member of this fault set as it has a northwest–southeast orientation and extends northwest towards the Hutton Field.

All the faults in the field were reactivated periodically during the rotation of the fault block including the east–west accommodation faults which die out to the west of the accumulation.

Deformation in the crestal area of the field and its structural style is dominated by local processes, prompted by rotation of the Brent Field fault block and the opening of the Viking Graben. At the Brent Group level collapse of the gravitationally unstable fault scarps by subaqueous failure led to tectonic, and in places sedimentological, reworking of the sediments. This subaqueous failure occurred along a series of listric faults soling out in the Cook Formation.

At the Statfjord Formation level the style of crestal deformation is broadly analogous to that seen at the crestal part of the Brent Group reservoir. The decollement horizon for the listric faults which have deformed the crestal Statfjord occurs near the base of the Statfjord Formation where sand rich intervals overlie shale dominated intervals.

Palaeontological evidence suggests that the gravitational induced crestal failure occurred at two main times, during the Bathonian along the whole crestal area, and during the early Cretaceous (pre-Santonian) on the Statfjord Formation crest in the southern part of the field.

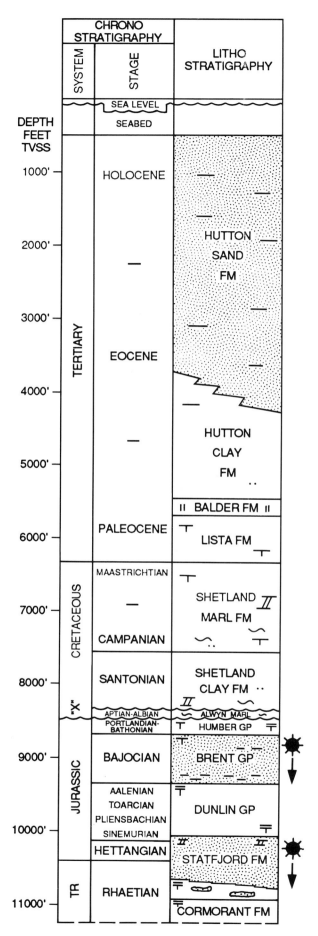

Fig. 4. Stratigraphic column for the Brent Field.

Geophysics

Seismic data over the Brent Field has been acquired at three different times.

(1) Wide spaced rectangular grid orientated east–west of 1970s vintage, augmented by traded and courtesy data. The oldest data set was shot in 1966. The seismic grid has north–south and east–west orientated lines with spacing varying from 3–7 km.

(2) A close spaced (200 m) grid oriented east–west recorded from 1980 to 1982 was processed as a pseudo-3-D dataset with sampling 25 m × 200 m.

(3) A 3-D grid recorded in 1986, spaced 12.5 m east–west by 37.5 m north–south but processed 25 m × 37.5 m.

The interpretation of the 1970 seismic outlined the north–south trend of the Tertiary Basin. Intermediate levels within the prospective Palaeocene interval were examined for appreciable closures, but none could be proven owing to the absence of reliable correlatable reflectors.

An important objective of the discovery well was a deep seated feature beneath the unconformity at an estimated depth of 8850 ft TVSS, based on a two-way travel time of about 2.4 seconds (Fig. 5).

The 1982 pseudo-3-D grid was an improvement over the various vintages that comprised the older 2-D seismic dataset, allowing reservoir horizons to be mapped. However it still proved insufficient to sample the faulting along the crestal and graben areas of the field.

The 1986 3-D seismic survey extends over the entire block with extensions into Blocks 211/24 to the north and 211/30 to the east and 3/4 to the south and was designed to image the complex fault pattern that had been seen by the 1982 seismic survey but was as yet unresolved on the crest of the field and the east flank. The seismic resolution has been improved by some 400% in the north–south direction and 40% in the vertical sense compared to 1982 2-D data. It enabled accurate interpretation of the west flank but, as there was little or no impedance contrast, the slump block and reworked sediments on eastern flank were not sufficiently imaged.

During 1988 reprocessing tests on selected 1986 seismic lines resulted in a significant improvement in data quality at the crestal area of the Statfjord Formation and at the crestal Brent Group and slump block levels. Reprocessing of the entire 1986 3-D seismic data was therefore initiated in January 1989.

A comparison of the 1970s 2-D seismic and 1986 3-D seismic is shown in Fig. 5.

Trap

Both the Statfjord Formation and Brent Group reservoirs are contained in a simple dip closed structure, dipping 8° to the west, with truncation unconformity traps (Fig. 2). The lower reservoir, the Statfjord Formation, is additionally partly trapped by fault juxtaposition of Lower Cretaceous mudstones east of the culmination. The upper seal is provided by a series of mudstones or chalky sediments overlying composite unconformity surfaces. In comparison to other oil fields in the area the Brent Field structure is relatively simple and lateral closure is provided by non-reservoir juxtaposition across E–W and NW–SE-trending faults (Fig. 3).

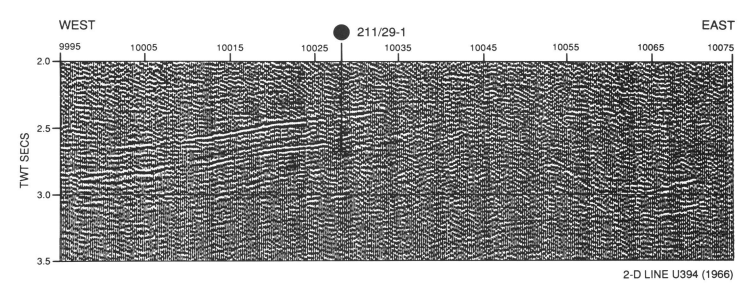

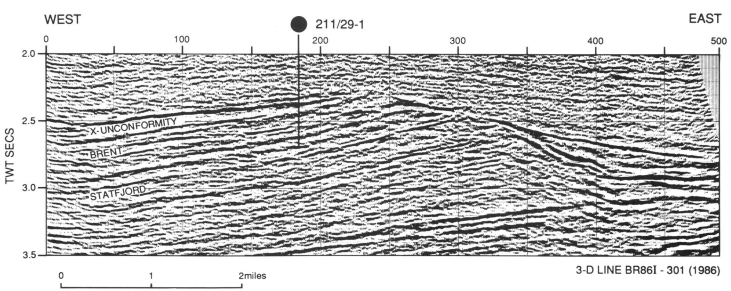

Fig. 5. Comparison of original seismic line through discovery well and latest seismic line.

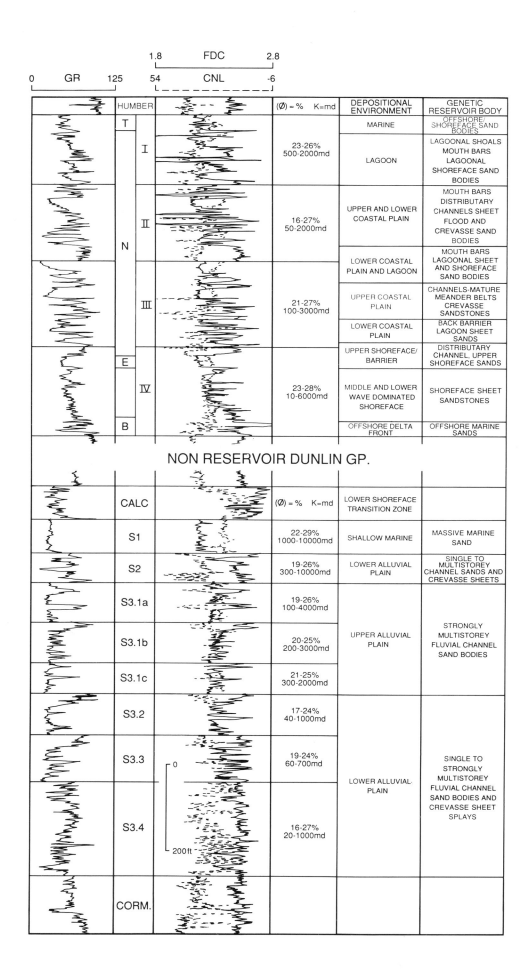

Fig. 6. Type log of the Brent and Statfjord reservoirs.

The trap was formed by Mid-Cretaceous times, long before oil migrated into the structure during the Eocene. Overpressures probably developed gradually during Tertiary burial of the structure and reservoirs are now overpressured by approximately 1800 psi.

Reservoir

Brent Group

The Brent Group reservoir varies between 780 and 850 ft in thickness and the producing horizons are all fine- to coarse-grained sandstones.

The stratigraphic subdivision of the Brent Group into five formations for regional correlation is too simplistic for field management purposes.

Instead the Brent Group reservoir is divided into four main genetic reservoir bodies or cycles based on sedimentological analysis of cores and logs. These Cycles I–IV are further divided fieldwide into 16 reservoir units. The abundance of coals and the use of pressure data to help define significant barriers to vertical pressure communication are used to establish the subdivisions and Cycles I, II and III represent a 'layer cake' sequence. The relationship of the cycles to the formations is shown in Figs 6 and 7.

Statfjord Formation

The Statfjord Formation is most fully developed towards the axis of the Viking Graben and reaches a thickness of 1000 ft in the Brent area.

The Statfjord Formation has been subdivided into eight units for reservoir management purposes: Unit S1, 2, 3.1a, 3.1b, 3.1c, 3.2, 3.3 and 3.4 as shown in Fig. 6. The boundaries of the units are taken at locally important shales which act as vertical permeability barriers. Field performance data indicates that only two of these boundaries are field-wide flow barriers; these are the boundary between Unit S2 and S3.1a and the boundary between S3.1c and S3.2. These two boundaries define the three parts of the reservoir which are managed separately during production and which can be correlated across the whole field.

Reservoir Quality (Table 2)

The reservoir quality of the Brent Group cycles is generally very good (Figs 6 and 7). The distribution of good quality sandstones and vertical permeability profiles are very variable and strongly lithofacies controlled. Five major rock types define the Brent Group reservoir: sands, micaceous sands, sand-dominated mudstones, mud-dominated sands and muds including coals. The reservoir rock characteristics range in porosity from 16–29% and the horizontal permeability from 10–6000 md in the Etive Formation.

The Broom, Etive, Ness and Tarbert Formations are dominantly subfeldspathic arenites or quartz arenites. The average feldspar content is 4.6%, average mica content is 2.4%, and the remainder of the framework minerals are mostly quartz. The Rannoch and some Ness Formation lagoonal deltaic sands contain significant amounts of muscovite (average 10.5%).

The main cement in the Brent Group is authigenic quartz (3.4% of the bulk volume). Carbonate cementation has been observed in the form of concretions in the water leg of the Rannoch Formation. Highly micaceous parts of the Rannoch Formation severely reduce permeability to less than 1 md. The Broom Formation also contains carbonate bands, which are not widely correlatable.

The Statfjord Formation reservoir Units S2 and S3 consist mainly of a complex alternation of fluvial sandbodies and intercalated floodplain shales. The reservoir rock has porosities ranging

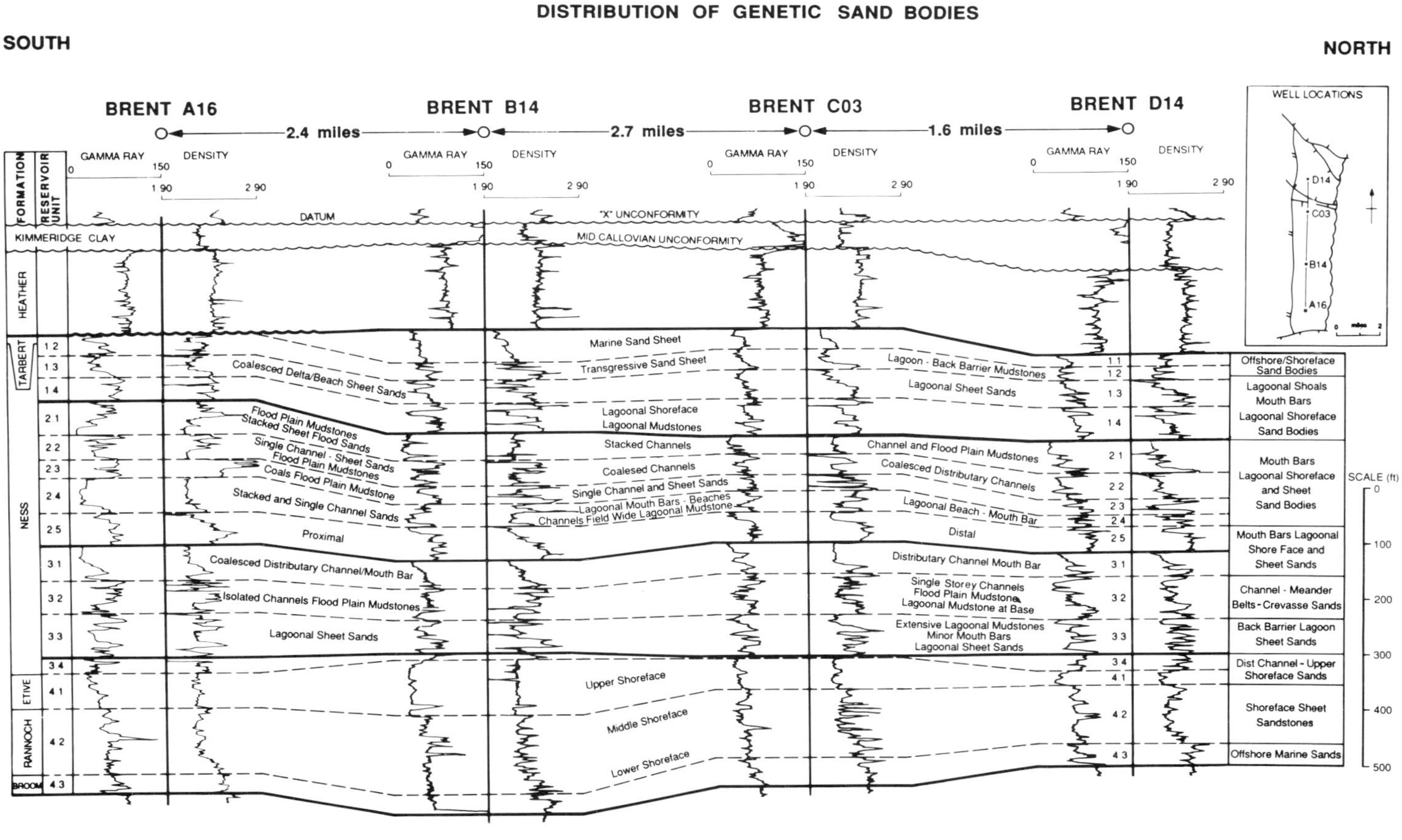

Fig. 7. Brent reservoir stratigraphic cross-section.

Table 2. *The reservoir quality per reservoir unit are given in following table:-*

FORMATION		RESERVOIR UNIT	NET SAND ISOCHORE (ft)	NET/GROSS RATIO (%)	AVERAGE POROSITY (%)	AVERAGE HORIZONTAL PERMEABILITY (md)
	Tarbert	B1.1	15	91	26	550
	Tarbert/Ness	B1.2	25	78	26	740
	Ness	B1.3	27	52	25	770
		B1.4	29	46	22	190
		B2.1	14	54	22	410
		B2.2	36	76	26	1300
BRENT		B2.3	10	29	20	100
GROUP		B2.4	22	71	24	460
		B2.5	12	22	25	880
		B3.1	42	80	26	1700
		B3.2	47	52	26	2100
		B3.3	37	52	25	1250
		B3.4	5	26	22	280
	Etive	B4.1	31	100	29	3800
	Rannoch	B4.2	104	87	23	200
	Rannoch/Broom	B4.3	12	22	16	10
	Nansen	S1	95	100	24	3200
	Eiriksson	S2	39	41	23	2900
STATFJORD		S3.1a	64	83	20	620
FM		S3.1b	97	82	22	2100
		S3.1c	66	63	21	1100
		S3.2	78	74	21	750
		S3.3	43	31	20	340
	Raude	S3.4	36	27	20	1150

from 16–26% and horizontal permeabilities from 20–10 000 md in Unit S2. This sequence is transgressively overlain by an even better quality and rather homogeneous, shallow marine sand: Unit S1, which has an average thickness of about 90 ft and porosities varying from 22–29% with horizontal permeabilities of 1 000 to 10 000 md.

The Statfjord Formation reservoir has a higher feldspar content than the Brent Group Reservoir (13.5% of the framework). This leads to higher secondary porosities. The overall diagenetic scheme is similar to that of the Brent Group.

Reservoir parameters are shown with the type logs in Fig. 6 and in Table 2.

Source

There are two main source areas for the hydrocarbons in the Brent Field: the Viking Graben and the East Shetland Basin. Both contain the same source beds, although they differ in thickness. The dominant oil source rock is the Kimmeridge Clay Formation which has a maximum thickness of 1600 ft in the East Shetland Basin and is over 3300 ft thick in the Viking Graben. The Kimmeridge Clay Formation has an average total organic carbon (TOC) of 5.6% (maximum 12.5%) in the East Shetland Basin (Goff 1983). The immature organic material is classified as Type II (degraded liptinite) kerogen and the non-soluble matter is dominantly sapropel (80%) with subordinate (20%) humic/coaly material (Brooks & Thusu 1977). The potential yield of the Kimmeridge Clay Formation has been calculated at 80 litres of oil per cubic metre of rock (Karlsson 1986). Diagenetic studies suggest that the Brent Field structure was filled by early Eocene times (Sommer 1978).

The Brent Group coals and vitrinite rich mudstones provide significant amounts of dry gas. Some of the organic rich lagoonal mudstones are also oil prone and other mudstone sequences in the area provide minor amounts of dry gas. The Dunlin Group contains up to 2% TOC in the Drake Formation but has <30% sapropelic material and therefore has limited oil potential. The Heather Formation contains 1–2% TOC in the form of vitrinite and inertinite and is capable of producing lean dry gas. The Cretaceous Cromer Knoll and Shetland Groups contain up to 2% TOC, mostly inertinite, and because of their extreme thickness in the Viking Graben they are capable of significant amounts of gas generation.

The Brent Group depth (8250 ft TVSS in the Brent Field) is at the point of oil threshold with a vitrinite reflectance value of $R_0 = 0.5\%$ (at 80°C). For the Type II kerogen that dominates the Kimmeridge Clay, oil is produced in a depth range of 10 650–15 000 ft TVSS, with peak oil generation at $R_0 = 0.7\%$ at a depth of 10 670 ft TVSS (at 120°C). Wet gas is produced at depths between 15 000 and 18 000 ft TVSS and dry gas only below the latter depth. Peak gas generation from Brent type coals is probably at an R_0 value of 1.5% which corresponds to a depth of 16 500 ft TVSS in the Viking Graben. The source rocks are therefore at all ranges of maturation depths and the vitrinite reflectance values and kerogen colorations reflect this.

There are two vertical migration routes into the fault block crest, up the steep faulted scarp and up the shallower dip slope of the block. Lateral migration paths in the Alwyn to Statfjord area can be estimated from the nature of the hydrocarbons and the oil–water contact depths in the fields.

The oil probably migrated through the Kimmeridge Clay and Heather Formation via oil-wet kerogen laminae and water-wet silty and sandy laminae through microfractures at low displacement pressure of between 20 and 50 psi (Goff 1983). Oil saturation in the expelled compaction fluids was at least 30%. The oil migrated into the nearest adjacent high porosity and permeability sequences (the Brent Group and Statfjord Formation).

Towards the north the reservoirs became less gaseous. The gas condensate accumulations in North Alwyn Field become gas cap reservoirs in the Brent Field and high GOR oil reservoirs in the Statfjord Field. Also along this trend the oil water contacts become progressively shallower and these hydrocarbon accumulations are filled to their spill points, either structurally dip closed or at sand to shale juxtapositions across faults. Therefore it is possible that gas migration is still taking place today in both reservoirs in the Brent Field (Bath *et al.* 1980).

Hydrocarbons

Both the Brent Group and Statfjord Formation reservoirs contain a reservoir fluid whose properties vary with depth. The variation can be explained by the action of gravity or possibly by recent migration of gas.

The oil of the Brent Group reservoir has an average density of 54.7 lbm/ft^3, a viscosity of 0.26 cP and an initial GOR of 2.58 Kscf/stb. The saturation pressure is 4961 psia at 8800 ft TVSS.

The composition of the Brent Group reservoir crude is: $C_1 = 0.55$; $C_2 = 0.08$; $C_3 = 0.06$; $C_4 = 0.03$; $C_5 = 0.02$; $C_6 = 0.02$; $C_{7+} = 0.22$; $NO_2 < 0.01$; $CO_2 < 0.02$.

The Statfjord Formation reservoir contains a supercritical fluid. Extensive fluid sampling has established that the hydrocarbon fluid column varies continuously with depth without a distinct gas/oil contact. The fluid column changes from a dew-point fluid at the top to a bubble-point fluid at the bottom, without a distinct phase change boundary. The transition occurs over an interval of about 100 ft from 9100 ft to 9200 ft TVSS.

The oil of the Statfjord Formation reservoir has an average density of 53 lbm/ft^3, a viscosity of 0.23 cP and an initial GOR of 2.17 Kscf/stb. The saturation pressure is 5170 psia at 9400 ft TVSS.

The composition of the Statfjord Formation reservoir crude is: $C_1 = 0.59$; $C_2 = 0.08$; $C_3 = 0.06$; $C_4 = 0.03$; $C_5 = 0.02$; $C_6 = 0.02$; $C_{7+} = 0.18$; $NO_2 < 0.01$; $CO_2 < 0.01$.

Reserves

At 1/1/89 total oil/condensate-initially-in-place was estimated to be 3500 MMBBL and the gas-initially-in-place 6700 TCF, of which 5400 TCF is solution gas and 1300 TCF free gas.

The ultimate oil/condensate and solution/free gas recovery are expected to be as follows:

	MMBBL Oil	TCF Gas
Brent Group reservoirs:	1220	2490
Statfjord Formation reservoirs:	95	1215
Total	1815	3705

Brent Field Data Summary

	Brent Reservoir	Statfjord Reservoir
Trap		
Type	Unconformity: tilted fault block	
Depth to crest	8240 ft TVSS	9000 ft TVSS
Lowest closing contour	9300 ft TVSS	10 700 ft TVSS
Gas/oil contact	8560 ft TVSS	9100 ft TVSS
Oil/water contact	9040 ft TVSS	9690 ft TVSS
Gas column	320 ft	100 ft
Oil column	480 ft	590 ft
Pay zone		
Formation	Brent Group	Statfjord Formation
Age	Middle Jurassic	Lower Jurassic/Triassic
Gross thickness (average/range)	810 ft; 780–850 ft	850 ft; 800–1000 ft
Porosity (average/range)	21%; 16–28%	23%; 16–29%
Permeability (average/range)	650 md; 10–6000 md	500 md; 20–10 000 md
Hydrocarbons		
Oil gravity	54.7 lbm/ft^3	53.0 lbm/ft^3
Oil type	Low sulphur light crude	Low sulphur light crude
Gas gravity	0.74	0.76
Saturation Pressure	4326–5750 psia	4400–5625 psia
Gas/oil ratio (average)	1.58 KSCF/STB	2.17 KSCF/STB
Condensate yield	138 BBL/MSCF	268 BBL/MSCF
Formation volume factor (average)	1.80 BBL/STB	2.04 BBL/STB
Formation water		
Salinity	25 000 ppm NaCl eqv	24 000 ppm NaCl eqv
Resistivity	0.236 ohm m at 77°F	0.270 ohm m at 77°F
Reservoir conditions		
Temperature	204°F	218°F
Pressure	5785 psia	6020 psia
Pressure gradient in reservoir	0.274 psi/ft (oil)	0.256 psi/ft (oil)
Field size		
Area	30 sq miles	
Drive mechanism	Water injection and gas injection	
Recoverable oil/condensate	1815 MMBBL	
Recoverable sol./free gas	3705 TCF	
Production		
Start-up date	November 1976	
Development scheme	4 platforms: export by tanker & pipeline	
Production rate (1988)	334 000 BOPD + 500 MMSCFD	
Cumulative production to end 1988	1205 MMBBL + 1786 TCF	

Besides the references quoted a substantial part of this article is based on Shell UK in-house study reports viz the 1990 Production Development Book (Vol. I), the Lithostratigraphic Nomenclature and 3-D Seismic Survey Interpretation Report.

The authors thank Shell UK, Expro and Esso Exploration and Production UK for giving permission to publish this paper.

References

BATH, P. G., FOWLER, W. N. & RUSSELL, M. P. 1980. The Brent Field, a Reservoir Engineering Review. *Proceedings of the European Offshore Petroleum Conference and Exhibition*, Paper EUR 164.

BROOKS, J. & THUSU, B. 1977. Oil Source Identification and Characterisation of Jurassic Sediments in the Northern North Sea. *Chemical Geology*, **20**, 283–294.

DEEGAN, C. E. & SCULL, B. J. 1977. *A proposed standard Lithostratigraphic Nomenclature for the Central and Northern North Sea*. Report of the Institute of Geological Sciences, 77/25.

GOFF, J. C. 1983. Hydrocarbon Generation and Migration from Jurassic Source Rocks in the East Shetland Basin and Viking Graben of the Northern North Sea. *Journal of the Geological Society, London*, **140**, 445–474.

JENSE, A. G. C., DING, C. N. & SMITH, I. F. 1988. Reservoir Management in Brent Field. *Seminar on Reservoir Management in Field Development and Production*. Norwegian Petroleum Society, Stavanger.

KARLSSON, W. 1986. The Snorre, Statfjord and Gullfaks Oilfields in the Habitat of Hydrocarbons on the Tampen Spur, Offshore Norway. *In* SPENCER, A. M. *et al.* (eds) *Habitat of Hydrocarbons on the Norwegian Continental Shelf*, Graham & Trotman, 181–197.

SOMMER, F. 1978. Diagenesis of Jurassic Sandstones in the Viking Graben. *Journal of the Geological Society, London*, **135**, 63–67.

The Cormorant Field, Blocks 211/21a, 211/26a, UK North Sea

D. J. TAYLOR[1] & J. P. A. DIETVORST[2]

[1] *Esso Exploration and Production UK Ltd, 21 Dartmouth Street, London SW1H 9BE, UK*
[2] *Al Furat Petroleum Co., PO Box 7660, Damascus, Syria*

Abstract: The Cormorant Oilfield is located approximately 150 km northeast of the Shetland Islands in Blocks 211/21a and 211/26a of the UK sector of the North Sea, in water depths of 500–550 ft. The field was discovered in 1972 by exploration well 211/26-1 and consists of four discrete accumulations spread along a major, north–south trending fault terrace. Hydrocarbons are produced from Middle Jurassic (Bajocian) sands of the Brent Group, which was deposited in a wave-dominated delta system. The reservoir has a typical gross thickness of 250–300 ft, locally increasing to 550 ft over faults active during sedimentation. Reservoir porosity varies from 16–28%, with average permeabilities ranging from tens of md to 1300 md. The accumulation contains under-saturated 34–36° API oil which was initially overpressured by some 1000–1270 psi. The stock tank oil initially in place and ultimate recovery are estimated at 1568 MMBBL and 623 MMBBL, respectively, reflecting a recovery factor of 39%. The reserves are produced through crestally-located wells supported by down-dip water injectors, and exported via two fixed platforms and an underwater manifold centre. To date, 59 wells have been drilled and 324 MMBBL (52%) of the estimated reserves have been produced.

The Cormorant Oilfield includes four discrete, but closely associated, oil accumulations in the UK sector of the northern North Sea, located approximately 150 km northeast of the Shetland Islands in water depths of 500–550 ft. The field straddles Licence Blocks 211/21a and 211/26a (Fig. 1), covers an area of 12 000 acres and is operated by Shell UK Exploration and Production on behalf of the Shell/Esso Joint Venture.

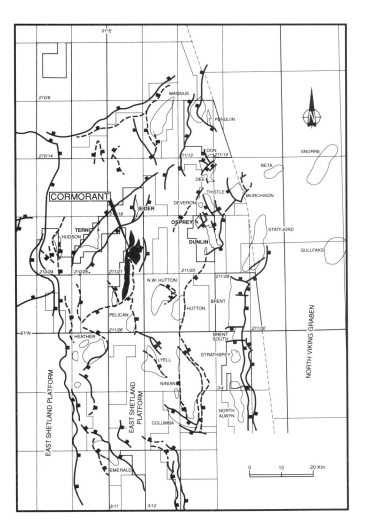

Fig. 1. Location of the Cormorant Oilfield in the East Shetland basin.

Four discrete accumulations make up the Cormorant Oilfield: Blocks I, II, III and IV (Fig. 2). The blocks are combined structural–stratigraphic traps lying along the crest of a large north–south striking fault terrace in the East Shetland Basin. Hydrocarbons are found in both Middle Jurassic and Triassic reservoirs. The Middle Jurassic reservoir occurs in the deltaic-shallow marine Brent Group sediments, and is found in all four blocks. The Triassic reservoirs occur within the alluvial deposits of the Cormorant Formation and are of secondary importance, being oil-bearing only in Block II and the southern part of Block I. A summary of data is given in the Cormorant Oilfield Data Summary at the end of the report, and an additional account of the development and geological histories of the field may be found in Howe (1991).

As with other Shell/Esso oilfields, the Cormorant Oilfield is named after a sea bird. Common around most of coastal Britain, the Cormorant is a skilled fisher, diving up to 30 ft.

History

Blocks 211/21 and 211/26 and several other Blocks were awarded to the Shell/Esso group as part of licence P232 in the Fourth Round allocation in 1972. In 1977 South Cormorant was designated P232 and North Cormorant and the rest of the original licence was re-designated P258. In March 1978 the unprospective parts of the original licence were subject to mandatory relinquishment and the retained portions in which Cormorant lies were named 211/21a and 211/26a. In June 1979, North Cormorant was designated as Production Licence P258, and the remainder of the original licence as P296. These licences expire in 2018.

The Cormorant Oilfield (Block I) was discovered in 1972 by Shell/Esso well 211/26-1. The well was located to test a large Palaeocene anticlinal structure in a near crestal position and a secondary pre-Cretaceous structure.

Water-bearing sandstones were encountered at the Palaeocene level, but the well penetrated and cored 236 ft of oil-bearing Middle Jurassic Brent Group sandstones (123 ft net pay). The oil–water contact was not seen. No reservoir sands were encountered in the underlying Dunlin Group, and the Lower Jurassic Statfjord Formation was found water-bearing. However, hydrocarbon indications were noted in the basal conglomerate of the underlying Triassic Cormorant Formation and in fractured Basement, but did not flow during testing. The Brent Group sandstones were production tested over four intervals at rates up to 7800 BOPD of 36° API oil.

The field was subsequently appraised with eight wells drilled between 1973 and 1977, proving the remaining accumulations,

Blocks II, III and IV. The wells 211/26-2 and -3, drilled in Block II, also showed the Triassic Cormorant Formation to be oil-bearing.

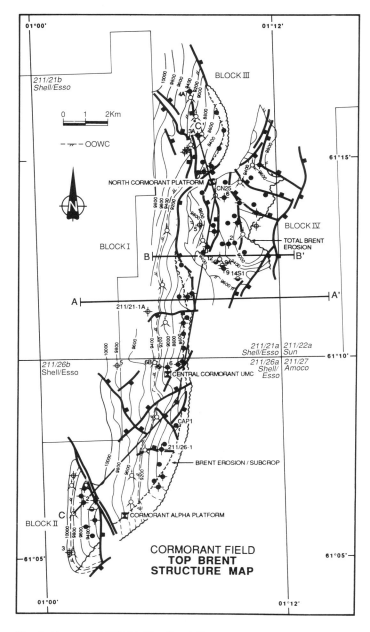

Fig. 2. Cormorant Oilfield; Top Brent Group structure map.

The elongate shape of the field (25 km north–south), led to the formulation of two separate development schemes, one for South Cormorant and one for North Cormorant, to recover the areally extensive reserves optimally. The South Cormorant Annex B was submitted to the Department of Energy in 1976, describing a platform sited to develop Block II and the southern part of Block I, and laying provision for a future sub-sea development to produce the Central Cormorant area. The latter scheme envisaged the suspension and tie-back of an appraisal well as a satellite producer, followed by the installation of an underwater manifold from which further production and injection wells would be drilled. The Cormorant Alpha concrete gravity platform was installed in May 1978 and production began in December 1979. The appraisal well 211/26-7 was drilled, completed, and tied back to Cormorant Alpha in 1980 as satellite well CAP1, and brought on stream in January 1981. The Annex B for the development of North Cormorant, involving Blocks III and IV and the northern part of Block I, was submitted in November 1978. The 40-slot steel platform, North Cormorant, was installed in August 1981 and linked by pipeline to Cormorant Alpha. First production was in February 1982.

The South Cormorant Annex B was updated and re-submitted in 1981, describing the sub-sea development for the Central Cormorant area which lay outside the drilling reach of the two platforms. This involved the siting of a 9-slot underwater manifold centre (the UMC), linked to production facilities on Cormorant Alpha. The UMC was installed in August 1982, with first production in May 1983. Drilling was completed in 1987 and comprised five template wells and four satellite wells linked into the UMC.

Oil from all facilities is evacuated via Cormorant Alpha into the Brent system, and then on to the crude terminal at Sullom Voe in the Shetland Isles. Sales gas is evacuated via the western leg of the Far North Liquids and Associated Gas System (FLAGS).

Production in all blocks was envisaged from rows of up-dip producing wells, supported by down-dip water injectors. In Blocks I, II and III this development concept has been largely maintained. However, in the structurally more complex Block IV, early development drilling indicated reservoir compartmentalization by sealing faults. It was therefore necessary to site producer/injector pairs within each fault compartment. Reflecting this conceptual change in development strategy, a revised North Cormorant Annex B was submitted to the Department of Energy in 1988.

A total of 33 production and 23 water injection wells have been drilled from the platforms, the UMC and related satellite locations. The peak average production of 155 000 BOPD was achieved in 1986. The average production for 1988 was 81 000 BOPD from North Cormorant and 41 100 BOPD from Cormorant Alpha (including the UMC).

Field stratigraphy

A schematic stratigraphic framework for the East Shetland Basin, applicable to the Cormorant area, is shown in Fig. 3.

Basement comprises pre-Devonian garnet–mica schists and weathered amphibolites, and has been encountered in five wells. The oldest sediment package overlying the Basement is the Triassic Cormorant Formation. This comprises 1000–1500 ft of red-brown, silty claystones interbedded with fine- to medium-, occasionally coarse-grained sandstones deposited in a distal alluvial plain environment. Although the Cormorant Formation is typically argillaceous, carbonate cemented, and therefore usually a poor reservoir, oil has been produced from the formation during tests in Block II.

The Cormorant Formation is overlain by the thin (10–30 ft), calcareous sandstone of the Lower Jurassic Statfjord Formation. The sandstone marks a return to marine conditions, which were maintained during the deposition of the overlying 150–200 ft package of marine shales and siltstones of the Dunlin Group.

A widespread fall in sea level during the Middle Jurassic led to the emergence of the Shetland Platform, and the establishment of a structural high in the Central North Sea area to the south. Erosional products from this area were deposited in the wave-dominated Brent delta system, which prograded in a north or north-easterly direction during the Middle Jurassic (Budding & Inglin 1981). The resulting sedimentary package is typically 250–300 ft thick in the Cormorant area, increasing locally to 550 ft. Deltaic sedimentation was terminated by a major marine transgression during the Bathonian. The onset of the transgression was marked by the deposition of the Tarbert Formation sandstones, followed by the shales and thin limestones of the Humber Group. The maximum thickness of the Humber Group over the Cormorant Oilfield is 740 ft, but this reduces to as little as 10 ft over the crests of the fault blocks, owing to the development of condensed sequences and the presence of local disconformities.

The deposition of the Jurassic sequence coincided with the further opening of the Viking Graben during the regional, Kimmerian tectonic event. This event culminated in the Late Jurassic, and was responsible for the break up of the East Shetland Basin into a series of north–south oriented fault terraces one of which runs through the Cormorant area. Additional sets of faults, oriented NW–SE and NE–SW, were also active, and are partly related to the reactivation of similarly oriented Caledonian faults in the Basement (Speksnijder 1987). Interference between these three fault sets divided the Cormorant fault terrace into the four constituent blocks of the Cormorant Oilfield. The four blocks developed a gentle westerly tilt during extension, prompting both the erosion of the Brent Group in the crestal areas of the blocks, and the thinning of the Humber Group sequence described above. The initiation of faulting during the early–mid-Jurassic in the Cormorant area is reflected in the dramatic thickness variations within the Brent Group observed between blocks.

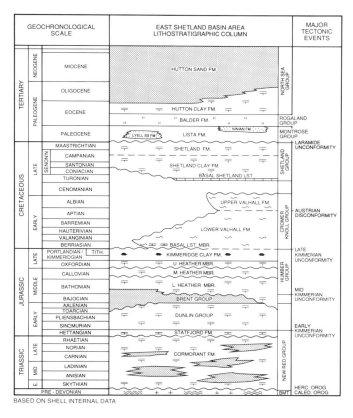

Fig. 3. East Shetland Basin area, generalized lithostratigraphic column.

The onset of regional basin subsidence is marked by the deposition of the Lower Cretaceous Cromer Knoll Group, which onlaps the locally eroded Jurassic sequence, defining a Base Cretaceous ('X') Unconformity. Lithologically the Cromer Knoll Group comprises limestones and marls which thicken and become more shaley off-structure. This Group is overlain unconformably by 3500–4000 ft of limestones and marls of the Upper Cretaceous Shetland Group. Tertiary to Recent sediments are represented by a 4500 ft succession of largely unconsolidated marine claystones and sandstones of the Montrose, Rogaland and Nordland/Hordaland Groups.

Geophysics

The Cormorant Oilfield is covered by four 3-D seismic surveys and a variety of 2-D data (Fig. 4). The first 3-D survey was shot in 1979 over North Cormorant, and the second, over the central Cormorant area, in 1981. In 1984, a higher quality 3-D survey was shot over the structurally more complex North Cormorant area, and this survey has been subsequently reprocessed. The latest survey was shot in 1988, covering the southern part of Block I and the whole of Block II, as part of the 3-D seismic survey of the Pelican prospect to the south (Fig. 1). The seismic data quality is variable, but has improved significantly since the acquisition of the original 2-D data (Fig. 5).

Six horizons are distinguished for field-wide mapping purposes: top Balder Formation (D10), the Base Tertiary Unconformity (D12), the Base Cretaceous ('X') Unconformity, top and base Brent Group, and top Basement. Additionally, a Lower Cretaceous ('K') Unconformity is mapped to the east of the main structures, and Triassic and intra-Brent reflectors may be locally mapped. Strong reflectors generally coincide with sharp, lithological contrasts and hence the top Balder, X, K and basement surfaces are excellent reflectors. The other reflectors are more variable in quality, the base Brent becoming locally indistinct, partly due to the more gradational nature of the Rannoch–Broom–Dunlin boundaries on a gross scale, and partly due to masking by residual water-bottom multiple energy from the strong X reflector.

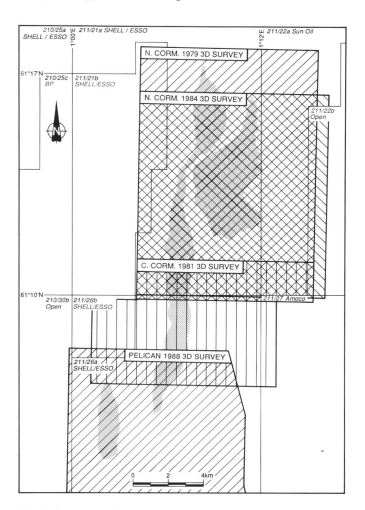

Fig. 4. Cormorant Oilfield; seismic coverage map.

Trap

Traps in the Cormorant Oilfield are formed from a combination of structural and stratigraphic features. On a kilometre scale, the geometry of the traps is structurally controlled by the regional fault sets described above, defining the four westward-tilted Cormorant fault blocks. However, as a result of the crestal erosion associated with the rotation of the fault blocks, the reservoir is typically truncated and sealed by an unconformable contact with the Heather Formation, rather than by fault juxtaposition. Where fault

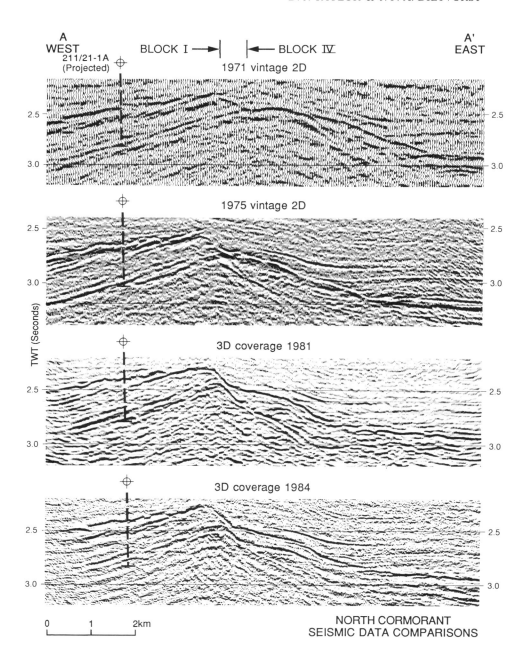

Fig. 5. North Cormorant Oilfield area, comparison of original 2-D and modern 3-D seismic through well 211/21-1A.

juxtaposition occurs, the reservoir is sealed against shales and limestones of the Humber Group and shales and marls of the Lower Cretaceous.

Blocks I and III are combined structural/stratigraphic traps, lying along the dissected axis of the large, N–S oriented fault terrace. The blocks are separated by a NW–SE trending fault with a throw of approximately 400 ft to the SW, and the crests of the structures lie at 8600 and 8200 ft TVSS in Blocks I and III, respectively. The Brent Group dips at approximately 10° to the west in both blocks.

Block II is a smaller accumulation, located SW of Block I. The reservoir is fault-bound to the east and northeast, eroded to the southeast, and dip-closed to the west. The Brent Group dips at 12° to the west, and reaches a crest at 9200 ft TVSS.

Block IV is a more isolated structure, down-faulted to the east of Blocks I and III. Unlike Blocks I, II and III, Block IV is not a simple westward-dipping structural/stratigraphic trap. The western boundary of the Block is a fault, crescentic in plan view, the fault dip decreasing steadily as it penetrates Basement. The maximum throw along the fault is approximately 1000 ft. The Block rises to a crest at 8500 ft TVSS, and is down-faulted on a network of N–S,

NE–SW and NW–SE faults, to produce a faulted dome-like structure (Fig. 6). In addition, the Brent Group is eroded along the eastern margin of the Block, such that the extent of the reservoir to the east, southeast and northeast is defined by a combination of structural, stratigraphic and dip closures.

Original oil–water contacts (OOWCs) differ in each block. Blocks I and III have OOWCs of 9250 ft TVSS and 9200 ft TVSS, respectively. Internal faults do not appear to be completely sealing, but may act as local transmissibility barriers. Block II has an OOWC at 9830 ft TVSS and production history has indicated that sub-seismic transmissibility barriers (possibly small sealing faults) have restricted communication between the crestal and down flank areas. Oil/water contacts for the Triassic Cormorant Formation in Block II have not been interpreted. The deepest oil-down-to level observed to date is at 11 037 ft TVSS. In Block IV, three OOWCs (9600, 9800 and 10 000 ft TVSS) have been inferred from log and RFT pressure data, and reflect fault compartmentalization within the structure.

The spillpoints for the Cormorant blocks are unknown, as the OOWCs in each block ultimately terminate against a major boundary fault. At these terminations, the sealing properties and exact juxtaposition geometries along the boundary faults are not known.

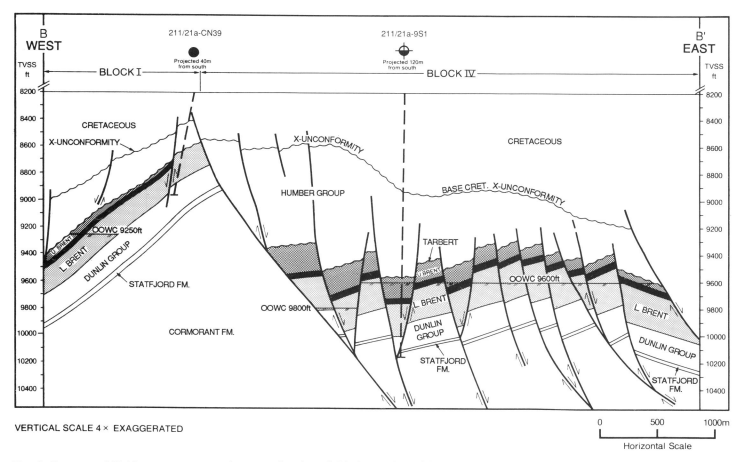

Fig. 6. Cormorant Oilfield; east–west structural cross-section through Blocks I and IV of the Cretaceous to Triassic interval.

Reservoir

The main reservoir occurs in the Middle Jurassic (primarily Bajocian) marine and non-marine deltaic sediments of the Brent Group. A smaller accumulation is also present in the Triassic reservoirs in Block II, and the southern part of Block I.

The *Brent Group* reservoir is divided into five, regionally correlatable formations: Broom, Rannoch, Etive, Ness and Tarbert. The sequence is further sub-divided within the Cormorant area for reservoir management purposes, and a summary of the main reservoir properties in these sub-units is given in Figs 7 and 8 for Blocks I/II and III/IV, respectively. A porosity and permeability profile through the Brent Group reservoir is illustrated in Fig. 9 with a type log of the field from the Block IV area.

The Brent Group varies in thickness from 250–300 ft in Blocks I, II and III to 300–550 ft in Block IV (Fig. 10). This variation is largely attributed to faulting and block rotation during sedimentation. The *Broom Formation* is considered to represent a suite of deltaic fans prograding into a shallow marine environment. The *Rannoch*, *Etive* and *Ness Formations* depict a transition from shoreface, through coastal barrier to delta plain environments, produced by the north to northeasterly passage of a wave dominated, coastal delta complex. The *Tarbert Formation* is interpreted to be the erosional product of the post-Ness marine transgression involving the deposition of reworked Brent Group sediments.

Lithologically, the Brent Group sandstones are fine- to very coarse-grained, variably sorted sandstones. The Tarbert and Ness Formations are dominantly quartzose whereas the Etive, Rannoch and Broom Formations are sub-arkosic. Mica and authigenic kaolinite are the main minerals that affect reservoir quality and are particularly common in the Rannoch Formation. Leaching of detrital feldspar played an important role in determining reservoir quality by creating secondary porosity and releasing silica. Reservoir quality deteriorates below the OOWC throughout the field due to the late-stage growth of fibrous illites, which seriously reduce reservoir permeability although not greatly affecting porosity. This is particularly pronounced in the deeper parts of Block IV and in all areas in Block II prompting, in the latter case, the location of some water injection wells in the oil column in order to achieve adequate rates of injection.

The *Cormorant Formation* ranges in thickness from 950–1030 ft in Block II, and is broadly correlatable into five units. The sediments are predominantly alluvial red-beds, containing thin channel deposits in which the sandstones are extensively carbonate cemented, have limited areal extent and restricted connectivity. Compared to the Brent Group reservoir, the average Cormorant Formation reservoir porosities and permeabilities are poor, typically 12–20% and zero to 600 md, respectively.

Source

The oil in the Cormorant Oilfield was derived from the organic-rich Kimmeridge Clay Formation. This formation thickens (up to 350 ft) and becomes mature for oil generation off structure in the deeper parts of the half graben. Hydrocarbon generation began at approximately 65 Ma (Palaeocene) and peaked between 50 and 40 Ma.

Hydrocarbons

Production rates are highest from the Ness and Etive Formations, productivity indices (PIs) ranging from 5–40 BOPD/psi. The Tarbert Formation is more variable in reservoir quality and thickness, resulting in lower PIs (typically 3–10 BOPD/psi). The Rannoch and

FORMATION	MEMBER/UNIT		GENERAL CHARACTERISTICS	SEQ.	DEPOSITIONAL ENVIRONMENT	GROSS ISOCHORE	SANDY BODY DISTRIBUTION	IMPERMEABLE LAYER EXTENT	OIL LEG BLOCK I/II	n/g	ø %	K_h (air)* mD
TARBERT			m-c grained, poorly sorted, poorly stratified sandstone. Blocky profile.		Transgressive marine sheet sand	Up to 30 ft BLOCK II	Not present in Block I only present downdip in NW part of Block II	Local mud drapes and calcite	II		16	100-1000
NESS	UPPER		Interbedded f-m gr sands, shales and coals in coarsening-up, fining-up and random sequences. Sands grouped in three packages - only basal package in N and E		LOWER DELTA PLAIN channels, splays, mouth bars	0-55 ft BLOCK I 0-70 ft BLOCK II	Upper sands 2-5 ft thick, network connectivity. Middle sands: 5-10 ft thick channels, 1-2 km wide; and 2-5 ft splays etc - network. Basal sand 5-15 ft thick, field-wide extent	Two block-wide shale intervals. Other shales kms-wide	I II	0.52 0.8	20 18	4-500 100-1000
	MID-NESS SHALE		Laminated shale with two 2-4 ft thick sands in middle		OPEN LAGOON	20-25 ft BLOCK I, II		Field-wide	—	—	—	—
	LOWER	A	Single f gr sand, occasionally with thin coal		MOUTH BAR/LAGOONAL BEACH	40-60 ft BLOCK II 9-36 ft BLOCK I	Block II, undifferentiated. Block-wide sand body, splits into two in north.	Coals: 10s-100's m Doggers: m's-10s m	II I	0.90	21 28	100-1000 30-1500
		B	Interbedded f-m gr sand, shale and coal in fining-up and coarsening-up sequences		LOWER DELTA PLAIN channels, splays, mouth bars	16-45 ft BLOCK I	Channels 15-25 ft thick, 1-2 km wide, E-W. Splays, etc (Interchannel) 2-6 ft thick sands. network connectivity	Coals: km's wide Interchannel shales: 100's-1000's m Channel shales: m's-10's m	I	0.63	24	30-1100
		C	Block-wide laminated shale with coal at base		OPEN LAGOON	23-30 ft BLOCK I		Block-wide	—	—	—	—
		D	Single f gr sand		BACK-BARRIER flood-tidal delta?	0-33 ft BLOCK I	Dies out rapidly northward. Two sands in CA12 and CA29	Doggers: m's-10's m	I	0.82	25	275-1600
		E	Bioturbated/rooted shale with coals. Thin minor sands		BACK BARRIER & ABANDONED CHANNEL	0-23 ft BLOCK I		Southern half of block	—	—	—	—
ETIVE			f-m gr sandstones in fining-up sequences (channels) and mixed sequences (barrier). Shale layers and clasts in channels. Micaceous layers near base in barrier		DELTA FRONT barrier and inlet channels	14-68 ft BLOCK I 60-80 ft BLOCK II	NW - SE channels occupy most of block, with thin 'abandoned channel' in UMC area. Barrier remnants in south and middle, full barrier in north	km-wide shale in southern channel area Channel shales: m's-10's m Doggers: m's-10's m	I II	0.95 >0.95	24 20-25	275-3000 100-1000
RANNOCH		A	Coarsening-up sequence of vf-f gr micaceous sandstones, siltstones and shales		DELTA FRONT middle-lower shoreface	40-60 ft BLOCKS I + II	Field-wide sandstone with laterally variable grain size distribution	Doggers: m's-10's m Shales over most of block at base	I	0.85	20	4-300
		B			upper offshore				I	0.63	19	1-250
		C							II	<0.4	17	1-50
BROOM			Crs gr, often pebbly and slightly muddy sand. Coarsening-up or blocky profile		OFFSHORE SHEET SAND	7-24 ft BLOCKS I + II	Block-wide. Thins onto crest of block	Irregular carbonate cementation: m's-100's m?	I II	0.62 0.6-0.9	19 16	3-700 1-100

* core data, arith. avg.

Sequence Notation: shale, very fine to fine sand, fine to medium sand, fine to coarse sand, coarsening upwards, fining upwards, homogeneous

Fig. 7. Cormorant Oilfield; Blocks I and II, reservoir characteristics.

FORMATION	RESERVOIR UNIT	GENERAL CHARACTERISTICS	SEQ.	ENVIRONMENT	GROSS THICKNESS (FEET)	SAND BODY DISTRIBUTION	AVERAGE SAND BODY (OIL LEG) RESERVOIR PROPERTIES
TARBERT	BLOCK Ic	Absent, due to erosion or non-deposition.		ESTUARINE / SHALLOW MARINE	0 - 100FT	LATERALLY CONTINUOUS IN BLOCK III DISCONTINUOUS IN BLOCK IV	EXCELLENT BLOCK III : ø = 23%, K_H = 280 - 1100mD BLOCK IV : ø = 22%, K_H = 0.5 - 1000mD CALCITE CEMENTATION LOCALLY PRESENT
	BLOCK III AND BLOCK IV	In Block III 5 - 18ft thick, in Block IV up to 100ft thick medium - coarse grained homogenised sands with occasional gravel layers, mudclasts and carbonaceous debris; increased shale content where reworked Upper Ness Shales. GRADATIONAL BOUNDARY					
UPPER NESS	BLOCK Ic, BLOCK III → BLOCK IV	Interbedded sands, shales and coals (in Block III coarsening upward to a blocky sand package with a shale break in the middle and at the top). In Block IV Upper Ness is thickened due to activation of main boundary fault displacing Block IV relative to main Cormorant structure. SHARP EROSIONAL BOUNDARY		MOUTH BAR	35 - 120FT	MOUTH BAR FAN SHAPED BODIES E-W EXTENT 1-2KM N-S EXTENT 2-4KM	BLOCK Ic : ø = 20%, K_H = 0.1 - 1000mD BLOCK III : ø = 22%, K_H = 0.3 - 4000mD BLOCK IV : ø = 20%, K_H = 0.1 - 2000mD K_H HIGHER AT OF MOST BODIES
MID NESS	FIELD WIDE	Shale + minor ripple (wave + current) laminated sands. Bioturbated Extensive back barrier lagoon. GRADATIONAL BOUNDARY		LAGOON	15 - 20FT	—	VERTICAL TRANSMISSIBILITY VERY LOW
LOWER NESS	BLOCK Ic, BLOCK III → BLOCK IV	Interbedded sands, shales and coals — shales: grey laminated muds, wave rippled, slumped bioturbated — sands: medium grained, cross bedded to fine grained micaceous, bioturbated, ripple. Main reservoir sands are in channels; crossbedded at the base passing into rippled sands capped by deformed silts, muds and coals. Thickness 70ft in Block IV and Block I, decreasing to 20ft in Block III. SHARP BOUNDARY		DELTA PLAIN and BACK BARRIER	20 - 70FT	ELONGATE CHANNELS FAN SHAPED MINOR MOUTH BARS SPLAYS	BLOCK Ic : ø = 19%, K_H = 1 - 2000mD BLOCK III : ø = 20%, K_H = 0.1 - 1000mD BLOCK IV : ø = 18%, K_H = <0.1 - 10000mD
ETIVE	FIELD WIDE	Fine to, medium grained sands generally cross bedded and horizontally stratified at base, partly stratified mottled or rippled sands above, mud, coal clasts and clay drapes occasionally present. GRADATIONAL BOUNDARY		UPPER SHORE FACE AND FORESHORE AREAS (BEACH, DUNE, CHANNEL / TIDAL INLET	30 - 100FT	LATERALLY EXTENSIVE	BLOCK Ic : ø = 26%, K_H = 100 - 10000mD BLOCK III : ø = 29%, K_H = 50 - 10000mD BLOCK IV : ø = 26%, K_H = 10 - 10000mD CALCITE CEMENTATION PRESENT
RANNOCH	FIELD WIDE	Bioturbated argillaceous and micaceous silts passing upward into horizontal / low angled stratified micaceous silts and fine sands with common calcite cemented horizons. SHARP BOUNDARY		MIDDLE - LOWER SHORE FACE	60 - 100FT	LATERALLY EXTENSIVE INHOMOGENEOUS DISTRIBUTION OF GRAIN SIZE	BLOCK Ic : ø = 16%, K_H = <0.1 - 100mD BLOCK III : ø = 20%, K_H = <0.1 - 1000mD BLOCK IV : ø = 20%, K_H = 1 - 1000mD CALCITE CEMENTATION THIN LATERALLY EXTENSIVE CALCITE STREAKS VERTICAL TRANSMISSIBILITY
BROOM	FIELD WIDE	Clay rich and burrowed, faintly laminated or cross bedded coarse grained sandstone. SHARP BOUNDARY		OFFSHORE SHALLOW MARINE	10 - 35FT (BLOCK Ic, III) 15 - 55FT (BLOCK IV)		BLOCK Ic : ø = 18%, K_H = 0.1 - 1000mD BLOCK III : ø = 20%, K_H = <0.1 - 1000mD BLOCK IV : ø = 20%, K_H = <0.1 - 10000mD VARIABLE EXTENSIVE CALCITE CEMENTATION

LEGEND: SEQUENCE NOTATION — COARSENING UPWARDS / FINING UPWARDS — MAIN SHALE LAYERS (schematic)

Fig. 8. Cormorant Oilfield: Blocks III and IV, reservoir characteristics.

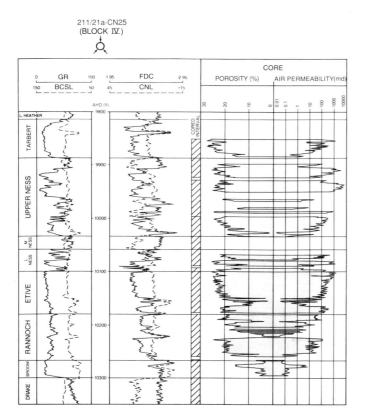

Fig. 9. Cormorant Oilfield; type log, well 211/21a-CN25.

Broom Formations also have relatively low production potentials, typically less than 10 BOPD/psi, combined. Individual wells can initially be produced at rates up to 36 000 BOPD.

The reservoir fluid properties vary from block to block across the field. The oil is highly under-saturated with gravity varying from 34–36° API, bubble points from 1040–2970 psi, gas/oil ratios from 224 to 770 SCF/BBL and viscosity from 0.56 to 1.18 cP. Reservoir pressures were originally 1000–1270 psi over hydrostatic pressure with the maximum overpressure in Block IV. Other reservoir properties are given in the data summary.

Reserves and resources

The stock tank oil initially in place for the Cormorant Oilfield is estimated to be 1568 MMBBL and ultimate recovery (UR) is estimated to be 623 MMBBL oil/condensate and 231 BCF gas, representing an overall recovery factor for the field of 40%. Total cumulative production at 1 January 1990 was 324 MMBBL (52% of UR).

Aquifer support is weak and water injection is required to maintain reservoir pressure and optimize sweep efficiency. In Blocks I, II and III, a line of injectors is situated close to the projected OOWC, supporting rows of crestal producers. In Block IV, owing to fault compartmentalization, dedicated pairs of injector/producer wells are required within individual sub-blocks.

Production problems consist primarily of sand production, scale build up, zonal water breakthrough and reservoir souring. Sand production is a problem in Blocks III and IV, where the reservoir sands are relatively unconsolidated. Barium sulphate scale develops in the producers when injected sea water mixes with formation

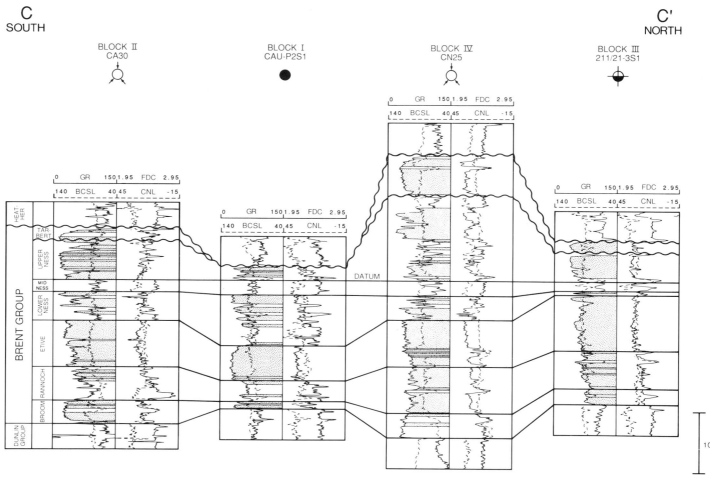

Fig. 10. Cormorant Oilfield; north–south stratigraphic cross-section.

water. Water breakthrough occurs preferentially in the reservoir units with higher permeability (e.g. the Etive Formation) and these zones need to be isolated in order to produce the other reservoir units. Hydrogen sulphide (H_2S) content has increased gradually over the field life, requiring treatment on the platforms before export of hydrocarbons.

The data and interpretations presented in this paper are the results of many years of study by Shell and Esso staff. The authors wish to thank the management of Shell UK Exploration and Production and Esso Exploration and Production UK Ltd for permission to publish this paper.

Cormorant Oilfield data summary (Brent Group Reservoirs only)

Trap
Type	Structural/stratigraphic, tilted fault block
Depth to crest	8200–9200 ft TVSS
Lowest closing contour	10 000 ft TVSS
Hydrocarbon contacts	OOWCs 9200–10 000 ft TVSS
Oil column	630–1500 ft

Pay zone
Formation/Group	Brent Group
Age	Mid-Jurassic (Bajocian)
Gross thickness (range)	300 ft (250–550 ft)
Net/gross ratio (range)	0.75 (0.4–1.0)
Cut off for N/G	13% porosity (= 1 md air permeability)
Porosity (range)	21% (16–28%)
Hydrocarbon saturation	0.70 (average)
Permeability (range)	50–1300 md air
Productivity index	9–60 BOPD/psi

Hydrocarbons
Oil gravity	33.8–36.0° API
Oil type	under-saturated oil
Viscosity (at Pi)	0.56–1.18 cP
Gas gravity	0.838–0.885 (air = 1)
Bubble point	1040–2970 psia
Gas/oil ratio	224–770 SCF/BBL
Formation volume factor, oil	1.147–1.423

Formation water
Salinity	14–18 g/l
Resistivity	0.12–0.17 ohms at datum (8690–9500 ft TVSS)
	NB: varies from Block to Block

Reservoir conditions
Temperature	195–225°F at datum (8690–9500 ft TVSS)
Initial reservoir pressure	4825–5265 psia
Pressure gradient in reservoir	0.32 psi/ft

Field size
Area	12 000 acres
Recovery factor, oil	40%
Drive mechanism	Weak aquifer, water injection
Recoverable oil/condensate	623 MMBBL
Recoverable gas	231 BCF

Production
Start-up dates	Cormorant Alpha in December 1979, North Cormorant in February 1982, UMC in May 1983.
Development scheme	Cormorant Alpha and North are platform developments and the UMC is an underwater manifold linked to Cormorant Alpha.
Peak production rate	155 000 BOPD (1986)
Cumulative production to date	324 MMBBL (1 January 1990)
Number/types of wells	4 exploration
	13 appraisal
	33 production
	19 injection
Average production rate in 1989	70 100 BOPD
Secondary recovery method	Water injection

References

BUDDING, M. C. & INGLIN, H. G. 1971. A reservoir geological model of the Brent sands in southern Cormorant. *In*: ILLING, L. V. & HOBSON, G. D. (eds) *Petroleum Geology of the Continental Shelf of North-West Europe*. Heyden & Sons, London, 236–344.

HOWE, B. K. 1991. The Cormorant Oil Field. *In*: *American Association of Petroleum Geologists Memoir*, in press.

SPEKSNIJDER, A. 1987. The structural configuration of Cormorant Block IV in the context of the Northern Viking Graben structural framework. *Geologie en Mijnbouw*, **65**, 357–379.

The Deveron Field, Block 211/18a, UK North Sea

R. R. WILLIAMS

BP Exploration, Farburn Industrial Estate, Dyce, Aberdeen AB2 0PB, UK

Abstract: The Deveron Field lies to the west of the Thistle structure and is a satellite development to the main Thistle Field. It holds separate field status but utilizes Thistle platform facilities. Although Deveron has a similar structural framework and stratigraphic sequence to the Thistle Field, subtle differences in reservoir character and behaviour are evident. The field was discovered by the first well on Block 211/18 but its small size prevented development until installation of the Thistle production facilities. The Deveron appraisal and development wells have all been drilled from the Thistle platform and so far no injection has been required. The current STOIIP is estimated at 41 MMBBL of which 20 MMBBL will ultimately be recovered. Cumulative production to the end of 1988 was 9.96 MMBBL from 3 production wells.

The Deveron Field, situated 1.7 miles west of the Thistle Alpha platform, lies in 530 ft of water, 130 miles NE of the Shetland Islands. Located on the western margin of the North Viking Graben to the W of the main Thistle structure, the field covers an area of 3.3 km² (815 acres) in Block 211/18a. The field lies within a gently eastward dipping, tilted fault block of Mesozoic age (Fig. 1). Almost the entire Brent Group sand sequence is hydrocarbon bearing, and the structure is fault sealed to the north and west and dip closed to the south and east. There are two dominant fault trends on Deveron and as in the Thistle structure, the N–S-trending faults exert the greater structural influence.

The field is named after a Scottish river, the Deveron, which rises in the foothills of the Grampian Highlands and flows east before heading northwards into the Moray Firth.

History

Licence P236 was awarded in the 4th round of UK Offshore Licensing in March 1972 to the Halibut Group led by the Signal Oil Company. The group, which varies slightly from the Thistle Field

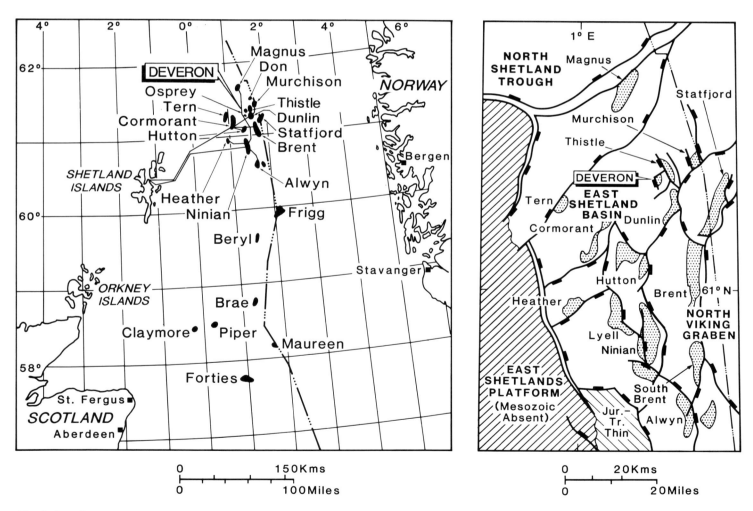

Fig. 1. Location map.

partnership, is now operated by BP Exploration on behalf of Britoil plc, and current participating companies are:

BP Exploration (Britoil plc)	15.95%
Deminex UK Oil and Gas Ltd	42.5%
Santa Fe Minerals (UK) Inc	22.5%
Arco British Ltd	10.0%
Premier Oil Exploration Ltd	8.05%
Monument Petroleum Mitre Ltd	1.0%

Oil was discovered by the 211/18-1 well drilled in 1972 on a Base Cretaceous structure. Although the field was discovered by the first well drilled on the block its small size and uncertain producibility prevented early development, as it was significantly below the threshold of economic viability for an independent facility. The following year the Thistle Field was discovered on a major structural feature to the east and oil was subsequently produced via the Thistle platform. Additional well slots were incorporated in the platform design for use either within the field or for satellite production.

The objective of well 211/18-1 was the Middle Jurassic Brent Group with a secondary objective in the Statfjord Formation. A DST in the Brent sands flowed 5293 BOPD of 38° API oil; the Statfjord had residual oil shows. After drilling well 211/18-1, a 2-D seismic survey was shot in 1979 and the area remapped. This data was used to determine the location of an appraisal well to be drilled as a long, highly deviated well from the Thistle platform. The objectives of this well were to determine the character and productivity of the Brent in this area and to establish an OWC for the accumulation. Well A44, drilled in 1982 was put on an extended (90 day) production test between April and August 1983.

After Development and Production Approval in 1984, the field came on production in 1984 through this one well. A further development well (A48) was drilled from the platform in 1985, however, this did not encounter hydrocarbons so the subsequent sidetrack A48Z became the second development well. In 1988, well A51 was drilled to provide a drainage point in the south of the field.

The Deveron Field holds separate field status and the separate PRT boundaries are illustrated in Fig. 2. Early in the development phase of Thistle, arrangements were negotiated between the Deveron participants and the Thistle Field Partnership Group in the form of a Combined Production Agreement to allow the use of Thistle Field and Brent Pipeline System Facilities by the Deveron Partnership. The proximity of Deveron to the Thistle platform provided an opportunity to recover economically reserves from a relatively small accumulation using only a small proportion of existing Thistle production facilities.

Oil production from Deveron is co-mingled with Thistle production and exported to Sullom Voe through the Brent Pipeline System. Deveron production is carefully allocated on the basis of frequent well tests and sampling. Deveron gas is also co-mingled with Thistle production.

The original development plan, based on a reservoir simulation model which included a history match of A44 test production, suggested the presence of a strong aquifer. It was envisaged that 3 production wells, including the appraisal well, would provide adequate drainage. Provision for water injection was made should reservoir pressure support be required. However the natural aquifer has so far maintained reservoir pressure.

Field stratigraphy

The general stratigraphic sequence of the Deveron Field is typical of this area and is outlined in the Thistle Field paper. However on Deveron the relatively straightforward layered sequence of the Brent Group identified on Thistle, although present, is difficult to correlate directly between wells. Of all the Deveron wells, only one, 211/18-A48Z, appears to be unfaulted in the reservoir interval. This

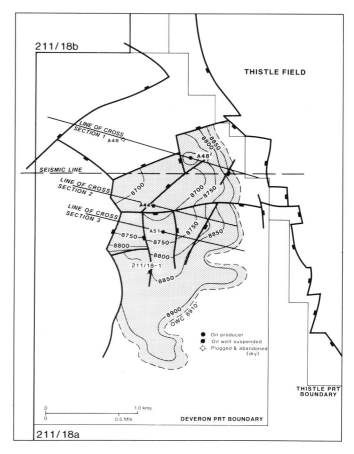

Fig. 2. Top Reservoir Structure Map.

also results in apparent wide variation in gross reservoir sand thickness and individual reservoir unit thicknesses across the field. As well A48Z penetrated the only complete Brent reservoir section, this is designated the type section for the field (Fig. 3).

Log correlation with wells on the Thistle Field indicates that the Broom, Rannoch, Etive, Ness and Tarbert formations of the Brent Group are all present on Deveron, however recent biofacies analysis suggests that in Deveron the Upper Brent lithostratigraphic/biostratigraphic relationship is more complex. Only in the southern part of Deveron are assemblages indicative of the Tarbert Formation present. Elsewhere, the entire D sand reservoir unit is interpreted as Ness Formation. There is therefore some evidence for areal diachroneity of the Upper Brent lithostratigraphic reservoir units.

Geophysics

The background to seismic data acquisition is similar to that outlined in the Thistle Field paper. Following initial acquisition of speculative and traded seismic data, an extensive grid of 2-D data became available from 1973, 1974, 1977 and 1979 surveys. In 1983 a 3-D survey was shot comprising 85 seismic lines orientated E–W and at 100 m spacing, amounting to 750 km (466 miles) of data across the Thistle and Deveron structures.

The data quality over most of the area is good. However, on the main Deveron structure, the fault pattern is extremely complex and is characterized by an ENE–WSW-trending cross fault system which offsets the main western boundary fault of the field. The 3-D database has a good broad band frequency content which overcomes early problems of interpretation below the Base Cretaceous reflector. The Base Cretaceous and Top Brent Horizons have been interpreted over the whole field, but the base reservoir envelope has

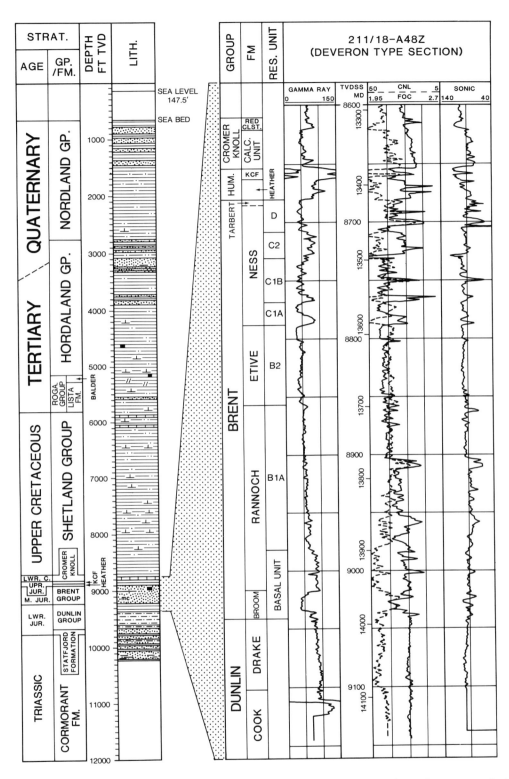

Fig. 3. General stratigraphic log. Type section, well no. 211/18-13.

only been interpreted in the north as the base reservoir reflection degrades in the south of the field (Fig. 4).

The reprocessing of the 1983 seismic data set over Thistle Field was extended to cover Deveron. This data has been interpolated to provide a normal line spacing of 33.3 m (109 ft). Improved signature deconvolution, short period multiple rejection and more detailed velocity analyses have significantly improved the quality of this data.

Trap

The Deveron structure can be identified at Base Cretaceous level as an eastward dipping, tilted fault block, downthrown to the west from the structurally higher Thistle structure. A similar structural configuration to the Thistle Field is mapped with a major N–S-trending normal fault forming the trap to the west with fault seal to the north and dip closure to the south and east. The crest of the structure at 8680 ft on the western margin of the field is 200 ft lower than that of Thistle. Mapped closure indicates the trap is full to spill point with an 8910 ft OWC approximately 400 ft higher than the Thistle contact. The structure is sealed by siltstones and claystones of the Lower and Upper Jurassic with the organic rich clays of the top sealing Humber Group also providing the source material.

Similar trends identified on Thistle of marked thinning of the reservoir sequence onto the crest of the structure are typically seen. Seismic evidence and well data indicate thinning of the gross

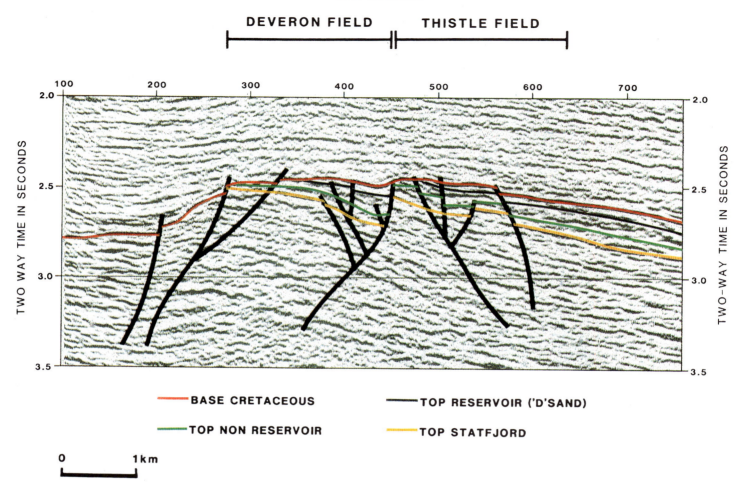

Fig. 4. 3-D seismic line across Deveron.

reservoir isochore towards the west and north involving some truncation by erosion along the crest of the structure. The field is subdivided areally into three distinct fault blocks each having undergone uplift and tilting resulting in erosion and non-deposition of the reservoir on the crest of the field.

As on Thistle, there are two dominant structural trends, the N–S trend exerting the greater structural influence and control on sedimentation. The major N–S fault, downthrowing to the west forms the western boundary of the field. The ENE–WSW- or E–W-trending faults have only limited throw and affect the structure in a more subtle style. The main fault in this E–W trend effectively dissects the field and forms the boundary between the Central and Southern Fault Blocks. Although the cross faults have little seismic expression, sealing potential is locally significant, illustrated by the ENE–WSW fault separating the dry Northern Fault Block from the Central Fault Block (Fig. 2).

Reservoir

In common with the Thistle Field, reservoir continuity is considered to be good, however the relationship between lithofacies and biofacies is more complex on Deveron. Correlation across the field is also difficult due to the combined effects of faulting, condensing and truncation of units, and the problems of interpretation of stratigraphically onlapping sandstones. Despite these correlation difficulties, the reservoir subdivisions identified on Thistle (see fig. 7 of that paper) can be extended to Deveron. The overall similarity of the Brent sequence suggests a similar depositional history for these sandstones and similar poroperm character.

Gross reservoir isochores reflect general thickening to the south and east away from the crestal part of the field. It is interpreted that active footwall uplift affected this crestal area causing thinning of the Brent Group either by deposition of condensed sequences, stratigraphic onlap, or by erosion. It is likely that a combination of these factors influenced depositional thickness.

The timing of this uplift resulted in more obvious structural influences affecting the Upper Brent sequence with Upper Brent zero sandstone edges located progressively further eastward.

The analogy with Thistle for Upper Brent sedimentation implies that structural controls exerted a stronger influence on D and C sandstone thickness. Similar trends are in evidence of D and C sandstone thinning and onlapping the crestal areas, and of significant syndepositional thickening of downfaulted section against N–S-trending normal faults. This thinning and thickening across the faults results in severe compartmentalization of upper reservoir sandstones with inherent problems of pressure support in isolated fault blocks. Well correlations show normal Brent stratigraphy until C sandstone deposition. As on the western margins of Thistle, incomplete C sandstone intervals occur on upthrown fault blocks. On Deveron, the lowermost C1A sandstone is absent in crestal areas and is interpreted to onlap these features.

Field-wide, Lower Brent reservoir thicknesses vary significantly with little apparent geographic control on trends. Early uncertainties as to lower reservoir thickness particularly in the NW of the field led to an interpretation of significant depositional thinning of the B2 and B1/A sandstones in comparison with Thistle. However, recent mapping and interpretation attributes all anomalously thin intervals of lower reservoir sandstones, particularly the B1/A, to missing section through faulting. New well data confirms that throughout the field, thinning of the lower reservoir is not significant. The current model for the field does not reflect significant depositional thinning of the lower reservoir sandstones, but allows for some degree of thinning of these sandstones towards the west of the field.

Source

In common with the Thistle Field, the highly organic Kimmeridge Clay Formation is the source rock for the Deveron crude. It is considered unlikely that sourcing is from within Block 211/18 but that stratigraphic juxtaposition of source and reservoir has occurred downflank along marginal faults. It is suggested that a migration path exists directed northwestwards towards the Deveron structure, from the Thistle Field spill point in the south.

Hydrocarbons

The Deveron crude is a highly undersaturated, low sulphur content oil with an API gravity of 38°. A well head sample obtained from A44 indicated a lower saturation pressure and lower GOR than Thistle. The reservoir is only slightly overpressured, its initial reservoir pressure being 5000 psig at a datum of 8800 ft TVSS. In common with Thistle, the reservoir contains an undersaturated oil with no free gas cap.

Test production from the appraisal well and early reservoir modelling indicated the presence of a large aquifer supporting Deveron production. This large natural aquifer has so far maintained reservoir pressure with no indication of any requirement for injection support. The strong pressure support from the SE indicates that the Deveron Field behaves in an analogous way to the Eastern and Southern Fault Blocks of the Thistle Field which received substantial pressure support from a natural aquifer before water injection was required. It is considered that the Deveron Field aquifer may be part of the same aquifer system as the Thistle Field Southern Fault Block.

All the Deveron wells are supported to some degree by this aquifer drive. Wells A44 and A48Z are producing formation water with a salinity of 23 500 ppm total dissolved solids. Water breakthrough has not yet occurred at well A51.

Reserves

The current estimate of initial oil in place is 41 MMBBL with ultimate recovery of 20 MMBBL, a recovery factor of almost 50%. The field came on stream via the Thistle platform in 1984 and cumulative oil production to the end of 1988 was 9.96 MMBBL with daily oil production averaging 4800 BOPD. Results from a remapping exercise similar to that completed on Thistle indicate that the initial oil in place will increase.

The principal development problems are similar to those experienced on Thistle and relate to the highly faulted, heterogeneous nature of the reservoir sandstones. These are described in detail in the Thistle Paper. An additional structural constraint on production relates to the isolation of individual fault blocks through partial or complete seal on faults, especially in the D sandstone. On Thistle this is well illustrated by the apparent isolation of the NE high pressure D sandstone fault block where erosion in the north and faulting in the east, west and south causes pressure isolation of this area. On Deveron it is more common for D sandstone units to become progressively condensed onto structural highs, leading to stratigraphic pinch out on the upthrown blocks. D sandstone deposition on structural highs is restricted to localized downfaulted areas. Well A51 was drilled at a crestal location to provide a D/C sandstone drainage point in the southern part of the field. It was perforated as a D sandstone producer and initially performed well before production steadily declined. At this point an artificial lift trial was initiated to evaluate well performance. The absence of lithological barriers to limit areal sweep within the D sandstone suggests that the poor well performance is related to structural controls.

The reservoir pressure was drawn down considerably when production began from well A51 indicating very limited pressure support from the typically strong natural aquifer. It is considered that the poor pressure support is a result of the well location within a small partially isolated fault block. The well drainage area is constrained by a stratigraphic pinch out to the south, fault seal of condensed D sandstone against Heather shale to the east and west and partial seal across the major E–W cross fault to the north.

This clearly illustrates that fault constrained areal sweep does appear to affect the D sandstone performance on the Deveron structure more severely.

The author wishes to thank BP Exploration and the Deveron partners (Deminex, Santa Fe, Arco, Premier and Monument) for permission to publish this paper. The author wishes to thank P. D. Begg for the selection of the seismic section, its interpretation and his comments on the geophysical content of this article.

Deveron Field data summary

Trap	
Type	Tilted and rotated fault block
Depth to crest (Top Brent Reservoir)	8650 ft TVSS
Oil/water contact	8910 ft TVSS (original)
Gross oil column	260 ft
Pay zone	
Formation	Tarbert, Ness, Etive, Rannoch formations of Brent Group
Age	Middle Jurassic, Bathonian–Bajocian
Gross thickness	450 ft
Net/gross ratio	12–100%
Porosity	16–30%, average 24%
Hydrocarbon saturation	average 0.7
Permeability	100 md–3-D
Hydrocarbons	
Oil gravity	38.0° API
Oil type	Light, low sulphur oil
Gas/oil ratio	150 SCF/BBL
Bubble point	650 psig
Formation volume factor	1.12 RB/STB
Formation water	
Salinity	23 500 ppm total dissolved solids
Resistivity	0.264 ohm m at 77°F
Reservoir conditions	
Temperature	220°F at 8800 ft TVSS
Pressure	5000 psig at 8800 ft TVSS in 1984
Field size	
Area	512 acres
Gross rock volume	49 723 acre ft
Recovery factor	48%
Recoverable hydrocarbons	20 MMBBL
Drive mechanism	Natural depletion with aquifer support
Production	
Start-up date	September 1984
Development scheme	Production wells drilled from Thistle 'A' platform
Number/type of wells	3 Production wells
Production rate: oil	4800 BOPD
water	15 500 BWPD
Cumulative production to end 1988	9.96 MMBBL
Secondary recovery method(s)	None

The Don Field, Blocks 211/13a, 211/14, 211/18a, 211/19a, UK North Sea

D. MORRISON[1], G. G. BENNET[2] & M. G. BAYAT[3]

BP Exploration, 301 St Vincent Street, Glasgow G2 5DD, UK
[1] 1 Ryat Drive, Newton Mearns, Glasgow G77 6SU, UK
[2] BP Exploration, Fairburn Industrial Estate, Dyce, Aberdeen AB2 0PB, UK
[3] BP Exploration, Britannic House, Moor Lane, London EC2Y 9SU, UK

Abstract: The Don Oilfield is located towards the western margin of the Viking Graben. It lies within four UKCS Blocks, 211/13a, 211/14, 211/18a, and 211/19a, some 15 km (9 miles) north of the Thistle Field. The field is structurally complex and consists of two discrete accumulations (Don NE and Don SW) which are separated by a WNW-ESE fault. The oil is trapped in sandstones of the Middle Jurassic Brent Group at depths of between 11 000 and 11 500 ft SS. Reservoir quality is variable with the Etive and Ness formations which contain the most productive intervals.

The Field is structurally complex and reservoir quality is highly variable. In order to minimize the impact of these uncertainties a phased development strategy has been adopted. This approach ensures the systematic reduction of risk and allows for flexibility to modify development plans as additional reservoir performance data is acquired.

The Don oil field is located mainly in UKCS Block 211/18a, but extends northwards into Blocks 211/13a and 211/14 and east into Block 211/19a. It lies some 150 km (93 miles) northeast of Shetland and 15 km (9 miles) north of the Thistle Field and it is one of the most northerly fields to be developed in the UKCS (Fig. 1). In this area, water depths are in excess of 500 ft. The field is structurally complex with the oil being trapped in Middle Jurassic Brent Group sandstones.

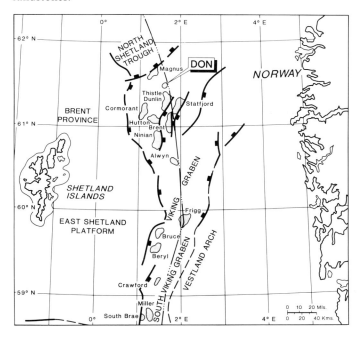

Fig. 1. Regional setting of the Don Field.

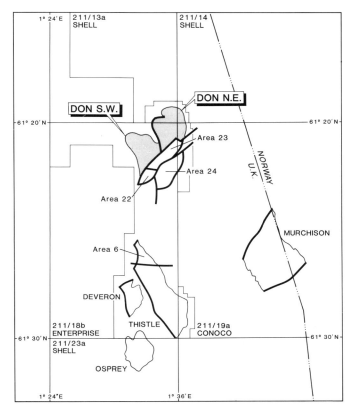

Fig. 2. Don Field location map.

The present field development area can be divided into two main fault blocks, Don SW and Don NE (Fig. 2). Don SW lies entirely within Block 211/18a but Don NE extends into Shell/Esso's Blocks 211/13a and 211/14 and also into the Britoil (BP)/Chevron/Conoco Block 211/19a. Following unitization and the acquisition of Shell, Esso and Conoco's interests by Britoil (BP), equity interests in Don NE are:

Britoil plc (BP)	55.247%
Deminex	23.245%
Santa Fe	12.306%
Arco	5.470%
Monument Oil & Gas	0.547%
Chevron	3.185%

History

UKCS Block 211/18 was awarded as Licence P236 in the fourth round of UKCS licensing in 1972 to a group with Signal Oil & Gas Co Ltd as operator. The current equities in 211/18a are:

Britoil plc (BP)	24.0%
Deminex	42.5%
Santa Fe	22.5%
Arco	10.0%
Monument Oil and Gas	1.0%

From Abbotts, I. L. (ed.), 1991, *United Kingdom Oil and Gas Fields,
25 Years Commemorative Volume*, Geological Society Memoir No. 14, pp. 89-93

A total of seven appraisal wells have been drilled on the main field, which was discovered in 1976 by well 211/18-12. This was drilled on a northerly dipping fault block (Don SW) and encountered a completely oil bearing Brent reservoir section. Well 211/18-13, also drilled in 1976, tested the Don NE fault block and again found an oil bearing Brent reservoir. The next two appraisal wells 211/18-15 and 211/13a-7, (drilled in 1977 and 1978) were both unsuccessful, penetrating the reservoir below the oil–water contact but in 1980 well 211/18a-21 was drilled as a successful appraisal well on the Don SW structure. This was followed in 1984/85 by the Shell-operated well 211/13a-8 which penetrated a thick immoveable oil column, but was sidetracked updip and successfully tested oil from the Don NE area.

In addition, wells 211/18a-22, 23 and 24 penetrated oil-bearing Brent Group sands in three fault blocks on the SE flank of the field. These may form additional targets for later phases of the development.

A NW–SE-orientated grid of 3-D seismic data was acquired over the Don area in 1983 and this forms the primary seismic data set on which the current reservoir mapping is based. This survey was reprocessed in 1985.

Field stratigraphy

The stratigraphy of the Don area is similar to that of much of the North Viking Graben. The deepest wells in the area have penetrated interbedded sandstones and shales of the Triassic Cormorant Formation. These are overlain by Statfjord Formation sandstones which, in this area, consist of medium to coarse sandstones 80–90 ft thick. These are in turn overlain by the marine siltstones and claystones of the Dunlin Group.

In the Don area, the sands of the Broom Formation are thin or absent resulting in shales of the Lower Rannoch Formation directly overlying the Dunlin Group. Details of the Brent Group stratigraphy are discussed below in the section describing the reservoir. Overlying the Brent Group sandstones are the siltstones and claystones of the Humber Group (Heather plus Kimmeridge Clay Formations). This interval varies considerably in thickness reflecting the intense tectonic activity that affected the area during Mid-Late Jurassic times (Badley *et al.* 1988).

The overlying Cretaceous section consists of a monotonous sequence of grey calcareous claystones with occasional limestones in the lower part. Claystones also predominate in the Tertiary sequence with the only sandstones encountered to date lying within the Miocene.

Geophysics

Two 3-D seismic data sets cover the Don Field. The original 1976/77 survey comprises east–west orientated lines 100 m apart with a 50 m CDP spacing. These data have been superseded by the 1982/83 survey which consists of 181 northwest–southeast-orientated lines (75 m spacing) totalling 1883 km of data. Data quality shows considerable improvement over the original survey, principally due

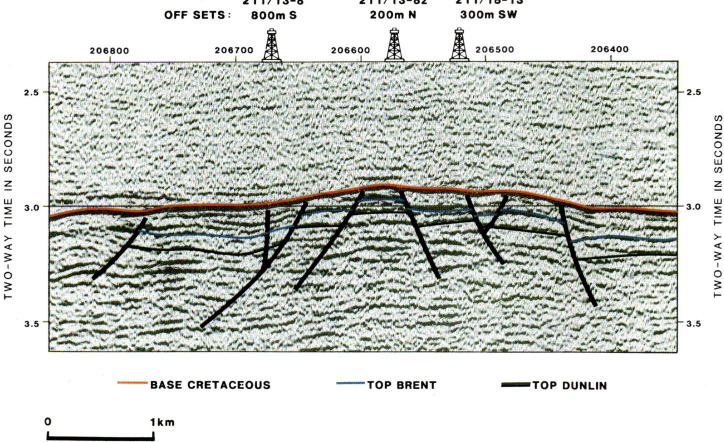

Fig. 3. Northwest–southeast seismic section through Don NE.

to the 12.5 m CDP spacing and improved frequency content of the data (Fig. 3).

The principal disadvantage of the 1982/83 survey was the poor migration of the cross lines. With the advent of better processing techniques, the data set was interpolated and remigrated in 1985 by Shell.

The mapping of the field can be split into two parts: Don NE, where the top Brent horizon can be mapped directly from the seismic data, and Don SW where the Top Heather reflection is mapped and the Heather formation isochore is added to create a Top Brent surface.

Trap

The hydrocarbons of the Don Field are trapped within two dipping fault blocks formed during the major Late Jurassic extension of the North Viking Graben (Fig. 4).

Two basic fault trends are developed in the Don area, these being NE–SW and NNW–SSE. Relative timing of these is difficult to assess, although clearly both systems were active during the deposition of the Kimmeridge Clay Formation as shown by the large variations in thickness of this formation in the wells.

There is no clear evidence from the well control to suggest that there was synsedimentary faulting during deposition of the Brent Group sandstones.

Different hydrocarbon water contacts have been encountered in the two main fault blocks and, together with the different hydrocarbon properties, indicate that the fault planes are acting as seals despite sand to sand juxtaposition. (The Don NE hydrocarbons are lighter and have a higher GOR than those encountered in Don SW.) The sealing capacity of the more minor faults within the reservoir is currently unknown and remains one of the major uncertainties in the development of the field.

Reservoir

The reservoir sequence in the Don Field is the Middle Jurassic age Brent Group (Fig. 5).

The sediments of the Brent Group in the Don Field were deposited at or near the northerly limit of the Brent 'province' and

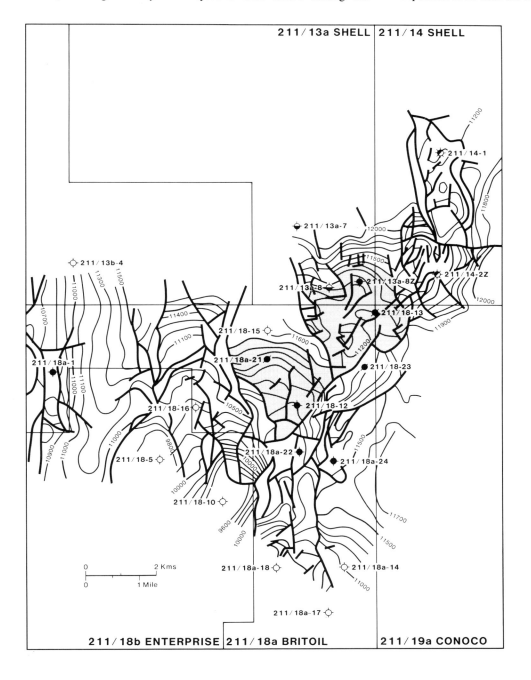

Fig. 4. Top Brent depth map.

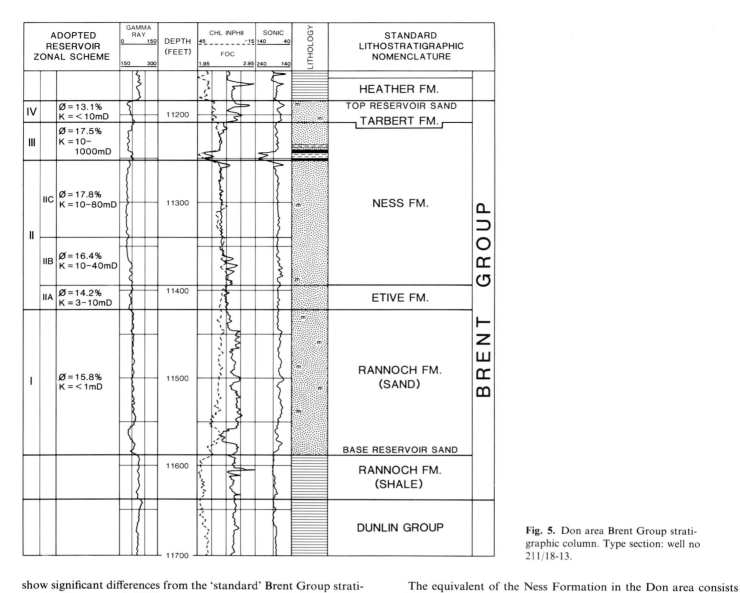

Fig. 5. Don area Brent Group stratigraphic column. Type section: well no 211/18-13.

show significant differences from the 'standard' Brent Group stratigraphy (Brown et al. 1987).

The Broom Formation, where seen, varies from 12–25 ft in thickness. No core material is available over the interval but on the basis of cuttings and comparison with Thistle Field to the south, it is likely to consist of poorly sorted, argillaceous and bioturbated shallow marine sandstones.

The overlying Rannoch Formation consists of a basal Rannoch Shale (30–50 ft in thickness) overlain by the Rannoch Sandstone (145–172 ft thick). These sandstones are fine- to medium-grained and micaceous with well developed coarsening upwards units. Low angle cross bedding is the dominant sedimentary structure and, together with wave ripples and rare burrows, suggest deposition in a lower-middle shoreface environment. The high content of detrital mica renders the Rannoch sandstones virtually unproductive, with permeabilities rarely in excess of 5 md.

The Etive Formation comprises fine- to coarse-grained sandstones with only minor amounts of mica and detrital clay. It is 30–100 ft thick and represents a continuation of the coarsening upwards sequence established in the underlying Rannoch Formation. The sandstones are massive or locally cross bedded with rare argillaceous laminae and wave rippled horizons. Rootlets and, locally thin, coal horizons are present in the upper part of the formation. Sedimentary structures and association with the adjacent formations suggest deposition in a middle/upper shoreface to backshore setting. Reservoir quality in the Etive Formation is generally poor to fair with porosity in the range 14–15% and permeabilities generally less than 10 md.

The equivalent of the Ness Formation in the Don area consists of 180–200 ft of fine- to medium-grained sandstones contrasting strongly with the 'standard' Ness sequences of interbedded fluvial sandstones, shales and coals typical of much of the Brent province. Correlation with wells to the south shows that this distinctive 'sandy' Ness Formation is laterally equivalent to, and interdigitates with, typical Ness Formation sediments (Brown et al. 1987). Lithologically, these Ness Formation sandstones closely resemble those of the underlying Etive Formation but the presence of several upwards fining and upwards coarsening units is believed to reflect a complex history of minor transgressive events. The only example of the 'standard' Ness Formation in the Don Field occurs near the top of the formation where a laterally persistent non-marine shale associated with carboniferous material and rootlet beds is present.

The sandstones of the Ness Formation include the most productive reservoir horizons within the Don area but their quality is highly variable. Porosity commonly ranges from 12–18% and permeabilities are generally less than 100 md but values as high as 25% and 1000 md have been encountered.

The overlying Tarbert Formation ranging in thickness from 21–65 ft, consists of very micaceous, strongly laminated sandstones. The high mica content readily distinguishes it from the underlying Ness Formation. These are interpreted as shallow marine sheet sands and are generally of poor reservoir quality with permeabilities of less than 10 md.

The well data shows that reservoir quality is generally poorer in Don SW than in Don NE. This difference in reservoir quality may in part be related to the presence of faulting in well 211/18-12 but the controlling factors on the areal distribution of reservoir quality are currently not well understood.

Production tests from wells 211/13a-8 and 8Z established the presence of several hundred feet of immobile oil column. Well-8 produced oil-free water from intervals which had oil saturations as high as 40%. Consequently by using all relevant data from logs, core and production test results, 'effective oil–water contacts' were established at 11 409 ft SS and 11 454 ft SS for Don NE and SW respectively.

Source

The source rocks for the Don Field are the organic-rich shales of the Kimmeridge Clay Formation which are regionally present in the North Viking Graben. The absence of hydrocarbons in wells that lie updip and to the southwest of the Don accumulation suggests that migration took place from mature source kitchen areas to the ENE of the field. Analysis of the oils within the separate Don fault blocks indicates that there are no major chemical differences between the oils, suggesting that the observed difference in PVT properties is related to variable migration routes, rather than different source kitchens for the oils.

Hydrocarbons

The Don crude is a light, low sulphur content oil with API gravity ranging from 42.1° in Don NE to 38.6° in Don SW. The reservoir is overpressured with an original pressure of 7460 psi at a datum depth of 11 500 ft and contains an undersaturated crude with no gas cap. Formation water has a salinity of 17 000 ppm dissolved solids and a resistivity of approximately 0.303 ohm m at 73°C.

Reserves

The Don reservoir is highly faulted and the sealing capacity of these faults is uncertain. Reservoir continuity is also doubtful, both because of the faulting and also because of its variable quality. These crucial uncertainties in the description of the reservoir have underlined the need for a flexible phased development plan which is capable of adjustment as additional drilling, production and injection data become available.

A phased sub-sea development plan has been proposed for the Don Field. Phase one (for which Annex B approval has been granted) is based on an initial assessment of the more prolific Don NE in the vicinity of wells 211/18-13 and 211/13a-8Z, involving two producers supported by an injector.

The plan envisages the re-entry of the well 211/18-13 and its completion as a producer. This would provide the necessary data to establish if this well is in a separate fault compartment to well 211/13a-8Z and the likely size of its drainage volume. The second producer will then be drilled close to well 211/13a-8Z and pressure data from this well, coupled with interference testing, will establish the degree of communication between these two wells. Both wells will be tied to a manifold located close to the 211/18-13 location and tied back to the Thistle platform.

It is planned to develop this field by water flooding as very little natural water influx is expected. This will be achieved by the drilling and completion of a downdip injector about one year after the first oil date. Injection water will be supplied from the Thistle platform. Initial production (governed by pipeline capacity) will be in excess of 11 000 BOPD. Reservoir performance and drilling data from the initial phase one development will be used to optimize the development plan which may lead to a phase two commitment. Phase two is currently envisaged to involve further development of Don NE and the initial development of the poorer quality reservoir of Don SW. Further development phases are possible.

The estimated reserves for a stand-alone phase one development are 24 MMBBL. However, there is some uncertainty in forecasting the reserves of this limited scheme. This is mainly due to the possibility of a limited drainage area for the two producers and a lack of support from the single injector mainly due to faulting.

The adoption of this phased sub-sea development plan for the field provides for systematic reduction of risks and retains a necessary degree of flexibility to respond to reservoir performance.

The authors wish to thank BP Exploration and the Don Partners (Deminex, Santa Fe, Arco, Monument and Chevron) for permission to publish this paper. The contents are based largely on the Annex B document and the authors wish to acknowledge the work of A. W. Mitchell (geologist), A. W. P. Thomson (reservoir engineer), J. R. Hook (petrophysicist) plus all members of the Don Project team.

Don Field data summary

Trap		
Type		Structural rotated fault blocks
Depth to crest		10 900 ft TVSS
Hydrocarbon contacts		11 409 ft TVSS (Don NE)
Oil column		500 ft
Pay zone		
Formation		Brent Group sandstones
Age		Middle Jurassic
Gross thickness (average)		420 ft
Net/gross ratio (range)		0.41–0.55
Porosity (average)		16.5%
Hydrocarbon saturation (average)		70%
Permeability:	Don NE	5–50 md
	Don SW	2–40 md
Productivity index:	Don NE	8–15 BOPD/psi
	Don SW	4–6 BOPD/psi
Hydrocarbons		
Oil gravity		38.6–42.1° API
Bubble point		1228–2940 psia
Gas/oil ratio		335–870 SCF/BBL
Formation volume factor		Don NE 1.47RB/STB
		Don SW 1.25RB/STB
Formation water		
Salinity		17 000 ppm
Resistivity		0.303 ohm m at 73°F
Reservoir conditions		
Temperature		265°F
Pressure		7220 psi at 11 200 ft ss
Pressure gradient in reservoir		0.36 psi/ft
Field size (Don NE only)		
Area		1390 acres
Gross rock volume		236 000 acre ft
Hydrocarbon pore volume		20 600 acre ft
Recovery factor: oil		22%
Drive mechanism		Waterflood
Recoverable hydrocarbons: oil		24 MMBBL
Production		
Start-up date		1989
Development scheme		Sub-sea tied back to Thistle Platform

References

BADLEY, M. E., PRICE, J. D., RAMBECH DAHL, C. & AGDESTEIN, T. 1988. The structural evolution of the northern Viking Graben and its bearing upon extensional modes of basin formation. *Journal of the Geological Society, London*, **145**, 455–472.

BROWN, S., RICHARDS, P. C. & THOMSON, A. R. 1987. Patterns in the deposition of the Brent Group (Middle Jurassic) UK, North Sea. *In*: BROOKS, J. & GLENNIE, K. (eds) *Petroleum Geology of North West Europe*. Graham & Trotman, 899–913.

The Dunlin Field, Blocks 211/23a and 211/24a, UK North Sea

A. BAUMANN & B. O'CATHAIN

Shell UK Exploration and Production, 1 Altens Farm Road, Aberdeen AB9 2HY, UK

Abstract: The Dunlin Oilfield is located in the East Shetland Basin, 160 km northeast of the Shetland Islands. It lies in UK Blocks 211/23a and 211/24a in about 500 ft of water. The field was discovered in June 1973 by well 211/23-1. The oil accumulation is trapped, in a north–south oriented, tilted fault block at the western margin of the Viking Graben, at a depth of about 8500 ft TVSS. The reservoir is contained in the formations of the Middle Jurassic Brent Group. In the Dunlin area they form a 450 ft thick sequence of sands and intercalated minor shales, which has been deposited by a shore face and delta system prograding northwards across the Viking Graben. The seal is formed by the shales of the Middle/Upper Jurassic Heather Formation. Reservoir properties of the Brent sands are fair to good with porosities of up to 30% and average permeabilities in the range from 10 to 4000 md. Development of the field is carried out from a single platform, from which production started in 1978. To date 40 development wells have been drilled and the total cumulative production amounts to 282 MMBBL of an ultimate recovery of 363 MMBBL.

The Dunlin Oilfield is located 160 km northeast of the Shetland Islands in the East Shetland Basin in a water depth of 495 ft. The field extends to two licence Blocks, 211/23a and 211/24a (Fig. 1). It consists of a main accumulation covering an area of 16.5 sq. km and a small satellite structure to the south-west, covering 1.5 sq. km.

The Dunlin Oilfield is a structural trap formed by a north–south oriented, tilted horst block. The reservoir is contained in the sandstones of the Middle Jurassic Brent Group, which are sealed by the shales of the Upper Jurassic Humber Group.

In the tradition of Shell's northern and central North Sea operations, the Dunlin Oilfield was named after a sea bird.

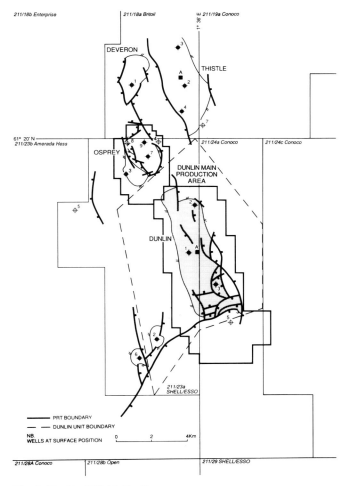

Fig. 1. Dunlin Oilfield situation map.

History

Blocks 211/23 and 211/24 were awarded in the Fourth Round and the Licence P232 issued in March 1972. The Dunlin Oilfield was discovered in June 1973 by well 211/23-1, which was drilled jointly by the Shell/Esso and Conoco/Gulf/BNOC (presently Chevron/OMV/Oryx) groups to test a structural high within Blocks 211/23 (Shell/Esso) and 211/24 (Conoco/GULF/BNOC). The discovery well encountered a gross reservoir sequence of 420 ft, which was fully oil-bearing. Prior to the installation of the platform, a total of eight appraisal wells were drilled in the Dunlin area which further defined the accumulations and confirmed the extent of the field.

Based on the results of the discovery well, the Shell/Esso and Conoco/GULF/BNOC groups agreed to a unitized development of the field with Shell appointed as operator. The Dunlin Unitization and Operating Agreement was finalized in April 1979. An equity re-determination became effective on April 1, 1987, which established the Shell/Esso share at 56.9% (formerly 70.7%) and the Conoco/Gulf/BNOC share at 43.1% (formerly 29.3%). The total stock-tank-oil-initially-in-place is determined at 827 MMBBL.

The Dunlin Oilfield was declared commercial in November 1976. It is being developed from a single concrete gravity platform with 48 conductor slots, installed in June 1977. Drilling started in November 1977 and by the end of 1988 a total of 40 development wells had been drilled in the main accumulation. Production commenced in August 1978, and crude is evacuated through the Brent System pipeline to Sullom Voe, Shetland Islands. The cumulative stabilized oil exported from the field at the end of 1989 was 282 MMBBL.

Field stratigraphy

Dunlin Oilfield lies in the East Shetland Basin in the northern North Sea, between the East Shetland Platform and the Viking Graben. The sedimentary record of the Dunlin area starts in the Triassic with the deposition of the fluviatile sediments of the Cormorant Formation, which were deposited from the high areas of the East Shetland Platform and the Fenno-Scandian High into the opening Viking Graben (Fig. 2). The terrestrial sands and siltstones of the Cormorant Formation are overlain by the shallow marine sands of the Upper Triassic Statfjord Formation. Further deepening of the Viking Graben/East Shetland Basin areas is reflected by the open marine shales and siltstones of the Lower Jurassic Dunlin Group. In the Middle Jurassic, this marine period was followed by a shallowing and the progradation of a deltaic system across the Viking Graben and East Shetland Basin in a northward direction. This delta deposited the sand/shale sequence of the Brent Group, in which the oil accumulation of the Dunlin

Oilfield is trapped. At the beginning of the Late Jurassic the progradation of the Brent delta was halted by a deepening and transgression, which led to the deposition of the marine Humber Group. The Upper Jurassic Humber Group comprises two formations; the Heather Formation at the base, consisting of shales and siltstones, and the overlying Kimmeridge Clay Formation, which is made up of organic-rich shales.

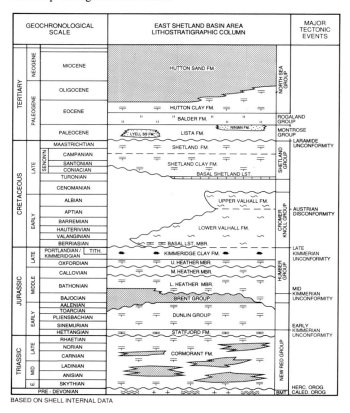

Fig. 2. Generalized stratigraphy of the East Shetland Basin.

At the end of the Mid-Jurassic and through the Late Jurassic, major periods of tectonic activity took place in the formerly rather inactive Viking Graben area. This resulted in a marked horst/graben structuration and the formation of numerous fault blocks in the East Shetland Basin, which were tilted to the west and differentially uplifted. These tectonic activities are evidenced by partial erosion of the Brent Group on the crestal parts of the higher fault blocks, and by thickness variations of the Heather Formation and the Kimmeridge Clay Formation. A relative fall in sea level, accompanied by widespread tectonic activity brought Humber Group deposition to a close. This event is represented by the Base Cretaceous 'X' Unconformity, which is thought to reflect the Early Alpine Cimmerian orogenic phase and forms a seismic marker of regional extent.

Above the 'X' Unconformity, the Lower Cretaceous Cromer Knoll Group was deposited as an onlap sequence during a marine transgression. These were succeeded in Late Cretaceous times by a monotonous sequence of Shetland Group Claystones.

Since Cretaceous times, only minor structural movements have occurred, and from the top Cromer Knoll ('K' Unconformity) to seabed the stratigraphy of the Dunlin area is characterized by a layer-cake sequence of marine shales and sands.

Geophysics

Delineation of the Dunlin Oilfield and development planning were based on a dense 2-D seismic grid acquired in 1974. In 1977–1979 one of the first North Sea 3-D seismic surveys was carried out over Dunlin, which resulted in a much improved definition of the field (Fig. 3). In 1983, re-processing of this survey was carried out, followed by a full-scale seismic interpretation and mapping of four main horizons. The most prominent is the Base Cretaceous 'X' Unconformity. This strong regional seismic marker was mapped at a high amplitude peak, generated by a marked decrease in acoustic impedance from the Lower Cretaceous Marls to the Upper Jurassic Kimmeridge Clay Formation. The top of the Brent Group horizon is an erosional unconformity, which also corresponds to a peak on this polarity of seismic. The acoustic impedance contrast across the horizon varies according to the preservation of the overlying Heather Formation, and the degree of truncation of the Brent Reservoir. The Base Mid Ness Shale Member reflector was mapped at a high amplitude peak. The Base Brent Group Horizon is often poorly defined. It corresponds to a low amplitude trough generated by a small increase in acoustic impedance where Broom Formation sandstones overlay Dunlin Group Shales. It was mapped at a zero crossing below a peak on zero phase reflectivity data, although residual water bottom multiple energy from the 'X' Unconformity reflector interferes with it in places.

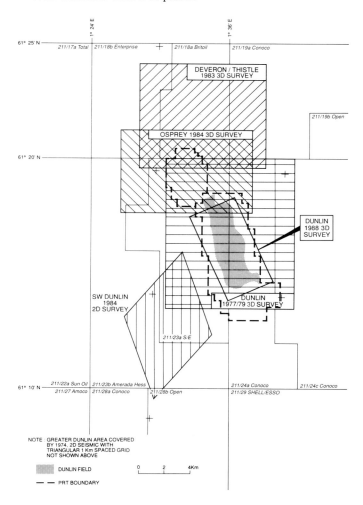

Fig. 3. Dunlin area seismic coverage map.

In 1988, 1600 line-km of new 3-D seismic was shot over the Dunlin Oilfield in order to provide an improved planning base for further reservoir management of the field and to identify the scope for additional development opportunities particularly in structurally complex areas. A comparison of the initial 1972 2-D seismic data and the 1988 3-D seismic data, through the discovery well 21/23-1, is shown in Fig. 4.

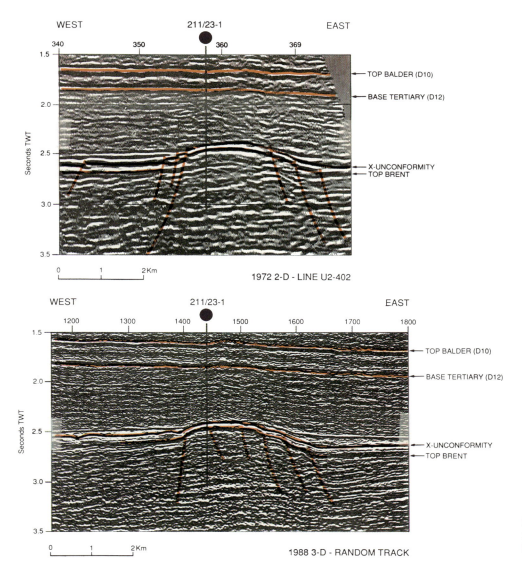

Fig. 4. Comparison of 1972 2-D and 1988 3-D seismic data through the Dunlin Oilfield discovery well 211/23-1.

Trap

The Dunlin Oilfield is a complex structural trap formed by extensional horst/graben type faulting and fault block rotation. It is subdivided into several blocks separated from each other by major faults (Fig. 5). The top seal is provided by the shales of the Heather Formation. The majority of the reserves are contained in a central horst block measuring approximately 6 × 2 km. It is bounded to the east, south and west by faults and to the north by dip closure (Figs 6 and 7). In this main block, the Brent reservoir dips to the north and west, and erosional truncation is observed in the east and southeast. Six smaller, fault-bounded blocks occur to the south and southeast of the main block.

The structural crest of the field lies at 8500 ft TVSS. Towards the north, where dip closure delineates the Dunlin accumulation, the spill point is at a depth of approximately 9450 ft TVSS. Two separate oil-water contacts have been encountered in the main block at 9205 ft TVSS and 9165 ft TVSS in the upper and lower part of the Brent reservoir respectively, separated by a field-wide sealing shale. The faulted nature of the southern and eastern flanks has led to a variety of oil-water contact depths in these sub-blocks.

Reservoir

The Dunlin oil accumulation is contained in the sands of the Middle Jurassic (Bajocian) Brent Group. The Brent Group has been extensively cored in 17 wells, giving good control on reservoir geometry, facies distribution and reservoir properties. The Brent Group is subdivided into five formations based on depositional facies. These are from bottom to top the Broom, Rannoch, Etive, Ness and Tarbert Formations (Fig. 8).

The Ness Formation consists of three members, the Lower, Middle and Upper Ness. For the purpose of reservoir subdivision, the Tarbert and Upper Ness form one unit, the 'Upper Reservoir'. It is separated from the 'Lower Reservoir' by the field-wide, sealing shale unit of the Middle Ness. The Lower Reservoir comprises four units, the Lower Ness, Etive, Rannoch and Broom. With the exception of the Tarbert and the Lower Ness, all units are continuous across the field and have a good log correlatability (Fig. 9).

The basal *Broom Formation* consists of poorly sorted, very coarse-grained and pebbly sands. It forms a regionally extensive, sheet-like deposit which across the Dunlin horst block shows a thinning from 45 ft in the west to 15 ft in the east, averaging 30 ft. Based on sedimentological evidence and regional isochore distribution, the Broom Formation is interpreted as being deposited as an easterly prograding series of fan deltas, into a shallow marine environment, prior to main Brent Sands progradation from the south. The heavy mineral assemblage of the Broom Formation points to a sediment source different from the overlying formations. The average net oil sand porosities range from 24–26%, and average (horizontal, air) permeabilities range from 0.5–1000 md.

The *Rannoch Formation* consists of fine micaceous sands which were deposited on the foreshore and shoreface of the coastal front of the Brent delta by onshore sediment transport. Four facies associations have been identified, labelled A to D from top to

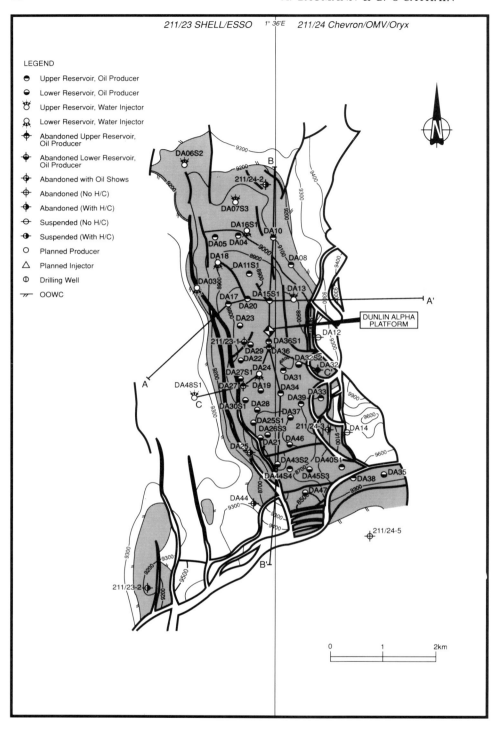

Fig. 5. Dunlin Oilfield, Top Brent structure map.

bottom. The Rannoch A and B are continuous sands and the Rannoch C and D intercalated sand/shale sequences. These facies types are related to water depth and hence wave energy, representing the transition from a distal, deep water, low energy environment (D) to a shallow water, higher energy environment (A). The Rannoch has a wedge-shaped geometry, thinning across the main Dunlin fault block from 170 ft in the west to 60 ft along the eastern boundary fault. Oil sand porosity averages vary from 16.9–25.0%, with average (horizontal, air) permeabilities ranging from 0.1–1000 md.

A common feature of the Rannoch Formation are carbonate-cemented layers ('doggers'), which are up to a few feet thick but have a limited lateral extent. In Dunlin, they comprise an average 12% of the gross thickness of the Rannoch Formation.

The coarse, well sorted sands of the *Etive Formation* are interpreted as being mainly upper shoreface and coastal bar deposits cut by channels and distributary outlets which formed along the coastal fringe of the Brent delta. The Etive Formation averages 90 ft in thickness, with a thickest development of up to 140 ft in the northern part of the main block and gradual thinning to 50 ft in the south. The Etive is the most prolific producing reservoir of the field with very good porosity and permeability characteristics. Porosity averages range from 24–30%, and average (horizontal, air) permeabilities ranging from 100–10 000 md.

The *Ness Formation* formed in a delta plain environment and is subdivided into three members in Dunlin (Lower, Middle and Upper Ness) which represent different sub-environments.

The *Lower Ness Member* is interpreted as representing a back barrier environment immediately landward of the Etive barrier system. It is thin or absent (generally less than 25 ft) in thickness and is composed of alternating sands and shales with occasional *in situ* coal layers. Average porosities vary between 16 and 30%,

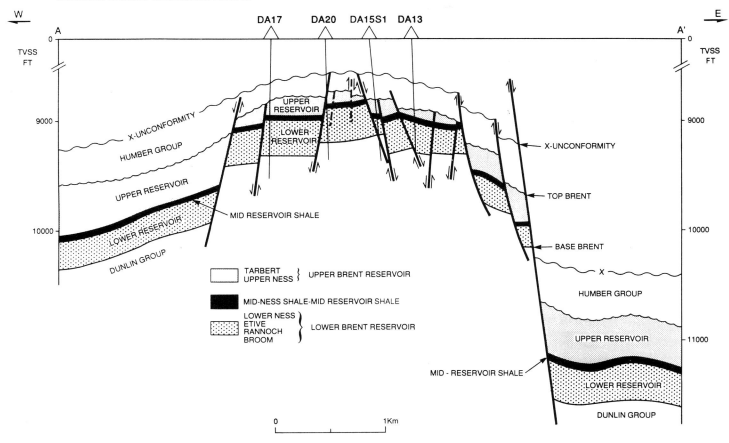

Fig. 6. Dunlin Oilfield, structural cross-section west–east.

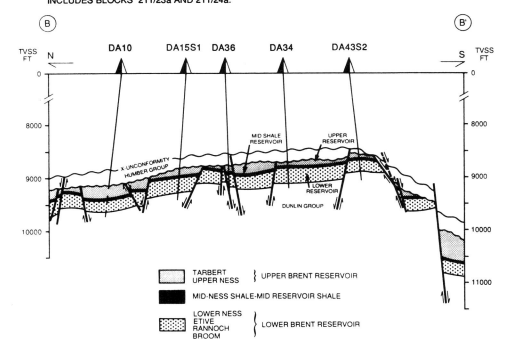

Fig. 7. Dunlin Oilfield, structural cross-section north–south.

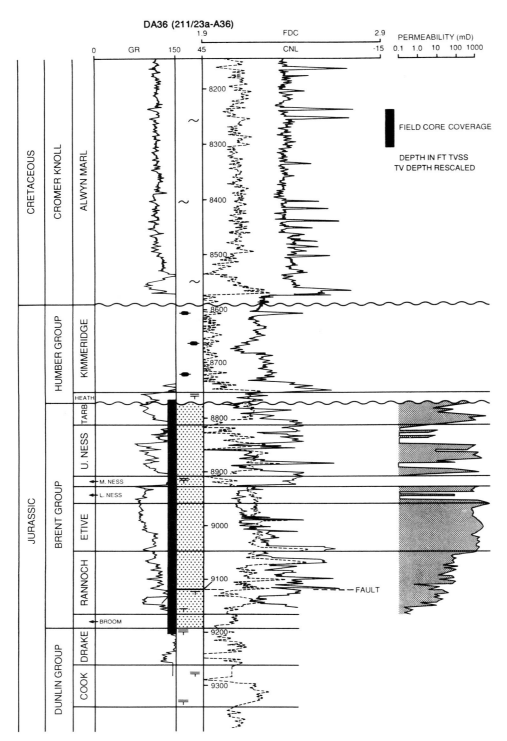

Fig. 8. Dunlin Oilfield, reservoir type log (Well DA36).

with average (horizontal, air) permeabilities ranging from 1.0–10 000 md.

The *Mid Ness Member* is a restricted environment lagoonal shale, on average 30 ft thick and with an even thickness distribution. It forms a significant vertical pressure barrier, effectively subdividing the Brent Group into an Upper and Lower Reservoir.

The *Upper Ness Member* is regarded as a sequence of upper delta plain deposits capped by lagoonal shale. At the base of the member is a coarsening-upwards sand sheet which is interpreted as a mouth bar deposit. Over the main fault block the thickness of the Upper Ness varies between 150 ft in the west and northwest and 60 ft near the eastern boundary, averaging 120 ft. Average porosities range between 10 and 30%, with average air permeabilities in the range 0.1–10 000 md.

The *Tarbert Formation* consists of medium- to coarse-grained, well sorted sands of marine origin. They form the basal part of the transgressive cycle overlying the deltaic sequence of the Brent Group. Its distribution across the field appears to be controlled by faulting and subsequent erosion at the top of the reservoir. Thickness varies from zero to 140 ft, with the thickest developments in the northwest of the main fault block and the eastern flank.

Porosity averages of the Upper Ness/Tarbert unit are between 17 and 28%, and average (horizontal, air) permeabilities range from 10–10 000 md.

Source

The hydrocarbons of the Dunlin Oilfield are derived from the organic-rich Upper Jurassic Kimmeridge Clay Formation, the

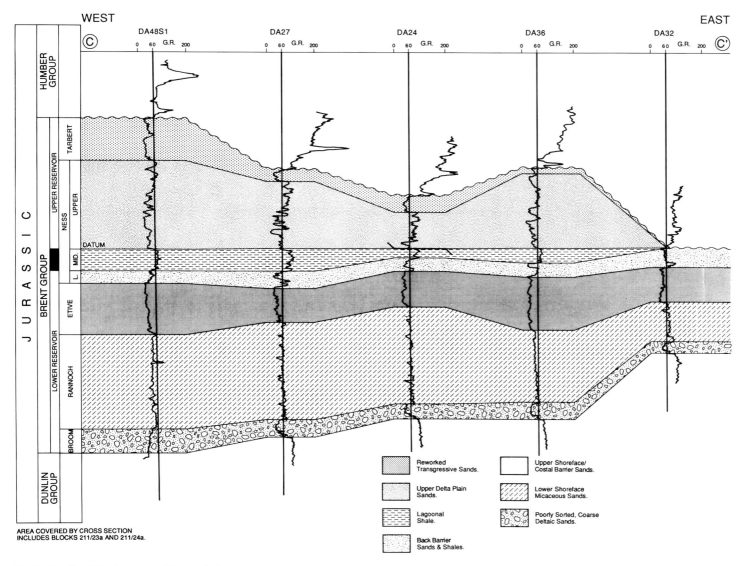

Fig. 9. Dunlin Oilfield stratigraphic correlation.

dominant source rock of the North Sea. In the Viking Graben area deposition of the Kimmeridge Clay Formation occurred during rapid subsidence. In the East Shetland Basin, syndepositional extensional tectonic activity resulting in a horst/graben structuration is reflected by thickness variations of the Kimmeridge Clay Formation. Kitchen areas are the Viking Graben system itself, where oil generation started in the early Tertiary, and the half-graben areas of the East Shetland Basin, where maturity was reached in the mid- and late Tertiary. Light oil generation is restricted to the central areas of the Viking Graben. The main migration paths from the kitchen areas into the fault block traps of the East Shetland Basin are provided by regional, large fault systems.

Hydrocarbons

The Dunlin crude has a specific gravity of 0.845–0.872 (at 60°F and 14.7 psia), which corresponds to 36.0–30.8° API gravity. The oil is undersaturated and has a paraffin wax content of 19 %W. The bubble point is 960 psia at 210°F. The solution GOR ranges from 102–277 SCF/BBL, averaging 220. Oil viscosity at initial reservoir conditions varies from 0.55–2.9 cP.

Initial reservoir pressure was 6020 psig at 9000 ft TVSS, and the reservoir temperature at this depth is 210°F. Formation water salinity is 24 000 ppm NaCl equivalents.

Reserves/resources

The current (1 Jan. 1990) expectation Stock Tank Oil Initially In Place (STOIIP) is 827 MMBBL with 182 BCF of solution gas. Oil produced to date is 282 MMBBL; gas produced is 62 BCF. The expected Ultimate Recovery is 363 MMBBL, giving an average oil recovery factor of 44%.

Oil production has been steadily declining since the field came on stream in 1978. Production peaked in early 1979 at approximately 126 000 BOPD and by 1989 average production had declined to 38 400 BOPD with an average watercut of 73%. Production is expected to last into the next century.

Material balance calculations indicate the existence of early aquifer support to both Upper and Lower Reservoirs, with a minimum expected connected volume of 60 and 40 million reservoir barrels, respectively. The Upper Reservoir initially received strong support from the north and west of the main block. This aquifer influx has declined, and the reservoir is now dependent on water injection. The Lower Reservoir received less support, and high initial offtake rates led to an early requirement for water injection.

Good productivities can be obtained from the Etive, Tarbert and Upper Ness Sands (PIs ranging from 10 BOPD/psi to 200 BOPD/psi). The Rannoch and Broom sands are much less prolific, with PIs in the range 1–5 BOPD/psi. Individual wells can produce at gross rates in excess of 30 000 BOPD. The Upper and Lower Reservoirs have been developed separately, with initial development wells being perforated across the majority of producing sands to maximize production. In recent wells, a selective perforation policy has been adopted in order to drain the less well-supported sands. In particular, adequate drainage of the Rannoch Sands require dedicated producers with much closer well spacing than is necessary to drain the Etive.

Production problems are caused primarily by scale build-up, zonal water breakthrough, and CO_2 corrosion. Barium sulphate scale develops in producing wells when injected seawater mixes with formation water, causing tubing bore restrictions. Water breakthrough occurs preferentially in high permeability zones, requiring isolation of these zones in order to produce oil from other reservoir units. Carbon dioxide corrosion causes short tubing life, requiring workovers of corroded wells.

Dunlin is a mature field with water ingress into most areas. The highly permeable basal Upper Ness and the base Etive are flushed throughout the field, and the Tarbert is largely flushed in the north.

This paper has been compiled from various sources within Shell UK Exploration and Production Ltd. The interpretations expressed reflect the results of studies carried out by Shell Expro in its role of operator of the Dunlin Field. The consent to publish given by BP Exploration plc, Chevron UK Ltd, Conoco UK Ltd, and Esso Exploration and Production Ltd is gratefully acknowledged.

Dunlin Oilfield data summary

Trap
Type	Structural; tilted fault block
Depth to crest	8500 ft TVSS
Lowest closing contour	9205 ft TVSS
Hydrocarbon contacts	Upper Reservoir: OOWC 9205 ft TVSS
	Lower Reservoir: OOWC 9165 ft TVSS
Oil column	705 ft

Pay zone
Formation/Group	Brent Group
Age	Middle Jurassic (Bajocian)
Gross thickness	170–650 ft, average 450 ft
Net/gross ratio	up to 1.0
Cut off for N/G	12 porosity units
Porosity (averages)	16% (L. Ness)–30% (Etive)
Hydrocarbon saturation	0.35 (Broom)–0.85 (Tarbert)
Permeability	up to 10 000 md
Productivity index	12–68 (Upper Reservoir) BOPD/psi
	42–110 (Lower Reservoir) BOPD/psi

Hydrocarbons
Oil gravity (average)	35° API
Oil type	Paraffinic, undersaturated
Viscosity (at p_i)	0.55–2.9 cP
Bubble point	960 psia at 210°F
Gas/oil ratio	220 SCF/BBL
Formation volume factor	1.13

Formation water
Salinity	24 000 ppm NaCl equivalents
Resistivity	0.110 ohm m at 187°F

Reservoir conditions
Temperature	210°F at 9000 ft TVSS
Pressure	6020 psig at 9000 ft TVSS
Pressure gradient in reservoir	0.35 psi/ft

Field size
Area	18 sq. km
Gross rock volume	878 341 acre ft
Hydrocarbon pore volume	120 166 acre ft
Recovery factor, oil	44%
Drive mechanism	Waterflood
Recoverable hydrocarbons	Oil 363 MMBBL
	NGL 6 MMBBL
	Gas 182 BCF

Production
Start-up date	August 1978
Development scheme	Platform-based, production into Brent System
Peak production rate	126 000 BOPD (1979)
Cumulative production to date	180 MMBBL (1 Jan. 1990)
Average production in 1989	38 400 BOPD
Number/type of wells	1 exploration
	7 appraisal
	41 development oil producers
	8 development water injectors
Secondary recovery method	Water injection

The Eider Field, Blocks 211/16a and 211/21a, UK North Sea

M. D. WENSRICH[1], K. M. EASTWOOD[2], C. D. VAN PANHUYS[3] & J. M. SMART[4]

[1] *Esso Exploration and Production, UK Ltd, 21 Dartmouth Street, London SW1H 9BE, UK*
[2] *Shell UK Exploration and Production Ltd., Shell Mex House, Strand, London WC2R 0DX, UK*
[3] *Al Furat Petroleum Co., PO Box 7660, Damascus, Syria*
[4] *Badley, Ashton & Associates Ltd, Winceby House, Winceby, Horncastle, Lincolnshire LN9 6PB, UK*

Abstract: The Eider Oilfield is located some 160 km northeast of the Shetland Islands mainly in Block 211/16a, part of Production Licence P296, with the southern tip extending into Block 211/21a. The discovery well was drilled in 1976 in a water depth of about 530 ft. Hydrocarbons are trapped at a depth of 8750 ft TVSS in an easterly dipping fault block that is part of the Tern–Eider horst in the East Shetlands Basin. The reservoir is the Middle Jurassic Brent Group, deposited in a wave-dominated delta system. It has an average gross thickness of 259 ft and an average net thickness of 210 ft. Porosities average 23% and permeabilities 375 md. The expected STOIIP and Ultimate Recovery are estimated at 204 MMBBL and 85 MMBBL respectively, representing a recovery factor of 42%.

The field is planned to be developed by five producers drilled along the crest of the structure. Pressure maintenance and sweep (secondary recovery) will be provided by four down-flank water injectors. The Eider production facility is a satellite platform of the neighbouring North Cormorant and Tern platforms and is designed for future unmanned operation. Production started in November 1988 and averaged some 54 000 BOPD. Water injection began in June 1989, using water, piped at injection pressure, from the Tern platform. First stage processing of the crude oil is carried out on the Eider platforms and then it is piped to the North Cormorant platform for further processing. From there it is evacuated via the Cormorant Alpha platform into the Brent System pipeline for export to the Sullom Voe terminal.

The Eider Oilfield is located 160 km to the northeast of the Shetland Islands in the East Shetland Basin (Fig. 1). Water depth averages 530 ft. The major part of the Eider Oilfield lies within Block 211/16a with the southern tip in Block 211/21a (Fig. 2). Both blocks are within Production Licence P296. The field area covers some 3.5 sq. km and contains 85 MMBBL of recoverable oil. The oil is trapped in the Middle Jurassic Brent Group reservoir within an easterly dipping horst block. The field, which was named after the eider duck, *Somateria mollissima*, is operated by Shell with Esso as a 50% co-venturer.

History

Licence P232 was issued on 16 March 1972. It comprised 9 concession blocks in the Northern North Sea, in which Brent, Cormorant, Dunlin, Osprey, Tern, Eider and Don Fields were subsequently discovered. Following years of exploration and appraisal, these fields were defined and the licence re-negotiated; Blocks 211/16a and 211/21a now form part of Licence P296 issued on 20 June 1979 to Shell UK Ltd as the operator and with Esso UK Ltd as a 50% co-venturer (Fig. 1).

The Eider oilfield was discovered in July 1976 by well 211/16-2 (Fig. 2). The well was drilled on a seismically defined structural high and found Middle Jurassic Brent Group sandstones to be oil bearing with an oil-down-to of 8880 ft TVSS. Appraisal well 211/16-4, drilled 2 km to the south of the discovery well, found the Brent Group sandstones oil bearing with a probable oil/water-contact (POOWC) at 9050 ft TVSS. Following a detailed seismic interpretation of Block 211/16a and the shooting of several additional seismic lines, well 211/16-5 was drilled in 1978 to appraise a separate fault block to the north west of the main block. The Brent Group sands were encountered at 8985 ft TVSS, 65 ft above the interpreted POOWC in 211/16-4, but were water bearing. Appraisal well 211/16-6, drilled 2.5 km north of the discovery well, confirmed the structural interpretation and observed an OOWC at 9033 ft TVSS. In 1981 a pilot 3-D seismic survey was shot across the central part of the field and full field 3-D seismic coverage was obtained in 1983.

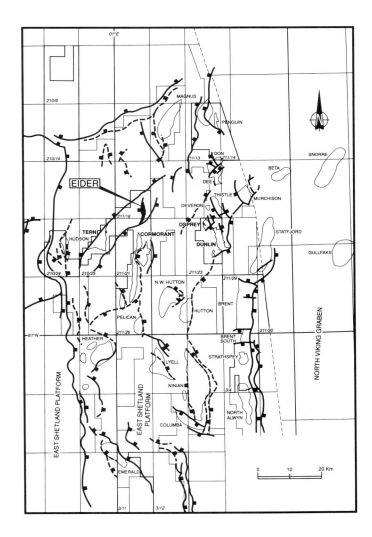

Fig. 1. Location map of the Eider Oilfield in the Northern North Sea.

From Abbotts, I. L. (ed.), 1991, *United Kingdom Oil and Gas Fields, 25 Years Commemorative Volume*, Geological Society Memoir No. 14, pp. 103–109

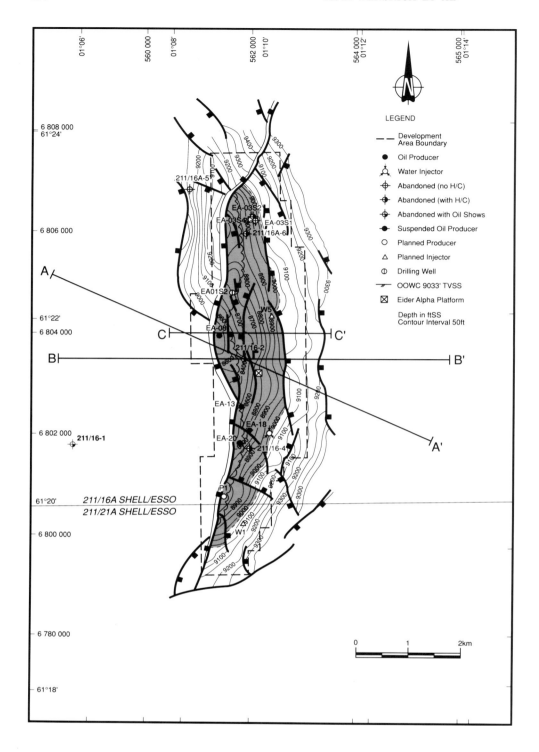

Fig. 2. Eider Oilfield: Top Brent Group structural contour map.

Reserves were estimated to be 85 MMBBL and it was concluded that development using a conventional, stand-alone production platform was only marginally economically attractive. Other development options were studied and a minimum facility satellite concept was found to offer a profitable means to develop this relatively small field. Government approval of the proposed development was obtained on 24 October 1985 and the field is now being developed with a single steel platform which will be largely unmanned during operation and remotely controlled from North Cormorant. The first stage gas/oil separation and crude dehydration to less than 5% BS&W is carried out on the Eider platform and the high vapour crude oil is then evacuated to the North Cormorant platform for second stage separation. From there it is exported via the Cormorant Alpha platform through the Brent pipeline system to the Sullom Voe terminal.

Production wells are being drilled along the crest of the structure, close to the western boundary fault. A secondary recovery scheme (pressure maintenance and sweep) comprises the drilling of a row of water injection wells along the oil/water-contact near the eastern boundary fault. Development well drilling started in September 1988 and first oil was produced in November 1988. Five development wells have been drilled to date (June 1989), four producers and one injector.

Field stratigraphy

The oldest sediments in the Eider area are those of the Triassic Cormorant Formation which rest on pre-Devonian granitic or metamorphic basement (Fig. 3). During the Devonian, Carboniferous and Permian, the Eider area formed part of the Shetland Platform. With the onset of rifting during the Triassic the area became the site of continental deposition of some 1000–1500 ft of red-brown silty claystones with occasional interbedded sandstones.

During the Early Jurassic the relatively thin (10–20 ft) calcareous sands of the Statfjord Formation were deposited as a marine transgressive lag. Continued rifting of the Viking Graben resulted in progressive subsidence of the area and the deposition of the marine shales of the Dunlin Group which are some 175 ft thick.

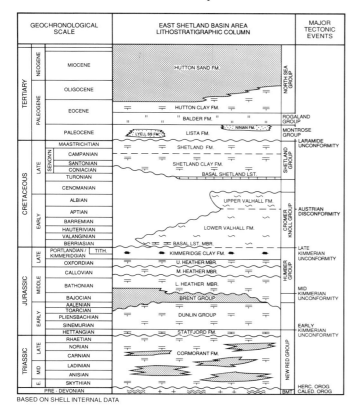

Fig. 3. Generalized stratigraphy of the East Shetland Basin.

During the Middle Jurassic a wave-dominated delta system prograded in a generally northern direction across the East Shetland Basin and Viking Graben. Its deposits formed the Brent Group which is typically 270–300 ft thick in the Eider area, compared to the 600 ft thick development in the Brent area. This marked reduction in thickness is ascribed to syn-depositional positive movements on the Tern–Eider horst. Delta progradation was halted by a renewed marine transgression and the marine shales of the Humber Group (Heather and Kimmeridge Clay Formations) were deposited during the upper Middle and Late Jurassic. Disconformities separate the Upper Heather Member from the Lower and Middle Heather Members below and the organic-rich Kimmeridge Clay Formation above. These disconformities are a result of the tilting of the Eider horst block to the northwest caused by tectonic movement along the Eastern Boundary Fault of the Tern–Eider horst during the Mid–Late Kimmerian orogenic phase. The Jurassic sediments are truncated by the Base Cretaceous 'X' Unconformity which reflects the culmination of this tectonic phase.

The Lower Cretaceous Cromer Knoll Group marls wedge out towards the west because of non-deposition/condensation (seismic onlap onto the Base Cretaceous 'X' Unconformity) during the next major tectonic event (Austrian phase). This tectonic phase reversed the Tern–Eider horst tilt towards the east, mainly via a process of intense faulting and re-activation of pre-existing Caledonian trends. The Upper Cretaceous Shetland Group completely covered the Tern–Eider horst with some 3000 ft of marine shales and marls. The Alpine tectonic phase resulted in the Laramide Unconformity which marks the end of the Cretaceous. A 4500 ft succession of Tertiary to Recent marine sandstones and claystones of the Montrose, Rogaland and Nordland/Hordaland Groups overlie the Cretaceous. The widespread and readily recognizable tuffs of the Balder Formation occur at the top of the Rogaland Group.

Geophysics

Prior to the 3-D survey, on which the present maps are based, seismic coverage consisted of a 0.5×1.0 km grid of lines shot in 1975 and 1976 with local infill surveys shot in 1977 and 1978. Seismic quality was good but line density did not allow detailed fault interpretation.

A 3-D seismic survey comprising 198 E–W oriented lines 75 m apart was recorded over the Eider Oilfield by Western Geophysical in two parts in 1981 and 1983. Data quality is considered reasonable and represents a significant improvement over the data initially used to define the structure (Fig. 4a and b).

In total five horizons (Base Tertiary Unconformity, Base Cretaceous Unconformity, Top and Base Brent Group and an intra-Triassic horizon) were mapped and tied to existing well data. The most prominent horizon is the Base Cretaceous Unconformity ('X') reflector which is expressed as a well developed negative (black) loop that is the result of the impedance contrast between the more dense sediments of the Cromer Knoll Formation that overlie the Kimmeridge Clay Formation.

A marked decrease in acoustic impedance from the marls of the Lower Heather Member to the porous sandstones of the Brent Group gives rise to a strong, negative reflection (black loop) which characterizes the Top Brent Group sandstones.

The quality of the reservoir sandstones deteriorates from the top towards the base of the Brent Group resulting in an insignificant acoustic impedance contrast between the Broom Formation at the base of the Brent Group and the shales of the Dunlin Group below. Hence there is no significant seismic marker which defines the Base of the Brent Group horizon.

Trap

The Eider Oilfield is a north–south-trending fault trap located within the NE–SW-trending Tern–Eider horst—a major structural feature some 40 km long and 6 km wide. The Eider accumulation is some 7 km long and 1.5 km wide, dips approximately 8 degrees to the east, and is bounded to the west by a major N–S-trending fault with up to 500 ft of throw (Fig. 5). The throw is well in excess of the total thickness of the Brent Group (290 ft) and thus juxtaposes these reservoirs with the shales of the Humber Group providing an effective seal.

The eroded structural crest of the reservoir is penetrated as high as 8458 ft TVSS giving a hydrocarbon column of 575 ft.

Reservoir

The Middle Jurassic Brent Group is divided into the Broom, Rannoch, Etive, Ness and Tarbert Formations (Fig. 6). The Rannoch Formation is further subdivided into the Rannoch Shale Member overlain by the Rannoch Sand Member. The Ness Formation is subdivided regionally into three members but in the Eider Oilfield its Lower Ness member is only locally present and the Etive Formation is in general directly overlain by the Mid Ness Shale Member.

The Tarbert Formation, the Upper Ness Member of the Ness and the Etive Formation constitute the three main reservoir units in the Eider Oilfield. Variations in facies and reservoir quality within the Etive Formation and Upper Ness Member of the Ness Formation allow some further subdivision.

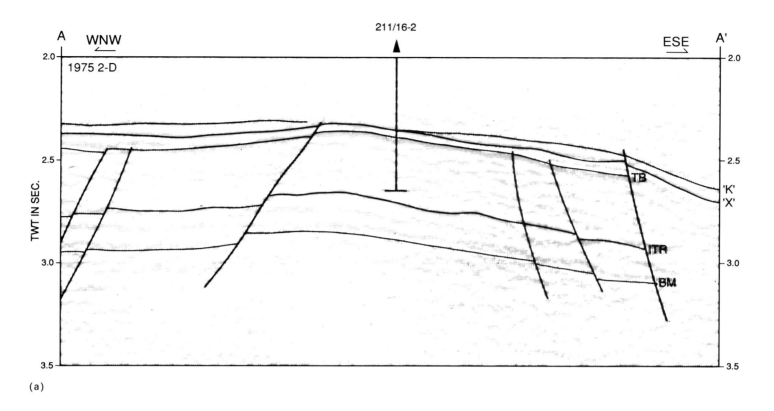

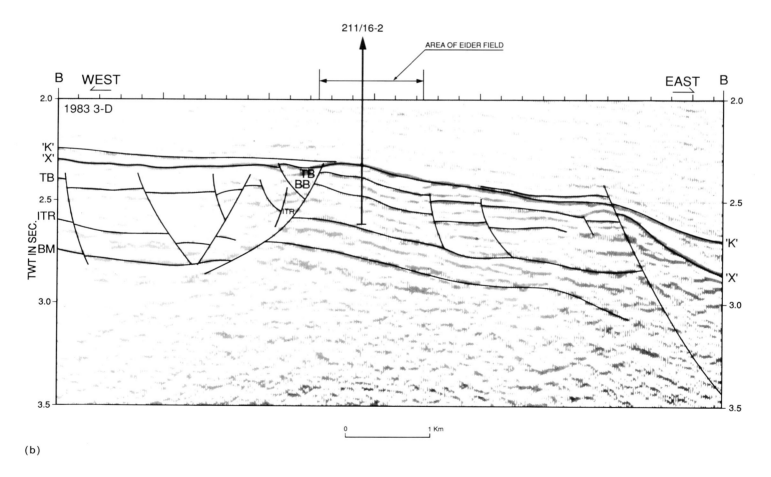

Fig. 4. (a) Eider, 1975 2-D seismic section through the discovery well 211/16-2. K, basal Shetland Limestone reflector; X, Base Cretaceous unconformity reflector; TB, Top Brent Group reflector; BB, Base Brent Group reflector; ITR, intra-Triassic reflector; BM, top basement reflector. (b) Eider, 1983 3-D seismic section through the discovery well 211/16-2. K, basal Shetland Limestone reflector; X, Base Cretaceous unconformity reflector; TB, Top Brent Group reflector; BB, Base Brent Group reflector; ITR, intra-Triassic reflector; BM, top basement reflector.

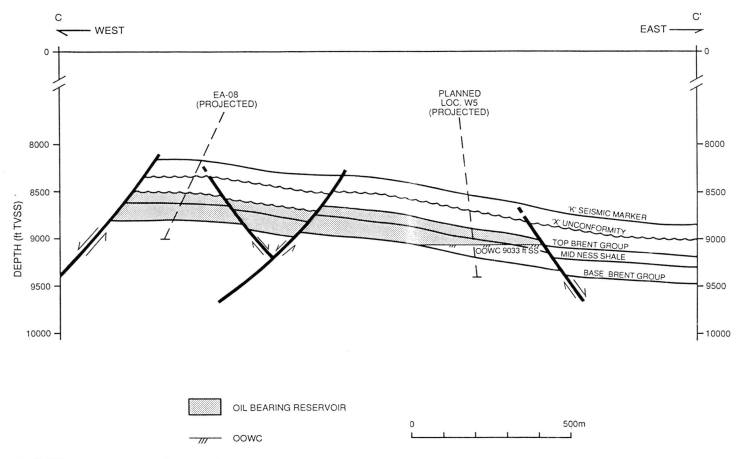

Fig. 5. Eider, east–west structural cross-section.

The *Broom Formation* is composed of very coarse-grained and pebbly, poorly sorted feldspathic sandstones with widespread calcite and siderite cementation. It is interpreted as representing a series of deltaic fans prograding into a shallow marine environment. This unit is relatively thin (9–12 ft) and is a poor reservoir because of the high matrix clay content and carbonate cement. Average porosities range from 6–18% and permeabilities up to 10 md.

The *Rannoch Shale Member* at the base of the Rannoch Formation is a bioturbated silty mudstone, which rapidly passes up into a siltstone to very fine sandstone that gradually coarsens upwards. The sequence is interpreted as an offshore mudstone overlain by lower shoreface sediments representing the onset of a northerly prograding coastal barrier complex. The unit is 44–57 ft thick and reservoir quality is poor to nil. Vertical permeabilities are reduced by mica rich laminae and the unit acts as a seal to the Broom Formation. Average porosities are in the range of 5–18% and permeabilities range up to 10 md.

The *Rannoch Sand Member* is a silt to fine-grained micaceous, moderately to well-sorted sandstone with carbonate cemented layers that form permeability baffles. The sandstones are horizontally to low-angle laminated and hummocky cross stratified towards the top of the unit and were deposited in a middle shoreface setting above storm wave base. This unit is 46–59 ft thick and is of poor to moderate reservoir quality. Average porosities are in the range of 5–24% and permeabilities range up to 100 md.

The lower part of the *Etive Formation* is composed of fine- to medium-grained and moderately well-sorted sandstones. Cross-bedded micaceous layers and minor ripple cross-laminae are present and bioturbation is lacking. Carbonate cementation is common at the base of the unit which was deposited in an upper shoreface setting by inlet channels cutting through and reworking the coastal barrier. Reservoir quality of the cross-bedded sandstones is very good although vertical permeabilities are much reduced by mica and clay layers. Unit thickness varies from 16–38 ft, porosities range from 16–28% and permeabilities range from 100–1000 md.

The upper part of the Etive Formation is composed of fine grained, moderately well-sorted sandstones, predominantly massive bedded but with some horizontal and ripple cross-lamination in its lower part. The unit represents an emergent coastal barrier comprising foreshore deposits that are overlain by dune sandstones capped by marsh sediments and coals. Reservoir quality is good and the thickness varies from 21–39 ft. Average porosities range from 16–27% and permeabilities range from 10–1000 md.

The Etive Formation is overlain by the lagoonal mudstones and silty–sandy layers of the *Middle Ness Shale Member*. This unit averages 10 ft in thickness and is generally non-reservoir (although some discontinuous sandstones may be of reservoir quality locally). It probably forms an extensive but discontinuous vertical transmissibility barrier in the field.

The entire *Upper Ness Member* varies in thickness from 35–51 ft. The lower part of this member coarsens upwards from very fine to medium to coarse sandstones. The sediments change from ripple cross-laminated in the lower part to cross-bedded in the upper, coarser part. Carbonaceous fragments and clay drapes are common. The unit represents shoaling of the lagoon (Middle Ness Shale Member) and the development of a lagoonal delta which is thickest in the southern part of the field, thins and becomes finer grained northwards. It is overlain by a sequence of fluvial channels which become abandoned upwards into a laterally extensive marsh in which thin coals have developed. Reservoir quality is very good although vertical permeability is reduced by clay drapes in the lower part. Average porosities range from 15–28% and permeabilities range from 20–6000 md.

The upper part of the Upper Ness Member consists mainly of irregularly laminated and bioturbated mudstones which coarsen upwards into fine-grained sandstones which are wave ripple-

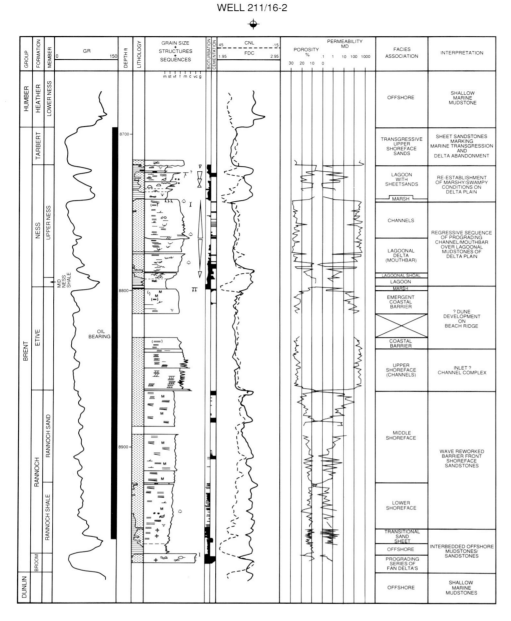

Fig. 6. Eider, type log and reservoir geological summary well 211/16-2.

laminated and bioturbated. The mudstones represent a return to lagoonal conditions and the sandy intervals may be minor mouth bar deposits. Reservoir quality is very poor and the sandy intervals are probably of limited lateral extent. Average porosities range from 5–22% and permeabilities range up to 30 md.

The *Tarbert Formation* is a fine- to coarse-grained moderately well-sorted sandstone and typically has a basal conglomerate lag about 1 ft thick. Carbonate cemented intervals are locally present. The unit represents the marine transgression which marked the end of the Brent Group delta progradation. The unit is eroded at the crest of the field and up to 78 ft thick on the flank; reservoir quality is excellent but vertical permeability may be impaired by carbonaceous layers and clay drapes. Average porosities range from 22–30% and permeabilities range from 100–5000 md.

Source

The source rock is formed by the organic rich shales of the Upper Jurassic Kimmeridge Clay Formation. Peak generation of hydrocarbons occurred during the Miocene, and oil migrated from the main mature oil 'kitchen' in the east to the fault trap, which was sealed by Upper Jurassic and Cretaceous claystones and marls.

Hydrocarbons

Productivity indices (PIs) for the various reservoirs are based on well tests from two appraisal wells (211/16-4 and -6) and one development well (EA13). Total PIs for the wells are expected to be in the range of 20–40 BOPD/psi. The average Tarbert reservoir contribution is expected to be some 24 BOPD/psi with the Upper Ness and Etive reservoirs having PIs of 10 BOPD/psi each.

The Eider crude oil has an API gravity of 34°, a GOR of 205 SCF/BBL and a sulphur content of 0.6% w/w. Initial reservoir pressure and temperature at datum (8750 ft TVSS) were 5020 psia (1150 psia above hydrostatic) and 107°C. The oil is undersaturated and slightly waxy with a viscosity at reservoir conditions of 1.18 cP. Formation water salinity is 19 000 mg/L NaCl equivalent.

Reserves

The estimated stock-tank-oil-initially-in-place of the Eider Oilfield is 187 MMBBL and the Ultimate Recovery is expected to be 85 MMBBL, giving a recovery efficiency of 46%. The initial development plan called for seven oil producers and seven water injection wells; well productivity has, however, been found to be better than expected and currently five producers and four injectors are envisaged for initial development.

Production during 1989 averaged some 39 000 BOPD dry oil. Water injection commenced in one well on 26 June 1989. The amount of gas associated with the Eider crude oil is too low to justify gas export facilities. Gas separated on the Eider platform during first stage separation will be used for power generation while remaining solution gas is transported in the crude oil to the North Cormorant platform. Available excess power generated on the Eider platform will be exported to the North Cormorant platform via a sub-sea cable.

The authors wish to thank those individuals who contributed to the internal Shell documents that were used to compile this paper. Permission from the managements of Shell UK Exploration and Production Ltd and Esso Exploration and Production UK Ltd is gratefully acknowledged.

Eider Oilfield data summary

Trap
Type	Structural; fault and dip closure
Depth to crest	8458 ft TVSS
Lowest closing contour	9033 ft TVSS
Oil/water contact	9033 ft TVSS
Oil column	575 ft

Pay zone
Formation/group	Brent Group
Age	Middle Jurassic
Gross thickness	290 ft (average)
Net/gross ratio	0.81 (average)
Cut off for N/G	1 md measured permeability (corresponding to approx. 11% porosity)
Porosity	23% (average)
Hydrocarbon saturation	0.75 (average)
Permeability	375 md (average)
Productivity index	40 BOPD/psi

Hydrocarbons
Oil gravity	34° API
Oil type	Undersaturated, slightly waxy, trace sulphur
Bubble point	1050 psia
Gas/oil ratio (range)	205 SCF/BBL (180–210 SCF/BBL)
Formation volume factor	1.10 (at original pressure)

Formation water
Salinity	19 000 ppm NaCl eq.
Resistivity	0.11 ohm m at 106°C

Reservoir conditions
Temperature	107°C at 8750 ft TVSS
Pressure	5020 psia at 8750 ft TVSS

Field size
Area	3.5 sq. km
Gross rock volume	278 million cu. m
Hydrocarbon pore volume	39 million cu. m
Oil recovery factor	42%
Drive mechanism	Water injection
Recoverable oil	85 MMBBL

Production
Start-up date	12 November 1988
Development scheme	1 satellite platform for unmanned operations, exporting into Brent System.
Peak production rate	57 000 BOPD (June 1989)
Cumulative production	15.2 MMBBL (1 January 1990)
Number/type of wells	1 exploration
	2 appraisal
	5 production planned
	4 injection planned
Secondary recovery method	Water injection

The Emerald Field, Blocks 2/10a, 2/15a, 3/11b, UK North Sea

D. M. STEWART & A. J. G. FAULKNER

Sovereign Oil and Gas plc, The Chambers, Chelsea Harbour, London SW10 0XF, UK

Abstract: The Emerald Oil Field lies in Blocks 2/10a, 2/15a and 3/11b in the UK sector of the northern North Sea. The field is located on the 'Transitional Shelf', an area on the western flank of the Viking Graben, downfaulted from the East Shetland Platform. The first well was drilled on the structure in 1978. Subsequently, a further seven wells have been drilled to delineate the field.

The Emerald Field is an elongate dip and fault closed structure subparallel to the local NW–SE regional structural trend. The 'Emerald Sandstone' forms the main reservoir of the field and comprises a homogeneous transgressive unit of Callovian to Bathonian age, underlain by tilted Precambrian and Devonian Basement horst blocks. Sealing is provided by siltstones and shales of the overlying Heather and Kimmeridge Clay Formations. The reservoir lies at depths between 5150–5600 ft, and wells drilled to date have encountered pay thicknesses of 42–74 ft. Where the sandstone is hydrocarbon bearing, it has a 100% net/gross ratio. Porosities average 28% and permeabilities lie in the range 0.1 to 1.3 darcies. Wireline and test data indicate that the field contains a continuous oil column of 200 ft. Three distinct structural culminations exist on and adjacent to the field, which give rise to three separate gas caps, centred around wells 2/10a-4, 2/10a-7 and 2/10a-6. The maximum flow rate achieved from the reservoir to date is 6822 BOPD of 24° API oil with a GOR of 300 SCF/STBBL. In-place hydrocarbons are estimated to be 216 MMBBL of oil and 61 BCF of gas, with an estimated 43 MMBBL of oil recoverable by the initial development plan.

Initial development drilling began in Spring 1989 and the development scheme will use a floating production system. Production to the facility, via flexible risers, is from seven pre-drilled deviated wells with gas lift. An additional four pre-drilled water injection wells will provide reservoir pressure support.

The Emerald Field is located in UKCS Blocks 2/10a, 2/15a and 3/11b in the northern North Sea some 85 miles northeast of Lerwick and 19 miles south of the Heather Field (Fig. 1), in 500 ft of water. The field occurs as a structural culmination on the eastern edge of the 'Transitional Shelf', a fault terrace downthrown from the East Shetland Platform, on the western flank of the Viking Graben (Wheatley *et al*. 1987). Production is from transgressive shallow marine sandstones of mainly Early to Middle Callovian age, referred to informally as the Emerald Sandstone. The structure has an area of closure of 3650 acres and is estimated to contain 216 million stock tank barrels of 24° API oil in place.

The field name was selected to complement the company logo of a Sovereign's crown. Several precious stones were considered, and a green emerald was thought most appropriate for an oil field.

History

Block 2/10 was awarded to Sovereign's former company, Sieben's Oil and Gas Ltd, in 1972. In 1978, following partial relinquishment, Chevron, Britoil and ICI farmed into Block 2/10a and Chevron became operator. In 1980 Blocks 2/15 and 3/11b were awarded, with Sovereign as operator of Block 3/11b. Areas of both blocks were subsequently relinquished at the end of the initial licence term in 1986. Substantial changes in licensees occurred between 1987, when Sovereign became operator, and 1989. Following unitization agreements, the Emerald Field interests are now fixed without further redetermination as follows: Sovereign Oil and Gas plc (Operator 30%), DNO Offshore plc (14%), Nedlloyd Energy (UK) Ltd (10.5%), Midland and Scottish Energy Ltd (44.23%) and Westburne Oil Ltd (1.27%).

The first well on the Emerald structure, well 2/10a-4, was drilled in 1978. The well was drilled to test a closed Jurassic fault block structure identified on seismic acquired in 1977. The primary objectives were Jurassic and possible Triassic sandstones with the Palaeocene sandstone offering a shallower secondary objective. This well was located at the highest point on the structure and encountered gas in a 42 ft thick Emerald Sandstone section. Oil and gas were also tested from two intervals within the Palaeocene. The first well to encounter oil in the Emerald Sandstone, well 2/15-1, was drilled in 1981. On test the well flowed at a combined maximum rate of 6000 BOPD from three test intervals within the Emerald Sandstone. Well 3/11b-3 was drilled in 1982 to determine whether the field extended into Block 3/11b. The well encountered 47 ft of oil-bearing sandstone. However, the following wells, 3/11b-4 and 4z, drilled on a separate basement level structure to the SSE of the Emerald Field, found the Heather Formation directly overlying the Devonian with the Emerald Sandstone absent. Well 2/10a-6 was drilled on a separate fault-controlled structure to the west of well 2/10a-4. This well encountered 74 ft of Emerald Sandstone, the greatest thickness encountered to date, and penetrated an oil–water contact higher than that of the Emerald Field, although the gas–oil contact appears to be the same.

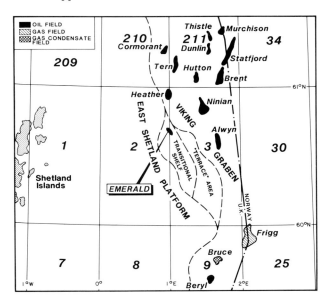

Fig. 1. Location map.

Following initial feasibility studies, wells 2/10a-7 and 7z were drilled on a cost sharing basis by the three licence groups in 1984. Well 2/10a-7 tested the NW–SE trending horst to the northeast of well 2/15-1 and found the Emerald Sandstone to be gas bearing, thereby demonstrating the presence of a second gas cap in the field. The well was then sidetracked into the oil zone.

The final appraisal well 3/11b-5, was completed in 1985. This was drilled to the southeast of well 3/11b-3 and used primarily for extended testing of the reservoir to provide reliable fluid parameters and reservoir flow characteristics.

Following the drilling and testing of well 3/11b-5, Sovereign undertook detailed development and feasibility studies which have enabled Sovereign to prepare an economically viable plan of development for the field. This work culminated in the submission of the Annex B to the Department of Energy in 1987. Approval was granted by the Minister of State on 29 January 1989.

The Emerald field is being developed using a floating production facility (FPF) tied back to subsea production and water injection wells by a flexible riser system. The crude oil will be exported from the FPF to a permanently moored storage tanker and then by shuttle tanker to port. Initially seven producer and four water injector wells are to be drilled. The wellheads are located either as part of a central cluster, directly below the FPF, or at two satellite locations. All the producing wells will be deviated at angles in excess of 80° through the Emerald Sandstone in order to improve productivity. The producing wells will be gas lifted and reservoir pressure maintained by water injection through the four water injector wells.

Pre-drilling of the four water injectors and seven production wells began in April 1989. One near-horizontal well has been tested and flowed at rates of up to 6822 BOPD without gas lift. Further wells are expected to be drilled once production commences, which is anticipated in 1991.

Field stratigraphy

The general stratigraphic sequence encountered by wells drilled on and adjacent to the field is illustrated in Fig. 2.

The Basement in the field area comprises Devonian Old Red Sandstone and Precambrian gneiss. The Devonian sediments, which exhibit low grade metamorphism, are found only in the southern part of the field. Across the field area there is no evidence of Carboniferous, Permian and Triassic rocks. The Basement surface was eroded and deeply weathered prior to Jurassic deposition.

A thin veneer of Lower Jurassic sediments overlies Basement in the northern part of the field (Wells 2/10a-4, 6 and 8). This interval consists of marginal marine deposits comprising calcareous sandstones, siltstones and shales, and often includes conglomeratic sequences. This unit is probably the lateral equivalent of the Dunlin Group.

In the Viking Graben the latter part of the Middle Jurassic was a period of major transgression, the initial phases of which are represented by the marine sandstones of the Tarbert Formation. By late Bathonian times the transgression had extended onto the Transitional Shelf and became fully developed during the Callovian, when the Emerald Sandstone was deposited as a sublittoral sheet-sand. Tectonic activity was largely absent on the shelf throughout this period and the Emerald Sandstone maintains a fairly consistent thickness of 42–64 ft across the field. Such variations in thickness as there are, can be attributed to deposition over minor irregularities on the unconformity surface. However, the sandstones are thinly developed or absent over former palaeohighs such as those around wells 3/11b-4 and 4z. The results of the initial development wells have supported the geological model of the Emerald Sandstone as a continuous body over the field area.

The overlying Heather and Kimmeridge Clay Formation represent continuing deposition during later phases of the transgression with no indication of a break in sedimentation. These Upper Jurassic formations comprise argillaceous and silty facies which form the seal to the underlying reservoir.

In common with much of the North Sea, a significant period of

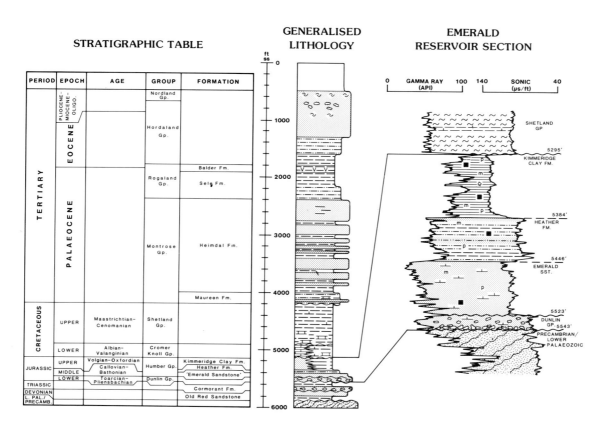

Fig. 2. Generalized stratigraphy.

faulting and erosion took place before Lower Cretaceous deposition. This was accompanied in the Emerald area by southwestwards tilting. Over the Transitional Shelf further tilting and intermittent fault movement took place during Early Cretaceous times. Lower Cretaceous limestones, marls and claystones were deposited in restricted marine conditions in a depocentre on the west of the Transitional Shelf while on the eastern margin, adjacent to the Viking Graben, more open marine conditions prevailed.

During the Late Cretaceous, the platform remained essentially stable. Upper Cretaceous sedimentation progressively overstepped the area from the Viking Graben. The Tertiary was the period of maximum subsidence and general regional tilting to the northeast. Marine conditions continued and a thick sequence of sandstones and shales was deposited.

Geophysics

Seismic surveys

The total seismic database over the Emerald Field comprises 658 km from ten different surveys. The initial interpretation of the Emerald Field was carried out using pre-1983 conventional seismic data. However, the data were unable to resolve adequately the Basement horizon and minor fault correlations. In 1983, 126 km of high-resolution seismic data were acquired over the whole field, with 650 m spacing between dip lines. These lines enabled much more accurate well prognoses to be generated than had been possible with previous surveys. Subsequently it was decided to acquire a detailed high-resolution survey in 1984 with spacing between dip lines of 325 m. The 1984 data set of 195 km has largely been used for the current interpretation and has allowed a detailed reinterpretation of the field without the need for older data.

This survey greatly improved the definition of the Basement reflector on which interpretation of the Top Reservoir configuration is based, and also increased confidence in the identification of fault compartments.

Seismic interpretation

Four principal horizons have been picked over the whole field; these are Top Basement, Base Cretaceous, Base Tertiary and Top Palaeocene. Other events were picked locally but relate primarily to the Palaeocene interval.

Depth conversion of the four horizons was carried out from velocity maps using a three-phase, layer-cake model. Velocity maps were used in preference to velocity/depth functions, as most of the velocity variations are caused by lithological variations which can only be accommodated by contoured maps.

As it was not possible to pick consistently the Top Emerald Sandstone event over the field, the Top Reservoir map was constructed by subtracting the combined Lower Jurassic and Emerald Sandstone isopachs from the Top Basement depth map. The Lower Jurassic isopach was based on the assumption that the specific occurrences seen in wells 2/10a-4, 6 and 8 and 2/15-1 form part of a continuous erosional remnant preserved beneath the Emerald Sandstone in the northern part of the field. The Emerald Sandstone isopach was constructed assuming that the reservoir was part of a continuous sheet-sand and that subtle variations in reservoir thickness are due to irregularities in the basement erosional surface.

Trap

The field is an elongate northwest–southeast-trending dip and fault closed structure located on the Transitional Shelf at the western edge of the Viking Graben. The Emerald Sandstone reservoir was deposited over the Basement and subsequently faulted. The fault blocks are disrupted by a number of en-echelon faults subparallel to the main graben edge faults, with throws usually in excess of 50 ft (Fig. 3). Late Jurassic siltstones and shales of the Heather and Kimmeridge Clay Formations overlie the reservoir and provide an effective vertical seal.

The trap developed during the main period of faulting that occurred during the latest Jurassic–early Cretaceous and was coin-

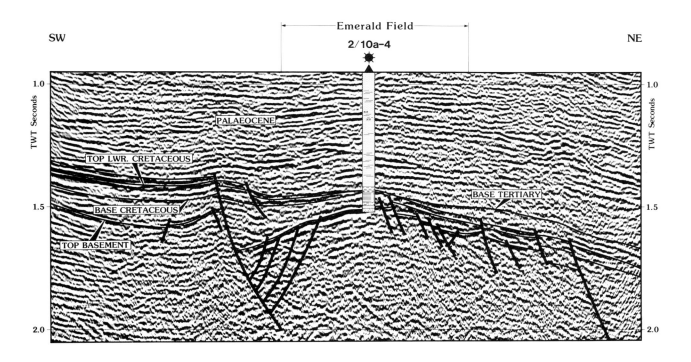

Fig. 3. Seismic section illustrating structure across the Emerald Field.

cident with the maximum phase of graben development. The trap was modified slightly by tilting to the northeast during the Tertiary.

The Emerald Field is subdivided into three main structural elements (Fig. 4). To the northwest is the north–south horst block on which well 2/10a-4 was drilled. To the east is a NW–SE elongate horst tested by wells 2/10a-7 and 7z. Separating the two horst blocks is a structurally lower region which extends almost the length of the field and forms the main 'oil fairway'. In the region of well 2/15-1, it forms a broad terrace broken by minor faults into small horsts and grabens. Towards the southeast, through wells 3/11b-3 and 3/11b-5, it becomes a horst feature bounded by the graben edge fault on its eastern flank. The structure is filled to spill point at 5580 ft and a common gas–oil contact of 5378 ft has been assumed, based on pressure data.

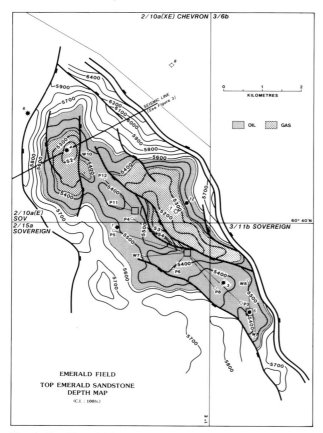

Fig. 4. Top Emerald Sandstone structure map across the Emerald Field.

Reservoir

The transgressive nature of the Emerald Sandstone is demonstrated by the presence of basal conglomerates and their abutment against a basal unconformity in most wells. The overall upward increase in the diversity of trace fossil associations and gradual increase in marine faunal elements in the overlying Heather Formation also indicates a general transgressive regime and progressive increase in water depth.

The Emerald Sandstone overlies a thin basal conglomerate and shows an overall fining upward tendency grading from medium to very fine sand and silt. It consists dominantly of quartz and orthoclase grains with minor mica and a low clay content. Shell debris, principally belemnite guards and bivalve fragments, is common. Bioturbation is extensive throughout and has destroyed most of the primary sedimentary structures. The sands appear to be sourced from dynamically metamorphosed argillaceous rocks of Caledonian or older age with some contribution from reworked sediments. Moderately common carbonized driftwood and plant fragments suggest close proximity to a vegetated land area.

Regional interpretation indicates that a NW–SE trending shoreline may have existed to the west and southwest of the field.

Within the Emerald Sandstone, two units have been recognized. The Lower Emerald Sandstone was deposited in a shallow nearshore environment below wave base, under moderate to low hydrodynamic energy conditions. There is no evidence of emergence or the presence of beach or shoreline deposits which were presumably reworked as the transgression advanced. The Upper Emerald Sandstone represents deposition under lower shoreface to offshore conditions of lower energy, occasionally interrupted by storms which deposited beds of coarser-grained material, mainly shelly debris. The base of this unit is usually marked by a thin, poorly sorted sandstone with granules, probably representing a lag deposit, and often shows as a heavy mineral effect peak on the gamma ray log. Petrographic analysis of this interval indicates an increase in heavy minerals and basement lithic fragments and is associated with increased levels of uranium on the spectral gamma ray log.

Biostratigraphy has yielded no specific ages from within the Emerald Sandstone. Based on the age of the overlying Heather Formation and the tentative ages that have been determined, the Emerald Sandstone is predominantly Early to Middle Callovian in age but possibly ranges from latest Bathonian to Earliest Oxfordian. A typical well section is illustrated by well 3/11b-3 (Fig. 5). The Upper and Lower Emerald Sandstone units cannot be distinguished palaeontologically but are correlated across the field area utilizing log characteristics together with sedimentological and petrographic data (Fig. 6). Across the field the thickness of Emerald Sandstone remains fairly constant, averaging 50 ft (varying from 42 ft to 64 ft) and exhibits little variation in composition, diagenesis and consequently reservoir quality. This reflects the uniform nature of deposition of the Emerald Sandstone and suggests that only minor variation in basement topography affected the sandstone thickness rather than any syndepositional faulting.

A similar diagenetic sequence is recorded in both sandstone units. Following deposition and intense bioturbation, local pyrite and carbonate cementation occurred. Before significant burial compaction, there was a phase of secondary quartz and feldspar cementation which helped to preserve the open framework of grains. The main diagenetic event, possibly during the early Palaeocene, was the formation of authigenic kaolinite and leaching of feldspars and micas. A minor phase of illitization of kaolinite occurred locally before oil migration commenced.

The Lower Emerald Sandstone unit is more porous and permeable than the Upper Emerald Sandstone unit owing to the upward decrease in grain size; the alteration products of feldspar (mainly kaolinite) have been trapped within the finer grain framework. The porosity averages 30.4% with a geometric mean permeability of 933 md in the lower unit; whereas in the upper unit the porosity averages 26.9% with a geometric mean permeability of 159 md.

Source

The Emerald Field crude oil is sourced primarily from the Kimmeridge Clay Formation on the flanks of the Graben area to the east, with possibly some additional sourcing from Middle Jurassic shales. The oil contains a dominant biodegraded component (oil gravity 24° API) mixed with small amounts (1–10%) of lighter gravity oils and gas.

Biodegradation is thought to have taken place in the late Palaeocene when meteoric invasion of the aquifer occurred during oil migration. The light hydrocarbons and gas represent a migration phase post-dating the cessation of biodegradation, probably due to further subsidence of the reservoir below the limit for bacterial activity and the associated increase in maturity of the source area.

Hydrocarbons

The maximum single flow rate derived from an Emerald Field appraisal well was 4680 BOPD from well 3/11b-3 on a 52/64 in.

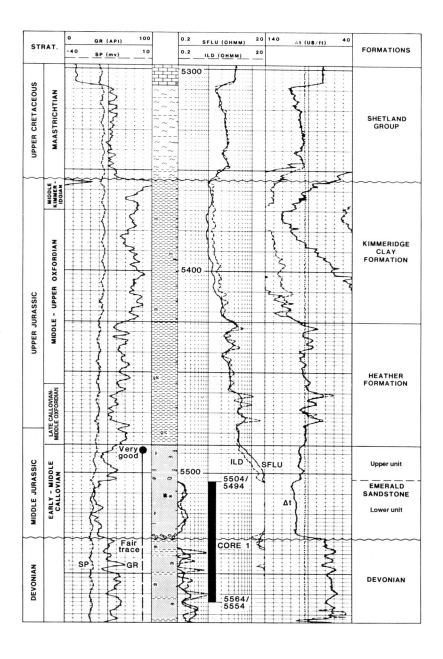

Fig. 5. Typical Jurassic well section.

choke, although initial results from the near-horizontal wells indicate the potential for flow rates of up to 10 000 BOPD using gas lift. The crude oil is saturated, with a gravity of 24° API, and a GOR of 300 SCF/STBBL at a reservoir pressure of 2425 psia. The reservoir temperature is 140°F and the reservoir is normally pressured.

Reserves/resources

Geological modelling associated with various feasibility studies has determined the in-place hydrocarbons of the Emerald Field to be 216 MMBBL of oil and 61 BCF of gas, with initial estimates of recoverable oil to be 43 MMBBL.

Expected initial daily production from the seven producing wells is likely to be as high as 40 000 BOPD. Production is due to begin in 1991 and last for at least six years.

The paper is based on the results of contributions made by many people from Sovereign who have worked on the Emerald Field over the past few years and is published with permission of Sovereign Oil & Gas plc, the Operator, and its partners: Midland & Scottish Energy Ltd, DNO Offshore plc, Nedlloyd Energy (UK) Ltd and Westburne Oil Ltd.

Reference

WHEATLEY, T. J., BIGGINS, D., BUCKINGHAM J. & HOLLOWAY, N. H. 1987. The geology and exploration of the Transitional Shelf, an area to the west of the Viking Graben. *In*: BROOKS, J. & GLENNIE, K. V. (eds) *Petroleum Geology of North West Europe*, **2**. Graham & Trotman, London.

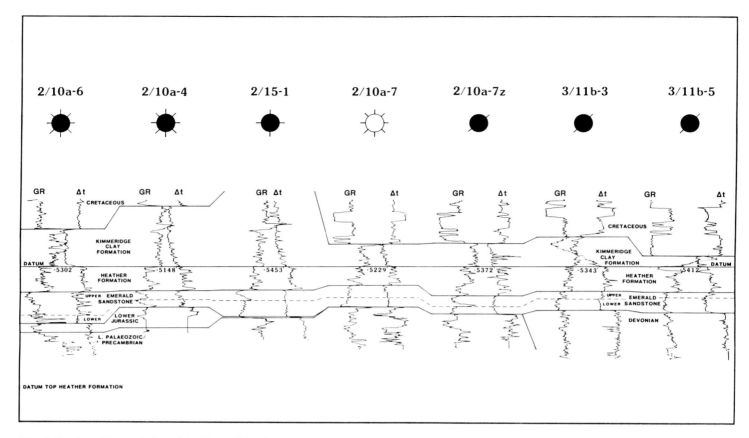

Fig. 6. Stratigraphic correlation of the Emerald Sandstone.

Emerald Field data summary

Trap
Type | Structural
Depth to crest | 5100 ft TVSS
Lowest closing contour | 5580 ft TVSS
Hydrocarbon contacts
Gas column | 5378 ft TVSS
Oil column | 5580 ft TVSS

Pay zone
Formation | Emerald Sandstone
Age | Bathonian–Callovian
Gross thickness (average/range) | 50 (42–64) ft
Net/gross ratio (average/range) | 1
Porosity (average/range) | 28.2 (27–33)%
Hydrocarbon saturation | 84%
Permeability (average/range) | 0.6 (0.1–1.3) darcies
Productivity index | 3–11 BOPD/psi (vertical well)

Hydrocarbon
Oil gravity | 24° API
Oil type | Sweet saturated black oil
Gas gravity | 0.653
Bubble point | 2425 psia
Gas/oil ratio | 300 SCF/STBBL
Formation volume factor | 1.137 RBBL/STBBL

Formation water
Salinity | 52 028 mg/l TDS
Resistivity | 0.14 ohm·m at 70°F

Reservoir conditions
Temperature | 140°F
Pressure | 2425 psia at 5378 ft TVSS
Pressure gradient in reservoir | 0.38 psi/ft

Field size
Area | 3650 acres
Gross rock volume | 170 000 acre-feet
Hydrocarbon pore volume | 48 000 acre-feet
Recovery factor | 20%
Drive mechanism | Waterflood
Recoverable hydrocarbons:
oil | 43 MMBBL

Production
Start-up date | 1991
Development scheme | Floating production system
Production rate | 40 000 BOPD
Cumulative production to date | 0
Number/type of wells | 7 producers, 4 water injectors
Secondary recovery method | Water injection

The Frigg Field, Block 10/1 UK North Sea and 25/1, Norwegian North Sea

J. BREWSTER

ELF UK, 197 Knightsbridge, London SW7 1RZ, UK
Present address: SNEAC(P), 26 avenue des Lilas, BP65, 64018 Pau Cédex, France

Abstract: The Frigg Field was the first giant gas field to be discovered in the northern North Sea. Its position on the boundary line between the UK and Norway called for international cooperation at an early stage in development. The Lower Eocene reservoir sands have extremely good poroperm characteristics but the heterogeneities within the sands control the water influx from the immense Eocene and Palaeocene aquifers below.

Frigg Field straddles the UK Norwegian median line between latitude 59° 48'N and 60°N and longitude 1°57'E and 2°15'E (Fig. 1). The field is 180 km east of the Shetlands and 390 km NE of Aberdeen on the UK side and 185 km WSW of Bergen and 215 WNW of Stavanger on the Norwegian coast. The water depth is 100 m. The field covers an area of 100 km² and consists of a gas pool up to 163 m thick overlying an oil disc which is up to 10 m thick. The reservoir is within the Lower Eocene sandstones of the Frigg Formation (Deegan & Scull 1977). Frigg is the name of the Norwegian goddess of married love and the hearth. She was the wife of Odin.

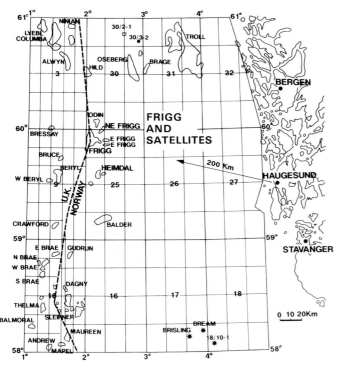

Fig. 1. Index map of the North Sea Frigg Field.

History

In the British Sector (Fig. 2), block 10/1, which contains most of the western portion of the field, was awarded to Total and Elf as part of licence P118 in 1970. On the Norwegian side of the median line, Frigg lies predominantly within Block 25/1, which was awarded in the second licensing round in 1969 as part of production licence 024, to the Petronord Group of Elf, Total and Norsk Hydro. The licence was awarded with the understanding that the Norwegian government would have the option to participate if a commercial discovery was made, and Statoil now have a 5% share in the licence.

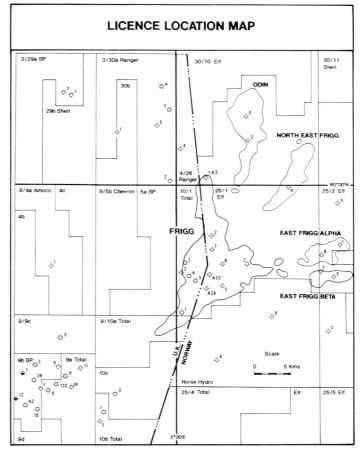

Fig. 2. Production licences.

Exploration history (Fig. 3)

Exploration in the area by the Pentronord Group had started in 1965 when a seismic grid was acquired with a spacing of 22 km by 15 km. In 1966, the initial interpretation of this survey outlined a structure at the Top Chalk reflector. In 1968, the Petronord Group acquired a further 700 km of seismic data in the area, and used this data for the licence application. Following the award of the Norwegian second round, a further seismic grid of 547 km with a 5 km spacing was shot in 1969. In 1970, the Frigg structure was remapped using all available data at the Top Chalk.

Following the agreement for Norwegian State participation in PL.024 and the UK award of P118, the first well was Norwegian 25/1-1 drilled in 1971, targeted at the crest of the Lower Tertiary structure. The well was drilled by the semi-submersible Neptune P81 and completed on 22 July.

From Abbotts, I. L. (ed.), 1991, *United Kingdom Oil and Gas Fields,*
25 Years Commemorative Volume, Geological Society Memoir No. 14, pp. 117–126

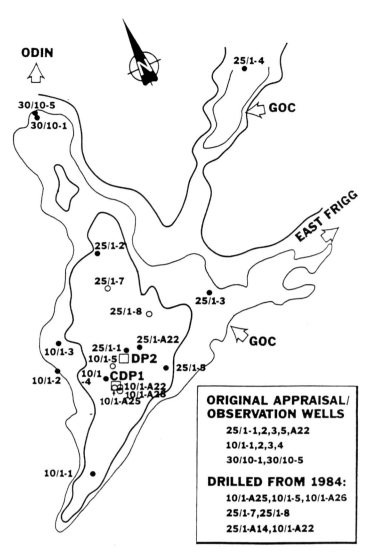

Fig. 3. Exploration and appraisal wells.

The prognosed Tuff Marker (Balder Formation) proved to be the top of the Lower Eocene gas-bearing sands at 1818 m msl. The gas-bearing reservoir was 128.8 m thick and was underlain by a 6.5 m oil column. The sands below the oil were water bearing. The well reached TD in Triassic strata with non-commercial shows in the Middle Jurassic.

A single DST of an 8 m zone within the gas-bearing interval produced 673 000 sm^3 of gas on a 7/8 in. choke. The pressure build-up on shut-in was immediate and stable at 196.7 bars at 1880 m below mean sea level (msl).

Appraisal history (Fig. 3)

Following the well results, a further 77 km of seismic data were acquired on the western flank of the structure in the UK blocks 10/1 and 9/5.

Between summer 1971 and summer 1973, five appraisal wells were drilled and a further 331 km of seismic data were acquired. Well 25/1-2 tested the northern extension of the field, and it encountered 44.4 m of gas-bearing sands underlain by 10.6 m of oil with the gas–oil contact (GOC) at 1948.1 m msl. Well 25/1-3 tested the eastern extension of the field and encountered 20.5 m of gas-bearing sand underlain by 9.6 m of oil with the GOC at 1947.8 m msl. Well 10/1-1 proved the southwestern extension of the field with 92.5 m of gas-bearing sand underlain by 10.3 m of oil. Well 10/1-2 on the western flank was a dry hole with very poorly developed Frigg Sands, and thus delimited the edge of the deposit and the field, and 30/10-1 on the northern flank proved 1.8 m of gas-bearing sand underlain by 8.5 m of oil with a GOC at 1947.1 m msl. The nearby well 30/10-5 was drilled in 1974 as a Jurassic exploration well, but it confirmed the 30/10-1 results in the Eocene.

During this appraisal phase, the field was continuously remapped by all the partners in both the UK and Norway. Estimates of gas in place varied from 320×10^9 sm^3 to 207×10^9 sm^3. The governments and oil companies involved agreed to appoint an independent expert to estimate the size of the field and the ownership. The expert considered the seismic coverage insufficient and, in 1973, a further seismic survey of 1870 km on a 1 by 1 and 1 by 2 km grid was acquired. The seismic survey was interpreted by a second expert in 1974, and an official map of the top reservoir was produced. The first expert deferred their conclusions on reserves and ownership, however, until a further 3 appraisal wells had been completed in areas of dispute between the various partners. These three wells were completed between 1975 and 1976. Well 25/1-5, on the eastern flank, encountered 62.2 m of gas and 7.4 m of oil, 10/1-3 on the western flank encountered 19.9 m of gas, and 10/1-4, in the crestal area, proved 128 m of gas. Using the 1973 seismic data and the results from the wells, the expert concluded in 1977 that the OGIP (original gas in place) was 268×10^9 sm^3 and that 39.18% belonged to the UK sector. The field was unitized in 1978.

During this period of intense activity in the early 1970s, the Norwegian gas field satellites to Frigg were also discovered; they all share the same aquifer.

Development (Fig. 4)

The field was declared commercial in 1972 and the development plan was finalized in 1974 following the signing in 1973 of a sales contract with the British Gas Council for the gas from the British sector of the field and the agreement of the Norwegian government for the sale of the Norwegian gas. The concept of both British and Norwegian involvement continued throughout the development plan. There are two pipelines to St Fergus, one theoretically for British gas the other for Norwegian. There are two drilling and production platforms, CDP1 in the UK sector and DP2 in the Norwegian sector (DP1 sank approaching location and was never used). CDP1 is a concrete drilling and production platform with drilling slots for 24 wells in two equal clusters, and DP2 is an eight-legged, piled steel platform with a similar well configuration to CDP1. These platforms were installed in 1975 and 1976 respectively. The flare platform was installed in 1975. Two processing platforms, TP1 and TCP2, were installed in 1976 and 1977, and the quarters platform QP was the first in the field in 1975. Treated gas flows through the two 32 in. sea-lines to St Fergus in Scotland via MCP-01, the manifold and compression platform, operated by Total, which is 186 km from Frigg and 1974 km from St Fergus. The pipelines were installed between 1974 and 1976 and hooked-up during 1977 and 1978.

Because of the high permeability of the sands (1-4D), the low viscosity of the gas and the possibility of an active water drive, the production wells were drilled in the apex of the structure and cover an area of approximately 5 km^2 compared to the field extent of more than 100 km^2. Production drilling started in 1976 on CDP1 and in 1977 on DP2, and was completed in 1978 and 1979 respectively. The well-top reservoir locations are approximately 250 m apart. To prevent near-well-coning and late-stage work-overs, the production wells only penetrated the top 60 m of reservoir and they were completed with screens in order to control sand production.

Commercial production started on 13 September 1977 from CDP1 and on 9 September 1978 from DP2. Production rates during the plateau period were up to 64×10^6 sm^3 in the winter and between 21 and 36×10^6 sm^3 during the summer.

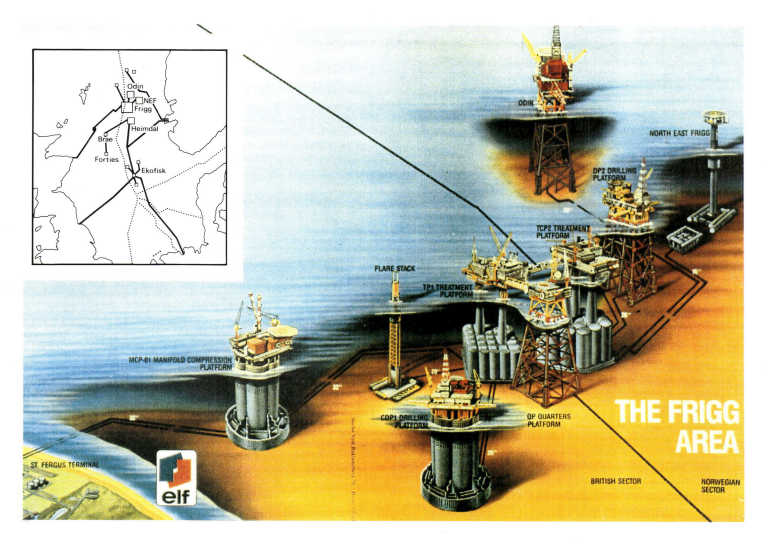

Fig. 4. Frigg installations.

Field stratigraphy

General stratigraphic sequence

The general stratigraphic sequence is shown in Fig. 5. The discovery well, 25/1-1, was used by Deegan & Scull (1977) as the type well for the Frigg Formation. The palynological zonation scheme employed by Elf is also shown in Fig. 6. This scheme has been detailed by Heritier et al. (1980), Revoy (1984), Conort (1986) and Mure (1987).

Geological history

The structural grain of the Frigg area was established during the Caledonian orogeny with lines of weakness of NW–SE and NE–SW. The succeeding tectonic events were all influenced by these basement trends. During the Devonian and Carboniferous, strike-slip wrench movements along the Caledonian trends continued. Extensional rifting episodes began during the Permian, with marginal evaporites overlain by siltstone redbeds during the Lower and Middle Triassic. The Frigg Field lies in the axial part of the Viking Graben, which developed in the Late Triassic (Early Cimmerian) as a series of N–S or NE–SW trending sub-basins separated by transverse, active ridges generally oriented NW–SE. These ridges have associated, small, pull-apart depressions, which complicate the well-to-well correlation of the Upper Triassic/Jurassic sequence. The field lies close to the intersection of the axial part of the graben and one of the main NW–SE wrench systems. Thick deposits of marine shales of the Dunlin Group provide the source rocks for the oil and gas of the field.

In the Frigg area there is a marked unconformity at top Dunlin, and the Broom, Rannoch and Etive Formations are not present. Thick sequences of Ness and Tarbert (or their equivalents) are found in some areas, suggesting that the underlying NW–SE trends were still influencing deposition.

The main Cimmerian rifting phase started during the Bathonian with the sequences generally thinning onto the basin margins. The final rifting phase occurred during the Upper Jurassic, with the reactivation of faults, large-scale subsidence in the basin axis and uplift on the flanks, as well as block rotation. During this period of the Late Jurassic, marine shales were deposited throughout the Frigg area, although they thin markedly onto the crestal areas of some of the fault blocks.

During the Early Cretaceous, the area was partially emergent and there are only sporadic occurrences of Lower Cretaceous sequences in the Frigg area. This is linked to the tectonic history, which reverted to a compressional phase following the Jurassic. During the Late Cretaceous, the area lay on the northern margin of the Chalk Basin, with pure chalks deposited in the Maastrichtian and interbedded chalks and marls throughout the rest of the Upper Cretaceous.

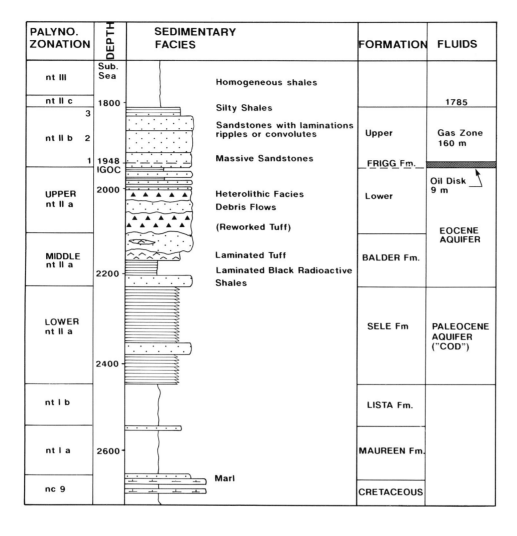

Fig. 5. General stratigraphic section.

There is a structural high beneath the Frigg Field at top-chalk level (Brewster & Jeangeot 1987). The structure is elongated generally north–south, and probably results from east–west compressional folding during the early-mid Tertiary.

The Maureen Formation (Nt 1a lower) overlies the chalk, and its base is a key seismic marker. In the Frigg area, the Maureen Formation is predominantly marine shale, but it contains sands up to 70 m thick.

The Lista Formation (Nt 1b) is also predominantly marine shale in the Frigg area but it contains thin sands up to a few metres thick, which are laterally persistent and are equivalent to the Heimdal Sands of Heimdal Field, 36 km south in Norwegian Block 25/4. Development drilling on the Heimdal Field established that there was pressure communication through the various Palaeocene sands.

The overlying Sele Formation (Nt 11a lower) is a more complex clastic sequence of thick sands and interbedded shales. The sands are informally named the "Cod Sands" and they constitute a major aquifer to the Frigg Field and its satellites.

The Balder Formation (Nt 11a middle) is unusual at Frigg Field. In the surrounding areas and in most of this part of the North Sea, the well data and consistent seismic response suggest a normal sequence of thin, interbedded, tuffaceous silts and shales with occasional sand bodies, but in the Frigg area the Balder Formation is extremely heterogeneous. It varies in thickness from less than 10 m to more than 80 m, and in addition to the usual interbedded shales and tuffaceous siltstones, it contains massive, irregular debris flows with very coarse, angular clasts of Sele and Lista-age material as well as reworked tuffs and siltstones. In some wells, massive, grain-flow sands have been recorded on logs and cores. The seismic response of the Balder beneath Frigg changes polarity and amplitude, and it appears that active erosion and redeposition in the Frigg area affected the deposition of the Balder. This marked heterogeneity allows for large areas of good pressure communication between the Palaeocene aquifers of the Maureen, Sele and Lista and the Lower Frigg Formation. The extent of this pressure communication, and the resulting very active aquifer and water drive, was not anticipated during the planning of the production. It has resulted in a complex sweep pattern, which has been determined by a number of wells and a 3-D seismic compaign described by Thouand et al. (1987) and outlined below.

The Frigg Formation (Nt IIa–c) overlies the Balder. It can be conveniently divided into upper and lower members on the basis of both palynology and lithostratigraphy. The Lower Frigg (Nt 11a upper) is a very irregular sequence of shales (both in situ and reworked), debris flows, grain flows and channelized sands. The sequence varies rapidly in thickness and continuity, and is difficult to correlate from well to well. The upper boundary is an unconformity, which can be recognized on seismic and which palynologically corresponds to a change from a predominantly terrestrial assemblage to a predominantly marine assemblage. The section immediately below the unconformity usually consists of a final series of debris flows with thin coals and abundant lignite. This Lower Frigg Member constitutes the local Frigg Aquifer. It is restricted to the Frigg area and the Frigg fan, and is coincident with the 'unusual' Balder facies below. The origin of this lower part of the Frigg Formation is problematical. It does not fit neatly into the submarine-fan model of Heritier et al. (1980), nor does it fit the deltaic model of the earlier workers and the expert appointed to determine the reserves. The predominantly non-marine palynological assemblage suggests a terrestrial source, and the very coarse, angular clasts of the debris flows and steep dip angles suggest local

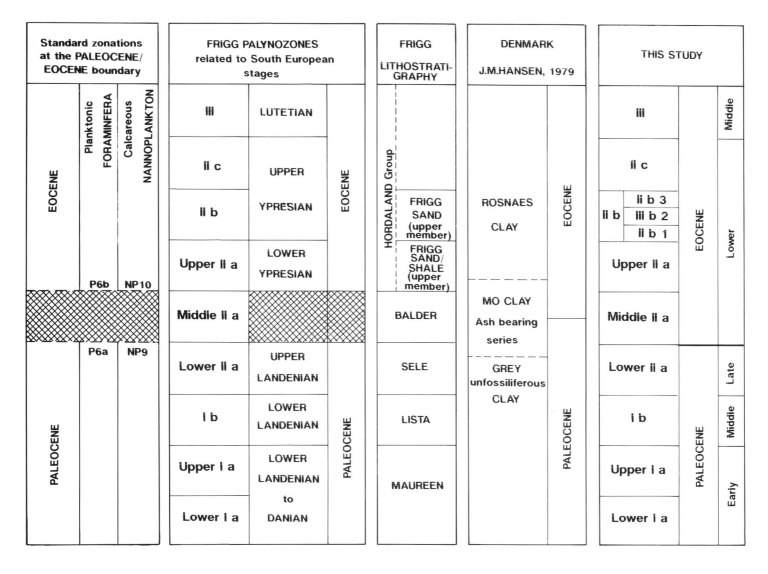

Fig. 6. Biostratigraphy, lithostratigraphy and chronostratigraphy.

deposition possibly associated with syn-depositional faulting. The association with the disturbed Balder sequence and the possible faulting (Brewster & Jeangeot 1987) suggest that the Frigg area was a local high at the end of the Palaeocene, and was an area of active tectonism during this period.

The upper part of the Frigg Formation (Nt II b–c), which constitutes the bulk of the gas-bearing reservoir, does fit the submarine-fan model of Heritier et al. It is a sequence of massive (decametric) channel sands with associated levee and lobe deposits. The flank areas of the fan, and hence the field, show decimetre- and metre-scale fining-up sequences and incomplete Bouma sequences, whereas the central parts of the field, which were the main channel areas, consist of clean, structureless sands with occasional thin, apparently discontinuous, carbonate-cemented layers as well as thin silts and shales, especially towards the top. The sequences thin towards the top, and the fine sands and thin shale layers are more extensive and continuous in the northern parts of the fan.

The reservoir is capped by brown-red, silty shales overlain by green, soft, pyritic clay. This Middle Eocene sequence is overlain by brown, gumbo clays ranging from Middle Miocene to Early Oligocene. Overlying the gumbo clays are green-brown, silty mudstones of Late–Middle Oligocene, which are overlain by a predominantly sandy Miocene–Pliocene sequence.

Geophysics

Seismic surveys available, interpretation methods (Fig. 7)

The early geophysical surveys were outlined above. Following the start of production in 1977, the interaction of the satellite pools to Frigg production prompted a time-lapse seismic study between Frigg and NE Frigg. A series of lines between Frigg and NE Frigg were shot, and then reshot in 1979, 1981 and 1983. The results are discussed by Revoy (1984) and demonstrate clearly the migration of gas towards Frigg from NE Frigg due to the pressure depletion of the main Frigg gas pool. In 1985, when the active water drive had been established and the heterogeneous water encroachment had been proved by appraisal wells, a complete 3-D grid of 4000 km was acquired, with a further 1000 km of peripheral 2-D.

Since then, further small 2-D surveys have been acquired to study particular areas of the field and to better define Jurassic exploration possibilities.

From the earliest surveys it was apparent that a prominent 'flat spot' corresponds to the initial gas–oil contact. The intersection of the flat spot with the top reservoir reflection has provided a series of depth control points around the edge of the field which define the

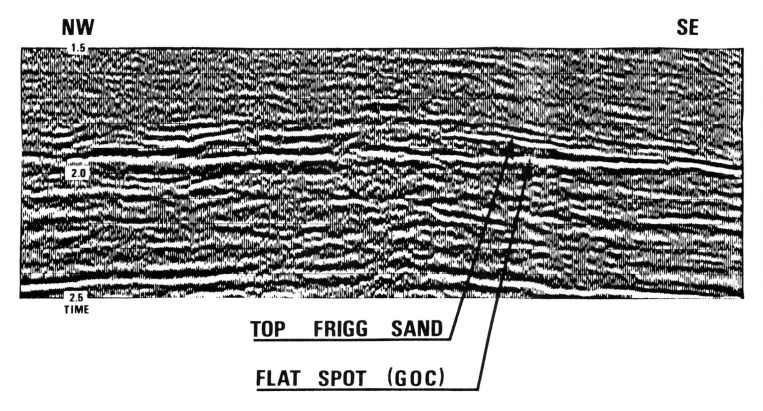

Fig. 7. Early seismic section across Frigg Field. Line 73-F-9. SSL Survey 1973.

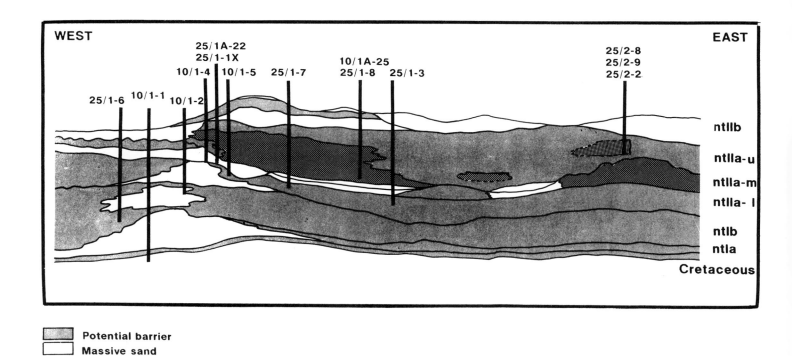

Fig. 8. General stratigraphic model.

accumulation and also control the structural map away from well control. The reflection from the top of the Frigg Formation is believed to be produced not by the contrast between the gas sands and the overlying shale but rather from a series of thin limestones and compact shales immediately above the top of the reservoir.

The acquisition of a post-production 3-D survey has led to an intensive study campaign on the use of seismic to monitor the water-rise. The results have largely proved to be semi-quantitative, due partly to the high residual gas saturation left behind the advancing water front, partly also the problem of differentiating the responses due to lithological contrasts from those of the gas/water interface, and also to the problem of comparing quantitatively the amplitudes of different vintages of seismic data. As the water rise is, to a large extent controlled by the geometry and continuity of the shale/silt barriers, separating lithological responses from fluid response is difficult, and in some cases impossible.

Geophysical definition of the trap

The latest geophysical maps of the Frigg reservoir are derived from the 1985 3-D seismic grid and its 2-D extension. Smoothed and migrated stacking-velocity values adjusted to the well data, and the initial flatspot-seismic marker intersection at the field limits, were used to depth-convert the map. The depth conversion of the underlying seismic events was performed using interval velocities derived from the well data. The final depth map of the top reservoir was contructed by adding the isopach map of the layer between the top reservoir and the seismic marker to the map of the seismic marker.

Seismic stratigraphic analysis of the intra-reservoir events, combined with the palynological analysis, provided the basic framework and geometry of the channel and lobe areas of the Upper Frigg Formation gas reservoir. Similar analysis in the aquifers provided indication of potential flow direction and communication between the various aquifer layers (Fig. 8).

Seismic monitoring of the water encroachment. The changes in the initial reflection of the gas–oil contact with time has been closely studied during the production life of the field by qualitative and attempted quantitative analysis using the Gassman equation approach. The first studies were published by Revoy (1984), who showed by a time-lapse technique that pressure depletion of the main Frigg Field accumulation was causing a gas tongue to move downdip from the NE Frigg satellite towards the main field.

Further work on quantitative analysis used the 1973 (pre-production) survey as a reference; by homogenizing, as far as possible, the amplitude of the post-production surveys of 1985 and 1987 and then comparing the amplitude response using rock and fluid properties derived from cores, logs, tests and production, indications of the vertical rise in the gas–liquid contact between the surveys were obtained. Results were only semiquantitative because of problems comparing the amplitudes and positioning of different surveys, and also to the interference effects of the lithology contrasts, which are difficult to predict precisely away from the well control.

Trap

The structural map is directly related to the depositional topography of the Frigg Formation. The morphology is typical of a 'birdsfoot delta' or a submarine fan with a feeder channel to the south west. The field covers slightly more than 100 km² and has 170 m of vertical closure. The reservoir is full to spill-point and has a gas–oil contact at 1948 m below msl and an oil–water contact at 1955 m msl. The reservoir is sealed by Middle Eocene marine shales with thin stringers of tight limestone common just above the reservoir.

Reservoir

Age, lithology, correlation, geometry

Since 1984, a number of deep wells from both the platforms and remote locations have been drilled to monitor the water encroachment into the field. Detailed sedimentological and palynological studies on the cores and sidewall cores from these wells have been revised to refine the reservoir subdivision and correlation as shown in Fig. 9. Dipmeter responses from non-cored wells calibrated to the cored wells provided a more detailed lithological breakdown and correlation of the original exploration and appraisal wells. The main results are presented in Fig. 6. Sedimentological facies analysis suggests that the sands of the Upper Frigg Formation generally prograde east and north from a narrow feeder channel connected to the Shetland platform.

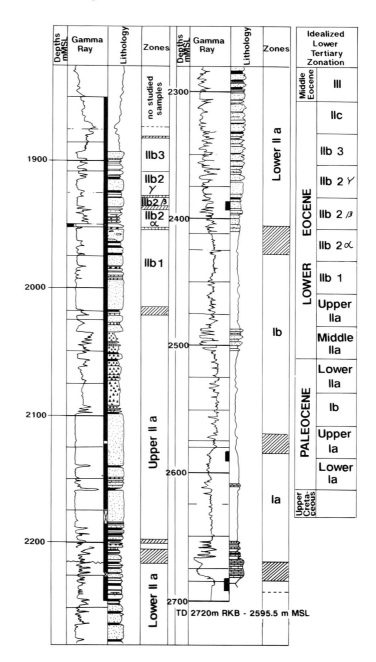

Fig. 9. 25/1-7 palynological summary.

Deposition, facies (Fig. 10)

In order to tie seismic data to the wells, the sedimentological description was simplified to the following three subdivisions: 'massive facies' representing sands with no continuous laminations affecting fluid flow, 'alternating facies' representing sands and silts with laminated micaceous sands and shales which do act as barriers to vertical water rise, and 'Breccia facies' representing the chaotic slumps, debris flows and grain flows often encountered in the Lower Frigg Formation and also present in the northern extremity of the field.

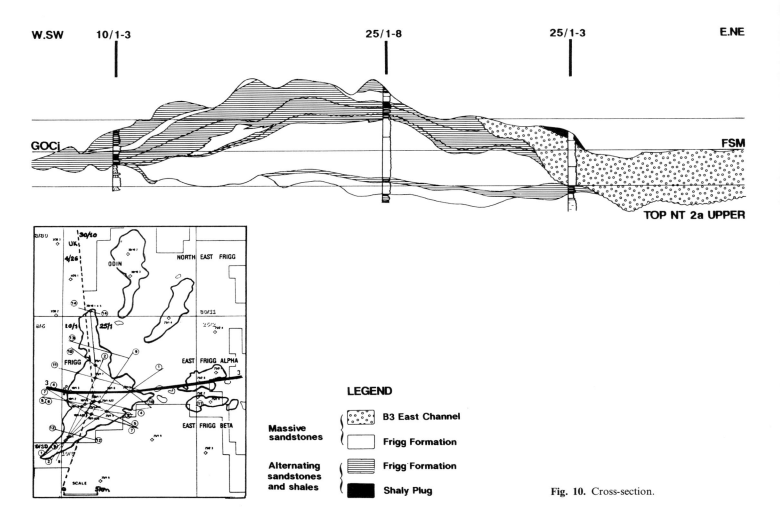

Fig. 10. Cross-section.

The main stratigraphic zone in the gas pool is the NtIIb sequence, which is up to 220 m thick and contains several channel systems. Within this zone and, for the purpose of reservoir modelling, seven 'massive facies' sands and six 'alternating facies' were delineated. The seismic resolution varies according to lithology so that within a 'massive facies' sand sequence, a shale as thin as 5 m can be discriminated whereas within an 'alternating facies' interval, sands must exceed 10 m before they can be resolved. The individual units were assigned petrophysical parameters derived from the wells, extrapolated on the basis of trend analysis and seismic data.

The flank wells show that the gross poroperm characteristics of the Upper Frigg Formation deteriorate away from the axial parts of the fan. But this is due to the decrease in thickness and number of the sands rather than a general increasing shale content within the reservoir layers.

Overlying the NtIIb2 sequence in the area between the Frigg Field and its satellites is a younger channel sequence (NtIIb3) which is upto 110 m thick. This channel system has cut down into the underlying sands and provides an area of good hydraulic communication between Frigg and its satellite fields to the east. From the well control, the petrophysical properties of this sequence appear to be as good as the best sands in the NtIIb2 sequence below.

Reservoir units, poroperm parameters

By and large, the reservoir units correspond to the depositional units outlined above. The appraisal wells drilled between 1984 and 1987 provided the key pressure profiles linked to core and log material, which clearly demonstrated the effect of the fine-grained units on the dynamic behaviour of the field during production (Fig. 11). Pressure barriers in excess of 8 bars were recorded across 'alternating facies' less than 10 m thick. The palynological stratigraphy linked to the seismic stratigraphy was used to map the reservoir layers throughout the field.

In the 'massive facies' sands, poroperm properties are homogeneously excellent with permeabilities up to 4 darcies and porosities up to 32%. The overall net-to-gross ratio deteriorates from axis to flank and, less markedly, from south to north, reflecting the general progradation direction. In the crestal areas net-to-gross ratio exceeds 95%. In the finer grained, laminated sequences of interchannel and channel abandonment deposits, which can act as local barriers to flow, the vertical permeability is reduced to less than 1 md.

Source

The Frigg Field gas was derived from local Lower and Middle Jurassic shales, which have present day vitrinite reflectance values that vary from 1.5 to 1.8. Migration of the gas into the field probably occurred through the pronounced 'gas chimneys' visible on seismic. These are areas of apparently disturbed acoustic velocity probably caused by high gas saturation and which extend virtually to the sea bed in some instances. Migration of the gas into Frigg probably occurred during the Oligocene and Miocene.

The oil in Frigg was also derived from the same source rocks. It was generated earlier and migrated earlier than the gas. The oil was then biodegraded into its present form. Upper Jurassic (Kimmeridge Clay) source rocks are present in the Frigg area but are unlikely to be the major source of the Frigg hydrocarbons as the pristane/phytane ratio of the oils differs from that of the Upper Jurassic extracts but matches that of the Middle and Lower Jurassic shales.

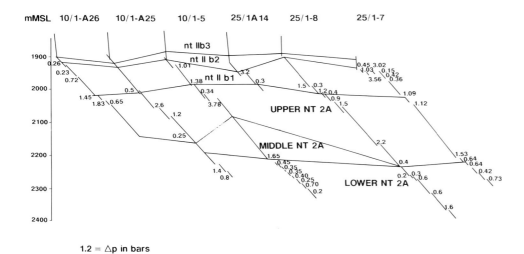

Fig. 11. RFT profiles from appraisal wells on Frigg.

1.2 = △p in bars

Hydrocarbons

Test results

The gas-bearing sands were tested in five of the exploration and appraisal wells: 25/1-1, 25/1-3, 25/1-5, 10/1-1 and 10/1-3. Measured flow rates were up to 750 000 sm³d. The pressure gradient from these wells was 0.01409 bar/m and initial static reservoir pressure was 196.9 bars at 1900 m below msl. Permeability estimates from test analysis were impossible because the extremely small drawdown, and instantaneous build-up was beyond the resolution of the gauges available at that time. Later measurements on production tests indicate effective permeabilities in excess of 1 D and up to 3.2 D.

Fluid types

Gas samples were taken from all the tests, and also FIT samples were gathered from the exploration and appraisal wells, but the PVT results are questionable because of poor sampling conditions and the very lean nature of the condensate. The definitive samples were measured from one of the production trains during the start-up period, and the composition was: N_2 0.3599%, CO_2 0.2902%, C_1 95.5225%, C_2 3.6887%, C_3 0.0379%, iC_4 0.0085%, nC_4 0.0047%, iC_5 0.0041%, nC_5 0.0004%.

Molecular weight was 16.28, density 0.5809, gas compressibility 0.8673, specific gravity 0.01382 g/cm², Formation Volume Factor 0.00508 vol/vol, viscosity 0.0205 centipoise, condensate content 4.3 g/m³, and condensate density 796.57.

The oil disc was tested in well 25/1-3 and the fluid properties were determined to be:
23° API naphthenic, Bo 1.148 m³/m³, specific gravity 0.8347 g/cm², solution GOR. 60.97 Sm³/Sm³, viscosity 4.83 centipoise, compressibility 9.14×10^{-5} m³/m³/bar.

Reservoir conditions

The initial static reservoir pressure was 196 bars at 1900 m below msl. All exploration and appraisal wells showed a common gradient both in the gas and water phases throughout the field. Initial pressures in the aquifer were based on fewer data points but, before production, there were no discernible pressure barriers or changes in gradient throughout the aquifer. Single-point, permanent pressure sensors in both the 'Cod' and Frigg Formation aquifers showed that the aquifer pressure decline was slower than predicted during the early phase of production and that the field was experiencing a very active, natural, water drive.

Reserves/resources

Recovery efficiency, drive mechanism, production problems

The recovery efficiency at Frigg is a function of Original Gas in Place (OGIP), residual gas saturation (gas left behind the water front), the sweep efficiency, the abandonment pressure and the potential for gas migration into Frigg from the satellite fields following Frigg production.

The drive mechanism is now known to be a strong, natural, edge and vertical water drive combined with limited pressure depletion.

There have been no serious production problems at Frigg. The main concern originally was possible sand production. Consequently the wells were completed with screens, and production rates were set in order to limit the through-screen velocity. Late-stage completions with normal perforated liners have also been successful.

Recoverable reserves, resources and their formation

Estimates of OGIP ranged from 207 to 329×10^9 Sm³.

Residual gas saturation was difficult to determine from the early well data as there were few suitable cores for special core analysis. The results on samples reconstituted from bagged sand suggested an Sgr of 19%. Later results from cores recovered using rubber sleeves and fibreglass core barrels gave values of Sgr of 29% for the average field porosity of 29% (deLeebeeck 1987).

The pressure decline at Frigg has been less than anticipated before production, and abandonment pressure will probably be less than 20 bars less than the original static pressure.

The sweep efficiency on Frigg should approach 97-98% as the production wells are crestal and the structure simple. Sweeping of the individual closures well be enhanced by the edge-water drive and by at least one high deviation production well designed to drain a structural high 2 km from the platform. The remote appraisal wells, which have been relogged on a yearly basis since 1985, show that no free gas remains beneath the shale barriers encountered in the wells. These shale barriers are extensive (kilometric) and give rise to a pressure barrier.

The volumetric estimate of original gas in place (OGIP) was redetermined by detailed analysis of the 1985 3D survey and a complete re-evaluation of the petrophysical parameters. The final estimate is 235×10^9 Sm³ $\pm$ 14×10^9 Sm³.

The present figure for recoverable reserves is between 175 and 178×10^9 sm³ giving a recovery factor of approximately 75%.

Frigg Field data summary

Trap
Type	Stratigraphic
Depth to crest	1785 m msl
Lowest closing contour	1955 m msl
Gas column	160 m
Oil column	9 m

Pay zone
Formation	Frigg
Age	Lower Eocene
Gross thickness	170 m (av 55m)
Net/gross	0.95
Cut off for N/G	$S_w < 60\%$
Porosity (av, range)	29%, 27–32%
Hydrocarbon sat	91%
Permeability (av, range)	1500 md, (900–4000)
Max well rate	$3 * 10^6$ Sm^3/w/d

Hydrocarbons
Gas gravity	0.581
Condensate yield	4.3 Sm^3/10^6 Sm^3 gas
Gas expansion factor	197
Oil gravity	0.835 g/cm^3
Oil type	Naphthenic
GOR	61 Sm^3/Sm^3

Formation water
Salinity	63 000 ppm

Reservoir conditions
Temperature	61°C
Pressure	198 bar abs
Pressure gradients	
gas	0.013 bar/m
oil	0.083 bar/m
water	0.103 bar/m

Field size
Area	100 km^2
Gross rock volume	4.79×10^9 m^3
Hydrocarbon pore vol	1.16×10^9 m^3
Recovery factor gas	75%
Recovery factor oil	0
Drive mechanism	Natural water drive and depletion
Gas in-place	235×10^9 Sm^3
Recoverable gas	$175-178 \times 10^9$ Sm^3

Production
Start-up date	13/9/77
Development scheme	Platforms and pipelines
Production rate	45×10^6 Sm^3/day (av. plateau)
Cumulative production	163×10^9 Sm^3 April 89

This paper has drawn heavily on the references given below, and acknowledgements are due to the exploration and reservoir teams of Elf. The opinions expressed herein are those of the author and not necessarily those of the operator EAN or of any of the partners in the Frigg Field. Permission to publish this paper was granted by the partners: namely, Elf Aquitaine UK plc, Total Oil Marine plc, Elf Aquitaine Norge A.S., Total Marine Norsk A.S., Norsk Hydro A. S., and Den Norske Stats Oljeselskap A. S. I. ¥xnevad provided the drafting.

References

BREWSTER, J. & JEANGEOT, G. 1987. The Production Geology of Frigg Field. *In*: KLEPPE, J. *et al.* (eds) *North Sea Oil and Gas Reservoirs.* Norwegian Institute of Technology, Graham and Trotman, London, 75–88.

CONORT, A. 1986. Habitat of Tertiary Hydrocarbons, South Viking Graben. *In*: SPENCER, A. M. *et al.* (eds) *Habitat of Hydrocarbons on the Norwegian Continental Shelf.* Graham and Trotman, London, 159–170.

DEEGAN, C. E. & SCULL, B. H. 1977. *A proposed standard stratigraphic nomenclature for the central and northern North Sea.* Report of the Institute of Geological Sciences Bulletin No. 77/25 N.P.D.no.1.

deLeebeeck, A. 1987. The Frigg Field reservoir: characteristics and performance. *In*: KLEPPE, J. *et al.* (eds) *North Sea Oil and Gas Reservoirs.* Norwegian Institute of Technology, Graham and Trotman, London, 89–100.

HERITIER, F. E., LOSSEL, P. & WATHNE, E. 1980. Frigg Field—large submarine-fan trap in Lower Eocene Rocks of the Viking Graben, North Sea. *Memoir of the American Association of Petroleum Geologists*, **30**, 59–79.

MURE, E. 1987. Frigg. *In*: SPENCER, A. M. *et al.* (eds) *Geology of the Norwegian Oil and Gas Fields.* Graham and Trotman, London, 203–213.

REVOY, M. 1984. Frigg Field; production history and seismic response. *In*: *Offshore Northern Seas Conference Proceedings—Reduction of Uncertainties by Innovative Reservoir Geo-Modelling.* Norsk Petroleums Foretning, Article G5.

THAUAND, A., BREWSTER, J. & MURE, E. 1986. Frigg Field. Mid life appraisal programme. *Proceedings of the Offshore Northern Seas Conference, Stavanger, August 1986.*

The Heather Field, Block 2/5, UK North Sea

BRIAN PENNY

Unocal UK Ltd, 32 Cadbury Road, Sunbury-on-Thames, Middlesex TW16 7LU, UK

Abstract: The Heather Oil Field is located at the western margin of the East Shetland Basin, in the northern North Sea, and is operated by Unocal UK Ltd. Oil is trapped in Middle Jurassic Brent Group sandstones below 9500 ft TVSS in a highly-faulted reservoir that has undergone considerable diagenetic modification. The field was declared commerical in 1974, began producing in 1978 from a single platform and has recovered 84 MMBBL of oil to date through primary and secondary recovery methods. Ultimate reserves are estimated to be 100 MMBBL of oil.

The Heather Oil Field is located in Block 2/5, 90 miles northeast of the Shetland Isles at the western margin of the East Shetland Basin (see Fig. 1). Oil is produced from a single reservoir, the Middle Jurassic Brent Group (Gp), between 9500 ft and 11 800 ft TVSS. The productive area of the field, which is a tilted fault block trap, is 10 500 acres. The field was named Heather in 1974 by a consortium led by Union Oil Company of California (Unocal). Oil is exported from the Heather Alpha platform, which stands in 470 ft of water, to the Sullom Voe terminal on the Shetland Isles via the Ninian Pipeline System. Initial production was in October 1978 and to date 84 million barrels of oil have been recovered. Recoverable reserves are estimated to be 100 MMBBL of oil.

History

Block 2/5 (licence P242) was awarded in the fourth round of licensing, in 1972, to a consortium comprising Unocal Exploration and Production Co. (UK) Ltd (31.25%, operator), Skelly Oil Exploration (UK) Ltd (31.25%), Tenneco (Great Britain) Ltd (31.25%) and Det Norske Oljeselskap (UK) Ltd (6.25%). A reconnaissance seismic survey, shot in 1971, had identified a large Mesozoic structure in the northeastern quadrant of the block. Subsequently, a 2 km grid was shot in 1972 and the structure defined as a northwesterly tilting fault block with a major displacement in the southeast along a fault trending NE–SW (see Figs 2 and 3).

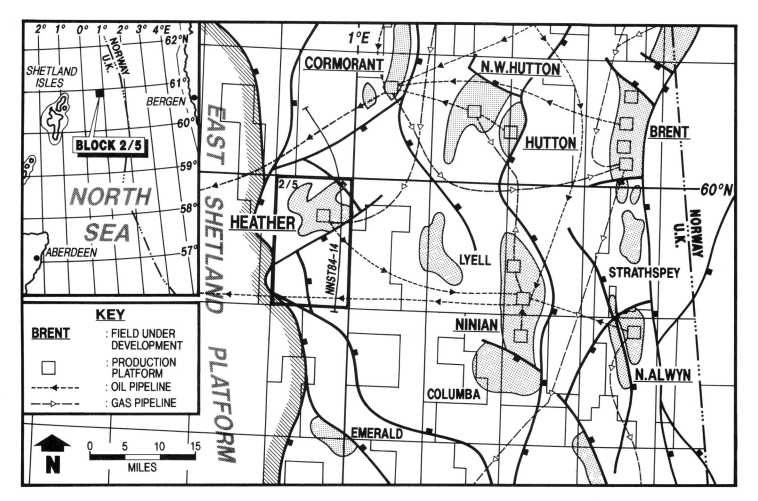

Fig. 1. Heather Field location map showing East Shetland Basin tectonic features and production infrastructure.

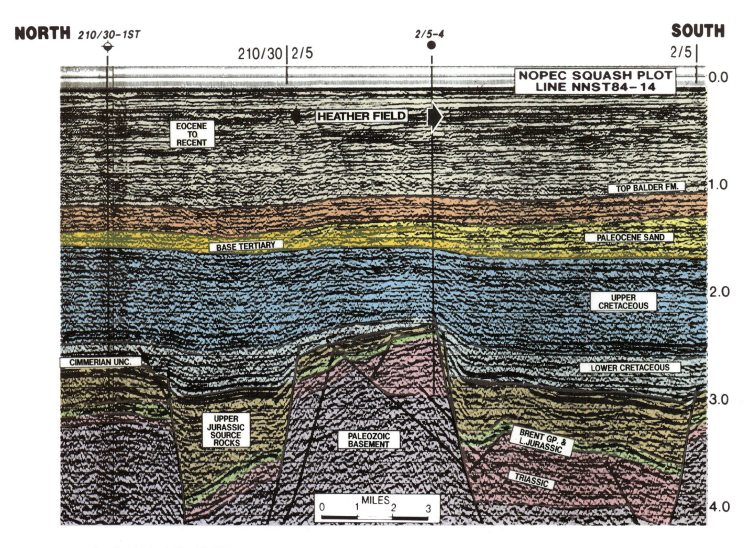

Fig. 2. Interpreted regional seismic line NNST84-14 across the Heather structure; see Fig. 1 for location.

The Heather Field discovery well, 2/5-1, was spudded in August 1973, its objective being the 'Dogger' Sandstone of Middle Jurassic age. The well encountered the anticipated pay sandstones at 10 070 ft TVSS on the flank of the fault block (see Fig. 3). A total reservoir section of 198 ft was drilled, of which 98 ft were net pay. The well flowed at a combined rate of 8950 BOPD from two zones. Wells 2/5-2 and 2/5-3, drilled early in 1974, also appraised the flanks of the field, the latter being dry and defining a 'highest known water' of 10 885 ft TVSS. Well 2/5-2 had proved 'oil down to' 10 807 ft TVSS. Reservoir conditions on the crest of the structure were determined by well 2/5-4. It encountered a thin Brent Gp section of 125 ft below 9506 ft TVSS, but confirmed a minimum oil column of 1300 ft. Well 2/5-4 also encountered oil in the Triassic Cormorant Formation (Fm) but four drill stem tests only recovered small amounts of oil and water.

This marked the end of the first phase of appraisal. For further details the reader is directed to Gray & Barnes (1981). At the end of 1974 the Heather Field was declared commercial and a tubular steel platform, with two drilling rigs and facilities able to handle 60 000 BOPD, was commissioned. The development plan envisaged 25 development wells, seven being peripheral water injectors serving 18 oil producers. This was based on an oil-in-place figure of 375 MMBBL with a closure of 5576 acres, reserves being estimated at 150 MMBBL.

The Heather Alpha platform was installed in May 1977 and the first development wells, H1 and H2, were spudded in July 1978. The first oil was exported through the Ninian Pipeline System in October 1978 and, ten years later, 80 MMBBL of crude oil had been recovered. This statement does not reveal the fluctuating fortunes of the field's development. During those ten years, reserve estimates had fallen to as low as 65 MMBBL, only to be steadily revised upward to 100 MMBBL today.

The early development of the Heather Field did not go according to plan. It soon became apparent that permeability barriers within the reservoir would not allow an efficient peripheral pressure maintenance system. In addition, five of the first ten development wells intersected unmapped faults. In 1980, however, several events coincided which rapidly increased the understanding of heterogeneities within the field. First, a 3-D seismic survey, shot in 1979, was interpreted and showed the Heather feature to be a mosaic of contiguous fault blocks. The new interpretation immediately increased confidence in locating well patterns. Secondly, the assumed fieldwide oil/water contact at about 10 850 ft TVSS was disproved by well H11 which encountered water at 10 400 ft TVSS in a fault block containing the dry appraisal well, 2/5-3. Thirdly, oil was discovered down-flank on the 'North West Heather' prospect by well 2/5-12A. The well flowed at a maximum rate of 5820 BOPD from the Brent Gp below 11 539 ft TVSS. Formation pressures recorded were found to be on the extrapolated Heather oil gradient and, therefore, indicative of a continuous oil column. Further appraisal of the North West Heather accumulation by wells 2/5-13ST and 2/5-14ST extended the oil column to 12 300 ft TVSS and confirmed pressure communication with the main producing area in the Heather Field.

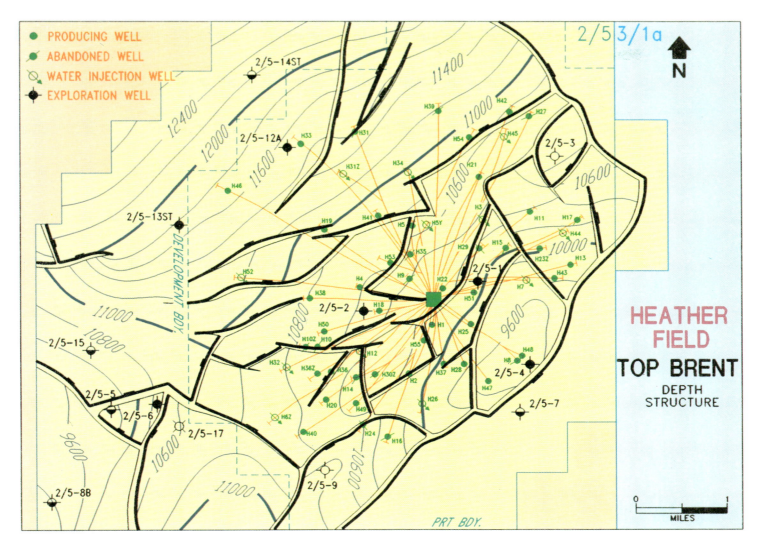

Fig. 3. Simplified Top Brent Group structure map of the Heather Field.

The appraisal of the northwestern flank of the Heather Field, below 10 850 ft TVSS, was successful and a second 3-D seismic survey was shot in 1981. The Heather Field, incorporating the North West Heather accumulation, has subsequently been divided into nine contiguous fault blocks and developed on a block-by-block basis. It now covers an area of approximately 10 500 acres. Reserves have been exploited by drilling highly-deviated wells, using non-toxic, oil-based muds coupled with modern bits and downhole motors. Well 2/5-12A was twinned in 1983 by well H33 and a further North West Heather step-out, well H46, reached a total measured depth of 19 646 ft, horizontal departure from the platform being 14 180 ft (2.7 miles). During the mature phase of development, the reservoir simulation model has been used continuously as a predictive tool with considerable success.

Peak production of 38 000 BOPD (monthly average) was attained in mid-1982. Since then there has been a slow decline to the current production of 13 000 BOPD, 18 500 BWPD. A total of 60 development wells have been drilled, including redrills, from 43 slots; three slots were added in 1988 to the original 40. In order to meet demands for both fuel and start-up gas, a 6 in gas import line was installed in 1987 which connects Heather to the western leg of the FLAGS system (see Fig. 1). There are currently 29 oil producing wells and 12 water injection wells. All producing wells are gas-lifted.

Of the original partners in the Heather Field, only Unocal and DNO remain. Skelly was bought by Getty Oil prior to development of the field and Getty was subsequently taken over by Texaco Britain Ltd in 1984. In 1988, BG United Kingdom, Inc. (British Gas) took over Tenneco's interest in the field.

Field stratigraphy

The stratigraphic sequence encountered during the exploration and development of the Heather Field comprises mainly clastic lithologies ranging in age from Triassic to Recent (see Fig. 4). A sheared gabbro of assumed Palaeozoic age has been encountered below 13 000 ft TVSS in well 2/5-4, above which are 3700 ft of Triassic red beds consisting predominantly of oxidized and weathered ferruginous sandstones deposited in a continental environment at a time when the northern North Sea basin was becoming established. These deposits are assigned to the Cormorant Fm.

A very thin sandstone unit (less than 15 ft thick) overlying the Cormorant Fm is possibly the Nansen Member of the Statfjord Fm, above which are marine shales and siltstones of the Lower Jurassic Dunlin Gp, which are 140 to 160 ft thick across the Heather Field. During earliest Middle Jurassic (Aalenian) times, coarse clastic sediments belonging to the Brent Gp were deposited into a shallow marine sea in the Block 2/5 area, probably from a provenance to the west on the East Shetland Platform. Overlying this formation are regressive, deltaic sandstones, siltstones and carbonaceous shales of Aalenian–earliest Bajocian age (Eynon 1981; Richards *et al.* 1988). Completing the Brent Gp sequence are early Bathonian transgressive marine sandstones, indicative of delta

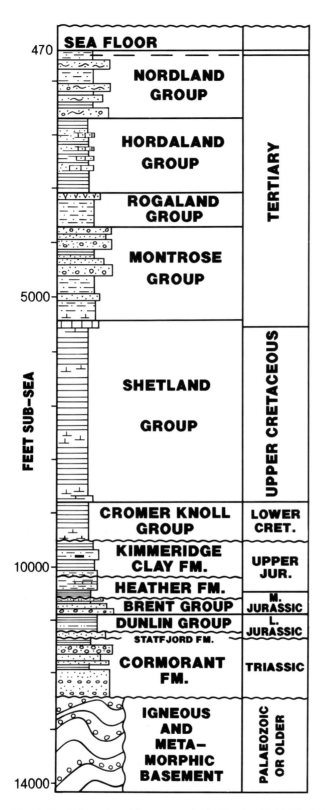

Fig. 4. Typical stratigraphic sequence in the Heather Field, Block 2/5.

anaerobic conditions (Barnard & Cooper 1981). These two formations make up the Humber Gp which varies between 250 ft and 1500 ft over the Heather structure, due to mid–late Jurassic structural growth, and thickens in the hanging wall regions where it has become a rich oil source (see Fig. 2).

Above the Kimmeridge Clay Fm are the limestones, claystones and characteristic red and green marls of the Cromer Knoll Gp. These marine Lower Cretaceous deposits range in thickness from 500–1000 ft over the Heather Field. Relatively deep marine conditions continued into the Upper Cretaceous and 3000–3500 ft of monotonous grey claystones of the Shetland Gp, capped by thin 'Danian' limestones, were deposited.

The Tertiary succession over the Heather Field cosists of 5500–6000 ft of predominantly clastic sediments deposited as a result of continued thermal subsidence. Early in the Tertiary, late Palaeocene sands were shed from the uplifted Shetlands region into the basin to the east. Faults at the margin of the East Shetland Basin became active and a coarse clastic wedge, up to 1600 ft thick, was accommodated, thinning northwards and eastwards across the Heather Field (see Fig. 2). These are the sediments of the Montrose Gp. They are overlain by clays, siltstones and tuffaceous deposits of the Rogaland Gp. The Eocene–early Oligocene is represented by clays and thin sands of the Hordaland Gp and the mid-Oligocene to the present by the sandy deposits of the Nordland Gp. Eocene to Recent sediments over the Heather Field are approximately 3500 ft thick.

Geophysics

Prior to development of the field in 1976, a N–S/E–W seismic grid, spaced at 0.5 km, was recorded. The survey tied the six wells in the area. Data quality was improved by reprocessing in 1983.

The structural complexity of the Heather Field was exemplified by the poor results of early development drilling in 1978. Therefore, in 1979, a 3-D seismic survey was acquired over the field. The subsequent extension of the field offstructure to the northwest in 1980 led to a further 'NW Heather' survey in 1981, which overlapped the 1979 survey by 1.5 km. Both surveys were oriented NW–SE with a line spacing of 75 m.

Shot point spacing was 25 m. The two surveys together cover an area of 110 km². Data quality for both surveys was good to fair, although interference from peg-leg multiples has hindered interpretation, particularly in flanking areas.

The 3-D data enabled the structure of the Brent Gp to be interpreted to a level of detail required for reservoir management. In 1986/87, the current Top Brent Gp depth map (Fig. 3) was constructed using a LANDMARK workstation. A major benefit of using the LANDMARK workstation was the ability to utilize instantaneous phase attribute sections. These improved the definition of minor faulting and helped to distinguish between primary and multiple energy. The primary structural interpretation was made at Base Brent Gp/Top Dunlin Gp level where a decrease in acoustic impedance over much of the field was represented by a strong seismic peak. The Top Brent Gp horizon is not an obvious seismic reflector over Heather, especially in crestal areas where the Jurassic units thin. The Top Brent Gp map was constructed by applying an average interval velocity of 12 500 ft/sec to the Brent Gp isopach and subtracting the time isopach from the Base Brent Gp horizon. Fine adjustments were then made manually to this pseudo-Top Brent Gp horizon, using the LANDMARK horizon editing facility, and the Top and Base Brent Gp horizons depth-converted. Depth conversion made use of average velocity maps, constructed to both horizons, that were based on all available well and velocity survey data.

A revised reservoir simulation model, based on this map, has proved reliable during the recent drilling of seven development wells. However, fault mapping has not resolved all the permeability barriers that are recognized through history-matching production data.

abandonment. The gross thickness of the Brent Gp, which forms the Heather Field reservoir, varies between 125 ft and 369 ft (see Fig. 5).

The continued drowning of the Brent delta led to the establishment of an open marine environment in late Middle Jurassic times. Shales, siltstones and thin limestones of the Heather Fm were laid down unconformably above the Brent Gp. During Upper Jurassic times, the dark grey brown, organic, radioactive shales of the Kimmeridge Clay Fm were deposited in a restricted basin under

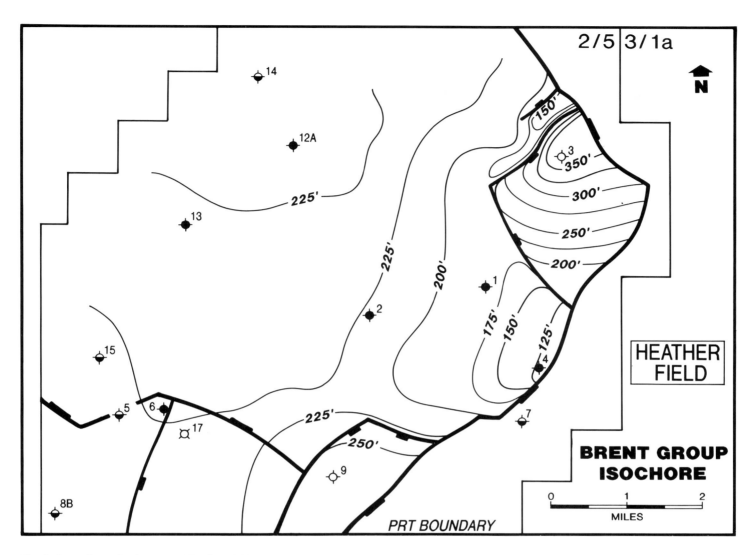

Fig. 5. Brent Group isochore map, Heather Field.

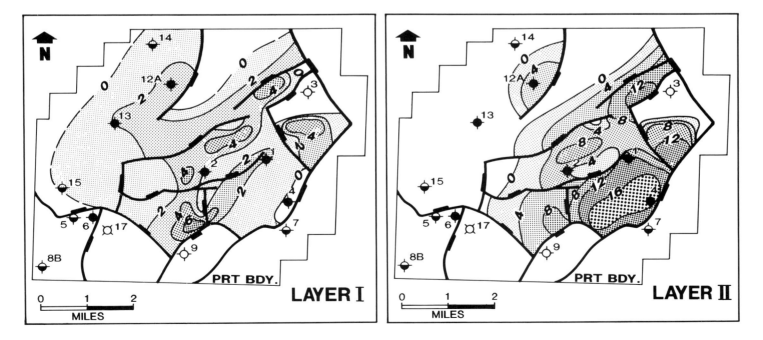

Fig. 6. Hydrocarbon-pore-feet maps for reservoir layers I and II depicting the overall distribution of producible oil in the Heather Field.

Trap

The Heather Field trap is a tilted fault block closure (see Fig. 2). In gross terms, it is a rectangular block, 7 miles long from northwest to southeast and 5 miles wide from northeast to southwest that tilts northwestwards at about 7°. This major structure formed during mid–late Jurassic times as a result of regional tensional forces related to the development of the North Viking Graben. The major bounding faults of the structure trend SW–NE and SE–NW. At the southeastern and northeastern margins of the trap, displacement at Brent Gp level is 3000 to 5000 ft. The southwestern boundary of the field is along a fault zone where the counter-regional dip of the Heather structure abuts the easterly, basinward dip of the West Heather structure (2/5-8B; see Fig. 3). Throw along this fault zone varies from less than 50 ft in the southeast to 1000 ft in the west, adjacent to the East Shetland Platform. At its northwestern extremity the structure forms the hanging wall block of another major SW–NE fault zone in block 210/30 and N–S-trending basin-margin fault system (see Fig. 1).

During the major period of structural growth, in the mid–late Jurassic, the shales and siltstones of the Heather Fm were laid down and sealed the reservoir sands of the Brent Gp beneath, both vertically and laterally. Where fault movement continued into the early Cretaceous period, the shales and marls of the Cromer Knoll Gp form the lateral seal to the trap. At the southwestern margin of the field the bounding fault acts as a lateral seal.

The complexity of the trap today is due in large part to fracturing of the block during middle and late Jurassic times. Small faults have offset the reservoir units and larger faults have broken the field up into nine discrete blocks. Many of the faults trend SW–NE and are related to Caledonide basement lineaments but, especially in the west of the field, a strong E–W trend is evident. Less apparent, but effective in creating impermeable barriers, is a N–S or NW–SE cross-trend (see Fig. 3). Pressure communication between the eight producing fault blocks is variable but they have generally been developed as separate entities. Faulting within some blocks has impaired the sweep efficiency of the water flood and infill drilling has been necessary.

Figure 6 depicts the distribution of oil within the two major reservoir layers of the Brent Gp (see Fig. 7). The preservation of porosity and permeability is considered to be strongly related to the early pooling of oil above local spill points synchronous with severe diagenesis (Glasmann et al. 1989). This had led to the bulk of producible oil being located in the crestal regions of individual fault blocks and is particularly evident in layer II, which contains 75% of Heather Field oil-in-place. The relative timing of diagenesis and oil entrapment has proved to be critical in creating a highly productive core in the highest regions of the structure (see Fig. 6). Peripheral regions, although accounting for 80% of the total field area, contain only 33% of estimated recoverable reserves. The distribution of oil within layer I is complicated by local erosional and depositional phenomena but is particularly significant on the northwestern flank. In the extreme northwest, layer I is oil-saturated down to 12 300 ft TVSS and is currently productive in well H46 at 11 660 ft TVSS (see Fig. 3).

At the end of the Oligocene a physico-chemical equilibrium is thought to have been reached whereby low permeabilities and high capillary pressures led to the cessation of both oil migration and diagenesis. As a result, high irreducible water saturations, coupled with an effectively immobile oil fraction, have been encountered at

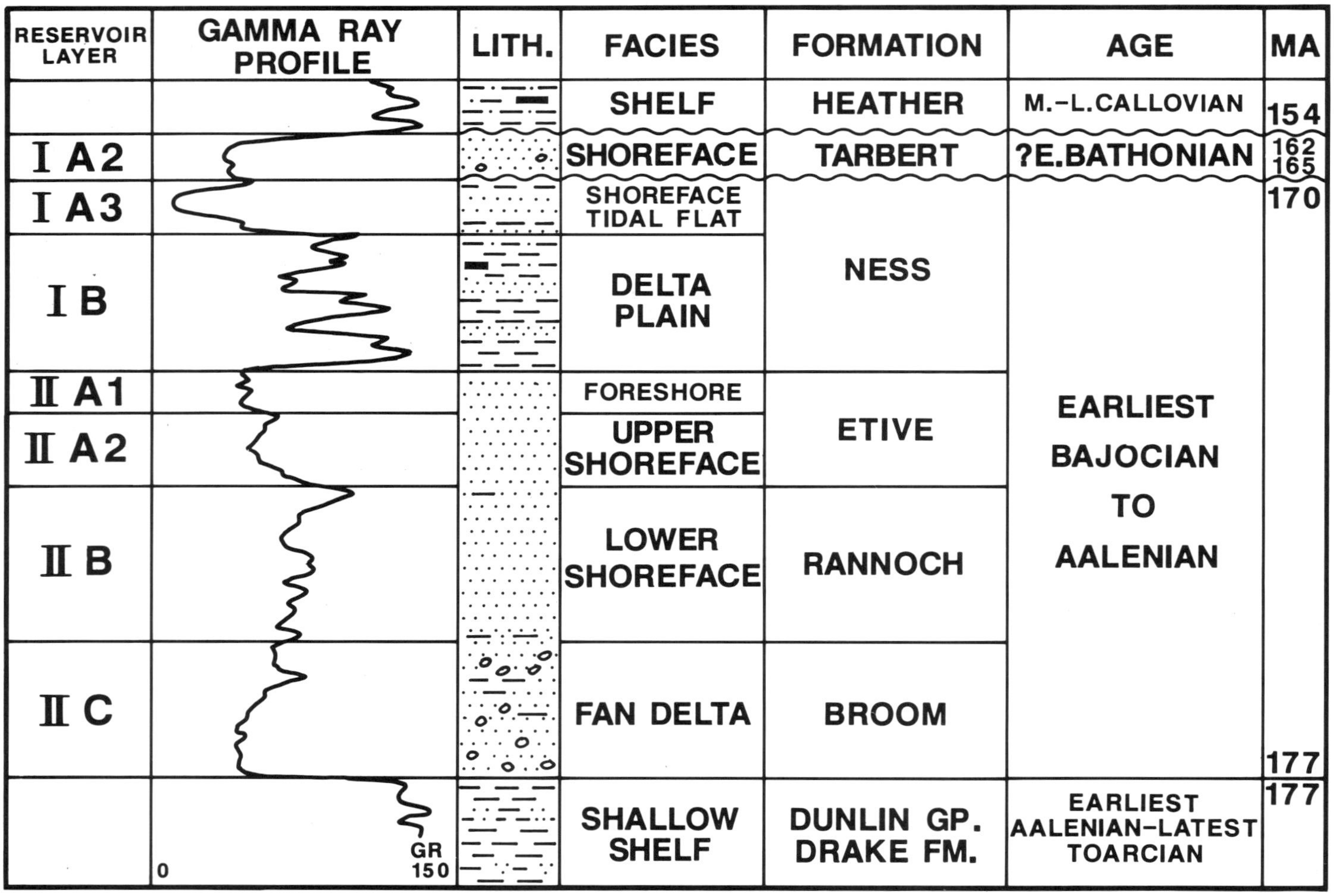

RESERVOIR LAYER	GAMMA RAY PROFILE	LITH.	FACIES	FORMATION	AGE	MA
			SHELF	HEATHER	M.-L. CALLOVIAN	154
I A2			SHOREFACE	TARBERT	?E. BATHONIAN	162 / 165
I A3			SHOREFACE TIDAL FLAT			170
I B			DELTA PLAIN	NESS	EARLIEST BAJOCIAN TO AALENIAN	
II A1			FORESHORE			
II A2			UPPER SHOREFACE	ETIVE		
II B			LOWER SHOREFACE	RANNOCH		
II C			FAN DELTA	BROOM		177
			SHALLOW SHELF	DUNLIN GP. DRAKE FM.	EARLIEST AALENIAN-LATEST TOARCIAN	177

Fig. 7. Summary of the Brent Group sedimentology, reservoir zonation and chronostratigraphy in the Heather Field.

depth and no free water level has been established for the Heather Field as a whole. The one exceptional area is the easternmost fault block in the field, which contains well 2/5-3 (see Fig. 6), where an oil/water contact has been established in the region of 10 400 ft TVSS. Pressure communication between this block and the central area of the field is minimal, although they have similar initial formation pressures of 4950 psig at 10 250 ft TVSS.

Reservoir

The Heather Field reservoir is the Middle Jurassic Brent Gp which is composed of sandstones with subordinate shales, siltstones and coaly deposits. The sequence can readily be divided into the formations of Deegan & Scull (1977; see Fig. 7). Reservoir quality in the Heather Field is controlled by diagenesis and sedimentary facies. Diagenesis has played a major role in altering reservoir quality, but is mostly responsible for permeability and porosity heterogeneities within individual facies. Sandstone thickness variations across the field (see Fig. 5) are due to syn-depositional faulting, post-depositional erosion, structural growth and differential compaction.

The lowermost reservoir zone in the Heather Field, layer IIC, is equivalent to the Broom Fm. It is characterized by coarse-grained sandstones deposited as delta-front splays and shallow shelf turbidites in the subaqueous portions of overlapping fan-deltas. The unit is commonly bi-lobed, 40–70 ft thick and thickens to the southwest where a third lobe is present. Layer IIC is overlain by the fine-grained, micaceous sandstones of layer IIB which is equivalent to the Rannoch Formation. These were deposited when delta-front sands, associated with the progradation of a major fluvio-deltaic system in Aalenian to earliest Bajocian times (Eynon 1981; Richards et al. 1988), were reworked into lower and middle shoreface sequences by wave and basinal processes. Continued progradation led to the deposition of the fine- to medium-grained, massive, upper shoreface/foreshore sands of layer IIA, (Etive Fm equivalent) and the delta plain and lagoonal deposits of layer IB (Ness Fm equivalent). Layers IIA and IIB are 30–40 ft and 50–60 ft thick respectively across the Heather Field, suggesting that a consistent rate of delta advance was achieved in this area (Budding & Inglin 1981).

Layer IB is commonly 40–65 ft thick but has been eroded on the crest of the structure. The sand bodies are thin and discontinuous but sand/shale ratios may be high. The sandstones of layer IB were deposited in distributary channel, distributary mouth-bar and crevasse splay environments. Younger sandstones, associated with the drowning of the delta, were deposited as stacked shoreface sequences. These lie within the layer IA reservoir zone which has been divided into four sub-units. The most important of these are layers IA2 and IA3, which are present over most of the field but are separated by an unconformity. Layers IA1 and IA4 are only locally developed. The layer IA3 unit contains tidal flat and shoreface facies and has been assigned to the upper part of the Ness Fm. The generally fine-grained, quartz-rich sandstones of IA3 can be correlated by their low gamma-ray count. This zone is a thin (less than 20 ft thick) but significant reservoir and was possibly a major conduit for oil migrating into the Heather structure. During early Bathonian times the more micaceous and bioturbated, transgressive sandstones of layer IA2 (Tarbert Fm equivalent) were deposited as the delta was abandoned. Layer IA varies between 0–60 ft in thickness.

Reservoir quality can be ranked by facies. From best to worst are: (1) upper shoreface; (2) foreshore; (3) fan-delta; (4) lower shoreface; (5) distributary channel; (6) distributary mouth-bar; (7) crevasse splay; (8) tidal flat. Generally the high-energy marine sandstones have the best porosity and permeability, layers IA and IIA being the most productive reservoir zones in the Heather Field. The primary depositional characteristics of the reservoir facies have been modified considerably by diagenesis (Glasmann et al. 1989).

Average porosity and permeability values for individual reservoir zones are not considered useful. In summary it can be said that in crestal regions of the field, and to some extent at the crests of individual fault blocks where early entrapment took place, the reservoir sandstones reach their fullest potential for preserving their original porosity and permeability. Calcite cementation is particularly pervasive in layers IIB and IIC and in some parts of the field may completely obliterate porosity in layer IIA2, the most productive reservoir zone. Pore-filling kaolinite is also present, particularly at depth where it replaces potassium feldspar and creates a microporous network of low permeability. The most damaging effects on reservoir quality, however, have been due to a relatively late episode of quartz and illite precipitation which has greatly lowered porosity and permeability, especially in the less quartzose sandstones on the flanks of the field.

Source

Geochemical correlation studies have established that the Kimmeridge Clay Fm is the main source of the Heather Field oils. Biomarker analyses indicate minor variations in the maturity of oils across the field. These are indicative of continuous migration of oil into the structure as thermal maturation of the Kimmeridge Clay Fm source beds increased. The oil is generally of low maturity (R_o equivalent 0.65–0.70); however, it is more mature in NW Heather.

Maturation modelling of the source kitchens to the north and south of the Heather structure (see Fig. 2), suggests that the Kimmeridge Clay Fm passed into the oil window during the Lower Palaeocene. K-Ar age-dating of illites in the Brent Gp (Glasmann et al. 1989) indicates that oil migration into the trap began in the Middle Eocene (45 Ma) and that the structure was full, or had reached a diagenetic equilibrium, by the end of the Oligocene (25 Ma). Migration into the trap is considered to have been mainly from the northwest.

Hydrocarbons

Heather Field crude is an undersaturated, low sulphur oil varying in gravity between 32° and 37° API. Initial pressure across the field is 4950 psig at 10 250 ft TVSS and the pressure gradient in the oil column is 0.3 psi/ft. Variations in the composition and characteristics of reservoir fluids have been noted. In the central and crestal producing area of the Heather Field the bubble point pressure (2370 psig), gas/oil ratio (700 to 800 SCF/STB), formation volume factor (1.36 STB/RB) and CO_2 content (2–3 mol.%) are relatively consistent. In flanking fault blocks to the southwest and northeast, however, lower gas/oil ratios are recorded with corresponding bubble point pressures in the region of 1930 psig. In the case of the easternmost fault block (wells H13 and H23Z; see Fig. 3) this is related to a consistently low CO_2 content of 1.44–1.47 mol.%. This relationship is not repeated in the southwestern fault block, however, where well H20 oil has a CO_2 content of 6.86 mol.%.

Elsewhere, as in well H21 and the NW Heather wells 2/5-12a and H33 which exhibit more gaseous compositions, high carbon dioxide content appears to be related to a relatively high gas/oil ratio. Oils from the NW Heather wells have recorded values of 1280 SCF/STB and 8.1–8.6 mol.% CO_2. The high CO_2 values are principally from layer I oils and may be indicative of charging by a more mature oil phase with a higher gas/oil ratio at a late stage in the filling of the Heather Field. A corollary to this may be that the low gas/oil ratios exhibited by the partially-filled easternmost fault block are a result of early filling and subsequent isolation synchronous with continued migration of oil into the main part of the structure from the northwest, via NW Heather. An analysis of biomarker data has not been able to add more detail to this simple model.

Initial flow rates of some early development wells in structurally high regions of the field improved on those achieved during the testing of the exploration wells. Wells H1 (layers I and II) and H2 (layer II only) produced at rates of 9300 and 7200 BOPD respectively, and well H8, on the crest of the structure, at 12 000 BOPD.

Uncontaminated formation water has been sampled from layer II in well H11. Total dissolved solids measured were 24 370 milligrams per litre and the resistivity, at 60°F, was 0.326 ohm.m. Samples of formation water that have been contaminated with injection water to varying degrees have been recovered and analysed across the field. These suggest that the field Rw may vary from 0.29 ohm.m, offstructure, to 0.46 ohm m, at 60°F, at the crest.

Reserves

Heather Field recoverable reserves are currently estimated at 100 MMBBL. Oil-in-place, using a 10% porosity cut-off, is currently estimated at 438 MMBBL. Recovery factors in the two structurally highest fault blocks, to the southeast of the Heather Alpha platform, average 37%. Contiguous fault blocks offstructure exhibit recovery factors of 16–18%, whilst 10–13% is estimated for the northwestern flank. Some of the oil-in-place on the flanks of the Heather Field is effectively immobile, but contributes to the dynamics of the reservoir through its expansion in response to production.

The recovery of oil from the Brent Gp has been hindered by the extremely heterogeneous nature of the reservoir. Secondary recovery through water flooding has been successful, especially in areas of better reservoir quality, but, even here, the strong vertical permeability contrast between layers has led to problems of early water breakthrough. Because of this problem offstructure, where layer IA3 is vastly more permeable than other reservoir zones, a pilot scheme to produce below the bubble point pressure was agreed with the Department of Energy in 1982. A water flood has subsequently been implemented.

The author would like to thank Unocal, Texaco, DNO and British Gas management for their permission to publish this paper. However, the interpretations and opinions expressed in this paper are entirely those of the author. The work of colleagues in Sunbury, Aberdeen, and at Unocal's Research Centre in Brea, California, during a major re-evaluation of geological and geophysical data in 1987, is hereby acknowledged.

References

BARNARD, P. C. & COOPER, B. S. 1981. Oils and source rocks of the North Sea area. *In*: ILLING, L. V. & HOBSON, G. D. (eds) *Petroleum Geology of the Continental Shelf of North-West Europe*. Institute of Petroleum, Heyden, London, 169–175.

BUDDING, M. C. & INGLIN, H. F, 1981. A reservoir geological model of the Brent Sands in Southern Cormorant. *In*: ILLING, L. V. & HOBSON, G. D. (eds) *Petroleum Geology of the Continental Shelf of North-West Europe*. Institute of Petroleum, Heyden, London, 326–334.

DEEGAN, C. E. & SCULL, B. J. (compilers) 1977. *A proposed standard lithostratigraphic nomenclature for the Central and Northern North Sea. Report of the Institute of Geological Sciences*, No. 77/25; *Bulletin of the Norwegian Petroleum Directorate*, No. 1.

EYNON, G., 1981. Basin development and sedimentation in the Middle Jurassic of the Northern North Sea. *In*: ILLING, L. V. & HOBSON, G. D. (eds) *Petroleum Geology of the Continental Shelf of North-West Europe*. Institute of Petroleum, Heyden, London, 196–204.

GRAY, W. D. T. & BARNES, G., 1981. The Heather oil field. *In*: ILLING, L. V. & HOBSON, G. D. (eds) *Petroleum Geology of the Continental Shelf of North-West Europe*. Institute of Petroleum, Heyden, London, 335–341.

GLASMANN, J. R., LUNDEGARD, P. D., CLARK, R. A., PENNY, B. K. & COLLINS, I. D. 1989. Isotopic evidence for the history of fluid migration and diagenesis: Brent Sandstone, Heather Field, North Sea. *Clay Mineralogy*, **24**, 225–284.

RICHARDS, P. C., BROWN, S., DEAN, D. M. & ANDERTON, R., 1988. A new palaeogeographic reconstruction for the Middle Jurassic of the Northern North Sea. *Journal of the Geological Society, London*, **145**, 883–886.

Heather Field data summary

Trap
Type	Tilted fault block
Depth to crest	9450 ft TVSS
Lowest closing contour	n/a
Hydrocarbon contacts	Free water level not seen
Gas column	n/a
Oil column	2350 ft +

Pay zone
Formation	Brent Group Sandstones
Age	Middle Jurassic (Aalenian–Bathonian)
Gross thickness (av./range)	210 ft/125–369 ft
Sand/shale ratio (av./range)	0.85/0.70–1.0
Net/gross ratio (av./range)	0.54/0.25–0.96
Cut off for N/G	10% φ; 35% V Clay
Porosity (av./range)	10%/0–27%
Hydrocarbon saturation (av./range)	65%/20–85%
Permeability (av./range)	20 md/0.1–2000 md
Productivity index	0.2–15.0 STBD/PSI

Hydrocarbons
Oil gravity	0.85 gm/cc (32° to 37° API)
Oil type	low sulphur (0.7%)
Gas gravity	c. 0.91 (air = 1.0)
Bubble point	1930 to 3890 psig
Dew point	n/a
Gas/oil ratio	450–1280 SCF/STB
Condensate yield	n/a
Formation volume factor	1.27–1.78 STB/RB

Formation water
Salinity	22 000 ppm NaCl equivalent
Resistivity	0.326 ohm m at 60°F

Reservoir conditions
Temperature	227–242°F
Pressure	4950 psig at 10 250 ft TVSS
Pressure gradient in reservoir	0.29–0.31 psi/ft

Field size
Area	10 500 productive acres
Gross rock volume	2 200 000 acre.ft
Hydrocarbon pore volume	77 000 acre.ft
Recovery factor: oil	0.23
Recovery factor: gas	n/a
Drive mechanism	water flood
Recoverable hydrocarbons:	
oil	100 MMBBL
NGL/condensate	10 MMBBL
gas	61 BCF (no sales)

Production
Start-up date	10th October 1978
Development scheme	single platform: individual development of fault blocks
Production rate	38 000 BOPD peak monthly rate (1982)
	13 000 BOPD current (June 1989)
Cumulative production to date	84 MMBBL (crude shipped)
Number/type of wells	41 wells
	29 oil producers
	12 water injectors
Secondary recovery method	water flood

The Hutton Field, Blocks 211/28, 211/27, UK North Sea

D. BRIAN HAIG

Conoco (UK) Ltd, Park House, 116 Park Street, London W1Y 4NN, UK

Abstract: The Hutton oil field is situated in the East Shetlands Basin in the UK North Sea on the western side of the Viking Graben. It straddles UK Blocks 211/27 and 211/28. The field was discovered in July 1973 by the Conoco well 211/28-1a and is operated by Conoco (UK) Limited. The structure comprises a series of southwesterly dipping tilted fault blocks. The reservoir sandstones are Middle Jurassic in age and were deposited as a result of deltaic progradation across the Hutton area. The oil bearing Brent Group sandstones vary in thickness from 150 ft to 380 ft with average porosities of 22% and permeabilities of 500–2000 md in the producing zones. Original recoverable reserves have been estimated at 190 MMBBL, of which the field has produced 107 MMBBL to date. The field is developed via a tension-leg platform, the first of its kind in the world.

The Hutton Field is situated in the East Shetlands area of the UK North Sea approximately 480 km northeast of Aberdeen (Fig. 1).

It straddles the Conoco/Chevron/Oryx block 211/28 (licence No.P204) and the Amoco/Mobil/Enterprise/Amerada/Texas Eastern block 211/27 (licence No.P473).

The field measures some 9 km north–south by 3 km east–west and the mean water depth at the platform location is 486 ft. It is a medium size oil field with approximately 550 MMBBL of oil originally in place.

The oil is found within the Brent Group reservoir sandstones which were deposited during a phase of coastal deltaic progradation during the Middle Jurassic.

The main field structure comprises three distinct fault blocks downthrowing to the northwest. Several subsidiary fault blocks on the eastern margin of the field and which are downthrowing to the east and southeast are also considered to be oil bearing.

Field development is via a centrally located Tension Leg Platform (TLP), the first of its kind in the world.

The field is named after James Hutton (1726–97) the Scottish geologist who originated many of the fundamental principles of modern geology notably his plutonic theory for the origin of the Earth.

History

The Hutton Field was discovered in July 1973 by the Conoco/Gulf/NCB well 211/28-1a within the UK production licence area P204. A further five appraisal wells were drilled to delineate the field: 211/28-2, 211/28-3, and 211/28-4 within the P204 area and 211/27-1a, 211/27-2 within the Amoco/Mobil/Gas Council/Texas Eastern/Amerada block 211/27 (licence Area P184 (now P473)).

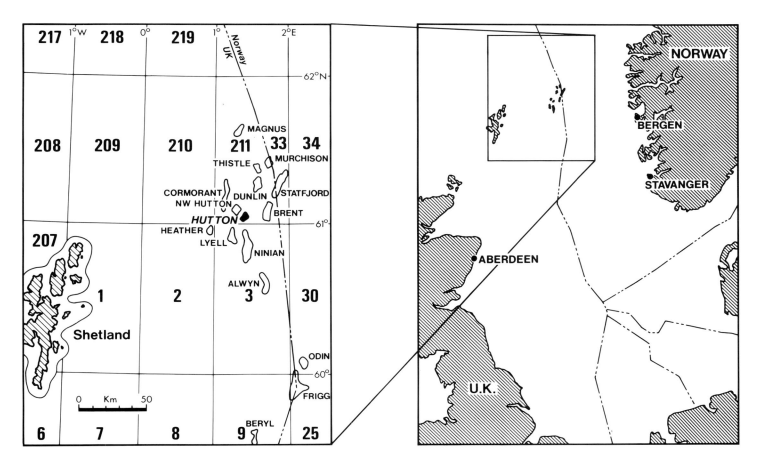

Fig. 1. Location of the Hutton Field relative to other major fields in adjacent blocks.

The Hutton Field Unitization and Unit Operating Agreement was signed in May 1980. The following are the present day licence holders:

P204 Licence Group, Block 211/28

Conoco (UK) Ltd (Operator)	33.33%
Chevron (UK) Ltd	33.33%
Oryx Energy (UK) Ltd	33.33%

P473 (formerly P184) Licence Group, Block 211/27

Amoco (UK) Exploration Company	25.77%
Enterprise Oil plc	25.77%
Mobil North Sea Ltd	20.00%
Amerada Hess Ltd	18.08%
Texas Eastern North Sea Inc.	10.38%

A three dimensional seismic survey was shot across the field in June 1980. During the same month an Annex B proposal was submitted to the Department of Energy to develop the field using a centrally located Tension Leg Platform (TLP) (Fig. 2).

Primary development commenced in May 1981 with the installation of a 32 slot template through which the first 10 development wells were drilled. These were drilled from a semi-submersible in order to enable production to commence as quickly as possible after installation of the TLP. To maximize the early production rate all of these 10 wells were completed as producers.

The platform was installed in August 1984 with production commencing the following month. A further 24 wells have subsequently been drilled from the TLP, including sidetracks. The status as of April 1989 was 16 producing wells, 12 water injectors and 6 abandoned wells.

The first Equity redetermination commenced in January 1985. Prior to its completion, however, all Unit members agreed that the following tract participation should be adopted with effect from 1st January 1987 and be retained throughout the life of the field:

Block 211/28 (P204) — 66.5% (Original Unit Share = 60%)
Block 211/27 (P473) — 33.5% (Original Unit Share = 40%).

Field stratigraphy

The Hutton Field is located in the East Shetland Basin on the western edge of the Viking Graben (Fig. 3). This basin was established during the late Permian/early Triassic and sedimentation into it continued through the Mesozoic and Tertiary. The sediments deposited during the Middle and Late Jurassic include the Brent Group sandstones, which form the reservoir for the Hutton field and the Kimmeridge Formation shales which comprise the prolific source rock for this and the other fields of the North Viking Graben. The Brent Group represents the deposition of the single cycle of deltaic sedimentation which prograded northwards in the Viking Graben.

The Hutton structure comprises a series of south-westerly dipping fault blocks ('A', 'B' and 'C' from west to east) uplifted from the main Graben area (Fig. 4).

The Brent Group in Hutton is subdivided into the following units, each representing a distinct phase of deposition (Fig. 5).

(1) The *Basal Sand Unit* (Broom) is interpreted as a sublittoral sand sheet deposited by storm generated currents. It represents the offshore equivalent of an estuarine/channel shoal system and is genetically separate from the overlying deltaic sequences.

(2) The onset of deltaic progradation is marked by the '*Mica Sand Unit*' (Rannoch) which is a sequence of micaceous silts and sandstones deposited in progressively shallower water on the lower and middle shoreface of the delta front.

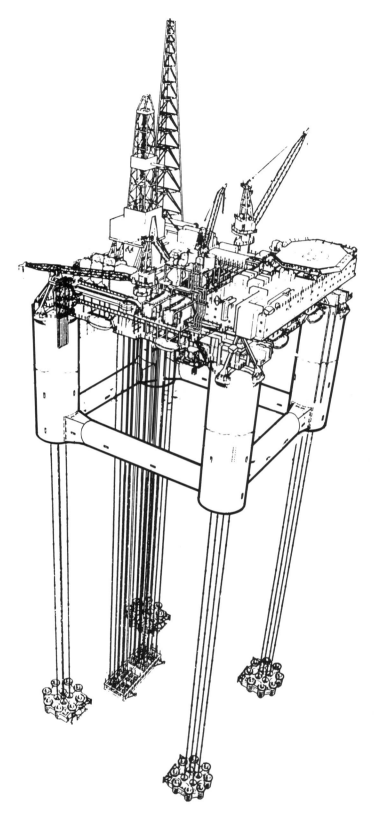

Fig. 2. Tension leg platform (TLP), Hutton Field.

(3) The *Massive/Mica Sand Unit* (Etive/Rannoch) is a complex unit occurring between the underlying middle shoreface sands of the Mica Sand Unit and the overlying upper shoreface/foreshore sands of the Massive Sand Unit. It comprises a series of coarse grained ebb-tidal delta mouthbar and interbar sandstones deposited by relatively high current energies.

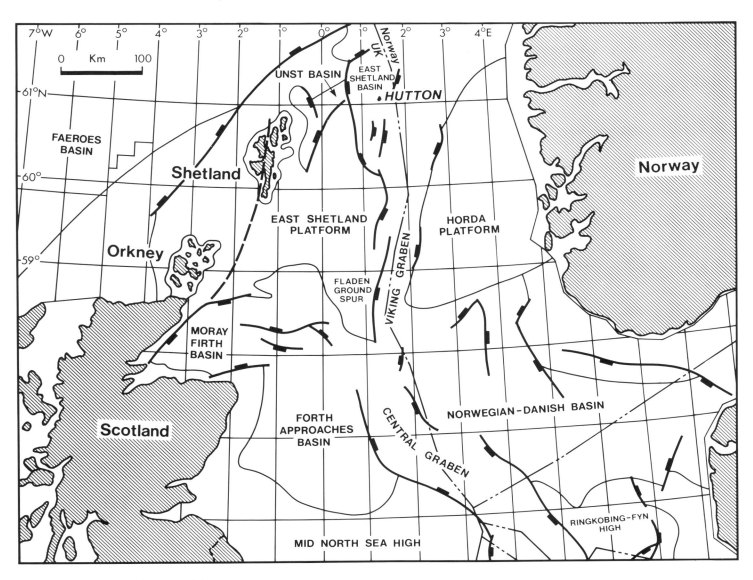

Fig. 3. Structural setting of Hutton Field.

(4) The *Massive Sand Unit* (Etive) comprises sandstones deposited as tidal channel facies and transitional upper shoreface, foreshore and backshore facies. Its thickness distribution is controlled by varying rates of progradation and aggradation across the field caused by changes in sea-level at the time of deposition.

(5) The Delta plain sediments of the *Middle Shaly Unit* (Ness) comprise a series of lagoonal and bay sandstones, mudstones, silts and coals deposited behind the barrier complex of the Massive Sand Unit. These sediments are frequently cut by fluvial channel deposits formed within narrow meander belts that have progressed across the field in a predominantly south–north direction. The Mudstones of the *Mid Shale Member* (Mid-Ness) represent a period of delta abandonment and flooding of the delta plain resulting in the deposition of a thick mudstone which is correlatable across the whole field and divides the Middle Shaly Unit into Lower and Upper units.

(6) The final abandonment of the delta system is marked by the deposition of the transgressive sands of the *Upper Sand Unit* (Tarbert).

At the end of Brent deposition the most structurally elevated parts of the Brent Group were further uplifted and eroded. They were overlain by the transgressive Mid–Upper Jurassic Humber Group which forms both the seal and the source for the Brent reservoir. This truncation is confined to the eastern edge of the field and across the whole of the 'C' fault block.

Many of the eastern margin fault terraces were partially downfaulted prior to this erosion and the successive thickening of Brent into these terraces is due to preservation of the uppermost units of the Brent Group.

Faulting during the Late Jurassic/Early Cretaceous created the three main fault blocks within the Hutton field. These downthrow progressively to the northwest.

Geophysics

The 211/27 and 211/28 Licence blocks were awarded in the UK Fourth Round licensing in 1971/72. Interpretation of the reconnaissance survey data revealed a closed structure which was confirmed by the Hutton discovery well 211/28-1a drilled in July 1973. Since then the following non-regional seismic surveys have been acquired.

1975	HUT survey	
1977	C211/28	Conoco Group
1977	AUK 77 survey	Amoco Group
1978	AUK 78 survey	Amoco Group
1980	3-D survey	Hutton Unit

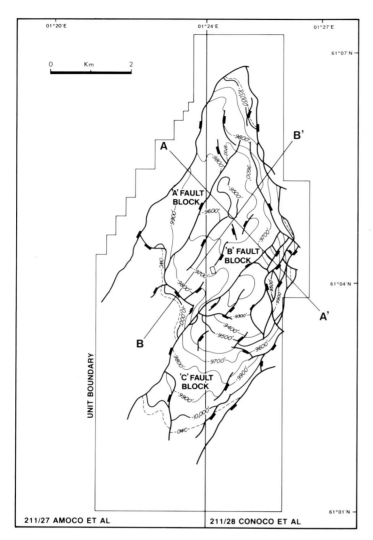

Fig. 4. Top Brent Group structure map. Contour interval is 100 ft.

The current 3-D survey, shot in 1980, covers an area of 50 km² and was shot on a NE–SW heading with a 25 m (in-line) and a 75 m (cross-line) spacing. From this data the following horizons are mapped:

(1) Base Cretaceous Unconformity;
(2) Top Brent Group;
(3) Top Dunlin Group.

Seismic data quality below the Base Cretacous Unconformity is reliable on the crest of the Hutton structure but becomes poorer across the adjacent flank areas (Figs 6 and 7).

The Hutton field is characterized by a large number of small-scale (<50 ft) intra-Brent Group faults which play a significant part in reservoir performance. These faults are below the resolution of the current seismic data.

VSP data has therefore become an important tool identifying these small faults and in improving the seismic interpretation at and near existing wells. Such data are currently available on more than 50% of the existing field wells.

Trap

The Hutton Field is divided into three major southwesterly dipping fault blocks ('A', 'B' and 'C' from north to south) outlined by SW–NE trending normal faults downthrowing to the northwest (Fig. 4)

These fault blocks are bounded in the east by a major north–south-trending fault complex which downthrows the Brent Group by approximately one thousand feet into the main graben area. Relative displacement has resulted in the most southeasterly block ('C') being structurally highest (Fig. 8).

Closure to the southwest is defined by structural dip (Fig. 9).

Within the major fault complex on the east of the field are numerous minor fault terraces which are considered to have considerable hydrocarbon potential although they are untested as yet.

Minor fault movement following the emplacement of oil has resulted in the formation of two oil–water contacts. The 'A' and 'B' fault blocks share a common contact at −10 020 ft TVSS. The 'C' fault block has its contact at −10 090 ft TVSS and has a much longer oil column than either of the other two blocks (910 ft ('C' block) 530 ft 'B' block) and 270 ft 'A' block)). The oil–water contact within the fault terraces of the eastern margin is as yet unknown but may be deeper than either of the main field contacts.

Reservoir

The Middle Jurassic Brent Group reservoir in the Hutton Field can be subdivided into six major reservoir units. A brief description of the reservoir qualities of each of these is given below. The average reservoir parameters shown in Fig. 10 and quoted below correspond to average values found in the main field hydrocarbon-bearing reservoir sands.

(1) Basal Sand Unit

This unit consists of a series of medium- to coarse-grained, poorly sorted sandstones commonly containing mud clasts, heavy minerals (e.g. garnet, hematite) and laterally extensive calcareous 'doggers'. Porosities range from 17–23% with permeabilities from 60–250 md. Lateral continuity of the sandstones is good and they exhibit good communication between wells.

(2) Mica Sand Unit

This unit consists of fine-grained, micaceous siltstones and sandstones. Reservoir quality is variable with average porosities of 20% within the net intervals but with generally low permeabilities (10–40 md). The best quality sandstones are found on the crest of the field.

(3) Massive/Mica Sand Unit

The reservoir quality of this unit is governed by its depositional environment. In the south and east of the field it is a medium–coarse-grained tidal-distributary channel sand with excellent porosity (25%) and permeability (2–3 Darcies). In the centre and west of the field the facies changes to medium-grained offshore barrier-bar sands with good porosity (22–25%) and permeability (1–2 Darcies). To the north of the field this unit thins and becomes more micaceous with porosities of 18–20% and permeabilities of 300–1000 md. The Massive/Mica reservoir sandstones are capped by a sequence of fine-grained micaceous siltstones. These siltstones vary in thickness (2–10 ft) and become an important vertical permeability barrier in the central regions of the field.

(4) Massive Sand Unit

This unit also displays variable reservoir quality across the field dependent upon its depositional environment. In the south and central regions of the field it consists of a thick (50–75 ft) section of

HUTTON FIELD

AGE		FORMATIONS	LITHO-LOGY	DEPOSITIONAL ENVIRONMENT	OPERATOR'S BRENT GROUP NOMENCLATURE	
UPPER JURASSIC	HUMBER GROUP	KIMMERIDGE FM.		OFFSHORE MARINE		
		HEATHER FM.				
MIDDLE JURASSIC	BRENT GROUP	TARBERT FM.		DELTA ABANDONMENT SHALLOW MARINE	UPPER SAND UNIT	
		UPPER NESS FM.		DELTA PLAIN DISTRIBUTARY CREVASSE AND LEVEE SYSTEM	MIDDLE SHALY UNIT	UPPER
		MID NESS		DELTA PLAIN BAY		MID-SHALE
		LOWER NESS FM.		DELTA PLAIN LAGOON SHORELINE CUT BY DISTRIBUTARY FLUVIAL CHANNEL SYSTEM		LOWER
		ETIVE FM.		DELTA FRONT UPPER SHOREFACE, FORESHORE, BACKSHORE AND TIDAL CHANNEL	MASSIVE SAND UNIT	
		ETIVE/RANNOCH FM.		DELTA FRONT NEARSHORE BAR AND EBB TIDAL MOUTHBAR	MASSIVE/MICA SAND UNIT	
		RANNOCH FM.		DELTA FRONT MIDDLE AND LOWER SHOREFACE	MICA SAND UNIT	
		BROOM FM.		PRE-DELTA SUB-LITTORAL SHEET SANDS	BASAL SAND UNIT	
LOWER JURASSIC	DUNLIN GROUP	DRAKE FM.		OFFSHORE MARINE		
		COOK FM.				
		BURTON FM.				

Fig. 5. Lithostratigraphic section of the Hutton Field.

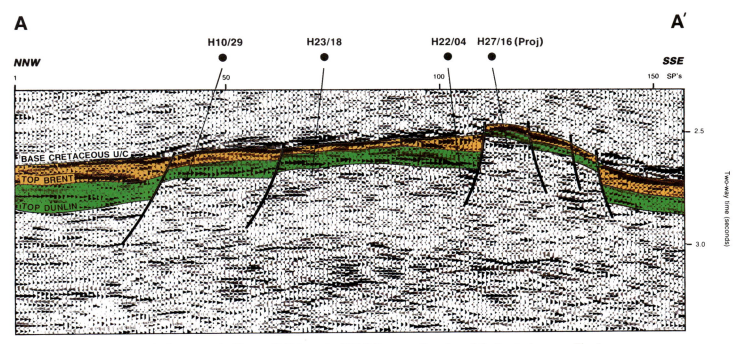

Fig. 6. Strike-orientated seismic line across the Hutton Field from the 1980 3-D survey. Location of the line is shown on Fig. 4.

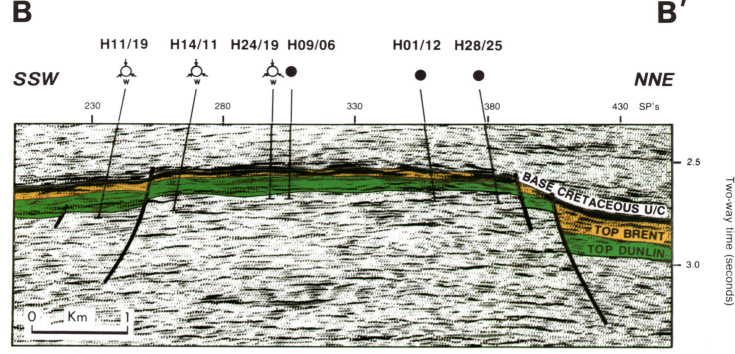

Fig. 7. Dip-orientated seismic line across the 'B' fault block in Hutton Field. Location of the line is shown in Fig. 4.

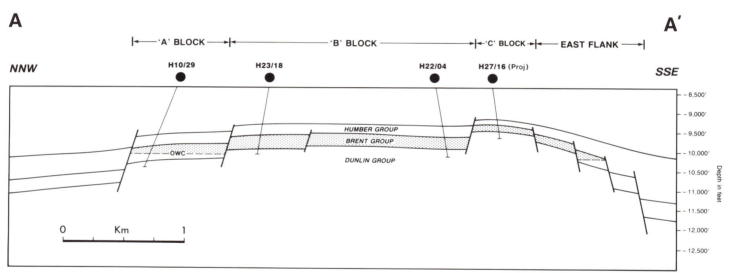

Fig. 8. Cross-section A–A' showing main structural elements of the Hutton Field. Location of the section is shown on Fig. 4.

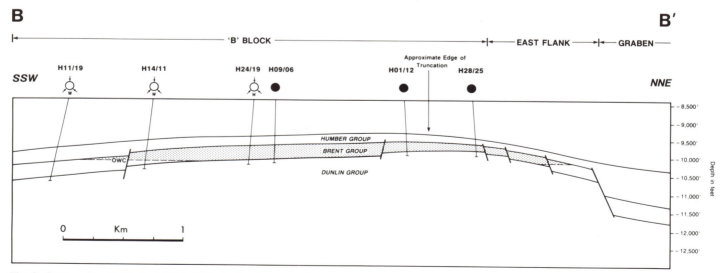

Fig. 9. Cross-section B–B' along 'B' fault block showing truncation edge. Location of this section is shown on Fig. 4.

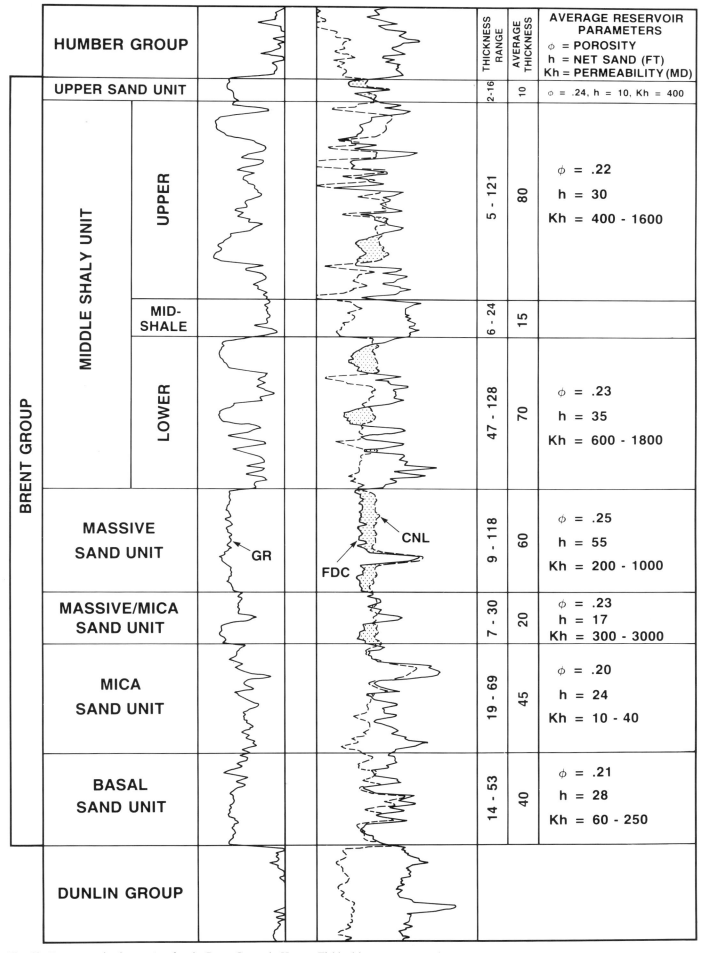

Fig. 10. Representative log section for the Brent Group in Hutton Field with average reservoir parameters.

fine–medium-grained generally featureless sandstones representing deposition of a thick barrier bar complex with foreshore and backshore sands. Porosities vary from 20–25% with permeabilities from 200–1000 md. Towards the north and east of the field the Massive Sand Unit thins as a result of rapid delta progradation northwards across the Hutton area. The Massive Sand Unit is represented by tidal channel sandstones which have excellent porosities (25%) and permeabilities (1–3 Darcies). The upper part of the thick Massive Sand Unit found in the south passes laterally northwards into a sequence of delta plain depositions similar to the overlying Middle Shaly Unit.

(5) Middle Shaly Unit

This unit comprises sequences of lagoonal siltstones, sandstones, coals and claystones cut by a series of fluvial channels. The best reservoir quality is found within these channels with porosities of 25% and permeabilities of 1600–1800 md. These channels are confined to narrow meander belts that cross the Hutton area in a predominantly south–north direction. Communication from west to east across the field is dependent upon the nature of the interchannel sediments. In the Lower Middle Shaly Unit these sediments are predominantly lagoonal sandstones and siltstones with porosities of 20–25% and permeabilities of 600–800 md. West to east communication at this level is therefore good. The interchannel areas of the Upper Middle Shaly Unit, however, are a series of lagoonal claystones and siltstones. West to east communication is therefore poor at this level. A thick (10–20 ft) correlatable claystone, the *Mid-Shale Member* which divides the Middle Shaly Unit into Upper and Lower units is a non-reservoir unit.

(6) Upper Sand Unit

Deposition of this unit marked the abandonment of the delta system. It was deposited as a transgressive sheet sand across the field. The sandstones are medium- to coarse-grained and poorly sorted with modest porosities (20–24%) and relatively low permeabilities (400 md).

The Brent Group sequence was followed by deposition of transgressive Mid–Upper Jurassic Humber Group shales. The transgressive eroded the uppermost units of the Brent Group in the east of the 'A' and 'B' blocks and across the entire 'C' block.

The phase of late stage secondary diagenesis has resulted in the creation of quartz overgrowths and pore filling illite and kaolinite. This secondary diagenetic phase has been confined to sandstones below the present oil–water contact.

Source

The source and the seal for the hydrocarbon bearing Brent Group sandstones in the Hutton field are the overlying Upper Jurassic Kimmeridge Fm clays. During the Late Cretaceous–Middle Tertiary the thick sequences of organically rich Kimmeridge Fm clays in the graben to the east matured and the oil that was produced migrated into the Hutton Field. It is commonly considered that the oil migrated into the field from the south. However, migration from the east via the flanking terraces is also considered likely.

Hydrocarbons

A drill stem test in the Brent Group Massive Sand Unit of the 211/28-1a(ST) discovery well flowed 4506 BOPD through a 1/2 in choke. The first development well H01, drilled prior to platform installation was tied back, perforated and brought on stream on 6 August 1984. Initial production rate was 6000 BOPD from the Basal and Mica Sand Units. The field as a whole reached peak production of 107 000 BOPD in December 1984. However, owing to the limited aquifer support and the lack of immediate water injection, this rate could not be sustained. Subsequently the production rate declined to alarming levels until sufficient water injection was able to restore an average production rate of 85 000 BOPD. Current production rate is 50 000 BOPD (31 March 1989).

Two separate oil–water contacts are present in the Hutton Field (−10 020 ft TVSS for 'A' and 'B' blocks and −10 090 ft TVSS for 'C' block). The hydrocarbons are, however, similar in both these accumulations being of low sulphur content with no H_2S and with a gravity of 34.5° API. There is no gas cap on Hutton and the gas/oil ratio is low.

Reserves

Original oil in place for the Hutton Field is estimated to be 550 MMBBL. The recoverable reserves have been calculated with the main field fault blocks only. It is estimated that a further 60 MMBBL may exist within the as yet undrilled downfaulted fault terraces on the eastern margins of the main field.

Production from the Hutton Field commenced in August 1984 with the tie-back of the 10 predrilled wells. Delays in injection support and unexpected reservoir quality and performance, however, resulted in lower than expected production rates. Cumulative production from the Hutton Field reached 100 MMBBL in November 1988. Current rates (March 1989) are 50 000 BOPD from 16 wells with 12 wells injecting water (Fig. 11).

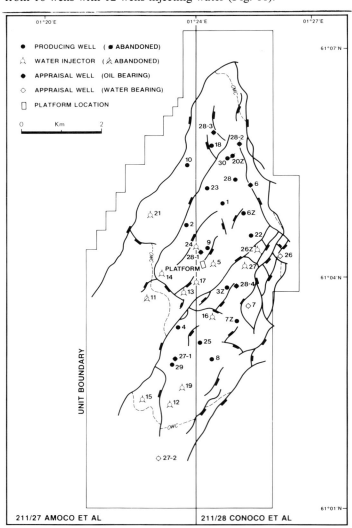

Fig. 11. Main structural elements, hydrocarbon limits and well locations of Hutton Field.

The author would like to thank the Hutton Group of companies for their cooperation and permission to publish this report. Particular thanks go to J. Smith and P. McCloskey of Conoco UK Ltd for their contributions. Special thanks to the Conoco UK Exploration draughting group for their help in draughting the text figures.

Hutton Field data summary

Trap
Type	Tilted fault block
Depth to crest	9750 ft TVSS ('A' Fault Block)
	9490 ft TVSS ('B' Fault Block)
	9180 ft TVSS ('C' Fault Block)
Oil–water contact	10 020 ft TVSS ('A' and 'B' Fault Blocks)
	10 090 ft TVSS ('C' Fault Block)
Oil column	270 ft ('A' Fault Block)
	530 ft ('B' Fault Block)
	910 ft ('C' Fault Block)

Pay zone
Formation	Brent Group
Age	Middle Jurassic
Gross thickness (average: range)	(275 ft : 150–400 ft)
Net/gross ratio (average: range)	(0.595 : 0.375–0.784)
Porosity (average: range)	(22% : 17.4–26.3%)
Hydrocarbon saturation (average: range)	(67% : 47.9–87.2%)
Permeability	(100–1500 md, 4500 md maximum)
Productivity index	15 BBL/day/psi

Hydrocarbons
Oil gravity	34.5° API
Oil type	Low sulphur crude
Bubble point	715 psig ('A' and 'B' Fault Blocks)
	800 psig ('C' Fault Block)
Gas/oil ratio	140 SCF ('A' and 'B' Fault Blocks)
	160 SCF ('C' Fault Block)
Formation volume factor	1.12 ('A' and 'B' Fault Blocks)
	1.125 ('C' Fault Block)

Formation water
Salinity	14 000 mg/l
Resistivity	0.3556 ohm m at 60°F

Reservoir conditions
Temperature	225°F/107°C
Pressure, initial	6300 psi
Pressure gradient in reservoir	0.338 psi/ft

Field size
Area	6600 acres
Gross rock volume	848×10^3 acre.ft
Hydrocarbon pore volume	80.2×10^3 acre.ft
Recovery factor	34%
Drive mechanism	Water injection, aquifer support
Initial recoverable oil	190 MMBBL

Production
Start-up date	Drilling May 1981, Production 1984
Development scheme	Tension leg platform, 32 slots
Production rate	Peak 107 000 BOPD (1984)
	March 1989 49 400 BOPD
Cumulative production (31 March 1989)	107.2 MMBBL

References

BADLEY, M. E. & PRICE, J. D. *et al.* 1988. The structural evolution of the Viking Graben and its bearing upon extensional modes of basin formation. *Journal of the Geological Society, London*, Vol. 145, pp. 455–472.

BROWN, S., RICHARDS, P. C. & THOMSON, A. R. 1987. Patterns in the deposition of the Brent Group (Middle Jurassic) UK North Sea. In BROOKS, J. and GLENNIE, K. (eds) *Petroleum Geology of Northwest Europe*, Graham & Trotman, pp. 899–913.

BUDDING, M. C. & INGLIN, H. F. 1981. A Reservoir Geological Model of the Brent Sands in Southern Cormorant. *Petroleum Geology of the Continental Shelf of North-West Europe*, pp. 326–334.

Northwest Hutton Field, Block 211/27, UK North Sea

L. H. JOHNES & M. B. GAUER

Amoco (UK) Exploration Co., Amoco House, Westgate, Ealing, London W5 1XL, UK

Abstract: The Northwest Hutton oil field lies in the East Shetland Basin entirely within UK Block 211/27. It was discovered by the Amoco Group 211/27-3 well in April 1975 and is operated by Amoco (UK) Exploration Company. The structure is a series of tilted fault blocks. The reservoir sandstones are the Brent Group Sandstones of Middle Jurassic age deposited in a progradational deltaic environment. The Brent Group sandstones vary in thickness from 320–500 ft over the field. Average porosity ranges from 20% at the crest of the structure to 13% on the western flank with permeability decreasing from an average of 267 md to 1 md. Recoverable oil is estimated to be 145 MMBBL. A total of 26 producing and 9 water injection wells have been drilled from a single steel jacket platform. Production peaked in May 1983 at 86 680 BOPD.

The Northwest Hutton oil field is located 80 miles northeast of the Shetland Islands in 475 ft of water within UK Block 211/27 (Fig. 1). The field has an area of 13 440 acres in which recoverable oil is estimated to be 145 MMBBL. The structure is formed by a series of tilted fault blocks and is fault bounded on the north and east. The reservoir is composed of Middle Jurassic Brent Group sandstones and was formed by the northeastwards progradation of a large deltaic system. The name of the field is derived from its location with respect to the Conoco Hutton Field, named after James Hutton (1726–97) the Scottish geologist who defined some of the fundamental principles of modern geology.

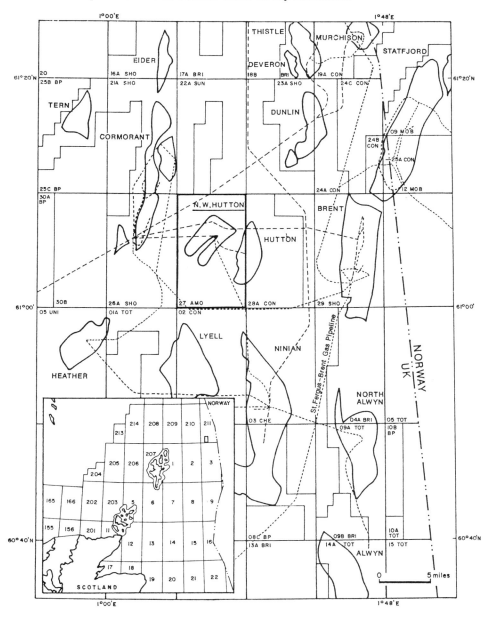

Fig. 1. Northwest Hutton Field location map.

History

Block 211/27 was awarded to the Amoco operated Group as licence P.184 during the UK 4th offshore licensing round in March 1972. The members of the Amoco Group are:

[1] Amoco (UK) Exploration Company	25.77%	(Operator)
[2] Enterprise Oil plc	25.77%	
Mobil North Sea Ltd	20.00%	
Amerada Hess (UK) Ltd	18.08%	
Texas Eastern North Sea Inc	10.38%	

The Northwest Hutton Field was discovered in April 1975 by the 211/27-3 well which encountered a 396 ft reservoir section at a depth of −11 100 ft TVSS. The discovery well was drilled to test the crest of a tilted fault block structure which was appraised by the drilling of a further eight appraisal wells in the period up to 1983. The 211/27-5 well was drilled downdip on the western flank of the structure and gave the first evidence of truncation of the Brent Group by the mid-Jurassic unconformity within the field. A downdip deterioration in reservoir quality was also seen in this well in comparison with the 211/27-3 well. The next appraisal well 211/27-6 was drilled in a central location and tested oil down to −11 895 ft TVSS compared with oil down to −11 998 ft TVSS in the 211/27-4a well indicating that the field did not have a single oil water contact. Wells 211/27-7 and 211/27-8, both dry holes, defined the eastern and western limits of the field.

The appraisal wells 211/27-9ST and 211/27-10 were both drilled in the established area of the field. The final appraisal well 211/27-11 was intended to test the southern limit of the field on the west flank and was drilled in 1983 after the field came on stream. An oil–water contact was established by this well at −12 432 ft TVSS.

Development of the Northwest Hutton Field began in October 1979. The first seven development wells were drilled through a 20 slot seabed template, prior to the installation of the fixed platform in 1981. The pre-drilling of the first production wells, which were tied back to the platform, enabled production to commence immediately the wells had been completed and perforated. This was the first time this technique had been used in the North Sea resulting in a considerable time saving in the initial phase of field development.

The platform is of conventional steel design with the capacity to handle production of 100 000 BOPD and 70 000 BWPD. The field came on production in April 1983 and peaked in May 1983 at 86 680 BOPD when the last of the predrilled wells was completed. Further development drilling took place from the platform with 26 production and 9 water injection wells drilled in total. Secondary recovery commenced in February 1984 with water injection in a line drive pattern and gas lift was initiated in October 1984.

Field stratigraphy

The Northwest Hutton Field is situated in the East Shetland Basin which forms a broad relatively shallow terrace within the Viking Graben. The oldest sedimentary rocks found within the basin are of Devonian age (Deegan & Scull 1977), lying unconformably on a basement of gneiss, granodiorites and granite. The Devonian is overlain unconformably by the Permo-Trias, the Carboniferous apparently being absent in the East Shetland Basin. The oldest rocks penetrated in the Northwest Hutton Field are Triassic red beds of the Cormorant Formation (Morton et al. 1987) (Fig. 2). Fluvial conditions, though probably in a less arid environment, continued with deposition of the Lower Jurassic Statfjord Formation. The end of fluvial deposition was marked by a regional marine transgression with thick marine mudstones of the Dunlin Group being deposited unconformably on the Statfjord Formation. The Middle Jurassic Brent Group Sandstones were deposited by a major deltaic system which prograded from southwest to northeast across the area (Richards et al. 1988). The Upper Jurassic is represented by marine shales as are the Lower and Upper Cretaceous. Sedimentation in the Tertiary was again dominated by the deposition of claystones.

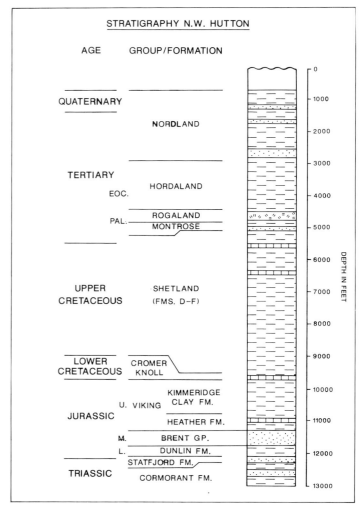

Fig. 2. Stratigraphic column, Northwest Hutton Field.

The East Shetland Basin is dominated by two major normal fault systems, a NW–SE-trending system that appears to have been initiated in the Permo-Trias and an older Caledonide NE–SW-trending system (Challinor & Outlaw 1981; Eynon 1981; Threlfall 1981). North–south faults are also evident which appear to be related to the earliest phase of rifting in the Permian to Early Triassic (Badley et al. 1988). Rifting of the North Sea in the Permo-Triassic to early Jurassic led to the formation of the Viking Graben and led to reactivation of the two main fault systems during the Jurassic and Cretaceous. Interaction of the two systems resulted in the formation of complex tilted fault block structures. Movement of the faults acted as a major control on sedimentation during the Jurassic and early Cretaceous. Faulting was dominantly normal but an element of wrench faulting is evident on the NE–SW and NW–SE-trending faults.

A second phase of rifting during the Upper Jurassic to Early Cretaceous gave rise to further faulting. Synsedimentary fault movement which began during the Middle Jurassic (Brown et al. 1987), is indicated by growth across faults and resulted in broad thickness variations in the individual formations. Further movements on these faults took place in the Late Cretaceous, the

---
[1] Formerly Amoco (UK) Petroleum Ltd.
[2] Enterprise Oil plc replaced Gas Council (Exploration) Ltd, following sale of the latter company's oil assets in 1984.

Palaeocene and the Miocene, accompanied by regional subsidence which has continued to the present.

Geophysics

Northwest Hutton Field was discovered with the 211/27-3 well in 1974. The block was loosely covered by several vintages of 2-D lines, 1972 to 1974. The unmigrated 24-fold and 30-fold data was of sufficiently good quality to define a structural high at the Top Kimmeridge, with suggestions of a tilted fault block beneath. Following the discovery, additional 2-D surveys were added each year from 1975 to 1978. With increased fold and migration the resolution of the Jurassic reflectors and faults improved. In 1979 a 3-D survey of 1752 line kilometres was shot by GSI using airguns. The survey was reprocessed by Digicon in 1984 to improve data quality by removing a seabed multiple which was obscuring reflections below the Base Cretaceous Unconformity. A second 3-D survey was carried out in 1984 by SSL using waterguns over the southern portion of the block, and overlapping the earlier 3-D survey. This later survey of 3182 line kilometres was processed by SSL on completion of the survey.

Interpretation of the 3-D surveys was carried out largely on a SCITEX RESPONSE 800 workstation. Identification and mapping of the Top Brent Sandstone is made difficult by the lack of a strong or consistent seismic reflector as a result of the variable and generally small sonic and density contrasts between the top Brent Sandstone and the overlying Viking Group Shales. Although attenuated, a water bottom peg leg multiple 200 m below the strong Base Cretaceous Unconformity reflector can occasionally interfere with the Top Brent Sandstone seismic pick. In the western part of the field, where the Upper Jurassic section is greater than 200 m thick the Top Brent Sandstone reflector is more consistent. Identification of major faults is generally not in doubt, however correlating across large faults can be problematic because of character changes associated with the Top Brent Sandstone reflector. Conversion from time to depth was accomplished by generation of an apparent velocity map to the Top Brent Sandstone, with contours based on stacking velocities and tied to well control.

The Northwest Hutton Field is mapped as a series of northeast to southwest oriented tilted fault blocks (Figs 3 and 4). In the east and centre of the field the fault blocks are relatively flat lying with dips of 6° to the west, increasing to westerly dips of 9° in the west of the field. The field is bounded to the west by a syncline with an axis running northeast to southwest. A major fault zone bounds the field to the north with a combined maximum throw down to the north of 800 ft.

Trap

The Northwest Hutton Field comprises a series of complex tilted fault blocks formed in the Middle to Upper Jurassic by interaction between the major northeast–southwest and northwest–southeast fault trends. Differential subsidence during the Upper Jurassic to Lower Cretaceous led to the development of structural dip of the field to the south and west (Fig. 5). The productive area of the field is bounded by three structural elements:

(i) a major NE–SW-trending fault on the eastern side which downthrows Brent Group Sandstones against Lower Jurassic Dunlin Group Shales;
(ii) a major NW–SE fault downthrowing to the north which juxtaposes the Brent Group against Upper Jurassic Shales on the northern side;
(iii) structural dip to the south and west.

The field is further divided into four distinct productive areas, the Eastern, Central, and Inner and Outer Western Lobes, bounded by northeast–southwest-trending faults parallel to the eastern boundary faults.

The trap is sealed vertically by shales of the Heather and Kimmeridge Clay Formations of Middle to Upper Jurassic age and the whole structure is draped by thick Cretaceous shales which create a regional seal.

No single oil–water contact is found in the Northwest Hutton Field, with progressively higher contacts from $-12\,934$ ft TVSS in the west to $-11\,930$ ft TVSS in the east.

Reservoir

The reservoirs of the Northwest Hutton Field are the Middle Jurassic Brent Group Sandstones of Bajocian–Bathonian age and can be subdivided into five formations (Deegan & Scull 1977) (Fig. 6). Two depositional systems are recognized: a lower system rep-

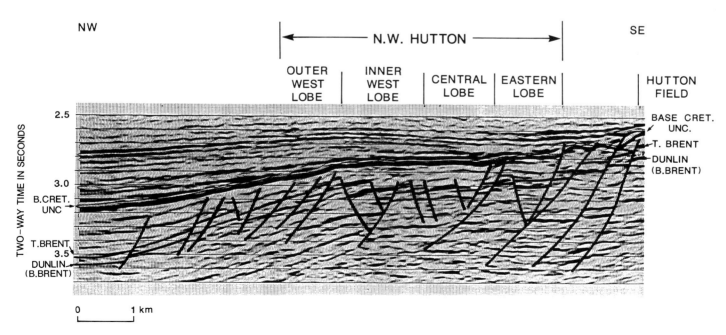

Fig. 3. NW–SE seismic line (Track 175).

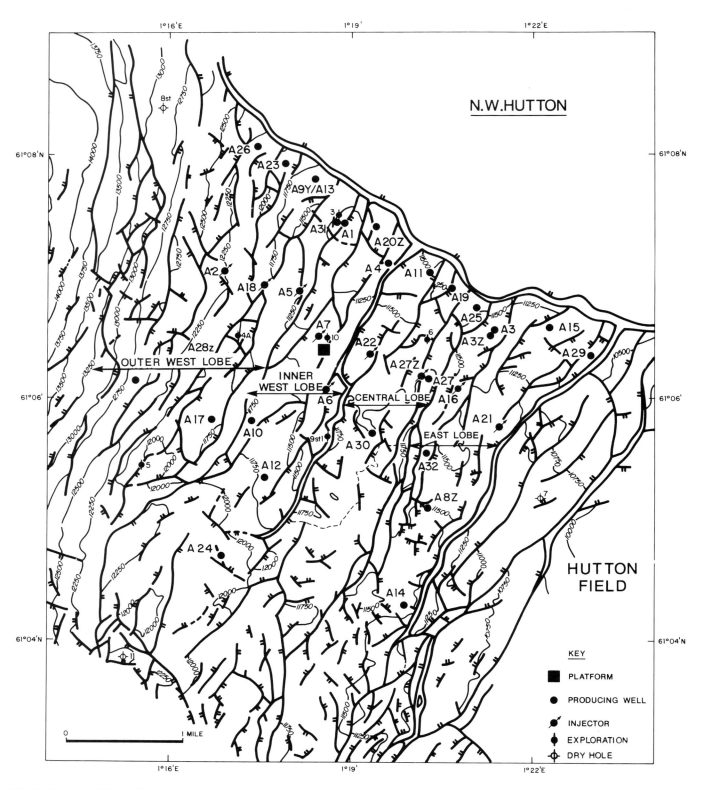

Fig. 4. Northwest Hutton Field Top Brent Group Sandstone structure map.

resented by the coarse basal sandstones of the Broom Formation deposited by an easterly prograding fan delta system (Brown *et al.* 1987) and an upper system composed of a regressive to transgressive clastic wedge comprising the Rannoch to Tarbert Formations.

Broom Formation

The Broom Formation is the basal member of the Brent Group and comprises a coarse-grained, poorly sorted, often bioturbated, subarkose of Bajocian age. The sandstones rest on the underlying Dunlin Group shales with either a sharp disconformable contact or a transitional boundary. The lower section of this unit is rarely burrowed and bioturbated sandy claystones are present. The overlying beds are fine- to medium-grained with horizontal to low angle cross stratification.

The upper part of the formation comprises interbeds of the previous lithology and micaceous siltstone to very fine sandstone with lenses and bands of burrowed sandy claystone. The sandstones are predominantly cemented by silica but locally early calcite

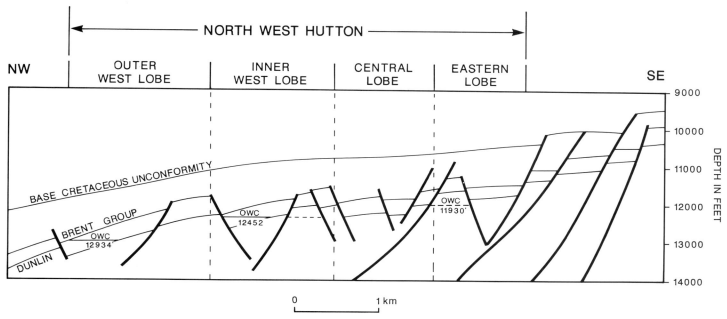

Fig. 5. Structural cross-section along Track 175.

cementation has totally destroyed porosity. The Broom Formation comprises a continuous sheet sandstone body across the field varying in thickness from 30 ft in the east to 60 feet in the west. No marine macro- or micro-organisms have been found in these sandstones, and the depositional environment as a sub-littoral sheet sand has been inferred from the common burrowing at the top of the sandstone and its sheet like distribution throughout the field.

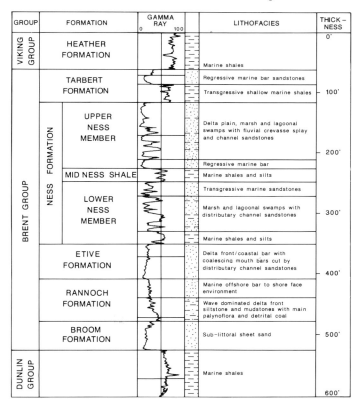

Fig. 6. Brent Group Sandstones lithofacies.

Rannoch Formation

The lower part of the formation consists of mudstones with thin interbedded micaceous sandstones which are commonly bioturbated. The formation coarsens and thickens upwards to very fine-grained sandstones often with low angle cross stratification and micaceous laminae. Bioturbation decreases upwards and locally thin intervals 1–6 ft thick are calcite cemented.

The Rannoch Formation represents the progradation of the Brent Delta across the Northwest Hutton area, the basal shaly unit being a low energy marine sequence characteristic of a wave dominated delta front. The upper part of the unit which comprises well sorted sandstones with wave lamination, scours, and hummocky cross stratification (Richards & Brown 1986), suggests deposition in a marine offshore bar to shoreface environment. The sandstones of the Rannoch Formation show good lateral continuity throughout the field ranging from 30–80 ft thick.

Etive Formation

The coarsening upward sequence of the Rannoch Formation is continued into the Etive Formation, with grain size typically grading from fine to medium grained. Within the Etive Formation local carbonaceous rip up clasts, wood and plant fragments occur while burrows are rare to absent. The presence of these indicates proximal deposition at the river mouths of the delta where wave conditions were sufficient to redistribute part of the sand along depositional strike, but not strong enough to winnow out the carbonaceous mud clasts and plant debris. Locally distributary channel sandstones have cut into the top of the barrier bar deposits giving rise to a fining upward sequence. The channel sequence has a sharp erosive base containing rip up clasts and is succeeded by horizontal to cross stratified medium-grained sandstones. These sandstones fine upwards to a fine-grained ripple-laminated sandstone, with carbonaceous shale laminae capped by rootlet layer with coal indicative of a channel abandonment facies. The thickness of the Etive Formation ranges from 25–80 ft over much of the field but increases to over 150 ft in the extreme southwest.

Ness Formation

In the Northwest Hutton Field the Ness Formation is subdivided into upper and lower members by the laterally extensive Mid Ness Shale Member.

Lower Ness Member. The main sandstone units present in the Lower Ness Member coarsen upwards from very fine ripple-stratified sandstone to cross-stratified fine-grained sandstone capped by a root mottled layer and a thick coal bed. Carbonaceous plant and woody material is common with burrows present locally but restricted to the lower part of the sequence. These crevasse splay sands, while individually non-persistent, may coalesce laterally to form sheet sandstones infilling large areas of the lagoon/bay.

Also present in the Lower Ness Member are sandstones interpreted as distributary channel sandstones. The base of these sandstones is erosional with rip-up carbonaceous mud clasts overlain by coarse-grained cross-bedded sandstones fining upwards to fine-grained ripple-laminated sandstones. At the top of the sequence are burrowed rootlet layers representing channel abandonment. The channel sandstones have limited lateral continuity but appear to be interconnected with the laterally equivalent crevasse splay sandstones and may incise into the crevasse splay sandstones.

The deltaic deposits of the Lower Ness Member are overlain by a transgressive marine sandstone. This sandstone, which is medium- to coarse-grained and cross-bedded, has excellent lateral continuity and can be traced across the whole field.

Mid Ness Shale Member. The shales of the Mid Ness Shale Member are wavy to lenticular bedded with wave ripple cross lamination, are locally burrowed and contain marine palynomorphs. The presence of these structures and their widespread areal extent, suggests an overall transgressive phase of deposition in a shallow marine environment.

Upper Ness Member. Deposition of the Mid Ness Shale Member was ended by a regressive event resulting in the deposition of a marine sandstone 20 ft thick at the base of the Upper Ness. The sandstone coarsens upwards from ripple-laminated fine-grained sandstones to a flat-bedded fine- to medium-grained sandstone capped by a rootlet layer and thin coals. The regression represents renewed progradation of the delta across the shallow marine Mid Ness deposits. Overlying the regressive marine sandstones are a series of delta plain sediments. Depositional facies comprise fine- to coarse-grained distributary channel sandstones 10–30 ft thick and fine-grained ripple-laminated crevasse splay sandstones with interbedded silty claystones and coal. The channel sandstones have a poor lateral continuity and a strong unidirectional trend in a NE–SW direction, parallel to the major fault trends, is inferred from log correlation. Interconnection of the channel sandstones occurs where they overlap or cut into the crevasse splay sandstone bodies.

Tarbert Formation

The Ness Formation is overlain by a transgressive marine shale representing the end of deltaic deposition and the onset of marine conditions. The shale averages 20 ft thick and is present over the whole of the field except where truncated by post Brent Group erosion. Overlying the shale is a regressive marine sandstone, with an erosive base which coarsens upwards with an associated decrease in bioturbation.

Ripple and wave cross stratification is common. Grain size is variable, ranging from very fine to very coarse. The Tarbert sandstone is a laterally continuous sheet sand.

The distribution of reservoir properties in the Northwest Hutton Field is strongly influenced both by depositional environment and by diagenetic factors which become more important with depth.

Reservoir quality in the Broom Formation is highly variable; permeability averages 17 md, and ranges from 10–100 md, while porosities range from 13–23%, averaging 14%. The basal shaly facies of the Rannoch Formation forms an extensive permeability barrier between the Rannoch and Broom Formations. Within the Rannoch Formation, porosity and permeability range from 10–20% and 0.1–100 md respectively. Locally, porosity is destroyed by carbonate cemented concretions. Porosity and permeability increase vertically from 5–18% and 0.1–100 md respectively in the coastal barrier bar facies of the Etive Formation. An upwards decrease in porosity and permeability is associated with the distributary channel sandstone facies of the Etive Formation, from 20–23% and 200–1000 md respectively at the base, to 17–20% and 10–15 md respectively at the top of the unit. A similar porosity–permeability relationship is seen in the distributary channel sandstones of the Ness Formation, while permeabilities of 1 md and porosities of 10–15% are typical of the crevasse splay lobe sandstones. The regressive and transgressive marine sandstones of the Ness Formation are of relatively uniform reservoir character, with an average porosity of 20% and permeabilities of 100–250 md. The Tarbert Formation shows a variation in permeability from 14–300 md although porosity is relatively uniform, averaging 20% across the field.

A diagenetic control on reservoir quality with depth is evident in all formations of the Brent Group. Porosity decreases from an average of 20% at the east of the structure to 12.7% on the western flank. The porosity reduction can be attributed to the increasing degree of quartz overgrowth cementation with increasing depth (Scotchman *et al.* 1989). Permeability also shows a decrease with depth from 296 md to 1 md, with a marked decrease in permeability occurring below 12 000 ft TVSS. The primary diagenetic control on permeability is illite cementation with a distinct increase in percentage of illite taking place at 12 000 ft TVSS.

The subdivision of the Northwest Hutton field into four lobes is primarily based on the presence of sealing faults and the variable oil–water contacts (Fig. 5). The identification of faults taken to be sealing is by a combination of well performance, pressure depletion and water injection response. The deterioration of reservoir quality with depth is most marked in the two western lobes where the deepest oil–water contacts are found.

Source

The hydrocarbon source rock for the Northwest Hutton Field is the Late Jurassic to Early Cretaceous Kimmeridge Clay Formation which exceeds 1200 ft in thickness. Within the vicinity of Northwest Hutton Field the formation has a total organic carbon content of 4.5–6.5%. Hydrocarbon generation occurred to the southwest of the field and maturation modelling indicates that peak oil generation was attained in the early Tertiary (Goff 1983). Lateral migration across faults allowed the hydrocarbons while moving structurally up dip to migrate into the stratigraphically lower Brent Group reservoir.

Hydrocarbons

The 211/27-3 discovery well tested oil at rates of between 6540 and 8160 BOPD with 2 million cu.ft of gas per day. The oil is 37° API gravity and under-saturated with a GOR of 800 BBL/SCF in the Outer West Lobe and 600 BBL/SCF in the East Lobe of the field. The reservoir is overpressured with an initial reservoir pressure of 7563 psi in the Outer West Lobe and 7315 psi in the East Lobe at 11 500 ft TVSS. Formation water is of relatively low salinity ranging from 19 700–25 600 ppm of total solids. Reservoir temperature is 245°F at the crest of the structure increasing to 260°F on the western flank.

Reserves

The Northwest Hutton Field contains an estimated 145 MMBBL of recoverable oil. Development drilling commenced in 1979 with 7 wells being drilled through a seabed template and subsequently tied back to the fixed platform. Production start-up was in April 1983 and peaked in May 1983 at 86 680 BOPD on completion of the last of the pre-drilled wells. The reservoir pressure and solution gas flow declined rapidly with initial production and water injection began in February 1984. A line drive pattern was selected with wells placed in the centre of the field with an orientation to permit sweep in a direction parallel to the northeast–southwest-trending fault system. A peripheral flood pattern was discounted because of the permeability degradation on the flanks of the field. Gas lift was initiated in October 1984. Oil production is taken via a 20 inch pipeline to the Cormorant A platform joining the Brent system flowline to Sullom Voe. Gas output is by a 10 inch spur line to the WELGAS link to Brent Field and the FLAGS system.

Amoco and the partners in the Northwest Hutton Group (Amerada Hess, Enterprise Oil, Mobil North Sea and Texas Eastern North Sea) are thanked for their permission to publish this work.

NW Hutton Field data summary

Trap	
Type	Tilted fault block
Depth to crest	11 000 ft TVSS
Oil–water contact	12 932 ft TVSS
Oil column	1932 ft
Pay zone	
Formation	Brent Group
Age	Middle Jurassic (Bajocian–Bathonian)
Gross thickness	320 to 500 feet
Net/gross ratio (average, range)	45%, 25–60%
Porosity	16%, 8–24%
Hydrocarbon saturation	45%, 20–90%
Permeability	0.1–2000 md
Hydrocarbons	
Oil gravity	37° API
GOR	600 BBL/SCF (Eastern fault block)
	800 BBL/SCF (Western fault block)
Formation volume factor	1.38 STB/RB (Eastern fault block)
	1.42 STB/RB (Western fault block)
Formation water	
Salinity	23 000 ppm
Resistivity	0.28 ohm m at 75°F
Reservoir conditions	
Temperature	245°F at 11 000 ft
Pressure	7315 psi (Eastern fault block) at 11 500 ft
	7563 psi (Western fault block) at 11 500 ft
Pressure gradient	0.64–0.66 psi/ft
Field size	
Area	13 440 acres
Gross rock volume	5 376 000 acre ft
Drive mechanism	Primary, solution gas
Recoverable oil	145 MMBBL
Production	
Start-up date	April 1983
Development scheme	Single steel platform
Production rate	Peak rate 86 680 BOPD (May 1983)
Production wells	25
Injection wells	9
Secondary recovery method	Water flood

References

BADLEY, M. E., PRICE, J. D., DAHL, C. R. & AGDESTEIN, T. 1988. The structural evolution of the northern Viking Graben and its bearing upon extensional modes of basin formation. *Journal of the Geological Society, London*, **145**, 455–472.

BROWN, S., RICHARDS, P. C. & THOMPSON, A. R. 1987. Patterns in the deposition of the Brent Group (Middle Jurassic) UK North Sea. *In*: BROOKS, J. & GLENNIE, K. W. (eds) *Petroleum Geology of North West Europe*. Graham & Trotman, 899–914.

CHALLINOR, A. & OUTLAW, B. D. 1981. Structural evolution of the North Viking Graben. *In*: ILLING, L. V. & HOBSON, G. D. (eds) *Petroleum Geology of the Continental Shelf of North West Europe*. Heyden, London, 104–109.

DEEGAN, C. E. & SCULL, B. J. 1977. *A proposed standard lithostratigraphical nomenclature for the Central and Northern North Sea*. Report of the Institute of Geological Sciences, 77/25.

EYNON, G. 1981. Basin development and sedimentation in the Middle Jurassic of the Northern North Sea. *In*: ILLING, L. V. & HOBSON, G. D. (eds), *Petroleum Geology of the Continental Shelf of North West Europe*. Heyden, London, 196–204.

GOFF, J. C. 1983. Hydrocarbon generation and migration from Jurassic source rocks in the E. Shetland Basin and Viking Graben of the northern North Sea. *Journal of the Geological Society, London*, **140**, 445–474.

MORTON, N., SMITH, R. M., GOLDEN, M. & JAMES, A. V. 1986. Comparative stratigraphic study of Triassic–Jurassic sedimentation and basin evolution in the northern North Sea and north-west of the British Isles. *In*: BROOKS, J. & GLENNIE, K. W. (eds) *Petroleum Geology of North West Europe*. Graham & Trotman, London, 697–709.

RICHARDS, P. C. & BROWN, S. 1986. Shoreface storm deposits in the Rannoch Formation (Middle Jurassic), North West Hutton Oilfield. *Scottish Journal of Geology*, **22**, 367–375.

——, ——, DEAN, J. M. & ANDERSON, R. 1988. A new palaeogeographic reconstruction for the Middle Jurassic of the northern North Sea. *Journal of the Geological Society, London*, **145**, 883–886.

SCOTCHMAN, I. C., JOHNES, L. H. & MILLER, R. S. 1989. Clay diagenesis and oil migration in Brent Group Sandstones of N. W. Hutton Field. *Clay Minerals* **24**, 339–374.

THRELFALL, W. F. 1981. Structural framework of the central and northern North Sea. *In*: ILLING, L. V. & HOBSON, G. D. (eds) *Petroleum Geology of the Continental Shelf of North West Europe*. Heyden, London, 98–103.

The Magnus Field, Blocks 211/7a, 12a, UK North Sea

M. SHEPHERD

BP Exploration, Aberdeen Operations, Dyce, Aberdeen AB2 0PB, UK

Abstract: Magnus is the most northerly producing field in the UK sector of the North Sea. The oil accumulation occurs within sandstones of an Upper Jurassic submarine fan sequence. The combination trap style consists of reservoir truncation by unconformity at the crest of the easterly dipping fault block structure and a stratigraphic pinchout element at the northern and southern limits of the sand rich fan. The reservoir is enveloped by the likely hydrocarbon source rock, the organic rich mudstones of the Kimmeridge Clay Formation.

The Magnus Oil Field is located 160 km (100 miles) NE of the Shetland Isles in Blocks 211/12a and 211/7a of the UK sector (Fig. 1). Magnus, situated near the northwestern boundary of the East Shetland Basin, is the furthest north of the producing fields in the North Sea (Latitude: 61° 37′ N). The reservoir (plan area: 33 sq km /8154 acres) lies at a depth of approximately 3000 m (9800 ft) subsea and comprises Upper Jurassic submarine fan sandstones in an eastwards dipping fault block (Fig. 2). The field is named after Saint Magnus of the Orkney Isles.

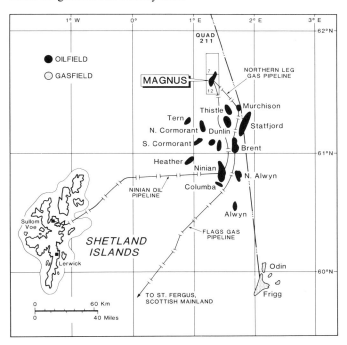

Fig. 1. Location of the Magnus Field.

History

UK licence Blocks 211/12 and 211/7 were awarded as 100% interest blocks to BP in the fourth round of offshore licences in 1972. In 1989, BP assigned certain fixed interests to four partner companies. The current licence interests are BP Petroleum Development Ltd 85%, Repsol Exploration (UK) Ltd 5%, Sun Oil Britain Ltd 5%, Brasoil UK Ltd 2.5%, Goal Petroleum plc 2.5%.

The rig SEDCO 703 spudded the discovery well 211/12-1 in March 1974. The well programme was designed to test a pre-Cretaceous fault block play in an area of the North Sea where recent major oil finds in the Middle Jurassic Brent sandstone had been announced. The well encountered 63 m (207 ft) net oil-bearing Upper Jurassic sandstone at a top reservoir depth of 2896 m (9501 ft) sub-sea. The prospective Brent Group sandstones were water bearing.

The subsequent appraisal stage saw six wells drilled in a period between 1975 and 1978. It became apparent that although the field was large with likely recoverable reserves of over 400 MMBBL, it would be difficult to develop. The oil field was shown to be long and thin (14 km (8.7 miles) north–south, 3.5 km (2.2 miles) east–west) and given a maximum platform drilling radius of about 3 km (1.7 miles), a two platform development was indicated. However, the sea depth in the Magnus area is 186 m (610 ft) and with the cost of constructing two jackets for this depth of water, the plan would have been uneconomical. Ultimately the decision was made to install a single fixed steel platform in the centre of the field with the northern and southern extremities of the field to be developed by satellite sub-sea wells. This development plan was subsequently approved by the Department of Energy in December 1978.

The Magnus platform jacket was installed in April 1982. Oil production started in August 1983 with water injection commencing in July 1984. Oil is produced from nine platform wells and two subsea wells. Pressure maintenance is currently achieved by water injection into six platform wells and three subsea wells.

Field stratigraphy

The generalized stratigraphic column for the Magnus Field area is illustrated in Fig. 3. The lithostratigraphic scheme used is that defined by Deegan & Scull (1977). The section above the Upper Jurassic reservoir is a 3 km (10 000 ft) thick Cretaceous to Recent, mudstone-dominated sequence with occasional sandstone horizons encountered. Shallow gas-bearing sandstone units up to about 30 m (100 ft) thick are present within the Pleistocene and Pliocene interval. Oil-bearing sandstones are occasionally found in the mudstone dominated Palaeocene section, with the oil trapped in a large anticlinal dome structure (136 sq km (52 sq mile) closure). The sandstones drilled so far have been thin (usually less than 1 m (3 ft) thick) with low permeabilities (less than 20 md).

The Upper Jurassic is dominated in the northern North Sea area by a marine mudstone sequence. The grey calcareous mudstones of the Heather Formation are succeeded by the dark grey-brown to black carbonaceous mudstones of the Kimmeridge Clay Formation. The Magnus Field reservoir comprises a small sand-rich submarine fan locally developed within the Kimmeridge Clay Formation. The reservoir sediments consist of fine- to medium-grained sub-arkosic sandstone with interbedded mudstones and siltstones.

The Middle Jurassic Brent Group is sandstone dominated in the upper part with an 87 m (285 ft) thick sandstone unit present in well 211/12-1. Core derived sandstone porosities are moderate and permeabilities are poor. The average porosity is 16.6% and the arithmetic mean permeability is 1.4 md for 211/12-1, the result of significant cementation in the sandstones. The absence of oil in the Brent Group in Magnus is probably due to the lack of a trapping mechanism. In particular, the presence of a lateral seal is in doubt.

From Abbotts, I. L. (ed.), 1991, *United Kingdom Oil and Gas Fields, 25 Years Commemorative Volume,* Geological Society Memoir No. 14, pp. 153–157

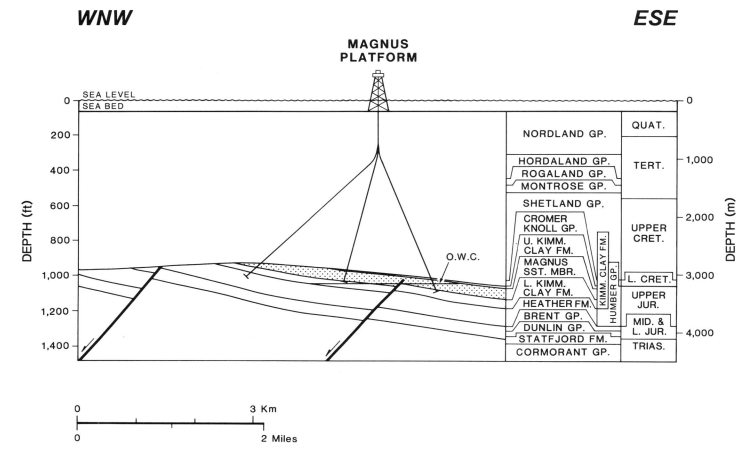

Fig. 2. Schematic geological cross-section of the Magnus Field.

Geophysics

Seismic acquisition and processing has improved markedly since the early 1970s when the first regional seismic lines were shot over the Magnus area. Early seismic sections showed very few obvious features below the Base Cretaceous reflection, to the extent that seismic correlation from well control was difficult at that time.

The current geophysical interpretation of the Magnus Field is based on a 3-D seismic survey acquired by GECO in 1983. The survey comprises 3238 sail-line kilometres with a line spacing of 75 m. The objective of the survey was to provide a detailed record of the reservoir geometry for reservoir management purposes. This was achieved. Four major seismic reflections (base Cretaceous, top reservoir, top Lower Kimmeridge Clay Formation, top Heather Formation) provide the framework for the definition of gross reservoir intervals (Fig. 4).

The gathering of production data has shown the Magnus Field to comprise several pressure regimes confined by both extensive intra-reservoir mudstones (restrictions to vertical flow) and faults (potential restrictions to lateral flow). With an increasing need for a better definition of intra-reservoir units and faults, the 3-D seismic survey was reprocessed in 1988. A new seismic interpretation is currently being developed, and it is already apparent that significantly more detail can be determined within the reservoir sequence using this reprocessed data.

Trap

The Magnus Field is located near the NW margin of the East Shetland Basin. Whereas the pre-Cretaceous structural style in the main basin area is dominated by N–S trending, down-to-the-east faults, the NW edge shows a NE–SW trending, down-to-the-northwest fault pattern. The N–S trend is consequent on the E–W tensional regime that gave rise to the North Sea graben system. The NE–SW fault system is more characteristic of the NE Atlantic continental margin fault pattern.

The Magnus structure is bounded to the NW by the Magnus High Boundary Fault. The early Cretaceous fault or fault zone is a notable feature on regional seismic sections. The top Jurassic horizon is downfaulted from a depth of about 3000 m (9800 ft) on the Magnus High to approximately 7000 m (23 000 ft) in the North Shetland Trough on the NW side of the fault (Evans & Parkinson 1983).

The main episode of faulting ceased in early Cretaceous times. Subsequent tectonism has been marked by subsidence and passive sediment infill of the basin.

The Magnus Field structure comprises an ESE dipping fault block bounded by major normal faults. The crest of the fault block is eroded and unconformably overlain by Cretaceous mudstones.

The trapping mechanism consists of a combination of stratigraphic pinch-out and reservoir truncation by unconformity. To the north and to the south of the field, the reservoir is limited by the depositional extent of the Magnus submarine fan sandstones. Reservoir seal is provided both laterally and vertically here by Kimmeridge Clay mudstones. At the crest of the fault block the Upper Kimmeridge Clay mudstones are missing due to erosion and the vertical seal is provided by Cretaceous mudstones.

The crest of the field is at 2800 m (9186 ft) subsea with the downdip limit of the field to the east defined by the oil–water contact of 3150 m (10 334 ft) subsea in the north and 3166 m (10 387 ft) subsea in the south. The stratigraphic trapping element of the reservoir results in some uncertainty as to whether the field is full to spill.

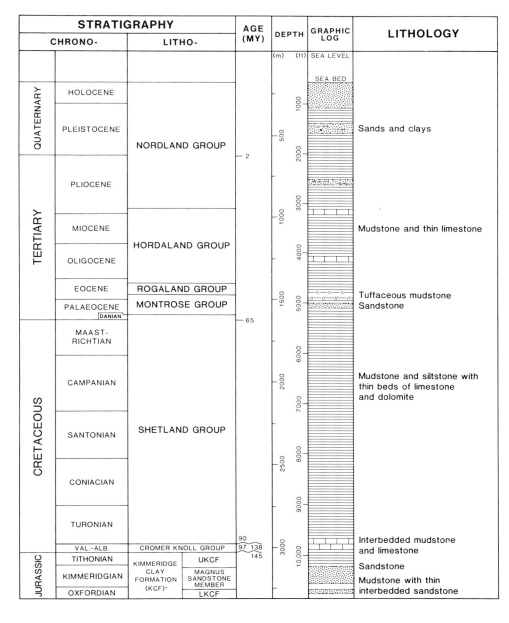

Fig. 3. Magnus Field generalized stratigraphic column.

Reservoir

The Upper Jurassic reservoir section has been assigned to the Kimmeridge Clay Formation (Oxfordian to Kimmeridgian age). For the purposes of reservoir description, the Kimmeridge Clay Formation has been subdivided into three units. The Magnus Sandstone Member is the sandstone dominant reservoir interval and sustains most of the production from the field. The underlying Lower Kimmeridge Clay Formation is mudstone dominated but with specific intervals of thin sandstones contributing some oil production. The Upper Kimmeridge Clay Formation is the highest unit and this mudstone interval acts as a reservoir seal over all but the crestal area of the reservoir.

In the crest of the field, a thin (less than 13 m (43 ft) thick) mid-Cretaceous sandstone interval unconformably overlies the eroded Magnus Sandstone Member. The unit is seen in only four wells and comprises less than 2% of total field oil in place.

Cores from the Upper Jurassic reservoir sequence show sedimentary associations and structures consistent with a submarine fan environment of deposition. For example, thick massive ungraded sandstone beds are interbedded with marine ammonite bearing mudstones. Bioturbation and cross bedding structures are rarely observed. Partial or occasionally complete classical Bouma sequences are commonly recorded in core.

The Magnus Sandstone Member is dominated by the stacked accumulation of thick-bedded fine- to coarse-grained sandstone units (individually about 2 m (7 ft) thick, but occasionally up to 7 m (23 ft) thick). The sand was probably deposited by high density turbidity currents.

An intra-Kimmeridgian faulting episode resulted in major change to the topography of the receiving basin for the submarine fan sediments. A north–south fault zone bisecting the Magnus Field was reactivated at a very early phase in the deposition of the Magnus Sandstone Member. An accommodation syncline (3 km N–S, 1.5 km E–W) to the west of the fault zone, ponded up to 160 m (525 ft) of a sand dominated sequence. Elsewhere within the field, where the basin topography is more subdued, the Magnus Sandstone is thinner; less than 100 m (320 ft).

Occasional interbedded finer-grained sandstone and mudstone units are seen within the Magnus Sandstone Member and are thought principally to represent deposition from a combination of low density turbidite flows and background hemipelagic mud deposition. Three of these fine-grained intervals appear to be present over a large area of the field. Four major sandstone sequences are compartmentalized with respect to fluid flow and pressures by these fine-grained units.

The Lower Kimmeridge Clay Formation is a low net-to-gross reservoir sequence (average N:G 23%) dominated by well lami-

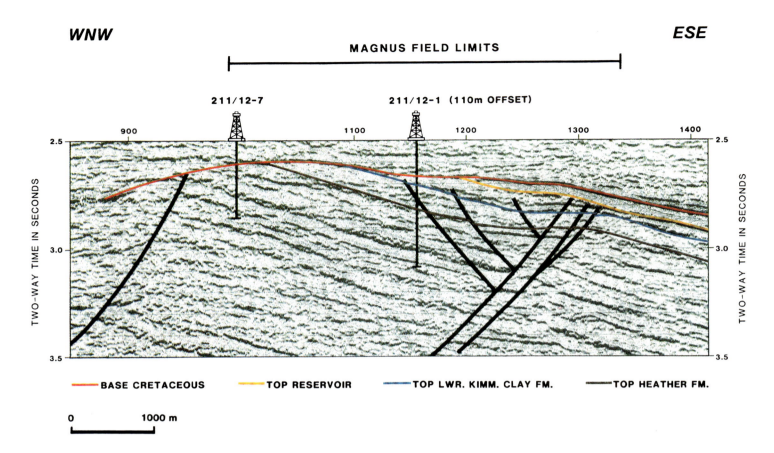

Fig. 4. Seismic section through Magnus Field acquired as part of the 1983 3-D survey shot by GECO and processed by Western Geophysical.

nated, hemipelagic/turbiditic mudstones. Sandstone zones are locally commonly developed, particularly towards the top of the sequence. These sand-prone zones comprise high density turbidite sandstones, interbedded with hemipelagic/turbiditic mudstones, debris flow horizons and slump sheets.

The Lower Kimmeridge Clay Formation is thought to represent a near slope sequence characterized by the influx of dilute muddy turbidites but with intermittent episodes of sand deposition.

The ungraded nature of the sand bodies suggests that the sand was deposited rapidly, probably due to the loss of energy at a break in slope. The localized presence of slump and debris flow units may have resulted in a somewhat irregular basin topography. It is likely that sand deposition was preferentially concentrated in the localized topographic lows between these units. Syn-sedimentary faulting may also have taken place, and this will have accentuated basin irregularities.

There are marked areal and depth controls on porosity and permeability values within the reservoir sandstones. Porosities vary from 24% at the crest to 18% along the downdip eastern margin. Permeabilities show a dramatic decrease from an average of 950 md at the crest to less than 20 md along the eastern flank. Current data indicates that the volume of diagenetic cement increases significantly with depth and that this is the main influence on porosity and permeability variation. The principal diagenetic mineral phases include quartz, dolomite, calcite, illite, kaolinite, feldspar overgrowths and pyrite.

The presence of filamentous illite below the oil–water contact is thought to have resulted in a reduction of permeabilities in the aquifer (Heaviside *et al.* 1983; Pallatt *et al.* 1984). The reduction in permeability is consequent on the large volume of bound water associated with the very high surface area of the filamentous illite ribbons.

Source

The reservoir sandstones of the Magnus Field are enveloped by the likely source rock, the organic-rich mudstones of the Kimmeridge Clay Formation. The main source area lay to the east of the field although some oil was probably also sourced from the north. The main phase of oil generation is thought to have commenced circa 75 Ma BP (Maastrichtian–Campanian).

Hydrocarbons

The crude in the Magnus Field is a light, low sulphur (0.28%), medium wax (4–6% by weight) oil with an API gravity of 39°. The initial gas–oil ratio is 750 SCF/STB and the average formation volume factor at reservoir conditions is 1.43. The crude is undersaturated with no gas cap present. Formation waters are very fresh with salinities averaging about 20 000 ppm total dissolved solids.

The datum reservoir pressure is 6653 psig at 3050 m (10 006 ft) subsea. This is about 2000 psi overpressured compared to a normal hydrostatic gradient. The formation temperature is 240°F at 3050 m (10 006 ft) subsea.

Reserves/resources

The Magnus Field initial oil in place is calculated as 1665 million stock tank barrels and ultimate recoverable reserves currently estimated as 665 million stock tank barrels. The 40% recovery factor is dependent on a water flooding scheme. The Magnus reservoir has a low energy drive mechanism and this would give a recovery factor of only 8% under natural depletion. Pressure support is maintained by a row of sea-water injection wells along the east flank of the field. The injection wells are generally perforated in the oil leg as the permeabilities in the water leg are generally too low (less than 15 md) for adequate injection rates. These wells are initially put on production to maximize downdip oil recovery prior to being turned around to injection.

The Magnus Field is into the sixth year of oil production and is currently on a production plateau of 139 300 BOPD. Individual well production rates are up to 30 000 BOPD. The field-wide water injection rate is currently about 190 000 BWPD. There are no significant production problems as yet. Water production from the producer wells is presently low (1988 average: 2600 BWPD), with only one well to date showing significant sea-water breakthrough and its associated scaling problems.

The reservoir management strategy is to maximize the economic recovery of the field, to ensure good sweep of the reservoir and to implement incremental reserves projects.

One particular area under review at the moment is the low net to gross Lower Kimmeridge Clay Formation reservoir. Currently the interval is produced under natural depletion only. If it can be established that the Lower Kimmeridge Clay Formation sandstones are of a sufficient lateral extent to justify a water flooding project, then recovery from the interval could be substantially improved.

This paper uses material from previous written material on the Magnus Field including Atkinson 1985, De'Ath and Schuyleman 1981 and Shepherd et al. 1989. The author wishes to thank the Magnus Field Partners, Repsol Exploration (UK) Ltd, Sun Oil Britain Ltd, Brasoil UK Ltd, and Goal Petroleum plc, for permission to publish this paper.

References

ATKINSON, J. P. 1985. The use of reservoir engineering in the development of the Magnus oil reservoir. *Proceedings of the Offshore Europe 85 Conference*, SPE 13979.

DE'ATH, N. G. & SCHUYLEMAN, S. F. 1981, The geology of the Magnus Field. *Proceedings of the Second Conference of Petroleum Geology of the Continental Shelf of NW Europe*. Insititute of Petroleum Geology, London, 342–351.

DEEGAN, C. E. & SCULL, B. J. 1977, *A proposed standard lithostratigraphic nomenclature for the Central and Northern North Sea. Report of the Institute of Geological Sciences*, 77/25.

EVANS, A. C. & PARKINSON, D. N. 1983. A half-graben and tilted fault block structure in the Northern North Sea. *In*: BALLY, A. W. (ed.) *Seismic Expression of Structural Styles*. Studies in Geology Series **15/2**, AAPG, Tulsa.

HEAVISIDE, J., LANGLEY, G. O. & PALLATT, N. 1983. Permeability characteristics of Magnus reservoir rock. *8th SPWLA London Chapter European Evaluation Symposium Transactions*, London.

PALLAT, N., WILSON., J. & MCHARDY, W. 1984. *The relationship between permeability and the morphology of diagenetic illite in reservoir rocks.* SPE of AIME Unsolicited Papers No SPE-12798.

SHEPHERD, M., KEARNEY., C. J. & MILNE, J. H. 1989. Magnus Field, East Shetland Basin, Northern North Sea, UK. *In*: BEAUMONT, E. A. & FOSTER, N. H. (eds) *Atlas of Treatise of Petroleum Geology*, AAPG, Tulsa, (in press).

Magnus Field data summary

Trap
Type: Combination stratigraphic pinchout/truncation unconformity trap
Depth to crest: 2800 m (9186 ft) subsea
Lowest closing contour: 3166 m (10 387 ft) subsea
Hydrocarbon contacts; 3150 m (10 335 ft) subsea (in the north), 3166 m (10 387 ft) subsea (in the south)
Gas column: No gas column present
Oil column: 350–366 m (1148–1201 ft) vertical closure

Pay zone
Formation: Kimmeridge Clay Formation
Age: Jurassic, Oxfordian–Kimmeridgian
Gross thickness: 0–250 m (0–820 ft)
Net/gross ratio: Magnus Sandstone Member 0.76 average; Lower Kimmeridge Clay Formation 0.23 average
Cut off for N/G: Net sand defined where log porosity is greater than 14% and the volume of shale (Vshale) is less than 33%
Porosity: range 18–24%, average 21%
Hydrocarbon saturation: 0.83
Permeability: range 0–950 md, geometric mean is 82 md
Productivity index: up to 200 STB/D/psi

Formation water
Salinity: 20 000 ppm total dissolved solids
Resistivity: 0.12 ohm m at 240°F

Reservoir conditions
Temperature: 240°F at 3050 m (10 006 ft) TVD subsea
Pressure: Initial datum pressure is 6653 PSI at 3050 m (10 006 ft) subsea
Pressure gradient in reservoir: 2.2 PSI/m (0.67 PSI/ft)

Field size
Area: 3300 hectares (33 km^2; 8154 acres)
Gross rock volume: 4.01×10^9 cubic metres (43.16×10^9 cubic ft)
Hydrocarbon pore volume: 0.38×10^9 cubic metres (4.09×10^9 cubic ft)
Recovery factor, oil: 40%
Recovery factor gas: Not applicable
Drive mechanism: Oil and rock expansion, limited water drive; water injection required for pressure maintenance
Recoverable hydrocarbons: 665 MMBBL
Recoverable oil: 665 MMBBL

Production
Start-up date: August 1983
Development Scheme: Single fixed steel platform. Production and injection wells are a combination of deviated wells from the platform and subsea wells.
Production rate: 139 300 BOPD (plateau)
Cumulative production to date: 244 million stock tank barrels (end 1988).

The Miller Field, Blocks 16/7B, 16/8B, UK North Sea

S. K. ROOKSBY

BP Exploration, 301 St Vincent Street, Glasgow, UK

Abstract: The Miller Oil Field is located on the western margin of the South Viking Graben in UKCS Blocks 16/7b and 16/8b. The oil is trapped in Upper Jurassic turbidite sands shed from the Fladen Ground Spur via the Brae complex submarine fan systems. The reservoir sands are of good quality with an average porosity of 16% and permeabilities occasionally in excess of 1 Darcy. The trap is formed within a subtle structural–stratigraphic combination. Overlying slow velocity Lower Cretaceous sediments produce a time flat which, after depth conversion, produces a 3-way dip closed feature. The trap is completed by stratigraphic pinchout of the reservoir sands to the northwest. The most recent (1985) seismic data allow the top reservoir reflector to be picked directly, which was not the case during the exploration and appraisal phase, when only the Top Kimmeridge Clay seismic pick could be made. The estimate of recoverable hydrocarbons is currently 300 MMBBL of oil and 570 BCF of gas. Development drilling commenced early in 1989. No results are yet available.

The Miller Oil Field is located 140 miles (225 km) northeast of Cruden Bay in Scotland, in the UK sector of the North Sea. The areal extent of the field is c. 45 km^2 (11 120 acres), shared evenly between blocks 16/7b and 16/8b. Water depth is 360 ft (110 m). The reservoir is composed of submarine fan sandstones of the Brae Formation deposited during the Late Jurassic. The field is named after the Scottish geologist Hugh Miller (1802–1856).

History

Licences P340 16/7b (BP operator) and P341 16/8b (Conoco operator) were re-awarded in the UK 7th round of licensing in 1980, subsequent to earlier compulsory relinquishments of third round licences by Pan Ocean/Marathon and Shell respectively. Current equity holders in Miller are BP 40% (operator), Conoco 30%, Enterprise 18% and Santa Fe 12%.

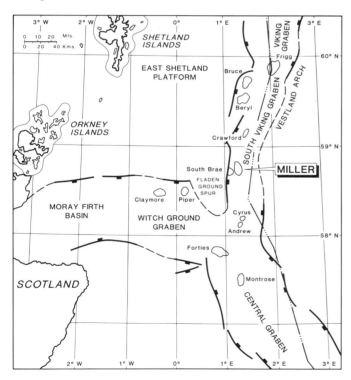

Fig. 1. Miller Field regional setting.

At the time of licensing, neither part block had been drilled. The early exploration was based on three 2-D seismic surveys acquired by both operators between 1981 and 1983, totalling about 600 km.

The first well spudded was 16/7b-20 by BP in November 1982, shortly followed by Conoco's 16/8b-2 in February 1983. Both wells encountered a gross pay of around 60 m (200 ft) in the Upper Jurassic Brae Formation, which flowed significant amounts of oil and gas on test. Data indicated a similar fluid type, a common oil–water contact and a common pressure regime. Results indicated a major oil accumulation. Between February 1983 and July 1984, wells 16/7b-21, -23, -24, -25, -26, 16/8b-3 and -5 were drilled in the Miller area to delineate the field. Between October 1984 and 1986, three further wells, 16/7b-28, 16/8b-6 and -7 were drilled to delineate field extent and the oil–water contact. The wells showed pressure depletion indicating pressure communication between Miller and the producing South Brae field and helped define the northern extent of the field.

The Development was approved by the Department of Energy in 1988.

Geological setting

The Miller Field is located along the western margin of the South Viking Graben, adjacent to the Fladen Ground Spur (Fig. 1). It lies in a north–south oriented half-graben which is fault-bounded to the west by the Devonian basement of the Fladen Ground Spur, gradually rising to the east towards the Vestland Arch. The reservoir is composed of Late Jurassic submarine fan sandstones within the Brae Formation which were derived from the Fladen Ground Spur to the west. The field straddles the boundary between Blocks 16/7b (BP) and 16/8b (Conoco Group) and lies to the east of Block 16/7a (Marathon Group) which contains the North, Central and South Brae Fields (Fig. 2). A generalized lithostratigraphic column is shown in Fig. 3.

Rifting of the South Viking Graben occurred between Late Permian and Triassic times and was reactivated during the Late Jurassic to Early Cretaceous. This later tectonism was initiated in the Late Oxfordian, peaking in the Kimmeridgian and waning through the Portlandian to Ryazanian. Faulting generally becomes younger towards the west, along the western margin of the graben. Zones of en-echelon faults developed, with the majority of these downthrown to the east. Figure 4 provides an illustration of the regional setting of the Miller Field. The thick Jurassic wedge within the graben can be clearly seen, juxtaposed against the fault-bounded Fladen Ground Spur to the west.

The Kimmeridge Clay Formation progressively onlaps the eastern margin of the basin reflecting the eustatic sea level rise during Late Jurassic times. During the period of highest sea level (Portlandian to Ryazanian) the Kimmeridge Clay Formation covered the Miller Field reservoir sands and onlapped the Devonian basement of the Fladen Ground Spur. This mudstone forms the seal over all of the field.

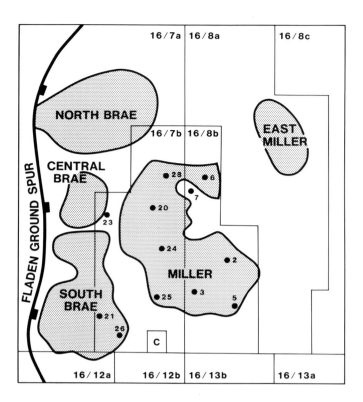

Fig. 2. The Brae–Miller Fields complex.

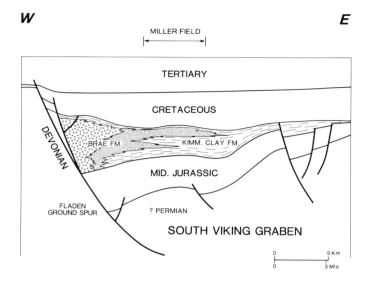

Fig. 4. Generalized E–W geological cross-section.

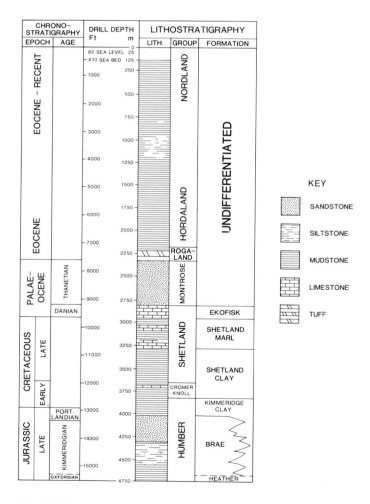

Fig. 3. Miller area generalized stratigraphy.

Geophysics

The definition of the Miller Field structure was primarily based upon three 2-D seismic surveys conducted by BP and Conoco between 1981 and 1983. These surveys total approximately 600 km (370 miles) and provide a grid spacing of about 0.5 km to 1 km over Blocks 16/7b and 16/8b with a number of well ties. All three surveys have been reprocessed to zero phase in order to maximize resolution at the reservoir level. Migrated stacks at 20 cm s^{-1} were used to map the Miller Field structure.

In addition to the 2-D seismic data-base a 3-D survey covering all of Blocks 16/7b, 16/8a and 16/8b as well as much of Blocks 16/7a and 16/8c was acquired post Annex B. Final 3-D migrated data were available in 1987 and have been used to remap the field for development purposes. This survey comprises 12 400 km of subsurface coverage (6200 sail line kilometres), of which about one-third is directly relevant to the Miller structure. This survey is being used to optimize the final positions of the development wells.

The 3-D data were also corrected to zero phase for interpretation. The Top Reservoir reflector was identifiable from these data. Therefore, the majority of the field can be mapped directly. As previously described the time maps give little indication of the structural closure. Given the current well control the problem of depth conversion is less significant than at discovery.

The low relief of the field, 110 m (361 ft) means accuracy in depth conversion is very important for both Volumetrics and Reservoir Development.

A representative seismic section over the crest of the field is shown in Fig. 5.

Trap

The areal extent of the Miller Field is defined by a combination of structural and stratigraphic trapping. In the south western part of Block 16/7b, the Miller Field is bounded by the intersection of the top reservoir structure and oil–water contact, where a structural saddle creates separation from the South Brae Field. A structural cross-section showing the South Brae to Miller relationship is seen in Fig. 6. The field is similarly bounded to the south, east, northeast, and to some extent to the north. Stratigraphic separation occurs to the north-west of the field where reservoir quality deteriorates towards the margin of the submarine fan system.

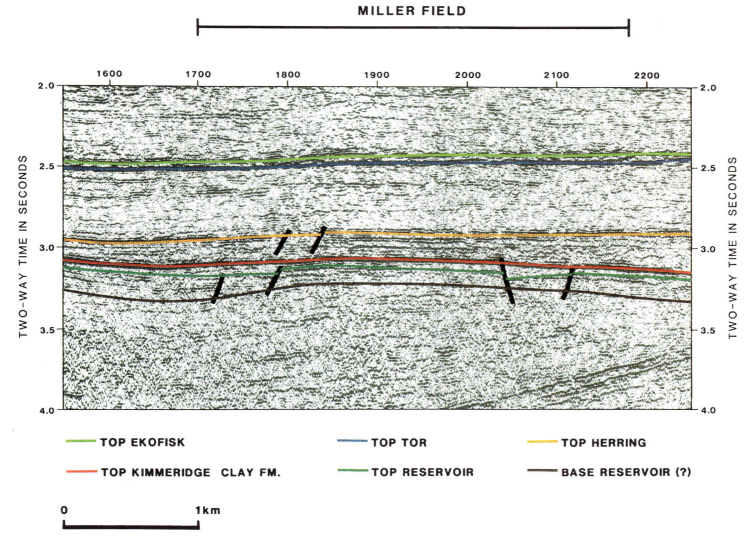

Fig. 5. Seismic section through Miller Field.

Reservoir

The Brae Formation forms the reservoir succession in the Brae and Miller Fields. It is divided on lithological grounds into three lithostratigraphic members: (1) Upper Member; (2) Middle Member (the main Miller reservoir); (3) Lower Member. The Lower Member is composed of apron fringe conglomerates deposited into a rapidly subsiding basin in Late Oxfordian to Early Kimmeridgian times.

With decreasing subsidence, areally extensive submarine fan systems sourced from the west were established during Early Kimmeridgian to Early Portlandian times. These sediments comprise the Middle Member and form the bulk of the Miller reservoir.

The Upper Member is also made up of submarine fan sandstones and shales, but with a northerly source. This member was deposited in Early to Late Portlandian times.

The Brae Formation has been subdivided, chronostratigraphically, into three reservoir units in the Miller area. These chronostratigraphic Units also relate, in part, to the Brae lithostratigraphic members. Unit 3 comprises the Lower Member of the Brae Formation. It represents much of the reservoir interval of the Central Brae Field in Block 16/7a.

Unit 2 is approximately equivalent to the Middle Member of the Brae Formation and forms the bulk of the Miller Reservoir and the bulk of the South Brae reservoir. The top of the Middle Member is diachronous. The top Miller reservoir is equivalent to the top Middle Member sandstones, and therefore must also be partially diachronous.

Unit 1 is equivalent to the Upper Member of the Brae Formation, and is thought to form the North Brae Field reservoir.

Reservoir quality varies with facies distribution throughout the Brae Formation, although within the Miller Field the facies and the reservoir quality are remarkably consistent. Reservoir distribution of Units 1 to 3 are shown schematically in Fig 7.

Primary intergranular porosity predominates within the sandstones of the Miller Field. Secondary porosity after feldspar dissolution provides a minor contribution to overall porosity. Microporosity developed between fibrous illite clays is rare. The predominance of quartz cements and volumetrically low proportions of authigenic illite within the sandstones indicate that the pore throats will be clean and the pore surface area of the sandstone should be low.

The Miller Field is not structurally complex. The top Unit 2 structure has the form of a southeast plunging nose, projecting off

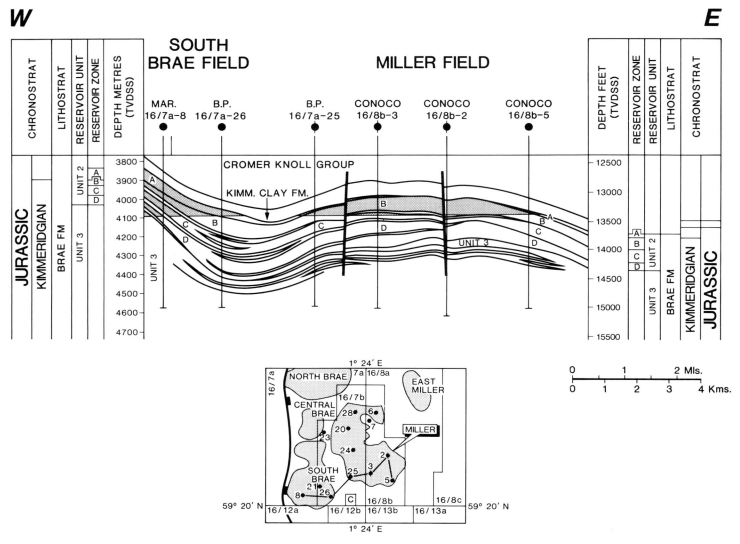

Fig. 6. Structural cross-section from South Brae to Miller Field.

the Fladen Ground Spur, dipping below the OWC to the south, southwest, east and north. Shale-out of reservoir sandstones provides a stratigraphic boundary to the northwest. The maximum hydrocarbon column is approximately 110 m (361 ft), much less than the 450 m (1476 ft) observed at South Brae.

Faulting is not extensive and maximum throws reach 130 m (328 ft). Most faults in the Miller Field have a NW-SE orientation, oblique to the main graben margin faults. Regional trends suggest that these faults are deep-seated and were possibly reactivated as synsedimentary features, but by Unit 2 time had probably ceased to exert any major control over sedimentation. However, later reactivation post-dating the Kimmeridge Clay Formation has occurred and today the main body of the Miller Field overlies a NW-SE trending horst block, with a small graben partially separating the northernmost portion of the field.

Faults do not appear to act as significant barriers to fluid transmissibility within the reservoir over most of the field. This is suggested by the pressure depletion observed within the oil column throughout the Miller Field in wells drilled since South Brae production began. A horizontal permeability barrier does exist, however, within Unit 2 Layers B and C, as these layers are undepleted at wells 16/7b-28, 16/8b-6 and 16/8b-7. This barrier is stratigraphically defined, where the fringes of the two submarine fan systems overlap and is a thin but extensive mudstone horizon (Fig. 7).

Relationship of the Miller Field to neighbouring fields

The separation between the Miller Field and the neighbouring hydrocarbon accumulations of South Brae, Central Brae, North Brae and those of Block 16/8a is both structurally and stratigraphically defined.

The South Brae and Central Brae Fields have an oil-water contact at 4095 m (13 435 ft) subsea, while the Miller Field has an oil-water contact at 4090 m (13 418 ft) subsea. Although a variation about these mean values occurs in each field, the difference of 5 m (16 ft) is statistically valid and significant 'blue water' exists between the fields. However, the similarity of these oil-water contacts may indicate that fluid communication over geological time may have been established. Late stage tectonic events may have contributed to the current variation seen in oil-water contacts, creating the present separation of the oil accumulations, but leaving a common aquifer. However, the mechanism controlling the observed oil-water contacts is poorly understood.

Source

The Kimmeridge Clay Formation, which was deposited in a deep-water anoxic environment, provides the source for the Miller Field hydrocarbons. It laterally interfingers with Brae Formation reser-

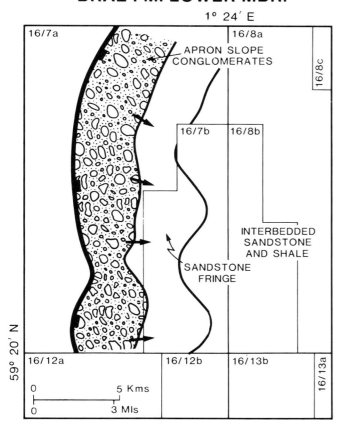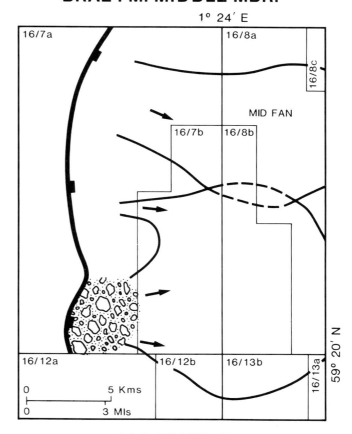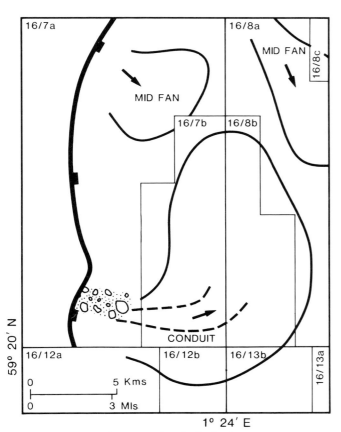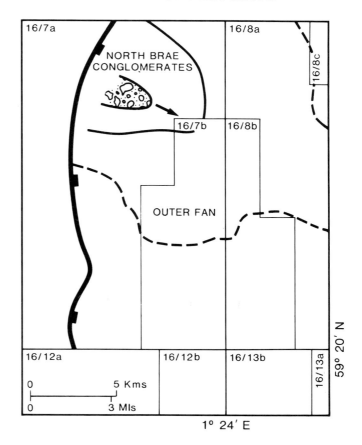

Fig. 7. Brae Formation reservoir unit distribution.

voir sandstones and directly overlies the main reservoir in the southern part of the field. Hydrocarbon generation and migration is believed to have initiated in the deeper parts of the basin, to the north, south and east of Block 16/8b and has continued to the present.

Hydrocarbons

The majority of hydrocarbons in the Miller reservoir are contained in Unit 2 of the Brae Formation. The Unit 2 accumulation extends from a crestal high of 3980 m (13 058 ft) subsea to an oil–water contact of 4090 m (13 418 ft) subsea.

The reservoir hydrocarbon is an undersaturated, CO_2-rich, sour, volatile oil with a reservoir density and viscosity of 623 kg m^{-3} and 0.20 cP respectively (39° API). The oil has a gas/oil ratio of 1813 SCF/STB and no gas-cap is present in the Unit 2 accumulation. High concentrations of barium ions have been seen in the water samples and it is likely that barium sulphate scaling will be a significant production problem.

Reserves

The aquifer surrounding the Miller reservoir is of limited size and does not provide significant pressure support. Production from South Brae has resulted in a decline in Miller reservoir pressure. Therefore, peripheral water injection will be employed to maintain production.

The Miller development is planned using 30 wells: 21 producers and 9 injectors. A total of ten wells (five producer and five injectors) are to be pre-drilled through a template prior to platform installation. The first of these was spudded on 8 March 1989.

First oil is planned in early 1992 with a plateau rate of 113 000 BOPD. The ultimate recovery is expected to be 300 MMBBL of 39° API oil, and 0.57 TCF of associated gas.

Many thanks to the efforts of my geology and geophysics colleagues who completed the reservoir descriptions for development planning, namely T. Kirkham, K. Mills, S. Adams, the contributions of the reservoir engineer, L. Westwick, and the Miller Field Partners: Conoco, Santa Fe and Enterprise in giving their permission for this paper to be published.

Miller Oilfield data summary

Trap	
Type	Combination; Structural 3 way dip & stratigraphic
Depth to crest (U Jurassic)	3980 m (13 058 ft)
Oil–water contact	4090 m (13 418 ft) ± 3m
Gross oil column	c. 110 m (361 ft)
Pay zone	
Formation	Brae Formation sandstones
Age	Late Jurassic (Kimmeridgian/ Portlandian)
Thickness	125–250 m
Net/gross	Average 0.8, range 0.4–0.98
Porosity	Average 16%, range 12–23%
Hydrocarbon saturation	85–90%
Permeability	11–1200 md
Hydrocarbons	
Oil gravity	0.83 gm cm^{-3} at 60°F/1.0 bar (38.5° API)
Oil type	Significant CO_2 and H_2S, undersaturated, volatile
Gas/oil ratio	1813 SCF/STB
Bubble point	4900 psia
Formation volume factor	1.97 rb/STB at 7250 psia
Formation water	
Resistivity	0.032 ohm m at 250°F
Salinity	70 000 ppm
Reservoir conditions	
Temperature	250°F
Pressure	7250 psig (original)
Pressure gradient in reservoir	0.27 psi/ft
Field size	
Area	45 km^2 (11 120 acres)
Recoverable oil and gas	300 MMBBL/0.57 TCF
Drive mechanism	Aquifer plus water injection
Production	
Anticipated first oil	1Q 1992
Anticipated daily production	113 000 BOPD
Development scheme	Steel jacket, drilling and export facilities for oil and gas
Number/type of wells	30 wells, 21 producers, 9 injectors

The Murchison Field, Block 211/19a, UK North Sea

JOHN WARRENDER

Conoco (UK) Ltd, Rubislaw House, Anderson Drive, Aberdeen AB2 4AZ, UK

Abstract: The Murchison oil field forms part of the Brent oil province in the East Shetland Basin, northern North Sea. The field, which straddles the UK–Norway international boundary, was discovered in 1975 and began production with Conoco (UK) Ltd as Operator, in 1980. Like many oil accumulations in the East Shetland Basin the trap consists of a northwesterly dipping rotated fault block of Jurassic–Triassic age sourced and sealed by unconformable Upper Jurassic shales. The productive reservoir consists of Middle Jurassic Brent Group sandstones which represent the south to north progradation of a wave/tide influenced delta system. The Brent Group on Murchison has an average thickness of 425 ft with average porosities of 22% and permeabilities in the 500–1000 md range in producing zones. The maximum hydrocarbon column thickness is approximately 600 ft. The oil is undersaturated and no gas cap is present. Recoverable reserves are 340 MMBBL from a total oil in place figure of 790 MMBBL. Oil production which is via a single steel jacket platform peaked at 127 000 BOPD in 1983 and currently averages 45 000 BOPD. Economic field life is expected to be at least 20 years.

The Murchison Field is located in the East Shetland Basin, northern North Sea at approximate latitude 61° 23′ N, longitude 1° 43.5′ E, 120 miles northeast of the Shetland Islands (Fig. 1). The field straddles the UK–Norway international boundary with the greater portion in the UK Block 211/19a and the lesser portion in Norway Block 33/9. Water depth is −512 ft BMSL. In the context of the North Sea the field is of medium size with an areal closure of approximately 7 square miles and contains 790 million barrels of oil in place. The productive reservoir consists of coastal deltaic sandstones of the Middle Jurassic Brent Group which lie between the marine shales of the Lower Jurassic Dunlin Group and the marine, organic-rich shales of the Upper Jurassic Humber Group. The trap is structural comprising a single, northwesterly dipping rotated fault block which has been sourced and sealed by the overlying Upper Jurassic shales. The field is named after the Scottish geologist Sir Roderick Impey Murchison (1792–1871), who is best known for his contribution to Palaeozoic stratigraphy.

History

Block 211/19 was awarded in 1970 to a partnership consisting of Conoco–NCB (subsequently BNOC)–Gulf as part of the UK 3rd Licensing Round. The adjacent Norwegian block 33/9 was awarded in 1973 to a consortium with Mobil as operator.

The field was discovered by the Conoco–NCB–Gulf well 211/19-2 which reached TD in August 1975. The well encountered 414 ft of gross Brent Group with 300 ft of net sandstone which was oil-bearing to the base of the reservoir. Underlying Statfjord and Triassic sandstones which were also penetrated by the well were water bearing. Two drill stem tests confirmed reservoir productivity. Delineation wells 211/19-3 and 211/19-4 proved the continuity of sandstone quality and thickness and established the main field oil–water contact. Exploration well 211/19-5 located approximately 3 miles southwest of the discovery well in a block downthrown from the main field found the Brent Group to be water bearing above the

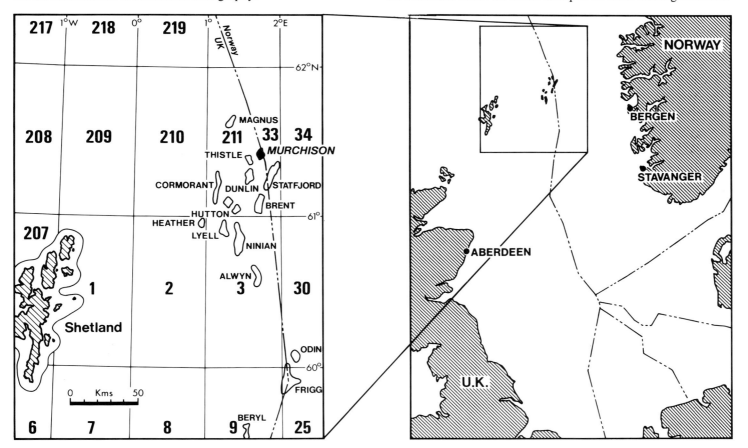

Fig. 1. Location map of the Murchison Field relative to other discoveries in adjacent blocks.

established OWC and confirmed the structural limits of the field.

The field was declared commercial in January 1976 and the decision to develop was made in September the same year. Conoco as operator of UK Block 211/19 submitted a development plan to the Department of Energy on behalf of interest holders Conoco–NCB–Gulf. As a portion of the field is in Norwegian waters a unitization agreement was drawn up between the UK (P104) licence holders and the Norway (PL037) licence group. Current unit participants are as follows.

UK Licence P104-Block 211/19

Conoco (UK) Limited (operator)	33.333%
Oryx UK Energy Company (formerly Britoil Limited)	33.333%
Chevron (formerly Gulf Oil Corporation)	33.333%

Norwegian Licence PL037-Block 33/9

Den norske stats oljeselskap AS	50.000%
Mobil Exploration Norway Incorporated	15.000%
Conoco Norway Incorporated	10.000%
AS Norske Shell	10.000%
Esso Exploration and Production Norway Incorporated	10.000%
Saga Petroleum AS	1.875%
Amoco Norway Oil Company	1.042%
Amerada Petroleum Corporation of Norway	1.042%
Enterprise Oil Norge Ltd	1.042%

Agreed equities prior to development were UK P104 group 83.75%, Norway PL037 group 16.25%. Two equity redeterminations have taken place in 1982–3 and 1984–6. The current equity shares are UK P104 group 77.8%, Norway PL037 group 22.2%.

Following the decision to develop the field preliminary engineering feasibility studies were initiated in late 1975 and completed in September 1976. Major design and service contracts were awarded in November 1976 and platform fabrication began in July 1977. The platform, a single steel jacket structure with 27 well slots was installed in August 1979 and drilling operations began in June 1980. Production started in September 1980. All of the original slots have been used and 6 new slots were installed in 1990, totalling 33. To date 30 wells and 10 sidetracks have been drilled, 40 well locations in total. Current status in February 1991 is 19 producing wells, 10 water injectors and 1 dual completion.

A detailed development plan was prepared in 1988 to optimize remaining recovery to the end of field life. The development plan leads to improvement in recovery from all areas of the field and includes further sidetracks, further perforations and recompletions and process modifications to handle greater volumes of produced and injected fluids.

Field stratigraphy

The Murchison Field forms part of the Brent oil province which is located in the East Shetland Basin at the northwestern end of the Viking Graben (Fig. 2). Field stratigraphy and the development of the Murchison structure have been largely controlled by the regional tectonics of the Viking Graben, the differential subsidence

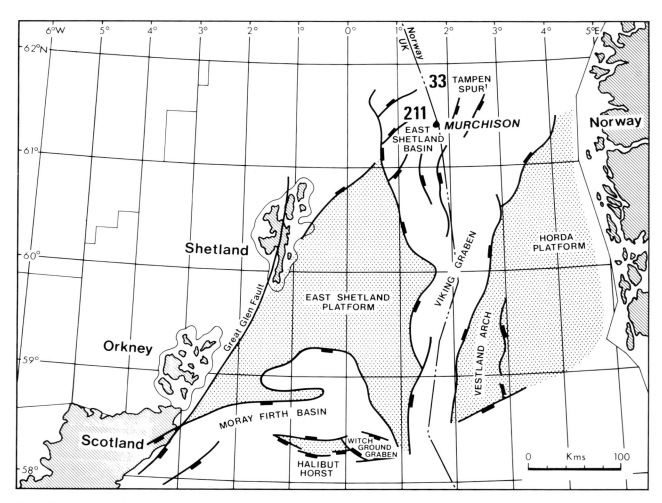

Fig. 2. Regional geological setting of the Murchison Field.

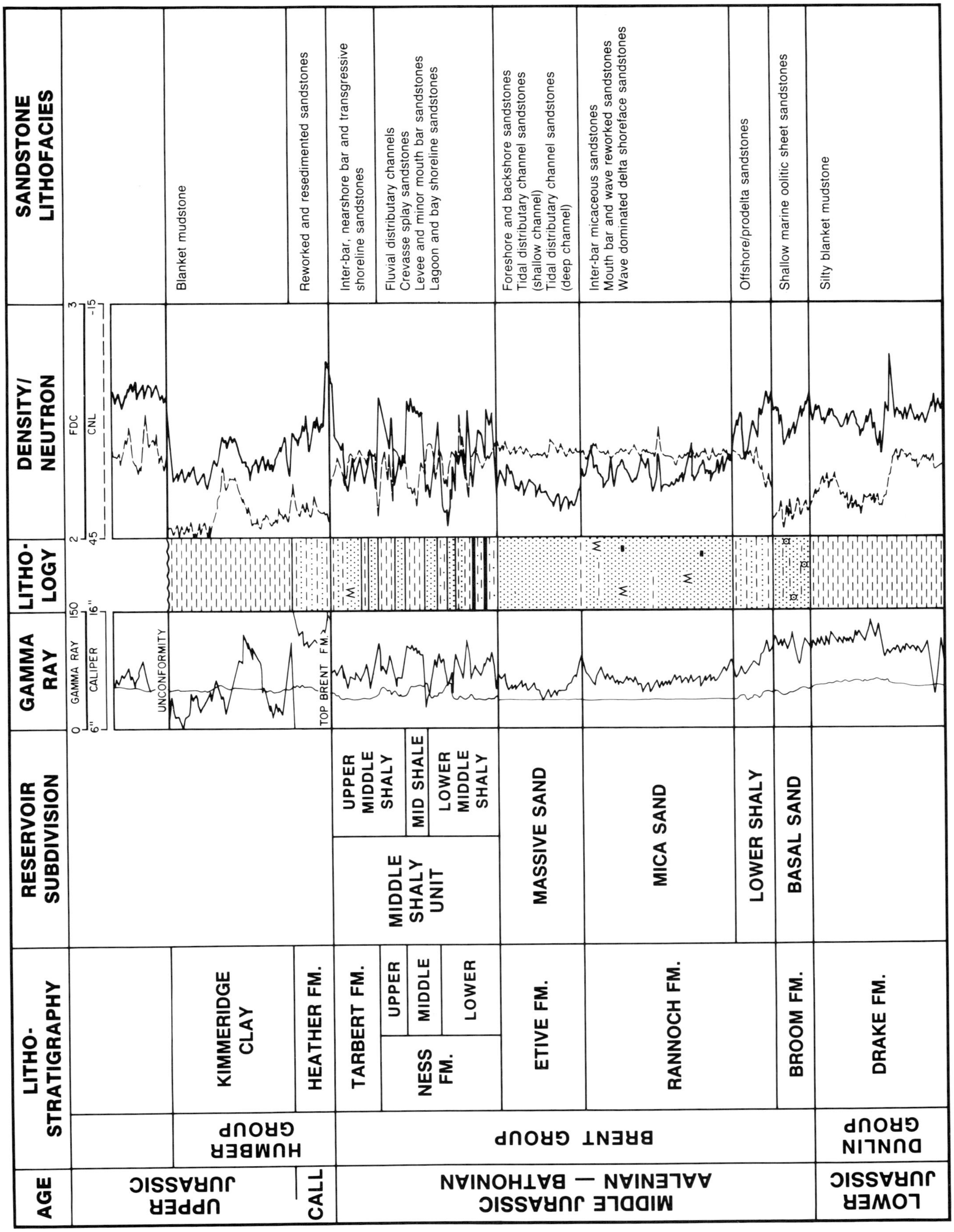

MURCHISON FIELD 167

AGE	LITHO-STRATIGRAPHY		RESERVOIR SUBDIVISION	GAMMA RAY	LITHO-LOGY	DENSITY/NEUTRON	SANDSTONE LITHOFACIES
UPPER JURASSIC	HUMBER GROUP	KIMMERIDGE CLAY					Blanket mudstone
CALL		HEATHER FM.					Reworked and resedimented sandstones
MIDDLE JURASSIC AALENIAN — BATHONIAN	BRENT GROUP	TARBERT FM.	UPPER MIDDLE SHALY / MID SHALE / LOWER MIDDLE SHALY (MIDDLE SHALY UNIT)				Inter-bar, nearshore bar and transgressive shoreline sandstones
		NESS FM. (UPPER / MIDDLE / LOWER)					Fluvial distributary channels / Crevasse splay sandstones / Levee and minor mouth bar sandstones / Lagoon and bay shoreline sandstones
		ETIVE FM.	MASSIVE SAND				Foreshore and backshore sandstones / Tidal distributary channel sandstones (shallow channel) / Tidal distributary channel sandstones (deep channel)
		RANNOCH FM.	MICA SAND				Inter-bar micaceous sandstones / Mouth bar and wave reworked sandstones / Wave dominated delta shoreface sandstones
			LOWER SHALY				Offshore/prodelta sandstones
		BROOM FM.	BASAL SAND				Shallow marine oolitic sheet sandstones
LOWER JURASSIC	DUNLIN GROUP	DRAKE FM.					Silty blanket mudstone

Fig. 3. Representative stratigraphic section for the Murchison Field.

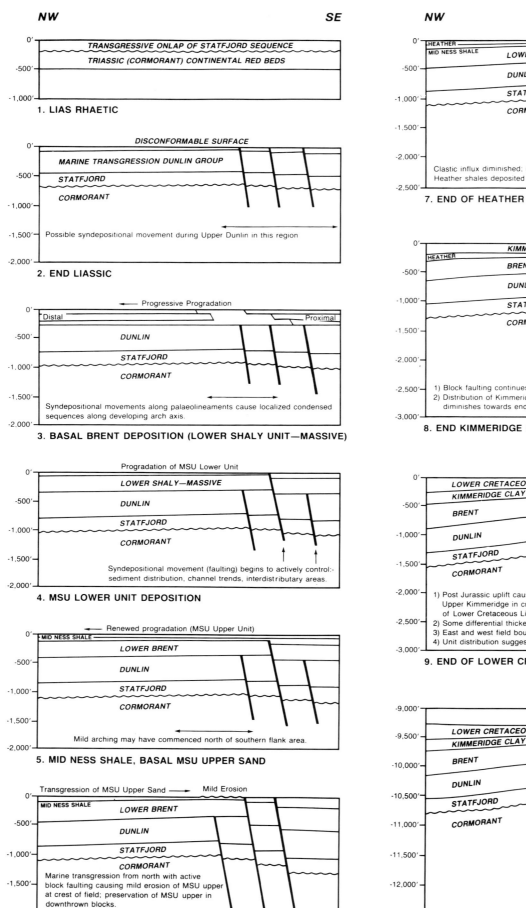

Fig. 4. Geological history of the Murchison Field.

history of the East Shetland Basin and global fluctuations in sea level. A generalized stratigraphic sequence and geological history of the field, reflecting these controls, are given in Figs 3 and 4.

The oldest rocks recognized in the Murchison Field are continental beds probably belonging to the Triassic Cormorant Formation. As the East Shetland Basin subsided, transgression from the Boreal Sea to the north resulted in progressive onlap of the Statfjord Formation during the early Jurassic. In Murchison two units of the Statfjord Formation are preserved, comprising the Eiriksson Member (estuarine sandstones and shales) and the Nansen Member (shallow marine sandstones). The basal Statfjord unit, the Raude Member marks continental deposition typical of the more elevated parts of the East Shetland Basin and is not present at Murchison. The Statfjord Formation is overlain conformably by marine shales of the Dunlin Group which occur throughout the remaining Lower Jurassic sequence.

During pre-Aalenian times preceding Brent Group deposition, NE–SW trending zones of weakness paralleling Caledonide palaeolineaments began to be established providing sites for later extensional fault movements which created the Murchison South Flank. However, tectonism affecting the Murchison area did not actively control sediment distribution at this time.

The Middle Jurassic Brent Group represents the south to north progradation of a sand dominated wave/tide influenced delta system. Deegan & Scull (1977) defined five widely distributed sub-units which from base to top are: Broom, Rannoch, Etive, Ness and Tarbert. The development of the Brent delta is represented by five main depositional units (see Fig. 3):

- pre-Brent delta sublittoral sands (Broom Formation);
- delta front (Rannoch and Etive Formations);
- delta plain (Ness Formation);
- delta abandonment (Tarbert Formation);
- post-Brent delta reworked sands ('Heather Formation' sand unit).

Early delta progradation was accompanied by mild tectonism along the established palaeolineaments, controlling sediment distribution to a limited degree. Significant control on deposition however, is not evident until Lower Ness Formation deposition where syndepositional fault movements controlled facies boundaries and separated slowly from more rapidly subsiding parts of the Murchison area. The major faulting which affected the South Flank and ultimately created the Murchison structure commenced during the late Bathonian (Tarbert Formation). Rotation of the main Murchison block occurred with relative uplift of the crestal parts and large scale differential subsidence in the South Flank area. At this time the deltaic cycle terminated as regional tilting in the East Shetland Basin allowed the passage of transgressive Tarbert Formation shoreline sands from the North over the delta.

Syndepositional faulting continued into the Upper Jurassic accompanied by the deposition of the thick transgressive Heather Formation and local erosion of the Brent Group in the crestal parts of the field. Deposition of the Heather Formation was followed by onlap of the Upper Jurassic Kimmeridge Clay Formation, differentially deposited (like the Heather) in the downthrown blocks. The distribution and thickness variations in the Kimmeridge Clay Formation suggest this episode of tectonism became considerably less active towards the end of the Jurassic.

Movement on faults was rejuvenated during post Jurassic to mid Cretaceous times associated with Late Cimmerian tectonism. As the East Shetland Basin continued to subside, differential uplift in the crestal area of the field caused minor subaerial erosion of the Upper Kimmeridge and the deposition of an incomplete Lower Cretaceous sequence. Early Cretaceous (Neocomian) sediments are absent and Barremian–Aptian sediments unconformably overlie the Jurassic. New faults became active during late Cimmerian movements, notably the east and west bounding transtensional faults, and reactivation of the South Flank faults with accompanied block rotation occurred to produce the Murchison structure, much as it is in its present form. Sequence distribution suggests that much of this tectonic activity ceased by the Mid Cretaceous.

A late Cretaceous sequence of Senonian open marine claystones and thin limestones and sandstones lie unconformably on Aptian sediments. These in turn are overlain by several thousand feet of Tertiary open to shallow marine sediments laid down in the subsiding Viking Graben.

Geophysics

The earliest geophysical map of the Murchison Field was prepared in 1971 following the UK 3rd Licensing Round award of block 211/19 to the Conoco–NCB–Gulf partnership. The map was based on a 3 km seismic grid. Infill seismic data which was shot in 1972 led to the mapping of a small four-way closure over the Murchison area with some indication of a continuous southern bounding fault. Regional seismic mapping in 1973 added weight to the trapping potential of the Murchison structure (Davies & Watts 1977).

Following the award of Norwegian block 33/9 in 1973 with Mobil as operator, two surveys were shot between 1973 and 1974 over both UK and Norwegian parts of the field. This data provided the most detailed maps prior to the drilling of the 211/19-2 discovery well and led to the identification of the east and west bounding faults (Engelstad 1986).

Further Conoco and Mobil operated surveys shot in 1976 and 1977 were combined with the early data and reprocessed by SSL to improve data quality within the Brent Group.

One further survey was acquired by the Unit in 1982 and the processed dataset was used in the 1984/86 equity redetermination.

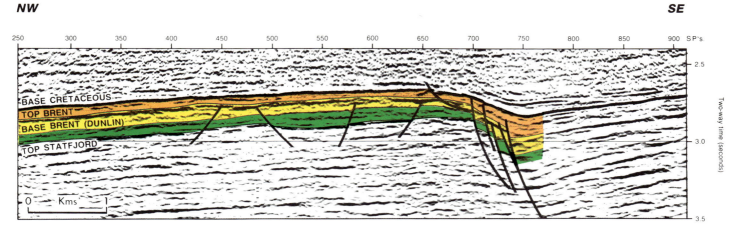

Fig. 5. Dip oriented seismic line across the Murchison Field from the 1977 CBG/211 survey.

Good data quality in the flank areas remained elusive, particularly at the Top Dunlin Group reflector.

In order to address these interpretational problems the owners elected to shoot a three-dimensional survey over the entire field. This took place between 1984 and 1985. 2600 km of 3-D data was acquired by GECO using a 25 m line spacing. The processed data which were interpreted between 1986 and 1987 form the basis for current field maps.

Seismic surveys available in order of completion are:

Reconnaissance survey	1971
Infill survey	1972
MNG (Mobil Group)	1973
MNG	1974
CNG (Conoco Group)	1974
MNG	1976
MNG	1977
CBG/211	1977
C/211-82	1982
3-D (Murchison Unit)	1985.

Trap

The structural configuration of the Murchison Field is that of a northwesterly dipping, rotated fault block bounded on three sides by normal or transtensional faults (Fig. 6).

The main axis of the field is approximately NW–SE, defined by two parallel field bounding faults, each with a strike length of about 3 miles. Throws on these faults range up to 500 ft and limit the hydrocarbon accumulation to the west and east. To the south a further bounding fault trends NE–SW and dips to the SE with a throw of up to 800 ft. This fault, approximately 3 miles in length connects the east and west bounding faults and limits the southerly extent of the field.

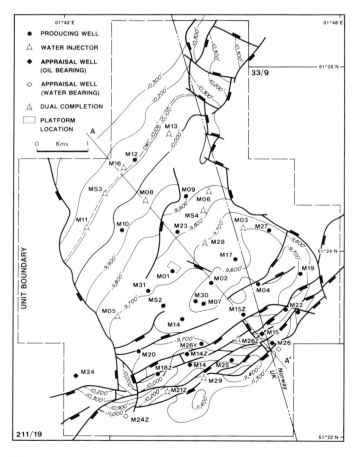

Fig. 6. Top Brent structure map showing wells drilled up to March 1989.

The South Flank is modified by a complex series of fault block terraces with similar orientations to the south bounding fault, which are progressively downthrown to the south. Throws on these faults vary between 50 ft and 250 ft. The crest of the field, a linear NE–SW trending high marks the intersection between the South Flank fault terrace and the main gently northwesterly dipping Brent block (see Fig. 7). Depth to the crest averages −9650 ft TVSS, with a maximum oil column thickness in the main field of approximately 600 ft.

To the northwest the field is delineated by an oil–water contact at −10 106 ft TVSS (Fig. 7). In the South Flank lowest known oil occurs considerably deeper at −11 142 ft TVSS. Pressure data indicate that there is reservoir communication with the main field across South Flank faults.

The vertical extent of the Murchison Field is limited by the draped Upper Jurassic Heather and Kimmeridge Clay Formations which truncate the Tarbert Formation and provide the regional seal.

Reservoir

The productive reservoir of the Murchison Field consists of Brent Group sandstones of Middle Jurassic (Aalenian–Bathonian) age.

The Brent section has been divided into zones and field correlations applied on the basis of reservoir characteristics rather than the lithogenetic Brent Group classification of Deegan & Scull (1977) or Morton & Humphreys (1983). Six units have been identified which from base to top are Basal Sand (rare), 'Lower Shaly Unit', 'Lower Mica Sand Unit', 'Upper Mica Sand Unit', 'Massive Sand Unit' and 'Middle Shaly Unit'. The 'Middle Shaly Unit' is further subdivided into a 'Lower' and 'Upper Member' separated by the 'Mid Shale'. The equivalence of these units to the Brent Group classification is shown in Fig. 3. A brief geological description of these intervals including facies relationships and reservoir properties is given below (see also Fig. 8).

'Basal Sand Unit' (Broom Formation)

The base of the Brent Group sequence in Murchison is represented by the Basal Sand Unit. It is a medium- to coarse-grained bioturbated and burrowed sandstone containing sideritized–kaolinitized ooliths and ranges from a few inches to several feet in thickness. It is interpreted as a pre-delta shallow water sublittoral sheet sand. The Basal Sand has no reservoir potential at Murchison.

'Lower Shaly Unit' (Rannoch Formation)

Overlying the Basal Sand Unit the Lower Shaly Unit comprises mudstones and siltstones with numerous wave-rippled and burrowed fine-grained sandstone lenses and laminae. These lenses and laminae increase proportionally towards the top of the unit where it passes into the base of the Lower Mica Sand. The unit which has an average thickness of 40 ft in the main field is the product of mud and fine sand deposition in front of the advancing Brent Group delta. The Lower Shaly Unit is not of reservoir quality at Murchison.

'Lower Mica Sand Unit' (Rannoch Formation)

The pro-delta mudstones and siltstones of the Lower Shaly Unit pass gradationally into a composite coarsening upward sequence of fine-grained sandstones which comprise the Lower Mica Sand Unit. This facies records the progradation and aggradation of the wave dominated delta shoreface. It has an average thickness of about 40 ft, 30–40% of which comprises sandstones of some reservoir

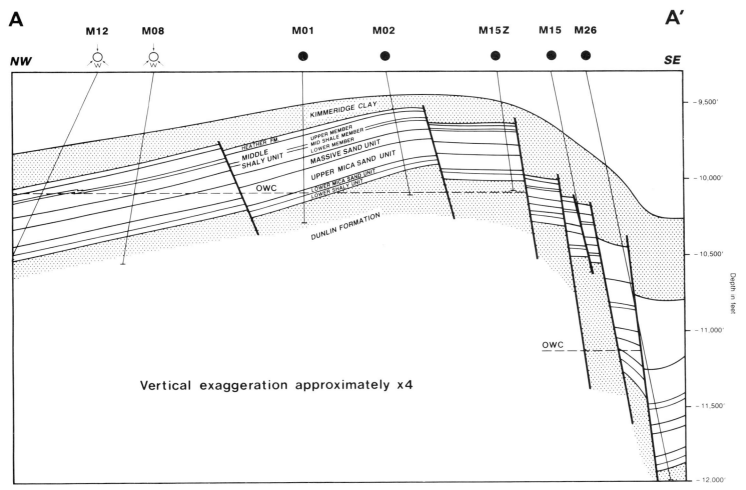

Fig. 7. Schematic geological cross-section through the Murchison Field.

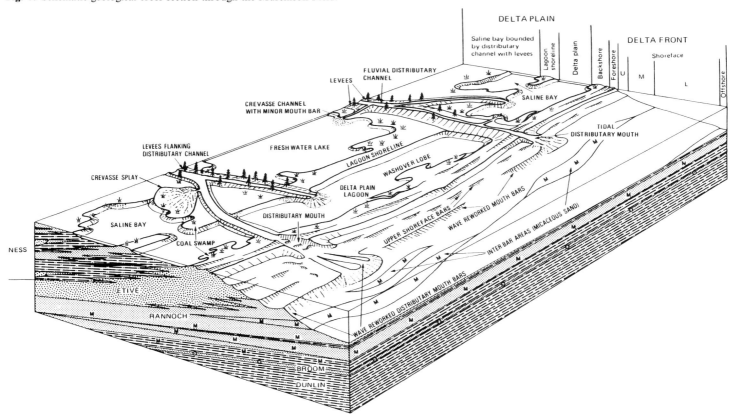

Fig. 8. Block diagram illustrating inferred facies relationships in the Brent reservoir. (From Robertson Research International Report on Sedimentology of the Murchison Field, 1986.)

potential. Average porosities are 15–16% and permeabilities are low. The unit is by far the poorest of the contributing reservoir intervals and contains approximately 4% of total oil in place.

'Upper Mica Sand Unit' (Rannoch Formation)

The coarsening upward sequence of the Lower Mica Sand Unit persists into the Upper Mica Sand Unit. Locally very coarse or pebbly sandstones are present. The sandstones may be cross-bedded, hummocky or planar laminated and are laterally variable. They represent middle shoreface mouth bar sandstones which are partially wave reworked. Thin, highly micaceous fine-grained sandstones are also present towards the top of the unit. These sandstones are planar laminated and represent low wave energy areas between the reworked mouth bars. Calcareous concretionary beds are common within this unit and may act as vertical permeability barriers. The Upper Mica Sand has an average thickness of about 140 ft of which 80–90% is considered of reservoir quality. Porosities range from 17–23% with permeabilities in the oil leg of 250–1000 md. Approximately 32% of total oil in place is contained within the unit.

'Massive Sand Unit' (Etive Formation)

This unit comprises clean cross-bedded or structureless sandstone bodies with sheet geometries and maintains the same characteristics throughout the Murchison Field. Two distinct facies types are recognized. The lower part of the Massive Sand consists of fining upward sandstone sequences, cross-stratified and current-rippled with mud drapes and partings. The upper part comprises clean fine- and medium-grained sandstones with heavy mineral concentrations and locally thin argillaceous sandstones and coals. The entire sequence is interpreted as a delta front shoreline facies, moving upward from a more distal, tidally influenced distributary channel sequence to a succession of delta front foreshore and backshore sands. Total thickness of the Massive Sand varies from 60–100 ft with a net/gross ratio of over 90%. Permeabilities in the oil leg are high, averaging 1500 md in the channel sequence with occasional very high permeability streaks of up to 10 Darcies. The Massive Sand is highly productive on Murchison and contributes to about 30% of oil in place.

'Middle Shaly Unit, Lower Member' (Lower Ness Formation)

This unit consists of a variety of sedimentological types characteristic of a non-marine delta top sequence sheltered by a coastal back-barrier bar. Fine- to medium-grained sandstones represent laterally coalescing distributary channels with crevasse splay, levee and minor mouth bar sandstones. Three distinct phases of channel activity are recognized. The sandstones have a sheet geometry and are locally isolated from each other vertically by coals and shales deposited as channel abandonment facies, and as interdistributary swamps and bays of the coastal back barrier lagoon. The lateral continuity of the sandstones and the nature of the depositional environment give the Lower Middle Shaly Unit excellent reservoir properties. The unit is variable in thickness from less than 25 ft over the crest to over 150 ft. Porosities are generally high (18–25%) and permeabilities are in the 500–100 md range. The unit contains approximately 23% of total field oil in place.

'Middle Shaly Unit, Mid Shale' (Mid Ness Shale Member)

Continental progradation ceased temporarily with the deposition of the Mid Shale, interpreted as a huge interdistributary bay deposit laid down during a period of minimal sand influx. The unit has an average thickness of about 15 ft, is correlatable field-wide and is a key marker in the Upper Brent Group. It has no intrinsic reservoir properties but acts as a fluid/pressure barrier between the Upper and Lower Members of the Middle Shaly Unit.

'Middle Shaly Unit, Upper Member' (Upper Ness/Tarbert Formations)

The Upper Member of the Middle Shaly Unit comprises two different facies types. A lower unit which corresponds to the Upper Ness Formation consists of a coarsening upwards sequence of micaceous, argillaceous sandstones with a sheet geometry. These sheet sandstones are cut by ribbon geometry channel sandstones with are in turn overlain by mudstones and thin argillaceous sandstones. The channels represent highly energetic fluvial systems which formed major multi-storey sand bodies in the subsiding and by now increasingly marine influenced delta. The Upper Unit which corresponds to the Tarbert Formation marks a complete change in sediment style and records the passage of transgressive shoreline sands from the north across the subsiding delta. Consisting of fine- to medium-grained sandstones and siltstones from the shallow marine nearshore bar and inter-bar the system records the end of the deltaic cycle.

Over the crest of the field the Upper Middle Shaly Unit is thinned by post-Brent Group erosion. The reworked Upper Middle Shaly sands were redeposited as Heather Formation sandstone which occurs as a continuous sheet over the truncated Brent Group sediments in the crest and flanks of the field. For practical reservoir description purposes the Heather Sandstone is included as part of the Middle Shaly Unit–Upper Member in the operator's zonation. Reservoir quality is very variable with porosities ranging between 11% and 23% in the sandstone units. The best permeabilities are found in channel sandstones where they commonly exceed 1500 md. The unit has an average thickness of about 50 ft with an estimated 11% of total field oil in place.

Source

The source and seal for the Murchison oil accumulation is the Upper Jurassic Kimmeridge Clay Formation. This formation consists of black shales, silty shales, brown oil shales and thin limestones deposited in a low energy anoxic marine environment. The shales are organic rich with total organic content ranging from 5–10%.

The favoured model for oil migration envisages primary migration from overpressured Kimmeridge Clay Formation along high lateral fluid pressure gradients to the Brent Group reservoir which is at most only slightly overpressured. The model proposes southward movement of hydrocarbons from the source rock across the stratigraphically lower Heather Formation to the reservoir situated updip on the inclined flank of the tilted Murchison fault block. Migration from downfaulted Kimmeridge Clay to the southeast of the South Flank via the fault system or along a path related to the Jurassic/Cretaceous unconformity surface is also likely (Brown 1984).

The Upper Jurassic is mature over much of the East Shetland Basin area. Gentle subsidence of the basin during the late Cretaceous to Tertiary brought the shales to maturity during the Palaeocene. Oil migration into the established Murchison trap occurred simultaneously and continued throughout the Tertiary period. Peak oil generation occurred during the late Eocene.

Hydrocarbons

The reservoir hydrocarbon is a low sulphur crude oil with a specific gravity of 37° API. The oil is undersaturated with an initial solution

GOR of 524 SCF/STB providing good well flow characteristics. Bubble point pressure is 1850 PSIG and is unlikely to be reached under future field operating conditions (Massie *et al.* 1985). Pressure trends in the reservoir show an initial decrease from 6300 psig when the accumulation was supported by the aquifer only to 4500 psig when injection support commenced in 1981. The reservoir pressure has remained relatively constant since the middle of 1982 as a result of balancing total fluid production with water injection.

Reserves

The field has a total STOOIP of 790 MMBBL with current booked reserves of 340 MMBBL which represents a recovery efficiency of 43%. Aquifer influx and injected water are the main drive mechanisms.

The original development strategy was a Lower Brent line drive flood with 3 rows of production wells and 2 rows of water injectors. The production wells were perforated initially in the lower productivity 'Mica Sand Unit' sandstones to encourage sweep of these sandstones, minimize water override through to the 'Massive Sand' and thus maximize field recovery. 'Lower Middle Shaly' sandstones, generally isolated from the underlying Massive/Mica complex were developed comingled with 'Mica' production. The 'Upper Middle Shaly Unit' was originally used for gas storage while the gas export line was built. Development of the 'Upper Middle Shaly' reserves began in 1986/87 initially using dual completions.

Perforation in the high productivity Lower Brent sandstones gave the field a high initial production rate. From Setember 1980 when the field was brought onstream production had built up to 75 000 BOPD by April 1981. At this time water injection commenced. Oil production peaked in February 1983 at 126 700 BOPD. Throughout 1983/84 production was kept at a plateau of 105 000 BOPD. The field came off plateau production in January 1985 and the oil rate has fallen steadily as the water cut has increased, to a current rate of 40–50 000 BOPD. The water production rate overtook the oil production rate in September 1986 and is currently averaging 130 000 BOPD. Reservoir simulation model runs indicate that the field will produce oil economically beyond 1998.

Oil production is via the Dunlin and Cormorant Fields, then by Brent Field pipeline to the Sullom Voe terminal on the Shetland Islands. Gas sales which started in 1983 go via the Northern Leg Gas pipeline and FLAGS system to St Fergus in Scotland.

This paper has been compiled from various sources within Conoco (UK) Ltd, in addition to the references cited. The ideas expressed have also been drawn from the work and experience of a number of the author's colleagues, in particular W. T. Gans, Conoco International, D. B. Haig, Conoco (UK) Ltd, London and A. J. Machell, Conoco Norway Inc., Stavanger. The author also acknowledges Conoco (UK) Ltd, and Murchison Unit members for permission to publish this paper.

Murchison Field data summary

Trap
Type	Tilted fault block
Depth to crest	−9500 ft TVSS, average −9650 ft TVSS
Lowest closing contour	−10 450 ft TVSS
Hydrocarbon contacts	−10 106 ft TVSS Main Field, −11 142 ft TVSS South Flank
Gas column	—
Oil column	600 ft maximum

Pay zone
Formation	Brent
Age	Mid-Jurassic
Gross thickness (average; range)	425 ft; 250–720 ft
Net/gross ratio (average; range)	0.7; 0.45–0.95%
Porosity (average; range)	22%; 16.5–26.5%
Hydrocarbon saturation (average; range)	70% 65–82%
Permeability	500–1000 md; maximum 10 d

Hydrocarbons
Oil gravity	37° API
Oil type	low sulphur crude
Bubble point	1850 psig at 230°F
Gas/oil ratio	524 SCF/STB
Formation volume factor	1.313

Formation water
Salinity	14 000 mg/litre
Resistivity	0.312 ohm m at 60°F

Reservoir conditions
Temperature	230°F
Pressure, initial	6300 psig
Pressure gradient in reservoir	0.31 psi/ft

Field size
Area (Top Brent)	4200 acres
Gross rock volume	1125×10^3 acre/ft
Hydrocarbon pore volume	5825×10^6 CuF. (Res. Vol.)
Recovery factor-oil	43%
Drive mechanism	water injection, aquifer support
Initial recoverable hydrocarbons	
Oil	340 MMBBL
Gas	138.3 BCF

Production
Start-up data	Drilling June 1980, production September 1980
Development scheme	Single platform, 27 slots + 3 subsea completions
Production rate	Peak 127 000 BOPD (1983); January 1991: 39 100 BOPD
Cumulative production (31 January 1991)	288.0 MMBBL
Number/type of wells	30; 19 producers, 10 water injectors, 1 dual completion
Secondary recovery method(s)	—

References

BROWN, S. 1984. The Jurassic, *In*: GLENNIE, K. W. (ed.) *Introduction to the Petroleum Geology of the North Sea.* Blackwell Scientific Publications.

DAVIES, E. J. & WATTS, T. R. 1977. The Murchison Oil Field. *In: Mesozoic Northern North Sea Symposium,* (Norsk Petroleumsforening), Article 15.

DEEGAN, C. E. & SCULL, B. J. (compilers) 1977. *A proposed standard lithostratigraphic nomenclature for the Central and Northern North Sea.* Bulletin Norwegian Petroleum Directorate No.1; Institute of Geological Sciences Report, 77/25.

ENGELSTAD, N. 1986. Murchison, *In*: Spencer *et al.* (eds) *Geology of the Norwegian Oil and Gas Fields.* Norwegian Petroleum Society.

MASSIE, I., BEARDALL, T. J., HEMMENS, P. D. & FOX, M. J. 1985. Murchison—A Review of the Reservoir Performance During the First Five Years. *60th Annual Technical Conference and Exhibition of the Society of Petroleum Engineers, Las Vegas, September 22–25,* 1985.

MORTON, A. C. & HUMPHREYS, B. 1983 The Petrology of the Middle Jurassic Sandstones from the Murchison Field, North Sea. *Journal of Petroleum Geology*, **5**, 245–260.

The Ninian Field, Blocks 3/3 & 3/8, UK North Sea*

E. J. VAN VESSEM & T. L. GAN

Chevron UK Ltd, 2 Portman Street, London W1H 0AN, UK

Abstract: The Ninian Field, located in the northern North Sea, lies in the East Shetland Basin on the west side of the Viking Graben. The field straddles Blocks 3/3 and 3/8 and is developed under a unitization agreement with Chevron UK Limited as operator. The structure is a westward tilted fault block. The estimated original oil-in-place contained in the marine–deltaic sandstones of the Middle Jurassic Brent Group, is 2920 MMBBL, of which an estimated 35 to 40% is recoverable. The oil is a paraffinic–naphthenic type with an API gravity of 36°. The field development consists of three fixed platforms with a total of 109 drilling slots. The natural drive in the field is negligible so that water flooding is required. The production of the Ninian Field started in December 1978 and reached a peak of 315 000 BOPD in the summer of 1982. At the end of 1988 over 811 MMBBL had been produced.

The Ninian Field is located in the northern North Sea, on the continental shelf between the United Kingdom and Norway. The nearest landfall is the Shetland Islands which are roughly 90 miles to the southwest (Fig. 1).

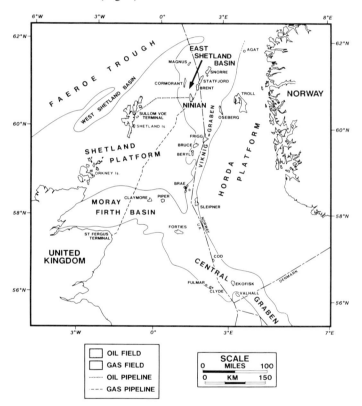

Fig. 1. Northern North Sea salient features map and location of Ninian and other selected fields.

The field extends across Blocks 3/3 and 3/8. The approximate centre of the field is defined by coordinates 60° 50′ N, and 1° 28′ E. Water depths in the area range from 440–490 ft.

The Ninian structure is a westward tilted fault block modified by subsequent faulting. The hydrocarbons occur in the Middle Jurassic marine–deltaic sandstones of the Brent Group. The proven productive area is about 22 000 acres with estimated ultimate reserves of 1045 MMBBL.

In the immediate vicinity there are several other important oil fields of which the Brent and Statfjord Fields to the northeast are the largest. Other producing fields in the area are North Alwyn, Hutton, Dunlin, Cormorant and Heather (Fig. 2).

The field was named after Saint Ninian, the patron saint of the Shetland Islands.

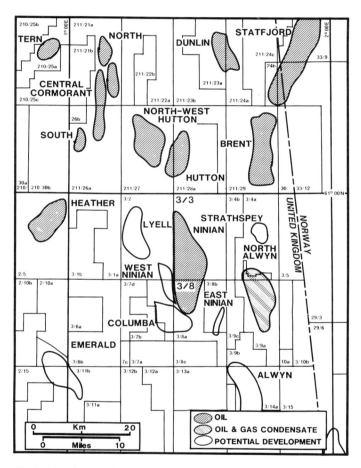

Fig. 2. Location of oil fields in the Ninian area.

History

The exploration licences covering the Ninian Field Blocks 3/3 and 3/8 were awarded by the British Government in the fourth licensing round in March 1972. Original participants in Licence P-202, which includes block 3/3, were: Burmah Oil (North Sea) Ltd (30%), Imperial Chemical Industries Ltd (26%), California Oil Company Ltd (24%), Murphy Petroleum Ltd (10%) and Ocean Exploration Company Ltd (10%).

---

* An expanded version of this paper appeared in the AAPG Treatise of Petroleum Geology Atlas of Oil and Gas Fields.

The original participants in Licence P-199 which comprises Block 3/8 were: BP Petroleum Development Ltd (50%), Ranger Oil (UK) Ltd (20%), Scottish Canadian Oil and Transportation Company Ltd (7%), London and Scottish Marine Oil Company Ltd (15.5%), Cawoods Holdings Ltd (3.75%) and National Carbonising Ltd (3.75%).

Over the years partners and licence interests have changed. The current licence holders and their equity shares in the Ninian Unit are as follows:

Britoil plc	21.38%
Chevron UK Ltd	17.10%
Enterprise Oil plc	18.53%
London and Scottish Marine Oil Company Ltd	17.25%
Murphy Petroleum Ltd	7.12%
Ocean Exploration Company Ltd	7.12%
Ranger Oil (UK) Ltd	11.50%

The late sixties discovery of oil in the central North Sea, followed by the discovery of the Brent Field in the northern North Sea in 1971, demonstrated the potential of the Jurassic in terms of source rock and reservoir.

The attractive key elements for drilling Ninian were good quality reservoir sandstone and the large size of the structure adjacent to a graben where source rock was likely to have generated hydrocarbons prior to trap formation.

Initial seismic mapping showed that the structure straddled the boundary between Blocks 3/3 and 3/8. Early exploration proceeded separately with BP as operator for Block 3/8 and Burmah operating Block 3/3. Both licence groups drilled their first wells simultaneously in late 1973–early 1974.

The discovery well, 3/8-1 (Fig. 3), was spudded on 16 September 1973 and was completed on 12 March 1974. The well bottomed in the Triassic at a total depth of 10 570 ft (10 475 ft TVSS). The lithological sequence encountered consists of about 6100 ft Tertiary sands and clays, 2700 ft of Cretaceous marls, limestones and claystones, 630 ft of Jurassic clay and sandstones, and 600 ft of Triassic sandstone, siltstone and claystone. The objective Middle Jurassic reservoir has a gross pay of 193 ft. Six open hole formation interval tests yielded 34° API oil.

The first well in Block 3/3 spudded on 4 October 1973 and drilled to a total depth of 10 699 ft (10 616 ft TVSS) in the Triassic. The well was completed on 10 April 1974. The primary objective, the Middle Jurassic sandstones, has a gross pay of 330 ft. Four separate well tests were conducted and a maximum flow of 7388 BOPD (34 to 35° API gravity) on a 0.5 in choke was recorded on an upper sandstone of the Middle Jurassic section. The overall lithology and thickness recorded are very similar to those of well 3/8-1. The secondary objective in the Palaeocene was found to be water wet.

Six subsequent appraisal wells established the extent of the field and showed it to have giant status with recoverable reserves in the order of 1000 MMBBL. Well and seismic data indicated that the larger portion of the reserves was contained in Block 3/3.

The joint development of the Ninian Field in Blocks 3/3 and 3/8 was initially led by Burmah, until Chevron acquired the operatorship in March 1975. BP is the operator for the construction and operation of pipelines and for securing terminal facilities. The construction of the production facilities was also started in 1975.

The field is developed with three platforms. The Central Platform is a gravity type concrete structure, which at the time of construction was the largest man-made structure in the world; the Southern and Northern Platforms are steel-piled jacket structures. All platforms have fully integrated production, injection, drilling and accommodation facilities. The total number of available drilling slots is 109. The Central and Southern Platforms were installed in 1978 and the Northern Platform followed a year later. First production started in December 1978 and at the end of 1988 the field had produced over 811 MMBBL. Average daily production reached a peak of 315 000 BOPD in the third quarter of 1982. Daily production in January 1989 averaged 130 000 BOPD.

The oil and natural gas liquids are transported via pipeline to Sullom Voe terminal (Shetland Islands) and the associated gas was used to power the platform. Excess gas was taken via the FLAGGS (Far North Liquid and Gas Gathering System) pipeline to St Fergus terminal in Scotland. Due to declining gas production this stopped in 1988.

Field stratigraphy

Ninian Field is located in the East Shetland Basin (Fig. 1) on the west side of the Viking Graben. This basin developed as part of the northwest European intracratonic basin complex under the influence of an extensional stress system related to the Mesozoic disintegration of the Pangaea megacontinent. The East Shetland Basin is bounded by the north–south trending Viking Graben, formed during Permo-Triassic times as part of the rift system through the North Sea. The term 'graben' is somewhat misleading as the Viking Graben is a half graben developed on a major set of easterly dipping faults with associated trends of westward tilted rotational fault blocks (Gibbs 1987).

Sedimentation in the basin lasted from Permo-Triassic until Recent times, with interruptions during the Jurassic and Early Cretaceous due to Kimmerian tectonic activity. This resulted in periods of erosion and marine transgression, of which the most regionally significant is the Base Cretaceous unconformity.

The Ninian structure, a gently west-dipping fault block, gradually developed during Mid–Late Jurassic times with the strongest accentuation occurring in Late Jurassic–Early Cretaceous times. The Middle and Upper Jurassic stratigraphic section shows the presence of unconformities. On the flank of the field they are small but become more significant in the crestal area. This is exemplified by the Base Cretaceous unconformtiy. On the flank the angularity of the unconformity increases from slight to pronounced near the crest where Cretaceous rocks may rest directly on those of Early Jurassic or Triassic age.

The Ninian Field stratigraphy is illustrated by the generalized composite stratigraphical column in Fig. 4.

The deepest well in the Field, 3/3-4A (Fig. 3), bottomed at 14 434 ft TVSS in crystalline basement. Overlying the basement is the Cormorant Formation, a series of fine-grained terrigenous clastic red beds of Triassic age.

The continental red bed facies is replaced in the Lower Jurassic by the marine shale facies of the Dunlin Group. In between the two is the transitional Statfjord Formation containing potential reservoir sandstones of fluviatile-marginal marine nature.

The Middle Jurassic Brent Group is the prime reservoir and largely represents a progradational deltaic sedimentation cycle. Five sub-units (Formations) are generally recognized (Deegan & Scull 1977): they are, in ascending order, Broom, Rannoch, Etive, Ness and Tarbert. The Tarbert sediments, representing renewed marine sedimentation, are either absent or thin over Ninian.

The Upper Jurassic section consists of claystones of the Heather and Kimmeridge Clay Formations. The Heather Formation unconformably overlies the Brent Group. The Heather and Kimmeridge Clay Formations are separated by an unconformity in the structurally higher parts of Ninian. The claystones of the Kimmeridge Clay Formation have a high organic content and are excellent oil-prone source rocks.

The Cretaceous rocks separated from the Jurassic by the Base Cretaceous unconformity, consist largely of fine calcareous mudstones, shales and occasionally thin limestones of the Shetland Group. In the Ninian area, Cretaceous sediments of pre-Barremian age are generally absent and the Lower Cretaceous Cromer Knoll Group is only thinly developed.

Cenozoic sediments reflect the overall subsidence and infilling of the basin. Deposition over the Cretaceous–Tertiary boundary was apparently continuous. The influence of the Early Tertiary Lara-

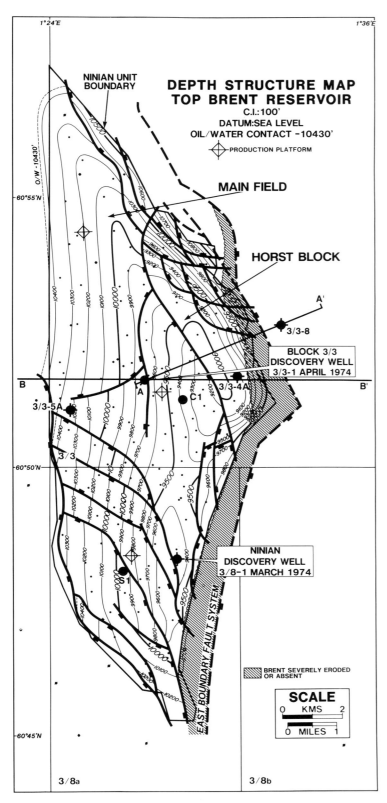

Fig. 3. Ninian depth structure map of the top Brent reservoir. A–A' and B–B' indicate the locations of the seismic profiles of Fig. 5.

mide tectonic phase that affected the basin sedimentation more severely elsewhere, is here restricted to the sudden introduction of coarser clastics in the Lower Palaeocene Montrose and Rogaland Groups.

These sandstones are interpreted as submarine fan deposits. The Rogaland sandstones are locally well developed in the Ninian area and have good reservoir potential.

The Eocene, Oligocene and Miocene rocks are predominantly argillaceous with interbedded siltstone and thin sandstones, while the Plio-Pleistocene sequence consists of massive, generally coarse sandstones, gravel and claystones.

Geophysics

Prior to discovery the seismic data coverage consisted of lines of various origins, an example of which is shown in Fig. 5 (line B-B') together with a modern 3-D line. The post-discovery seismic coverage comprises four different vintage surveys. The earliest field survey, acquired in 1974, is a 1 km grid of 24-fold data. In a 1978 resurvey of the area, a 0.5 km grid of 48-fold data was recorded. To solve the complexities of the faulted eastern part of the field, two overlapping 52-fold 3-D surveys with 50 m line spacing were shot in 1981. These surveys yielded 530 3-D migrated profiles with a total

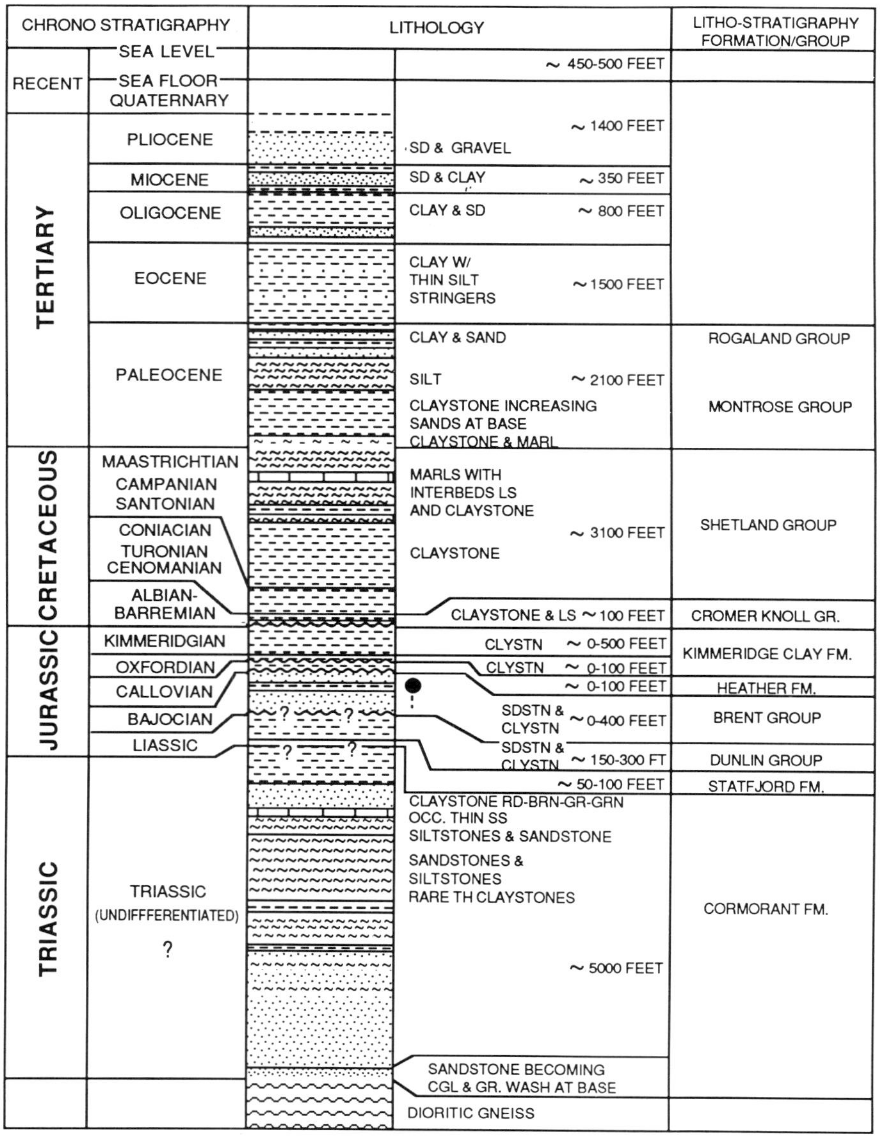

CHRONO STRATIGRAPHY		LITHOLOGY	LITHO-STRATIGRAPHY FORMATION/GROUP
RECENT	SEA LEVEL — SEA FLOOR — QUATERNARY	~ 450-500 FEET	
TERTIARY	PLIOCENE	SD & GRAVEL ~ 1400 FEET	
	MIOCENE	SD & CLAY ~ 350 FEET	
	OLIGOCENE	CLAY & SD ~ 800 FEET	
	EOCENE	CLAY W/ THIN SILT STRINGERS ~ 1500 FEET	
	PALEOCENE	CLAY & SAND	ROGALAND GROUP
		SILT ~ 2100 FEET	
		CLAYSTONE INCREASING SANDS AT BASE	MONTROSE GROUP
		CLAYSTONE & MARL	
CRETACEOUS	MAASTRICHTIAN CAMPANIAN SANTONIAN	MARLS WITH INTERBEDS LS AND CLAYSTONE	SHETLAND GROUP
	CONIACIAN TURONIAN CENOMANIAN	~ 3100 FEET CLAYSTONE	
	ALBIAN-BARREMIAN	CLAYSTONE & LS ~ 100 FEET	CROMER KNOLL GR.
JURASSIC	KIMMERIDGIAN	CLYSTN ~ 0-500 FEET	KIMMERIDGE CLAY FM.
	OXFORDIAN	CLYSTN ~ 0-100 FEET	
	CALLOVIAN	~ 0-100 FEET	HEATHER FM.
	BAJOCIAN	SDSTN & CLYSTN ~ 0-400 FEET	BRENT GROUP
	LIASSIC	SDSTN & CLYSTN ~ 150-300 FT	DUNLIN GROUP
		~ 50-100 FEET	STATFJORD FM.
TRIASSIC	TRIASSIC (UNDIFFFERENTIATED) ?	CLAYSTONE RD-BRN-GR-GRN OCC. THIN SS SILTSTONES & SANDSTONE SANDSTONES & SILTSTONES RARE TH CLAYSTONES ~ 5000 FEET	CORMORANT FM.
		SANDSTONE BECOMING CGL & GR. WASH AT BASE	
		DIORITIC GNEISS	

Fig. 4. Stratigraphic column, Ninian Field area, based on wells 3/3-4A and 3/3-5A. The Brent Group is the pay zone (after: Albright *et al.* 1980). For well locations see Fig. 3.

coverage of approximately 5300 km. The most recent survey was a 1983 2-D–3-D reconnaissance survey acquired to detail the western boundary of the field. This survey was recorded 60-fold with a 200 m line spacing.

The strongest and most regionally continuous reflector near the Brent Group is from the Base Cretaceous unconformity. The high reflectivity at this level is caused by the high impedance contrast between the thin Lower Cretaceous limestones and the underlying low velocity Kimmeridge Clay Formation.

Below the Base Cretaceous unconformity weaker reflectors are identified as the Base Kimmeridge Clay Formation, the Base Brent Group, and the Top Cormorant Formation. The quality of the seismic data is fair to good; however, over the crestal and eastern part of the field, identification of the reflectors is difficult because of decreasing thickness of individual units and increased structural complexity.

The Ninian structure (Fig. 6) is divided into two components, the west-dipping flank, constituting the main part of the field, and a horst block which forms the eastern part of the Ninian structure. Figure 6 also shows the thinning of the Heather and Kimmeridge Clay Formations towards the East Boundary Fault System. The Heather Formation is completely truncated in the horst block area. In the main field, the mappable reflector nearest to the Brent Group is the Base of the Kimmeridge Clay Formation. Top Brent cannot be mapped because the Kimmeridge/Heather interface has a stronger reflectivity than the Heather/Brent contact. In addition the Heather is too thin for the top to be resolved on seismic data. Over the horst block both the Heather and Kimmeridge Clay Formations become too thin to be resolved seismically, so that in this area the Base Cretaceous unconformity is the mappable horizon nearest to the Brent reservoir.

Trap

The two structural components of Ninian, the west-dipping fault block, or main field, and the horst block (Figs 3 & 6), have different trapping styles.

The main field is a simple structural trap. The eastern closure is against the west-bounding horst block fault, where the throw varies from several hundred feet in the north to some 200 ft or less in the south. Generally the Brent Group reservoir is juxtaposed against claystones of the Jurassic Dunlin Group. The result of pulse tests suggests that the northern part of the fault is sealing. However, pressure and production data from the horst block indicates that some communication exists with the main field, suggesting that the southern area of the fault is only partially sealing.

The westward down-dip extension of the field is in the northwest determined by the oil–water contact and, in the southwest by a series of faults. The combined throws of these faults result in

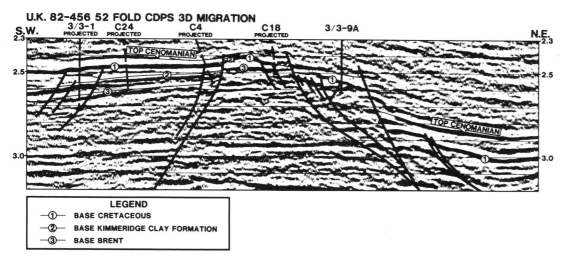

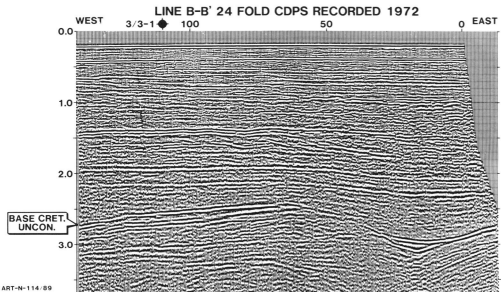

Fig. 5. Examples of an original seismic line (1972) and detail from a modern 3-D line. For location see Fig. 3.

complete reservoir separation between Ninian and the hydrocarbon accumulations of the adjacent Columba and West-Ninian structures (Fig. 2). All indications are that these faults are sealing. Claystones of the Upper Jurassic Heather and Kimmeridge Clay Formations provide the vertical seal in the main field.

The horst block, forming the eastern part of the field, is structurally complicated and is separated from the main field by the earlier described westerly downthrown normal fault. A major easterly downthrown normal fault system forms the eastern boundary with greater than 1500 ft of throw.

The horst block forms the structural crest of Ninian, which at the level of the Base Kimmeridge Clay Formation, marks a reversal in dip to the east. In the western part of the horst block at the base of the Brent Group, a similar change from west to slight east dip is observed. The structure in the eastern part of the horst block is complex owing to a closely spaced series of down-stepping faults. The dip of beds in these small fault blocks is difficult to discern even on the modern 3-D seismic data. In addition, well data are sparse here, while the effect of incomplete sections caused by faulting and possible subtle variations in facies and stratigraphic thickness further impede an integrated structural-stratigraphical interpretation. The dip at the Base of the Brent Group in the eastern part of the horst block is most likely to the west. Thus, the trap in the western part of the horst is a combination of faulting and structural roll over, while in the eastern part numerous small unconformity traps appear to be present. Lateral and top seals in the horst block are provided by the Kimmeridge Clay Formation and the claystones of the Cromer Knoll or Shetland Groups.

The Ninian Field is about 22 km long trending from north to south and, on average, 4 km wide with a productive area of about 22 000 acres. The oil–water contact is taken to be at 10 430 ft TVSS indicating the trap is not filled to spill point.

Reservoir

In the Ninian area there are three potential reservoirs of which only the Middle Jurassic Brent Group is oil-bearing; both the Palaeocene Rogaland Group and Early Jurassic Statfjord Formation are water bearing.

By the end of 1988, 136 wells had been drilled within the field limits. For most wells, a standard set of wireline logs is available consisting of density, neutron, gamma-ray, laterolog and/or induction and sonic logs. Forty nine wells have been cored for detailed stratigraphical, sedimentological and petrological analysis. Integration of core studies and log response enhances reliability of the stratigraphical correlation of uncored wells.

A brief summary of facies and reservoir characteristics of the Brent Group component formations (Deegan & Scull 1977; Vollset & Doré 1954) is given below. The Chevron zonal subdivision is shown in brackets. Figure 7 illustrates a typical reservoir profile.

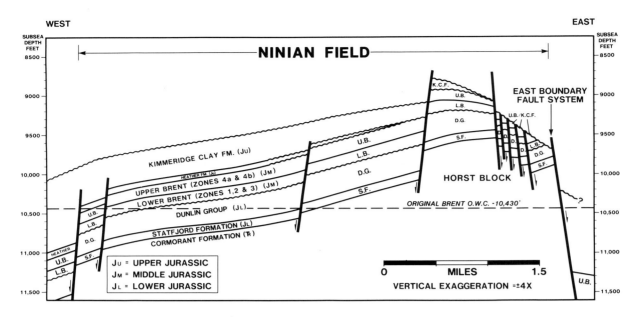

Fig. 6. Diagrammatic west–east structural cross-section typical for the central-northern part of the Field.

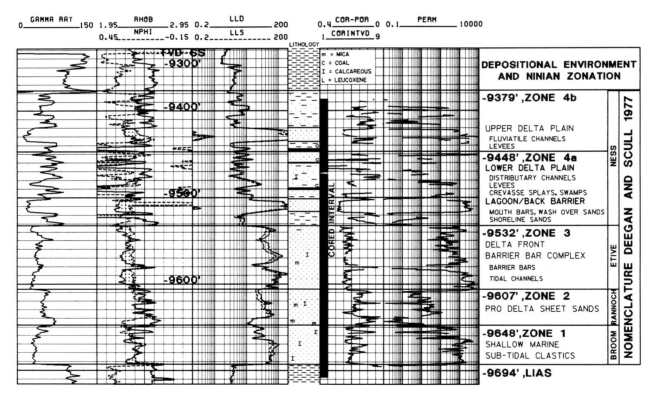

Fig. 7. Profile of well 3/3-C1, showing log and core derived data in combination with the zonation and the interpretation of environmental deposition. For well location see Fig. 3.

Broom Formation (Zone 1)

Lithology. Medium- to coarse-grained, feldspar-rich sandstones, often argillaceous, with a general fining upwards trend. The basal part is often pebbly and arkosic. Incidental occurrence of heavy minerals and calcite cementation is present in the lower part of the Formation.
  Depositional environment. Shallow marine to tidal flat.
  Average reservoir properties. Net pay 40 ft, porosity 19% and permeability 100 md.

Rannoch Formation (Zone 2)

Lithology. Very fine-grained sub-arkosic sandstones, rich in detrital mica (mostly muscovite and biotite), argillaceous and locally cemented by calcite. This unit is characterized by an overall coarsening upward trend with decreasing mica and argillaceous content.
  Depositional environment. Shallow marine pro-delta sheet sands with micaceous deposition concentrated through current and wave winnowing.
  Average reservoir properties. Net pay 30 ft, porosity 16% and permeability 25 md.

Etive Formation (Zone 3)

Lithology. Dominantly medium- to coarse-grained, massive and homogeneous sandstones, locally cemented near the base by concretionary calcite. Mica appears as a minor constituent. The lower part of the Etive Formation consists of stacked, fining-upward sequences of coarse- to fine-grained sandstones. These units are gravelly to pebbly at the base, grading upwards into micaceous laminae with occasionally finely divided wood debris. Locally there is a persistent 5–10 ft interval with heavy minerals, mainly zircon and ilmenite (or its weathering product leucoxene). The upper part of the Formation consists of mostly fine- to medium-grained sandstones with argillaceous partings and some bioturbation. Often these sandstones are capped by a coal or rootlet bed.

Depositional environment. The variations in lithology suggest various environments which, to some extent, are all related to a regressive prograding barrier bar complex comprising tidal inlet fill facies, high or low energy barrier shore facies and back barrier sand apron facies.

Average reservoir properties. Net pay 40 ft, porosity 20% and permeability 500 md.

Ness Formation (Zones 4a and 4b)

Lithology. The Ness lithology is diverse and consists of varying proportions of interbedded fine- to coarse-grained sandstones, siltstones, claystones and coals. The lower part (Zone 4a) is often dominated by carbonaceous siltstones and claystones, interbedded with coals and fine- to medium-grained sandstones, frequently micaceous or carbonaceous. The upper section (Zone 4b) consists mostly of sequences of fining-upward sandstones, interbedded with claystones. The sandstones are pebbly at the base and silty to shaley at the top with rootlet zones.

Depositional environment. The lower and upper delta plain environment. The lower basal part of the Ness Formation consists of lagoon to back barrier facies with mouth bar, washover and shore line sands. There is a progressive change to domination by a fluvial environment with stacked channel sequences and related levee, crevasse splay and inter-distributary swamp deposits.

Average reservoir properties. In Zone 4a: net pay 40 ft, porosity 17%, permeability 150 md; in Zone 4b: net pay 50 ft, porosity 20% and permeability 1500 m.

Tarbert Formation

Lithology. Glauconitic, argillaceous, micaceous, fine- to medium-grained sandstones.

Depositional environment. Transgressive marine. In Ninian thin Tarbert sequences are recognized with some confidence in only two cores. For practical reasons, these minor remnants are included in Zone 4b.

Source

The Kimmeridge Clay Formation, largely Oxfordian–Kimmeridgian in age, is the principal source rock in the East Shetland Basin. The kerogen is type II from marine sediments with autochthonous organic matter, derived from a mixture of phytoplankton, zooplankton and micro-organisms, deposited in an anoxic environment. In the Ninian area the total organic carbon (TOC) of this unit ranges from three to nine percent. The yield of the source rocks is ten kgs/tonne.

In the Dunlin and Heather Formations and shales of the Brent Group, TOC values may reach several percent. However, the kerogen type in these shales is dominantly inertinitic with little oil generating potential.

The claystones of the Kimmeridge Clay Formation in the Ninian Field are insufficiently mature to have generated oil. The burial history diagram for well 3/3-8 (Fig. 8), which is a deep well on the east side of the east boundary fault (Fig. 3), shows an onset of oil generation of approximately fifty million years ago (Early Tertiary). In still deeper areas around Ninian oil generation is expected to have started earlier, probably during the Late Cretaceous. The formation of the trap in Late Jurassic–Early Cretaceous times clearly pre-dates the time of oil migration.

Oil migration to the Ninian structure has to rely on contact between the source rock and the reservoir, or on flow along fault planes or unconformity surfaces. Both the Base Cretaceous and the Base Upper Jurassic unconformities and the east boundary fault system may have served as migration paths.

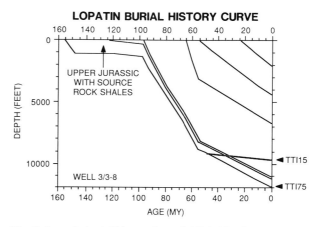

Fig. 8. Lopatin burial history for well 3/3-8. The theoretical maturity values (TTI) of 15 and 75 correspond to the onset of oil and peak oil generation respectively. For well location see Fig. 3.

Hydrocarbons

The initial reservoir pressure in Ninian was 6490 psi at a datum of 9750 ft TVSS. The oil saturated interval has, on average, a pressure gradient of 0.335 psi/ft, which is consistent with the measured fluid density from pressure–volume–temperature (PVT) analysis. All intervals were originally pressured above the normal hydrostatic gradient. The estimated reservoir temperature at 9750 ft TVSS is 215°F.

Ninian oil is a paraffinic–naphthenic type, of 36.4° API gravity, with 0.39 weight percent sulphur and a viscosity of 0.91 cp at a saturation pressure of 1330 psi. The original gas/oil ratio (GOR) was 301 SCF/STB and the formation volume factor (FVF) was 1.20 reservoir BBL/STB.

The water salinity is 35 000 ppm (TDS) giving a resistivity at reservoir conditions of 0.11 ohm m^2/m.

The test data for the initial wells were already mentioned in the History section. More wells in the field were tested as Ninian was appraised, with results broadly in line with those of the 3/3-1 well. The first producing well, 3/8-S1, had an initial flow rate of 25 000 BOPD.

Reserves

The Ninian Field has little natural drive and reservoir pressure is maintained by water injection. The injection capacity for Zone 4b is much greater than that of the underlying Zones 1 through 4a. A pressure barrier separates Zone 4b from the underlying zones. Hence, for proper pressure maintenance the interval comprising Zones 1 through 4a and the Zone 4b interval were developed as separate reservoir units.

Production difficulties in the field are twofold. First, variation in primary depositional conditions determines reservoir quality, distribution and continuity of the reservoir. Secondly, faults create complications as they act as barriers to flow, and through the juxtaposition of different reservoir zones, can cause unexpected cross-flow patterns in the reservoir.

The field will be produced entirely through water flooding. Estimated recoverable reserves are 1045 MMBBL. The recovery efficiency was estimated earlier to be 39%. However, the oil in place figure was revised upward in 1989 to 2920 MMBBL and an adjustment to the recovery factor may follow, when a current simulation study is concluded.

The authors have drawn freely on the experience of many earth scientists who over the years have contributed to our understanding of the Field. We would like to extend a general acknowledgement to all of them for their contributions.

The views presented in this paper are those of Chevron and do not necessarily reflect those of the Ninian partners.

Chevron management and Ninian partners are thanked for giving permission to publish this paper.

Ninian Field data summary

Trap
Type	Structural trap/structural truncation unconformity traps
Depth to crest	8946 ft TVSS
Oil/water contacts	10 430 ft TVSS
Oil column	167 ft (average)

Pay Zone
Formation	Brent Group
Age	Middle Jurassic, Bajocian–Bathonian
Gross thickness (average/range)	215 ft/150–370 ft
Net/gross ratio (average)	0.7
Cut off for N/G	1 md
Porosity (average/range)	Upper Reservoir: 20%/16–25% Lower Reservoir: 19%/14–24%
Hydrocarbon saturation	60–65%
Permeability (average/range)	Upper Reservoir: 700 md/20–2850 md Lower Reservoir: 250 md/15–1300 md

Hydrocarbons
Oil gravity	36.4° API
Oil type	Paraffinic-naphthenic
Bubble point	1330 psi
Gas/oil ratio	301 SCF/STB
Formation volume factor	1.20 BBL/STB

Formation water
Salinity	35 000 ppm TDS
Resistivity	0.11 ohm m^2/m (reservoir conditions)

Reservoir conditions
Temperature	215° F at 9750 ft TVSS
Pressure	6490 at 9750 ft TVSS
Pressure gradient in reservoir	0.335 psi/ft

Field size
Area	22 000 acres
Gross rock volume	4 760 000 acre-ft
Hydrocarbon pore volume	744 BBL/acre-ft
Recovery factor—oil	35–40%
Drive mechanism	Water injection
Recoverable hydrocarbons:	
oil	1045 MMBBL
NGL	30 MMBBL
gas	150 BCF

Production
Start-up date	23 December 1978
Initial rate first well	25 000 BOPD
Development scheme	3 platforms with a total of 109 slots; oil and NGL transported via pipelines to Scotland. Dual reservoir completion, with variable number of producing or injection wells.
Production rate	Peak rate 315 000 BOPD (3rd quarter 1982); January 1989 130 000 BOPD
Cumulative production	811 355 336 BBL (to 31 December 1988)
Secondary recovery method	Water injection

References

ALBRIGHT, W. A., TURNER, W. L. & WILLIAMSON, K. R. 1980. Ninian Field, U.K. sector, North Sea, *In*: HALBOUTY, M. T. (ed.) *Giant Oil and Gas Fields of the Decade, 1968–1978*. AAPG Memoir **30**, 173–194.

DEEGAN, C. E. & SCULL, B. J. 1977. *A Standard Lithostratigraphic Nomenclature for the Central and Northern North Sea*. Institute Geological Sciences, Report 77/25; (also Norwegian Petroleum Directorate. Bulletin 1, 1–36).

GIBBS, A. D. 1987. Deep Seismic Profiles in the Northern North Sea, *In* BROOKS, J. & GLENNIE, K. W. (eds) *Petroleum Geology of North-West Europe*. Graham & Trotman, 1025–1028.

VOLLSET, J. & DORÉ, A. G. 1984. A Revised Triassic and Jurassic Lithostratigraphic Nomenclature for the Norwegian North Sea. *Norwegian Petroleum Directorate, Bulletin*, **3**, 1–40.

The Osprey Field, Blocks 211/18a & 211/23a, UK North Sea*

JOHN W. ERICKSON[1] & C. D. VAN PANHUYS[2]

[1] *Esso Exploration and Production, UK Ltd, 21 Dartmouth Street, London SW1H 9BE, UK*
[2] *Al Furat Petroleum Co., PO Box 7660, Damascus, Syria*

Abstract: The Osprey Oilfield is located 180 km northeast of the Shetland Islands in Blocks 211/23a and 211/18a in the UK sector of the northern North Sea. The discovery well 211/23-3 was drilled in January 1974 in a water depth of 530 ft. The trap is defined at around 8500 ft TVSS by two dip and fault closed structures, the main 'Horst Block' and the satellite 'Western Pool'. The hydrocarbons are contained in reservoir sandstones belonging to the Middle Jurassic Brent Group which was deposited by a wave-dominated delta system in the East Shetlands Basin. The expected STOIIP and ultimate recovery are estimated at 158 MMBBL and 60 MMBBL of oil respectively, which represents a recovery factor of 38%. The 'Horst Block' contains 85% of the reserves with an OOWC about 150 ft shallower than in the 'Western Pool'. Reservoir quality is excellent, with average porosities varying from 23–26% and average permeabilities varying from 35–5300 md. The development plan envisages eleven satellite wells, six producers and five water injectors, closely clustered around two subsea manifolds. First production is expected in late 1990/early 1991. The wet crude oil will be piped to the Dunlin 'A' platform for processing and from there to the Cormorant Alpha platform into the Brent System pipeline for export to the Sullom Voe terminal.

The Osprey Oilfield is a relatively small accumulation, with an expected ultimate recovery of 60 MMBBL of oil. It is situated in the northern North Sea some 180 km northeast of the Shetland Islands and 7 km northwest of the Dunlin Oilfield (Fig. 1). The field is located in an average water depth of 530 ft and covers an area of 4.5 sq. km in Blocks 211/23a and 211/18a.

The hydrocarbons are trapped in Middle Jurassic Brent Group sandstones which were deposited by a wave-dominated delta system in the East Shetland Basin.

Two distinct fluid contacts separate the main field 'Horst Block', a domed and faulted structure containing 85% of the reserves, from the structurally lower satellite 'Western Pool'. The two structures are fault and dip closed, and are not interpreted to be filled to spill point.

The Osprey oilfield is named after the large, fish-eating bird of prey known to ornithologists as *Pandion haliaetus*.

History

Licence P232 was issued on 16 March 1972. It comprised 9 concession blocks in the northern North Sea, in which the Brent, Cormorant, Dunlin, Osprey, Tern, Eider and Don Fields were subsequently discovered. Following years of exploration and appraisal, these fields were defined and the licence re-negotiated; Block 211/23a now forms part of Licence P296 issued on 20 June 1979 to Shell UK Ltd and Esso UK Ltd on an equal basis (Fig. 1). The northern part of the Osprey Oilfield extends into Block 211/18a, which forms part of Licence P236, held by the 'Halibut Group'. A Unitization Agreement was reached in 1988 between the co-venturers and the interests were distributed as follows: Shell UK Ltd 46.75% and designated operator for this venture; Esso UK Ltd 46.75%; Deminex UK Oil and Gas Ltd 2.7625%; Santa Fe Minerals (UK) Inc. 3.0875%; Arco UK Ltd, 0.65%.

The oilfield was discovered in January 1974 by well 211/23-3 which penetrated 62 ft of net oil sandstone over the Brent Group interval in the 'Western Pool' (Fig. 2). An OOWC was identified at 8918 ft TVSS in the Tarbert Formation. Well 211/23-4, located 2.5 km northeast of the discovery well was also drilled in 1974, but no hydrocarbons were found. A 3-D seismic survey was acquired during 1977–79 over the greater Dunlin area, including most of the Osprey field. In 1982 well 211/23a-7 evaluated the prominent 'Horst Block' between wells 211/23-3 and 211/23-4. This well penetrated 269 ft of net oil sandstones with an OOWC at 8764 ft TVSS in the Rannoch Formation. Pressure data from these wells suggested communication via the aquifer with the Dunlin and/or Thistle/Deveron Fields as well as a pressure differential within the Brent Group between the 'Upper Reservoir' (Tarbert Formation and Upper Ness Member) and the 'Lower Reservoir' (Lower Ness Member, Etive, Rannoch and Broom Formations). The well flowed on production tests at stable rates of 9400 BOPD and 8100 BOPD from the Upper and Lower Reservoirs, respectively.

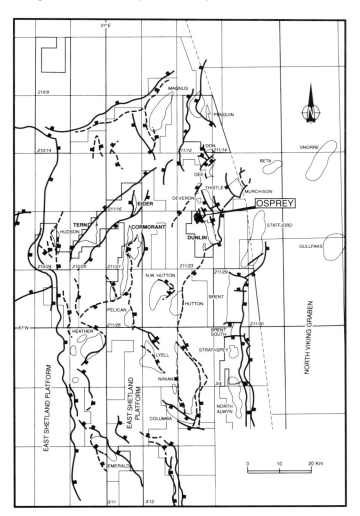

Fig. 1. Regional structural map and oil fields in the East Shetlands Basin, featuring the licence areas.

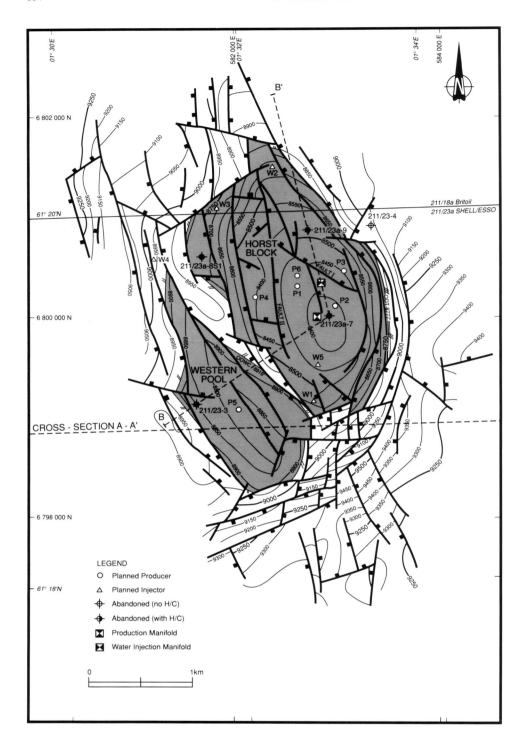

Fig. 2. Osprey Oilfield: Top Brent Group structural contour map.

Following the interpretation of a new 3-D seismic survey acquired in 1984, well 211/23a-8s1 was drilled in 1985 to evaluate the western flank of the field. The deeper oil/water contact of the 'Western Pool' was confirmed at 8918 ft TVSS as encountered in the discovery well.

The last appraisal well, 211/23a-9, was drilled in 1985 in the northern part of the 'Horst Block'. The well encountered oil down to 8777 ft TVSS in the Upper Ness Member, somewhat lower than the OOWC at 8764 ft TVSS established for the Lower Reservoir in well 211/23a-7. RFT pressure measurements confirmed the vertical segregation of the hydrocarbon accumulations in the Upper and Lower Reservoirs. The well flowed on production tests at stable rates of 3310 BOPD and 6590 BOPD from the Tarbert Formation and the Upper Ness Member, respectively.

Wells 211/23-3, 211/23-4 and 211/23a-7 also penetrated the Lower Jurassic Statfjord and Triassic Cormorant Formation which were encountered water-bearing.

The oilfield is to be developed from two subsea manifolds, one for oil production and one for water injection. Eleven wells, six oil producers and five water injectors, are planned to be drilled with subsea completions and hooked up to the manifolds. The wet crude oil will be piped to a dedicated processing module on the Dunlin 'A' platform and from there to the Cormorant Alpha platform into the Brent System pipeline for export to the Sullom Voe terminal. Injection water will be provided from the Dunlin 'A' platform to the Osprey water injection manifold.

Production is planned to start in January 1991. The plateau production rate of 25 000 BOPD is expected to be reached in the same year. Field life is estimated to be 16 years. Water injection at 70 000 BWPD is expected to maintain reservoir pressure at 4700–5000 psi.

Field stratigraphy

The Osprey Oilfield is located in the East Shetland Basin (Fig. 1) on the broad western flank of the North Viking Graben, a branch of the Mesozoic rift system which was initiated during Permian times. The Graben subsided rapidly during the Triassic and reached its maximum structural development by the end of the Jurassic. Tectonic activity during the Late Jurassic and Early Cretaceous produced the pattern of generally north–south trending rotated fault blocks characteristic of the East Shetland Basin.

The deepest formation penetrated in the Osprey area (Fig. 3) is the Triassic Cormorant Formation, comprising continental claystones and sandstones deposited on a distal flood plain. These are overlain by the thin (<100 ft) sandstones of the Lower Jurassic Statfjord Formation deposited in a coastal plain environment. Further subsidence then resulted in the deposition of the marine shales and siltstones of the Dunlin Group.

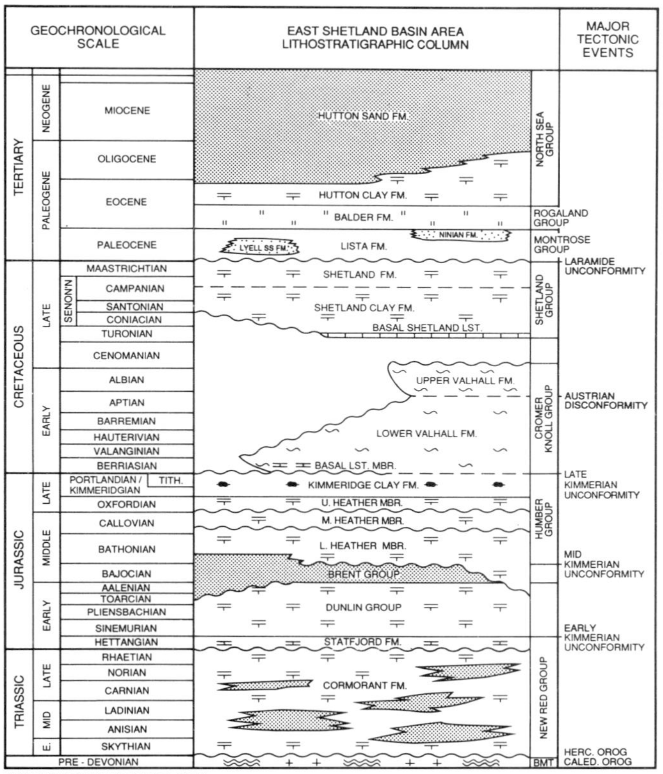

Fig. 3. Generalized stratigraphy of the East Shetland Basin.

During the Middle Jurassic an eustatic drop in sea level led to the emergence and erosion of the East Shetland Platform and other structural highs. The Brent Group sandstones form the reservoirs of the Osprey accumulation and were deposited as part of a major coastal plain and deltaic complex which prograded northwards along the axis of the Viking Graben.

The Tarbert Formation sandstones and shales comprise the erosional remnants of the Humber Group transgression. The Humber Group is composed of marine shales of the Heather and Kimmeridge Clay Formations. These shales provide the top seal to the Brent Group sandstones in this area. The Heather Formation ranges in thickness from 50–150 ft in the Osprey area. The overlying Kimmeridge Clay Formation is the source rock and consists of organic-rich shales and varies in thickness from 240 ft in the flank wells to 60 ft in the 'Horst Block' wells.

During the Early Cretaceous, the marls and limestones of the Cromer Knoll Group were deposited. These were followed in Late Cretaceous times by approximately 3000 ft of Shetland Group claystones. The succeeding Tertiary and Quaternary sediments are dominated by clays with only minor sand deposition, unlike other East Shetland Basin areas.

Geophysics

The discovery of the Osprey Oilfield in early 1974 was based on a 500 × 500 m 2-D seismic grid acquired during 1972. This grid spacing was reduced to approximately 250 × 200 m during 1973 and 1974 allowing a more detailed structural interpretation of the area (Fig. 4). The Osprey structure was partly, but not entirely, covered by the 3-D Dunlin seismic survey acquired from 1977–1979. The current seismic interpretation of the Osprey Oilfield is based on a 1600 km 3-D survey acquired in 1984. This data set covers 70 sq. km and consists of 180 east–west oriented lines spaced 37.5 m apart, with an in-line trace spacing of 25 m. The data shows good resolution and continuity of seismic events (Fig. 4). Three velocity intervals have been identified and mapped, calibrated and controlled by the exploration wells and neighbouring Dunlin seismic. Time to depth conversion has been performed using a 3-layer model. An average velocity map has been used to depth convert the surface to 'X' Unconformity interval and interval velocity maps have been used to convert the underlying Humber and Brent Group.

In total four horizons have been mapped and tied to existing well data. The most prominent horizon is the Base Cretaceous Unconformity ('X') reflector. It is characterized by a high amplitude peak produced by a large acoustic impedance contrast between the Basal Cretaceous claystones/marls and the underlying organic-rich shales of the Upper Jurassic Kimmeridge Clay Formation.

Over the 'Horst Block' area, the Top Brent Group horizon is calibrated at the upper zero crossing of a black loop. Over the western part of the field, this horizon is calibrated at the lower zero crossing of the same loop as a result of a change in character due to impedance contrast variations between the Brent Group and the overlying Humber Group shales.

The Base Mid Ness Shale Member reflector is taken at the peak of a black loop. The Base Brent Group horizon can be identified with confidence over the Horst Area, at a zero crossing below the black loop, but elsewhere the horizon is poorly defined and difficult to identify.

Trap

The hydrocarbons in the Osprey Oilfield are mainly trapped in two fault and dip closed structures, elongated in a NNW–SSE direction (Figs 2 and 4). Locally, where the Brent Group is truncated by erosion, hydrocarbons are stratigraphically trapped against overlying shales of the Humber Group. These shales also form the cap rock elsewhere in the field. Lateral seals are formed through juxtaposition across faults of Brent Group reservoirs with shales of the Humber Group and shales and marls of the Cromer Knoll Group. The formation of these traps took place during the Mid–Late Kimmerian and Austrian Tectonic events.

The structurally higher north–south trending 'Horst Block' (3 × 1.5 km) is bounded by sealing faults with minor dip closure and contains 85% of the hydrocarbons. At the structural culmination (8389 ft TVSS) the oil column is 388 ft and the OWC is in the Rannoch Formation. Reservoir pressure differences of up to 200 psi and different oil–water contacts suggest two separate reservoirs in the 'Horst Block'. The Upper Reservoir (Tarbert Formation, Upper and Middle Ness Member) has an ODT at 8777 ft TVSS and the Lower Reservoir (Lower Ness Member, Etive, Rannoch and Broom Formations) has a somewhat shallower OOWC at 8764 ft

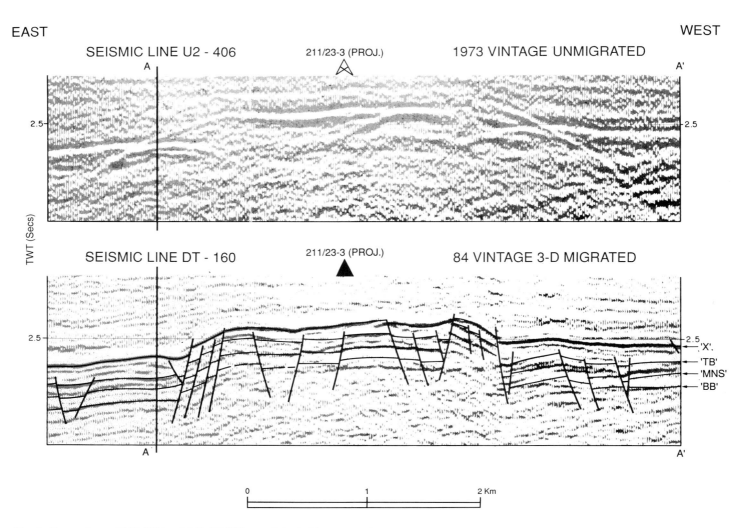

Fig. 4. Comparison of 2-D 1973 and 3-D 1984 Vintage Seismic Lines approximately through the Discovery Well 211/23-3. Four horizons have been mapped: 'X', Base Cretaceous Unconformity; 'TB', Top Brent Group; 'MNS', Base Mid Ness Shale Member; 'BB', Base Brent Group.

TVSS. The 'Horst Block' is dissected into four sub-blocks by easterly hading, NNW–SSE trending faults with throws up to 150 ft, which may constitute transmissibility barriers to flow.

The 'Western Pool' (2 × 1 km) is formed by a dome-shaped, partly fault, partly dip closed structure elongate in a NNW–SSE direction. The structure culminates at 8775 ft TVSS and contains a maximum oil column of 143 ft. It is down faulted and structurally separated from the 'Horst Block', forming a discrete accumulation with an OOWC at 8918 ft TVSS in the Upper Reservoir. The Lower Reservoir is expected to be entirely water-bearing. Neither accumulation is filled to its spill-point. There are well defined structural separations between Osprey and the neighbouring fields Dunlin, Thistle and Deveron.

Reservoir

The productive reservoirs of the Osprey Oilfield belong to the Middle Jurassic Brent Group. All wells have penetrated the complete Brent Group interval which varies in thickness from 560–590 ft. However, seismic evidence suggests that over the eroded northwestern part of the 'Horst Block' only some 375 ft may have been preserved. The five formations of the Brent Group from the base upwards are the Broom, Rannoch, Etive, Ness and Tarbert Formations (Figs 5 and 6). The Rannoch Formation is further subdivided into the Rannoch Shale Member and the overlying Rannoch Sand Member. The Ness Formation is subdivided into the Lower Ness, the Mid Ness Shale and the Upper Ness Members.

The *Broom Formation*, ranging in thickness between 30–50 ft, comprises predominantly poorly sorted, coarse-grained, coarsening upwards, bioturbated sandstones. It is currently interpreted as deposits of an easterly prograding series of deltaic fans into a shallow marine environment. The sandstones have moderate to good reservoir quality with porosities ranging between 16–32% and permeabilities between 10–1000 md.

The *Rannoch Shale Member*, thickening towards the SSE from 20–70 ft, is lithologically similar to the shales in the Dunlin Group, which are interpreted as offshore muds. It is essentially a non-reservoir interval with permeabilities normally less than 1 md.

The *Rannoch Sand Member*, thickening from 120–150 ft towards the north, represent, together with the Etive Formation, a coastal barrier complex, forming a prograding wave-dominated deltaic to coastal/inter-deltaic sequence. The Rannoch sandstones are generally well to moderately sorted, fine- to medium-grained, coarsening upwards sandstone sequences with micaceous laminae and tight carbonate cemented layers and nodules in the lower part. They have been deposited in a lower to middle shoreface environment. The reservoir quality is moderate to good, with porosities ranging from 20–30% and permeabilities between 10–300 md in the lower part and 100–1000 md in the upper part. The higher permeabilities in the upper part as compared to those in the lower part, are due primarily to increased grain size, and decreased occurrence of mica and carbonate cemented horizons.

The *Etive Formation*, increasing in thickness from 40–70 ft towards the south, comprises moderate to well sorted, medium- to coarse-grained sandstones, which tend to coarsen upwards in the

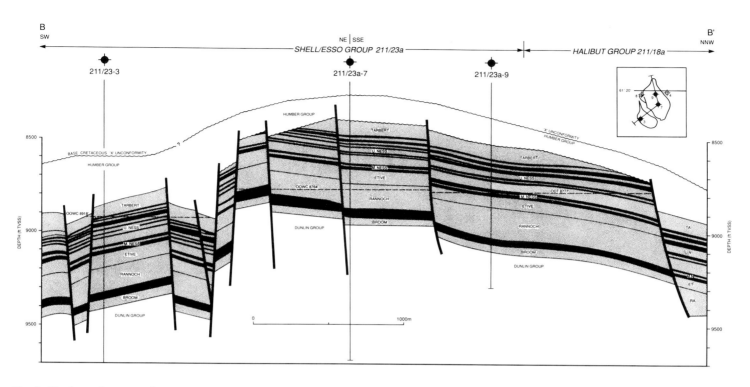

Fig. 5. North–south structural cross-section showing lithological boundaries and distribution of reservoir quality within the Brent Group.

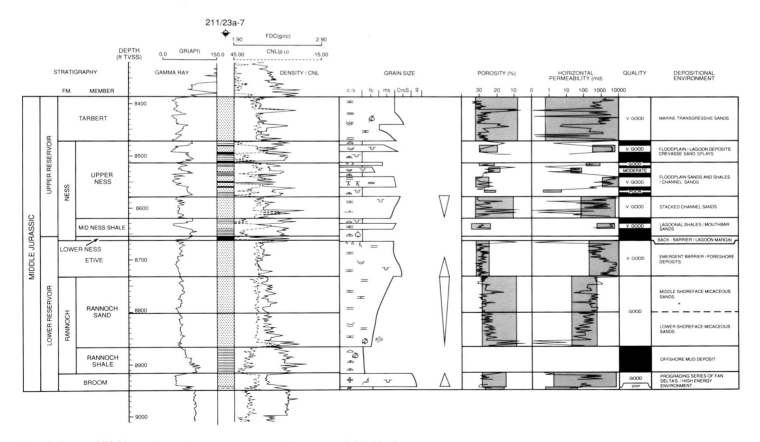

Fig. 6. Osprey Oilfield, type log and reservoir geological summary well 211/23a-7.

lower part and fine upwards in the upper part. They represent beach foreshore and emergent barrier deposits, such as dunes and interdunes. The sandstones are of good to excellent reservoir quality, with porosities ranging between 20–30%, and permeabilities between 100–7000 md. There is a general trend of decreasing permeability upwards corresponding to a similar trend of decreasing grain size.

Overlying the Etive Formation is a sequence of fluvial and lagoonal deposits, comprising the Lower Ness, Mid Ness Shale and Upper Ness Members of the *Ness Formation*.

The *Lower Ness Member* consists of an upper and a lower coal, separated by a shale or silty sandstone, representing a back barrier/lagoonal margin deposit. The interval is essentially non-reservoir, with permeabilities less than 1 md. The sequence is relatively thin (0–30 ft) on structure and it thickens to some 50 ft off-structure to the east.

The *Mid Ness Shale Member*, ranges in thickness between 40–50 ft, comprises a bioturbated lower and upper shale layer with a medium- to coarse-grained, prominently coarsening upwards sandstone interval between. The member represents essentially lagoonal deposits with a possible mouth-bar or wash-over sand intercalation and acts as an impermeable barrier between Upper and Lower Reservoir. The sandstone interval has very good reservoir properties, with porosities between 20–30%, and permeabilities between 100–5000 md. Although this sandstone interval is 'sandwiched' between two shales it is oil bearing within the 'Horst Block'.

The *Upper Ness Member* varies in thickness between 140–180 ft, but is interpreted from seismic to be eroded to only some 20 ft over the western flank of the 'Horst Block'. It consists of moderately to well sorted, fine- to coarse-grained sandstones deposited in a predominantly fluvial environment (stacked, fining-upwards channel sequences and crevasse/sheet splays) in the lower half of the sequence, and lagoonal/flood plain shales with thin sequences of coals and mouth-bar/wash-over sandstones with increasing bioturbation upwards in the upper part. Sorting is, however, poor at the erosive bases of individual channel sequences. Tight, carbonate-cemented horizons occur near the base of this member. The reservoirs are generally of moderate to very good quality, with porosities ranging between 10–32%, and permeabilities between 1–5000 md. Deterioration of the reservoir quality within this member is mainly due to the variations in grain size and argillaceous content as well as the presence of carbonate cemented zones.

The *Tarbert Formation* overlies the Ness Formation with an erosive contact and varies in thickness between 70–120 ft on the western and eastern flanks where it is fully preserved. It is interpreted to be eroded completely over the western part of the 'Horst Block'. It consists of moderately sorted, fine- to medium-grained, locally micaceous sandstones in a series of fining-upwards cycles, with tight carbonate cemented horizons developed in the basal part. It is interpreted as a transgressive marine sand deposit which marks the end of the Brent Group deltaic/coastal to inter-deltaic conditions in response to a progressive sea-level rise. The reservoir quality is good, with porosities ranging predominantly between 20–30%, and permeabilities between 100–1000 md. Reduction of the reservoir quality occurs locally due to the presence of micaceous horizons and the development of carbonate-cemented layers and nodules.

Source

The source rock is formed by the organic-rich shales of the Kimmeridge Clay Formation which have not reached maturity in the Osprey area. However, in the vicinity of the field, from the Cormorant syncline in the south to the Magnus syncline in the north, the source rock reached maturity in the Miocene, leading to generation and expulsion of hydrocarbons. By this time, the present structures had been formed, capped by sealing Upper Jurassic shales and Cretaceous claystones and marls.

Hydrocarbons

PVT analyses have been carried out on samples from wells 211/23-3, 211/23a-7 and -9. The PVT data indicate the Osprey crude oil to be highly undersaturated with a bubble point at 598–685 psi at 95–99°C and at a reservoir pressure of about 5000 psi. The gas/oil ratio for single stage separation to stock tank conditions varies from 89–109 SCF/BBL. Oil gravity ranges from 29.4–31.7° API.

Osprey Oilfield data summary

Trap	
Type	Structural, faulted dome
Depth to crest	8389 ft TVSS
Lowest closing contour	9300 ft TVSS
Hydrocarbon contacts	ODT (Horst Block, Upper Reservoir 8777 ft TVSS
	OOWC (Horst Block, Lower Reservoir) 8764 ft TVSS
	OOWC (Western Pool, Upper Reservoir) 8918 ft TVSS
Oil column	375 ft maximum, 'Horst Block'
Pay zone	
Formation/group	Brent Group
Age	Middle Jurassic
Gross thickness	375 ft (down to OOWC)
Net/gross ratio	0.70 (avg.)
Cut off for N/G (porosity)	12.1–16.6%
Porosity	25% (range 22–28%)
Hydrocarbon saturation	83% (maximum average)
Permeability	300–2100 md (average)
Productivity index	12–65 BOPD/psi (actual)
Hydrocarbons	
Oil gravity (range)	31° API (29.4–31.7° API)
Oil type	Highly undersaturated
Oil viscosity	2.5 cP
Bubble point	605 psig
Gas/oil ratio (range)	110 (89–109) SCF/BBL
Formation volume factor	1.09
Formation water	
Salinity	22 000 mg/l NaCl eq.
Resistivity	0.28 ohm m at 25°C
Reservoir conditions	
Temperature	101°C (214°F) at 8550 ft TVSS
Pressure	6000 psi initial
	current 4800–5000 psi at 8550 ft TVSS
Pressure gradient in reservoir	0.37 psi/ft
Field size	
Area	4.5 sq. km
Gross rock volume	270 million cu. m
Hydrocarbon pore volume	25 million cu. m
Recovery factor: oil	38%
Drive mechanism	Pressure support through water injection
Recoverable hydrocarbons:	
oil	60 MMBBL
gas	7 BCF
Production	
Start-up date	January 1991 (estimate)
Development scheme	Subsea development, 11 wells (planned) via Dunlin Oilfield. Evacuated via Brent System
Cumulative production	None (1 January 1990)
Production rate	25 000 BOPD (planned) (28 000 peak)
No. production wells	6 (planned)
No. injection wells	5 (planned)

The formation volume factor at reservoir conditions is between 1.08–1.10 and oil viscosity at reservoir conditions is 2.4–3.1 cP, which is somewhat higher than the 1 cP determined for the crudes in the nearby Dunlin, Deveron, and Thistle fields.

No H_2S was detected during the production tests and no major variations in properties are expected between the 'Horst Block' and the 'Western Pool'. A formation water sample from the Etive Formation in well 211/23a-9 yielded chloride concentrations of 13 630–13 700 mg/l with total dissolved solids of 23 407–23 745 mg/l. Based on the production test results in the appraisal wells 211/23a-7 and 211/23a-9, productivity indices (PI) of up to 80 and 48 BOPD/psi can be expected for the Upper and Lower Reservoir respectively. For planning purposes, however, a minimum initial average well productivity index of 25 BOPD/psi is assumed for both the Upper and Lower Reservoir.

Reserves

Limited aquifer support is anticipated and reservoir pressure will be maintained at 4700–5000 psi by voidage replacement through water injection. The field has been subdivided into five sub-blocks: four fault blocks on the 'Horst Block' and a fifth formed by the 'Western Pool'. The separating faults may form transmissibility barriers, hence the development plan envisages producer–injector pairs for the Lower and Upper Reservoirs in the fault blocks, with development planned to minimize well numbers depending on transmissibility information from the first phase of development drilling. The STOIIP is estimated at 158 MMBBL of oil with the 'Horst Block' containing 135 MMBBL (85% of the stock tank oil initially in place). A full field simulation study indicated that 60 MMBBL of oil is expected to be recovered. Drilling started in early 1990, and first oil is planned for late 1990 or early 1991 with a plateau production of 25 000 BOPD to be reached in the same year. The field life is expected to be some 16 years.

The authors wish to acknowledge the support and advice from Esso Exploration and Production UK Ltd staff and Shell Expro staff. They also wish to thank Shell Expro, Esso Exploration and Production UK Ltd, Deminex UK Oil and Gas Ltd, Santa Fe Minerals (UK) Inc., Arco UK Ltd and Monument Oil and Gas Ltd for permission to publish.

The Tern Field, Block 210/25a, UK North Sea

M. VAN PANHUYS-SIGLER[1], A. BAUMANN[2] & T. C. HOLLAND[2]

[1] *Department of Petroleum Geology, Marischal College, Aberdeen, UK*
[2] *Shell UK Exploration and Production (UEDN/422), 1 Altens Farm Road, Aberdeen, UK*

Abstract: The Tern Oilfield is situated 150 km northeast of the Shetland Islands in Block 210/25a in the UK sector of the northern North Sea. The discovery well 210/25-1 was drilled in 1975 in a water depth of about 541 ft. The trap is defined at around 8000 ft TVSS by a tilted horst-structure. The hydrocarbons are contained in reservoirs belonging to the Middle Jurassic Brent Group sands deposited by a wave-dominated delta system in the East Shetland Basin. Complex faulting of the structure is responsible for the division of the field into two areas with different original oil–water contacts: the Main Area of the field with an oil–water contact at 8260 ft TVSS, and the Northern Area with a possible oil–water contact at 8064 ft TVSS. Reservoir quality is good with average porosities ranging from 20–24% and an average permeability of 350 md. The expected STOIIP and ultimate recovery of oil are 452 and 175 MMBBL, respectively which represents a recovery factor of 39%.

The initial stage of the development plan calls for ten wells, five oil producers and five water injectors, to be drilled from a single platform, Tern Alpha. Development drilling started in February 1989 and first oil was produced on 2 June 1989. The oil is evacuated via the North Cormorant and Cormorant Alpha platforms into the Brent System pipeline for export to the Sullom Voe terminal.

To date, two producers have been drilled and total cumulative production is 6.4 MMBBL (1 January 1990). Ultimate recovery is estimated to be some 175 MMBBL.

The Tern Oilfield is situated 150 km northeast of the Shetlands in UK Block 210/25a in a water depth of 541 ft (Fig. 1).

The field has a triangular shape and is bounded to the west by the major NE/SW Tern–Eider Horst Boundary Fault and to the east by a zone of complex faulting. To the south the field is delimited by dip closure (Fig. 2).

The reservoir section consists of the Middle Jurassic Brent Group sands, which were deposited by a shore face/deltaic system prograding northwards across the Viking Graben and East Shetland Basin (Glennie 1986). The top seal is provided by the Upper Jurassic Humber Group shales (Fig. 3).

The reservoir is hydrostatically pressured and a strong aquifer response is not expected. The recoverable reserves are estimated at 175 MMBBL.

History

Licence P232 was issued on 16 March 1972. It comprised 9 concession blocks in the northern North Sea, in which Brent, Cormorant, Dunlin, Osprey, Tern and Eider Oilfields were subsequently discovered. Following years of exploration and appraisal, these fields were defined and the licence re-negotiated; Block 210/25a now forms part of Licence P296 issued on 20 June 1979 to Shell UK Ltd as the operator and with Esso UK Ltd as a 50% co-venturer (Fig. 1).

The earliest seismic was shot in 1972 with a line spacing of 1 km. The first exploration well 210/25-1, drilled in April 1975, discovered the Middle Jurassic Brent Group sands entirely oil-bearing (oil-down-to at 8227 ft TVSS). The Triassic Cormorant Formation and Lower Jurassic Statfjord Formation (Figs 3 and 4) were also found to be oil-bearing but later judged non-prospective. In the same year a 0.5 × 1 km seismic grid was shot over the Tern structure and well 210/25-2 was drilled in a downflank position to appraise the oil–water contact in the Main Area. The Brent Group was encountered fully water-bearing with a water-up-to at 8351 ft TVSS.

A third well 210/25-3, was drilled in January 1977 to appraise the Northern Area of the field. The Brent Group sands were encountered completely water-bearing and with a reservoir pressure some 500 psi higher than the main block, establishing the fault west of the well as a sealing boundary fault of the Tern accumulation (Fig. 2). The well was subsequently sidetracked to the NW and found an oil–water contact between 8064 and 8080 ft TVSS in the northernmost fault block of the field.

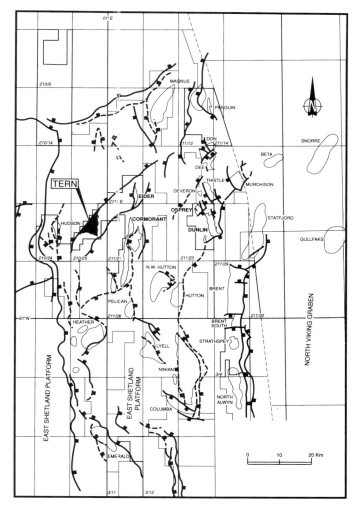

Fig. 1. Location map of the Tern Oilfield in the northern North Sea.

In 1980 a full field 3-D seismic survey was shot with a 75 m line spacing to allow a detailed fault interpretation. In the following years the Tern project was postponed because of high development costs associated with water injection and gas lift requirements; however, changes in fiscal policy led to further field appraisal. A

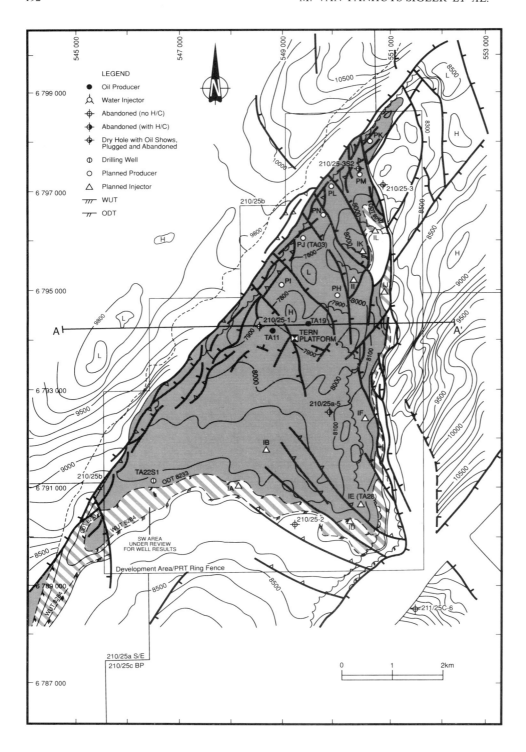

Fig. 2. Tern Oilfield, geometry and well locations: structure map of the top of the Brent Group Reservoir.

fourth well, 210/25a-5 was drilled in 1983 confirming oil-bearing Brent Group sands and established an oil-down-to at 8232 ft TVSS and a water-up-to at 8284 ft TVSS in the southeast of the field. The 3-D seismic was reprocessed in 1984. This resulted in a greater confidence in the reserve estimates for the field, which together with the development of a new and cheaper project design, led in 1985 to the decision to develop the Tern Oilfield. A primary development plan with 5 oil producers and 5 water injectors was defined to optimize early oil potential and information gathering in the Main Area. This constitutes an initial 'core' development which can be supplemented as understanding of the field increases.

The field will be developed from a single platform, Tern Alpha, installed in May 1988. Oil will be evacuated via the North Cormorant platform. Water injection is planned to maintain reservoir pressure and gaslift will be provided to sustain production rates.

Development drilling started in February 1989 and first oil was produced on 2 June 1989.

Field stratigraphy

The Tern area evolved geologically as part of the expanding Viking Graben Rift System, but with a strong Caledonian influence on the orientation of the bounding Tern Horst faults. In the early rift phase, approximately 1000 ft of sandstones, siltstones and shales of the Triassic Cormorant Formation were deposited in a distal flood plain environment (Fig. 3).

Progressive subsidence and inundation across the East Shetland Basin (Fig. 1) during the Lower Jurassic, resulted in the deposition of the thin (approximately 20 ft) calcareous sandstones of the

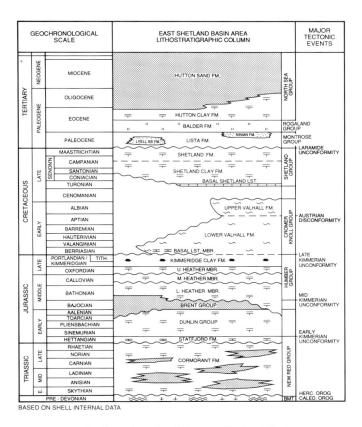

Fig. 3. Generalized stratigraphy of the East Shetland Basin.

Statfjord Formation, followed by the marine shales of the Dunlin Group, which vary in thickness from 150–200 ft.

During the Middle Jurassic a wave dominated delta system prograded in a northerly direction across the subsiding Viking Graben and East Shetland Basin (Glennie 1986). It formed the Brent Group which constitutes the main reservoir for the field. The Brent Group has been subdivided from base to top into the Broom Formation (shallow marine sands), the Rannoch and Etive Formations (shallowing-upwards shore face/coastal barrier system) and the Ness Formation (back barrier lagoon and delta plain deposits). The topmost Tarbert Formation consists of transgressive marine sands; in Tern, however, it is absent over most of the field due to erosion or non-deposition.

The Tarbert marks the onset of a major transgression, which halted the progression of the Brent delta system and finally resulted in the deposition of several hundred ft of marine shales and clays of the Upper Jurassic Humber Group. The Humber Group constitutes a regional seal.

The Late Jurassic saw the culmination of major tectonic events which are reflected by the occurrence in the Humber Group of disconformities and depositional thinning. They result from the tilting of the Tern–Eider horst block towards the northwest which was caused by a pronounced uplift of the southeast boundary area during the Late Kimmerian orogenic phase. The next major tectonic event, the Austrian phase, caused a reversed tilt movement of the Tern–Eider Horst towards the east, mainly via a process of intense faulting and reactivation of Caledonian trends. This is reflected by a wedging out of the Lower Cretaceous Cromer Knoll Group towards the west.

Continued subsidence during the Late Cretaceous and Tertiary resulted in the deposition of thick marls, shales and sands.

Geophysics

The pre-1980 seismic data base in the Tern Oilfield area consisted largely of a 0.5 × 1.0 km grid of conventional 2-D surveys mainly of 1972, 1975 and 1976 vintage. Seismic resolution of the fault pattern at Brent Group level, particularly along the eastern flank of the Tern Oilfield, was limited (Fig. 4), mainly due to coarse subsurface sampling. In 1980 a 3-D seismic survey was shot consisting of 128 lines oriented NE–SW, spaced 75 m apart and covering an area of about 110 sq. km. In 1984 the whole 3-D survey was reprocessed.

In total seven horizons were mapped and tied to existing well data. The most prominent horizon is the Base Cretaceous 'X' Unconformity reflector. It is characterized by a high amplitude peak produced by a large acoustic impedance decrease at the boundary between the Cretaceous claystones/marls and the organic-rich shales of the Upper Jurassic Kimmeridge Clay Formation.

The Top Brent Group horizon is the uppermost peak of a medium strength 'doublet' which corresponds approximately to the upper part of the Brent Group. It is generated by subtle changes in acoustic impedance from silts and claystones of the overlying Humber Group to the underlying sandstones and intercalated shales of the Ness Formation.

The Base Brent Group horizon is ill-defined, and difficult to correlate due to the low acoustic impedance contrast of the sandstones and shales in the lower part of the Brent Group with the underlying Dunlin Group shales.

The data quality in the Main Area of the field is generally very good in the shooting direction and large faults are well defined. The definition of the smaller NW–SE faults is limited in the area close to the North-Western Boundary Fault where the Humber Group becomes thin (Fig. 4).

Trap

The trap consists of a dip and fault closed structure which was shaped during the Kimmerian and Austrian tectonic events by multiphase faulting and tilting. The top seal is provided by Upper Jurassic shales of the Humber Group. The crest of the structure at the top Brent Group level is at a depth of about 7610 ft TVSS (Figs 3 & 5).

Two main areas can be distinguished, characterized by varying intensity of faulting and different oil–water contacts, which indicates the presence of one or several sealing faults. The Main Area of the field is relatively unfaulted in the southwest but its eastern side is traversed by N–S faults which form a series of stepped fault blocks. In the Northern Area, the reservoir is more shaly and dissected by numerous NW–SE faults.

Reservoir

The Tern oil accumulation is contained in the sandstones of the Middle Jurassic Brent Group which varies in thickness from 255–290 ft. It is subdivided into a Lower Reservoir, consisting of the Broom, Rannoch and Etive Formations and the Lower Ness Member, separated by the Mid Ness Shale Member from the Upper Reservoir (Upper Ness Member and Tarbert Formation). A porosity and permeability profile through the Brent Group reservoirs is illustrated in Fig. 6 with a type log of the field.

The basal *Broom Formation* has a thickness of 10–20 ft and consists of coarse to pebbly sands. It is continuous over the Tern Oilfield and is interpreted as a series of deltaic fans prograding into a shallow marine environment. The Broom shows strong variations in rock matrix properties and locally extensive carbonate cementation. Average porosities vary from 15–22% and permeabilities from 7–540 md.

The overlying fine-grained micaceous sandstones of the lower to middle shoreface *Rannoch Formation* form a laterally continuous, sheet-type deposit with hummocky cross-stratification. It has a maximum thickness in the central part of the field, where the best

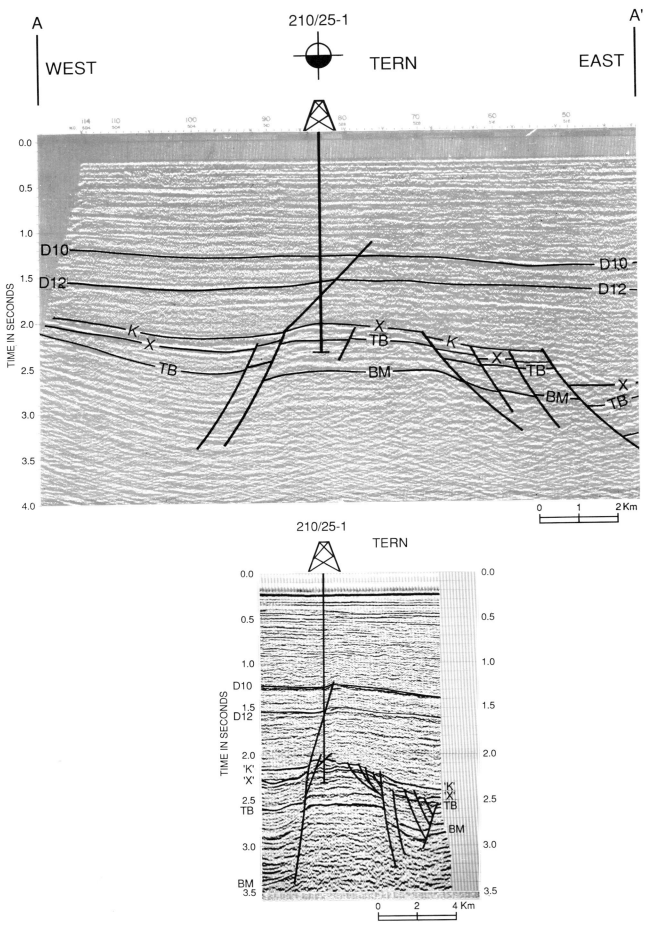

Fig. 4. Tern Oilfield: comparison of 1972 and 1988 seismic sections. D10, Top Balder Tuff Formation reflector; D12, Base Tertiary reflector; K, basal Shetland Limestone reflector; X, Base Cretaceous unconformity reflector; TB, Top Brent Group reflector; BM, top basement reflector.

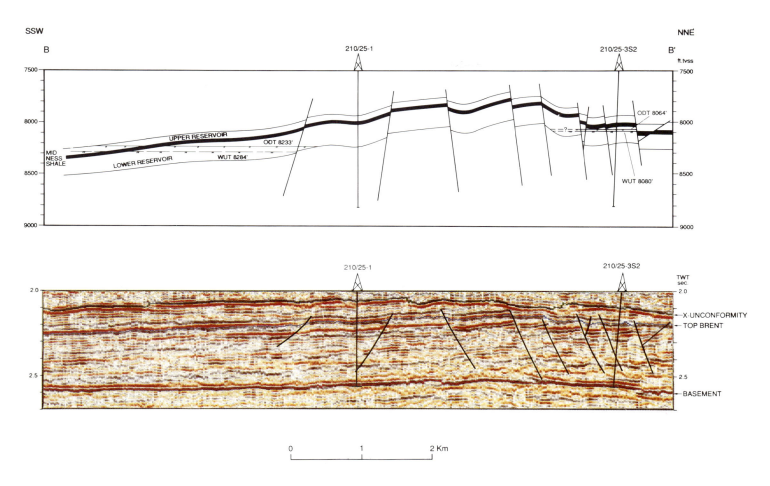

Fig. 5. Tern Oilfield: structural cross-section and seismic line through wells 210/25-1 and 210/25-3s2.

reservoir properties are also found. Average porosities vary from 19–23% and average permeabilities are in the range of 50–270 md. The highest values, up to 29% porosity and 620 md permeability, are found in an offshore bar facies. Calcite-cemented layers of up to a few feet thickness are common in the Rannoch Formation. These 'doggers' appear to be randomly distributed both laterally and vertically. They are non-net reservoir and on average comprise 14% of the gross thickness.

The upper shoreface and emergent coastal barrier *Etive Formation* forms a continuous deposit with a thickness of 30–80 ft. It consists of well sorted, medium- to coarse-grained sandstones. Reservoir properties are rather uniform, with an average porosity of 25.6% and an average permeability of 1000 md. Maximum thickness and best reservoir properties are found in the central part of the field. Towards the north, reservoir properties deteriorate with the appearance of a 15–20 ft thick sequence of non-reservoir sandy shales and coals at the base of the Etive. Throughout the field, minor carbonate-cemented layers are present.

The sediments of the back barrier/delta plain depositional environment of the *Ness Formation* are characterized by internal heterogeneity and strong variations in reservoir properties. The *Lower Ness Member* consists of a sequence of medium-grained tidal delta and washover sands with intercalated lagoonal and floodplain muds. It forms a wedge-shaped body with a maximum thickness of 60 ft in the south, wedging out in a northerly direction. Reservoir properties are vertically very variable. The average porosity and permeability are 24% and 1120 md, respectively, with the highest values found in the washover sands. The lagoonal *Mid Ness Shale Member* forms a barrier to vertical cross-flow and has a rather constant thickness of about 25 ft. The *Upper Ness Member* is a 40–70 ft thick, laterally continuous sequence of shales and sands. The base of the Upper Ness Member consists of 25–35 ft of very fine to coarse, ripple cross-laminated and bioturbated sands representing a lagoon-fill/mouthbar deposit. These sands are overlain by approximately 15 ft of floodplain and lagoonal muds with thin intercalated silts and very fine sands, which are essentially non-reservoir. The top of the Upper Ness is locally affected by erosion. Where present, it consists of tidal sand flat and channel deposits. The Upper Ness Member is characterized by strong vertical and lateral reservoir property variations. Average porosities range from 15–25%, permeabilities from 19–460 md.

The *Tarbert Formation* consists of fine- to coarse-grained shallow-marine sandstones with parallel and low-angle bedding and bioturbated silty and argillaceous sands. Reservoir properties are very variable; in the clean sands they can reach up to 25% porosity and 650 md permeability.

The Lower Reservoir is expected to form a continuous, single reservoir over the entire field. In the Upper Reservoir, however, lateral continuity is likely to be adversely affected by faulting, owing to its lower net/gross ratio and the presence of shales at the base and the top. In the central and northern part of the field, fault juxtaposition is expected to result in some communication between the Upper and Lower Reservoir.

Source

The hydrocarbons of the Tern Oilfield were derived from the organic-rich Kimmeridge Clay Formation, which in the deeper parts of the half grabens surrounding the Tern horst reached maturity in the Miocene.

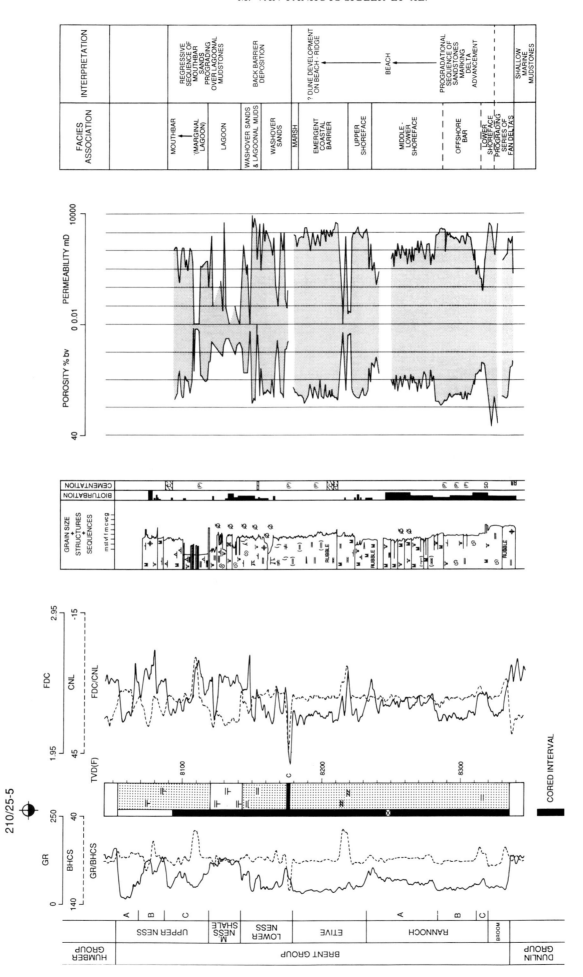

Fig. 6. Tern Oilfield, reservoir type log: Well 210/25a-5.

Tern Oilfield data summary

Trap
Type	Structural, fault and dip closure
Depth to crest	7800 ft TVSS
Lowest closing contour	8260 ft TVSS
Oil–water contacts:	
Northern area	8064 ft TVSS (estimated)
Southern area	8260 ft TVSS (estimated)

Pay zone
Formation/group	Brent Group
Age	Middle Jurassic
Gross thickness	284 ft
Net/gross ratio	0.62 (average)
Cut off for N/G	1 md measured permeability (corresponding to approx. 11% porosity)
Porosity (range)	22% (20–24%)
Hydrocarbon saturation	0.71 (average)
Permeability	350 md (average)
Productivity index	20 BOPD/psi

Hydrocarbons
Oil gravity	34° API
Oil type	Undersaturated, trace sulphur
Oil viscosity	1.26 cP
Bubble point	895 psia
Gas/oil ratio (range)	177 SCF/BBL (150–190 SCF/BBL)
Formation volume factor	1.13

Formation water
Salinity	21 000 ppm NaCl eq.
Resistivity	0.267 ohm m at 28°C; 0.106 ohm m at 93°C

Reservoir conditions
Temperature	93°C at 8000 ft TVSS
Pressure	3580 psia at 8000 ft TVSS
Reservoir oil gradient	0.34 psi/ft

Field size
Area	10.0 sq. km
Gross rock volume	866 × 10^6 m^3
Hydrocarbon pore volume	118 × 10^6 m^3
Oil recovery factor	39%
Drive mechanism	Gas lift and water injection
Recoverable oil	175 MMBBL

Production
Start-up date	2 June 1989
Development scheme	1 fixed platform, producing into Brent System
Peak production rate	52 000 BOPD (expected in 1992)
Cumulative production	6.4 MMBBL (1 January 1990)
Number/type of wells	1 exploration 3 appraisal 2 producers (1 January 1990)
Secondary recovery method	Water injection

Hydrocarbons

The Tern crude is highly undersaturated with a trace of sulphur. Its bubble point pressure is low at 895 psia and it has a GOR of 150–190 SCF/BBl. The oil gravity is 34° API. The oil from the Upper Reservoir is a slightly heavier and more viscous crude with a lower solution GOR than the oil from the Lower Reservoir. Water salinity is 21 000 ppm NaCl eq. The formation temperature at datum level (8000 ft TVSS) is estimated to be 200°F (93°C), with a temperature gradient of 13°F/1000 ft. The pressure regime is hydrostatic (3580 psia at datum level).

Reserves

Reserves are estimated at 175 MMBBL, which represent 39% recovery of the stock tank oil initially in place of 452 MMBBL. As no significant aquifer support is anticipated, pressure maintenance by water injection will be implemented from the outset. Furthermore, because the pressure regime is hydrostatic, gas lift is required to sustain adequate production rates. Transmissibility is expected to be reasonable in the Main Area of the field where faulting is less extensive than in the Northern Area where transmissibility and hence recovery efficiency is expected to be lower.

The last two wells (210/25-3s2 and -5) were production tested and based on these tests an average productivity index of 20 BOPD/psi can be expected if the producers are completed over the entire Brent sand sequence. The initial flow performance of the first oil producer was 23 000 BOPD. Water injection rates up to 16 000 BWPD per well are expected. Production started on 2 June 1989, and is projected to build to a peak of 52 000 BOPD in 1992. Field life is expected to be some 20 years.

The authors wish to acknowledge the support and advice from Shell UK Expro Staff. They also wish to thank Esso Exploration & Production UK as well as Shell UK Expro for their permission to publish this paper.

Reference

GLENNIE, K. W. (ed.) 1986. *Introduction to the Petroleum Geology of the North Sea*. Blackwell Scientific Publications.

The Thistle Field, Blocks 211/18a and 211/19, UK North Sea

R. R. WILLIAMS[1] & A. D. MILNE

BP Exploration, Fairburn Industrial Estate, Dyce, Aberdeen AB2 0PB, UK
[1] *Present address: 53 Albury Place, Aberdeen, UK*

Abstract: The Thistle Oilfield is located on the western margin of the North Viking Graben in UK Blocks 211/18a and 211/19. The field was discovered in 1973 by the Halibut Group, operated by the Signal Oil Company. The discovery well was drilled in 1973 and the field was declared commerical before completion of the first two appraisal wells. The extension of the field into Block 211/19 was confirmed by well 211/19-1 drilled in 1974 and the field was subsequently unitized.

The Middle Jurassic Brent Group forms the oil-bearing reservoir sequence. The structure is an easterly dipping, rotated fault block divided into five main compartments by N–S normal faults and E–W cross faults. A 60 slot drilling and production platform was installed in 1976 and production began in February 1978. It is now connected by pipeline to Sullom Voe and production through this link commenced in late 1978. Due to only limited aquifer support, injection wells have been drilled on the east flank of the field, in order to maintain reservoir pressure. In addition, several wells in partly isolated fault blocks are produced by gas lift completions. The current STOIIP is estimated at 794 MMBBL of which 396 MMBBL will ultimately be recovered. Cumulative production to the end of 1988 was 334 MMBBL from 34 production wells, with support from 13 injection wells.

The Thistle Oilfield lies in 530 ft of water, 130 miles northeast of the Shetland Islands on the western margin of the East Shetland Basin (Fig. 1). The field covers an area of 16 km² (c. 4000 acres) primarily in Block 211/18a with a small extension into Block 211/19. The trap is formed by an easterly dipping structure bounded by a major fault in the west and dip closed to the east and south. To the north an E–W fault separates the main Thistle Field from the Area 6 structure, an isolated, downfaulted, area which lies within the Thistle PRT boundary and so does not hold separate field status.

The reservoir sandstones of the Brent Group were deposited as a regressive sequence of sands overlain by a transgressive marine interval. The Brent Group is underlain by the Dunlin Group, an

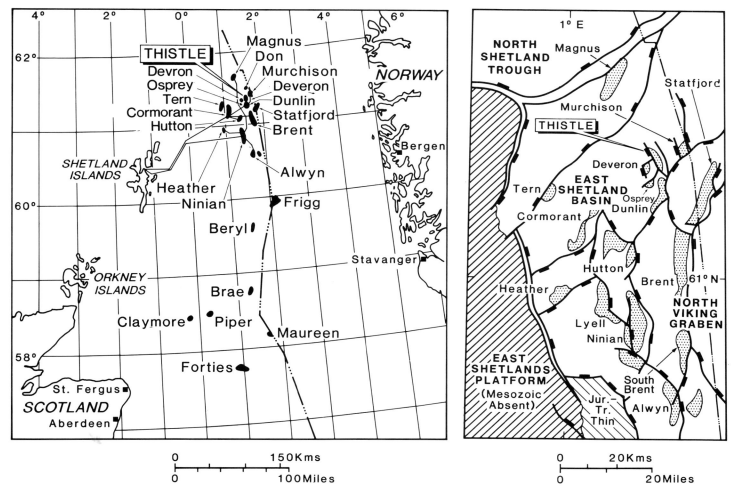

Fig. 1. Location map.

From Abbotts, I. L. (ed.), 1991, *United Kingdom Oil and Gas Fields,*
25 Years Commemorative Volume, Geological Society Memoir No. 14, pp. 199–207

open marine claystone dominated non-reservoir sequence, and is overlain by the Humber Group marine claystones which provide seal and charge for the reservoir (Fig. 2).

The field is named after a wild flower, *Onopordum acanthium*, the floral emblem of Scotland.

History

Licence P236 was awarded in the 4th round of UK Offshore Licensing in March 1972, to the Halibut Group led by the Signal Oil Company. The group is now operated by BP Exploration on behalf of Britoil plc, with present participating companies being:

BP Exploration (Britoil plc)	18.41%
Deminex UK Oil and Gas Ltd	42.5%
Santa Fe Minerals (UK) Inc	16.87%
Arco British Ltd	10.0%
Premier Oil Exploration Ltd	8.4%
Monument Petroleum Mitre Ltd	2.41%
Ultramar Exploration Ltd	1.41%

The extension of the field into Block 211/19 was proven in 1974, the licence P104 being operated by Conoco North Sea Inc. The field was subsequently unitized, and in June 1984 the P104 licence interests in Thistle were assigned to the P236 Halibut Group.

The field was discovered in 1973 with well 211/18-2 drilled on

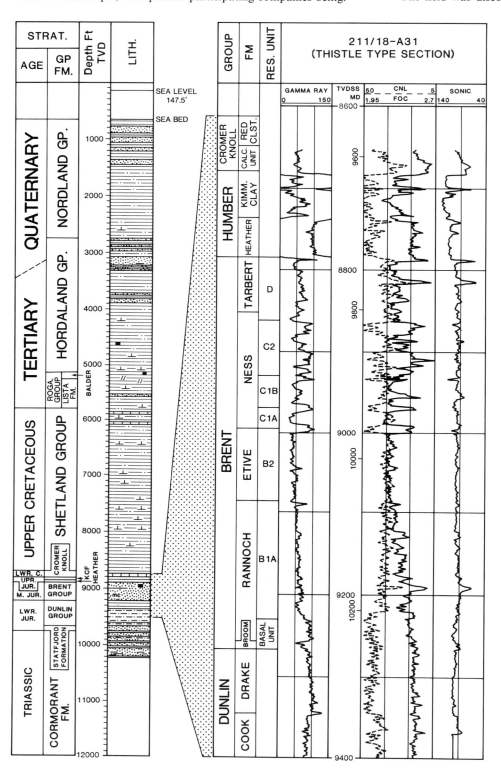

Fig. 2. General stratigraphical log.

a base Cretaceous structure. The find was confirmed by wells 211/18-3 and 211/18-4 in 1973 and three further appraisal wells were drilled after Annex B submission. Well 211/18-8 confirmed the northern extent in 1975 and 211/19a-7 in the southeast confirmed field extent and proved separation between Thistle and Dunlin in 1979.

The Thistle Field was declared commercial in 1973 and development plans were initiated at that stage.

The early development and production history of the reservoir has been well documented by Hay (1977) and Nadir & Hay (1978). The decision to proceed with development was followed by the drilling of four expendable appraisal wells. The single steel platform was installed on location in 1976. Development drilling from the twin derrick platform began in September 1977, with first oil produced in February 1978.

The platform initially produced oil through an offshore loading system (SALM: Single Anchor Leg Mooring) but by late 1978 production through to Sullom Voe commenced via a pipeline link to Shell's Dunlin platform. Until completion of the NLGP (Northern Leg Gas Pipeline) in 1983, associated gas was disposed of at the platform by flaring and re-injection but is now produced through a short spur tying into the NLGP and from there via FLAGS (Far North Liquids and Associated Gas System) to St Fergus.

The original development plan was based on the optimized reservoir simulation model and involved using all available well slots. The field was to be developed by simple line drive technique, natural water drive being augmented by downflank water injectors supporting crestal oil producers. The ultimate aim of the development plan was to waterflood successively the reservoir layers by sequential upward completions.

Field stratigraphy

The stratigraphy succession encountered on Thistle Field is presented in Fig. 2. The lithostratigraphic subdivisions proposed by Deegan & Scull (1977) have been adopted with only minor modification.

The Lower Jurassic sequence of coarse clastics of the Statfjord Formation is conformably overlain by the Dunlin Group claystones. During the Bajocian and Bathonian periods, the Brent Group sequence was deposited predominantly as sandstones with interbedded claystones and rare coaly horizons. The lithostratigraphic units Broom, Rannoch, Etive, Ness and Tarbert are roughly equivalent to the reservoir unit subdivision of Basal Unit (including the claystone at the base of the Rannoch), B1/A, B2, C and D sand intervals respectively. The Basal Unit is not considered reservoir as it is dominated by claystones and siltstones with only a thin basal sand.

The Brent Group is overlain by the Upper Jurassic, Humber Group sequence of Heather and Kimmeridge Clay Formations and is unconformably overlain by a very attenuated Lower Cretaceous sequence of Barremian limestones and Albian–Aptian red claystones. From the top of the Lower Cretaceous through to the Quaternary the sequence is dominated by claystones. Minor traces of tuffaceous siltstone mark the top of the Palaeocene and above this, the only major variation in the lithology is in the Upper Miocene with deposition of a thick, glauconitic sand.

Geological history

Crustal extension in the North Viking Graben was initiated during

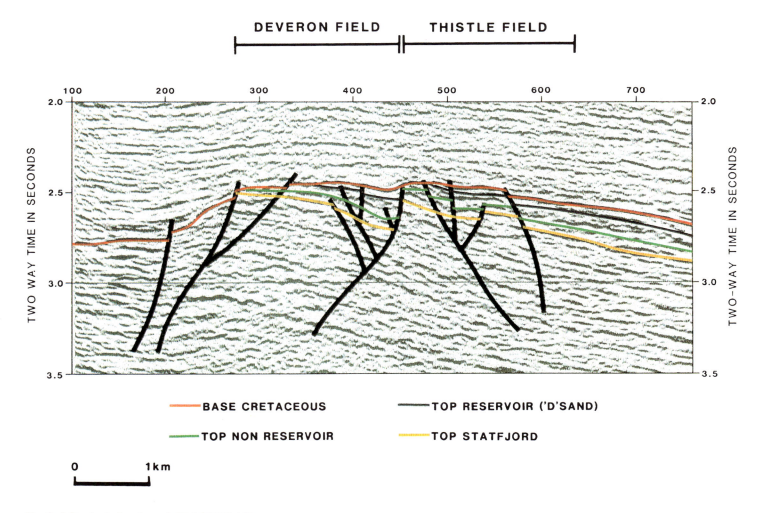

Fig. 3. 3-D seismic line through Well 211/18-A31.

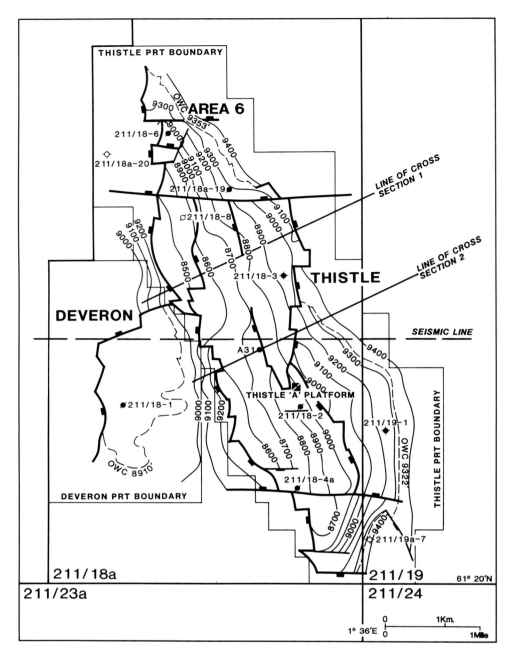

Fig. 4. Top reservoir structure map.

the Permian, reached a peak in the Jurassic, and continued on into the Cretaceous. Movements associated with rifting have therefore been active over a prolonged period resulting in repeated reactivation of faults.

Some of the faults initiated during the early rifting period continued to move sporadically throughout the Middle Jurassic. Deposition of the reservoir sequence thus occurred in an actively subsiding graben with evidence of contemporaneous synsedimentary faulting affecting the Upper Brent Group and Humber sequences. Considerable structural influence has therefore been exerted on the pattern of Brent Group sedimentation with evidence of thinning and/or onlap of sands onto structurally high areas.

Following deposition of the Upper Jurassic, early Cretaceous fault movement caused further uplift, tilting and erosion. The dominant N–S structural grain of the field which influenced sedimentation throughout the Jurassic was still active during the Lower Cretaceous. This end Jurassic activity resulted in non-deposition of early Lower Cretaceous sediments. At the end of the Lower Cretaceous, uplift resulted in an unconformable relationship between the Lower Cretaceous sequence and the overlying Upper Cretaceous Shetland Group. From the end of the Lower Cretaceous, widespread, rapid subsidence and deposition persisted until the late Tertiary.

Geophysics

Prior to the award of Block 211/18 in 1972, approximately 125 km of speculative and traded seismic data were available. Since then, an extensive grid of 2-D data was acquired. In addition, two 3-D seismic surveys were shot in 1976 and 1983. The 1983 3-D survey was shot specifically to enhance existing seismic data quality fieldwide. It comprises 85 sail lines at 100 m interval with a 12.5 m CMP interval amounting to 750 km of surface data coverage. However, the 1983 3-D survey was completed after installation of the Thistle 'A' platform and no undershooting of the platform was undertaken. To remedy the gap in the regular 3-D grid, an infill high resolution 2-D survey of 120 km was shot late in 1987.

In 1988, the 1983 3-D data were reprocessed in order to improve the seismic definition particularly below the Base Cretaceous unconformity. The data were interpolated to provide a line spacing of 33.3 m to aid in fault delineation. An example track from these data is illustrated in Fig. 3.

Early seismic data revealed the basic structural configuration of the field as an easterly dipping fault block, bounded to the west, north and south by major faults. During the exploration and appraisal stage, a structure map of the Base Cretaceous unconformity formed the basis for future development. Detailed mapping of the field was hampered by a lack of reasonable seismic data below the Base Cretaceous reflector and interpretations of Top Reservoir structure were largely based on well data and some seismic dip data. Top and Base Reservoir structure maps were constructed by the addition of isochores, interpolated from well data, to the Base Cretaceous structure surface. This was a major source of uncertainty in trap definition at this stage.

The current structural interpretation of the field is based on the 1983 3-D seismic data. This 3-D survey allows the reservoir envelope to be mapped directly from seismic data for the first time. Definition of the dominant faults which form the structural framework of the field is good, and the inferred E–W cross faulting can be identified by transverse jumps of the N–S fault trends and on time-slice data. This is illustrated by the western margin of the field where E–W cross faults can be identified on the original 100 m spacing cross tracks.

The structure maps interpreted from the 1983 3-D seismic data form the basis for current infill drilling and future field development planning.

Trap

The Thistle 'area' is part of a branched system of graben and horsts with associated minor fault blocks within the North Viking Graben. The regional pattern is of extensional normal faulting with many faults active throughout the Jurassic providing not only the trap geometry for migrating hydrocarbons but also a local and regional control of reservoir distribution.

The producing hydrocarbon accumulations within the Brent Province are dominantly structurally controlled and are typically found on the westward dipping flanks of tilted fault blocks. The Thistle structure, however, is an easterly dipping structure controlled by movement on a major N–S trending fault on its western flank. This fault downthrows to the west and numerous terraces step down into a half graben before rising farther to the west to form the smaller Deveron structure (Fig. 4). The Thistle structure is also fault sealed to the north and a combination of dip and fault closure seal it to the south. Three surfaces describe the geometry of the hydrocarbon-bearing reservoir; Top Brent Group reservoir, Base Brent Group reservoir and the hydrocarbon water contact. The 3-D seismic data now allows more confident mapping and this illustrates progressive thinning of the reservoir envelope onto the crest of the structure on the western flank (Fig. 5). The oil–water contact at 9322 ft TVSS is identified as the free water level. On current mapping the trap is assumed to be full to spill point and relies on Humber Group claystones to provide vertical and lateral

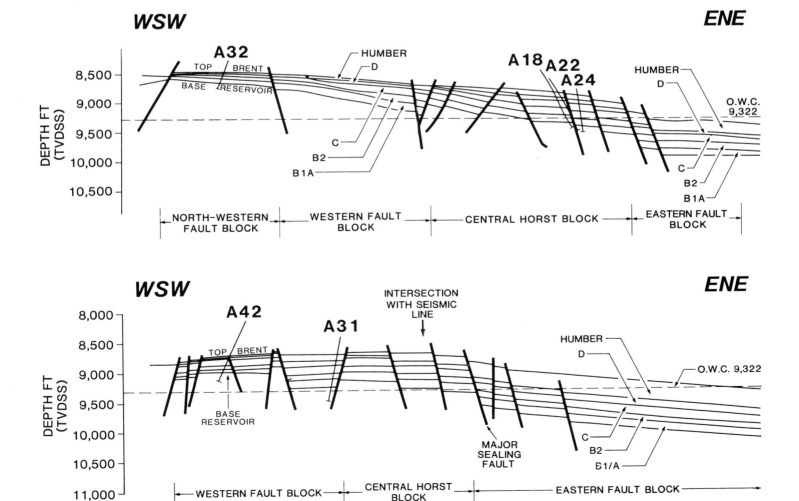

Fig. 5. Structural cross-section.

seals with the Rannoch Formation claystone providing a bottom seal.

The entire structure is heavily faulted and major faults severely restrict pressure support to the main field from the aquifer in the east. RFT measurements confirm that in the south a major fault separates the main structure from the Southern Fault Block. Analyses of formation water chemistry suggests that the Southern Fault Block is in communication with the Deveron and Dunlin aquifers but not the main Thistle aquifer to the east.

The fault pattern highlights the main structural grain trending N–S, however there is also an E–W trending structural component. Both these structural elements are considered part of the same tectonic regime and together they divide the field into discrete structural blocks (Fig. 6). The faults and structures which form the boundaries to these structural elements are inferred to have long histories of movement. Specifically the E–W faults appear to cross cut the more dominant N–S faults, resulting in an orthogonal fault configuration. It is thought likely that within the field the N–S faults have exploited an underlying E–W crustal weakness and followed the path of least resistance during their development. The E–W faults appear to act as transfer faults to the N–S normal faults with no significant strike slip component involved. Although these cross faults are difficult to detect directly from seismic data, the N–S trending structures are often dramatically offset across them.

Within the reservoir envelope, reservoir maps have been generated using 50 well track sections and 76 serial cross sections. From these, individual reservoir unit structure surfaces and isochores were derived and together these provide a robust, three dimensional, geological model of the reservoir (Fig. 5). Use of the constraints of the seismic reservoir envelope to define gross reservoir thickness has led to the modelling of significant Upper Brent Group thinning by erosion and/or non-deposition over the crest of the structure. In general, reservoir unit isochores show little structural control on lower reservoir thickness apart from slight thinning towards the western margin of the field. However, the effect of faulting is crucial on the upper reservoir in its control on both deposition and preservation of reservoir section.

Reservoir

The Thistle Field reservoir is formed by the sandstones of the Brent Group. The upper four Brent Group formations are identified as distinct reservoir units based on log correlation across the field and

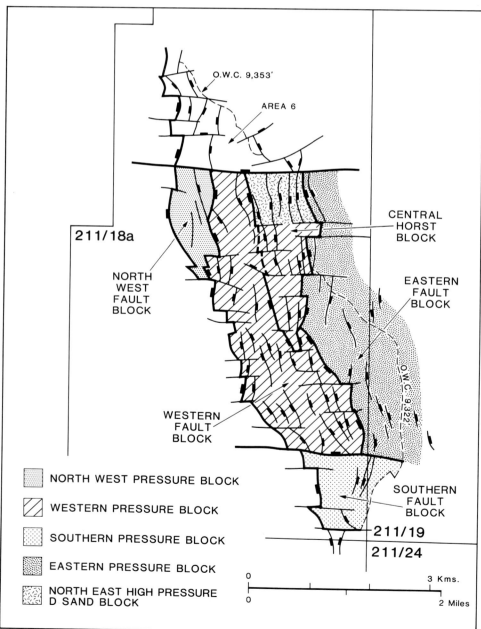

Fig. 6. Fault blocks.

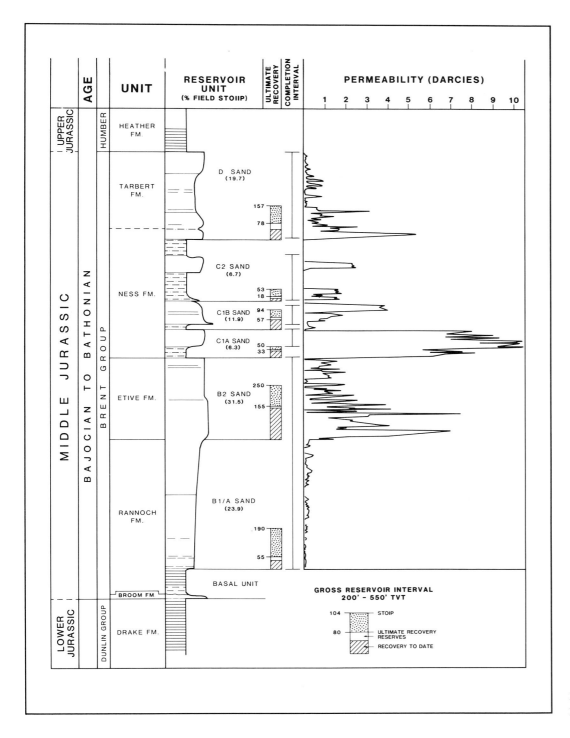

Fig. 7. Reservoir sand units and permeability distribution.

are termed B1A (Rannoch), B2 (Etive), C (Ness), and D (Tarbert) Sands (Fig. 7). The depositional environment of each unit controls reservoir character and each unit boundary represents a major change in environment, resulting in complex vertical variations in character.

Rannoch Formation

The Rannoch Formation is a very fine- to fine-grained sandstone with highly micaceous laminae, which becomes coarser and cleaner and improves in reservoir quality towards the top of the sequence. Gross thickness of the unit is typically 50 ft, but the unit thins to a minimum of 50 ft in the SW of the field. The unit has been interpreted as a prograding lower shoreface sand sequence deposited in a storm and wave dominated environment.

These micaceous, laminated sandstones have porosities averaging 22% with permeabilities ranging between 10–1000 md. The presence of pronounced but discontinuous micaceous laminae, together with preferential development of silica along these, results in a distinct contrast between horizontal and vertical permeabilities. On a larger scale, effective permeability is also controlled by bedform geometry, which comprises stacked lensoid units that are internally laminated and are truncated by successive layers. This heterogeneity results in extremely tortuous paths of fluid movement.

Doggers also occur apparently randomly in the upper part of the Rannoch Formation. Although important locally, these tight calcite cemented zones do not significantly affect permeabilities or reservoir performance on a large scale since they are volumetrically insignificant.

Etive Formation

The Etive Formation comprises a unit of upwards coarsening, massive, clean sandstones of sheet-like geometry, typically 90 ft in thickness. It directly overlies the Rannoch Formation and represents a prograding barrier bar environment with associated beach, back barrier and mouth bar facies. Despite their apparently massive character, the Etive sandstones are sedimentologically complex.

The Etive is an excellent reservoir, with porosities and permeabilities average 26% and 3000 md respectively. The basal and top boundaries of the Etive are tight streaks which constitute important vertical barriers across the field. Within the unit, porosity is partly depth dependent.

The largest proportion of original field reserves were present in the Etive Formation and early production was mainly from this unit. In consequence it watered out early in field life, which in turn has caused severe problems in producing the underlying poorer quality Rannoch.

Ness Formation

The Ness Formation comprises an alternating sequence of fine- to medium-grained sandstones and claystones with occasional thin coals. It is subdivided into three separate reservoir units, each comprising a sandstone underlain by a claystone. Reservoir quality of the lower two units is excellent. Porosity ranges between 25–30% and permeability 1–6000 md. The thickness of the respective sandstones ranges between 20–50 ft. The topmost sandstone in contrast is thicker (100 ft) but is heavily bioturbated and is of poorer quality (porosity 20%, permeability 0.2–1000 md). Throughout the Ness, facies is a major control on porosity.

The sandstone units of the Ness are difficult to correlate field-wide. This reflects the complex and varied delta top depositional environment.

The Ness Formation represents the culmination of the sustained regression that characterizes Brent Group deposition.

Tarbert Formation

The Tarbert Formation is a predominantly sandstone unit, interpreted as a marine sheet sand deposited after rapid flooding of the Brent deltaic/coastal facies. The unit ranges in thickness from 80 ft in the central part of the field, to zero over the crest and on other highs within the field, illustrating the strong structural control on Tarbert sand deposition and distribution. Locally a marked erosional surface within the Tarbert can be identified in cores.

The Tarbert is a compound stratigraphical unit, comprising a series of sub-units with distinct permeability and porosity characters controlled by mica distribution. As was the case in the Rannoch Formation, variation in mica content results in contrasts in permeability of several orders of magnitude.

Source

Geochemical analysis indicates that the Kimmeridge Clay with its high sapropelic kerogen content is the source rock for the Thistle crude. In the absence of significant expulsion in block 211/18, the Thistle oil is believed to have been sourced from the north and northeast. Migration probably took place in early Tertiary times.

Hydrocarbons

The Thistle crude is a light low sulphur content oil with an API gravity of 38°. The reservoir is overpressured by 2000 psi, its original pressure being 6060 psig at a datum of 9200 ft TVSS, and contains an undersaturated crude, with no free gas cap. Individual well GORs are in the range 250–350 SCF/BBL.

Formation water is relatively fresh with a salinity of 23 000 PPM dissolved solids and has neutral to slightly alkaline pH. Water resistivity is 0.253 ohm m at 77°F. The water chemistry is typical for the Brent Province. The OWC is at 9322 ft TVSS and the aquifer size is very large.

Along its eastern flank the Thistle Field is in communication with an active aquifer. However a major N–S fault extends the length of the field and prevents good aquifer support in the main field area.

The Southern Fault Block appears to have separate aquifer support, related to the Deveron and Dunlin aquifer system.

Reserves

The current estimate of initial oil in place is 794 MMBBL with ultimate recovery of 396 MMBBL, a recovery factor of 50%. The Field came on stream in 1978 and cumulative oil production to the end of 1988 was 334 MMBBL, with daily oil production averaging 50 000 BOPD. Field life is predicted to extend to 2004. Reserves (excluding Area 6) remaining in individual reservoir units are summarized in Fig. 7. A detailed model of distribution of reserves within the field is now available, leading to a more accurate definition of potential infill and sidetrack targets, which would result in ultimate recovery being greater than 396 MMBBL. The greatest proportion of reserves remain in the Rannoch and Tarbert units.

The principal development problems on Thistle relate to the highly faulted heterogeneous nature of the reservoir. Both structural and lithological factors control and restrict reservoir performance. Reference has already been made to the isolation of the main field from the active aquifer downflank to the east, due to the presence of the major N–S fault. Elsewhere within the field individual fault compartments exist which require artificial lift to compensate for poor pressure support in the reservoir. However, fault compartmentalization appears to be a problem mainly with production from the Tarbert.

The main uncertainty regarding potential for reserves growth relates to the Tarbert (and Ness) where erosion has been mapped, particularly in crestal areas. If erosion is not as severe as mapped then considerable potential may exist for increased oil in place.

The main lithological factors controlling reservoir performance relate to the behaviour of claystones in the upper reservoir as effective pressure barriers, and the permeability contrasts within the Tarbert and Rannoch. These vertical variations in reservoir character have significantly influenced perforation policy for these two sandstones.

Water coning is a dominant feature of Rannoch production behaviour, caused by early flooding of the Etive, resulting in turn in a distinctive early watercut in the Rannoch sandstone and high water production (over 90%) for long periods. Again perforation policies can considerably alleviate this problem, by initially perforating the lowest units to reduce downward coning of water from the Etive and maximizing recovery from the poorer quality lower sandstones. Simulation studies highlight the poor recoveries anticipated from the Rannoch, calculated at 23% in contrast to 55% for the other units. However, recent modelling indicates that improved recovery factors of 40% may be achieved by producing the wells for longer periods at high watercuts and by providing adequate drainage points. Predicted improvements in reservoir performance and extended well life may ultimately lead to a requirement for increased platform water handling capacity.

The author wishes to thank BP Exploration and the Thistle Partners (Deminex, Santa Fe, Arco, Premier, Monument and Ultramar) for permission to publish this paper. The author wishes to thank P. D. Begg and A. T. James for their constructive criticism and comments on the text.

Thistle Field data summary

Trap
Type	Tilted and rotated fault block
Depth to crest (Top Brent reservoir)	8500 ft TVSS
Oil/water contact	9322 ft TVSS (original)
Gross oil column	822 ft

Pay zone
Formation	Tarbert Ness Etive Rannoch formations of Brent Group
Age	Middle Jurassic, Bathonian–Bajocian
Gross thickness	200–550 ft
Net/gross ratio	20–100%
Porosity	range 16–30%, average 24%
Hydrocarbon saturation	average 0.78%
Permeability	range 40 md to 4 d

Hydrocarbons
Oil gravity	38.4° API
Oil type	Light low sulphur oil
Gas/oil ratio	290 SCF/BBL
Bubble point	960 psig
Formation volume factor	1.18 RB/STB

Formation water
Salinity	23 000 ppm total dissolved solids
Resistivity	0.253 ohm m at 77°F

Reservoir conditions
Temperature	220°F at 9200 ft TVSS
Pressure	6060 psig at 9200 ft TVSS

Field size
Area	3970 acres (16 km^2)
Gross rock volume	910 000 acre ft
Recovery factor	50%
Recoverable hydrocarbons	396 MMBBL

Production
Start-up date		February 1978
Development scheme		single fixed steel platform
Number/type of wells		single 34 production/13 injectors
Production rates (annual average)*		
oil	1982	124 000 BOPD
	1988	50 000 BOPD
water	1982	92 000 BWPD
	1988	207 000 BWPD
Cumulative production to end 1988		334 MMBBL
Drive mechanism		Local water drive
Secondary recovery methods		Water injection support, some gas lift

* Fluids production is facilities constrained

References

DEEGAN, C. E. & SCULL, B. J. (compilers) 1977. *A standard lithostratigraphic nomenclature for the Central and Northern North Sea.* Institute of Geological Sciences Rept. No. 77/25. HMSO, London.

HAY, J. T. C. 1977. The Thistle oilfield. *Mesozoic Northern North Sea Symposium.* Norwegian Petroleum Society, Oslo, Section 11.

NADIR, F. T. & HAY, J. T. C. 1978, Geological and reservoir modelling of the Thistle field. *Proceedings of the European Offshore Petroleum conference*, Society of Petroleum Engineers, London, **2**, 223–237.

The Central Graben and Moray Firth

The Arbroath and Montrose Fields, Blocks 22/17, 18, UK North Sea

R. CRAWFORD, R. W. LITTLEFAIR & L. G. AFFLECK

Amoco (UK) Exploration Company, Amoco House, West Gate, Ealing, London W5 1XL, UK

Abstract: The Arbroath Field was discovered by the Amoco operated group comprising Amoco (UK) Exploration Company, Gas Council (Exploration), Amerada Hess Ltd and Texas Eastern (UK) Ltd with the 22/18-1 well in May 1969 and was the first commercial oil field to be discovered in any sector of the northern North Sea. Two years later in 1971, the adjacent Montrose Field was discovered. Both Arbroath and Montrose are located 130 miles east of Aberdeen towards the northern end of the Central Graben area. The two fields have simple non-faulted, anticlinal structures separated by a structural saddle with only 15 ft of relief. The structures are a product of Alpine tectonism combined with differential compaction of the reservoir section. The hydrocarbons have a Kimmeridge Clay source and occur wholly within the Forties Sandstone interval. The oil is trapped by mudstones of the Sele Formation. The reservoir sandstones were deposited in a prograding submarine fan complex and have maintained an average porosity of 23% and a permeability of 80 md.

To date only the Montrose Field has been produced. The main recovery mechanism is natural bottom water drive supplemented by water injection. First oil was produced in 1976 via a tanker mooring system, replaced by a dedicated pipeline in 1984. The Arbroath Field is under development and first oil was produced in April 1990.

The Montrose and Arbroath fields are located approximately 130 miles east of Aberdeen in 300 ft of water within UK blocks 22/17 and 18 of the North Sea (Fig. 1). The two fields are separated by only 15 ft of structural relief in a saddle 500 yards wide; both have Palaeocene Forties Formation reservoirs and hence they are presented together. Both fields have anticlinal structures related mainly to Alpine tectonic deformation but with some overprint of differential compaction of the reservoir. Ultimate recovery is expected to be 98 MMBBL for Montrose and 102 MMBBL for Arbroath (production start up in April 1990). The fields are named after the maritime communities of the same names situated on the east coast of Scotland.

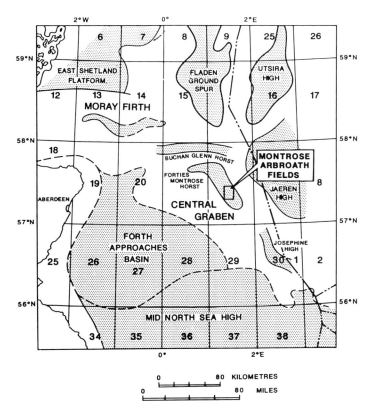

Fig. 1. Montrose/Arbroath Fields location map.

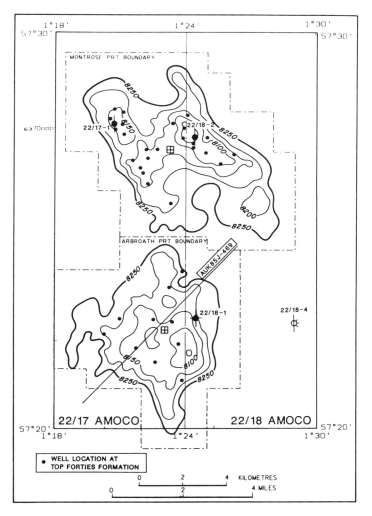

Fig. 2. Near Top Forties event structure map.

History

Montrose and Arbroath fields (Fig. 2) are wholly located within Blocks 22/17 and 22/18 awarded in the first Offshore Licence round,

September 1964, as parts of licences P019 and P020 respectively, to the Amoco operated partnership now consisting of:

Amoco (UK) Exploration Co.	30.77%
Enterprise Oil plc	30.77%
Amerada Hess Ltd	23.08%
North Sea Inc.	15.38%

Amoco (UK) Exploration Company was formerly Amoco (UK) Petroleum Ltd and Enterprise Oil plc replaced Gas Council (Exploration) Ltd in the government sale of the latter company's oil assets in 1984.

The early history of discovery and development of the Montrose Field, as the area comprising the Montrose and Arbroath structures was then known, is described by Fowler (1975). Early seismic reconnaissance work carried out in the northern North Sea (all of the area north of the Mid-North Sea High) prior to 1964 had indicated the presence of a large northwest-trending, low-relief, anticlinal structure with 30 square miles of closure in the base Tertiary interval in UK Blocks 22/17 and 22/18. This structure drapes over a deeper fault-bounded anticline thought to be the Upper palaezoic section.

At the time of award of the blocks the geology of this North Sea area was little known. Significant gas discoveries had been made onshore in the Netherlands and there were expectations of extending this Permian play to the offshore in the southern North Sea. In 1964 the hydrocarbon potential of the northern North Sea was considered uncertain and speculative although it was realized that the area had all the ingredients to be oil prospective. In the mid- to late 1960s exploration activity in the northern North Sea was concentrated in the Norwegian sector and despite some small discoveries of gas condensate there and of oil in the Danish sector, the results until late 1969 were disappointing. Only five wells were drilled in the United Kingdom offshore north of latitude 57°N when, in May 1969, the Amoco operated well 22/18-1 was spudded.

22/18-1 was drilled at shot point 317A on seismic line 480 by the semi-submersible rig Sea Quest, subsequent to the structural lead being further defined by 450 miles of seismic data acquired in 1967/68. This additional data had shown that the Palaeocene prospect had two culminations on its longer north–south axis. 22/18-1 was designed to test the southern feature (later to be named Arbroath) and its planned depth increased to test the fault-bounded anticline in the possible Upper Palaeozoic section beneath. The well discovered oil below 8156 ft subsea in clean Palaeocene sandstones some 380 ft thick. The oil column measured 100 ft to the oil–water contact at 8256 ft subsea. On a short-term test the well flowed 40° API gravity, low sulphur oil at a maximum rate equivalent to 2160 BOPD. This was the first commercial oil to be discovered in the United Kingdom northern North Sea.

The prospect was further delineated by the interpretation of an additional 160 miles of seismic data acquired in 1970 which indicated additional closure and the separation of the northern structure (Montrose) into two NNW–SSE trending structural culminations (Fig. 2). Exploratory well 22/18-2, drilled in 1971, tested the eastern lobe of the Montrose structure and confirmed its pressence when oil flowed at a rate equivalent to 4190 BOPD from a column 185 ft thick. Appraisal well 22/17-1, drilled the following year, tested the western lobe successfully, testing 1962 BOPD from an oil column 170 ft thick.

Montrose was developed in preference to Arbroath as it was interpreted to have greater reserves, a lower risk value and the best economics. It was also deemed that Arbroath would require further appraisal and that a delay in the development of Montrose was not necessary since Arbroath could easily be linked into the overall development.

The Montrose structure was declared commercial in 1972. It was developed utilizing an eight-legged conventional steel structure with 24 well slots and a three-legged flare support structure. The field was initially produced via a tanker mooring system replaced in 1984 by a pipeline. First oil was produced in June 1976. Production peaked in early–mid 1979 at 35 000 BOPD.

A template was placed over the Arbroath structure in 1979 and deviated appraisal wells 22/17-T1, 22/17-T2 and 22/17-T3 confirmed the extent of the Arbroath field.

Arbroath is to be developed using a single four-legged satellite platform to be in place by late 1989. This platform is linked to the Montrose facility by four pipelines: produced oil, produced gas, injection water and gas lift. Facilities at Montrose are to be modified to allow gas lift and injection water to be supplied to Arbroath via the appropriate pipelines. First oil from Arbroath was produced in April 1990.

Field stratigraphy

The Montrose and Arbroath fields are situated towards the south central part of the Forties–Montrose High (Fig. 1), a positive structural feature, 50 miles long within the Central Graben that extends from Block 21/10 and the Forties Field in the north to Block 22/24 and the Marnock discovery in the south. The high is a horst feature with an approximate NNW to SSE trend bounded on the east and west by normal faults and at its north and south ends by normal faults with an approximate east–west trend. The origin of this positive area can be probably related back to the Carboniferous (see below) where the horst area formed part of a larger positive Carboniferous block of Variscan tectonic origin. Development of the horst feature began in late Permian times and, along with the graben system development, reflects the early stages of the opening of the Atlantic. The north and south ends of the horst are defined by probable Carboniferous (Variscan) features rejuvenated in the late Permian. Movement on all bounding faults reached a maximum during the Triassic but continued into the Jurassic. At Montrose/Arbroath the complete absence of any Carboniferous and Jurassic section combined with a regional thinning seen in the Zechstein, Triassic, Cretaceous and Palaeocene sections suggests that the area was a dominant, positive feature from the Palaeozoic through to the Tertiary (Fig. 3).

The oldest rocks penetrated in the area belong to the Devonian System (Fig. 4) and are represented by a siltstone/shale sequence with infrequent intercalations of thin, fine sandstone beds. No Carboniferous strata have been identified and derived clasts of Carboniferous strata have not been recognized in the Permian siliciclastic section suggesting that the Montrose Forties Horst was a positive feature during Carboniferous times, resulting in non-deposition. The Permian is represented by rocks attributed to both the Rotliegendes Formation and the Zechstein Group. The Rotliegendes Formation comprises shales interbedded with tight and porous sandstones deposited in an, as yet, undetermined environment. The overlying Zechstein is represented by a carbonate section composed entirely of amorphous micritic dolomites sometimes vuggy with vugs open or fully dolomite cemented. The depositional environment of the Zechstein cannot be ascertained but the absence of any evaporites coupled with the presence of the sometimes thick, massive dolomite beds with thin shale beds suggest that the Montrose/Arbroath area was positive in Zechstein times.

The area remained positive throughout the Triassic and the Zechstein was subaereally exposed creating karstic porosity and microporosity in some areas with co-eval cementation elsewhere. The Triassic is thin (400 ft approximately) and represented by continental red beds of mudstone and siltstone. Tectonic activity at periods throughout the Triassic is evidenced by the presence of conglomerates, composed of boulders and pebbles of the same red beds, deposited along the Forties–Montrose High fault scarp. Sediments belonginng to the Jurassic System are absent in the Montrose/Arbroath area and Lower Cretaceous sedimentation is represented only by a thin sequence of Aptian to Albian chalk-marls. Chalk of the Upper Cretaceous is of the order of 1000 ft thick.

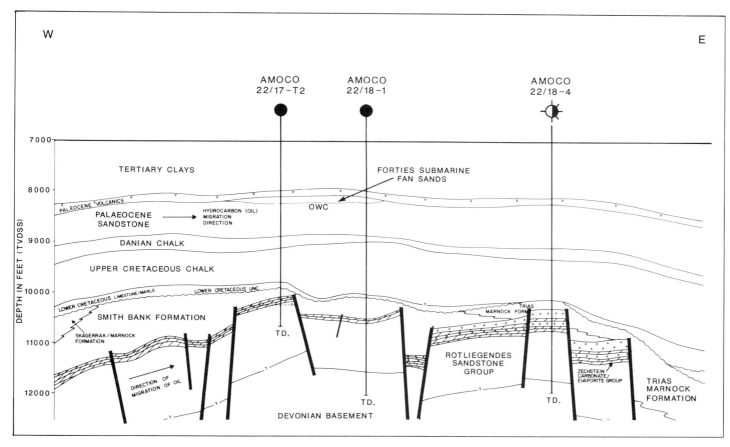

Fig. 3. Forties–Montrose High and Arbroath Field structural cross-section.

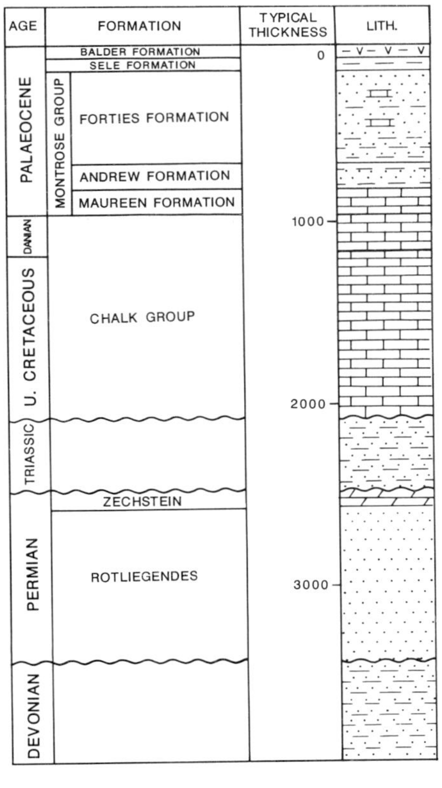

Fig. 4. Stratigraphic succession for the Montrose/Arbroath area.

Chalk sedimentation, represented by the Ekofisk Formation, continued into the Early Palaeocene. Much of the Palaeocene interval is represented only by mass flow deposits of the Montrose Group. These commence with resedimented chalk conglomerates/ breccias and sandstones of the Maureen Formation which are of variable thickness (up to 200 ft) and may be absent in some wells. The overlying Andrew Formation is approximately 100 to 350 ft. thick and represented by shales and turbidite sandstones. The Andrew Formation sequence shows a general coarsening, and sometimes thickening, upwards. Sandstone units reach a maximum thickness of approximately 100 ft where they probably represent several catastrophic events. The overlying Forties Formation also preserves a tendency to overall upward coarsening and thickening. The formation is 300 to 600 ft thick and is, in its lower 100 to 200 ft, composed of mainly mudstones with thin turbidite interbeds (the 'Forties Shale'). This is overlain by a turbidite sandstone sequence with beds reaching a maximum thickness of 60 to 70 ft (the 'Forties Sandstone').

Towards the close of the Palaeocene subsidence, previously centred on the Central Graben, became widespread. A marine transgression over the Moray Firth delta system cut off the sediment supply to the Forties submarine fan complex resulting in the deposition of grey to black, sometimes laminated, mudstones of the Sele Formation. Tuffaceous siltstones with mudstones of the Balder Formation record the close of Palaeocene deposition. The overlying Eocene to Recent deposits consist of undifferentiated mudstones and siltstones up to 8000 ft thick.

Geophysics

In 1985, 850 kilometres of seismic data, with a dominant northeast to southwest line orientation and 250 m line separation, were shot over the entire Arbroath Field and the majority of the Montrose Field to supersede all previous surveys over that area. Recent (1987) reprocessing of this data, including horizontal velocity analysis, enabled the top of the Forties Formation reservoir to be picked directly. Previous top reservoir depth maps had been constructed by isopaching downward from the top Balder Ash Formation or upwards from the first sufficiently continuous reflector beneath.

CDP gathers showed signs of increased amplitude with increasing shot to geophone distance in the zone some 0.025 s two-way time below the top Forties Sandstone. Although considered to be real data the inclusion of the further traces in the stack was believed to be detrimental to the resolution and detection of subtle events such as the top sandstone and consequently they were excluded from the reprocessing sequence.

Sandstone quality within the reservoir was seen to vary in some early wells and so analyses of range-dependent attributes (RDA) were conducted on the complete range of offset data. A credible basis for the prediction of sandstone quality was not obtained owing to inconsistent well ties. It was concluded that the definition of the lateral distribution of the uppermost shaley sandstones was beyond the resolution of the seismic data.

A study of the petrophysical and acoustic properties of the Forties Formation from borehole information to assess the feasibility of lithology calibration of micromodelled data was also attempted but concluded that shaliness and effective porosity cannot be distinguished from acoustic impedance alone within the formation.

Six well velocity surveys, including four vertical seismic profiles, were recorded in Montrose wells and, up to the end of the first stage of development drilling in Arbroath, eight VSPs and two checkshot surveys had been recorded from the eleven wells drilled. VSP character correlations to seismic sections were good.

The strongest and most continuous reflectors on the sections are identified as the Palaeocene Balder Ash/Sele Shale composite and the Top Maureen/Ekofisk event (Fig. 5) at respective crestal two way reflection times of 2.502 and 2.668 s. Upper zero crossings were correlated at each level.

The close proximity of the Balder Ash (36–80 ft thick) to the Sele Shale (62–84 ft thick) causes the individual seismic responses of each of these horizons mutually to interfere to produce a composite response which leaves the top Sele Formation response undefined. The top of the Forties Formation has been picked on the next upper zero crossing after that of the Balder Ash. Within the field boundaries the pick is unequivocal across varying lengths of line between sharply defiend zones of discontinuity. These interruptions are interpreted as small slump faults in some cases and facies boundaries in the Forties Sandstone in others.

The crest of the structure at top Forties Formation level is at a two-way section time of 2.527 s and has 0.036 s of one-way time relief to the closing contour. This equates with structural culmination in both Arbroath and Montrose at around 8030 ft subsea with structural closure at 8265 ft.

Within the Forties Formation and in the remaining pre-Forties Palaeocene beds, no field-wide continuous events are seen. However, gently tilted events at or near the oil–water contact have been recognized and followed over short segments of seismic lines. However, difficulties are met in tying these events from line to line. In places they are so weak and nebulous that only a phantom remains with which to create a convincing field-wide picture.

Other than the possibility of small faults as mentioned above no direct faulting on a regional or local scale is interpreted as causing localized displacements of the reservoir.

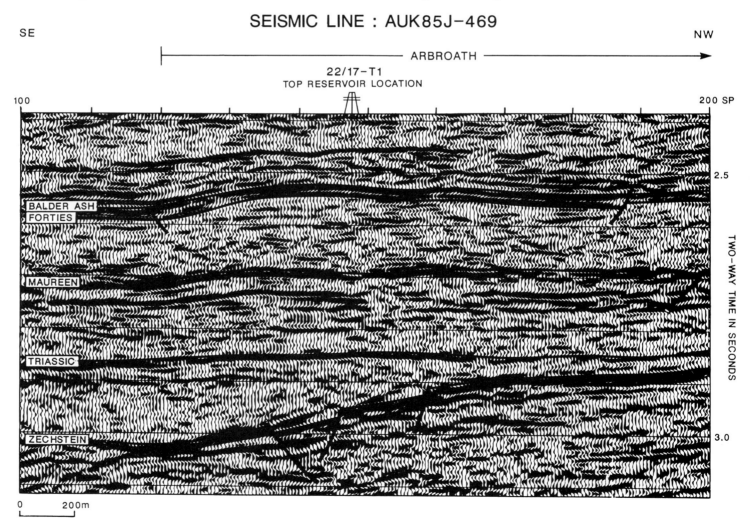

Fig. 5. Seismic line AUK 85J-469.

Trap

Oil in the Montrose and Arbroath fields is trapped by dip closure of the Forties Formation which is sealed by mudstones of the Sele Formation. The structures in both fields formed in early Eocene times and can be broadly defined as domal anticlines with gently dipping (1–2°) non-faulted limbs. Similar structures are mapped at the top Danian horizon indicating that the structures are probably of tectonic origin. Second-order relief has been added to the structures by the differential compaction of the Palaeocene sandstone/shale package. The structural highs occur at the places where the sandstone section is thickest whereas towards the field perimeters the sandstone section is often thinner and shale interbeds are common. At Montrose the two structural high trends are related to sandstone thick sections and these are separated by a shalier section in the structural low.

The structural features measure approximately 4 by 5 miles and show a maximum of 230 ft of relief. The oil–water contacts are variable, being deepest (8265 ft subsea at Arbroath, 8318 ft subsea at Montrose) where the sandstone package is thick and clean towards the centres of the fields. They are shallowest where the sandstones are argillaceous and of lower permeability, generally towards the field perimeters. The original reservoir pressures in both fields are of the order of 3700 psi indicating a normal pressure gradient. The structures are full to spill point.

Reservoir

The productive interval of the Montrose and Arbroath fields is wholly located within the upper sandstone-dominated section of the late Palaeocene Forties Formation (Fig. 6). Oil-bearing sandstones do not extend down into the mudstone dominated section of the lower Forties Formation and they are not known in the underlying Andrew Formation.

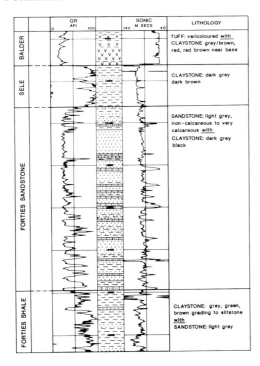

Fig. 6. Forties Formation reservoir stratigraphy.

The Forties Formation is represented by a sequence of sandstones and mudstones that show general coarsening and thickening-upward trends. The formation varies in thickness from 300 ft to 600 ft, thinning towards the edge of the field, and it can be divided into a lower mudstone unit (the Forties Shale) and an upper sandstone unit (the Forties Sandstone). The Forties Shale is on average 100 to 200 ft thick and is variable in character, ranging from a well defined massive mudstone sequence with increasing sandstone content to a sequence of sandstones and mudstones of equal proportions. The overlying Forties Sandstone is 200 to 400 ft thick and generally shows increasing sandstone content upwards. The thickest beds defined by shale breaks are of the order of 60 to 80 ft but these represent up to ten catastrophic events. In both fields, individual sandstone beds have sheet-like geometries though the thickness and lithofacies of each sheet is variable throughout its area. In general, the structural highs locate the positions where beds are thickest and where sandstones are stacked and represented by massive sandstones often with a granule/pebble layer at the base. Towards the field perimeters sandstone beds tend to be thinner, finely laminated, and separated by shales.

The reservoir sandstones are typically poorly sorted, very fine- to medium- or coarse-grained and friable to moderately cemented with angular to sub-angular grains. The average sandstone is composed of grains of monocrystalline quartz (75 to 86%) and feldspar (5 to 10%) with small quantities of mud clasts, mica, chert and occasionally glauconite. Heavy mineral traces include zircon and tourmaline but exclude epidote, this last feature suggesting provenance in the Orkney–Shetland Platform area (Knox et al. 1981). The feldspar component is dominated by plagioclase with minor amounts of orthoclase and microcline. Orthoclase is commonly highly altered, whereas plagioclase and microcline are fresh. Diagenesis is late and has resulted in the precipitation of quartz, calcite, kaolinite and chlorite cements. None of the cements is a significant porosity occluder, their effect being mainly to reduce permeability.

The sediments of the Forties Formation are interpreted to have been deposited by turbidity currents and to have accumulated in a prograding submarine fan setting. The base of the section is dominated by pelagic mudstones. Thin turbidite sands are gradually introduced, thickening and coarsening upwards to culminate in a sequence of major and minor stacked mid-fan channellized sandstones. Sandstone deposition is concentrated in the channel areas (now the zones of structural highs) where the turbidity currents were strongest.

The reservoir has been subdivided into seven layers (Montrose) and eight layers (Arbroath) for the purpose of estimating reserves and modelling reservoir performance. Layers are bounded by thin shale beds and have variable petrophysical characters associated with internal facies changes. In general porosities are commonly in the 23–25% range and show no significant change between layers. Permeabilities are commonly 70 to 90 md but range from less than 1 md to 2000 md and show a tendency to increase upwards through the section.

Source

Oil at Montrose and Arbroath has been typed to an Upper Jurassic Kimmeridge Clay source which is mature (present day vitrinite reflectances are greater than 1.3%) within the Central Graben area to the west of the fields in Blocks 22/16, 21 and 22. Here, the Kimmeridge Clay is 100 to 1000 ft thick and has an average TOC content of 8%, composed predominantly of amorphous, marine, type 1, kerogen. It is believed to have reached peak oil generation (VR = 1.0%) 10 to 20 million years before present. Migration through the Jurassic and Cretaceous sections is thought to have followed normal fault systems and over steepened beds which mark the edge of the Forties Montrose High. Once in the Tertiary section the oil followed a more gently inclined route through the Maureen and Andrew sections to be trapped below the Sele Formation in the Forties Sandstone. Lateral migration need only have been 6–10 miles.

Hydrocarbons

The Arbroath discovery well, 22/18-1, tested hydrocarbons from the Forties Sandstone at a rate of 2160 BOPD with a flowing bottom hole pressure of 2198 psig. The Montrose discovery well 22/18-2, and appraisal well 22/17-1, tested oil at rates of 4190 BOPD and 1962 BOPD with flowing bottom hole pressures of 2665 psig and 3354 psig respectively. Appraisal wells 22/17-T2 and 22/17-T3 on the Arbroath structure tested oil at rates of 3053 BOPD and 8554 BOPD with flowing bottom hole pressures of 2895 psig and 3227 psig respectively. Average formation pressures are 3744 psig for Montrose and 3700 psig for Arbroath indicating normal pressure gradients in the area.

The reservoir oil is light brown with a gravity of 38 to 42° API and a viscosity of 2.69 centistokes at 50°C. Sulphur content is low (0.16%), wax content is 8.78% weight and the pourpoint is 9°C. The oil pool is undersaturated and GORs range from 310 SCF/BBL to 640 SCF/BBL.

Formation waters recovered in drill stem tests and during isolated formation pressure tests show a significant difference in chemistry and resultant resistivity between the two fields. At Montrose total dissolved solids measure from 110 000 to 112 000 mg/l and resistivity measures 0.027 ohm m at reservoir temperature of 257°F. At Arbroath total dissolved solids measure 132 000 to 139 000 mg/l and resistivity measures 0.023 ohm m at reservoir temperature of 245°F.

Reserves/resources

Production of the Montrose Field began in June 1976. The Arbroath Field is under development and production is planned to begin in early 1990. The main recovery mechanism in the Montrose Field is bottom water drive. The field is underlain by a large aquifer which is in vertical communication with the oil column. The aquifer support is good in the western lobe of the field but is less effective in the more stratified eastern lobe. Initially full aquifer support was assumed; however, the reservoir pressure in each of the lobes began to decline with production and it became necessary to supplement this natural water drive with water injection into the aquifer.

Aquifer injection began in 1978 with injection into four wells, three in the east lobe and one in the west. Reservoir pressures have since been maintained above bubble point (2737 psig in the east lobe and 2348 psig in the west) and are currently in the 3000 to 3200 psig range. The waterflood was expanded in 1986 with the conversion of two wells to water injection (one in each lobe). At this time all six injectors were opened up into the oil column to increase injection and improve recovery. The ultimate field recovery is expected to exceed 90 MMBBL.

The main production problems in the field have been associated with oil transportation, scale formation and artificial lift. Initially oil transportation was via a tanker mooring system. This system suffered from periods of downtime due to bad weather and was replaced with a pipeline in 1984. Scale formation inhibited production as some wells began showing signs of injection water breakthrough, the injection water and formation water being incompatible. This situation has been effectively treated with scale inhibition squeezes and scale is no longer considered to be a problem.

The main production problem that still remains concerns the short runlives associated with the electric submersible pumping (ESP) system. Though some progress has been made with extending the runlives, the Montrose downhole environment with its deviated, hot, gassy and sandy wells is not ideally suited for ESPs.

The Arbroath Field development plan has incorporated the lessons learned at Montrose. The planned development utilizes thirteen producers and six injector wells with an additional two producers to be converted to injectors late in the field life. The injectors are peripheral and will provide both aquifer support and edge water drive. Injection is planned to begin with production start-up and reservoir pressure will be maintained above the bubble point of 1991 psig throughout the field life. Gas lift is planned for all producers. Ultimate recovery is expected to exceed 100 MMBBL.

Arbroath and Montrose Fields data summary

	Montrose	Arbroath
Trap		
Type	Domal Antiform	Domal Antiform
Depth to crest	8040 ft	8030 ft
Lowest closing contour	8250 ft	8250 ft
Hydrocarbon contacts	8250 ft variable	8250 ft variable
Gas column	none	none
Oil column	210 ft	220 ft
Pay zone		
Formation	Forties	Forties
Age	Palaeocene	Palaeocene
Gross thickness (average/range)	330/260–440 ft	330/260–440
Net gross ratio (average/range)	0.5/0.3–0.8	0.5/0.3–0.8
Porosity (average/range)	24%/3–30%	24%/3–30%
Hydrocarbon saturation	55%	55%
Permeability (average/range)	80 md/1–2000 md	80 md/1–2000 md
Hydrocarbons		
Oil gravity	40° API	38°–42° API
Bubble point	2348 psig (West) 2737 psig (East)	1991 psig
Gas/oil ratio	600 (West) 800 (East)	490
Formation volume factor	1.467 (West) 1.557 (East)	1.327
Formation water		
Salinity	111 000 mg/l	135 000 mg/l
Resistivity	0.027 ohm m at 257°F	0.023 ohm m at 245°F
Reservoir conditions		
Temperature	257°F	245°F
Pressure	3744 psig at 8500 ft	3700 psig at 8500 ft
Field size		
Area	9910 acres	7712 acres
Gross rock volume	748 000 ac.ft	555 000 ac.ft
Drive mechamism	water drive	water drive
Recoverable oil	98 MMBBL	102 MMBBL
Production		
Start-up date	1976	1990
Development scheme	Single steel platform	Satellite platform
Production rate	35 000 BOPD (peak)	35 000 + BOPD
Production wells	15	6
Injection wells	6	4
Secondary recovery method	Water flood	Water flood

The authors wish to thank Amoco (UK) Exploration Company and its partners Enterprise Oil plc, Amerada Hess Ltd and Texas Eastern (UK) Ltd for their permission to publish this paper. There have been more than 25 years of geological, geophysical and engineering investigation in the Montrose/Arbroath area. This paper represents only a brief summary of the results of this investigation and the authors wish to take the opportunity to thank all those who have previously worked the area without whose efforts this paper would not have been possible.

References

FOWLER, C. 1975. The Geology of the Montrose Field. *In*: WOODLAND, A. W. (ed.) *Petroleum and the Continental Shelf of Northwest Europe. Vol. 1. Geology.* Applied Science, Barking, 467–476.

KNOX, R. W. O'B., MORTON, A. C. & HARLAN, R. 1981. Stratigraphical Relationships of Palaeocene Sands in the UK Sector of the Central North Sea. *In*: ILLING, L. V. & HOBSON, G. D. (eds) *Petroleum Geology of the Continental Shelf of North-West Europe.* Heyden, London, 267–281.

The Argyll, Duncan and Innes Fields, Blocks 30/24 and 30/25a, UK North Sea

D. ROBSON

Hamilton Brothers Oil and Gas Ltd, Devonshire House, Piccadilly, London W1X 6AQ, UK

Abstract: The Argyll, Duncan and Innes Fields are situated in the UK Central Graben in Blocks 30/24 and 30/25a. They produce oil from Devonian, Rotliegendes, Zechstein and Jurassic reservoirs through a common floating production facility, the Deepsea Pioneer.

Argyll was the first UK North Sea oil field to come on stream, with first oil produced in June 1975. Duncan and Innes were added later and the estimated STOIIP for the three fields is approximately 270 MMBBL. To date over 90 MMBBL have been recovered, and the fields are now on decline.

The Argyll, Duncan and Innes fields are located in UKCS Blocks 30/24 and 30/25a, in the Central North Sea, 230 miles east of Edinburgh, and in 260 feet of water (Fig. 1). The Argyll Field was the first productive oilfield in the UKCS, with production start-up being in June, 1975. The fields are produced through a common floating production facility; the Deepsea Pioneer, and the oil exported via shuttle tanker and CALM buoy.

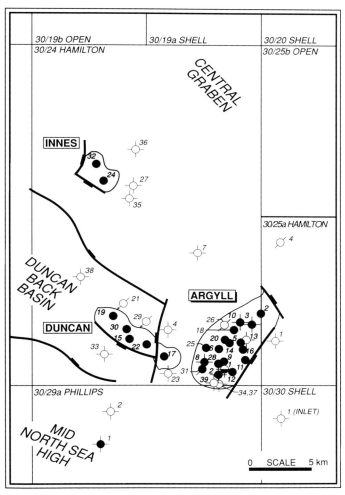

Fig. 1. Block 30/24 Field location map.

The Argyll Field has produced from Devonian, Rotliegendes, Zechstein and Jurassic reservoirs, whilst Duncan produces from the Jurassic, and Innes from the Rotliegendes. Argyll and Innes are produced by depletion, assisted by gas lift in the case of Argyll, whilst Duncan is produced using water injection. The three fields are estimated to have contained initially 270 MMBBL of oil, and have produced over 90 MMBBL of oil to date. The fields are now on decline, after achieving a peak production rate of some 40 000 BOPD in 1985. The geology of the Argyll Field was originally described by Pennington (1975) and the field development summarized by Bifani & Smith (1985).

The Argyll Field began the Hamilton Brothers scheme of naming fields after areas of Scotland or Scottish Clans. Duncan and Innes also follow this tradition.

History

Blocks 30/24 and 30/25 were awarded to the P073 licence group as part of the second UKCS licence round. The original licensees were Hamilton Brothers (operator), RTZ Corporation, Associated Newspapers Group and Kleinworth Benson Lonsdale Ltd. Following equity changes and company acquisitions, the current licence holders are as follows:

Hamilton Oil GB plc	28.8%
Hamilton Brothers Petroleum UK Ltd	7.2%
Elf Oil and Gas Ltd	25.0%
Texaco North Sea UK Ltd	24.0%
Blackfriars Oil & Gas (Ultramar)	12.5%
Renown Petroleum Ltd	2.5%

The first well on the licence, 30/24-1, was drilled by the rig 'Ocean Traveller' on a Tertiary prospect in 1969. The objective was dry, but the well encountered oil shows in Permian Zechstein carbonates. 30/24-2 investigated this interval further, the well being drilled on a horst feature identified at base Cretaceous level. The well penetrated an oil bearing Zechstein, and Devonian sequence. The Zechstein interval flowed 37° API oil at a rate of over 4300 BOPD on test after acidizing. This well, the Argyll discovery well, was followed by the 30/24-3 well which evaluated the structure to the north. 30/24-3 encountered a tight Zechstein sequence, but penetrated oil-bearing Devonian sediments. The 30/24-5 and -6 wells appraised the Zechstein reservoir and were suspended for future production. In addition to the Zechstein reservoir, these wells also encountered oil-bearing Upper Jurassic and Devonian sandstones. They were tied back as sub-sea production wells to the Transworld 58 (TW58), a converted semi-submersible drilling rig. Oil production began from the Argyll Field in June 1975 through the TW58 floating production facility, making Argyll the first productive oilfield in the UKCS.

Considerable uncertainty still remained as to the ultimate reserve in the Argyll Field, and further wells were drilled to add to production and the understanding of the accumulation. The 30/24-8 well delineated the western limit of the field, and proved a field-wide oil–water contact in the Zechstein. The Zechstein was the principal productive horizon during the first five years of field development and has produced the majority of the oil from the field. In 1979 the 30/24-11 well penetrated oil-bearing Rotliegendes

From Abbotts, I. L. (ed.), 1991, United Kingdom Oil and Gas Fields, 25 Years Commemorative Volume, Geological Society Memoir No. 14, pp. 219–225

sandstones in the central part of the field. This proved to be a prolific producing horizon and further drilling took place targeted at this reservoir. Additional potential from the Devonian was suggested by the 30/25a-2 well, drilled on the crest of the structure and the 30/24-25 well proved productive from this unit after fracture stimulation. The latest well drilled on the Argyll Field, 30/24-39 was targeted for Jurassic potential on the western flank of the field. This well was unsuccessful and the Argyll Field oil production is now declining. In total some 25 wells and sidetracks have been drilled to explore, appraise and produce the Argyll Field.

Early seismic mapping of Block 30/24 had identified Permian structures to the west of Argyll. The 30/24-4 well was drilled in 1972 on such a structure, but encountered water-bearing Zechstein carbonates. However, the well did penetrate a significant thickness of Upper Jurassic sandstones, with some poor oil shows. Following this well, attention was switched to the Argyll Field. In 1976 and 1978 new seismic data were aquired over the area which showed two separate structural highs, to the south and southwest of the 30/24-4 well. Well 30/24-15, drilled in 1980 on the larger of these two structures was the Duncan Field discovery well, penetrating oil bearing Upper Jurassic sands which flowed at over 9500 BOPD on test. The second, smaller structure was drilled in 1981 by the 30/24-17 well. This well also encountered productive oil-bearing Upper Jurassic sandstones and this accumulation was named East Duncan. Duncan was given separate field status, and a further eight wells have been drilled in the field, which began production in 1983.

Seismic surveys shot in 1980, 1981 and 1982 identified a prospect some 9 miles northwest of the Argyll Field. The 30/24-24 well was drilled in 1983 to evaluate the Jurassic, Zechstein and Rotliegendes potential of this prospect. The Jurassic was absent, the Zechstein non-reservoir but a thick sequence of oil-bearing Rotliegendes sandstone was encountered. These sandstones flowed at rates in excess of 6000 BOPD and have produced at over 10 700 BOPD. The accumulation was named Innes and production began in 1985. The 30/24-27 appraisal well was unsuccessful but the 30/24-32 well drilled in 1985 became the second producer in the field.

The Argyll, Duncan and Innes Fields are currently produced through a common floating production facility, the Deepsea Pioneer (DSP). However, this has not always been the case. Initially the Argyll Field was produced through the Transworld 58 (TW58) floating production facility. Following the discovery and initial testing on the Duncan Field, however, it became apparent that water injection would be required to recover the maximum reserves from this field. Initially Duncan was produced through the TW58, but in 1984 this was replaced by another floating production facility, the Deepsea Pioneer, located over a new Argyll sub-sea manifold. The DSP has facilities for water injection and for gas lift, and has been used to produce oil from the 30/24 fields since that date. The Innes field was initially produced through the TW58, following that vessel's move from Argyll/Duncan, with oil being exported from the TW58 via the DSP. At the end of 1986 however the TW58 was taken out of service, and the Innes Field produced directly through flowlines to the DSP.

Field stratigraphy

The oldest rocks encountered in the Argyll, Duncan and Innes Field areas are of Middle Devonian age (Fig. 2). These form a sequence of claystones and siltstones which are overlain by Middle Devonian limestones. Upper Devonian continental sediments overlie these carbonates, with red siltstones and fine sandstones grading upward into coarser clastic sediments of fluvial origin. A major angular unconformity exists at the top of the Devonian sequence. The Devonian is present beneath all of the fields, but the distribution of overlying sediments is controlled both by later erosive episodes, and the palaeotopography.

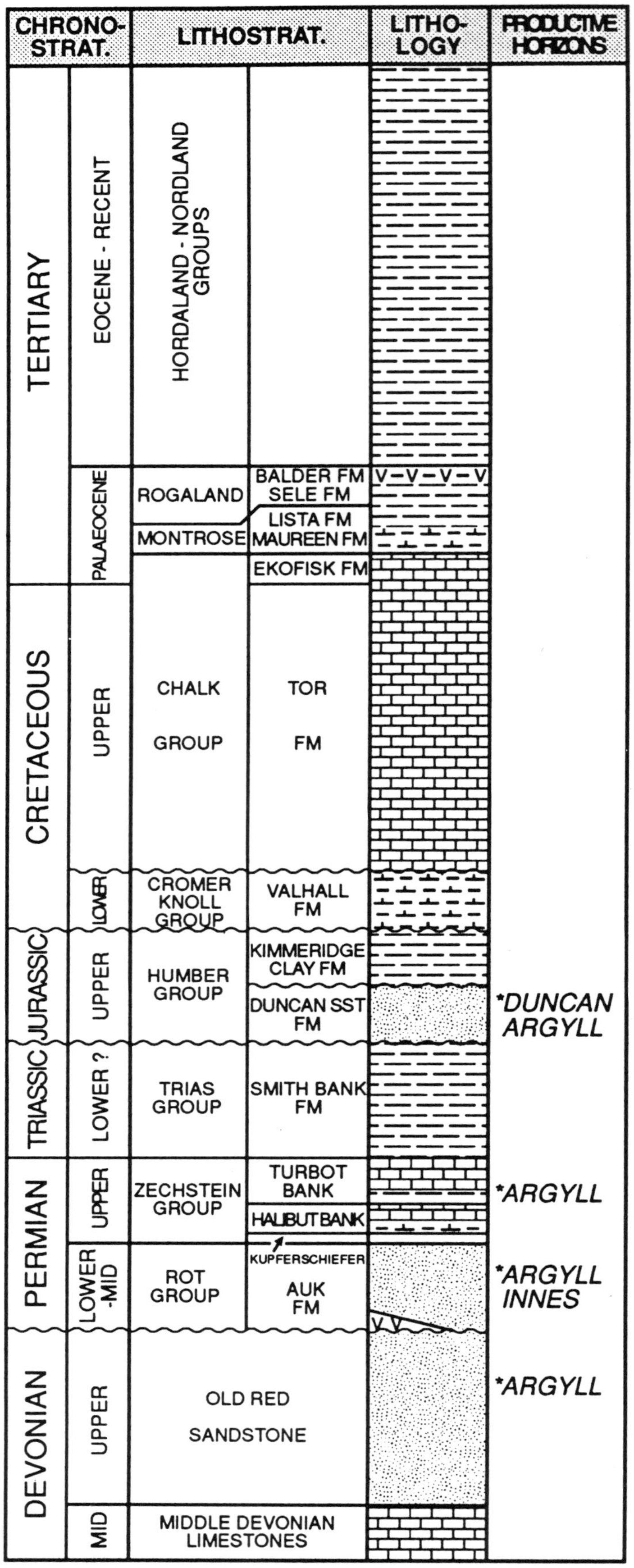

Fig. 2. Seismic stratigraphy.

Permian Rotliegendes sandstones assigned to the Auk Formation were deposited on the denuded Devonian surface. The areal distribution of this unit is limited and controlled by local palaeotopography. However, thick sequences of aeolian, fluvial and sabkha sediments are present in the central part of Argyll and in Innes. Locally an altered volcanic horizon occurs at the Devonian/Permian boundary. These rocks have been dated as Permo-Carboniferous in age, and it is likely that they are related to the thick sequences of volcanics which occur farther south.

A thin Kupferschiefer shale marks the onset of the Zechstein transgression, with suggestions of marine reworking in the uppermost parts of the Rotliegendes sequence.

Zechstein carbonates conformably overlie the Kupferschiefer. These sediments are of a marginal facies and consist of a lower dolomitic horizon, assigned to the Turbot Bank Formation and an overlying sequence of anhydrite, dolomite and claystone assigned to the Halibut Bank Formation. An organic-rich dolomitic interval known as the Sapropelic Dolomite separates the upper and lower parts of the Zechstein interval. The dolomite shows evidence of extreme karstification, with common vug and collapse breccias. Early Triassic claystones assigned to the Smith Bank Formation overlie the Zechstein sequence.

The Cimmerian tectonic events have produced major unconformities in the sequence. As a result no Lower or Middle Jurassic sediments occur in the fields, and erosion has resulted in removal of post-Devonian sediments on the crest of Argyll, and in severe truncation of the post-Devonian section in other areas.

Upper Jurassic sandstones onlap the high areas of Argyll and are present as thick sequences over Duncan. It is difficult to predict if the thickness variations in this unit result from depositional topography or subsequent erosion. However, these sands are of shallow-marine to shoreface origin and have been assigned to the Duncan Sandstone Formation, roughly equivalent to the Fulmar Sand Formation developed in the Fulmar and Clyde Fields to the northwest. These sandstones show an element of diachronism, ranging from Oxfordian to Volgian in age. They are overlain in Duncan by organic-rich claystones of the Kimmeridge Clay Formation, although these sediments are absent due to truncation over Argyll and the crestal parts of East Duncan.

Thin Lower Cretaceous (Cromer Knoll Group) sediments occur away from the higher areas and the Upper Cretaceous to Danian Chalk Group is omnipresent over all of the fields. Subsidence continued through Tertiary depositing thick sequences of claystones on top of the chalk.

Geophysics

Over 3000 km of 2-D seismic data have been collected over blocks 30/24 and 30/25a. These data have been used to define the Argyll, Duncan and Innes Fields, as well as several other prospects and leads on the blocks. The most extensive surveys were shot in 1982 and 1984.

The primary seismic events which are mappable over the area are top Chalk Group, base Cretaceous, top Zechstein and top Middle Devonian Limestones. In addition a top Duncan Sandstone pick and near top Rotliegendes can be made in some areas. The top of the Triassic and Devonian are very poorly defined.

The Argyll structure can be defined at base Cretaceous level as a NE-SW-trending high, generally plunging southwestwards. At top Zechstein level the structure is similar, but shows a greater degree of faulting. A major fault bounded high area occurs in the northeast of the field, with several southwestward throwing faults dissecting the field. A major horst feature extends across the central portion of the structure, making the field extremely complex in this area. At Middle Devonian level a broadly similar structure exists, but with an increased dip and slightly different orientation.

Duncan and East Duncan can be identified at base Cretaceous level as two highs separated by a major N-S fault which downthrows to the west. The top of the Duncan Sandstone can be seismically mapped, and at this level the main Duncan Field structure can be seen to dip northeastward, away from a major NW-SE fault system which marks the southern boundary of the field. East Duncan is a small dip-closed structure, separated by a structural saddle from the western flank of Argyll.

Innes is identified as a complex, faulted, northward dipping structure at top Rotliegendes level.

Trap

The 30/24 Fields are located on the southwestern flank of the Central Graben. The major graben edge faults are to the northeast of the Argyll Field, with Argyll being located on a NW-SE-trending Palaeozoic high feature, the Argyll Ridge. To the north the Argyll Ridge is offset to the west, where the same feature forms the Auk Ridge. Between the Argyll Ridge and the Mid-North Sea High is an extensional back-basin in which the Duncan Fields are located. This back-basin becomes continuous with the Central Graben to the north.

The Argyll Field hydrocarbons are trapped in a southwestward dipping tilted fault block, with dip closure to the south and east, and covering some 2800 acres, with the crest of the structure being at 8600 ft sub-sea (Fig. 3). The post Devonian reservoirs are also truncated towards the crest of the field. This provides an additional potential trapping mechanism for these units. However, the Devonian does not provide a bottom seal and therefore the accumulation relies primarily on the major graben edge faults in the northeast of the field, which downthrow non-reservoir lithologies against the Devonian (Fig. 4). These major NW-SE-trending faults appear to be pre-Permian in age and reactivated during the Jurassic. In addition to these faults a further set of NW-SE-trending faults dissect the field, and four major structural areas can be identified:

(1) the Crestal Area where Chalk rests on Zechstein or Devonian;
(2) the Central Graben, where thick Rotliegendes sandstones were deposited and where a thin Jurassic sequence is preserved;
(3) the Central Horst, where Chalk rests on Zechstein or Jurassic sediments;
(4) the West Flank, where Jurassic sandstones are preserved and the Chalk rests on Jurassic or Triassic strata.

From the fluid contacts, and pressure history of the field it is apparent that the majority of the field is in pressure communication, with essentially a common initial oil–water contact in the Permian and Devonian reservoirs at 9430 ft TVSS. Based on the top Zechstein map this would imply that the structure was full to spill point. However, the thin Jurassic reservoir has a somewhat higher contact and isolated fault compartments may occur.

The Duncan Fields, located in the back-basin to the west of Argyll, are somewhat less complex in terms of structure and faulting. The main Duncan structure is essentially a NW-SE-trending antiform (Fig. 5), dip-closed to the north and northeast and closed by a combination of dip and fault closure to the south. A major NW-SE-trending fault system cuts the southern flank of the antiform and forms the southern boundary of the field along most of its length. The structure plunges to the NW and is cut by several faults which appear to have a major effect on Jurassic sedimentation. To the SE the structure is closed from East Duncan by a major N-S-trending fault. The structure appears full to spill, with an oil–water contact at 9658 ft TVSS. Top seal is provided by Jurassic claystones or Cretaceous sediments, whilst Triassic claystones provide the bottom seal. East Duncan lies on the upthrown side of the major fault which throws the main Duncan sands against Triassic claystones. The structure is a four-way dip-closed dome, full to spill, with an oil–water contact at 9250 ft TVSS and the structural crest being at 9150 ft TVSS.

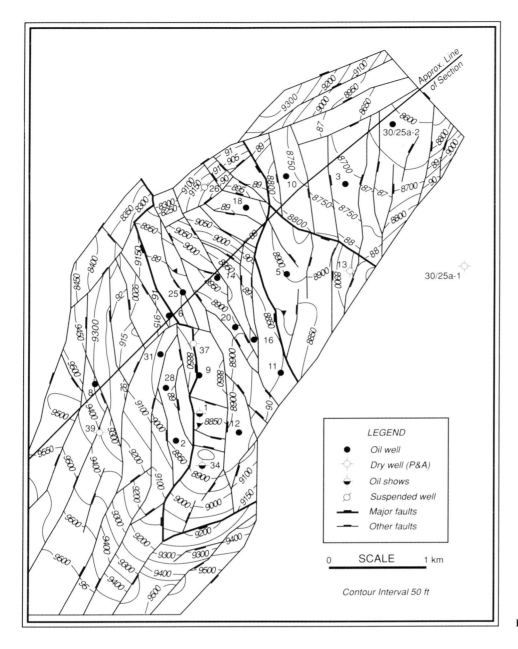

Fig. 3. Argyll Field structure map.

The Innes structure is a simple northward dipping fault block (Fig. 6). It is sealed by major faults downthrowing to the south, west and east and has an oil–water contact at 12 385 ft TVSS. Top seal is provided by tight Zechstein sediments and overlying Kimmeridge Clay. The productive area is approximately 500 acres and the maximum oil column reaches almost 400 ft.

Reservoir

Argyll

The major reservoir intervals in the Argyll field are the Zechstein, Rotliegendes and Devonian.

The Zechstein Group carbonates have excellent reservoir characteristics, primarily due to the presence of a system of vugular and fracture porosity developed as a result of post-depositional tectonic uplift and karstification. The sequence can be divided into two shoalling upward cycles; a lower marginal sequence (Halibut Bank Formation) representing an upward shallowing from the Kupferschiefer dolomitic shale, and an upper-shallow marine sequence (Turbot Bank Formation) with a basal sapropelic dolomitic claystone (Stinkschiefer). This organic-rich layer acts as a vertical permeability barrier and separates the upper, more permeable Zechstein from the lower, generally poorer quality unit. These two sequences are considered to represent the regional Zechstein Z1 and Z2 transgressive cycles. The Zechstein is almost fully dolomitized, with very little primary mineralogy remaining. It is interpreted as having been deposited in shallow-marine, lagoonal and subtidal environments.

The Lower Zechstein consists of thin lagoonal and shelf carbonates at the base, and a reef complex towards the top. The organic-rich Kupferschiefer marks the onset of the Zechstein transgression and represents deposition in a basinal, anoxic development. This is overlain by lagoonal and reef talus detritus, with carbonate breccia and conglomerates being present. An overlying reef complex with both domal stromatolites and algal laminates makes up the upper part of this unit.

The Upper Zechstein consists primarily of dolomite, with the sequence passing upwards from a basinal organic claystone, through shallow marine dolomitic claystones and into dolomites and oolitic grainstones, representative of a marginal to subtidal environment. At the very top of the sequence there is the suggestion of the base of a third cycle. No significant anhydrite occurs,

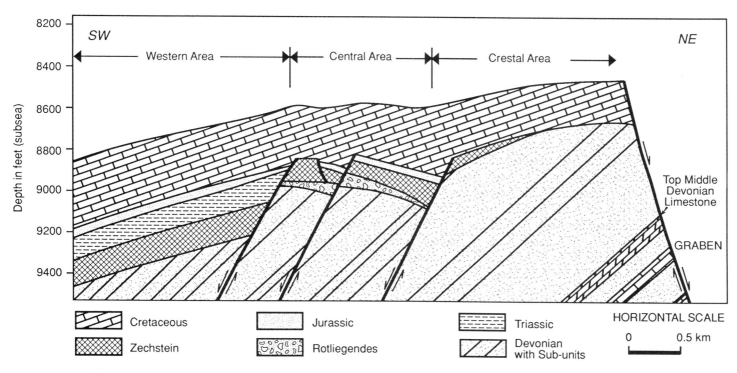

Fig. 4. Argyll Field schematic cross-section.

although this does occur in the same unit in the Duncan Field. Almost all primary sedimentary features have been destroyed by dolomitization and collapse brecciation, due to the dissolution of evaporitic material from the sequence. Generally the Zechstein thickens to the east and south and is absent due to truncation on the crest of the field.

The Rotliegendes reservoir is composed of good quality massive aeolian dune sandstones, restricted to the Central Graben of the Argyll Field. The sandstones are medium-grained, clean and display large-scale cross-bedding. They show a high degree of primary porosity and high net : gross ratios. They are interpreted to have been deposited as self dunes from southeasterly directed palaeowinds.

The Devonian reservoir has been recently re-examined and the age of the majority of the interval revised. Previously (Bifani *et al.* 1987) the majority of the arenaceous reservoir sequence below the Zechstein was interpreted to be of Permian (Rotliegendes) age. However, based on data from log correlation and from dipmeters, these sediments have now been assigned to the Devonian and divided into 25 stratigraphic units. A complex sequence of fluvial sands occurs, these being of variable age becoming older to the north east. Post-depositional faulting complicates correlation across the field, and different parts of the section appear to be present over the major fault blocks of the field. The sediments are interpreted to be of braided stream and ephemeral lake origin, deposited in an arid continental environment. Primary rock quality is poor to fair, generally being better in the stratigraphically younger parts of the sequence.

Upper Jurassic shallow marine sands are also oil-bearing in Argyll, although only a very small amount of production has been obtained from this unit. They occur on the western flank of the field where the sequence is similar to East Duncan; and in Central Argyll. In the latter area they are poorer in quality.

Duncan

The reservoir in the Duncan Fields are sandstones of Upper Jurassic age. The Duncan Sandstone Formation is dated as Oxfordian to Kimmeridgian and was deposited in a lower to middle shoreface environment as storm or sheet sandstones. The existing topography and syn-sedimentary tectonic activity appears to have been crucial in controlling sedimentation patterns, whilst the sequence is truncated by post depositional erosion. As a result the thickness of the reservoir varies across the field, from 4 ft to over 150 ft.

The reservoir can be sub-divided into five zones, based on log and core correlation, which reflect minor regressions and transgressions in the sequence. The main reservoir zone, Zone II, lies beneath a silty, non-reservoir zone which forms the uppermost part of the Duncan Sandstone. Zone II is of good quality, being composed of fine- to medium-grained cross-bedded sandstones, interpreted to have been deposited in a middle shoreface environment by storm currents as near-shore submarine dune features. This zone is almost entirely net sand and has porosities of up to 24%. The underlying, more argillaceous Zone III makes up the remainder of the productive reservoir in the field.

Innes

The Permian Rotliegendes Auk Formation comprises the reservoir sequence in Innes. Based on core and log correlation, three reservoir zones have been identified. The top and basal layers consist predominantly of fluvial and interdune sandstones, whilst the middle layer is a dune sand, exhibiting the best reservoir characteristics.

The thickness of the lower layer varies sympathetically with the underlying topography and the zone consists of debris flow conglomerates and sands of poor reservoir quality. The good quality aeolian dune sands which overlie these sediments have an average thickness of some 80 ft and show a gradual thickening towards the south and west over the field. These are capped by a unit of interdune and reworked dune sands, of moderate reservoir quality.

Source

The source rock for the hydrocarbons in all three fields is the Upper Jurassic Kimmeridge Clay Formation. This organic-rich claystone

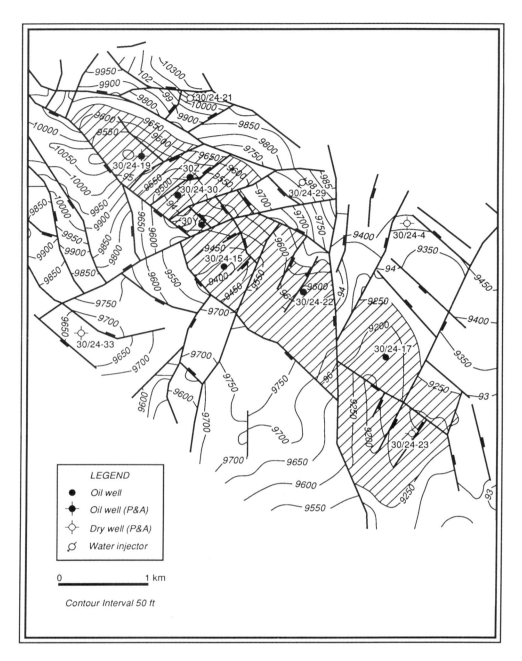

Fig. 5. Duncan Field structure map.

was deposited over the whole area but is not mature over the Argyll or Duncan fields. Geochemical data suggest a marginal-marine depositional environment for the source rock, with some terrestrial input. It is likely the oils have migrated from the deeper, more basinal Central Graben to the north of the fields. The migration route into the traps would have been relatively simple, given the amount of faulting and reservoir juxtaposition which is present. There is some evidence of the development of secondary porosity due to dissolution of cements by acidic pore fluids associated with hydrocarbon emplacement. Minor geochemical differences are apparent between the oils, this being due to differences in thermal history and migration pathway. In particular the Innes oil indicates a somewhat higher thermal regime, perhaps consistent with being deeper and closer to the potential migration routes from the graben.

Hydrocarbons

Oil is the primary hydrocarbon type produced from all of the 30/24 fields. All of the fields contain undersaturated black oil in their reservoirs and there is no evidence of any gas caps in any of the fields. However, there are variations in hydrocarbon composition and the amount of gas produced with the oil.

Argyll produces a low GOR 37° API crude from the major reservoir zones, with the bubble point being well below reservoir pressure. As such only a small amount of solution gas is produced from the Argyll field. Duncan produces a slightly higher GOR 38° API crude, but also has a relatively low bubble point. Innes, however, has a higher solution gas–oil ratio and bubble point pressure. As such a significant amount of gas is produced with the Innes 45° API crude. This gas together with the other gas production, is used for fuel and for the gas lift systems. The small remaining amount of gas is flared.

Formation waters in the field show an areal and stratigraphic variation. Of particular note is the difference in chemistry between the Duncan and East Duncan waters. This suggests that faults in the area seal separate aquifer systems.

Reserves and recovery

The 30/24 fields are estimated to have contained initially some 286

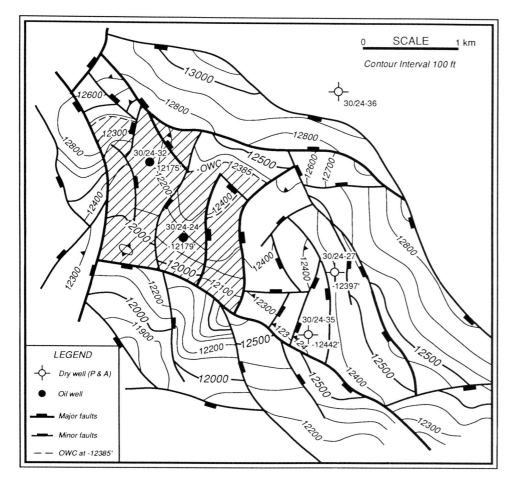

Fig. 6. Innes Field structure map.

MMBBL STOIIP. Of this, the vast majority (218 MMBBL) was reservoired in Argyll, whilst Duncan (49 MMBBL) and Innes (19 MMBBL) make up the remainder. To date over 90 MMBBL of oil has been recovered from the fields.

Well productivities and performance vary, both with the field and with the reservoir unit. The presence of an active aquifer in the area has assisted recovery, most notably in the Permian and Devonian reservoirs in the Argyll Field, and in the Jurassic of East Duncan. Water injection was required in Duncan to maintain reservoir pressure and to sweep the reservoir effectively. However, the current reservoir pressure in Argyll, Duncan and East Duncan is still above the bubble point. Innes, a small solution gas drive reservoir, has only limited aquifer support, and is currently producing with a reservoir pressure below the bubble point.

The most productive reservoir in the Argyll Field has been the Zechstein, where flow rates of clean oil in excess of 16 000 BOPD have been achieved. However, the nature of the fracture and vugular porosity in this unit makes its performance areally variable and difficult to predict. The Rotliegendes sandstones have also proved good producers, with rates up to 11 000 BOPD. Production performance from the Devonian reservoir has been somewhat less, although rates in excess of 5500 BOPD have been achieved. Water breakthrough has occurred in all of the reservoir zones, and several wells have been abandoned after watering out. Gas lift is currently being used to increase production from all of the remaining Argyll wells.

Duncan and East Duncan have produced at fairly steady rates from the Jurassic, with maximum single well rates up to 10 000 BOPD. Both fields are now experiencing water encroachment, but the reservoir drive mechanism of water injection, and aquifer support appears to have been effective in producing the maximum reserves from the field.

Innes initially began production from one well at a rate of 7000 BOPD. By the time production was switched to the Deepsea Pioneer the combined rate from the two production wells had fallen to 4450 BOPD. Production has since declined, although there is some evidence of limited natural pressure support.

The 30/24 fields are now declining in production, although even after 15 years on stream the current combined oil production through the Deepsea Pioneer is still in excess of 11 000 BOPD.

I would like to acknowledge Hamilton Brothers Oil and Gas and the Argyll/Duncan/Innes Partners for permission to publish this paper. In addition I would like to thank A. Yaliz, C. A. Smith and M. Boztas for critical review of the manuscript. The interpretations expressed in this paper are those of the author and are not necessarily those of the Argyll/Duncan/Innes partnership.

References

BIFANI, R. & SMITH, C. A. 1985. *The Argyll Field after a decade of production.* Society of Petroleum Engineers Paper 13987/12.

——, GEORGE, G. T. & LEVER, A. 1987. Geological and reservoir characteristics of the Rotliegendes sandstone in the Argyll Field. *In*: BROOKS, J. & GLENNIE, K. W. (eds) *Petroleum Geology of North West Europe.* Graham & Trotman, London, 523–531.

PENNINGTON, J. J. 1975. The Geology of the Argyll Field. *In*: WOODLAND, A. W. (ed.) *Petroleum and the Continental Shelf of North-West Europe, Volume 1, Geology.* Applied Science Publishers, London, 285–294.

The Auk Field, Block 30/16, UK North Sea

NIGEL H. TREWIN[1] & MARK G. BRAMWELL[2]

[1] *Department of Geology and Petroleum Geology, University of Aberdeen, Aberdeen AB9 2UE, UK*
[2] *Shell UK Exploration and Production 1 Altens Farm Road, Nigg, Aberdeen AB9 2HY, UK*

Abstract: The Auk field is located in Block 30/16 at the western margin of the Central Graben. Oil is contained in a combination stratigraphic and structural trap which is sealed by Cretaceous chalk and Tertiary claystones. An oil column of up to 400 ft is contained within Rotliegend sandstones, Zechstein dolomites, Lower Cretaceous breccia and Upper Cretaceous chalk. Production has taken place since 1975 with 80% coming from the Zechstein, in which the best reservoir lithology is a vuggy fractured dolomite where porosity is entirely secondary due to the dolomitization process and leaching of evaporites. Both Rotliegend dune slipface sandstones, and the Lower Cretaceous breccia comprising porous Zechstein clasts in a sandy matrix, also contribute to production. Poor seismic definition of the reservoir results in reliance on well control for detailed reservoir definition. The field has an estimated ultimate recovery of 93 MMBBL with 13 MMBBL remaining at the end of 1988.

The Auk field is situated in Block 30/16 of the Central North Sea about 270 km ESE from Aberdeen in 240–270 ft of water (Fig. 1). The field covers an area of about 65 km² and is a combination of tilted horst blocks and stratigraphic traps, located at the western margin of the South West Central Graben. The Auk horst is about 20 km long and 6–8 km wide, with a NNW–SSE trend. It is bounded on the west by a series of faults with throws of up to 1000 ft, and the eastern boundary fault has a throw of 5000 ft in the north reducing to zero in the south (Fig. 2). The horst is a westward tilted fault block in the north which grades into a faulted anticline in the south. The Auk accumulation is largely contained within Zechstein dolomites and is ultimately sealed by Cretaceous chalk which overlies the base Cretaceous erosion surface. An E–W cross-section of the field is illustrated by Fig. 3. Auk was the first of the alphabetical sequence of North Sea sea-bird names used for Shell/Esso fields.

History

Licence P.116 was part of the Third Round allocation to the Shell/Esso Licence Group in 1970. The Auk discovery well (30/16-1) was drilled in 1971 by the semi-submersible rig 'Staflo' and encountered oil-bearing Zechstein dolomites unconformably overlain by Cretaceous chalk. The well tested 5900 BOPD of 38° API oil. The initial target was Rotliegend sandstones, and the preservation of porous Zechstein carbonates on the eroded horst block was unexpected (Brennand & van Veen 1975).

Appraisal well 30/16-2 was drilled 4 km SE of the discovery and encountered productive Zechstein overlain unconformably by Triassic shales; this well also tested oil from the Rotliegend. 30/16-3 was drilled 2.5 km east of the discovery and again located Zechstein, but overlain by tight Lower Cretaceous conglomerate. In 30/16-4 chalk was found resting on wet and tight Rotliegend, the entire Zechstein having been eroded. As a result of the poor seismic control and unpredictable faulting and erosion of the Zechstein, initial reserves of the Zechstein were estimated with a wide spread of 30–100 MMBBL with an expectation of 60 MMBBL (Buchanan & Hoogteyling 1979).

The development decision was made in late 1972 when a 10-slot steel drilling/production platform with facilities for peak oil throughput of 80 000 BOPD was ordered. Oil loading was planned through an exposed location single-buoy mooring into two dedicated tankers, but only one was used after 1980. Since 1986 production has been via a pipeline link to Fulmar. The platform was installed in 1974 and production from the first well commenced in December 1975 at rates increasing to 35 000 BOPD (Buchanan & Hoogteyling 1979). Oil production from the field peaked at 70 000 BOPD in May 1977 and had declined to 14 000 BOPD net oil with 77% water cut in July 1986. Production rates through 1988 were 9500 BOPD net oil with 81% water cut.

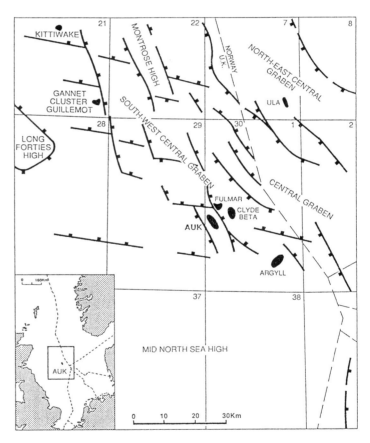

Fig. 1. Location map, Auk oilfield.

Field stratigraphy

The general stratigraphic sequence of the Auk field area is shown in Fig. 4. Unconformities are present in the area at base Devonian, base Permian, base Trias, sub-Lower Cretaceous, sub-Upper Cretaceous (chalk) and sub-Palaeocene. These produce a considerable variation in stratigraphy in different parts of the field and in adjacent areas which cannot be detailed in this brief account. Lower Palaeozoic basement was penetrated in 30/16-5 and consists of steeply dipping low grade metamorphic siltstones and claystones with extensive quartz veining.

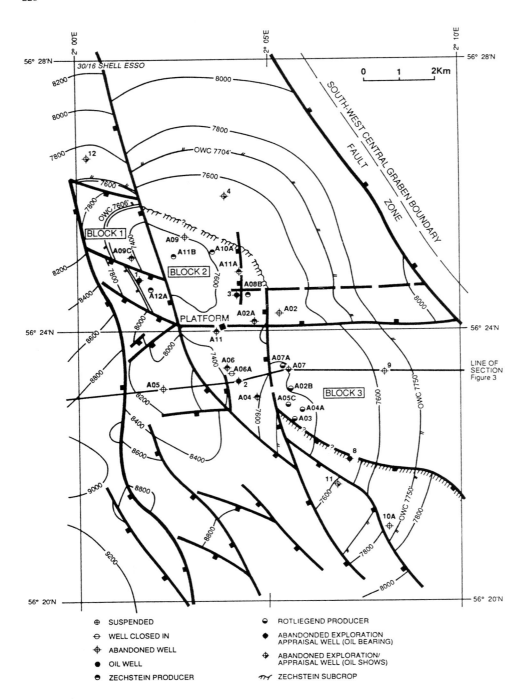

Fig. 2. Structure map of Auk Field area contoured at top reservoir (Zechstein + Rotliegend). The limit of Zechstein subcrop, generalized oil–water contacts, major faults and positions of blocks 1, 2 and 3 of the field are indicated.

Devonian

Two Devonian carbonate units each up to 100 ft in thickness and separated by 200 ft of red-brown claystone rest unconformably on the basement with a thin basal conglomerate of locally derived material. The carbonates are calcareous dolomites and dolomites which have yielded tabulate and rugose corals and are thus of shallow-marine origin. Similar coral-rich limestones in Argyll field are considered to be of Middle Devonian age (Pennington 1975). Succeeding the carbonates with apparent conformity are up to 3000 ft of porous (av. 17%) sandstones and shales of inferred Devonian age and probably of flood plain origin. These sandstones are thickest in the west of the field but are absent in the east due to pre-Permian erosion (Fig. 3).

Permian

The Permian Rotliegend sequence (Auk Formation of Deegan & Scull 1977) is about 900 ft thick and onlaps eastwards on to the Devonian substrate (Heward 1991). The base is locally marked by a conglomerate with basalt pebbles (30/16-1 and 30/16-8) and a possible in situ flow of porphyritic leucite–nepheline basalt is present in 29/25-1. The reflector associated with the base Permian becomes weak on the eastern part of the Auk horst which was at a higher structural level during basal Rotliegend deposition. The overlying Rotliegend sandstones commence with thin conglomerates and sandstones of alluvial origin possibly derived from the Mid-North Sea High to the south. These deposits are followed by extensive pinkish-red-brown dune sandstones with large scale cross-bedding which indicates that the wind came from the west. Interdune sandsheets are represented by horizontally laminated wind-rippled sandstone. The top part of the sequence (reservoir Unit 1 below) contains waterlain massive mass flow and stratified sandstones, intraformational conglomerates and thin lacustrine shales, which yield an Upper Permian flora. Pennington (1975) recorded a flora of Upper Permian age from the Rotliegend in the Argyll field, thus the convention of equating Rotliegend with Lower Permian may not be valid in this area. A wetter climate with dune

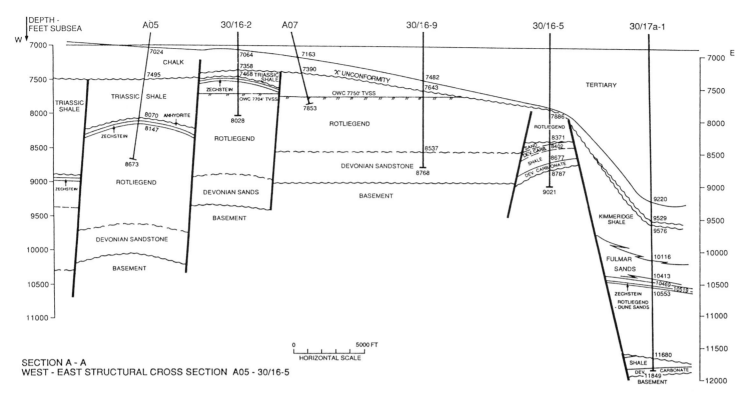

Fig. 3. Structural W–E cross-section of the Auk Field. The line of section is indicated on Fig. 2.

stabilization and reworking is favoured as an explanation for the waterlaid deposits in the Auk area. This contrasts with the usual 'Weissliegendes' interpretation in which dunes and sandstones suffered homogenization and deformation during the Zechstein transgression by rapid rise of the water table associated with expulsion of air trapped in the sands following rapid flooding of the dune field (Glennie & Buller 1983).

The sandstones show only minor reworking at the top prior to deposition of the Kupferschiefer, a 3–5 ft thick bituminous shale deposited below wave base in the restricted marine setting of the Zechstein basin. This horizon forms an important gamma-ray log marker throughout the area.

The main reservoir, the Zechstein dolomites, averages 28 ft in thickness and is probably the time equivalent of the Z1 carbonates of northern England. The Zechstein comprises the Argyll Formation of Deegan & Scull (1977). The dominant lithology seen in core is a laminated dolomicrite with sub-parallel organic-rich laminae on a millimetre scale. Solution vugs after evaporite minerals occur, some of which are confined by organic laminae. Ghosts of peloids, indeterminate fossil material, and scattered quartz grains occur in the dolomite. Pervasive brecciation is present, probably due to collapse of the more evaporite-rich lithologies.

The laminated lithology has been interpreted as stromatolitic algae (Brennand & van Veen 1975) and some domal stromatolites are present. A saline inter-to-supratidal sabkha is suggested as the depositional environment. The organic laminae could also be relics of planar algal mats of a sabkha environment. Rapid fluctuation of water depth in the Zechstein basin may have resulted in the exposure and dolomitization of the sequence prior to deposition of the second Zechstein cycle. This reservoir unit, discussed below in more detail, is overlain by another organic-rich shale with a highly gamma-ray response. In the SW part of the field (A05) 30 ft of anhydrite overlies the upper shale and is interpreted as the Z1 anhydrite.

In earlier interpretations an Upper Zechstein carbonate unit was considered to be present (Brennand & van Veen 1975; Buchanan & Hoogteyling 1979) and was interpreted as the collapse breccia residue of a higher Zechstein cycle. This unit is now thought to be a depositional breccia of Lower Cretaceous age and is considered below.

Trias

Red-brown to grey-green silty claystones of Lower Triassic (Scythian) age are thickest in the west of the field (500 ft in A05) but have been extensively eroded on the top of the block. These rocks are interpreted as the flood plain deposits of a fluviatile system and are probably a part of the Smith Bank Formation (Fisher 1986).

Cretaceous

The four Lower Cretaceous lithological units recognized in the field area are the Upper Carbonates breccia, a Hauterivian basalt flow, an Albian–Aptian conglomerate, and a marl. The Upper Carbonates breccia (previously interpreted as an Upper Permian collapse breccia) consists of Zechstein dolomite and possible Triassic clasts in a matrix which includes rounded (Rotliegend derived) sand grains and an open marine fauna of probable Neocomian age. This deposit only occurs between Zechstein and other Cretaceous deposits where the Trias is absent, and is nowhere overlain by Triassic sediments. Deposition of the breccia was associated with erosion, karstification and faulting of the horst in early Cretaceous times. A basalt flow occurs in the NE of the field (Wells A11A, A10A), and radiometric dating suggests a Hauterivian age (Heward et al. 1988). The Albian–Aptian conglomerate fringes the basalt flow (30/16-3; A11B) and contains rounded basalt and Zechstein pebbles in a matrix of sand (reworked Rotliegend). Shell fragments of an open marine fauna (Aptian–Albian age) are present and bivalve borings penetrate Zechstein clasts. This conglomerate is interpreted as the deposit of an early Cretaceous marine transgression. The marl overlies the Upper Carbonates breccia (30/16-1) and

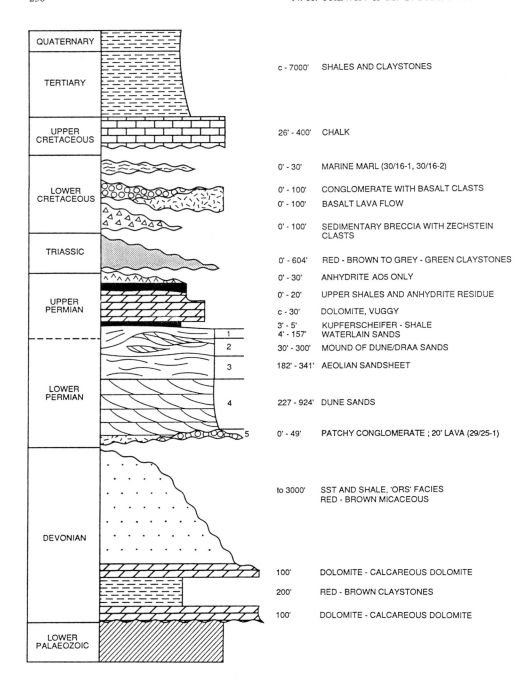

Fig. 4. Composite stratigraphic column for the Auk field area.

Trias (30/16-2), and is of marine origin, deposited in shallower conditions than equivalent Lower Cretaceous marls in the surrounding areas.

The Upper Cretaceous chalk rests unconformably on the eroded horst structure. Foraminiferal packstones of Coniacian to Santonian age grade up into Campanian foraminiferal wackestones. The chalk thins from 400 ft in 30/16-1 to only 26 ft in the east (30/16-5) (Figs 2 & 3).

Tertiary–Recent

Palaeocene claystones and shales of variable grey, green and red hues unconformably overly the chalk and comprise the Montrose Group and the Sele Formation of the Rogaland Group (Deegan & Scull 1977). The Balder Formation at the base of the Eocene is characterized by tuffaceous sediments. The Palaeocene is generally around 100 ft in thickness and the Eocene 600–900 ft. Thick Oligocene claystones are glauconitic and around 2500 ft thick, and the base Miocene is at about 3900 ft. Following rapid subsidence in the Palaeocene creating deep marine conditions, sedimentation rates increased dramatically in the Oligocene and appear to have remained high to the present day.

Geological history

The Lower Palaeozoic sediments were metamorphosed during the Caledonian orogeny and subsequently eroded to a peneplained surface on which marine Devonian limestones were deposited due to marine transgression from the south. Marine influence was short-lived and 3000 ft of continental (Old Red Sandstone) facies succeed the limestone.

The Hercynian orogeny resulted in differential uplift and erosion in the Auk area and Carboniferous rocks are absent. Tilting to the SW along faults active at the eastern margin of the future Auk horst is possibly related to formation of the Central Graben (Glennie 1986).

Rotliegend clastics resulting from denudation of the Mid-North Sea High were deposited on the eroded land surface. Initial deposition of thin waterlain conglomerates and sandstones was followed by a thick aeolian dune and sandsheet sequence which onlaps the horst. The top of the Rotliegend comprises water-lain massive and stratified sands which modified dune topography and were deposited in a wetter period when dunes became inactive. The Zechstein transgression rapidly and gently flooded the remaining topography of stabilized dunes and water-lain deposits with only minor reworking. Rapid deepening resulted in deposition of the Kupferschiefer below wave base in the restricted marine Permian Basin. The overlying Zechstein dolomites accumulated in a sabkha environment in the first Zechstein evaporative cycle. Post-Zechstein to Lower Trias erosion and leaching of the Zechstein took place prior to deposition of the Triassic mudstones and sandstones, which probably covered the whole area.

The Mid-Cimmerian phase of major rifting was instrumental in shaping the Auk area, resulting in faulting and erosion of much Trias with local exposure of Permian. The present day fault pattern was established at this time (Figs 1 & 2). The main faults, striking NNW–SSE, have a dextral transcurrent component, offset being observed at the level of the Zechstein subcrop against the base of the chalk. These faults also have a variable vertical throw, which is particularly important in the case of the eastern fault which has a throw of 5000 ft in the north reducing to zero in the south.

Jurassic sediments are not known to be preserved on the Auk horst, but thick Upper Jurassic is present in the adjacent Central Graben where it contains the major source rock and also the Fulmar reservoir sandstones. The Auk area remained high in the Lower Cretaceous and only local deposits of conglomerate, breccia and volcanics are preserved owing to continued faulting.

Late Cretaceous sea-level rise resulted in drowning of the Auk horst and deposition of the chalk which shows depositional thinning over the crest of the block. Rapid subsidence of the area took place in the early Tertiary and resulted in deposition of c. 7000 ft of mudstones. This phase of burial produced maturation of the Upper Jurassic source rocks in the Central Graben and migration of oil into the Auk structure, probably from Oligocene times to the present.

Geophysics

The Auk structure was originally defined on seismic (e.g. line UK2-242 (Brennand & van Veen 1975)) as a drape seal over the horst block. Subsequent coverage consists of a 1 × 1 km grid of 896 km conventional 2-D lines shot in 1976 and a 279 km infill grid of 1980. The reliable mappable reflections are top and base chalk, base Rotliegend and top Devonian limestone. At the time of writing two new test lines have been shot and give better reservoir definition (Fig. 5), and a 3-D seismic programme is now planned.

There is no reliable seismic event associated with the principal Zechstein reservoir, and where the chalk-Zechstein interval is less than 200–300 ft the top Rotliegend event is not visible. Base chalk is the nearest horizon that can be mapped and structure at top Zechstein is obtained by isopach extrapolations from base chalk. Over the crest of the structure control is insufficient to construct meaningful isopachs of Lower Cretaceous, Trias and Zechstein and maps rely on well control. The extent of the Zechstein subcrop is thus in doubt and estimated reservoir depths of the Zechstein can be 50–100 ft or more in error. Complications with the several unconformities and minor faults necessitate reliance on well information.

Reservoirs

The four reservoir units below are recognized in the Auk field, of which the most important is the Zechstein dolomite. The structural contours of the top of the reservoir (Fig. 2) refer to the top of the

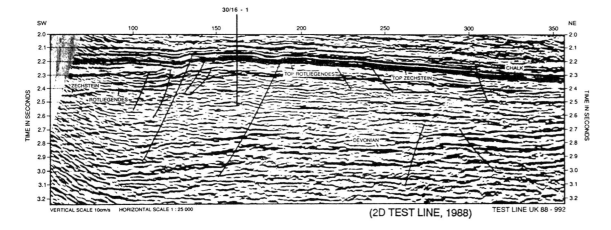

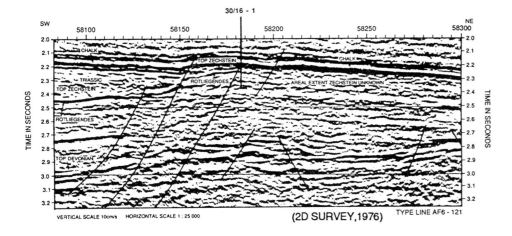

Fig. 5. Geophysical cross-sections of the Auk field through the discovery well showing the greatly improved quality of the 1988 test line over the original data.

Table 1. *Summary of Rotliegend units in the Auk Field, together with core analysis data and percentage of dune slipface lithofacies in each unit*

Unit	Description	% Slipface	Average porosity (range)	Average permeability. (range)
1 Flanking drape	Marine Kupferschiefer drapes relict topography. Waterlain mass-flow sands infill depositional lows. Intercalated lacustrine shales and limestones. Freshwater origin, 4–157 ft	0	19.9% (13–27)	3.6 md (0.2–8)
2 Mounded sheet	Aeolian wind-rippled sands with thin waterlaid conglomerates. Forms depositional mound in Block 3 which is 7 km wide across wind direction. 284 ft thinning to 32 ft at edges. Dune and draa, slipfaces oriented E. Fine-grained aeolian sands banked on northern flanks.	10	16.7% (11–22)	1.9 md (0.2–2.0)
3 Undulating sheet	Aeolian wind-rippled sandsheet with dune apron and dry interdune sands, stacked dune slipface sands oriented to the E. 182–342 ft.	40	20.8% (15–27)	10.7 md (1–125)
4 Wedge onlapping horst	Wedge shaped body of porous dune sands dominated by slipfaces to the SE. Onlaps the Devonian horst, 924 ft in W to 227 ft in E.	70	No core	No core
5 Patches	Localized fluviatile conglomerates which may fill topography on the unconformity, 0–49 ft	0	No core	No core

Summarized from Heward 1991.

Zechstein plus Rotliegend reservoirs. The line of erosional truncation of the Zechstein reservoir is also shown in Fig. 2.

Reservoir	Produced to date
U. Cretaceous, chalk	Waste zone
L. Cretaceous, Upper Carbonates	7 MMBBL Block 1, A09C, A12A
U. Permian, Zechstein	61 MMBBL
L.–U. Permian, Rotliegend	12 MMBBL

Rotliegend

Rotliegend in Auk can be divided into five units (Heward 1991) which possibly represent distinct phases of desert sedimentation and are summarized in Table 1. Lithologically the reservoir comprises seven genetic lithological types (Table 2) of which four are aeolian and three waterlain. These units are treated in more detail in Heward (1991).

The ranges of porosity and permeability for the main sand body types are summarized in Fig. 6. Reservoir units 1, 2 and 3 of Table 1 occur above the OWC in the Auk field. There is variation in the OWC depth in the field with oil staining in permeable sands to 7690–7720 ft in Block 3, around 7700 ft over much of Block 2 and shallowing northward to 7595 ft in 30/16-4. This feature may reflect gross reservoir capillary effects and pathways for oil entry to the reservoir. It is considered likely that the Rotliegend has been charged by downward movement of oil from the overlying Zechstein. Sandstones with poor porosity/permeability characteristics are frequently not oil-stained above the OWC and in cores oil-staining can be seen following permeable coarse laminae. The Rotliegend oil column is around 400 ft in Block 3, 300 ft in Block 2 and 75 ft in Block 1, but an interval of black bituminous oil 20–100 ft thick has been observed above the OWC in some wells which locally reduces the gross reservoir column.

Only the massive waterlain sands of Unit 1 and the dune slipface sands (Fig. 7) of Units 2 and 3 have reasonable reservoir characteristics. The waterlain sands are widespread in the north of Block 2 and south of Block 3, however, the best quality sands are lensoid and encased in poorer material. Short term production of only a few hundred STBPD has been achieved from these sands.

Table 2. *Characteristics of the four aeolian and three waterlain lithofacies recognized in the Rotliegend of the Auk Field*

Aeolian	
Dune slipface (Fig. 7)	Medium-grained, well-sorted, sandflow cross-stratified sandstones with sedimentary dips of 24–28°. High porosity.
Dune apron	Crs.–fine-grained wind-ripple laminated sandstone in mm-cm alternations of steepening-up dips of 5–25°. Upward increase in porosity.
Low-angled stratified	Crs.–fine-grained wind-ripple laminated sandstone in mm alternations, low dips. Low porosity.
Fine grained association	Fine–v. fine-grained, well sorted, mm–cm lamination. Wind-ripple laminated, grainfall and sandflow cross-stratified microporous.
Waterlain	
Fluvatile extraformational conglomerates	Polymict conglomerates with fine- to medium- sandstone matrix.
Mass flow sands and intraformational conglomerates	Fine–medium-grained massive and stratified sandstones with thin intraformational dolomitic conglomerates and shales. Fining-up sequences.
Lacustrine shales and dolomites	Laminated grey shales, dolomite.

Core analysis data from Heward 1991.

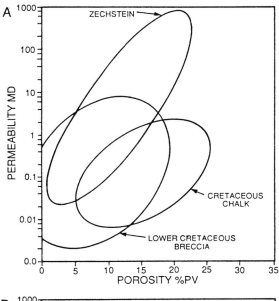
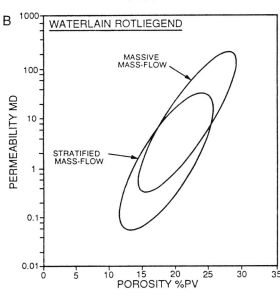
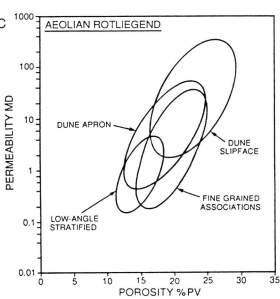

Fig. 6. Porosity/permeability data for reservoir units of the Auk Field.

The aeolian slipface sands occur in Units 2 and 3. In Unit 2, 50 ft of slipface sands correlate over more than 750 m in a north–south direction in the south of Block 3 and account for the bulk of the 12 MMBBL of Rotliegend production from Auk, with wells producing at 500–1700 STBPD.

Unit 3 contains the most porous and permeable sands above the OWC and these occur in the crest of Block 3 and in part of Block 2 around 30/16-2. Rapid water breakthrough affected these sands due to their proximity to the underlying aquifer of Units 3 and 4 which provides pressure support for the field. Future Rotliegend production depends strongly on sedimentological interpretation of the reservoir architecture and location of significant connected volumes of slipface sands.

The diagenetic history comprised an early environment-related phase with clay infiltration, formation of grain coatings and some dolomite. Periodic subaerial exposure from late Permian to early Cretaceous resulted in meteoric diagenesis with formation of kaolinite and quartz and some dissolution of unstable minerals. An unusual bladed tufty form of pore-bridging kaolinite is the subject of further investigation. Following late Cretaceous and Tertiary burial further dolomite and minor barytes cementation took place. The main reservoir characteristics are dependent on depositional factors modified (detrimentally) by diagenesis.

Zechstein

The Zechstein dolomite reservoir has a reasonably uniform thickness of 25–30 ft where present in the field area. It is absent from the NE of the field (Fig. 2) where it was probably eroded in Jurassic to Lower Cretaceous times.

The reservoir lies between the Kupferschiefer and the Upper Shale and is a fractured dolomite with secondary intercrystalline, mouldic and vuggy porosity (Fig. 7) (Lafferty 1987). The original sediment was a carbonate mud with intervals of millimetre-scale alternations of carbonate and organic laminae of algal origin. Evaporites (gypsum/anhydrite) grew in the soft sediment to produce a carbonate/evaporite mush. Early dolomitization took place very soon after deposition and destroyed much original texture. Only minor compaction had taken place prior to dolomitization, and Lafferty (1987) considered a likely mechanism to be 'seepage reflux' or 'evaporative pumping' thus clearly implying a shallow to supratidal environment at the time of dolomitization. Early dolomitization affected the carbonates whilst apparently retaining the evaporites, but some minor brecciation may be due to solution of thin evaporite seams. Dolomitization may also have been associated with a phase of subaerial exposure prior to Upper Shale deposition.

Burial, probably in the Triassic, resulted in stylolitization along organic-rich laminae, but evaporites remained. Subaerial exposure in Jurassic–Cretaceous times resulted in leaching of remaining evaporites, and possibly some carbonate, resulting in extensive vuggy porosity. Stylolite seams confine and bridge some vugs indicating that compaction predated leaching. Some parts of the rock were so extensively leached that mechanical collapse took place, and subsequent faulting resulted in extensive fracturing of the weakened rock. No primary matrix porosity exists in the dolomites, and connection of vugs is due to secondary intercrystalline porosity, fractures and probable solution channels. Core analysis data seriously underestimate the reservoir quality since only the tightest and least fractured material has survived coring. Vugs ranging up to centimetre size are difficult to evaluate in conventional core analysis. Frequently only rubble or very friable material is recovered by coring, therefore core-measured porosity values are thought to be unrepresentative. Permeability measurements on core plugs range from 0.02–620 md (av. 53 md), but formation permeability exceeds these values due to pervasive fracturing of the rock. Permeabilities of tens of Darcies have been measured during production and build-up tests.

The Zechstein reservoir appears to be in communication with the Rotliegend but pressure interference testing has proved inconclusive. Early water break-through and high water cuts with continued oil production are due to the inhomogeneities of the reservoir with a hierarchy of permeability features ranging from multidarcy (fractures, solution enlarged joints) to hundreds of millidarcies (vuggy friable dolomites) and less than 10 md (the tighter, vuggy, laminated dolomites). Thus, whilst water is produced via fractures, oil may still seep from the less permeable parts of the reservoir.

Upper Carbonates

The Lower Cretaceous Upper Carbonates reservoir is up to 100 ft thick and consists of a breccia dominated by dolomitic clasts in a sandy to marly matrix (Fig. 7). The reservoir is poorly defined having been cored in only three wells with very poor recovery. The reservoir is interpreted as a sedimentary breccia deposited as a result of local faulting and erosion. In this interpretation the distribution of the reservoir is considered to be structurally controlled and dependent on a source of suitable clasts (mainly porous Zechstein). Vuggy porosity in the clasts is inherited from the Zechstein and matrix porosity is also present. Red staining of matrix and clasts may be due to initial subaerial diagenesis. The reservoir exhibits considerable lithological heterogeneity, making meaningful petrophysical evaluation difficult. Core-measured porosities range from 0.5–25% (av. 10%), and permeabilities only average 2.5 md. However, reservoir porosity and permeability were probably significantly higher since core recovery was poor and probably not representative of the formation; fracture porosity is probably important in this reservoir. 7 MMBBL production has been assigned to this reservoir.

Chalk

The lowermost 30 ft of the chalk is an oil-stained foraminiferal packstone of Coniacian to Santonian age. Burrows and large fragile *Inoceramus* fragments are present, and this chalk is thought to be autochthonous rather than the result of re-deposition as in the case of other chalk reservoirs (Taylor & Lapre 1987). Porosity in the oil-stained zone ranges from 6.5–34% (av. 17%). Stylolites are present with large (> 1 cm) amplitudes. Permeability ranges from 0.01–4.7 md (av. 0.7 md); the highest values representing stylolite and fracture pathways. The oil is present in fine-grained matrix and in foraminiferal chambers and porosity is very poorly connected. No production can be sustained from the chalk and it is regarded as a 'waste zone' despite having an estimated 59 MMBBL oil in place.

Trap

The Auk field is essentially a stratigraphic trap within a tilted horst. The trap is formed by the erosional truncation of Zechstein dolomite, Rotliegend sandstones and Lower Cretaceous breccia by the sub-chalk unconformity. The basal 30 ft of the chalk is porous and contains some hydrocarbon, but it rapidly grades upwards into an effective seal. The western boundary of the field is faulted against Triassic shale.

The field is about 12 km long by 6 km wide, with a field crest at around 7337 ft and the OWC at a maximum depth of 7750 ft in porous and permeable rocks but rising to around 7600 ft in the north of the field. The irregularity of the OWC is thought to be due to downward charging from the Zechstein being restricted by impermeable cemented horizons and facies distribution in the Rotliegend.

Fig. 7. Typical core samples from the three main reservoir units of the Auk field. The inset photograph is an SEM photograph of a pore case of the Zechstein dolomite.

Source

The major source of the Auk oil is Upper Jurassic (Kimmeridgian) organic-rich shales in the adjacent Central Graben. The Auk oil may be due in part to migration from the Fulmar area which is filled close to spill point. The GOR in Auk is lower than in Fulmar and gas may have leaked from Auk to the overlying Tertiary where over-pressured gas is present in Oligocene sediments at around 4500 ft.

Maturation of Upper Jurassic source rocks probably took place from late Tertiary times and migration was controlled by the bounding fault system of the graben, and the permeable pathway provided by the Zechstein.

Hydrocarbons

The Auk field contains 38° API low sulphur (0.4%) oil with a GOR of 190 SCF/STB and a relatively high wax content (13%). Reservoir temperature is 215°F with an initial pressure of 4067 psia at 7600 ft TVSS. Viscosity at reservoir conditions is 1.24 cP. Water salinity is 105 000 ppm NaCl equivalent.

Auk Field data summary

Trap
Type — Combination stratigraphic and structural
Depth to crest — 7337 ft TVSS
Lowest closing contour — 7900 ft TVSS
Hydrocarbon contacts — Block 1 OWC 7606 ft TVSS
Block 2 OWC 7704 ft TVSS
Block 3 OWC 7750 ft TVSS
Oil column — 400 ft

Main pay zone

Formation	Zechstein dolomites	Rotliegend sandstones
Age	Upper Permian	Lower–Upper Permian
Gross thickness (average/range)	28 ft/25–30 ft	985 ft/485–1588 ft
Net/gross ratio	1.0	0.85/0.46–0.92
Cut-off for N/G	Gamma Ray, ±75°API	14–16% in-situ porosity
Porosity (average/range)	12%/2–26%	19%/11–27%
Permeability average (average/range)	53 md/0.1–10 000 md	5 md/0.2–125 md
Productivity index	50–150 BOPD/psi	1 BOPD/psi

Hydrocarbons
Oil gravity — 38°API
Oil type — low sulphur (0.4%)
Bubble point — 700 psi
Gas/oil ratio — 190 scf/bbl
Condensate yield — NA
Formation volume factor — 1.154 rb/stb

Formation water
Salinity — 105 000 ppm NaCl equivalent
Resistivity — 0.025 ohm m at 205°F

Reservoir conditions
Temperature — 215°F
Pressure — 4067 psi at 7600 ft TVSS
Pressure gradient in reservoir — 0.33 psi/ft

Field size
Area — 65 sq. km.
Gross rock volume — 1.4 MM acre-ft
Hydrocarbon pore volume — 529 MMBBL STOIIP
Recovery factor — 18%
Drive mechanism — natural water drive
Recoverable reserves — 93 MMBBl oil

Production
Start-up date — 1975
Development scheme — Central 10-slot steel production/drilling platform, oil evacuation via tankers till 1986, then via pipeline to Fulmar field.
Number/type of wells — 10 exploration/appraisal
9 development (producing)
10 development (suspended/abandoned)
Secondary recovery methods — none at present
Production rate — 9500 BOPD net oil (end 1988)
Cumulative production to date — 80 MMBBL

Reserves and resources

Owing to the complexity of the several reservoirs, lack of geophysical definition of the reservoirs, and doubts on the porosity and permeability distribution in the reservoirs, accurate reserve estimates are difficult to achieve. The (1988) revised ultimate recovery estimate was 93 MMBBL (13 MMBBL remaining); a figure approaching the original discovery high estimate. Assignment of oil between reservoirs is complicated by the fact that the Zechstein has produced more oil in parts of the field than the estimated STOIIP for this reservoir. Possible reasons for this are (1) that some migration from the Rotliegend to Zechstein has taken place, (2) that the extent of the Zechstein reservoir has been underestimated, and (3) the Zechstein reservoir parameters used in the calculations of STOIIP were pessimistic.

A good natural water drive exists in the field which is underlain by an excellent Rotliegend aquifer. Further production by re-drilling slots on the platform will concentrate on maximizing production from the slipface dune sands of the Rotliegend, and possible finds of attic oil resulting from better structural definition of the reservoirs by the current geophysical programme.

This paper has been compiled from numerous sources within Shell, review contributions by R. Buchanan, G. Stalker and V. C. Vahrenkamp were particularly useful. The work of A. P. Heward, Y. Jensen and J. van Limborgh formed the basis of the Rotliegend interpretation. Consent to publish given by the managements of Shell UK Exploration and Production and Esso Exploration and Production UK is gratefully acknowledged.

References

BIFANI, R. 1975. A Zechstein depositional model for the Argyll Field. *In*: TAYLOR, J. C. M. (ed.) *The role of evaporites in hydrocarbon exploration*. JAPEC course notes 39.

BRENNAND, T. P. & VAN VEEN, F. 1975. The Auk oil-field. *In*: WOODLAND, A. W. (ed.) *Petroleum Geology and the Continental Shelf of NW Europe*. Applied Science Publishers Ltd., 271–281.

BUCHANAN, R. 1979. *Auk area—a production geology report*. Shell UK Expro report.

—— & HOOGTEYLING, L. 1979. Auk field development: A case history, illustrating the need for a flexible plan. *Journal of Petroleum Technology*, 31, 1305–1312.

DEEGAN, C. E. & SCULL, B. J. 1977. *The standard lithostratigraphic nomenclature for the Central and Northern North Sea*. Institute of Geological Sciences, Report No. 77/25.

FISHER, M. J. 1986. Triassic *In*: GLENNIE, K. W. (ed.) *Introduction to the Petroleum Geology of the North Sea* (2nd Ed.). Blackwell, 113–132.

GLENNIE, K. W. 1986. The structural framework and pre-Permian history of the North Sea area. *In*: GLENNIE, K. W. (ed.) *Introduction to the Petroleum Geology of the North Sea* (2nd Ed.). Blackwell, 25–62.

—— & BULLER, A. T. 1983. The Permian Weissliegend of NW Europe. The partial deformation of aeolian dune sands caused by the Zechstein transgression. *Sedimentary Geology*, 35, 43–81.

HEWARD, A. P. 1991. Inside Auk—the anatomy of an aeolian reservoir. *In*: MIALL, A. D. & TYLER, N. (eds) *Three-dimensional facies architecture of clastic sediments*. Society of Economic Palaeontologists and Mineralogists, (in press).

——, JENSEN, Y. & VAN LIMBORGH, J. 1988. *Rotliegend reservoir of the Auk Field Central North Sea UK Sector*. Shell UK Expro Report.

LAFFERTY, K. 1987. *A diagenetic and depositional history of the Zechstein carbonates in Well 30/16-A06A of the Auk Field*. MSc Thesis, Univ. Aberdeen.

PENNINGTON, J. J. 1975. The geology of the Argyll Field. *In*: WOODLAND, A. W. (ed.) *Petroleum Geology of the Continental Shelf of NW Europe*. Applied Science Publishers, 285–291.

TAYLOR, S. R. & LAPRE, J. F. 1987. North Sea Chalk diagenesis: its effects on reservoir location and properties. *In*: BROOKS, J. & GLENNIE, K. W. (eds) *Petroleum Geology of North West Europe*. Graham & Trotman, 483–495.

The Balmoral Field, Block 16/21, UK North Sea

P. C. TONKIN & A. R. FRASER

Sun International E and P Company Ltd, 80 Hammersmith Road, London, UK

Abstract: The Balmoral oilfield operated by Sun Oil Britain Ltd, lies within UK blocks 16/21b and 16/21c, 140 miles off the northeast coast of Scotland. The field was discovered by the drilling of well 16/21-1 in 1975. Andrew Formation sandstones of Late Palaeocene age form the reservoir, which is sealed by Lista Formation claystones. The sandstones are of submarine fan origin sourced from the north and west of the area. The trap is structural, formed by the differential compaction of Tertiary sediments over a Palaeozoic structural high.

The upper section of the reservoir consists of two units of consolidated sandstone (units U and M) of channel-fill origin separated by a channel abandonment claystone (unit S1). Porosities for these sandstone units range from 17–28% and permeabilities are up to 3300 md. The lower section of the reservoir consists of friable sandstones (Unit F), characterized by grain-coating clays which have prevented consolidation. This unit is mainly of submarine fan lobe origin. Porosities range from 20–28% and permeabilities are up to 700 md.

Balmoral came on stream in November 1986. Recoverable reserves are estimated to be 68 MMBBL of undersaturated 39.9° API oil, and annual production remains at the 35 000 BOPD plateau rate.

The oil is produced from 12 wells with reservoir pressure maintained by the injection of water through a further six wells. These are all tied into a floating production vessel (FPV), the first such purpose-built production facility to be used in the North Sea. Production in Balmoral is expected to continue until the year 2001.

The Balmoral Field lies within blocks 16/21a, 16/21b and 16/21c, 140 miles from the northeast coast of Scotland in a water depth of 470 ft. The field is operated by Sun Oil Britain Ltd, a wholly owned subsidiary of Sun Company Incorporated.

Balmoral is elongated in a NW–SE direction, measuring 6.5 miles by 2.0 miles with a closure of 3240 acres. Reserves are currently estimated to be 68 MMBBL recoverable, in a sandstone reservoir of Late Palaeocene age. Hydrocarbons have accumulated at the structural crest of a high formed by drape and compaction over a deeper Palaeozoic structure.

Balmoral is named after the Scottish Royal Castle in Grampian. Other Sun Oil operated fields in block 16/21 (Glamis, Blair and Stirling) are also named after Scottish castles.

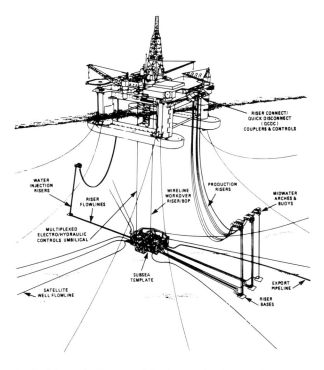

Fig. 1. Schematic diagram of floating production vessel.

History

Block 16/21 was awarded to a consortium headed by North Sea Sun Oil Co. Ltd as part of the Fourth Licensing Round in March 1972. Block 16/21, together with block 211/22 was awarded under licence P.201. The Balmoral Field was discovered in July 1975 with the drilling of well 16/21-1. Following the mandatory 50% relinquishment of the block in March 1978, North Sea Sun Oil Ltd retained 13 462 acres as the 16/21a block. The relinquished portion became block 16/21b.

In December 1980, during the Seventh Round, block 16/21b was awarded to a group headed by the then British National Oil Company, as part of licence P.344. Operatorship of 16/21b passed to ARCO British Limited in February 1986. ARCO further relinquished 50% of 16/21b in December 1986, retaining blocks 16/21b and 16/21c, and leaving block 16/21d as open acreage to the north. The Balmoral Field now straddles blocks 16/21a, 16/21b and 16/21c, with the common interests in the field covered by the Balmoral Field Unitized Area. The current equity for each group is shown in Table 1.

Table 1. *Balmoral Field: participants and equity*

16/21a Licence P.201	
Sun Oil Britain Ltd (Operator)	62.0%
Deminex UK Balmoral Ltd	15.0%
Clyde Expro plc	10.0%
LASMO (TNS) Ltd	8.0%
Clyde Petroleum (Balmoral) Ltd	5.0%
16/21b and 16/21c Licence P.344	
ARCO British Ltd (Operator)	35.0%
Summit North Sea Oil Ltd	40.0%
Goal Petroleum plc	25.0%

The 16/21-1 discovery well was drilled in July 1975 to test the upthrown fault block closure of a section presumed to be of Jurassic age. No hydrocarbon shows were encountered in the faulted section which actually proved to be of Devonian age, but a 78 ft oil column was encountered in late Palaeocene sandstones of the Andrew Formation. The 16/21-1 well tested at a maximum rate of 4049 BOPD of 39.4° API oil through a 1 inch choke.

Further delineation drilling was not undertaken until 1979, when the 16/21a-2 well was drilled approximately 1.4 miles to the northeast of the discovery well. It had been planned to test an apparently separate Palaeocene structure, but the well encountered an 88 ft oil column in the same Palaeocene sandstones and with an oil–water contact identical to that found in the 16/21-1 well. This indicated a continuous accumulation between the two wells.

In addition, hydrocarbons were discovered in fractured sandstones of late Devonian age. This second accumulation was subsequently named the Stirling Field.

Appraisal of Balmoral continued in late 1981, with drilling of the 16/21a-3 and 16/21a-3st wells. 16/21a-3 was abandoned due to deteriorating hole conditions after penetrating the Devonian. The sidetracked hole, however, tested the Palaeocene reservoir at a maximum rate of 5476 BOPD. The 16/21b-4A well was drilled by BNOC in 1982 near the southeastern end of the Palaeocene structure. The well encountered a short oil column that tested at a rate of 3540 BOPD. The well was later completed as the first water injector for the field. The 16/21b-5 well appraised the northwest extension of Balmoral. This well tested 5100 BOPD from the Palaeocene.

Well 16/21a-7 was drilled in 1983 as a Palaeocene and Devonian appraisal, and was subsequently sidetracked as 16/21a-7z to provide a second water injection well for Balmoral. Well 16/21b-9 was drilled by Britoil later in 1983 to test the southwestern-most extent of Balmoral, but encountered only oil shows in Lista Formation sandstones.

Following this initial exploration/appraisal phase of drilling, it was decided that Balmoral was commercially viable. A development plan was prepared for six water injectors and thirteen producers designed for a plateau production rate of 35 000 BOPD.

A new and innovative concept was applied in the Balmoral Field by the design of a purpose-built floating production vessel (FPV), the first of its kind to be used in the North Sea (Fig. 1).

Between January 1984 and September 1985, wells 16/21a-10z, 16/21a-11, 16/21b-14 and 16/21a-16 were drilled to provide the full complement of water injection wells. 16/21b-12 was drilled in 1984 as a producer to drain the northwest culmination of the field. Ten of the production wells, 16/21a-B1 to B10, were drilled as deviated wells from a central template location. Three remaining satellite producers, 16/21a-2, 16/21a-3st and 16/21b-12, were tied back to the production facility via flowlines. The FPV was installed in the field above the template in August 1986.

First oil was produced in November 1986, with the plateau production rate of 35 000 BOPD being reached by February 1987. The 16/21a-B9 well started to produce water at this time and was subsequently shut in. The oil is exported from Balmoral via a 14 inch line to the main Brae–Forties trunkline, and then to Cruden Bay, Scotland. Production is expected to continue to the year 2001.

Field stratigraphy

The oldest rocks identified in the Balmoral area are continental mudstones, sandstones and conglomerates of Devonian age. These were uplifted, tilted and eroded as a result of Hercynian movement. No sediments of Carboniferous, Permian or Triassic age are known in block 16/21.

The initiation of the Cimmerian orogeny in the Late Triassic, marked the onset of major rifting and the formation of the Witch Ground, Central and Viking Graben systems. A thin sequence of marine transgressive sandstones and conglomerates was deposited during Volgian time, and this constitutes the reservoir section for the Glamis Field, southwest of Balmoral. The Late Jurassic clastic sequence is overlain by the organic-rich Kimmeridge Clay Formation, the source rock for the hydrocarbons in the Balmoral, Glamis and Stirling Fields. Late Jurassic rocks are absent in the central Balmoral area, probably through non-deposition on the Fladen Ground Spur.

A thin sequence of claystones and marls of the Barremian Valhall Formation was deposited at the onset of the Cretaceous marine

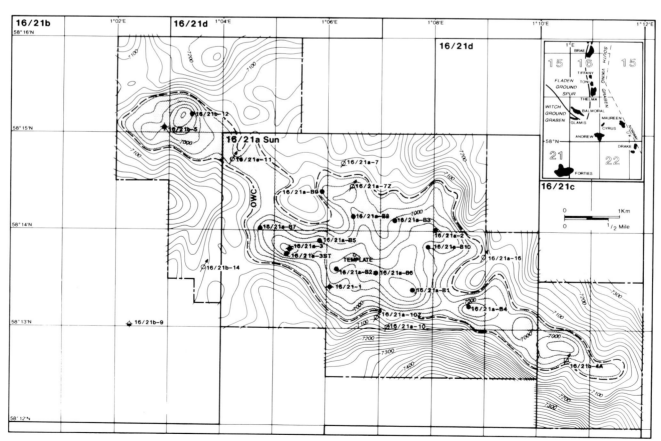

Fig. 2. Balmoral Field depth structure map. Top Andrew Formation (reservoir model version).

incursion. Deepening and extension of the Cretaceous sea led to the deposition of marls and chalks of the Flounder, Tor and Ekofisk Formations, which attain a combined thickness of up to 600 ft in the Balmoral area. In the axial zone of the subsiding basin, clays, sands and reworked chalk that now constitute the Maureen Formation were deposited.

With the cessation of chalk deposition during the early Palaeocene, there followed a rapid and continual influx of sands and deep water pelagic clays, which, by the end of the Cenozoic, had filled the subsiding graben areas with 7000–10 000 ft of sediment. The sands were sourced from areas to the north and west and locally from intra-basinal highs over salt domes and along fault scarps. The Andrew Formation comprises mostly submarine fan sands deposited downslope from a shelf area to the west.

In block 16/21, the thickness of the Andrew Formation increases from about 600 ft in the Balmoral area to more that 1000 ft in the south of the block. Subsequent compaction and drape of the Andrew Formation over a Palaeozoic structural high produced the broadly anticlinal Balmoral structure (Fig. 2).

The Lista Formation claystones and Forties Formation sandstones were deposited only thinly in block 16/21, the latter as submarine fan deposits sourced from the Orkney–Shetland platform.

In the axial part of the graben systems, mudstones of the Sele Formation were deposited in dysaerobic conditions. Deposition of tuffaceous claystones and mudstones of the Balder Formation followed at the end of the Palaeocene. This formation is a widespread and convenient Eocene–Palaeocene boundary marker.

The Hordaland Group, of Eocene to Late Miocene age, and the Nordland Group of Late Miocene to Pliocene age, were deposited as basinal muds with localized channel sands and turbidity flows. The two groups have a combined thickness of 5000–6000 ft. Depth of burial of the Kimmeridge Clay Formation was sufficient in deep graben areas for maturation and oil generation to commence during the Oligocene.

Geophysics

Until 1983 structural mapping was based on more than a dozen seismic surveys shot in the period 1971 to 1982, generally at a line spacing of a kilometre or greater. Poor quality of the Upper Palaeocene seismic reflections caused continual problems in structural delineation and depth prediction. Acoustic data from wells drilled after 1982 indicated that the primary cause of the problems was the relatively weak and gradual acoustic impedance increase associated with the Top Andrew Formation sandstone in this area. A strong velocity increase at the top of the sandstone is offset by a density decrease in nearly all of the Balmoral wells.

Prior to commencement of the development stage of the Balmoral project in 1983, steps were taken to improve the reliability of structural mapping by acquiring seismic data which would permit a less ambiguous interpretation of the top reservoir event. A test programme was carried out by three contractors using different energy sources—an airgun, a watergun, and a high pressure steam gun (Starjet). The objective was to compare results with particular reference to the quality of the Top Andrew reflection. Best results were obtained with the Starjet data, and the acquisition and processing contract for a detailed seismic program was duly awarded to Compagnie Générale de Géophysique. The Balmoral Detailed Seismic Survey of 1983 consisted of 488 km of 48-fold data with dip lines oriented NNE at an average spacing of 250 m.

Reflection quality at target level was considerably improved, though still poor in the eastern part of the field where scattering of seismic waves in the overlying Eocene channel sandstones is believed to affect adversely the stacking process. An interpreted seismic line through wells 1, B2 and B8 is shown in Fig. 3.

Trap

The Balmoral Field trap is of a structural nature, resulting from the drape of Andrew Formation sediments over a Palaeozoic structural high. Reactivation of earlier Hercynian or Cimmerian faults may have accentuated the structure, but there is no major fault offset at Palaeocene level (Fig. 4). The field is oriented WNW along the trend of the Palaeozoic structural high. It is approximately 6.5 miles long, 2 miles wide and has an areal closure of 3240 acres.

The structure is dip closed on all sides at Palaeocene level. As mapped, the field is not full to spill, but uncertainty in seismic depth conversion means that the possibility of a northwestern spill point cannot be excluded. The depth of the oil–water contact varies from 7033 ft ss in well 16/21b-12 to a maximum of 7065 ft ss in well 16/21a-B9. The field-wide average value is close to 7050 ft ss.

The Andrew Formation sandstones are overlain by Lista Formation claystones, which form an effective seal for the Balmoral accumulation.

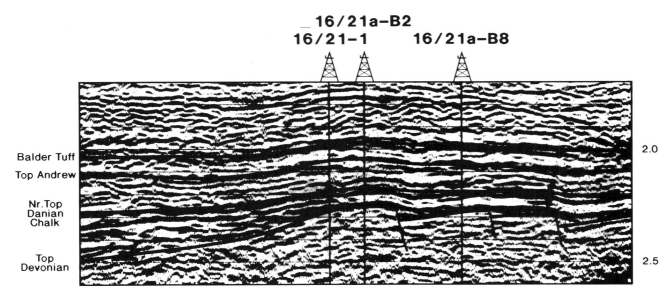

Fig. 3. Portion of seismic line SU 83-22 across Balmoral Field.

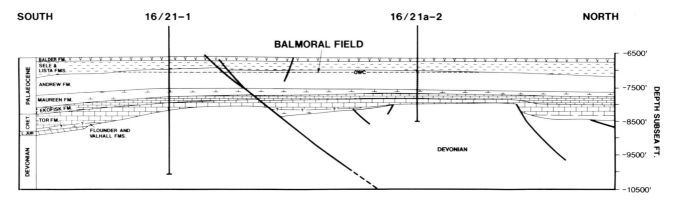

Fig. 4. Geological cross-section through Balmoral Field.

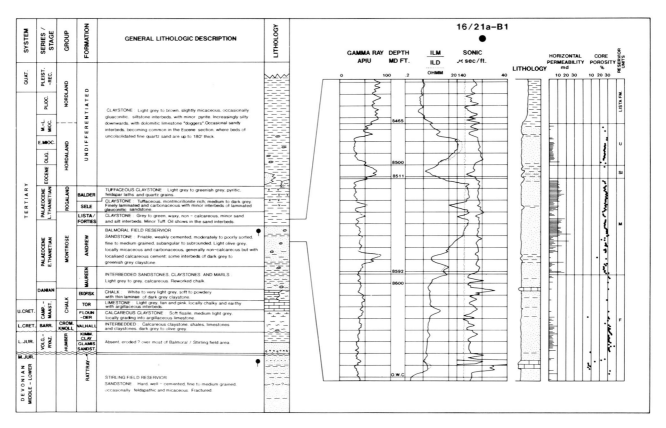

Fig. 5. Stratigraphy and reservoir units of Balmoral Field.

Reservoir

Within the regional stratigraphic framework formulated by Deegan & Scull (1977), the sediments of the Balmoral Field reservoir are assigned to the Andrew Formation of the Montrose Group (Fig. 5). This formation is of Early to Middle Thanetian age and the reservoir section falls within the Alisocysta Margarita dinocyst zone of Knox et al. (1981).

The Andrew Formation varies from 600–850 ft in thickness over the Balmoral area, and consists of amalgamated massive sandstone units with minor sections of interbedded mudstones and sandstones. The sandstones are fine- to medium-grained, poorly to moderately sorted, and contain up to 60% quartz with lithic fragments, feldspars, detrital clays and traces of mica making up the remainder of the composition. There is a uniform sonic log response with a correspondingly low gamma ray trace that enables the formation top and hence the top of the reservoir to be distinguished from the overlying Lista Formation claystones.

The Andrew Formation sandstones were deposited from a series of submarine fan flows, sourced from fluvio-deltaic sediments north and west of Balmoral. Periodically, during the late Palaeocene, large flows of clastic material, possibly triggered off by slope instability during periods of sea level low stand travelled into the Central Graben Basin as submarine turbidity currents. Basin slopes were about 1–2°. Sediments gradually filled the deeper basinal areas to the south of the Balmoral Field area forming a series of stacked lobes. These constitute the lower part of the Andrew Formation. Subsequent deposition of sands of the uppermost Andrew Formation was largely in the form of channel fills which cut down into pre-existing turbidic sediments. Moderate to poor sorting and the presence of angular fragments of carbonaceous debris in both the lobe and channel fill sands indicate a short transport distance, possibly only of the order of miles.

The waning phase of turbidity current deposition was characterized by channel abandonment facies, recognized as claystones with thin interbedded sandstone units. Following a period of much reduced sand input, during which Lista Formation claystones were deposited, the discontinuous sandstones of the Forties Formation were laid down. As basin subsidence continued in the main areas of deposition south of the Fladen Ground Spur, the Forties Formation depocentre was located south of Balmoral.

The benthonic foraminifera which have been observed in the Andrew Formation give evidence for an outer shelf environment, with oxygenated bottom waters. The end of local sandy sediment influx is characterized by the elimination of calcareous forams such as *Globigerina triloculinoides*, and the advent of agglutinating forams as the water depth increased. Water depths during this time were probably in the range 500–1000 ft.

With the renewed rise in sea level and the reduction of the clastic supply, the pelagic shales of the overlying Lista Formation were deposited. There are some thin sandstone units (up to 12 ft thick) within this formation, but they are calcite cemented and subsequently have low porosities and permeabilities. They are sometimes oil-bearing, but are not included as part of the Balmoral reservoir, as they are not commercially productive.

Four reservoir units are defined. From oldest to youngest, these are named Friable Sand (Unit F), Main Sand (Unit M), Shale Unit (S1) and Upper Sand (Unit U), (Fig. 6). The sedimentological work and reservoir subdivision is based on 1900 ft of core from 17 appraisal and development wells in the Balmoral area, as well as log information.

Unit F varies from 0–300 ft in thickness and consists of poorly consolidated fine- to medium-grained sandstones of lobe origin which contain appreciable amounts of detrital clay, mica, lignitic debris and grains of volcanic and tuffaceous material.

These sandstones are poorly consolidated due to a lack of quartz overgrowths. The formation of quartz overgrowths was prevented by an abundance of early diagenetic grain coating clays, chiefly chlorite with minor illite. The bound water in these grain-coating clays produces paths of high conductivity between sand grains, and this is the cause of anomalously low resistivities in the oil column where this unit occurs near the oil–water contact.

A feature of this reservoir unit is the presence of hard, calcite-cemented zones up to 5 ft thick, and of unknown lateral extent. These are loosely termed 'doggers' and are believed to be the result of early carbonate diagenesis produced from the breakdown of organic material.

Individual beds have sharp tops and bases, mudstone rip-up clasts, and occasionally show current ripple lamination. These features are consistent with periodic rapid influxes of sand resulting from turbidity currents, although complete Bouma sequences are never observed in core beyond amalgamated A-A-A cycles.

The Upper Sand (Unit U) and Main Sand (Unit M) comprise a series of stacked channel fill sands with minor channel abandonment claystones. The stacked sands are typically 1–2 ft in thickness, fining upwards above sharp bases and frequently containing mudstone rip-up clasts. Within individual beds, water escape structures (dish structures and fluid escape pipes) are common, indicative of rapid deposition. Additional structures include horizontal and inclined laminae, trough cross-bedding and current ripples.

Unit S1 consists of thin but laterally extensive claystones up to 10 ft thick, which separate Units U and M vertically. Thin sandstone injection structures occur in the claystones, with occasional thin interbedded fine-grained sandstones. The S1 claystones form local vertical permeability barriers, but there is sufficient channel downcutting from Unit U to Unit M to provide sand–sand contacts and a continuous hydrocarbon phase in the reservoir. The total thickness of Units U, S1 and M varies from 35 ft to 140 ft.

Porosities for reservoir units U and M vary from 17.4% to 27.7% and permeabilities are up to 3300 md. The Unit F friable sandstones show a similar porosity range, 20.4% to 28.4%, but permeabilities tend to be lower, varying from 20–700 md (Fig. 5). This is due to the occlusion of pores by detrital clays and chlorite grain coatings. In the friable sandstones especially, the lenses of calcite-cemented sandstone ('doggers') produce zones of very low permeability that act as localized barriers to vertical fluid movement.

For all reservoir units, the best reservoir properties occur in the central part of the field, and the lowest values for porosity,

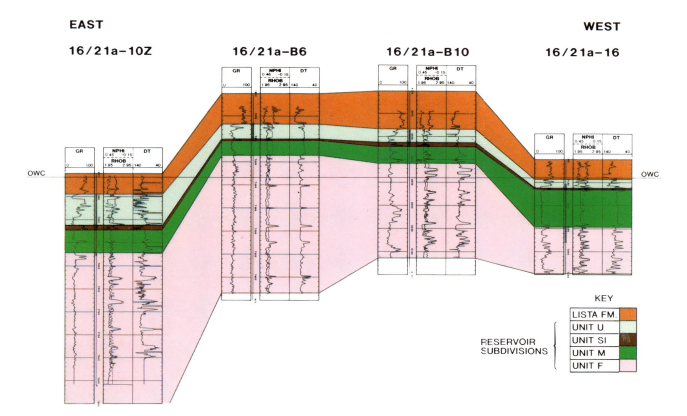

Fig. 6. Balmoral log correlation.

permeability and net to gross are found on the flanks of the structure. This is believed to be due to the deposition of the coarser and cleaner sand on the Balmoral 'high', where a decrease in turbidity current velocity was most marked. The structurally lower, flank areas of the field received finer, more argillaceous material, and consequently have lower porosities and permeabilities.

A similar observation has been made in the Frigg Field (Heritier *et al.* 1979).

Source

A Kimmeridge Clay Formation source is indicated by the presence of bisnorhopane in the pentacyclic triterpane biomarker fraction. Furthermore, high concentrations of rearranged steranes (diasteranes) indicte a clastic source, and the higher concentration of C35 hopanes relative to C34 hopanes is indicative of deposition of the source rock in an anoxic environment.

The drape and compaction concept for the origin of the Balmoral structure indicates that the trap must have been formed soon after the late Palaeocene deposition of reservoir sediments. Regional studies show that the Kimmeridge Clay Formation had reached maturity for oil generation as early as the Oligocene. The 16/21b-9 well, just over 1 mile southwest of Balmoral penetrated 22 ft of Kimmeridge Clay Formation. Whilst this proved to be immature for oil generation, a greater thickness of Kimmeridge Clay Formation occurs to the south and southwest of Balmoral, in the Witch Ground Graben. It is most likely that hydrocarbons migrated into the Balmoral structure from this direction up faults and through fractured chalks during the Oligocene.

Hydrocarbons

The Balmoral oil is undersaturated with an API gravity of 39.9° (0.826 g/cc) and an original gas/oil ratio of 366 SCF/STB at a reservoir pressure of 3145 psig. The average reservoir temperature is 207°F and formation waters have an equivalent salinity of 72 000 ppm NaCl.

Seventeen of the Balmoral wells were drillstem tested, typically at rates of 2500 to 3500 BOPD. Test results for each well are shown in Table 2.

Table 2. *Balmoral Field: production test results*

Well	Flow Rate
16/21-1	4049 BOPD maximum
16/21a-2	5200 BOPD maximum
16/21a-3,3st	5476 BOPD maximum
16/21b-4A	3541 BOPD maximum
16/21b-5	5100 BOPD maximum
16/21a-7	3082 BOPD maximum
16/21b-12	2408–3018 BOPD (increasing after acidizing)
16/21a-B1	3200 BOPD average
16/21a-B2	4437 BOPD average
16/21a-B3	2992 BOPD average
16/21a-B4	2936 BOPD average
16/21a-B5	2867 BOPD average
16/21a-B6	3153 BOPD average
16/21a-B7	2966 BOPD average
16/21a-B8	2944 BOPD average
16/21a-B9	3198 BOPD average
16/21a-B10	2366 BOPD average

Reserves/resources

Reservoir pressure is maintained by water injection, which commenced at the start-up of field production and has been increased to a peak injection rate of 52 000 BWPD. Gas lift facilities have been installed in all the template wells and three wells are currently on gas lift. This has nearly doubled oil production in these three wells. Early in the development plan, it was anticipated that sand production would be a potential problem. For this reason, each producer was gravel packed, with the gravel placed in the annulus between the production tubing and internal wire screens.

Production from Balmoral commenced in November 1986. The 16/21a-2 well had not been brought on at that time, pending completion of an extended well test (EWT) from the Devonian Stirling Field. After this EWT, the 16/21a-2 well was recompleted into the Palaeocene and commenced production at a rate of 3000 BOPD in May 1988.

Recoverable reserves are estimated to be 68 MMBBL of which 43 MMBBLS had been produced by April 1990. The average production rate is 35 000 BOPD (April 1990).

Currently water injection is being maintained at 57 000 BWPD, and the average water cut for all wells is 25.2%. Water injection is designed to account for voidage plus 10%.

Infill wells may be drilled in the future to maximize recovery.

Balmoral Field data summary

Trap	
Type	Structural; drape over horst block
Depth to crest	6905 ft ss at Top Andrew Fm.
Lowest closing contour	7110 ft ss at Top Andrew Fm.
Oil–water contact	7050 ft ss
Oil column	145 ft maximum
Pay zone	
Formation	Andrew Fm.
Age	Thanetian (Late Palaeocene)
Gross thickness	550–850 ft
Lithology	Sandstone
Porosity (average)	25%
Hydrocarbon saturation (average)	81%
Permeability	20–3300 md
Net to gross (average)	0.87
Hydrocarbons	
Density	39.9° API (0.826 g/cc)
Gas/oil ratio (initial)	366 SCF/STB
Formation volume factor	1.256 RB/STB
Bubble point (initial)	1453 psig
Formation water	
Salinity (average)	72 000 ppm NaCl equivalent
Resistivity (average)	0.039 ohm m at 194°F
Reservoir conditions	
Temperature	207°F
Pressure (initial)	3145 psig
Field size	
Net area	3240 acres
Recoverable oil	68 MMBBL
Recovery factor	45%
Production	
Start-up	November 1986
Development	12 producers, 6 water injection wells
Production rate	35 000 BOPD (average, April 1990)
Cumulative production	43 MMBBL (to April 1990)

This article is based on in-house and contractor studies, and the publications listed below have been used for additional information. The authors would like to thank Sun International Exploration and Production Company Ltd and its partners in the Balmoral Field, for giving permission to publish this paper.

The opinions expressed in this paper are those of the authors and are not necessarily those of the Balmoral owners.

References

DEEGAN, C. E. & SCULL, B. J. 1977. *A standard lithostratigraphic nomenclature for the Central and Northern North Sea*. Institute of Geological Sciences, Report 77/25.

HERITIER, F. E., LOSSEL, P. & WATHNE, E. 1979. Frigg Field—large submarine fan trap in Lower Eocene rocks of the North Sea Viking Graben, *Bulletin of the American Association of Petroleum Geologists*, **63**, 1999–2020.

KNOX, R. W. O'B, MORTON, A. C. & HARLAND, R. 1981. Stratigraphical relationships of Palaeocene sands in the UK sector of the Central North Sea, *In*: ILLING, L. V. & HOBSON, G. D. (eds) *Petroleum Geology of the Continential Shelf of Northwest Europe*, London Institute of Petroleum, 267–281.

The Beatrice Field, Block 11/30a, UK North Sea

V. STEVENS

BP Exploration, Farburn Industrial Estate, Dyce, Aberdeen AB2 0PB, UK
Present address: Stevens Associates, 78 Salisbury Place, Aberdeen AB1 6QU, UK

Abstract: The Beatrice Field was discovered in 1976 in Block 11/30a within the Inner Moray Firth Basin. The reservoir consists of multilayered Lower and Middle Jurassic sediments containing a STOIIP of 480 MMBBL of high wax crude oil. The reservoir sequence comprises 1100 ft of Hettangian (Lower Jurassic) to Callovian (Middle Jurassic) sandstones, siltstones and claystones. The Beatrice Field is co-sourced by a combination of Devonian and Jurassic rocks. The hydrocarbon trap comprises a tilted faultblock, the top of which is truncated by the main field boundary fault. The field has low energy, and the P_i of 2897 psi at 6500 ft TVSS, GOR 126 SCF/STBBL and P_b of 635 psi, together with the poor aquifer influx, necessitated the development of water injection from the start of production and use of electrical submersible pumps in all wells. Ultimate oil recovery is expected to be 146 MMBBL. The field has been developed with four platforms at three sites in 133 ft waterdepth. The crude is transported 42 miles by pipeline to the dedicated oil terminal at Nigg on the Cromarty Firth.

The Beatrice Oilfield is located within the Moray Firth in NE Scotland (Fig. 1) in UKCS Block 11/30a. It is the only producing oil field in the North Sea within sight of land, lying only 14 miles from shore, near Lybster in Caithness. It lies in water depth of *c.* 133 ft and occupies an area of 5800 acres. The development plan, which included environmental protection measures, was approved by the Department of Energy in May 1978. The Beatrice Field has total recoverable reserves of 146 MMBBL out of a total of 480 MMBBL STOIIP. The hydrocarbon trap consists of a tilted faultblock, truncated by the main field fault. The crest of the reservoir lies at a depth of 5900 ft TVSS, with dip closure to the north and west, and fault closure to the southeast (Fig. 2). Vertical relief from the OWC to the crest is 1100 ft. Total net sand within the gross reservoir section is 450 ft in thickness. The field was named Beatrice in 1977. The name is derived from Latin and means 'bestower of blessings'. The choice of the name was made by T. Boone Pickens, then president of Mesa Petroleum (UK) Ltd, after his wife Beatrice Louise Carr.

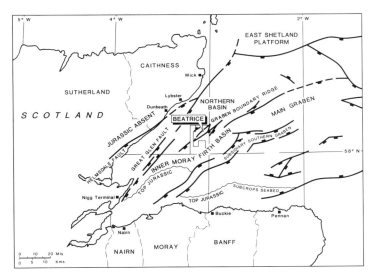

Fig. 1. Location map of Beatrice Field, showing main structural elements of the Moray Firth Basin.

History

Production licence P.187 on UKCS Block 11/30 was awarded to Mesa Petroleum (UK) Ltd in 1972 as part of the Fourth Round of licensing. After six years the licence area was halved to 27 000 acres. Block 11/30a was subsequently operated by the British National Oil Corporation (BNOC) who became Britoil. BP, who now own Britoil, operate the field on their behalf. The operator's interest is 28%, with partners Kerr-McGee Oil plc, 25%; Deminex UK Oil and Gas Ltd, 22%; Lasmo North Sea plc, 15%; and Hunt Oil UK Ltd, 10%.

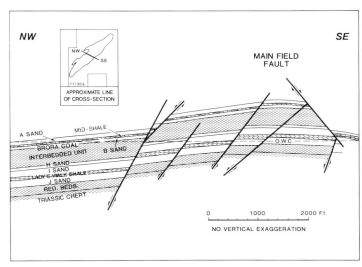

Fig. 2. Structural cross-section NW–SE across Beatrice Field.

The Beatrice discovery well, 11/30-1 was spudded in July 1976. The well penetrated the reservoir high on structure, close to the main field fault (Fig. 3). In August 1976 a major oil show was confirmed at a depth of 5933 ft TVSS in Upper Jurassic (Callovian) sandstones, with a gross oil column of 831 ft and a net pay of 304 ft. Drillstem tests on four different reservoir intervals produced an aggregate flowrate of 6066 BOPD of 38° API low GOR oil. A total of eight appraisal wells were subsequently drilled to define the extent of the Beatrice Field.

Approval for the Beatrice development plan was granted in August 1978. The crude oil was to be transported by pipeline to Nigg, on the Cromary Firth, and four platforms were eventually located at three sites (Fig. 4).

A site. This consists of two platforms linked by a bridge. The AD Platform is for drilling and accommodation, and has a 32-slot drilling capability with a main drilling rig and a small workover rig. The AP Platform is for production process facilities and for exporting the processed crude to shore. It is connected via 25 km of

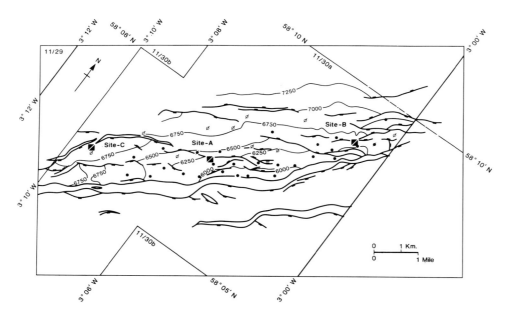

Fig. 3. Top reservoir structure map showing well locations.

submarine cable to Dunbeath where it receives electric power from the North of Scotland Hydro-electricity Board.

B Site. The 12 slot satellite B Platform is manned and it lies 3.5 miles east of the A Site, located over appraisal well 11/30-3. It exports its oil and water to the AP complex and receives treated seawater from the same for injection.

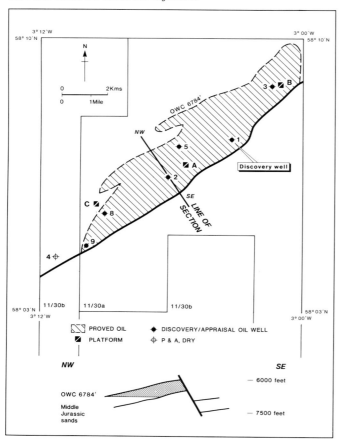

Fig. 4. Beatrice Field: offshore development facilities.

C Site. This small unmanned satellite platform injects water into the southwestern part of the field, to support the updip A Site producing wells. The total distribution of development wells drilled in the Beatrice Field by platform is: A Platform 19 producers, 8 injectors: the B Platform 8 producers, 2 injectors and the C Platform 2 injectors (Fig. 3).

Field stratigraphy

The basement rocks consist of pre-Devonian metasediments and granites. Sediments overlying the Caledonian Basement rocks in the Beatrice Field and in the surrounding Inner Moray Firth (IMF) Basin, range in age from Devonian Lower Old Red Sandstones to Lower Cretaceous Albian marine clays (Glennie 1986).

Devonian. The Caledonian Basement is overlain by the Devonian Old Red Sandstone, thick continental clastics derived from the erosion of the Caledonian mountains. In some areas sediments were deposited in a lacustrine environment, rich in organic matter. The Old Red Sandstone outcrops along the Moray Firth to the NW of the IMF Basin in Caithness, and to the south in Grampian, but it has not been recorded to date in Block 11/30a.

Carboniferous. Carboniferous sediments have not been reported in wells drilled in the IMF Basin, although possible Carboniferous basaltic volcanics, dated as 316 MY were penetrated by well 12/23-1. Carboniferous sediments were removed by Variscan uplift and erosion of the area.

Permo-Triassic. The Permian Rotliegend Sandstone overlies the Devonian and is the erosion product of Devonian and Caledonian/Variscan highlands. An arm of the Zechstein Sea entered the IMF Basin from the east, where thick salt sequences occur, whilst further west anhydrite deposits are encountered. Sedimentation of the reddish coloured mudstones and sandstones, continued from Rotliegend into Triassic times. A thin wedge of clastic Zechstein deposits is interpreted in well 11/30-2. Continental conditions persisted into the Triassic. Sandstones were deposited in a low lying desert environment (e.g. Smith Bank Formation). At the end of the Triassic sediment supply waned and a hardground developed, the Stotfield Cherty Rock Member, which forms a distinctive stratigraphic marker (Fig. 5).

Jurassic: (i) Dunrobin Group. Unconformably overlying the Stotfield Member is a series of multicoloured shales and sands of Rhaetian to Lower Pliensbachian age, which form the lower part of the Beatrice reservoir. The Dunrobin Group was deposited initially in an alluvial–lagoonal environment, and becomes progressively more marine upwards, marking the first phase of the Lower Jurassic marine transgression in Block 11/30a (Linsley *et al.* 1980).

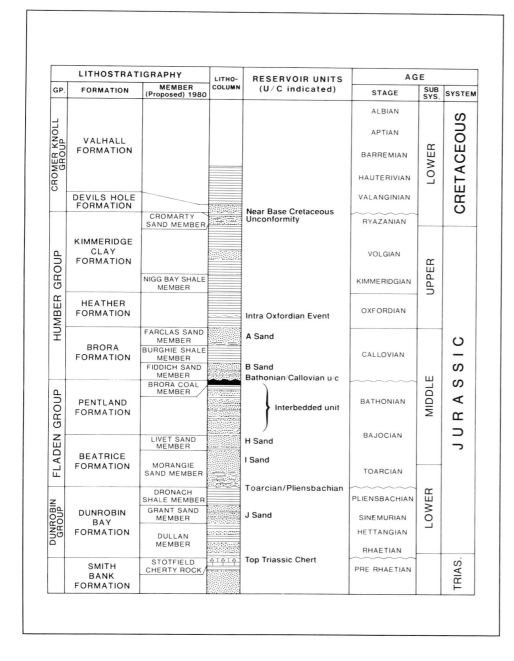

Fig. 5. Lithostratigraphic units of Beatrice Field.

(ii) Fladen Group. A regressive period lasting from Toarcian to Late Bathonian times. The base of the group is represented by a sequence of restricted marine and delta front sandstones and claystones and is overlain by a sequence of deltaic to lagoonal fluvial sandstones and claystones, capped by a distinctive coal horizon, the Brora Coal.

(iii) Humber Group. The Brora Coal is unconformably overlain by a cyclical marginal marine sequence of Lower Callovian sandstones and claystones. These sandstones form the upper part of the Beatrice Field. The base of the Humber Group marks the return to more marine conditions in this part of the IMF Basin. Above the Beatrice reservoir, further transgression during the Oxfordian led to the deposition of thick marine claystones and minor sandstones of the Heather Formation. This was followed by the deposition of organic-rich Kimmeridge Clay Formation. Towards the end of the Upper Jurassic the IMF Basin experienced tectonic uplift, with minor erosion on the structural highs. The Volgian Sandstone is representative of this process in the Beatrice area. Albian marine clays, which subcrop at the seafloor, are the youngest sediments of the Beatrice Field.

Geophysics

Available seismic surveys

The initial geophysical reconnaissance of the Moray Firth was carried out by The Institute of Geological Sciences (IGS) in 1966. In 1970 they carried out a further seismic investigation using a 10 km rectangular grid of traverse lines running NE–SW and NW–SE. One NW–SE traverse runs across the Beatrice Field. IGS carried out other additional seismic surveys in 1971, 1972 and 1974 (McQuillin & Bacon 1974; Chesher & Bacon 1975). Mesa Petroleum (UK) Ltd obtained lines over Block 11/30 in 1970, and they acquired the block in 1972 during the UKCS Fourth Round of licensing. The Beatrice discovery followed in 1976. In November 1976 a post-discovery seismic survey was conducted with a 1 km line spacing, and two further surveys followed in 1977 and 1979. A 3-D survey over the Beatrice Field was carried out in 1980 by GSI, in which 151 sail lines were shot at 100 m intervals, totalling 1124 km. This improved structural control and enhanced fault definition.

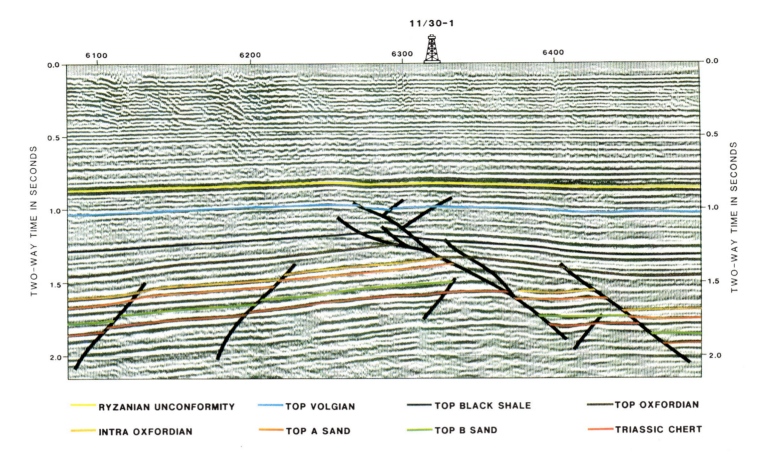

Fig. 6. Seismic section through well 11/30-1.

Geophysical definition of the reservoir

The current structural interpretation of the field is based on the 1980 3-D seismic survey. Figure 6 shows a seismic section through well 11/30-1. This survey enabled both Upper Callovian A Sand and Lower Bajocian H Sand to be mapped over the entirety of the field. However, the seismic response for the A Sand diminishes to the southwest of the field, where structural closure is difficult to define. Trial vibroseis was shot between the 11/30-4 and the 11/30-9 wells to resolve a possible shale-out of the A Sand in this area.

Trap

The Beatrice Field is situated in a structurally high, tilted fault block, trending in a NE–SW direction. It encloses a hydrocarbon bearing area of c.5800 acres above the field-wide oil–water contact of 6784 ft TVSS. Seismic and well evidence indicate that most of the fault activity responsible for the development of the Beatrice structure was confined to a relatively short period during the Upper Oxfordian. However, some thinning of strata across the structure is evident prior to Oxfordian times, indicating that structural movements had already started by that stage.

The Upper Oxfordian fault pattern responsible for the formation of the Beatrice structure is related to the initial structural development of the IMF Basin. This started with Caledonian movements, which occurred mainly along the Great Glen fault and associated faults. Late Caledonian shearing produced sinistral NE–SW aligned faults in the basement rocks prior to the deposition of the Devonian clastics in the IMF Basin. Subsidence and sedimentation ensued, both of which were greatest close to the Great Glen fault. Continental rifting was initiated in Permo-Triassic times; in the Moray Firth area this phase manifested itself in rejuvenation of Caledonian faults. Rogers *et al.* (1989) suggest that only normal fault movements took place in Mesozoic times, with no further associated lateral movement. The faulting of the Beatrice structure is normal, although controlled by trends of earlier strike-slip systems. The main field fault forms the southeastern boundary to the Beatrice Field, and runs in a NE–SW direction. It has a maximum throw of 1000 ft to the SE. The trend of the main field fault is slightly oblique to the Great Glen fault, and it is one of a number of faults splaying off the Great Glen fault which formed during the extensional phase of the IMF Basin's development (Chesher & Lawson 1983). Most fault activity ceased by Kimmeridgian times, except along the main field fault, where some movement continued until possibly as late as the Ryazanian (McQuillin *et al.* 1982). Tectonic inversion took place during the Upper Cretaceous, when tilting and uplift occurred, accompanied by some small scale faulting, which affected the Beatrice structure. Tectonic inversion has caused the removal of all sediments younger than Lower Cretaceous from the Beatrice area.

Reservoir

The Beatrice Field reservoir consists of Lower to Middle Jurassic shallow marine and coastal sediments. The gross reservoir section from the base of the J Sand to the top of the A Sand is 1100 ft, with an average net sand thickness of 450 ft. The Beatrice reservoir has been divided into eight lithostratigraphic units, which range in age from Hettangian to Callovian (Fig. 7). There are five sandstone layers (A, B, H, I and J), one interbedded unit and two shale horizons. The reservoir sandstones of the Beatrice Field are correlatable across the field, except within the Interbedded Unit. Recovery

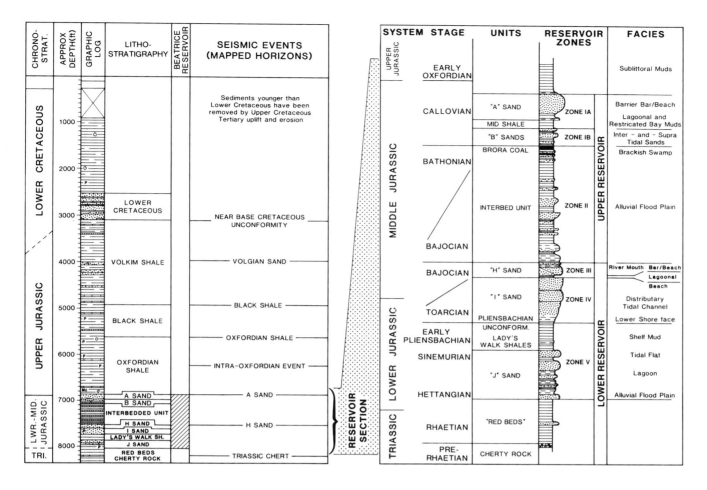

Fig. 7. Stratigraphic column. (Left) reservoir to seabed; (right) Beatrice reservoir section indicating zonation.

factor for the various reservoir sandstones is greatly affected by sand quality and continuity of the different units. In addition to the alphabetical sandstone notation, the reservoir is subdivided into six reservoir zones for reservoir simulation modelling. (Zones IA, IB, II, III, IV and V): each reservoir zone contains a recognized reservoir sand unit, the base of one zone forming the top of the next zone.

Zone V

Zone V is the lowermost of the reservoir zones of the Beatrice reservoir, it is made up of the J Sand and the non-pay Lady's Walk Shale.

The J Sand forms the lower part of Zone V and it is dated as Hettangian to Sinemurian in age. It has a gross thickness between 150–200 ft, with net/gross ratios of 15–65%, typically with 30–35 ft of net sand. The best quality reservoir sands occur in the upper J Sand unit. The J Sand has an average porosity of 15% and an average permeability of 400 md. The lower J Sand unit consists of fine-grained, moderately sorted alluvial flood plain/channel deposits. Sometimes the sandstones and siltstones may be calcareous with intervening mottled grey-green claystones. Bedding is rarely more than 5 ft thick. In the upper J Sand unit two sandstone horizons separated by a claystone occur. The sandstones are generally fine- to medium-grained and may contain rare shell fragments and display planar and ripple cross-bedded structures. Rootlet zones, rare plant material and desiccation cracks occur in the interbedded claystones which are black and organic-rich, containing an abundant flora and fauna, reflecting the change to fully marine conditions.

The Lady's Walk Shale forms the upper part of Zone V. The unit is Late Sinemurian to Early Pliensbachian. It comprises a series of transgressive calcareous shelf muds which overlie the J Sand sequence. The 100–120 ft of Lady's Walk Shale comprises dark grey, hard micaceous claystones which are variably silty, sandy or slightly calcareous. It contains bivalves, gastropods, crinoid fragments and occassional Early Pliensbachian ammonites.

Zone IV

The I Sand unit makes up the whole of Zone IV. The I Sand is dated as Uppermost Toarcian to Early Bajocian. It is usually between 120–170 ft thick and is the thickest of the reservoir sandstones in the Beatrice Field. The I Sand marks the change of environment from marine to regressive deltaic/barrier bar sandstones. The I Sand has an average porosity of 15%, and a permeability ranging between 0.1–500 md; a high permeability streak of *c.* 4000 md is sometimes present in the uppermost part of Zone IV. The net/gross ratio ranges from 60–80% (average 75%) and the sandstone coarsens steadily upwards from very fine-grained sandstone and siltstone immediately above the underlying Lady's Walk Shale, through to fine-, medium- then coarse-grained sandstone towards the top of the unit. The clay content decreases upwards, with the lower part of the I Sand having a high proportion of detrital clay and low permeabilities (<1 md). Clays are sparser in the upper part of Zone IV where they mostly comprise authigenic kaolinite.

Zone III

The H Sand unit comprises Zone III. It is of Early Bajocian age and it varies between 30–55 ft thick, and a net/gross ratio between 34–90%. The I Sand commences with a basal shale bed consisting of thin lagoonal delta top claystones, above this are river mouth sandbar deposits (Zone III sub-units do not correlate in detail between wells). The sandstones are fine- to coarse-grained quartzose sandstones with traces (<1%) of feldspar, becoming coarser and more friable upwards and with improving poroperm characteristics. Average porosity is 14% and average permeability 50 md, improving in the upper sands to between 500–1000 md. Quartz overgrowths are common and can reduce permeability and porosity.

Zone II

The Interbedded Unit is synonymous with Zone II and is of Bajocian to Bathonian age. It comprises an intercalated sequence of sandstones, siltstones and claystones. The Interbedded Unit is around 300 ft on the crest of the structure, thickening up to 440 ft down flank. It has a net/gross ratio of 6–21%. Within the Interbedded Unit there are four main sandstone horizons (Upper Sand, 40 Foot Sand, Middle Sand and Lower Sand) and two thinner sandstone horizons have also been identified (Top Sand and Lower Thin Sand). These form shoe-string sands with limited intercommunication and which are not correlatable across the whole field. The Interbedded Unit sandstones have an average net sand thickness of 80 ft with an average porosity of 16.5% and an average permeability of 320 md. The Interbedded Unit was laid down in a freshwater swamp, cut by alluvial channels. Most sandstone horizons have a sharply defined base, starting with a well sorted sandstone, which fines upwards into a mottled siltstone, topped by silty claystone. At the top of the Interbedded Unit is the Brora Coal which formed in an extensive, low lying swamp.

Zone IB

The B Sand, Zone IB is typically between 45 to 65 ft thick. It is dated palynologically as Early Callovian in age. Zone IB has a net/gross ratio of between 41–99% and a total average net sand thickness of 50 ft. It generally consists of three or more coarsening upward clastic cycles. A full cycle starts with a fine-grained, transgressive deposit of siltstone or fine sandstone rich in siderite and shell fragments. Above this, fine-grained sandstone occurs, which coarsens upwards into clean, medium-grained cross-bedded sandstone of a barrier bar or beach environment. These sandstones have good porosity and permeability characteristics in the crestal and northeastern area of the field, and deteriorate off structure. Rapid variations occur. For example well 11/30a-A4 has 20 ft net sand, whereas the nearby well 11/30-5 has no net sand.

Zone IA

Zone IA is the uppermost of the Beatrice reservoir zones. It comprises a lower unit called the Mid Shale and an upper pay zone called the A Sand, both of Callovian age.

The Mid Shale forms the lower part of Zone IA and is a claystone horizon of field-wide extent which separates the B Sand below from the A Sand above. It consists of 25–45 ft of fossiliferous shale with an argillaceous sandstone horizon locally developed in the northeastern part of the field, where up to 22 ft of net sand is observed. It is of Early Callovian age.

The A Sand forms the uppermost and most productive sandstone unit of the Beatrice Field, and it is Middle to Upper Callovian in age. It consists of a coarsening upwards sequence of deposits reflecting deposition in barrier bar and associated environments. The unit is on average about 100 ft thick, with a net/gross ratio of 41–99%, an average porosity of 16%, and permeability ranging between 50–500 md. The lower part of the A Sand unit represents the sandstone and siltstone deposited seaward of the main barrier bar complex. The upper A Sand is made up of well sorted beach and barrier bar sandstones, occasionally interrupted by claystone intercalations, and also containing some high permeability streaks (c. 2000 md). The uppermost part of the A Sand, and the top of the Beatrice reservoir is defined by a 2 ft thick, very fine-grained sandstone, with numerous belemnite guards. This layer forms a transitional zone to the shelf mudstones above the reservoir.

Source

Controversy has existed over the origin of oil at Beatrice, due to its unusual chemical composition when compared with other North Sea crudes. Three possible source rocks have been proposed: Upper Jurassic, Lower to Middle Jurassic and Devonian.

(i) Upper Jurassic. The Upper Jurassic Kimmeridge Clay Formation is unlikely to be the source rock at Beatrice, since it lacks thermal maturity in this part of the IMF Basin, late Cretaceous tectonic inversion halting the maturation process. In addition kerogen analysis indicated that these shales could not produce the long chain n-alkanes of the high wax crude oil at Beatrice (Pearson & Watkins 1983). Several source dependent bio-markers, abundant in Kimmeridgian bitumen, do not occur in Beatrice oil (Peters *et al.* 1989)

(ii) Lower to Middle Jurassic. The Middle Jurassic source beds frequently include oil shales, which are rich in spores and high hydrocarbons, and which can give rise to high wax crude oils. Greater depth of burial has occurred, so that the precursor hydrocarbons of the Beatrice crude would have reached maturity.

(iii) Devonian. The Devonian lacustrine shales of the IMF Basin have the required organic richness (TOC 0.4–1.0%) for potential oil generation (Trewin 1989). At several Devonian outcrops in Caithness, TOC levels up to 4.2% occur, and at Strathpeffer, the algal rich Devonian Foetid Beds have a TOC of 5.0% (Hall & Douglas 1981). Carbon isotope and biomarker evidence have given support to a significant Devonian contribution to Beatrice oil (Duncan 1986; Duncan & Hamilton 1988). The dominant organic components are amorphous organic matter of mixed algal and bacterial origin which predominantly yield the lighter fraction hydrocarbons (C17–C23), and are thus unlikely to have been the sole source for the high wax crude oil of Beatrice.

(iv) Co-sourcing. Beatrice is most likely to be sourced by a combination of both Devonian and Lower and Middle Jurassic source rocks. Peters *et al.* (1989) conducted a multi-parameter geochemical study on Beatrice source rocks and they concluded that Beatrice was co-sourced by 60% Devonian and 40% Lower and Middle Jurassic contribution.

(v) Migration and Timing. Devonian and Lower/Middle Jurassic hydrocarbons, generated in the deeper parts of the IMF Basin, to the east of the main field fault and the Great Glen fault, could readily migrate vertically into the sandstone horizons of the Beatrice structure. The migration path would necessarily be short, due to the waxy nature of the crude.

A phase of maximum burial occurred during the Mid-Cretaceous, maturing the deeper source rocks to the east of the major fault. Migration is considered to have taken place prior to the Late Cretaceous.

Hydrocarbons

Beatrice crude oil is classified as a low sulphur, high wax, high pour point, paraffinic crude. The waxy composition of Beatrice crude oil is atypical compared with other North Sea oils. It is not a heavy oil (38° API), however it has a high wax content (17%). This makes it highly viscous and, together with its high pour point (24°C) could have potentially adverse rheological properties in the reservoir, pipeline and production facilities. Treatment of the produced crude oil was originally considered to be essential and the oil was treated with Pour Point Depressants (PPDs) to keep it liquid whilst cooling in the submarine pipeline. Subsequent trials to transport the crude oil without using PPDs were successful providing the crude oil temperature was not allowed to fall below a critical temperature (13.3°C), below which high gel strengths develop. In the event of a shut down, rapid displacement of the pipeline to water can be effected with help from either the water injection system or cement pumps.

Tests of the appraisal wells indicated that the reservoir is normally pressured and little natural water drive would occur. Water injection was therefore started early in the development of the Beatrice Field, with facilities on the C Platform dedicated to water injection alone. Separate water injection is required for both the Upper Reservoirs (Zones IA, IB and II) and for the Lower Reservoirs (Zones III, IV and V) and wells are sited accordingly. The Upper and Lower Reservoirs are being concurrently developed. Crestal wells initially produce from the Lower Reservoir, and, when this waters out, the wells are recompleted as Upper Reservoir and/or comingled producers.

The reservoir pressure (initial pressure 2897 psig at 6500 ft TVSS) is not high enough for the wells to produce at an economic rate, especially when the watercut increases. The low gas/oil ratio of the crude (126 SCF/STBBL) and the low bubble point of 635 psig precluded gaslift as a viable option. Therefore, all producing wells are equipped with electrical submersible pumps (ESPs). Pumps are set close to the top of the reservoir and designed so that wireline access is possible for production logging, while the pumps are operating. Pump performance, when compared to other North Sea installations, has been good, since operating conditions are almost ideally suited to the use of ESPs (Kilvington & Gallivan 1983; Way & Hewett 1982). The reservoir is cool, there is little sand production and the low GOR and bubble point preclude gas locking problems. Pump runs in excess of four years have been achieved, and most average over one year.

Reserves

The ultimate recovery is estimated to be 146 MMBBL, out of a total STOIIP of 480 MMBBL, representing an overall recovery factor of 30%. However, the recovery in each reservoir unit varies between a low value of 12% in Zone V (STOIIP of 20 MMBBL), to a high value of 45% in Zone III (STOIIP of 33 MMBBL). Zone II has a value of 15% (STOIIP of 73 MMBBL) with part of its reserves residing in non-communicating sands which make a negligible contribution to recovery. The recovery in Zone IA, which has the largest STOIIP of all zones (250 MMBBL) will be 37%. The variation in the recovery factor is wholly dependent on the sand quality and continuity. Oil production started in September 1981 and averaged some 32 000 BOPD until July 1984, when the B Platform came on stream and the production went up to some 60 000 BOPD. The 1989 oil rate was 26 000 BOPD. Cumulative oil recovery at the end of 1989 was 109 MMBBL.

I would like to thank BP Exploration and the Beatrice Partners (Deminex, Hunt, Kerr-McGee and Lasmo) for permission to publish this paper. I would also like to thank P. Begg for his comments on the geophysical part of this paper, H. Allen and R. McQuillin for reading the draft manuscript, M. Littlewood for his comments on the use of PPDs and M. Miller for editing the final manuscript.

Beatrice Field data summary

Trap
Type	Tilted fault block
Depth to crest	5900 ft TVSS
Oil–water contact	6784 ft TVSS
Net area	5800 acres

Pay zone
Formations	Brora Fm, Pentland Fm, Beatrice Fm.
Age	Lower to Middle Jurassic (Hettangian to Callovian)
Gross thickness	1100 ft
Average net/gross ratio	41%
Porosity range	13–22%
Average water saturation	21.5%
Permeability range	1–4000 md

Hydrocarbons
Oil gravity	38° API
Wax content	17%
Viscosity	2.23 centipoise
Oil type	Waxy, high paraffin
Bubble point	635 psig at 80°C
GOR	126 SCF/BBL
Natural pour point	24°C
Formation volume factor	1.09

Formation water
Salinity	35 000 ppm chlorides
Resistivity	0.062 at 80°C
Other ions	Barium, strontium

Reservoir conditions
Average temperature	80°C
Average pressure	2897 psig at 6500 ft TVSS
Pressure gradient	0.33 psi/ft

Field size
Area	5800 acres
Recovery factor	30% overall
Recoverable oil	146 MMBBL
Drive mechanism	Water injection, artificial lift by ESPs on all wells

Production
Start-up date	15 September 1981
Development scheme	4 platforms at 3 sites
Production rate	c. 30 000 BOPD
Cum. production 1-1-90	109 MMBBL
Number/type of wells	27 oil production/12 water injection
Secondary recovery method	Water injection

References

CHESHER, J. A. & BACON, M. 1975. *A deep seismic survey in the Moray Firth.* Institute of Geological Sciences Report, No. 75/11, HMSO, London.

—— & LAWSON, D. 1983. *Geology of the Moray Firth.* Institute of Geological Sciences Report No. 83/5, HMSO, London.

DUNCAN, A. 1986. *Organic geochemistry applied to petroleum source potential and tectonic history of the Inner Moray Firth basin.* PhD Thesis, University of Aberdeen.

—— & HAMILTON, R. 1988. Palaeolimnology and organic geochemistry of the Middle Devonian in the Orcadian Basin. *In*: FLEET, A. J., KETTS, K. R. & TALBOT, M. R. *Lacustrine Petroleum Source Rocks.* Geological Society, London, Special Publication, **40**, 173–201.

GLENNIE, K. W. (ed.) 1986. *Introduction to the petroleum geology of the North Sea.* Blackwell Scientific, Oxford.

HALL, P. & DOUGLAS, A. 1983. The distribution of cyclic alkanes in two lacustrine deposits. *In*: BJORROY, M. (ed.) *Advances in Organic Geochemistry*. Proceedings of the 10th International Meeting in Organic Geochemistry, Bergen. Wiley, New York, 576–587.

KILVINGTON, L. J. & GALLIVAN, J. D. 1983. Beatrice Field: Electrical submersible pumps and reservoir performance 1981–1983. *Offshore Europe Proceedings*. SPE paper 11881.

LINSLEY, P. N., POTTER, H. C., MCNAB, G. & RACHER, D. 1980. The Beatrice Field, Inner Moray Firth, UK North Sea. *In* HALBOUTY, M. T. (ed.) *Giant Oil and Gas Fields of the decade 1968–1978*. Memoir of American Association of Petroleum Geologists, **30**, 117–129.

MCQUILLIN, R. & BACON, M. 1974. *Preliminary reports on seismic reflection surveys in sea areas around Scotland, 1969–1973*. Institute of Geological Sciences Report 74/12, HMSO, London.

——, DONATO, J. A. & TULSTRAP, J. 1982. Development of basins in the Inner Moray Firth and North Sea by crustal extension and dextral displacements of the Great Glen Fault. *Earth & Planetary Science Letters*. **60**, 73–80.

PEARSON, M. J. & WATKINS, D. 1983. Organofacies and early maturation effects in the Upper Jurassic sediments from the Inner Moray Firth basin, North Sea. *In*: BROOKS, J. (ed.) *Petroleum Geochemistry and Exploration of Europe*. Geological Society, London, Special Publication, **12**, 147–160.

PETERS, K. E., MOLDOWAN, J., DRISCOLE, A. & DEMAISON, G. J. H. 1989. Origin of Beatrice oil by co-sourcing from Devonian and Middle Jurassic source rocks, Inner Moray Firth, UK. *Bulletin of the American Association of Petroleum Geologists*, **73**, 454–471.

ROGERS, D. A., MARSHALL, J. A. & ASTIN, T. R. 1989. Devonian and later movement on the Great Glen Fault System. *Journal of the Geological Society, London*, **146**, 369–372.

TREWIN, N. H. 1989, Petroleum Potential of the ORS of Northern North Scotland. *Scottish Journal of Geology*, **25**, 201–225.

WAY, A. R. & HEWITT, M. A. 1982. Engineers evaluate submersible pumps in North Sea Field. *Pertroleum Engineering International*, **8**, 92–108.

The Buchan Field, Blocks 20/5a and 21/1a, UK North Sea

C. W. EDWARDS

BP Petroleum Development Ltd, Farburn Industrial Estate, Dyce, Aberdeen AB2 0PB, UK
Present address: Hillhead Croft, Chapel of Garcoch, Inverurie, Grampian AB51 9HE, UK

Abstract: The Buchan Oilfield is located in Blocks 21/1a and 20/5a of the UK North Sea, on the southern side of the Witch Ground Graben. The Buchan structure is a complex tilted and dissected fault block formed during Jurassic extension and rifting. The Upper Devonian–Carboniferous reservoir is composed of fluvial Old Red Sandstone facies sandstones sealed by Lower Cretaceous mudstones and contains a 585 m (1919 ft) thick oil column. Poor matrix porosities and permeabilities are enhanced by a pervasive fracture system, although faulting in the reservoir restricts communication between several of the nine producing wells. Hydrocarbon migration has occurred from a mature Upper Jurassic source rock north of the field into the structure across the flank faults. Production of the highly under-saturated oil is by depletion drive with some aquifer sweep to a floating production facility and onward transmission to the Forties Field by pipeline. Production commenced in 1981 and original recoverable reserves are estimated at 90 MMBBL of which 71.5 MMBBL have already been produced.

The Buchan Field is located 153 km (95 miles) NE of Aberdeen (Fig. 1) in water depths between 112–118 m (367–387 ft). It is estimated to contain 90 MMBBL of recoverable oil, of which 71.5 MMBBL had been produced by the end of 1989. The field came off the 'export pipeline capacity' plateau of 28 000 BOPD in December 1988. The field is unusual in the North Sea in producing from a fractured Upper Devonian–Lower Carboniferous reservoir, now incorporated in a tilted and eroded fault block structure, which was created during the late Jurassic.

The name for the field is derived from the adjacent Buchan district of northeast Scotland.

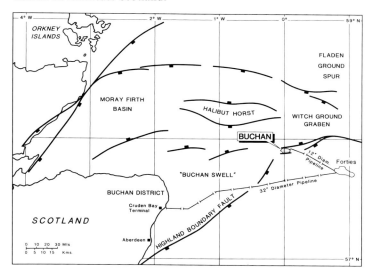

Fig. 1. Location map of the Buchan Field.

History

The field is predominantly contained within Block 21/1 which was awarded to Transworld Petroleum (UK) Ltd in July 1972 as part of the P.241 United Kingdom 4th Round licence. A small part of the field extends into Block 20/5 which initially formed part of licence P.294, awarded to Texaco Inc., also in a 4th Round allocation.

The Buchan Field discovery well, 21/1-1, was drilled in August 1974 at the 21/1-20/5 block boundary as a joint Transworld/Texaco well to test a complex horst structure with closure at the Base Cretaceous level which straddles the block boundary. The 21/1-1 well successfully penetrated and tested 477 m (1565 ft) of a highly over-pressured, fractured Upper Devonian–Lower Carboniferous reservoir.

Three appraisal wells, 21/1-2, 21/1-3 and 21/1-4 were drilled to delineate the highly faulted field, but all encountered major problems because of the overpressure. Transworld Petroleum (UK) Ltd submitted a proposed development and production programme (Annex B) to the Department of Energy in June 1977. In July 1977 BP Petroleum Development Ltd farmed-in to the licence, took over operatorship and a 54.167% interest in the Buchan Field. The present Buchan Unit participants and working interests are:

	% share
BP Petroleum Development Ltd	49.163191
BP Petroleum Development (Alpha) Ltd	12.706793
Monument Exploration and Production Ltd	12.706793
Goal Petroleum plc	2.269070
Transworld Petroleum (UK) Ltd	12.706793
Shenandoah Expro Ltd	0.907628
Brabant UK Ltd	0.302542
Clyde Petroleum (Buchan) Ltd	9.237190

The development scheme subsequently adopted by BP for the field consists of a sea-bed template, installed in August 1977, through which four wells (21/1-6, 21/1-7ST, 21/1a-9 and 21/1a-11) were drilled prior to the arrival of the floating production facility. Three satellite wells (21/1a-8, 21/1a-10 and 21/1-2ST) were tied back to the subsea manifold by flowlines. Well 20/5a-1, which was drilled by Texaco in 1981 as part of the September 1980 Unitization Agreement, was also tied into the manifold. A further well (21/1a-13), intended to drain the north-west flank of the field, was drilled in 1984 but failed to flow at economic rates and was abandoned. A successful infill well (21/1a-14) was drilled in 1987–88 to accelerate and increase oil recovery from the field.

The complexly faulted nature of the reservoir, its 'marginal' size and the uncertainty of the recovery factor in the fractured sandstone reservoir made the declaration of the commerciality of the field subject to rigorous technical and feasibility studies. Ultimately the preference was for production through a converted semi-submersible rig rather than a fixed platform. The perceived economics which dictated a floating rig were influenced by likely abandonment cost (Darnborough 1980). A Pentagone-type semi-submersible drilling rig (Drillmaster), was converted to the floating production facility, now called Buchan Alpha, during 1978–80. First oil was produced in May 1981 and the design throughput of 72 000 BOPD was first achieved in July 1981.

Initially hydrocarbon export was through a CALM (Catenary Anchor Leg Mooring) buoy and shuttle tanker, with associated gas flared at the platform. However, following an economic review this system was replaced in late 1986 by a 12 inch diameter oil pipeline to the Forties Field, 55 km (34 miles) ESE of the Buchan Field, for onward transmission to the Cruden Bay landfall. This Buchan–

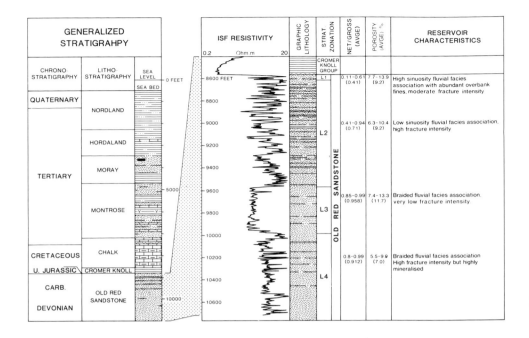

Fig. 2. Generalized stratigraphy of the Buchan Field based on well 21/1-6.

Forties line has a capacity of 28 000 BOPD, and although lower than the tanker option the average annual weather-related production downtime decreased from 45% to less than 20%.

Gas lift installation was undertaken on Buchan Alpha during 1985 and all producing wells have gas lift mandrels installed.

Field stratigraphy

The Buchan Field is situated on the crest of a tilted and eroded fault block located near the southern edge of the Witch Ground Graben. The stratigraphy of the field reflects its tectonic setting; a Lower Cretaceous to Recent succession lies unconformably on the Devonian-Lower Carboniferous reservoir sandstones over most of the field (Fig. 2).

The Buchan Field reservoir is composed of an 'Old Red Sandstone facies' alluvial sequence of sandstones and mudstones of Devonian–Lower Carboniferous age. The youngest palynological determination in the reservoir occurs in well 20/5a-1 and assigns rocks to the NC miospore zone (Neves et al. 1972) of latest Visean–earliest Namurian age. Deeper into the reservoir mid-Asbian (Visean, NM to TC miospore zones) and Fammenian ages (GM and VU miospore zones of Clayton et al. 1978) have been identified. Rare plant macrofossils (?Archaeosigillaria sp.) and fragments of pleuracanth cartilage have been recovered from core of probable Fammenian (VU miospore zone) age. The base of the sequence has not been penetrated in the field.

In the crestal part of the field only Lower Cretaceous and younger sediments are preserved above the Base Cretaceous unconformity that forms the top of the reservoir sequence, but down-flank, a thin Upper Jurassic sequence has been encountered in two wells. In the deep basins to the north and south of the field it is likely that thicker Upper and Middle Jurassic and older sediments are present. The Upper Jurassic Kimmeridge Clay Formation in well 21/1a-13 is only 10 m (33 ft) thick and overlies an 8 m (26 ft) thick sandstone of indeterminate but possibly Late Jurassic age. In well 21/1a-8, 17.5 m (57.4 ft) of Kimmeridge Clay Formation mudstones of Late Kimmeridgian age overlie the Late Fammenian reservoir.

The oldest Cretaceous sediments identified over the field are Hauterivian to early Barremian mudstones of the Valhall Formation. These are overlain by Rødby and Hidra Formation mudstones. A thin Lower Cretaceous mudstone sequence over the crest, 57 m (187 ft) in well 21/1-6, thickens dramatically downflank as onlap was maintained in post-Jurassic times. Mudstone lithologies persisted into the basal part of the Chalk Group Herring Formation before the sequence passes up into the micritic limestones and calcareous mudstones of the Herring, Flounder and Tor Formations. The base of this Turonian–Maastrichtian white limestones (Chalk Group) interval is particularly cherty and, together with three distinct red/brown mudstones (Hill & Smith 1979), provide consistent and useful field-wide lithological markers. The deposition of the Chalk Group persisted into the Danian. The presence of a thin sandstone at the base of the Ekofisk Formation represents the end Cretaceous regressive event.

The Tertiary (Thanetian and younger) Montrose and Moray Groups are composed predominantly of poorly sorted sandstones, siltstones and mudstones, and are representative of the various submarine fan systems which blanketed the area from a source area to the west (Stewart 1987). The Hordaland and Nordland Groups are composed predominantly of clays with minor sands.

Geological history

The Devonian–Carboniferous reservoir sequence was deposited in an alluvial environment but its overall position in a regional setting is less certain. Fluvial and marginal marine sediments are known in the Midland Valley of Scotland, and in northeast Scotland the Upper Old Red Sandstone is predominantly fluvial in origin (Mykura 1983). Within the Buchan Field reservoir an overall fining-upwards pattern has been recognized, superimposed on an upward transition from sandstone deposition in braided-stream to meandering-stream environments. Carboniferous mudstones of Visean age in well 21/1a-13 may represent an erosional remnant of a more extensive Carboniferous marginal marine sequence but generally little is known of deposition in the area between Old Red Sandstone times and the Late Jurassic. The Kimmeridge Clay Formation, seen only on the flanks of the field in wells 21/1a-8 and 21/1a-13, may also have been deposited over the crest and subsequently eroded prior to the early Cretaceous. During the Late Jurassic transgressive phase regional north–south extension created the predominantly east–west structural grain and the Buchan tilted fault block. Post-rift passive infill of the thermally-subsiding basin continued thereafter with the deposition, and draping of the structure, by Cretaceous–Recent sediments. Faulting largely ceased by the commencement of Chalk Group deposition although minor adjustments across major faults continued until Maastrichtian times.

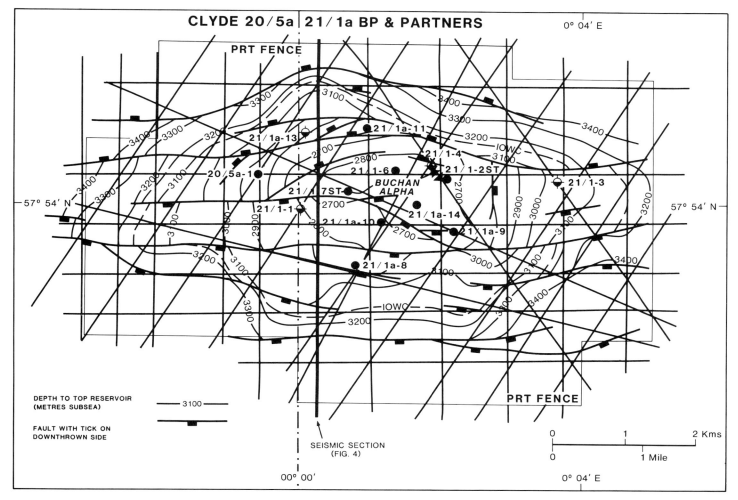

Fig. 3. Seismic coverage over the Buchan Field showing the location of line BPWE 87-310 (Fig. 4).

Geophysics

A total of eight seismic surveys were acquired over the field between 1970 and 1981. There is no uniform seismic coverage and some large gaps exist in which no data are present because lines have been shot in a number of different surveys with differing grid sizes and orientations (Fig. 3). Furthermore, significant mis-ties occur between surveys of various vintages with differing acquisition and processing parameters and major but unquantifiable inaccuracies of some of the navigation data. As a result the current seismic interpretation and mapping in the field is based only on data from the 1975, 1977 and 1978 surveys.

Despite several reprocessing trials the quality of the data is very poor. This is partly due to the age of the data, and partly results from the stratigraphy and complex faulting of the structure. Recently, however, a single line (Fig. 4), which forms part of a 1987 regional survey and crosses the western end of the structure, and a concurrent further reprocessing trial on one of the 1977 lines provided a dramatic improvement in the resolution of the Base Cretaceous unconformity. These two lines, plus two additional regional lines shot in 1982 and 1987, together with a better understanding of the regional structural development, have allowed a more confident interpretation of the field structure to be made.

The key seismic horizon mapped is the base Cretaceous reflector which represents Top Reservoir over the central part of the field. Due to a high degree of faulting, the unconformity is a difficult horizon to map with any confidence. To obtain a Top Reservoir map (Fig. 5) a small depth correction has been applied to the Base Cretaceous unconformity horizon on the flanks to account for the thin veneer of Upper Jurassic sediments known or surmised to be present. From reservoir dipmeter studies it is apparent that some faults have no topographical expression at the Top Reservoir horizon and yet offset lithofacies layers within the reservoir. However, no coherent reflectors can be mapped within the reservoir, a feature consistent with the poorly-defined internal lithological subdivision. With only three 'modern' lines there is still scope for an improved structural model, and an infill seismic grid was shot during 1989 to improve the data.

Trap

The Buchan Field is situated on the crest of a horst-like east–west elongate, tilted, fault-block structure, and is defined by four-way dip closure at the Base Cretaceous unconformity level. Seen in isolation, the structure resembles a simple horst with large-throw faults north and south of the oil-bearing reservoir (Fig. 6a). Regionally the structure is a culmination on a rotated fault block which can be traced eastwards into Block 21/2.

The areal extent of the field is 14.28 sq km (5.51 sq miles), delineated by the Top Reservoir surface and the original field-wide oil–water contact at 3165 m (10 384 ft) TVSS. The east–west extent of the field is about 6.8 km (4.2 miles) and north to south the maximum width of the field is about 3.5 km (2.2 miles). The crest of the structure is at 2580 m (8464 ft) TVSS giving a gross oil column of 585 m (1919 ft). Based on current mapping the field cannot be regarded as full to spill.

The throw on the major horst-bounding fault zones at the Base Cretaceous level is in excess of 800 m (2625 ft). Splay faults from the two major faults structurally isolate an elevated central zone from

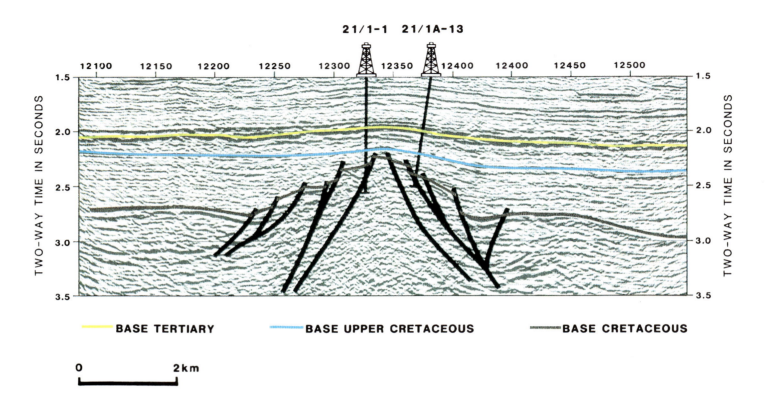

Fig. 4. Seismic line BPWE 87-310 aligned north–south across the Buchan structure showing the major faults on the flanks of the field.

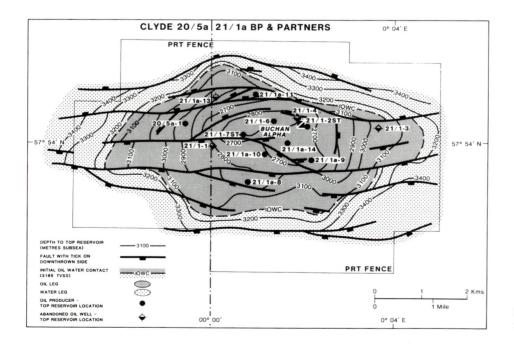

Fig. 5. Top reservoir structure map of the Buchan Field.

the northern and southern flank areas of the field. The subdivision into central and flank areas is also reflected in the distribution of porosity and permeability characteristics of the reservoir between the two areas.

Both the lateral and vertical seals for the reservoir are the mudstones of the Cromer Knoll Group Valhall Formation, although on the flanks of the structure the Kimmeridge Clay Formation mudstones would also seal the structure. These mudstones also form the transition zone from the normal pressure regime of the upper part of the section to the highly over-pressured regime of the reservoir. This pressure transition zone is only some 30 m (100 ft) thick (Hill & Smith 1979).

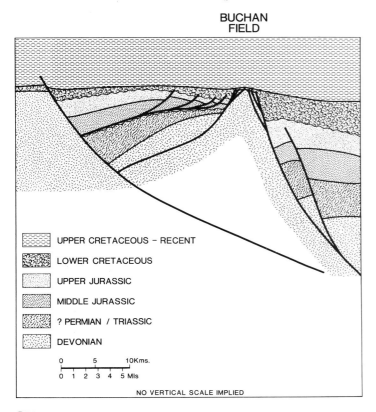

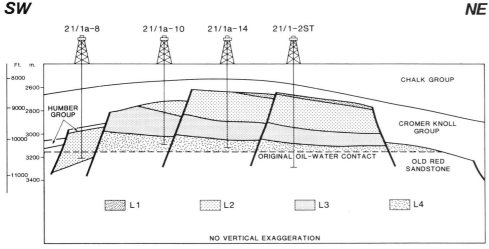

Fig. 6. (a) North–south cross-section through the field showing the position of the field in relationship to the major listric faults. (b) NE–SW cross-section showing the correlation of lithofacies zones through the field.

Reservoir

The reservoir in the Buchan Field is composed of sandstones of Upper Devonian–Carboniferous age. A thin 8 m (26 ft) fine-grained sandstone, possibly of Jurassic age, has also been encountered in one well (21/1a-13). Its areal distribution is uncertain, and the production rate from this sequence was low. The main reservoir is composed primarily of interbedded fluvial arkosic and sub-arkosic sandstones, with subordinate siltstones, mudstones and minor amounts of intra-formational mudstone and calcrete-pebble conglomerate. In the absence of persistent palynological material, correlation within the reservoir relies on electric log and sedimentological lithofacies correlation. A four-fold subdivision can be consistently identified over most of the field (Figs 2 & 6b). The four lithofacies zones correspond approximately with depositional environment changes within the sequence. A braided stream environment where clean, well sorted, fine-grained arkosic sandstones predominate (Zone L4), passes upwards through a low-sinuosity (Zone L3) to high sinuosity fluvial system (Zones L1 and L2) near the top of the reservoir section, in which siltstones and mudstones become more important.

Calcrete horizons, particularly within the upper zones, attest to pedogenic processes immediately after deposition. The upper zones have suffered differential erosion off the crests of some of the intra-field fault blocks.

There appears to be a consistent pattern of thinning from south to north in the lower zones (L3 and L4), and this is also reflected in a south to north or northwest decrease in net/gross ratio implying a sediment source or higher energy environment somewhere to the south. These trends, although less defined, are also seen in Zones L1 and L2. Mineralogically, the sediments are indicative of an acid-igneous source. This is consistent with speculation that a granite body in block 13/29 to the west-north-west of the Buchan Field may have provided sediment to the area (Richards 1985). However, rare coarse polycrystalline quartz grains showing highly sutured boundaries suggest that there may also have been a metamorphic source component. The reservoir quality has been adversely affected by calcite and dolomite cementation, and grain compaction due to depth of burial. Matrix porosities and permeabiities are generally low, typically 5–7% and 0.2–20 md respectively. Later differential solution, particularly of calcite and feldspar grains has produced thin, 0.5 m (1.6 ft), higher permeability streaks within some of the sandstones. However, the main factor controlling production within the reservoir appears to be at least three distinct systems of ramifying open fractures. Locally some of the fractures are partially or totally mineralized with barytes, galena, sphalerite, pyrite, calcite and silica. Production log data suggest that the fractures provide conduits for fluid flow within the reservoir, with production thought to rely largely on fluid expansion into the fractures. Intervals of significant production roughly coincide with mud loss zones encountered whilst drilling, which are interpreted as highly fractured permeable intervals.

The reservoir is highly over-pressured with an original reservoir pressure of 7506 psig at 2926 m (9600 ft) TVSS and this pressure will have assisted in maintaining open fractures within the reservoir.

Source

A mature source rock has not been identified in any of the wells within the field. However, from regional geochemical evidence, the source rock is likely to be the Kimmeridge Clay Formation which is known to be present in the syncline to the north of the field, and is in faulted contact with the flank of the Buchan Field structure at depth. Migration from this source kitchen would be across the fault and into the reservoir. South of Buchan, any Kimmeridge Clay Formation present will be shallower and may be immature for oil generation. The Kimmeridge Clay Formation was probably not buried sufficiently deep for the onset of oil generation to occur until early Tertiary time (*c.* 50 Ma).

Hydrocarbons

The discovery well on the field (21/1-1) flowed 4412 BOPD from a 106 m (347 ft) perforated interval. In contrast, the 21/1-3 well flowed only 784 BOPD from a 74 m (242 ft) net perforated interval and the 21/1a-13 well failed to flow. The poorer quality reservoir downflank is due primarily to a decrease in porosity with depth coupled with an increase in mineralization of fractures. Currently the wells near the crest of the field (21/1-6, 21/1-7ST and 21/1a-14) still produce virtually dry oil and contribute approximately 80% of the production. All the other wells cut variable amounts of an extremely saline formation water (116 528 ppm chlorides).

The 33.6° API crude oil from the Buchan Field has a comparatively high sulphur (0.8% w/w) and vanadium (23 ppm) content, and an extremely high asphaltene (2.7% w/w) content for a North Sea crude oil. It is not unduly waxy (5% w/w), nor is there any H_2S. The composition of the oil varies slightly across the field with oil in the flank well 21/1a-8 being somewhat heavier and with an appreciably higher sulphur content.

Reserves/resources

The original estimate of oil recovery by Transworld gave a range of 5–130 MMBBL, with the uncertainty due to a lack of understanding of how the matrix would recharge the fractures, the effects of aquifer break-through, and the degree of pressure support. The installation of gas lift and the pipeline, together with better than expected reservoir performance, has increased the operating efficiency of the field and has raised the reserves figure from 50 MMBBL at the time of initial development to the present day figure of 90 MMBBL (1989). The current decline in production on a well-by-well basis suggests that the economic life of the field will probably end beyond 1995.

The continuing revisions of the recovery estimate are indicative of the uncertainty in the characteristics of the reservoir performance and of the recovery mechanisms at work. In particular, the apparently random distribution of the fractures and higher permeability intervals, and the pervasive faulting, make coherent reservoir descriptions and predictions tentative.

Currently production is totally from the upper (L1 and L2) lithofacies zones. Zone L3, despite having an apparently higher matrix porosity than the zones above and below, has a lower fracture intensity, produces nothing and effectively acts as a seal to the L4 zone. The proximity of the L4 zone to the water leg has precluded its use as a producing interval where channelling of water through fractures could quickly kill any oil production. It is possible that locally the L4 zone may contain small volumes of oil trapped beneath the less permeable L3 zone.

Initially it was believed that the field was highly compartmentalized by the faulting and hence that with the exception of the central horst area all of the wells were in pressure isolation. In addition the field was modelled as a homogeneously-fractured dual porosity system.

The first PLT logs were run in the wells during 1985, and showed that the production was from thin discrete intervals with large intervals contributing little or no flow. In addition these logs showed that the productive intervals often had differing reservoir pressures and hence, when the wells were shut in, there was strong wellbore cross flow.

As a consequence of the new data from the production logs, the field is currently being simulated using a heterogeneous, multi-layered system with significant vertical permeability barriers which are thought to be due to shale and/or less fractured intervals.

Recent pressure data have indicated that there may be more extensive pressure communication between the wells than previously thought.

Expansion (depletion) recovery appears to be assisted by the lateral influx of water from the aquifer which is believed to be sweeping both the fractures and the highly permeable zones. The recovery factor is estimated to be between 15 and 18%.

Production and well monitoring facilities are limited by the floating production system and satellite wells which do not permit the collection of reliable data without operational or financial penalty.

There are currently no spare risers on the platform, thus making production from further infill wells only possible at the expense of an existing riser.

Similarly, deck space and weight restrictions will not permit the installation of water injection facilities for pressure maintenance. In addition, it is feared that water injection would force water through the fractures and high permeability channels preventing production, by depletion, of the oil from the low permeability matrix. This could lead to the early abandonment of the field as the production wells water out.

The Buchan Field Unit partners have given permission to publish this paper.

References

CLAYTON, G., HIGGS, K., KEEGAN, J. B. & SEVASTOPULO, G. D. 1978. Correlation of the palynological zonation of the Dinantian of the British Isles. *Coloquio Internationale Palinologia Leon*, **1**, 137–147.

DARNBOROUGH, E. 1980. The Buchan Field development. *Proceedings of the European Offshore Petroleum Conference and Exhibition*, London, October 21–24, 177–184.

HILL, P. J. & SMITH, G. 1979. Geological aspects of the drilling of the Buchan Field. *Offshore Europe 79 Conference and Exhibition*, Aberdeen, SPE 8153, 1–7.

MYKURA, W. 1983. Old Red Sandstone. *In*: CRAIG, G. Y. (*ed.*) *Geology of Scotland.* Scottish Academic Press, Edinburgh, 205–251.

NEVES, R., GUEINN, K. J., CLAYTON, G., IOANNIDES, N. S. & NEVILLE, R. S. W. 1972. A scheme of miospore zones for the British Dinantian. *Comptes Rendu 7th cong. Int. Strat. Geol. Carb.* (Krefeld, 1971), **1**, 347–353.

RICHARDS, P. C. 1985. Upper Old Red Sandstone sedimentation in the Buchan oilfield, North Sea. *Scottish Journal of Geology*, **21**, 227–237.

STEWART, I. J. 1987. A revised stratigraphic interpretation of the Early Palaeogene of the central North Sea. *In*: BROOKS, J. & GLENNIE, K. W. (eds) *Petroleum Geology of the North Sea.* Graham and Trotman, London, 557–576.

Buchan Field data summary

Trap
Type: Horst-like tilted fault block sealed by Lower Cretaceous mudstones
Depth to crest: 2580 m (9464 ft) TVSS
Lowest closing contour: Unknown
Hydrocarbon contacts: Original OWC 3165 m (10 384 ft) TVSS. There is no gas cap
Oil column: 585 m (1919 ft)

Pay zone
Formation: Upper Old Red Sandstone
Age: Fammenian–late Visean/?early Namurian (Devonian–Carboniferous)
Gross thickness: in excess of 665 m (2182 ft)
Net/gross ratio: field-wide average is 0.823 with a range from 0.115 to 0.956
Cut-off for net/gross: 50% *V* clay and 3.5% porosity
Porosity: average matrix porosity is 0.0885 with a range from 0.109 to 0.071
Hydrocarbon saturation: varies from 0.099 in the lowermost zone to 0.6771 in the upper zone of the reservoir
Permeability: Matrix permeability is generally low (0.1–2 md) but fractures pervade the reservoir and these improve average permeabilities to 38 md
Productivity index: 2–20 BPD/psi

Hydrocarbons
Oil gravity: 33.6° API
Oil type: Low sulphur: 0.81% (central) to 1.06% (flank), high asphaltene (2.7% wt), wax 4.5–5.5% w/w, undersaturated
Bubble point: 1271 psig at 222°F
Gas/oil ratio: (producing) 285 SCF/STBBL
Formation volume factor: 1.205 at 2860 m TVSS (volume-weighted average reservoir elevation)

Formation water
Salinity: 116 528 ppm NaCl equivalent.
Resistivity: 0.067 ohm m at 60°F

Reservoir conditions
Temperature: 222°F at 2926 m (9600 ft)
Initial reservoir pressure: 7506 psig at 2926 m (9600 ft) (original pressure)
Pressure gradient in reservoir: 0.35 psi/ft

Field size
Area: 14.28 sq km (5.51 sq miles)
Gross rock volume: 3.3811 billion cubic metres
Hydrocarbon pore volume: 466 MMBBL
Recovery factor: oil: 15–18%
Drive mechanism: pressure depletion/aquifer support
Recoverable hydrocarbons: oil: 90 MMBBL

Production
Start-up date: May 1981
Development scheme: Four template wells and five satellite locations connected to a floating semi-submersible production facility by subsea flowlines to a subsea manifold
Secondary recovery: Gas lift installation in 1985
Production rate: Restricted by Buchan–Forties pipeline capacity to 28 000 BOPD. Production declined below export capacity during December 1988
Cumulative production to date: 71.5 MMBBL to end December 1989

The Chanter Field, Block 15/17, UK North Sea

H. R. H. SCHMITT

Occidental International Exploration and Production Company, 1200 Discovery Drive, Bakersfield, CA 93389, USA

Abstract: The Chanter Field is located 11 km southeast of the Piper Alpha platform location in Block 15/17. The field was discovered by the 15/17-13 well, which tested 37.8° API crude from Galley sandstone turbidites and 52.1° API gas-condensate from the shallow marine sands of the Piper Formation. The Galley sandstone reservoir is overpressured, whereas the Piper sandstone reservoir is normally pressured. The bulk of the field occupies the culmination of a structural terrace on the downthrown side of the main E–W fault forming the northern margin of the Witch Ground Graben in this area. It has dip closure to the east, south and west and fault closure to the north. The 'main block' gas–water contact of 13 080 ft TVSS in the Piper sandstone and the oil–water contact of 12 240 ft TVSS in the Galley sandstone were deduced from wireline pressure measurements and have not been encountered in wells. Estimated proved reserves of 4.6 MMBBL of crude oil and condensate liquids and 20.8 BCF of gas will be produced through a cluster of wells tied back to the Piper Field by a subsea pipeline.

The Chanter Field is in UK North Sea Block 15/17, approximately 110 miles northeast of Aberdeen and 7 miles southeast of the Piper Alpha platform location (Fig. 1). The field lies in 473 ft of water and covers an area of about 500 acres, containing up to 17 MMBBL STOIIP and 98 BCF GIIP in two Upper Jurassic reservoirs. Galley sandstone turbidites within the Kimmeridge Clay Formation are oil-bearing, and shallow-marine Piper sandstones contain gas-condensate above dew point. Hydrocarbons are trapped in a south-tilted terrace with northern fault closure (Figs 1–3).

The name Chanter refers to the pipe on a set of bagpipes which is provided with finger holes and on which the melody is played.

History

Block 15/17 was awarded in the UK Fourth Round of Licensing in 1972, and is operated by Occidental on behalf of the following consortium:

Occidental Petroleum (Caledonia) Ltd	36.5%
Texaco Britain Ltd	23.5%
Lasmo North Sea plc	20%
Union Texas Petroleum Ltd	20%

However the Chanter discovery well, 15/17-13, and subsequent appraisal wells were drilled without Texaco's involvement. Equity in the Chanter Sole Risk Step Out Area (a two mile square centred on 15/17-13) is:

Occidental Petroleum (Caledonia) Ltd	51.6814%
Lasmo North Sea plc	28.3186%
Union Texas Petroleum Ltd	20%

The 15/17-13 discovery well, completed in September 1985, encountered 193 ft of pay in the Galley sandstone and 229 ft of pay in the Piper Formation. The well was drilled on the crest of the Chanter structure, a culmination on a structural terrace, immediately south of the major E–W faults delineating the northern margin of the Witch Ground Graben (WGG). The well was located north of a saddle separating Chanter from the 15/17-8A well drilled 2.5 km to the south in 1976 (Fig. 4). The Galley and Piper sandstones were both present in 15/17-8A but only traces of hydrocarbons were recorded. In 15/17-13 however, the Piper sandstone flowed 3850 BPD of 52.1° API gravity condensate with 24.5 MMSCFD of gas, and the more massive Galley sandstone flowed 9260 BPD of 37.8° API gravity oil. No oil–water or gas–water contact was observed.

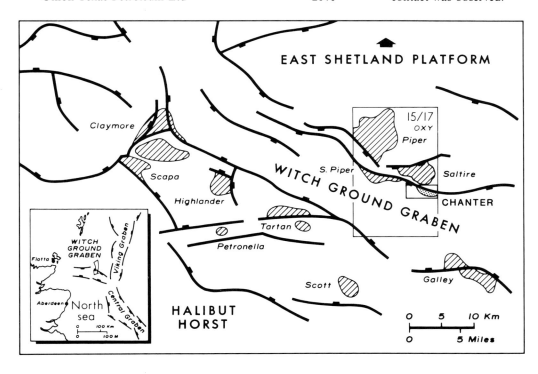

Fig. 1. Structural elements, Witch Ground Graben.

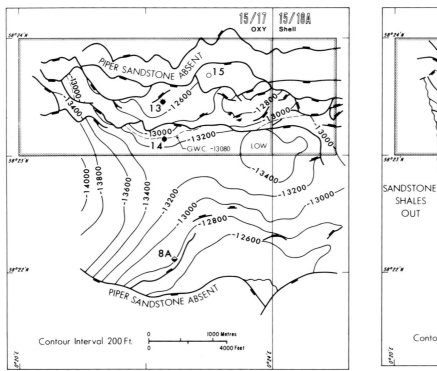
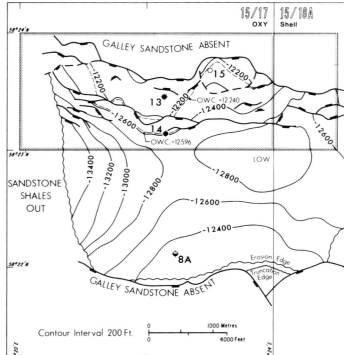

Fig. 2. Top structure maps for Piper (left) and Galley (right) sandstones.

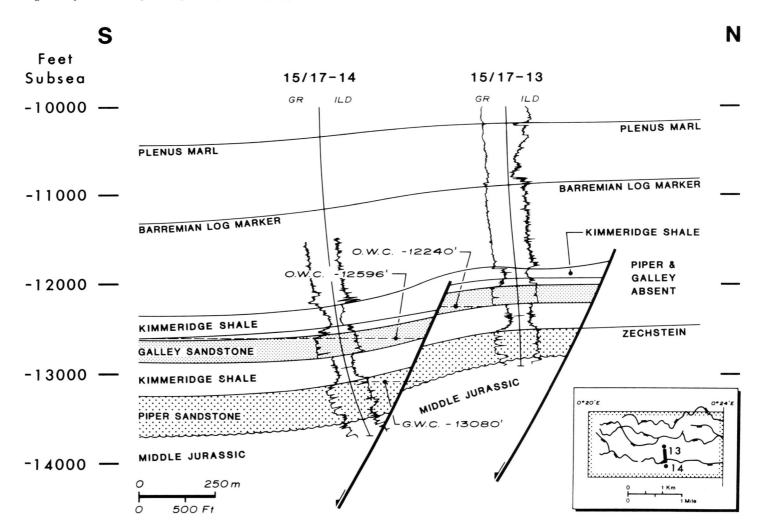

Fig. 3. Chanter Field S–N structural cross-section.

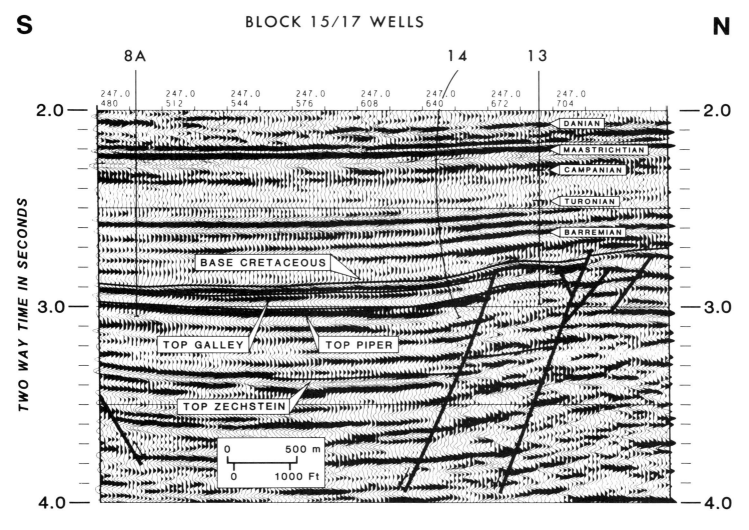

Fig. 4. Chanter Field S–N interpreted seismic line.

Following the 15/17-13 well, 15/17-14 was drilled approximately 600 m to the south. The well was intended to intersect the oil–water contact downdip in the Galley sandstone. The well penetrated a similar sequence to that in 15/17-13. The Galley sandstone flowed a maximum of 5800 BOPD but rapid depletion was observed, indicating a small accumulation. An oil–water contact was present at 12 596 ft TVSS but wireline pressure plots indicate that the oil columns in 15/17-13 and 15/17-14 are not in direct communication (Fig. 5). The Piper sandstone tested water and a gas–water contact of 13 080 ft TVSS was deduced from the wireline pressure plot from wells 15/17-13 and 15/17-14 (Fig. 5).

In 1986, the final Chanter appraisal well, 15/17-15, was drilled 800 m northeast of 15/17-13 in order to investigate a mapped upthrown terrace. After encountering a thin Galley sandstone, the well passed through a thin underlying Kimmeridge shale into Carboniferous clastics, the Piper sandstone being absent due to faulting and erosion. The Galley sandstone was wet and wireline pressure data from this well and 15/17-13 indicated that the main Chanter Field oil–water contact in the Galley sandstone is at approximately 12 240 ft TVSS.

The appraisal programme had thus established the size and extent of the accumulation. The decision was taken to develop the field using a cluster of wells at the 15/17-13 location and tied back to Piper Alpha via a 6 inch flowline. An Annex B was submitted to the Department of Energy in September, 1987 and approved in December. Oil production from the Galley sandstones was to have commenced late in 1988 but was prevented by the accident on Piper Alpha. Piper sandstone gas-condensate production was planned to commence between 1990 and 1992, depending on the Galley sandstone reservoir performance.

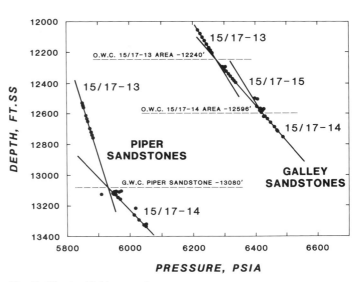

Fig. 5. Chanter Field reservoir pressures.

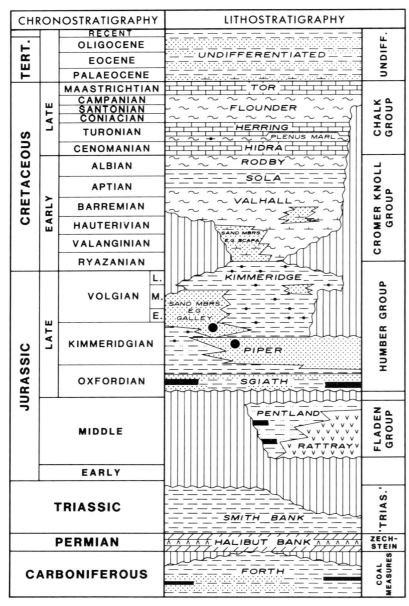

Fig. 6. Chanter stratigraphy.

Field stratigraphy

The stratigraphic sequence for Chanter is shown in Fig. 6. Two reservoirs are present in the Upper Jurassic Humber Group; the Galley sandstone within the Kimmeridge Clay Formation and the Piper sandstone (Fig. 7).

Extensional subsidence of the WGG Basin may have taken place during the Permian and Triassic, but the major phases of graben formation occurred in the Late Jurassic and Early Cretaceous. The graben probably developed as a reactivation of Hercynian features.

During the Middle Jurassic, the area lay on the northern flank of the exposed Central North Sea Arch (Boote & Gustav 1987). In early Late Jurassic times, the arch subsided and was transgressed by the Oxfordian Sgiath paralic facies and the Late Oxfordian–Early Kimmeridgian shallow-marine, wave-dominated deltaic facies of the Piper Formation (Harker et al. 1987). Although the Piper sandstone is broadly similar to that in Piper Field, the Sgiath in Chanter contains no reservoir sandstones comprising only clays and silts. During the Kimmeridgian, the area began to sag in response to incipient rifting. The basin deepened in the Late Kimmeridgian–Early Volgian, and the sand-dominant Galley turbidite system developed. The early synrift depression finally collapsed in the Middle–Late Volgian, forming a deep trough over the area of the WGG.

During the Volgian, erosion of the crestal and frontal faces of the emerging blocks immediately north of Chanter resulted in much of the earlier cover being sloughed off into the graben and redeposited as turbidites. However, a major part of the Galley sandstone was probably derived from a more distant source terrain, possibly on the East Shetland Platform, over 100 km to the north. As tectonism waned, the entire graben and surrounding platform was blanketed by Late Volgian–Early Ryazanian organic-rich black shales deposited under anoxic conditions. Further tectonism occurred during the Ryazanian when fault-block rotation led to thickening of the black shales adjacent to reactivated faults. By this time, the structure of the Chanter reservoirs was much as it is today. The graben was gradually infilled with hemipelagic shales and marls of the Cromer Knoll Group, and major tectonism effectively ceased by the end of the Early Cretaceous. The Cretaceous sequence in Chanter is more complete than in the Piper Field, since the formations below

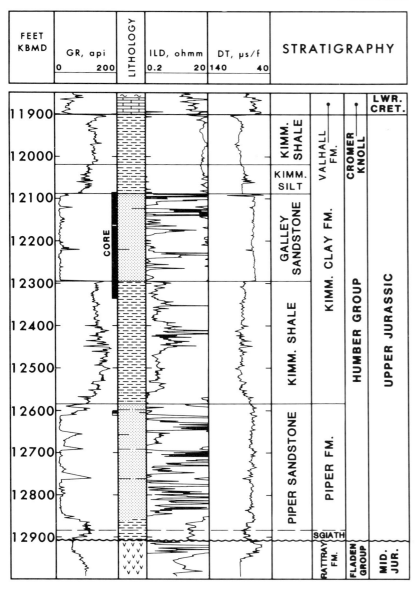

Fig. 7. 15/17-13 stratigraphic type well.

the Flounder onlap the Piper Field and are absent or thin therein.

Continuous subsidence and deposition of marine sediments took place through Late Cretaceous and Palaeocene times, accompanied by sedimentary compaction. The Upper Cretaceous is characterized by hemipelagic limestones, marls and chalks of the Chalk Group, but clastic sedimentation gradually returned in the Tertiary.

Geophysics

The seismic coverage within Block 15/17 consists of both 2-D and 3-D data (see Piper paper in this volume), but only 3-D data was used in mapping Chanter. The geophysical definition of both trap and reservoir is good (Fig. 4).

Trap

Chanter Field occupies the culmination of a fault-bounded structural terrace on the downthrown side of major E–W faults forming the northern margin of both the WGG and the field itself. To the north of the field, the Piper and Galley sandstones are absent along the fault scarps due to erosion in Late Jurassic to Early Cretaceous times. Along the western, southern and eastern margins of the field, there is a combination of dip and fault closure (Figs 2–3).

The vertical seal for the field is formed by shales of the Kimmeridge Clay Formation. Along the northern 'boundary' fault of the field, the Upper Jurassic reservoirs are juxtaposed against Triassic shales and Middle Jurassic claystones, siltstones and volcanics. These units form the updip lateral seal to the reservoir. In common with other fields in the area, faults seal only where total offset of the permeable section occurs. The E–W fault between 15/17-13 and 15/17-14 constitutes a partial barrier along the southern margin of Chanter Field (Fig. 3).

The main field fluid contacts of 12 240 ft TVSS in the Galley and 13 080 ft TVSS in the Piper, were deduced from wireline formation-pressure plots (Fig. 5) and are subject to inaccuracies associated with this procedure. Constraints on these fluid contacts are included in Table 1.

In the 15/17-14 area, a different Galley oil–water contact was intersected by the well at a depth of 12 596 ft TVSS. The deeper fluid contact is made possible by total offset of the Galley sandstones by faulting between 15/17-13 and 15/17-14 (Fig. 3).

The Piper sandstone in Chanter is filled to its spill point at 13 080 ft TVSS, as mapped at the eastern end of the reservoir. The Galley sandstone spill point of 12 240 ft TVSS is at the western end of the main part of the field juxtaposed against Piper sandstones across a fault. In the 15/17-14 block, a spill point of 12 596 ft TVSS is present across the fault running north of the well.

Table 1. Chanter Field volumetrics summary

Galley Sandstone
Most likely reserves 3.1 MMSTB

	Depth (ft TVSS)	STOIIP (MMBBL)	Solution gas (MMSCF)
Oil-down-to (15/17-13)	12 211	11.5	10.5
Pressure gradient intersection*	12 240	12.8	11.7
Deepest possible pressure gradient intersection*	12 255	13.3	12.2
Water-up-to (15/17-15)	12 291	–	–

Piper Sandstone
Most likely reserves 26.7–47.3 BCF gas, 5.4–7.3 MMBBL condensate liquids

	Depth (ft TVSS)	GIIP (BCF)	Residue gas (BCF)	Condensate liquids MMBBL
Gas-down-to (15/17-13)	12 769	26.4	21.3	4.6
Pressure gradient intersection*	13 080	95.4	76.7	16.6
Water-up-to (15/17-14)	13 088	–	–	–

* From wireline pressure measurements

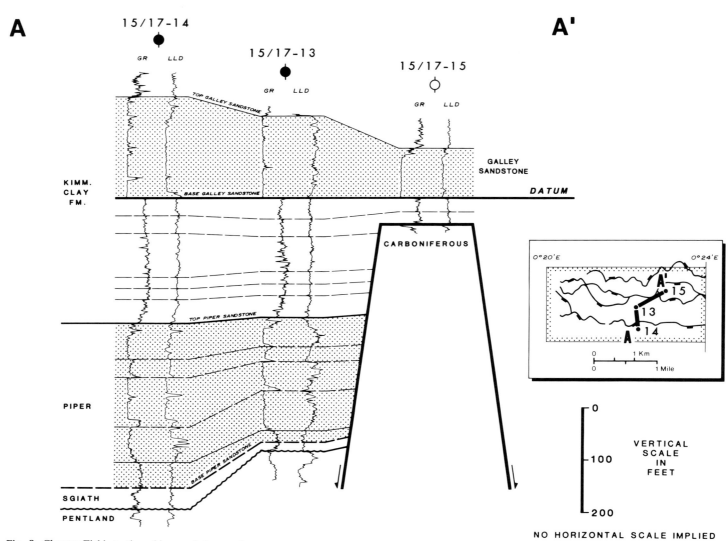

Fig. 8. Chanter Field stratigraphic correlation section.

Reservoir

As mentioned previously, two reservoirs are present, both within the Upper Jurassic Humber Group; the Piper sandstone and the Galley sandstone.

Piper sandstone

The Piper sandstone in Chanter comprises a series of stacked sand bodies deposited in a shallow-marine environment as part of a wave-dominated delta complex. The prograding sands were covered by transgressive regional shales and locally reworked by shelf processes. Three distinct facies were recognized in cores: laminated to massive cross-bedded sandstones, bioturbated units and a shale facies.

The laminated to massive, cross-bedded sandstones are generally medium- to coarse-grained, moderately to well sorted and locally burrowed, containing carbonaceous material, pebbly zones, erosional contacts, slump bedding and fossils. The clean nature of the sandstone and its fossil assemblage suggest a high-energy traction current origin for the sands in a shallow-marine setting. The highly bioturbated mudstone, siltstone and sandstone of the bioturbated unit are locally fossiliferous and occasionally contain faint remnants of sedimentary structures. These units were deposited at lower energy levels than the cross-bedded sands. The shale facies consists of highly fossiliferous, laminated to massive carbonaceous mudstone, with abundant sedimentary structures. This facies is interpreted as having been deposited during high sea level stands accompanying or following major transgressions.

The sandstones are moderately to well cemented with silica in the form of quartz overgrowths. Locally, there are moderate amounts of calcite, dolomite, clay and hematite. Clay content is typically less than 3% but is as high as 10% in the bioturbated units. Calcite is minor in 15/17-13 but has completely cemented-up some horizons below the gas–water contact in 15/17-14. The pore network comprises both primary intergranular and secondary dissolution porosity. Reservoir parameters are given in the data summary. Although the Piper sandstone has a higher average porosity than the Galley sandstone the average permeability is lower. This is due almost entirely to bioturbation and the resulting increase in clay content.

Correlation within the Piper sandstone is good (Fig. 8), and it is anticipated that it will prove to be a well organized layered reservoir as in Piper Field.

Galley sandstone

The Galley sandstone in Chanter is Early Volgian in age. It was deposited in an outer neritic or bathyal setting as channel fill, sand rich turbidites infilling a N–S trending trough extending from Block 15/12 down the eastern side of Block 15/17 and into the Galley Field in Block 15/23. The turbidites became progressively younger as they prograded southwards (Fig. 9).

The Galley and Piper sandstones have almost identical mineralogy, suggesting that they may have been derived from a similar source terrain or that the Galley sandstones were derived, at least in part, from Piper sandstones. The most northerly penetration of the Galley sandstones was in 15/12-1. The main source for these sands was even further north on the East Shetland Platform, with some more locally derived component originating from Piper sands eroded off nearby fault scarps.

The Galley sandstone in Chanter is a medium- to coarse-grained, moderately to well sorted, silica-cemented quartz arenite with little or no clay (<3%). Beds are massive or parallel laminated, with thicknesses of up to 12 ft but averaging 4 ft. Graded beds occur in the coarser units, and parallel laminae and pebbly intervals occur in the middle to lower parts of the sandstone. Shale rip-up clasts, carbonaceous material and displaced shallow-marine fossils are found throughout the sandstones, and disturbed bedding, stylolites and dish structures all occur locally. Burrows are extremely rare. Rare, thin, shale intercalations comprise less than 1% of the reservoir interval.

In 15/17-14 and 15/17-15, the massive Galley sandstones are overlain by interbedded sandstones, shales and silts. These are interpreted as channel-fill turbidites deposited in a system of lower hydraulic energy than that of the massive sands below.

Correlation within the Galley sandstone is poor (Fig. 8). Reservoir parameters are given in the data summary.

Source

The organic-rich shales of the Kimmeridge Clay Formation are the source of the Chanter hydrocarbons. These shales contain 4–16% TOC comprising predominantly amorphous marine kerogen. The likely migration path is out of the WGG and updip into the Chanter reservoirs.

Migration of hydrocarbons into Chanter took place in Late Cretaceous to Early Tertiary times. This coincides with the period

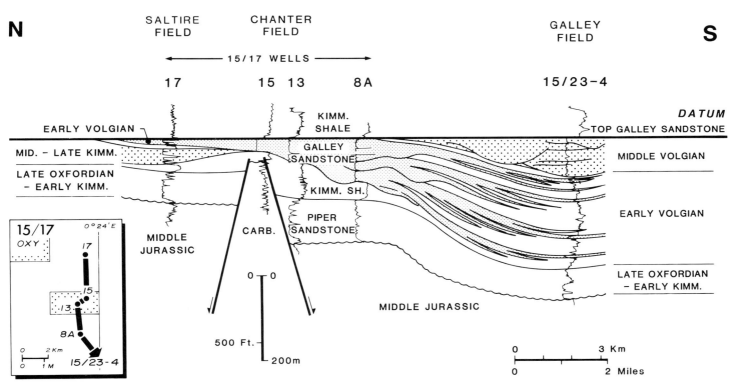

Fig. 9. Galley sandstone stratigraphic cross-section.

of most rapid burial of the Kimmeridge shales in this area. The gas condensate was a late-stage product of the maturation process, probably generated at a greater depth within the WGG.

Hydrocarbons

Test results were discussed in the preceding History section. The main characteristics of the hydrocarbons recovered from the Piper and Galley sandstones in the 15/17-13 area, and the formation water recovered from 15/17-14 are given in the data summary. No variations in the hydrocarbon properties with depth are anticipated, because of the limited areal extent and low structural relief of the reservoirs.

Reserves/resources

As mentioned previously, fluid contacts in the main part of Chanter Field (the 15/17-13 block) have not been encountered in wells but have been deduced from wireline pressure measurements (Fig. 5). Due to inaccuracies in this technique, a degree of uncertainty must be attached to fluid contacts derived by it. Table 1 summarizes information relating to fluid contacts within the main part of Chanter, and also illustrates STOIIP and GIIP values calculated using the different contacts. Most likely values for Galley STOIIP and Piper GIIP are 12.8 MMBBL and 95.4 BCF respectively.

Chanter's small size and a number of unfavourable geologic factors means that recovery efficiency for both reservoirs will be low. Aquifer influx into both reservoirs will be severely restricted by the E–W fault along the southern edge of the field, particularly in the Galley sandstones. However, cross-sectional studies indicate that limited cross-fault communication within the Piper sandstones may occur in the most permeable sandstone units. The combined effect of this could be severe pressure depletion of the Galley sandstone during oil production and early water breakthrough in the Piper sandstone during gas-condensate production.

Not only is the average permeability of the Piper sandstone lower than that of the Galley sandstone, the permeability range is far greater, being very low in the bioturbated units but much higher in the cross-bedded and laminated units. If the drive mechanism is one of water drive with aquifer influx, the low-permeability layers will not contribute significant reserves.

Estimated recovery efficiencies vary from 18–34% in the Galley oil reservoir. In the Piper reservoir, the estimated gas recovery factor is 34–60% and the liquids recovery factor, 32–43%. The variation is due to different degrees and directions of aquifer support. The relative quantities of gas and liquids recovered from the Piper sandstones are heavily dependent upon the production facilities they are processed through. Most likely reserves are 3.1 MMBBL of oil from the Galley, 26.7–47.3 BCF of gas from the Piper and 5.4–7.3 MMBBL of condensate liquids from the Piper. However, 'proved' reserves are much lower at 4.6 MMBBL of oil and condensate liquids and 20.8 BCF of gas.

The effect of uncertain aquifer support on field performance has been catered for by adopting a flexible development scheme. The principal theme is that additional wells (injectors and/or producers) can be drilled once the nature of the drive mechanism of both reservoirs becomes clear through production data.

Chanter Field data summary

	Reservoir	
Reservoir parameters	Galley sandstone	Piper sandstone
Porosity (%)	11.6	13.5
Average permeability (md)	250	80
Maximum net pay (ft)	193	229
Average water saturation (%)	12.9	18.0
Reservoir conditions		
Datum (ft TVSS)	11 935	12 446
Temperature (°F)	252	252
Pressure (psia)	6196	5860
Fluid contact (ft TVSS)	12 240 (OWC)	13 080 (GWC)
Hydrocarbons		
Oil/condensate gravity (°API)	37.8	52.1
GOR/GLR (SCF/STB)	914*	5727†
Bubble point/dew point (psia)	3305	5371
Formation volume factor	1.542	262.6 SCF/RCF
Viscosity (cp)	0.33	
Formation water		
Salinity (Cl ppm)	44 000	
Resistivity (ohm m)	0.0445 at 200°F	

* At 200 psi and 90°F
† Using Piper Alpha process

Permission to publish this paper was given by Occidental and its partners (Lasmo North Sea plc and Union Texas Petroleum Ltd) and is gratefully acknowledged. The interpretations expressed in this paper are entirely the responsibility of the author. In addition, I wish to thank the Occidental drafting department for their patient assistance.

References

BOOTE, D. R. D. & GUSTAV, S. H. 1987. Evolving Depositional Systems within an Active Rift, Witch Ground Graben, North Sea. *In*: BROOKS, J. & GLENNIE, K. (eds) *Petroleum Geology of North West Europe*. Graham & Trotman, 819–833.

HARKER, S. D., GUSTAV, S. H. & RILEY, L. A. 1987. Triassic to Cenomanian Stratigraphy of the Witch Ground Graben. *In*: BROOKS, J. & GLENNIE, K. (eds) *Petroleum Geology of North West Europe*. Graham & Trotman, 809–818.

The Claymore Field, Block 14/19, UK North Sea

S. D. HARKER, S. C. H. GREEN & R. S. ROMANI

Occidental Petroleum (Caledonia) Ltd, 1 Claymore Drive, Aberdeen, UK

Abstract: The Claymore Field is located in UK North Sea Block 14/19 on the southwest margin of the Witch Ground Graben. The principal structure is a southerly tilted and truncated fault block. The field is divided into three producing areas. Major production is from Upper Jurassic paralic sandstones of the Sgiath Formation and turbidite sandstones of the Claymore Sandstone Member of the Kimmeridge Clay Formation in the downflank Main Area. Minor production is from Permian carbonates of the Halibut Bank Formation and Carboniferous sandstones of the Forth Formation in the crestal Central Area. The Northern Area is a northerly plunging nose, extending grabenwards from the Claymore tilt block. Production in the Northern Area is from Lower Cretaceous turbidite sandstones of the Valhall Formation.

A small amount of oil was recovered on a wireline test in 1972 from Permian carbonates in the crestally located 14/19-1 well, in what is now termed the Central Area. In 1974 the Main Area was discovered by the southerly downdip well 14/19-2, and the Northern Area was discovered by the northerly downdip well 14/19-6A. Initial oil in place was 1452.9 MMBBL with currently estimated ultimate proved recovery of 511.0 MMBBL of oil. A 36-slot steel platform was installed in 1977. Two subsea water-injection templates were added in 1981 and 1985. Cumulative production to 6 July 1988 was 322.9 MMBBL of oil and daily production was 75 000 BOPD of oil from 28 producers, supported by 16 injectors.

Claymore is located on the southwest margin of the Witch Ground Graben, a northwest-trending arm developed off the Central-Viking North Sea graben system (Fig. 1). The Claymore Field lies in UK North Sea Block 14/19 in 360 feet of water, 112 miles (180 km) southeast of the Flotta Oil Terminal in the Orkney Islands. The field contained a total of 1452.9 MMBBL of oil originally in place in three field divisions (Fig. 2, Table 1): the Main Area (MAC); the Northern Area (NAC) and the Central Area (CAC). The reservoirs are in sandstones of Late Jurassic, Early Cretaceous and Early Carboniferous ages and carbonates of Late Permian age (Fig. 3). The trap is a combination of a southerly tilted truncated fault block of the Main and Central Areas, and a down-faulted, northerly plunging structural nose of the Northern Area (Figs 2 and 4). The field is named after the large two-edged sword of the Scottish Highlanders, known as the 'Claymore' (Gaelic: *Claidheamh*, sword plus *Mor*, large).

History

Exploration and appraisal

In March 1972, Occidental, as operator of a group then including Allied Chemical (North Sea) Ltd, Getty Oil Co. and Thomson North Sea Ltd, was awarded Block 14/19 in the UK Fourth Round of Licensing. The current Claymore Field operating group is listed in Table 1. An anticlinal feature at Tertiary level was mapped using the limited 2-D seismic control that had been shot by Occidental during 1970–1971.

The first well, 14/19-1, was drilled in mid 1972 in a near-crestal location and found a thick, though non-hydrocarbon bearing, Tertiary section. A thin Cretaceous sequence unconformably overlay a truncated Triassic with hydrocarbon shows detected in thin Lower Cretaceous sandstones, Permian carbonates and Carboniferous sandstones, down to a depth of 8650 ft subsea (Maher &

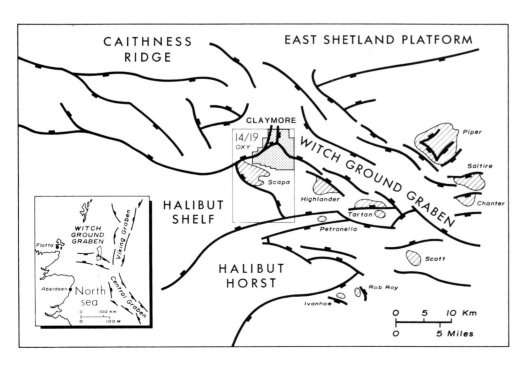

Fig. 1. Witch Ground Graben oilfields and fault trends map.

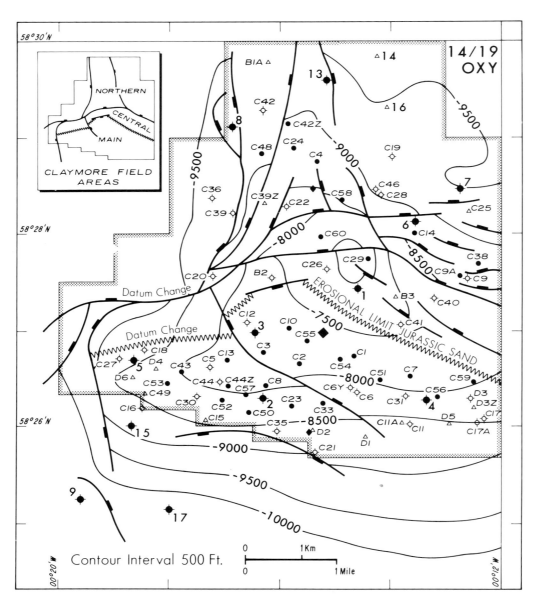

- ● APPRAISAL WELL
- ● PRODUCTION WELL
- △ INJECTION WELL
- ◇ ABANDONED WELL
- ■ CLAYMORE ALPHA PLATFORM
- ◆ SUBSEA INJECTION TEMPLATE

STRUCTURE CONTOURS
- NORTHERN AREA ON TOP LWR. CRET. SAND
- CENTRAL AREA ON BASE VALHALL
- MAIN AREA ON TOP JURASSIC SAND

Fig. 2. Claymore composite depth structure map.

Harker 1987). A Formation Interval Test at 7953 ft below Kelly Bushing (KB) recovered 8200 cc of 33° API oil from a Permian dolomite.

The 14/19-1 well discovered what was to become known as Central Area Claymore (Figs 2 and 4). The second well, 14/19-2, was drilled to the southwest and downdip of 14/19-1 in 1974. This well encountered 604 ft of oil-sand in the Upper Jurassic, above an oil–water contact at 8655 ft subsea, thereby discovering Main Area Claymore. The well tested 2017 and 1370 BOPD over two intervals of 240 ft and 150 ft (Harker & Maher 1988). The 14/19-3, 14/19-4 and 14/19-5 wells, all drilled in 1974, appraised the stratigraphic variation and updip erosional truncation of the Jurassic Sands. No further oil–water contact was found, but formation pressure gradients confirmed the depth at 8655 ft subsea. The 14/19-5 well also found oil in Lower Cretaceous sandstones, though no drill-stem tests were run.

Northern Area Claymore was discovered by the 14/19-6A well, drilled in late 1974, downdip to the northeast of 14/19-1, and it penetrated 194 ft of gross oil-sand in the Lower Cretaceous. A drill-stem test over the upper 57 ft recorded a flow of 5124 BOPD. Oil shows were also detected in Carboniferous sandstones, and a drill-stem test over 150 ft recorded a flow of 729 BOPD.

The 14/19-7 well was deviated downdip to the northeast from 14/19-6A, and found the Lower Cretaceous sandstones to be water-

Table 1. *Claymore Field summary; Location: UK North Sea Block 14/19*

	Field divisions			
	Main Area	Northern Area	Central Area	
Trap	Tilted fault block	Downthrown nose	Tilted fault block	
Reservoirs	Upper Jurassic Sands	Lower Cretaceous Sands	Permian Carbonates	Carboniferous Sands
Area (acres)	3505	3729	850	885
Oil originally in place (MMSTB)	1002.6	357.1	10.9	68.1
Recovery factor	32.7	45.3	45.0	5.0
Original recoverable reserves (MMSTB)*	337.5	164.2	4.9	3.4
Total oil originally in place		1452.9 MMBBL		
Total original recoverable reserves		511.0 MMBBL		

* Minor additional Claymore reserves assessed with oil originally in place of 14.2 MMBBL and original recoverable reserves of 1.0 MMBBL.

Operating Group:

Occidental Petroleum (Caledonia) Ltd (Operator), Texaco Britain Ltd, Union Texas Petroleum Ltd, Lasmo North Sea plc, Agip (UK) Ltd, Atlantic Resources (UK) Ltd, Ranger Oil UK Ltd, Coalite Oilex Ltd, Goal Petroleum plc, Nedlloyd Energy (UK) Ltd, Peko Oil North Sea plc, Pict Petroleum plc, Sovereign Oil & Gas plc, Total Oil Marine plc.

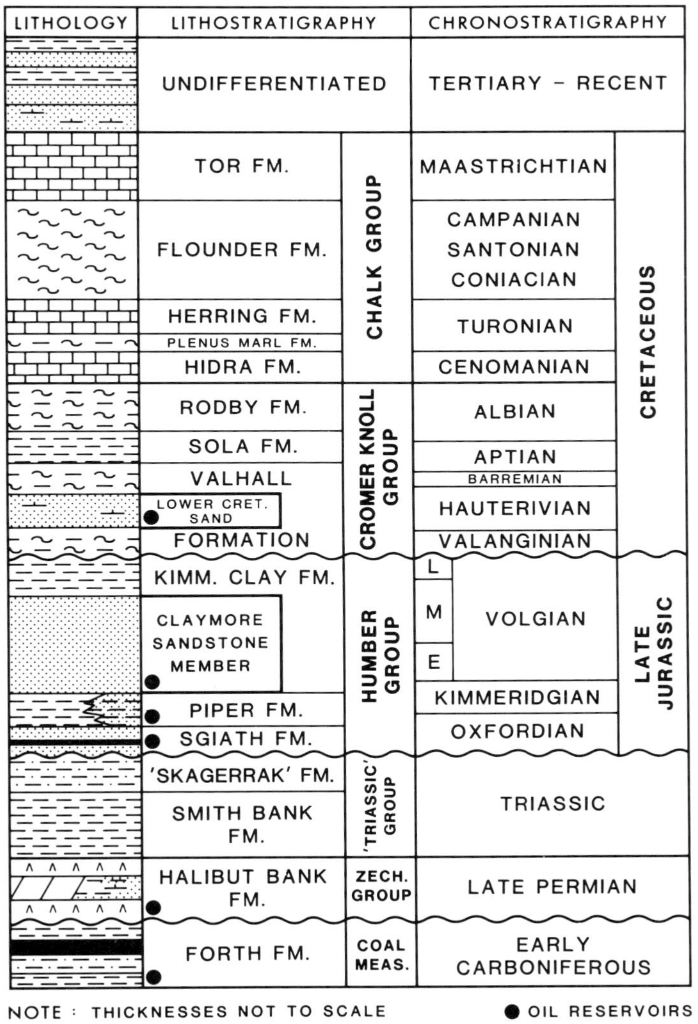

Fig. 3. Claymore stratigraphy.

bearing. The deeper Jurassic sandstones had oil shows, but were not tested. The 14/19-8 well, drilled in early 1975, was located on the northwestern flank of the Northern Area structural nose (Fig. 2). This well also found water-bearing Lower Cretaceous sandstones, but from formation-pressure gradients, the oil–water contact was determined at 9400 ft subsea.

Development

The decision to proceed with development of the Claymore Field was made in mid-1975, following the drilling of the eight exploration and appraisal wells described above. A 36-slot steel platform was designed, built and installed, so that development drilling commenced in May 1977. The first development well, C1 in Main Area, found the Upper Jurassic sands thin and truncated, unconformably overlain by Lower Cretaceous marls (Figs 2 and 4). Production wells C2, C3, C6, C7 and C8 were drilled into the central and eastern regions of the main area. The C4 well was drilled between 14/19-6A and 14/19-8 to appraise the development of the Northern Area Lower Cretaceous sandstones. The C5 well was abandoned before reaching the target. Seven production wells came on stream in November 1977 at a combined rate of 70 000 BOPD.

A significant proportion of the Claymore reservoirs is underlain by water (Fig. 4), and an early decision was made to proceed with flank water injection to flood the reservoir. In order to free platform slots for production wells, two water-injection templates were added. The Northern Template for Northern and Central Area injection was installed in 1981 and the Southern Template for Main Area injection in 1985. As of 6 July 1988, there were 28 producers, supported by 16 injectors with a daily oil production of 75 000 BOPD and cumulative oil production of 322.9 MMBBL.

Field stratigraphy

The stratigraphic sequence for Claymore is shown in Fig. 3 (see also Maher & Harker 1987). The oldest sediments penetrated in Claymore are Early Carboniferous Coal Measures of the Forth Formation. These consist of a thinly interbedded deltaic sequence of sandstones, shales and coals. Late Permian sediments, which unconformably overlie the Carboniferous, represent evaporitic deposits of the Zechstein Sea. The basal Kupferschiefer is succeeded by dolomites, dolomitic shales, limestones, anhydrites and thin carbonate-cemented sandstones. The Halibut Bank Formation comprises the Zechstein reservoir facies, near the crest of the Claymore structure. Fluvio-lacustrine sedimentation followed,

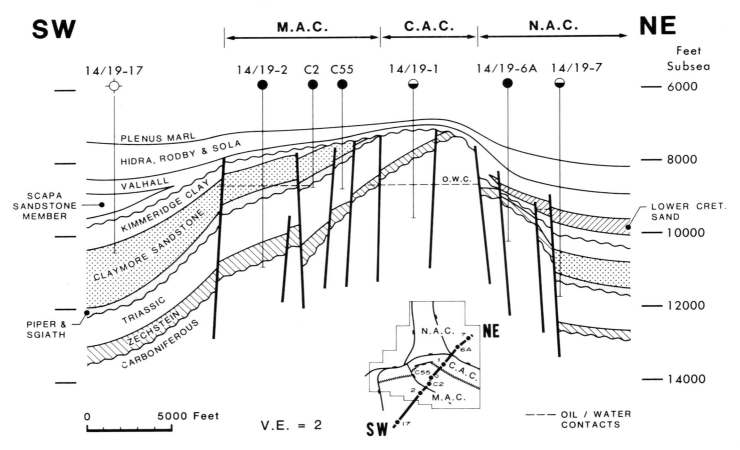

Fig. 4. Claymore schematic structural cross-section.

apparently conformably, with argillaceous Triassic redbeds of the Smith Bank Formation and fluvial silty sandstones of the 'Skagerrak' Formation.

No Early or Middle Jurassic rocks are preserved at Claymore. The Sgiath Formation represents an Oxfordian (Late Jurassic) brackish transgression into the Witch Ground Graben (Harker et al. 1987). This thin unit of paralic sandstones, carbonaceous shales and coals is succeeded by c.160 ft of the Late Oxfordian–Kimmeridgian Piper Formation. The Piper consists of mostly transgressive marine mudstones, except in the southeastern part of Claymore, where bioturbated argillaceous shelf sandstones are present (Harker & Maher 1987). A major rifting episode of the Witch Ground Graben began at the Kimmeridgian–Volgian boundary. Large volumes of sand-rich turbidites flooded into the Claymore area throughout the Early to Middle Volgian, forming the Claymore Sandstone Member of the Kimmeridge Clay Formation (Boote & Gustav 1987). The Mid–Late Volgian organic-rich argillites of the overlying Kimmeridge Clay, mark a return to a low-energy depositional environment.

The major Late Cimmerian earth movements strongly affected the Claymore region at the close of the Jurassic. Fault-block rotation and erosion affected the crest, removing much of the Jurassic and Triassic cover. The Early Cretaceous Valhall sediments mark a carbonate transgression and onlap, and cap the Claymore structure with sandy marls and limestones. Sandstone turbidites were deposited in the topographic lows around the Claymore structure. The northern Valhall sandstones comprise the reservoirs of the Northern Area, and the southern sandstones, the Scapa Field reservoirs (Fig. 4). Tectonic activity diminished during the later Cretaceous with hemipelagic post-rift deposition of marls, limestones and chalks. Tertiary sedimentation marked a return to clastic deposition with the accumulation of c. 6000 ft of sands and muds, accompanied by minor regional tilting to the ESE.

Geophysics

The Claymore structure was identified in 1972 on just a few of the regional seismic lines then available. In 1974, when the discovery well 14/19-2 was drilled, the seismic grid had a spacing of 1 by 2 km. This was infilled in stages to a final 2-D grid of approximately 500 by 500 m. A 3-D survey, with north–south orientated lines covering most of the 14/19 block, was shot in 1982, utilizing a line spacing of 75 m. The quality of this survey was superior to any of the previous 2-D data in terms of both resolution and structural definition. However, it still suffered from persistent multiples and the poor quality of mappable events in or near the Jurassic reservoirs. Recent attempts at reprocessing have been moderately successful, and at the time of writing (mid-1989) the entire 3-D survey was being reprocessed.

Current structure maps are based on a combination of seismic interpretation and well data. The top of the Lower Cretaceous sands of the Northern Area can be mapped successfully from the seismic data. In the Main Area, only the top of the Kimmeridge Clay Formation/Base Cretaceous Unconformity and the Top Zechstein can be mapped reliably (Harker et al. 1987). These two events clearly show the tilted fault block and the erosional truncation of the Jurassic section (Fig. 5).

Trap

The Claymore structure lies at the intersection of three major tectonic lineaments. These are the NE–SW Caledonian trend, the E–W Hercynian trend and the dominant NW–SE Cimmerian trend of the Witch Ground Graben. The Claymore trap is a combination of structural and stratigraphic elements, and covers an area of 8100 acres (Table 1, Figs 2, 4 and 5). The Main and Central Areas form

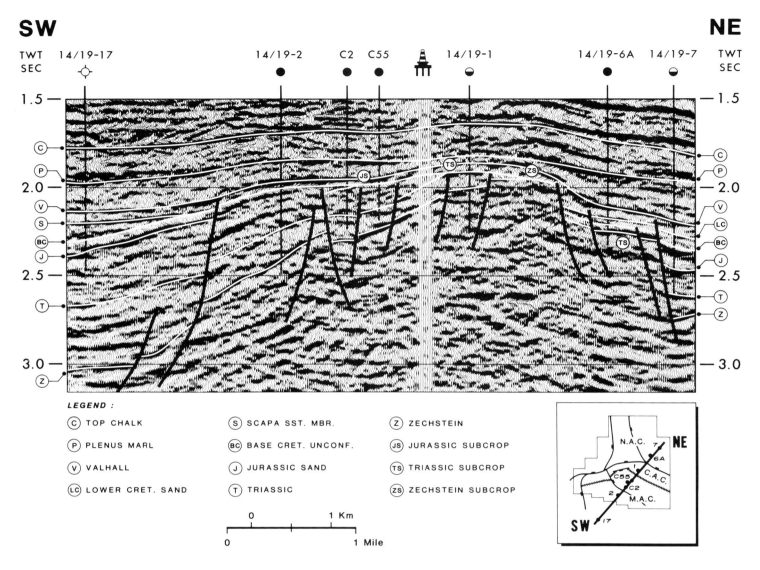

Fig. 5. Claymore SW–NE seismic line.

a triangular southerly tilted and truncated fault block. The Northern Area is a triangular, faulted, northerly plunging structural nose that is downthrown from the major Central/Main Area bounding faults.

Non-reservoir marls of the Valhall Formation form the major caprock for the truncated crestal Main and Central Areas at the Base Cretaceous Unconformity (Fig. 4). In addition, the Valhall marls form the updip and top seals for the onlapping Lower Cretaceous sandstones of Northern Area. Downdip, to the south of the crest of the Claymore structure, Triassic shales of the Smith Bank Formation form the top seal for the Zechstein reservoir of Central Area. Further downdip, the Kimmeridge Clay silts and mudstones form the top seal for the Claymore sandstones of Main Area.

The oil–water contact for Main Area is 8655 ft subsea; for Northern Area it is 9400 ft subsea. Neither of these areas nor Central Area is full-to-spill point. The determination of the Central Area oil–water contact is difficult. This is because of the poor quality reservoir rock and lenticular nature of the sand bodies present at depth in the Carboniferous, and the trend to impermeable anhydrite downdip in the Zechstein. However, with the structural continuity of the Main and Central Areas tilted fault block, and the general lack of good hydrocarbon indications any deeper, the Central Area oil–water contact is estimated at 8655 ft subsea (i.e., the same as Main Area).

Reservoir

A summary of Claymore reservoir parameters is presented in Table 2.

Main Area

An E–W correlation section across Main Area, including C2, the type well, is shown in Fig. 6 (TVT: True Vertical Thickness logs). The Oxfordian paralic Sgiath Formation varies in thickness from 21–121 ft. It thins towards the crest, which is evidence for the early existence of the Claymore structure (Harker & Maher 1988). The general sequence of a basal coarsening-upward sandstone, a coal bed and interbedded shales and fining-upwards sandstones, correlates well across the field. The Late Oxfordian to Kimmeridgian, marine-shelf facies of the Piper Formation has poor quality reservoir rocks only in the region of the 14/19-4 well in the southeast.

Over most of the field, the Piper consists of three easily correlatable coarsening-upwards, silty mudstone cycles. The Piper ranges in thickness from 48–188 feet and, over most of the field, forms a transmissibility barrier between the Sgiath Formation and the Claymore Sandstone Member.

The turbidites of the Late Kimmeridgian to Mid Volgian Clay-

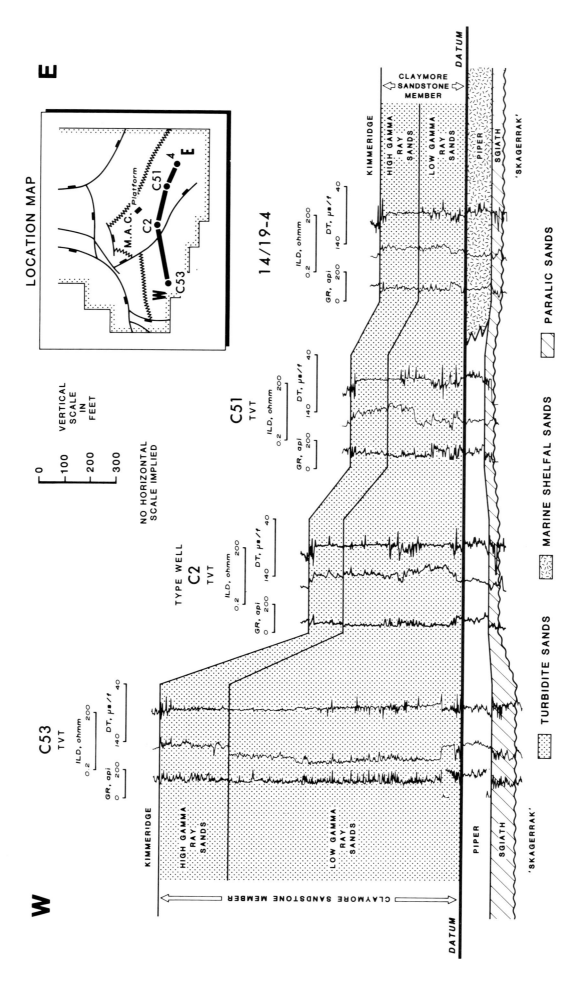

Fig. 6. Main Area Claymore stratigraphic correlation section.

Table 2. *Claymore reservoir parameters*

	Facies/ Lithology	Porosity %	Permeability md	Net/gross ratio
Main Area				
High Gamma Ray Sands	Turbidite Sands	20	10–400	0.90
Low Gamma Ray Sands	Turbidite Sands	21	20–1300	0.90
Piper	Marine Shelf Sands	18	0.2–13	0.50
Sgiath	Paralic Sands	19	30–600	0.50
Northern Area				
Massive Sand Unit	Turbidite Sands	24	50–4000	0.95
Lower Sand Unit	Turbidite Sands	20	20–1500	0.40
Central Area				
Halibut Bank	Carbonates	15	0.2–899	0.30
Forth	Deltaic Sands	19	0.2–100	0.20

more Sandstone Member are divided informally into the High and Low Gamma Ray Sands (Maher & Harker 1987). These contained 95% of the oil originally in place in Main Area. They thicken rapidly from east to west and downdip to a combined thickness of 1600 ft. The sands are fine-grained subarkoses with low authigenic clay content, but significant proportions of clay lithoclasts, micas and organic matter give rise to the distinctive high gamma radiation levels. The High Gamma Ray Sands are not as good reservoir rocks as the Low Gamma Ray Sands, being thinner bedded (0.5 versus 1.5 ft) with consequently more silty intercalations. The sediments are interpreted to have been transported by turbidity flows, with the Claymore structure acting as a sediment trap, ponding multitudes of low-volume sand flows, each of low lateral continuity. Thus correlation on a bed-by-bed basis is extremely difficult. The southerly tilting of the Claymore structure during turbidite deposition is indicated by dipmeter data and by the marked thickening, in particular of the Low Gamma Ray Sands, downdip from the crest (Harker & Maher 1988).

Northern Area

A NW–SE correlation section across Northern Area, including C28, the type well, is shown in Fig. 7. The principal informal twofold subdivision of the sandstones of the Hauterivian Valhall Formation into an upper Massive Sand unit (unit 1) and a Lower Sand unit (unit 2) is easily distinguished from their wireline log characteristics, and is easily correlatable fieldwide. The highly radioactive 'F' Marker above the reservoir interval forms a convenient datum for correlation. The Massive Sand unit is of better reservoir quality and contained 85% of the oil originally in-place in Northern Area.

The Lower Cretaceous sandstones are quartz arenites of variable grain size, commonly with a carbonate cement. The generally finer-grained Lower Sand represents a unit of many low-volume turbidite sand pulses into a regime of hemipelagic carbonate deposition. These sands are thought to have been derived principally from a Late Jurassic source, the eroded leading edge of the Claymore Structure, as they contain common reworked Late Jurassic palynomorphs. The Massive Sand unit is generally coarser-grained and thought to be derived from a Carboniferous source, as it contains abundant reworked Carboniferous palynomorphs. The Massive Sand probably consists of coalesced mass-flow sands over the entire Northern Area, although subsequent dewatering has obscured most bed boundaries.

The Lower Cretaceous sandstones have a depositional zero edge near the Main/Central Area bounding faults (Fig. 4). They thin over the structural nose and thicken up to 494 ft into the adjacent structural lows. A third informal Lower Cretaceous sandstone unit, known as the 'Stringers', is present only in the northwest flank, in the area of the B1A well. This unit consists of a highly radioactive mudstone sequence with thin interbeds of fine-grained turbidite sand. The 'Stringers' overlie and onlap the Massive Sand unit (Fig. 7).

Central Area

A log of the type well for Central Area, 14/19-1, is shown in Fig. 8. The Early Carboniferous sandstones of the Forth Formation are lenticular deltaic sand bodies. They are poorly sorted quartz arenites with pervasive ferroan dolomite cement. The sands are thinly interbedded with coals and overbank and prodelta shales. The succeeding Late Permian Zechstein evaporites are in reservoir facies only near the crest of the Claymore structure. The productive dolomites and limestones are progressively replaced downdip by anhydrite, which can be mapped using a change in seismic character. The productive area of the Zechstein consists of dolomites and limestones of the Halibut Bank Formation, some of which show evidence of debris-flow deposition. High permeabilities result from a combination of intergranular, vugular (leached) and fracture porosity. However, the Zechstein pay zones are generally thin, resulting in a low volume of oil in place (Table 1).

Source

The Late Volgian organic-rich mudstones of the Kimmeridge Clay Formation comprise the major source rock for Claymore. These mudstones are mature in the off-structure areas to the northwest, northeast and south of the Claymore high (Fig. 1). Maturity was initially reached in the Eocene, with maximum expulsion occurring probably in the Miocene. Migration paths were directly updip towards the structure in Jurassic sandstone carrier beds, and via fault conduits into older and younger reservoir rocks.

Hydrocarbons

Test results of the discovery wells and production volumes for Claymore are discussed in the History section above. The statistics of the reservoir conditions, hydrocarbons and formation waters are presented in Table 3.

Reserves/resources

The Claymore Field contained 1452.9 MMBBL of oil originally in place and estimated ultimate proved recoverable reserves of 511.0

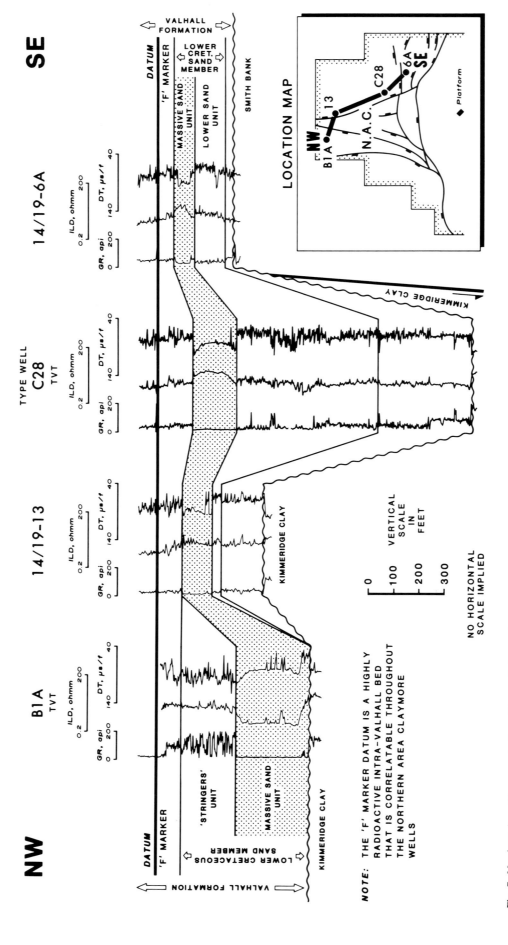

Fig. 7. Northern Area Claymore stratigraphic correlation section.

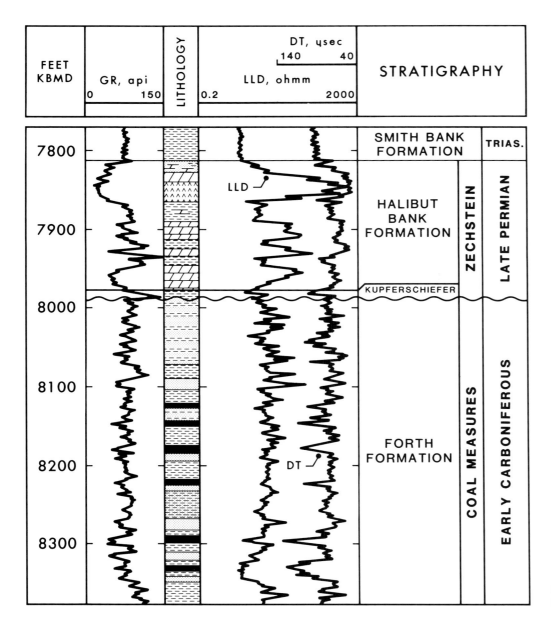

Fig. 8. Central Area Claymore type well 14/19-1.

Table 3. *Claymore fluid parameters*

	Main Area Jurassic	Northern Area Cretaceous	Central Area Perm.	Central Area Carb.
Reservoir				
Datum (ft ss)	8300	8900	8300	8300
Temperature (°F)	170	190	175	175
Pressure (psi)	3785	4080	3915	3915
Oil/water (ft ss)	8655	9400	8655	8655
Hydrocarbons				
Oil gravity (°API)	28.5	33.8	32.8	31.7
Gas/oil ratio (stf^3/STB)	120	408	325	302
Bubble point (psi)	680	1550	1194	1270
Formation volume factor	1.097	1.230	1.236	1.153
Viscosity (cp)	2.5–7.0	1.09	1.35	1.27
Formation waters				
Salinity (Cl ppm)	49 630	53 400	41 040	64 345
Resistivity (ohm-m)	0.045	0.039	0.053	0.036

MMBBL of oil. A maximum vertical oil column of 1003 ft has been proved in Main Area Claymore. The breakdown of these volumes into the Main, Northern and Central Area reservoirs is presented in Table 1. Aquifer support for Claymore is generally poor and is supplemented by flank water-injection drive. Production problems, particularly in Main Area, have been related to variable pressure depletion and uneven water advance and sweep across the field (Chen *et al.* 1987). These phenomena result from the effects of partial transmissibility barriers of stratigraphic and structural origin. Minor faults, with throws of only 30 ft, have been shown to be mineralized and able to withstand pressure differentials of 1000 psi within the reservoir (Harker *et al.* 1990). The correction of these problems is addressed by the integration of geological, geophysical and engineering disciplines. The combined knowledge is used to judiciously locate and selectively complete both production and injection wells.

The writers wish to thank Occidental Management and their Claymore operating group partners for permission to publish this paper. The interpretations expressed in this paper are entirely the responsibility of the authors. In addition, we wish to thank A. Downton and his drafting group for their aid in completing this paper.

References

BOOTE, D. R. D. & GUSTAV, S. H. 1987. Evolving depositional systems within an active rift, Witch Ground Graben, North Sea. *In*: BROOKS, J. & GLENNIE, K. (eds) *Petroleum Geology of Northwest Europe*, Graham & Trotman, 819–833.

CHEN, H. K., ROBINSON, T., HARKER, S. D. & MAHER, C. E. 1987. The Main Area Claymore Reservoir, a review of geology and reservoir management. *S.P.E Offshore Europe 87*, **2**, 16556/1-10.

HARKER, S. D. & MAHER, C. E. 1988. Late Jurassic sedimentation and tectonics, Main Area Claymore reservoir, North Sea. Giant Oil and Gas Fields. *S.E.P.M. Core Workshop* No. 12, 395–458.

——, GUSTAV, S. H. & RILEY, L. A. 1987. Triassic to Cenomanian Stratigraphy of the Witch Ground Graben. *In*: BROOKS, J. & GLENNIE, K. (eds) *Petroleum Geology of Northwest Europe*, Graham & Trotman 809–818.

——, MCGANN, G. J., BOURKE, L. T. & ADAMS, J. T. 1990. Methodology of Formation MicroScanner image interpretation in Claymore and Scapa fields (North Sea): *In*: HURST, A., LOVELL, M. A. & MORTON, A. (eds) *Geological Applications of Wireline Logs*. Geological Society, London, Special Publication, **48**, 11–25.

MAHER, C. E. & HARKER, S. D. 1987. Claymore Oil Field. *In*: BROOKS, J. & GLENNIE, K. (eds), *Petroleum Geology of Northwest Europe*. Graham & Trotman, 835–845.

The Clyde Field, Block 30/17b, UK North Sea

D. A. STEVENS[1] & R. J. WALLIS

BP Exploration, Fairburn Industrial Estate, Dyce, Aberdeen AB2 0PB, UK
[1] *Present address: Shell UK, Exploration and Production, 1 Altens Farm Road, Nigg, Aberdeen AB9 2HY, UK*

Abstract: The Clyde Field, which was discovered in 1978, is located on the SW edge of the North Sea Central Graben. The reservoir is developed with Late Jurassic shallow marine sands of the Fulmar Sand Formation. An estimated 408 MMBBL of oil is present (Annex B), of which 154 MMBBL is considered recoverable.

The structure of the Clyde Field takes the form of a rotated Jurassic fault block, truncated at its crest by a major unconformity. Oil is retained within a combination trap, sourced from Late Jurassic Kimmeridge Clay thermally matured in the highly productive basinal lows, adjacent to the field.

Reservoir sand quality is highly variable, ranging from excellent with permeabilities in excess of 1d, to poor with permeabilities of less than 1 md. The principal control on reservoir quality appears to be original depositional texture, although strong diagenetic effects are also present.

Production is from a single, centrally located, platform provided with thirty slots. Aquifer support is insufficient to maintain reservoir pressure at the current plateau production rate of 50 000 BOPD and so a programme of water injection has been implemented.

The Clyde field is located 290 km (180 miles) southeast of Aberdeen in the British sector of the North Sea (Fig. 1). It lies on the SW margin of the Central Graben within Block 30/17b. The block contains three known accumulations, of which only the largest structure (Beta) has so far been developed (Fig. 2). The field is operated by British Petroleum on behalf of Shell UK Ltd and Esso Exploration and Production UK Ltd. Britoil as part of British Petroleum holds 51% equity, Shell and Esso 24.5% each.

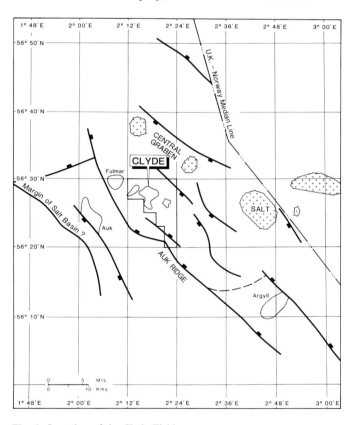

Fig. 1. Location of the Clyde Field.

The reservoir comprises Upper Jurassic shallow marine deposits in a combined structural and stratigraphic trap. STOIIP is currently estimated at 408 MMBBL, of which 154 MMBBL (38%) is considered recoverable. Production is via a single, centrally located, thirty-slot platform from which oil is evacuated via the Fulmar Field Floating Storage Unit, and gas by the Fulmar Field–St Fergus pipeline.

Prior to the incorporation of Britoil plc into British Petroleum plc, it was policy to name newly discovered Britoil fields after Scottish rivers. The name Clyde was selected not only in deference to alphabetical progression, but also in recognition of Britoil's close association with the Scottish City of Glasgow.

History

Block 30/17 was originally awarded to Gulf in 1964 following the 1st UK Licensing Round. Gulf then relinquished the eastern portion of the block that now forms part block 30/17b in 1970. The relinquished acreage was subsequently awarded to the BNOC/Shell UK/Esso Exploration and Production UK Ltd in 1977 during the 5th UK Licensing Round. BNOC was operator for the acreage under the privatized name of Britoil prior to acquisition by British Petroleum plc in 1988.

The initial seismic data set over block 30/17b comprised 363.5 km (226 miles), of 2-D data. Early interpretation revealed the existence of three structures at base Cretaceous level (Fig. 2). In March 1978 well 30/17b-2 was drilled as a crestal test of the largest of these structures, Beta, and discovered a 515 ft gross oil column within the Late Jurassic Fulmar Sand Formation. On test a maximum flow rate of 4974 BOPD was achieved from two horizons.

The 30/17b-2 discovery well was followed in December 1978 by appraisal well 30/17b-3 (Fig. 3). This was drilled to delineate the southern edge of mapped closure on the Beta structure and encountered a 179 ft gross oil column in Fulmar sands marginally thinner and poorer in quality than seen on the crest.

The next well, 30/17b-5, was drilled to evaluate the separate Gamma structure located 2 miles to the east of the main Beta accumulation (Fig. 2). The well, spudded in January 1979, encountered a 291 ft gross oil column in the Fulmar Sand Formation. However, reservoir quality proved to be relatively poor and pressure data suggested isolation from the main Beta accumulation.

In 1979, an extensive 3-D seismic survey grid was shot over most of block 30/17b to provide better definition of the fault pattern and more reliable correlation of intra-reservoir sands.

In July 1980, well 30/17b-7 was spudded to appraise the northeast of the Beta accumulation which penetrated Fulmar Sands uncomformably overlain by Late Cretaceous Chalk. A gross pay section of 431 ft was encountered in good-quality sands above an oil–water contact consistent with that in the nearby 30/17b-2

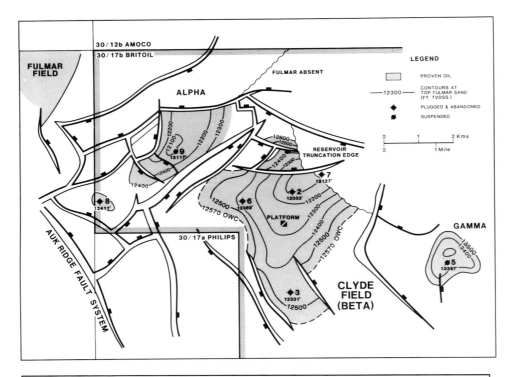

Fig. 2. Clyde oil accumulations and position of appraisal wells.

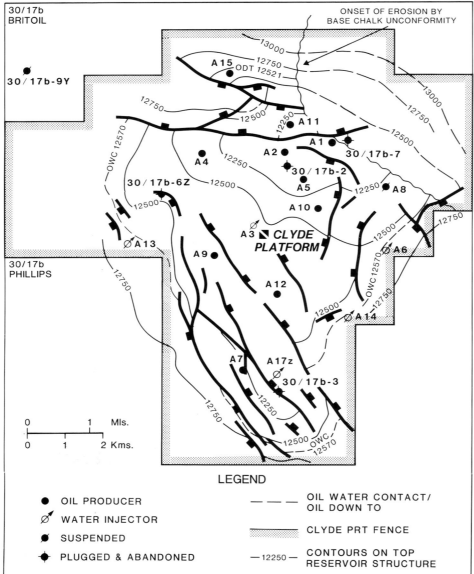

Fig. 3. Clyde Field structural elements, wells and platform locations.

discovery well. This well confirmed the existence of commercial reserves within the field and allowed development planning to proceed.

Wells 30/17b-8 and -9 were drilled in mid-1981 and mid-1983 respectively on separate culminations within the Alpha structure to the west of the main Beta accumulation. Although both encountered significant oil columns, reservoir quality and flow were considerably better in well 30/17b-9.

Department of Energy development approval for the Clyde Field was granted to the BNOC/Shell UK/Esso partnership in December 1982. Construction work on the platform started in January 1984, leading to jacket installation in July 1985 and topside addition in June 1986. After hook-up and commissioning, development drilling commenced on 2 January 1987, and first oil was achieved on 18 March 1987, six months ahead of schedule. To date 16 development wells have been drilled comprising 10 producers, 5 injectors and one suspended well (Fig. 3). Production has so far been restricted to the Beta structure. However, depending on the results of further appraisal drilling, it is also planned to develop the adjacent Alpha structure using Clyde facilities. No plans currently exist to develop the Gamma accumulation.

Field stratigraphy

The geological setting of the Clyde Field is best appreciated by reference to the sub-regional tectonic elements map, generalized stratigraphic column, and sections presented in Figs 1, 4, 5 and 6.

The dominant structural feature of the Clyde 'area' is a series of NW–SE trending en-echelon rotated fault terraces that step down from the Auk Ridge into the Central Graben to the east. Further into the basin faulting is synthetic to the deep structure and a number of halokinetic features are developed (Fig. 5). Two main phases of faulting are apparent on seismic sections, namely an Early Permian–Early Triassic phase and subsequent Late Jurassic to Early Cretaceous (Kimmerian) phase associated with the main

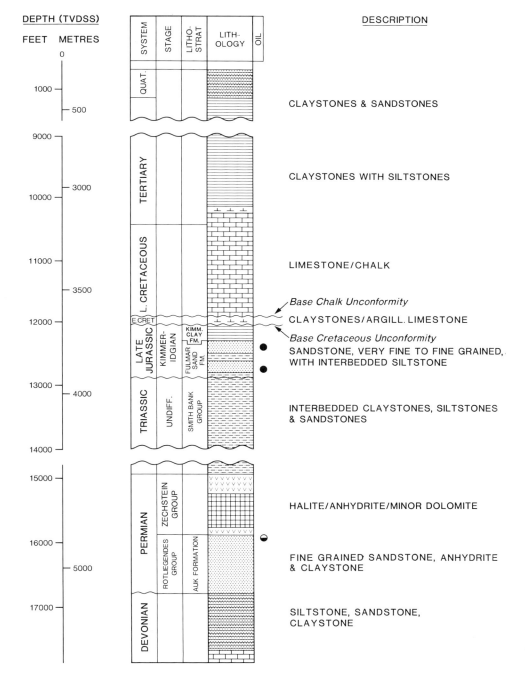

Fig. 4. Clyde Field generalized stratigraphic column.

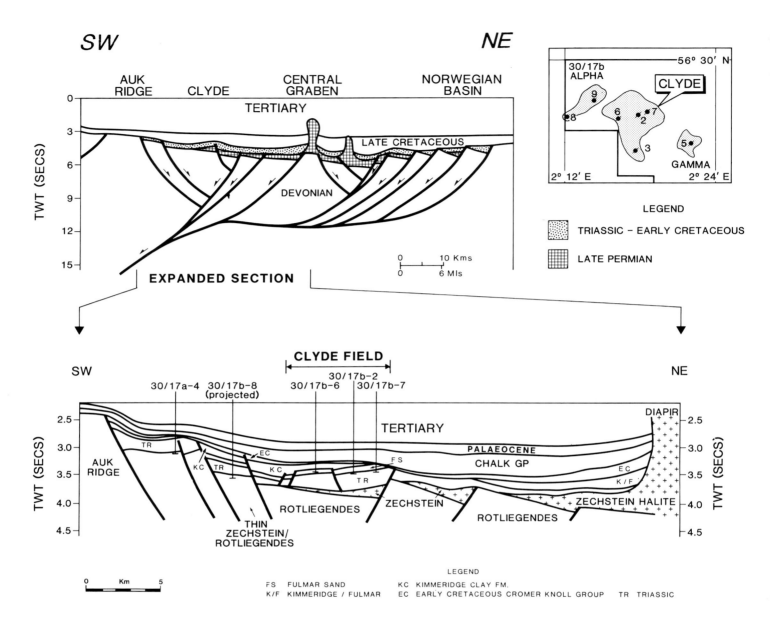

Fig. 5. Geological sections through Clyde Field.

period of basin extension. The majority of later movements appear to have occurred along reactivated Permo–Triassic structures, although a series of smaller, secondary, faults which generally detach at the level of the Zechstein evaporites are also developed locally.

The pre-Jurassic and Jurassic syn-rift sequence in the Clyde area consists of Devonian continental deposits overlain by Permian Rotliegendes Group sands and Zechstein Group evaporites of variable thickness. These are succeeded by continental red bed clastics of Triassic age unconformably overlain by Late Jurassic shallow marine sands belonging to the Fulmar Sand Formation. The overlying Kimmeridge Clay Formation and Early Cretaceous Cromer Knoll Group sediments comprise organically rich marine clays and argillaceous limestones respectively. The thickness of these post-reservoir intervals varies considerably over the block and it appears that syndepositional faulting was a significant local control on their distribution. The post-rift sequence comprises a thick sequence of onlapping Late Cretaceous Chalk, conformably overlain by Tertiary marine claystones.

Geophysics

The initial interpretation was based on 363.5 km (226 miles) of 2-D seismic data acquired as part of Shell, BNOC, and Spec surveys shot over block 30/17b between 1972 and 1978. The combination of these different surveys provided an areal coverage of approximately 1 km (0.6 mile) line spacing, sufficient to allow definition of the extent of the Clyde field and satellite structures.

In 1979 a 3-D seismic survey was shot over block 30/17b in order to better define fault trends and improve seismic resolution and correlation of the reservoir sands. A total of 205 lines was shot along a NE–SW orientation with a 75 m spacing, to form a complete survey of 2500 km (1553 miles). Reprocessing trials subsequently indicated that considerable improvement in data quality could be achieved, and on the basis of this the data was reprocessed in 1986 prior to the production of field development maps. Later development well results have indicated that the geometry of the intra-Fulmar sand bodies is considerably more complex than originally mapped and uncertainty remains over their

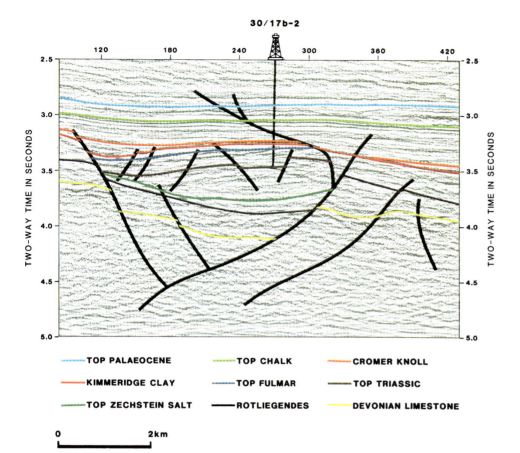

Fig. 6. Seismic section through Clyde Field.

geometry and interconnection. In particular, the dislocation of reflectors formerly interpreted as representing faulting of continuous sand horizons is now in many cases interpreted, by the operator, as the product of the lateral termination of individual, stacked, sand packages.

Trap

The Clyde structure comprises a rotated fault block underlain by a Zechstein salt wedge. The northeastern edge of the field has been truncated by crestal erosion beneath a major unconformity at the base of the Late Cretaceous Group (Figs 3, 5 and 6). Of particular interest is the lensoid geometry of the Jurassic–Triassic sediment package, and its relation to the present structural crest and underlying Zechstein wedge (Figs 5 and 6). The geometry has been interpreted as the product of tectonic inversion caused by halokinetic movements (Smith 1987, Fig. 7a), or gravitational sliding (Gibbs 1984, Fig. 7b). However, rotation of the terrace during late Jurassic extension and/or reverse faulting on the eastern boundary fault to the terrace (Fig. 5) can also account for the structure.

The Clyde Field is a combined structural and stratigraphic trap. Top seal is provided by the overlying Kimmeridge Clay Formation, and up-dip seal by the truncation of the Fulmar Formation beneath the Base Chalk unconformity. Down-dip, simple three way dip closure provides the trapping mechanism.

The oil–water contact for the main field has been identified at 12 570 ft TVDSS. This is higher than the currently mapped closure, indicating that the field is not full to spill point. There is no clear evidence for post-emplacement movement of the oil.

Reservoir

The reservoir of the Clyde Field is developed within the Fulmar Formation and is of Early Kimmeridgian age. The dominant

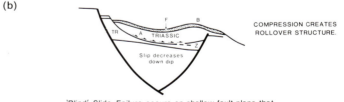

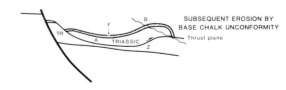

Fig. 7. Geological models describing the development of the Jurassic–Triassic sediment package in Clyde. (**a**) Halokinetic model (Smith 1987). (**b**) Gravity slide model (Gibbs 1984).

lithofacies are heavily bioturbated, arkosic, fine-grained, sandstones and siltstones deposited in a shallow marine shelf environment and probably derived from the adjacent Permo-Trias palaeotopography.

Appraisal and development drilling showed the field to comprise three coarsening upwards cycles termed the Upper, Middle and Lower Reservoir Units (Zones A, B and C respectively). The Upper Reservoir Unit thickens towards the west downflank from the crest; conversely the Middle and Lower Reservoir units thicken onto the crest of the field to the east.

The main characteristics of the reservoir zones are summarized below.

(i) *Upper/Middle Reservoir Sand Units* (Zones A and B). The sandstones within these zones exhibit relatively uniform porosities of between 20–25%, but highly variable permeabilities which can range from 0.1 md to over 1000 md. The higher permeabilities generally occur over intervals 20–30 ft thick, but production data indicate that they are variable in extent (Fig. 8).

(ii) *Lower Reservoir Sand Unit* (Zone C). These sandstones are of uniformly excellent quality across the field, with an average porosity of 23% and average permeabilities in the range of 400–500 md (Fig. 8).

Reservoir quality appears to be predominantly facies controlled (grain size and detrital clay content), with subordinate diagenetic control (dolomite and microcrystalline quartz cements).

The variation in facies, and hence reservoir quality, is believed to be palaeobathymetrically controlled. The Fulmar sandstones were deposited on a broad shallow marine shelf that deepened to the northeast. Each reservoir sand unit represents a prograde across this shelf subsequent to a transgression at the base of the reservoir unit. Poorest quality sands and silts were deposited in the deepest water at the lowest part of the reservoir unit; as the sands built up so reservoir quality increased until they were being deposited at or above medium wave base, where the best quality sands were deposited. No sands are believed to have been deposited above the mean low water line (foreshore to backshore).

Each coarsening upward prograde was deposited in successively more sediment starved, or deeper, waters. This resulted in the decreasing reservoir quality observed from the lower to the upper reservoir units.

Laterally, facies variation is caused by the complex interplay of shelf hydrodynamics, sediment supply and especially syndepositional fault movements/basin subsidence due to underlying salt movements. The Lower Reservoir unit prograded from a high in the west into deeper water filling up the lower topography in the east and hence thickening in that direction. A similar depositional pattern is found in the Middle Reservoir Unit. In the Upper Reservoir unit, salt removal from underneath the western part of the field created a relative low in the west and hence this unit thickens in that direction.

Siliceous sponge spicules are common, particularly in the upper reservoir units. Spicule dissolution frequently results in the creation

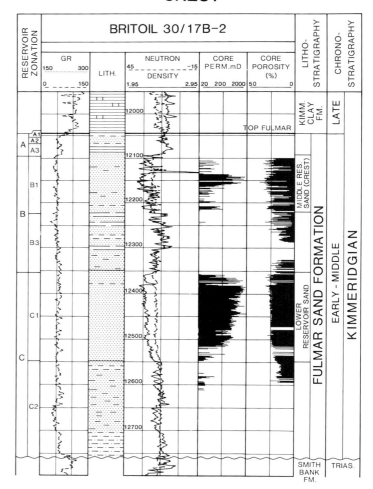

Fig. 8. Reservoir zonation and reservoir properties of the Clyde Field.

of discrete secondary porosity with little or no permeability enhancement. Silica reprecipitation occurs as nodular, horizon constrained or pervasive microcrystalline quartz with resultant high microporosity and low permeability. Silica is less commonly precipitated as syntaxial quartz overgrowths with good permeability characteristics occurring mostly in the coarser cleaner sandstones. Dolomite also constitutes a significant pore filling or replacive authigenic phase and does not appear to be lithofacies constrained.

Source

Clyde oil was sourced from the Late Jurassic Kimmeridge Clay Formation. The majority of the oil in Clyde was probably generated off structure in adjacent structural lows such as the highly productive Josephine Graben. The latter area generated oil in the Early to Middle Miocene when the Kimmeridge Clay Formation achieved a depth of burial of 9500–10 000 ft. The relative immaturity of Clyde oil reflects this early to middle generation phase, and suggests that later, more mature hydrocarbon phases bypassed the field (Robertson Research International 1988).

Hydrocarbons

Clyde oil is undersaturated in all reservoir zones. The Zones B and C have the same oil type with an API of 38°, GOR of 490 SCF/BBL, and a bubble point of 2370 psia. Zone A has an API of 37°, GOR of 340 SCF/BBL, and a bubble point of 1722 psia. The reservoir temperature is hot at 297°F, and the reservoir is overpressured with an initial pressure of 6458 psia at 12 300 ft TVSS.

The mobility ratio is very favourable, being slightly less than 1 and high recovery factors are expected in the good quality Zone C sands.

Zone C producers have a good production rate of over 23 000 BOPD with PIs in excess of 25 BOPD/psi. The Zone A and B producers have poorer performance due to the poor quality sand. Typical rates and PIs are 5000 BOPD with a PI of 3–4 BOPD/psi for Zone B and 2000 BOPD with a PI of 1–2 BOPD/psi for Zone A.

Reserves

Aquifer support in the Clyde field is insufficient to maintain reservoir pressure at the present production rate of 50 000 BOPD. Consequently, a programme of water injection has been implemented.

Recovery factors will vary considerably between the reservoir zones. Reservoir Zone C is expected to have a high recovery factor of around 55% due to good quality sandstone, favourable permeability distribution which improves up in the sequence, and good apparent lateral continuity. Water injection is proving successful in recouping and maintaining pressures.

Reservoir Zone B will have a lower recovery due to the poorer quality sandstone, less favourable permeability distribution, and restricted drainage area of the wells.

Reservoir Zone A possesses the poorest quality sandstone, and this makes pressure support difficult. The wells typically come on stream in excess of 3500 BOPD, but quickly decline to rates of less than 2500 BOPD. As a consequence, Reservoir Zone A production is being accelerated to maximize recovery over field life. This is being achieved by concentating on Reservoir Zone A development by drilling producer–injector pairs and by introducing a low pressure manifold to reduce the FTHP.

Artificial lift by gas is also being considered for both Zone A and B producers, along with fracturing of Zone A producers.

This paper represents a very brief synthesis of work conducted by a number of people over a number of years. However, we would particularly like to thank N. H. Allen, P. D. Begg, D. G. Barr, M. I. A. Pye, C. E. A. Reddick, R. Smith and G. Yielding for their technical contributions, and P. J. Turner for his comments on the manuscript. In addition, we would like to thank Shell and Esso representatives for their permission to publish this work and for many stimulating discussions over the last three years. Special appreciation is also extended to S. Gowland et al. of Geochem Poroperm Ltd for their contribution to our understanding of the sedimentological and petrographic history of the Clyde reservoir.

Clyde Field data summary

Trap
Type	Combination structural/stratigraphic
Depth to crest	12 050 ft TVSS
Oil–water contact	12 570 ft TVSS
Gross oil column	520 ft
Net area	2973 Acres

Pay zone
Formation	Fulmar Formation
Age	Early Kimmeridgian
Average gross thickness	650 ft
Net/gross range/average	50–80%/60%
Porosity range/average	5–30%/20%
Average water saturation	25%
Permeability range	1–1000+ md

Hydrocarbons
Oil gravity	37° API (Zone A), 38° API (Zones B/C)
Oil type	Light undersaturated oil
Gas/oil ratio	340 SCF/BBL (Zone A), 490 SCF/BBL (Zones B/C)
Viscosity	0.57 cP (Zone A), 0.49 cP (Zones B/C)
Bubble point (297°)	1722 psia (Zone A), 2370 psia (Zones B/C)
Formation volume factor (Boi)	1.24 (Zone A), 1.34 (Zones B/C)

Formation water
Water gradient	0.445 psi/ft
Salinity	110 000 ppm chlorides
Resistivity	0.016 ohm m at 297°F
Other ions	Suphate (300 ppm), sodium (55 000 ppm), potassium (3500 ppm), calcium (8000 ppm)

Reservoir conditions
Average temperature	297°F
Average pressure	6458 psia at 12 300 ft TVSS (initial)
Pressure gradient in reservoir	0.9 psi/ft

Field size
Area	2773 acres
Recovery factor	38% (average)
Recoverable oil	154 MMBBL
Drive mechanism	Natural depletion then water injection

Production
Start-up date	First production 8th March 1987
Development scheme	1 central platform
Number/type of wells	12 oil producers/11 water injectors planned
Average production rate	50 000 BOPD
Cumulative production	At end 1988: 30.7 MMBBL
Secondary recovery method(s)	Gas lift/jet pumping

References

GIBBS, A. D. 1984. Clyde field growth fault secondary detachment above basement faults in the North Sea. *Bulletin of the American Association of Petroleum Geologists*, **68**, 1029–1039.

SMITH, R. L. 1987. The structural development of the Clyde field. *In*: BROOKS, J. & GLENNIE, K. (eds) *Petroleum geology of North West Europe*. Graham & Trotman, London, 523–531.

ROBERTSON RESEARCH INTERNATIONAL LTD. 1988. *Central Graben North Sea: Stratigraphy, Structure and Petroleum Geochemistry of the Lower Cretaceous, Jurassic and Older Strata*. Robertson Research International Ltd in association with Scott, Pickford and Associates.

The Crawford Field, Block 9/28a, UK North Sea

A. YALIZ

Hamilton Brothers Oil and Gas Ltd, Devonshire House, Mayfair Place, London W1X 6AQ, UK

Abstract: The Crawford Field was discovered in 1975 in UK Block 9/28 and the first oil was produced in April 1989. The field has a complex structural history. The reservoir is located on a down-faulted, westward tilting faultblock along the western margin of the Viking Graben. The eastern margin of the faultblock is severely truncated at Base Cretaceous level. The main producing zones comprise Middle Jurassic (Brent Group equivalent) and Triassic (Skagerrak Formation) sandstones. The seal is formed by Cretaceous marls and limestones. Reservoir quality and thickness are extremely variable, and drainage areas are limited. The reservoir fluid is a medium gravity oil having a thin gas cap. Oil in-place is in the order of 130 MMBBL but recovery factors are expected to be low.

The Crawford Field is located in North Sea Block 9/28a, 180 miles northeast of Aberdeen in water depths of 385 ft. The field trends NE–SW on a terrace structure along the western margin of the Viking Graben (Fig. 1).

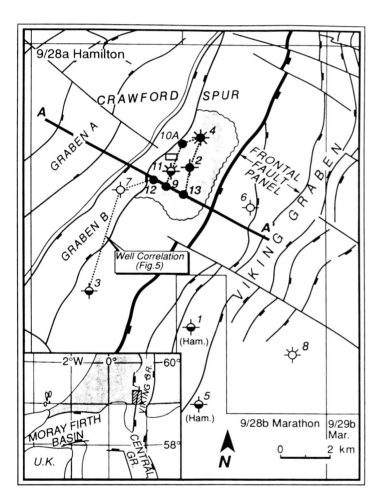

Fig. 1. Structural elements and well locations of the Crawford Field. The inset shows the field location. Line A–A' represents the cross-section in Fig. 4, and the dashed line represents the correlations shown in Fig. 5.

Oil is produced from Middle Jurassic and Triassic sandstones. The combined maximum thickness of the sediments reach 1000 ft in the western part of the field. However, the reservoir is much reduced along the eastern margins due to severe truncation at Base Cretaceous level.

Production of the Crawfield Field is based on a phased development plan. The oil is produced using the floating production platform, Northsea Pioneer.

In common with all other Hamilton Brothers Oil and Gas fields the Crawford Oilfield was named after a Scottish clan.

History

Block 9/28a was awarded in March 1972 as part of the Fourth Round Licence P209. At present the licence is held by the following group of companies:

Hamilton Oil GB plc	22.33333%
Hamilton Brothers Petroleum (UK) Ltd	14.00000%
BP Petroleum Development Ltd	30.33333%
Elf UK plc	23.33334%
Blackfriars Oil and Gas Ltd (Ultramar)	8.33334%
Renown Petroleum Ltd	1.66666%

The block is operated by Hamilton Brothers Oil and Gas Ltd.

The field was discovered in 1975 with well 9/28-2 which was drilled to test a terrace structure along the western margin of the Viking Graben (Fig. 1). The well proved producible oil in a 752 ft gross reservoir interval of Triassic age. Maximum flow rate was 2610 BOPD. The Rotliegendes was also tested but no hydrocarbons were produced to surface.

A second well, 9/28-3, was drilled in 1975 on the extreme southern edge of the prospect, but no Jurassic or Triassic rocks were penetrated. In 1977, a further appraisal well, 9/28-4, was drilled to the north of the discovery well and tested oil from both Jurassic and Triassic reservoir intervals with a maximum flow rate of 1750 BOPD.

Well 9/28a-7 was drilled in 1984 on the western side of a small graben structure (Graben B) to the west of the field (Fig. 1). No hydrocarbons were found and the well was plugged and abandoned. This was followed in 1987 by the successful 9/28a-9 well, located to the south of the discovery well. A Middle Jurassic reservoir interval tested a maximum of 3500 BOPD.

A draft Annex B was prepared in early 1988 and a further appraisal stage began with well 9/28a-10A in the northwest corner of the field. This well encountered highly productive Jurassic marine sandstones which flowed at a maximum rate of 7650 BOPD. The Triassic was non-productive.

Three further development wells followed in quick succession. Well 9/28a-11 had minor oil and gas shows and was suspended without testing. The Annex B document was updated and submitted for production consent in September 1988.

In the meantime well 9/28a-12 was drilled to the west of well 9/28a-9. The final well, 9/28a-13, located in the southeast part of the field proved a Cretaceous section immediately overlying Triassic reservoir with Jurassic missing. The Triassic was tested at a maximum rate of 6850 BOPD.

After completion of facilities installation, the first oil from the Crawford Field was produced on 1 April 1989 from Middle Jurassic and Triassic reservoirs.

Field stratigraphy

The oldest sediments penetrated in the Crawford area are of Lower Permian age (Fig. 2).

The field is located on the Crawford Spur which is a northerly extension of the Fladen Ground Spur (Fig. 1). The oldest sediments comprise Rotliegend aeolian and sabkha deposits which attain a thickness of at least 1250 ft (well 9/28-3). They are overlain by a Zechstein marginal marine clastic sequence which thickens towards the west.

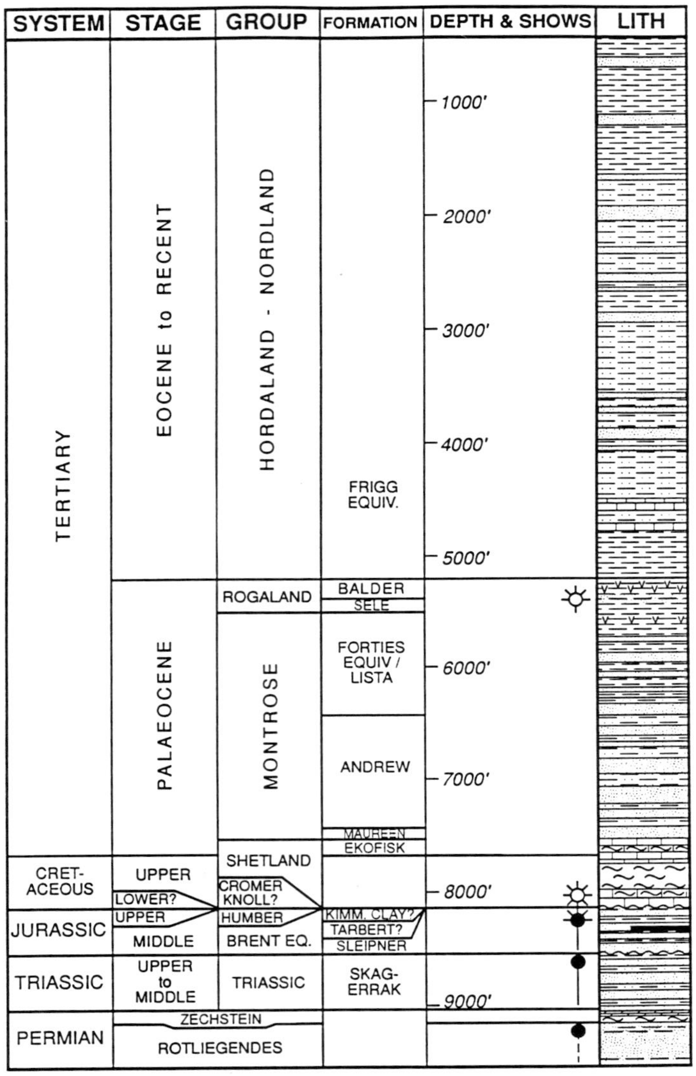

Fig. 2. Crawford Field generalized stratigraphic column.

The early Cimmerian extensional tectonic phase during the Late Triassic led to the formation of a number of small fault blocks. At this stage extensive Triassic erosion occurred over structurally high areas possibly followed by deposition of Lower Jurassic sediments of the Statfjord Formation, which were not preserved in the area due to subsequent uplift and erosion.

The Middle Jurassic sediments of the Crawford Field are assigned to the Brent Group equivalent. After the Lower Bajocian tectonic pulse the area was blanketed by coastal plain deposits (Sleipner Formation equivalent) with the main drainage taking place towards the nearby Beryl Embayment. During Upper Bajocian–Lower Bathonian a southwesterly marine transgression deposited marine facies (Tarbert Formation equivalent) which have been preserved only as a thin layer as a result of Cretaceous truncation.

During the Oxfordian, movement of the graben-edge faults caused the erosion of the Crawford Spur and the deposition of thick Upper Jurassic sediments in the Viking Graben. The Upper Jurassic was found only in well 9/28a-10A as a very thin dark shale of Lower Ryazanian/Upper Volgian age (Kimmeridge Clay Formation).

The Triassic and Jurassic are overlain unconformably by Lower or Upper Cretaceous sediments as a result of onlapping from the east and northeast.

During the early Tertiary the entire area was covered by complex marine deposits of the Ekofisk Formation equivalent, followed by the deposition of submarine fans of Palaeocene age. After the deposition of the Balder Formation the area was again dominated by marine sedimentation.

Geophysics

The seismic data base consists of 1600 km of data. The Crawford field mapping was based on the Hamilton 1980 and 1987 surveys. The re-processed AUK-77 data have also been used as a guide to pick the Top Zechstein event in the west. Further seismic lines totalling 80 km were shot in 1988. These are now being incorporated into the field interpretation.

Synthetic seismographs were produced for five wells and used in the calibration of seismic data. Six seismic markers were selected for interpretation: Top Balder, Near Top Danian, Base Cretaceous, Middle Jurassic Coals, Near Top Triassic, Top Zechstein.

Three of these horizons, Top Balder, Base Cretaceous and Top Zechstein, were mapped using average velocity maps for depth conversion. The average velocity maps were based mainly on pseudo-velocities computed using seismic times at well control points. The time and velocity maps were digitized, gridded and then combined to produce depth maps.

The Base Cretaceous is a well-defined event representing the top of the Middle Jurassic or Triassic reservoirs. The structural configuration of this unconformity surface is characterized by a steep ESE dip representing an erosional surface on westerly dipping fault blocks (Figs 3 & 4). The surface dips to the west into Graben B on the west flank of the field. The eastern edge of the field is defined by the Base Cretaceous unconformity which severely truncates the reservoir. Over most of the area, faulting at Base Cretaceous level is minor or absent.

The Top Zechstein is a strong reflector and defines the fault pattern affecting the Triassic and possibly the Jurassic sequences. It also provides constraints on the thickness of the Jurassic/Triassic section. The Zechstein map shows the Crawford Spur to be bounded by NNE–SSW-trending normal faults parallel to the South Viking Graben axis. The down-to-the-east faulting displays steep planes which become listric at depth. Complementary antithetic faults were formed resulting in the development of Graben B structure to the west of the field.

Trap

The Crawford Field is considered to be a down-faulted terrace on which the reservoir section has been preserved by pre- and early Cretaceous faulting. The field is interpreted as being completely closed with a dip and fault closure to the northeast and dip closure to the southeast. The boundary to the west is interpreted to be a sealing fault separating the field from Graben B.

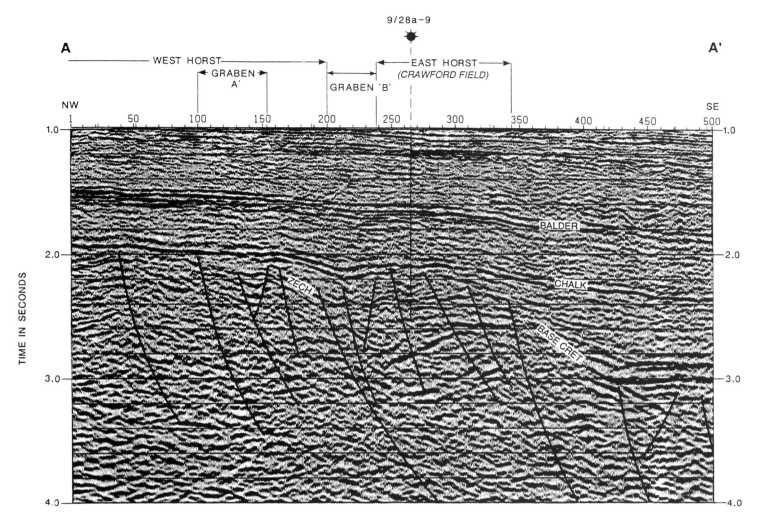

Fig. 3. A NW–SE seismic line across the Crawford Field.

The gross structure of the field can be described as a rotated west-dipping horst block bounded by a major set of faults trending in a NNE–SSW direction. The horst is truncated along its eastern edge. The vertical seal to the reservoir is provided by Cretaceous marls which overlap unconformably over the Triassic and Jurassic reservoirs.

The reservoir has an elongated geometry ($3\frac{1}{2} \times 1\frac{1}{2}$ miles) with its long axis parallel to the Viking Graben axis. The gross reservoir thickness varies from a few feet to over a thousand feet. The reservoir lithologies are Middle Jurassic and Triassic sandstones.

The field is split into distinct fault panels mapped at Top Zechstein level. The faults are thought to extend up into the reservoir, although very little faulting is seen at the Base Cretaceous level. However, from extensive fracture studies it is evident that the Triassic is heavily fractured.

There is no apparent consistency in the observed oil–water contacts between the wells and therefore no single field-wide contact can be defined. For example, in well 9/28-4 the contact is seen at 8366 ft SS whereas in well 9/28a-13 it is defined at 8762 ft SS. In the volumetric calculations different contacts were used within different fault blocks.

A gas–oil contact is also seen in three central wells at varying depths between 7993 ft SS and 8064 ft SS.

Reservoir

The Crawford reservoir consists of lithologically and depositionally distinct Middle Jurassic and Triassic intervals. The predominantly arenaceous red and green deposits of Triassic age have been assigned to the Skagerrak Formation and subdivided informally into 'lower' and 'upper' units. The Middle Jurassic interval consists of coastal alluvial plain and overlying marine deposits which have been assigned to Sleipner and Tarbert Formations, respectively.

Triassic

Lithostratigraphy. The Skagerrak Formation is subdivided into five facies types.

(1) *Flat-bedded fluvial channel-fill sandstones* are dominated by horizontal or low-angle lamination, with rare horizons of aligned intraformational granules. Grain size is predominantly fine to medium sand grade. Deposition probably resulted from ephemeral high power stream flows.

(2) *Dune-bedded fluvial channel-fills* are characterized by sandstones with planar and trough cross-stratifications. Granules of claystones and calcrete clasts frequently lag basal erosional surfaces, often showing pebble imbrication. The sandstones frequently show an upward fining grain-size from a pebbly base to fine sand grade at the top. Minor amounts of claystone occur locally between the channel-fill sandstones. There are also thin and very fine silty sandstone horizons rich in mica.

Channel-fill sandstones show considerable variation in bed thickness from well to well. The thickness of the stacked sequences also varies dramatically as in well 9/28a-10a (20 ft), 9/28a-9 (320 ft) and 9/28a-13 (600+ ft).

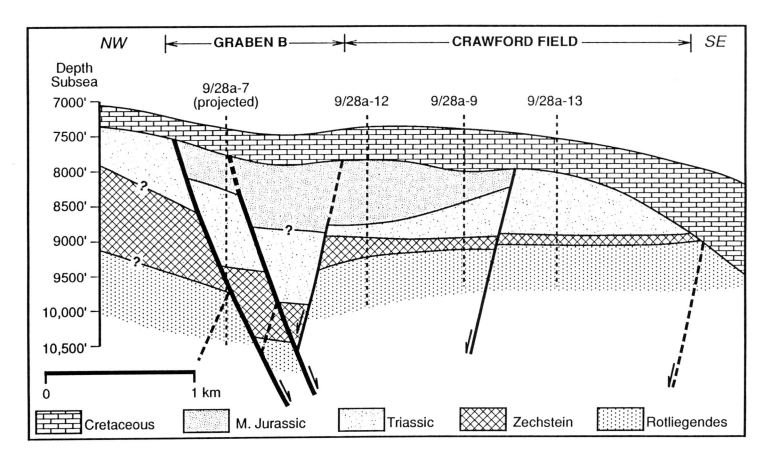

Fig. 4. Structural cross-section through Crawford Field, along line A–A' in Fig. 1.

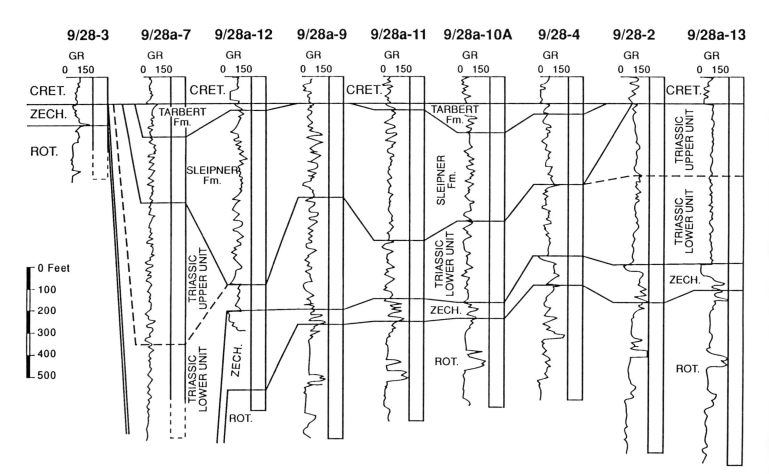

Fig. 5. Correlation of the Middle Jurassic, Triassic and Permian intervals of the Crawford Field wells.

(3) *Sheetflood Sandstones* form small scale (generally less than 1 ft thick) fining upward sequences with siltstones and argillaceous material at the top. They have sharp bases and the sedimentary structures are dominated by horizontal and sub-parallel lamination which grades vertically into silt drapes suggesting a relatively rapid waning in flow energy.

(4) *Lacustrine deposits* consist of sandstones, siltstones and claystones characterized by high levels of synsedimentary deformation and bioturbation.

(5) *Pedogenic calcrete deposits* are characterized by red/brown/green mottled argillaceous sandstones with abundant nodular carbonate cements separating early diagenetic precipitates.

The Skagerrak Formation has proved to be barren of any age diagnostic palynoflora, except in well 9/28a-4 where an Upper Triassic (Carnian) miospore was found.

Depositional model. There are significant variations in the total Triassic thickness from well to well, ranging from 119 ft in 9/28a-12 to 1520 ft in 9/28a-7 (Fig. 5). Because of this thickness variation and lack of biostratigraphic data the correlation of events within the Triassic has been unsuccessful. However, on the basis of sedimentological data the Triassic is divided into two units of contrasting depositional style and sedimentary characteristics.

The '*lower unit*' was deposited in a semi arid climate in a continental alluvial plain environment traversed by fluvial channels containing dune/bar scale bedforms. The sediment transport direction was towards the west-northwest. Interfluve regions were areas of sheetfloods formed only during sudden rainstorms. Areas of the floodplain receiving no sedimentation formed stable surfaces where mature calcretes were developed.

The '*upper unit*' is characterized by much coarser-grained channel-fill sandstones which were deposited on a braided plain. The channels were relatively deep with thick developments of dune-scale bar bedforms. The lateral channel migration was quite rapid and thus overbank deposits were rarely preserved. Dipmeter and core goniometry data again indicate a WNW-dipping palaeoslope.

The 'upper' unit has been preserved mainly on the eastern part of the field and attains a thickness of 363 ft in well 9/28a-13. The Triassic intervals seen in the western wells consist chiefly of the 'lower' unit, because the 'upper' unit has been eroded. Prior to this phase of erosion, the Triassic units were probably ubiquitous throughout the area.

Reservoir characteristics. Within the 'lower' Triassic unit the main potential reservoir sandstones are those of dune-bedded fluvial channel origin. However, laterally extensive mature calcretes with carbonate cements may give rise to horizontal stratification and compartmentalization of the sand bodies.

Log calculated porosities of the reservoir sandstones from the 'lower' unit vary from 19% to 26% but the net/gross ratios are extremely low. Permeability values vary even more widely from 0.01 md to around 2500 md.

The 'upper' Triassic unit lithofacies types are exclusively dune-bedded fluvial channel sandstones which, due to their stacked nature, form essentially a single sandstone body of considerable thickness. The size and distribution of the 'upper' unit reservoir is governed by the extent of post-Triassic erosion. Although the 'upper' unit is a much better reservoir than the 'lower' unit, reservoir heterogeneities may exist due to channel shaped surfaces, large-scale cross-stratification, relict clay plugs, channel linings, pore lining chlorites and extensive clay-filled micro-fractures.

The 'upper' unit is only preserved in wells 9/28-2 and 9/28a-13. Log analyses provide a range of porosities between 20% and 23%. Air permeabilities show a wide range of 0.07–2000 md, but most of the values exceed a few hundred millidarcy. However, the brine permeabilities are in the order of several tens of md.

Middle Jurassic

Lithostratigraphy. The Middle Jurassic sediments are subdivided into seven facies types, five in the Sleipner Formation and two in the Tarbert Formation.

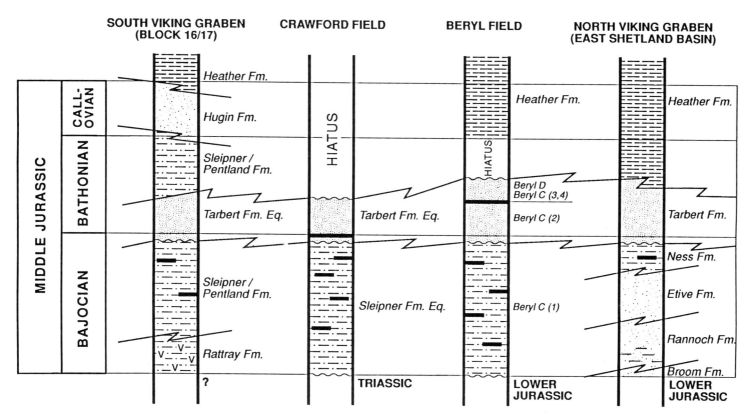

Fig. 6. Relationship of the Middle Jurassic stratigraphy of the Crawford Field to other areas in the Viking Graben.

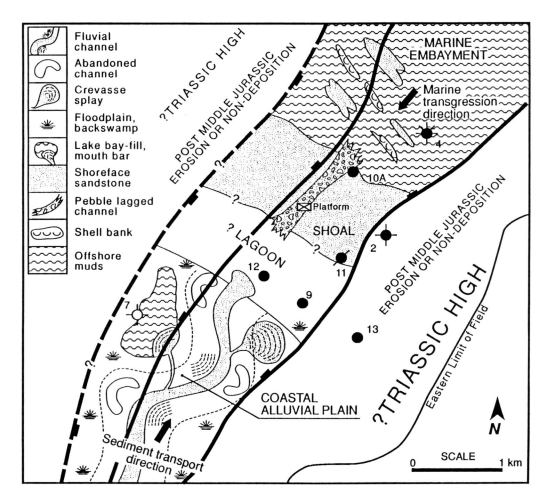

Fig. 7. Idealized Late Bajocian palaeogeographic model for the Tarbert and Sleipner Formations in the Crawford Field.

(1) *Fluvial channel-fill sandstones* show fine to coarse sand grade and generally have erosive bases. Sedimentary structures include trough and planar cross-stratification, clay, silt and coaly drapes.

(2) *Overbank fluvial sandstones* resemble the fluvial channel-fill sandstone facies but are more argillaceous. Crevasse splay sandstones and levee deposits are included in this facies.

(3) *Interdistributary bay-fill deposits* comprise lacustrine and saline bay-fill sediments. They are generally claystones or argillaceous siltstones.

(4) *Overbank/floodplain deposits* have formed within the floodplain or interfluvial areas between the fluvial channels as a result of 'fall-out' of fines from suspension in floodwaters.

(5) *Lacustrine/saline bay delta front/mouth bar deposits* comprise fine-grained sandstones, siltstones and claystones often showing low angle/parallel lamination.

(6) *Shoreface/inner shelf sandstones* have an exclusively sandy lithology and are occasionally pebbly. The sediments represent depostion in an open marine shoreface environment or high energy channels which cut through the shoreface deposits.

(7) *Low energy muddy shelf deposits* are predominantly argillaceous and formed in a low energy marine environment. Siltstones and very fine-grained sandstones have abundant bivalve shell fragments. Shell bank deposits are a minor part of this facies. The shell fragments have varying degrees of marine affinity.

Biostratigraphy. In constrst to the Triassic reservoir, the Middle Jurassic sediments are rich in palynoflora. Based upon quantitative palynological data 12 correlative datums have been found within the Brent Group equivalent.

The onset of sedimentation was initiated during Middle Jurassic with the development of a northward draining non-marine, fluvial depositional regime. The Sleipner Formation equivalent in the Crawford area represents various palaeoenvironmental changes in the fluviatile/delta-top settings during Early–Mid-Bajocian times. At the same time the Sleipner Formation/Pentland Formation accumulated in the S. Viking Graben. The distal equivalents of this drainage system are represented by the Broom and Ness Formations in the N. Viking Graben (Fig. 6).

The palynofloras within the Sleipner Formation indicate an initial colonizing macroflora dominated by small tree ferns with no coal seams. A change in sedimentation patterns accompanied the establishment of peat swamp conditions and large tree ferns. Adjacent areas were characterized by crevasse splay and interdistributary bay facies.

The upper part of the Brent Group is composed of marginal marine Upper Bajocian–Bathonian sediments (the Tarbert Formation equivalent) and contains very characteristic marine microplankton suites. This unit can be accurately equated with the Tarbert Formation of the Brent Group in the N. Viking Graben and the 'Beryl Formation'/'Middle Beryl Sandstone' of the Beryl Embayment (Fig. 6). Based upon microfaunal data the upper sub-unit of the Tarbert Formation equivalent was deposited under a more marine environment than the underlying lower sub-unit. The upper sub-unit sandstones probably represent shoreface and shallow, subtidal settings, with occasional storm deposits, and are considered to have had reasonable potential for lateral continuity. The lower sub-unit of the Tarbert Formation is a heterogeneous sequence deposited in a variety of settings within a restricted coastal lagoon environment. Salinities were highly variable and the sub-unit represents the initial deposits of the marine transgression as the latter moved south over the Crawford field area. The lower sub-unit probably passed northwards into more open marine influenced, shelf-type sands.

Depositional model. The Sleipner Formation was deposited over an irregular highly eroded Triassic surface, and infilled depressions in the eroded surface with 'valley-fill' type fluvial channel-fill sandstones. The depositional surface soon became a low-relief plain, gently dipping towards the north. Differential subsidence occurred across the area but sedimentation rates kept pace with subsidence. An idealized palaeogeographic reconstruction of the area during the Late Bajocian is shown in Fig. 7.

The lower part of the Sleipner Formation is represented by floodplain deposits with crevasse splay sandstones which were traversed by fluvial channels. In the middle parts of the formation the supply of clastic detritus was limited and extensive development of peat took place resulting in thick coaly horizons. Towards the end of the deposition of the Sleipner Formation fluvial conditions returned to the area.

Biostratigraphic evidence suggests a minor unconformity between the Sleipner Formation and the overlying marine sediments of the Tarbert formation.

The Upper Bajocian to Bathonian Tarbert Formation represents deposition in a radically different setting. These shallow marine sediments comprise high energy storm sandstones and conglomerates and low energy claystones and sandstones which may have an interdigitating contact trending NW–SE (Fig. 7). The orientation of lamination in the sediments indicate a gradient towards the offshore region in the north or northeast. It is highly probable that the marine transgression was from the north-northeast in the direction of the Beryl Embayment.

Reservoir characteristics. Both the Sleipner and the Tarbert Formations possess good/excellent reservoir sandstones.

The Sleipner Formation sandstone bodies show highly complex geometries of isolated ribbon and tabular sand shapes of fluvial channel origin. The pore network is essentially open and relatively well connected. However, net/gross ratios are often very low reflecting the isolated nature of channelized bodies. Log calculated porosities vary from 18% to 21%. Permeabilities are often very good to excellent and range from 0.01 md to several Darcies.

The Tarbert Formation sediments form more simplified, stacked, sheetlike sandstone bodies, interdigitating with offshore mudstones. Sands are cleaner, coarse-grained and show an open pore network. These high energy sandstones provide the optimum reservoir quality, forming tabular or sheet-like bodies. However, they probably pass towards the northeast into low reservoir quality offshore muds. In the reservoir sandstones, log porosities have a range of 20–22%, while permeabilities range from 0.01 md to 6000 md.

Source. The source of the hydrocarbons is thought to be the Kimmeridge Clay Formation situated in the deep S. Viking Graben to the east. Migration was probably along fault planes during Upper Cretaceous or Tertiary.

Hydrocarbons

Six wells have been tested in the field. Well 9/28-4 was productive from both the Middle Jurassic and Triassic reservoirs. Wells 9/28a-9, 9/28a-10A and 9/28a-12 produced from the Middle Jurassic and wells 9/28-2 and 9/28a-13 from the Triassic only.

Most of the tests were characterized by linear flow and depletion which indicated that the drainage areas were elongated sand bodies of limited volume.

The Jurassic reservoir fluid is a medium gravity black oil and initially near saturation pressure. A thin gas cap in pressure communication with the oil leg was seen in well 9/28a-9.

The Triassic reservoir fluid is more undersaturated and shows a possible variation of oil gravity with depth. No free gas was seen in the Triassic.

Formation water samples are only available from RFT fluids. Salinities are in the region of 70 000–75 000 ppm NaCl eq. and the formation water resistivity is calculated to be 0.043 ohm m at 185°F.

A summary of the PVT properties is given in Table 1.

Reserves/resources

Appraisal drilling showed that the quality and thickness of the Crawford reservoir were extremely variable. There is doubt whether a significant portion of the mapped volume can be drained by a reasonable number of wells. Because of this uncertainty a phased development plan has been adapted.

The fluid PVT tests showed that the oil was saturated at initial reservoir pressure. Since the wells are draining limited volumes, the producing gas–oil ratios are expected to rise quickly. The gas handling capacity of the floating production facility is the major limiting factor in liquid processing capacity.

Oil-in-place (STOIIP) calculations have given a probable figure of 130 MMBBL. The recoverable reserve estimates are somewhat more difficult to make as the Crawford accumulation is made up of many small, closed drainage volumes. However, it is estimated that ultimately around 9 MMBBL may be recoverable.

Table 1. *Crawford Field PVT data summary*

	9/28-4	9/28a-9
Formation	Triassic	Jurassic
Depth, ft	8602–8794	8162–8568
Res. pressure, psia	3811	3700
Temperature, °F	190	190
Sat. pressure, psia	3235	3690
Co. at Pbpm, 1/psia	8.82E-6	12.18E-6
Gor, SCF/STB	529	794
FVF, BBL/STB	1.317	1.410
Oil gravity, API	25.2	31.2
Oil vis. at Pbp, cP	2.06	0.66

The author wishes to thank the management of Hamilton Brothers Oil and Gas Ltd and the participating Partners for permission to publish this paper.

The author would also like to thank R. Ezzat, B. Morrow, A. Vaughan, A. Johnson and M. Taylor for their contributions at various stages of the development of the Crawford Field and A. Taylor for a critical review of the draft manuscript. Special thanks are also due to J. Fenton for information on biostratigraphy and to A. Canham for information on sedimentology.

The Cyrus Field, Block 16/28, UK North Sea

D. G. MOUND[1], I. D. ROBERTSON[1] & R. J. WALLIS[2]

[1] *BP Exploration, 301 St Vincent Street, Glasgow G2 5DD, UK*
[2] *BP Exploration, Farburn Industrial Estate, Dyce, Aberdeen AB2 0PB, UK*

Abstract: The Cyrus Oilfield is located in Block 16/28 of the UK sector of the North Sea approximately 250 km (155 miles) NE of Aberdeen and 55 km (34 miles) NE of the Forties Field. The trap consists of a broad, very low relief four-way dip closure developed over a deeper tilted fault block. The reservoir consists of submarine-fan sandstones of late Palaeocene age, belonging to the Andrew Formation. Provenance was to the NW resulting from the early Tertiary sea-level fall which exposed the East Shetland Platform. The reservoir has been sub-divided into two zones, an upper zone of interbedded sandstones and mudstones with net to gross ratios of 0.4 to 0.6 and sandstone porositites of 12% to 18%, and a lower zone of massive fine-grained sandstones plus subordinate thin shales and limestones, with net to gross ratios in excess of 0.9 and porosities averaging 20%. The reservoir is filled with undersaturated oil of 35° API and is normally pressured. The estimate of initial oil-in-place is 75 MMBBL. Development of the field is centred on the use of BP's SWOPS (Single Well Offshore Production System) vessel using two horizontal field development wells which feed into a single seabed template for offtake. Ultimate recovery from the field is estimated to be approximately 12 MMBBL.

The Cyrus Field is located in the UK Sector of the North Sea, approximately 155 miles NE of Aberdeen and 34 miles NE of the Forties Field (Fig. 1). It covers an area of 13 km² (3212 acres) in the NW corner of Block 16/28 in a water depth of 360 ft. Submarine fan sandstones of late Palaeocene age constitute the reservoir rocks. The field is named after St Cyrus, a martyred Egyptian physician beheaded in Alexandria, *c.* 303 AD.

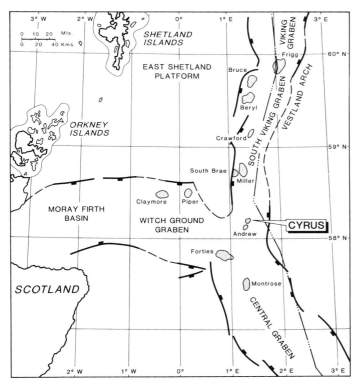

Fig. 1. Location of the Cyrus Oilfield.

History

Block 16/28 was awarded in June 1970 in the Third Round of UKCS licensing as part of Licence P092. BP is the operator of the block and licensee; there have been no relinquishments on the block.

There have been 13 wells drilled in the block to date (Fig. 2). The first of these wells was drilled in 1974 and discovered the Andrew Field in the SW of the block, encountering gas and oil-bearing sandstones of Palaeocene age and gas-bearing sandstones of early Cretaceous age occurring in four-way-dip closures over a salt diapir. The second and third wells in the east of the block failed to encounter hydrocarbons. The fourth well in the NW of the block was targeted at a low relief, four-way dip closure at the top of the Palaeocene sequence, with a secondary target at the Base Cretaceous unconformity. This was the Cyrus Field discovery well 16/28-4S which encountered a gross 30.5 m (100 ft) oil-bearing sandstone at the top of the Palaeocene strata. The well also encountered oil shows from Lower Cretaceous sandstones.

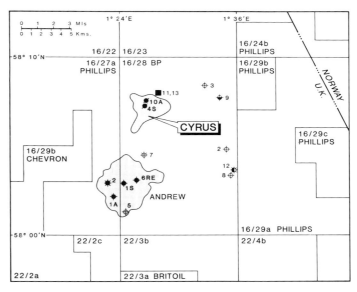

Fig. 2. Block boundaries, Field outlines and exploration drilling in the vicinity of Cyrus.

On the basis of the interpretation of a total of 660 km of seismic data acquired in 1977, 1981 and 1983, together with the single discovery well 16/28-4S, field development approval was granted at the end of 1984. It was proposed to develop the field via two vertical production wells tied back to a single seabed template with offtake using BP's innovative SWOPS (Single Well Offshore Production System).

The first of the two vertical development wells drilled in 1986/7 encountered a gross oil column of 32.4 m (106 ft), less than the 50 m (164 ft) which was prognosed and less than the 45 m (148 ft) deemed necessary for efficient production designed to avoid water-coning.

From Abbotts, I. L. (ed.), 1991, *United Kingdom Oil and Gas Fields,*
25 Years Commemorative Volume, Geological Society Memoir No. 14, pp. 295–300

Following a further field evaluation in 1987, benefiting from reprocessed seismic data and the additional well control, it was proposed to exploit the field by producing from two 'horizontal' wells. The first of these two horizontal wells was successfully drilled in 1988 (Fig. 3). The second horizontal well was drilled from the same surface location during late 1989.

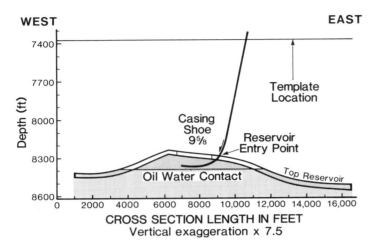

Fig. 3. Horizontal well layout in Cyrus.

Field stratigraphy

The Cyrus Field is located at the confluence of the Witch Ground, South Viking and Central Grabens of the Central North Sea (Fig. 1).

The generalized stratigraphic column is illustrated in Fig. 4. The lithostratigraphic nomenclature follows that of Deegan & Scull (1977).

Underlying the Cyrus Field is a rotated fault block containing Mesozoic and Palaeozoic age sediments. This is overlain by Lower Cretaceous mudstones and sandstones, and Upper Cretaceous mudstones, marls and chalk. The early Tertiary is represented by Palaeocene sandstones and shales of the Montrose Group which consists of the Maureen Formation, the Andrew Formation (containing the Cyrus Field reservoir) and the Forties Formation equivalent, which is represented in this area by a shale. Overlying the Montrose Group are the Early Eocene shales and tuffs of the Sele and Balder Formations which are very thin in this area.

Higher in the section, the Hordaland Group contains two sequences of local sandstone occurrence which appear as channelized features. The more important sequence contains large scale channels which are infilled with various amounts of sandstone; these are important for seismic depth conversion of deeper horizons. A further sandstone sequence which occurs near the top of the Eocene is more regional in extent. The remainder of the Hordaland and Nordland Groups (Fig. 4) consist of mudstones, clays and interbedded sands.

Geophysics

Field definition was originally based on interpretation and mapping of 30-fold seismic data acquired by S & A Geophysical in 1977 and processed by BP. This grid, orientated N–S and E–W had a line spacing averaging 750 m and totalled 167 km of field-specific data.

Following the successful 16/28-4S well, the field was remapped with the benefit of additional seismic data shot in 1981 and 1983. These 60-fold data were acquired by Western Geophysical and processed respectively by SSL and Merlin Geophysical. These lines were principally orientated NW–SE and NE–SW at line spacings varying from 750 m to 1 km. These data show a marked improvement on the poorly resolved 1977 data and were used as a basis for

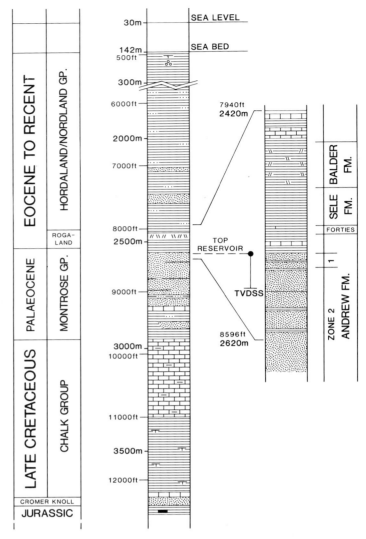

Fig. 4. Generalized stratigraphic column; Cyrus Field.

the 1984 mapping on which the field development was approved.

The latest remapping, following the disappointing results of the 16/28-10 well has benefited from an additional 100 km of 60-fold seismic data acquired by Western Geophysical in 1986 and processed by Merlin Geophysical. A selection of the 1981, 1983 and 1986 seismic data was reprocessed by BP, including the use of processing sequences designed to display the variations of reflection amplitude with source/streamer offset. These reprocessed data have enabled definition of a reflector corresponding closely to top reservoir, and more importantly provide better resolution of the overlying Eocene strata wherein lithological variations give rise to lateral variations in interval velocities (Fig. 5).

The reprocessed data along with the original 1981, 1983 and 1986 data provide the primary seismic database for the field interpretations described herein.

Trap

The Cyrus Field occurs in a four-way dip closure resulting from drape and compaction of the massive Palaeocene submarine fan systems of the Maureen and Andrew Formations over a deeper tilted fault-block of Jurassic age (Fig. 5). The fault block rotation is a likely response to salt movement connected to the diapirism in the SW of the block beneath the Andrew Field. Although some fault expression is seen in the deeper Palaeocene sequence, it dies out beneath top reservoir. There is potential for some small scale faulting related to compaction and slumping in the overlying

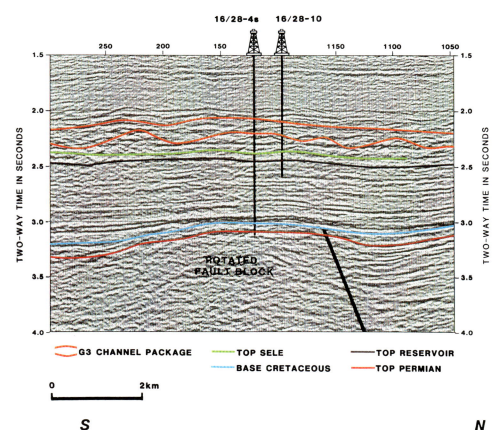

Fig. 5. Seismic section through the Cyrus Field.

Eocene strata, though this does not have a major impact on the geometry of the top reservoir.

The thickest oil column encountered in the Cyrus Field was 32.5 m (106.6 ft), found in the 16/28-10 well. Current mapping indicates a maximum gross oil column potential of 40 m (131 ft) at two locations; 750 m (0.46 miles) to the west of 16/28-10 and 1500 m (0.93 miles) to the southeast. This equates approximately to the perceived mapped closure at the 2551 m TVSS contour, coincident with the field oil–water contact (OWC). These culminations have a depth to crest of 2511 m (8238 ft) TVSS (Fig. 6).

The current depth structure map on top reservoir has been derived by mapping of the relevant seismic reflector (Fig. 5). The key problem in structural definition lies in the ability to define the effect of the lateral variations in lithology, and hence interval velocity, in the overlying Eocene strata. The presence of sand bodies (G3) in these strata with sigificantly higher interval velocities than the surrounding shale leads to a seismic 'pull-up' of up to 15 milliseconds of underlying reflectors as illustrated in Fig. 7. Mapping and calibration of the Eocene channel geometries to generate a 'pull-up' compensation surface has been used to correct the structure map on the top reservoir in two-way-time prior to depth conversion. The accuracy of the depth mapping is a function of the ability to map accurately the channel geometries on the seismic data and estimation of the quality of sand within these features.

The current depth map on top Zone 2 reservoir (Fig. 6) defines a spill point at its western extremity. The overlying shales of the Sele Formation in this area provide an adequate seal to the reservoir, which is similarly proven to the south over the Andrew Field reservoir.

Reservoir

The reservoir sequence in the Cyrus Field is Palaeocene age sandstones of the Andrew Formation. This is an extensive submarine-fan system, similar to, but pre-dating, the Forties Formation submarine-fan system. The provenance area for the fan was from the NW, due to early Tertiary sea level fall which exposed the East Shetland Platform. Regionally the Andrew Formation reservoir shales-out to the east approximately coincident with UKCS/NOCS boundary.

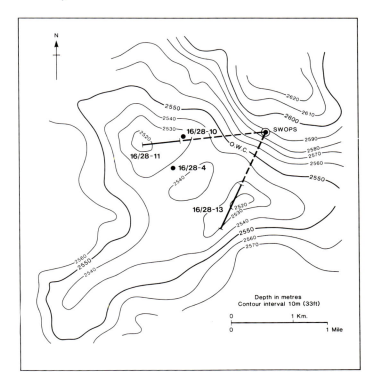

Fig. 6. Map of reservoir, top Zone 2.

The Cyrus Field reservoir has been sub-divided into two zones (Fig. 8).

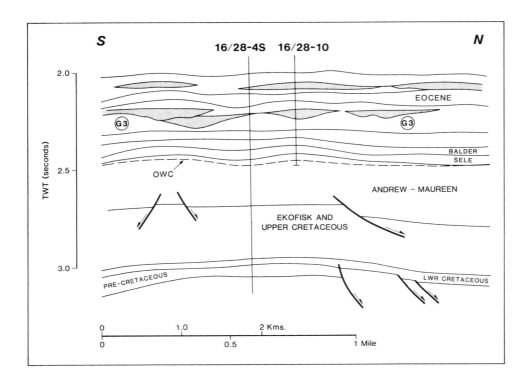

Fig. 7. Geoseismic section through Cyrus showing effects of Eocene sands on OWC reflection time.

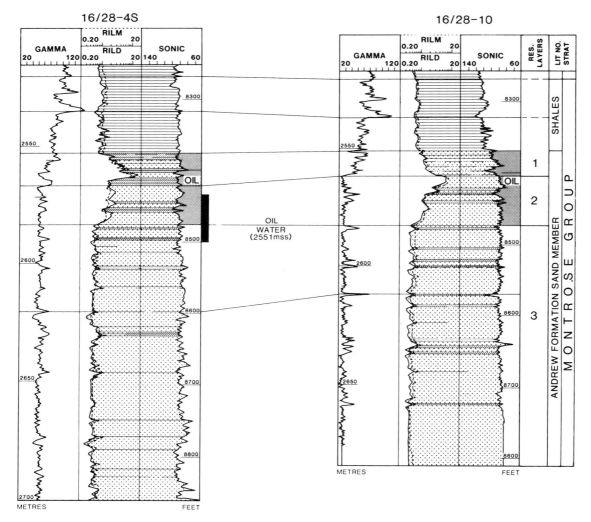

Fig. 8. Cyrus Field Palaeocene reservoir zones.

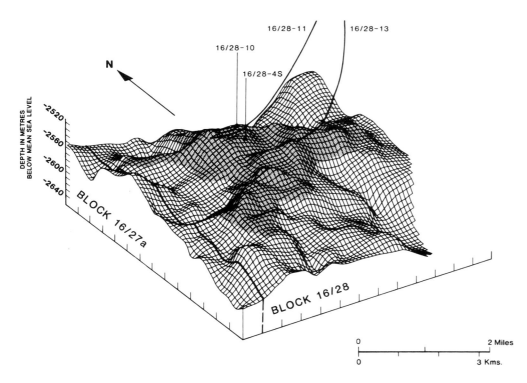

Fig. 9. Cyrus Field; top reservoir. A SW isometric view showing horizontal well trajectories.

Zone 1 occurs at the top of the reservoir where the massive sandstone units of Zone 2 fine upwards into an interbedded unit. This Zone 1 in turn passes into the Andrew Formation mudstone, which is regarded as indicative of waning-flow and ultimately fan-abandonment.

Zone 2 consists of a thick sequence of massive fine-grained sandstones with subordinate thin shales and limestone. The massive sandstones probably represent deposition in the channelized mid-fan environment of a submarine-fan complex.

The Zone 2 massive sandstones represent the main part of the reservoir and are both laterally and vertically extensive beyond the field limits.

Vertical thicknesses in the area exceed 250 m (820 ft) although the quoted parameters are based on analysis of the top 30 m (100 ft) only. The net/gross ratio is relatively uniform across the field generally exceeding 90%. Porosities average 20%, with a water saturation of c. 35%; average permeability is 200 md.

The Zone 1 interbedded unit comprises thin sheet or channel-like sandstones of thickness between 0.5 m and 2.5 m (1.6 ft and 8.2 ft), and of limited lateral extent, interbedded with mudstones. The gross thickness over the reservoir is 10–15 m (33–50 ft), thinning to the SW and thickening to the NE. The net/gross ratios from wells range from 0.39 to 0.64 illustrating the variable nature of this zone. Porosities range from 12–18% and permeabilities of 50 md are recorded. Average water saturations are 50%.

Source

The source rocks for the Cyrus Field are believed to be the organic-rich Kimmeridge Clay shales of Latest Jurassic to Earliest Cretaceous age which occur beneath the field and are regionally present across large areas of the South Viking Graben. The Kimmeridge Clay was mature for oil generation and expulsion during Oligocene times and the expected mechanism for migration through the Cretaceous section was both through faults (associated with diapirs) and through the Cretaceous section itself, even if unfaulted. Severe overpressures generated within the pre-Cretaceous section may be expected to have driven hydrocarbons vertically upwards, particularly in the deepest parts of the graben.

Once into the Early Tertiary sandstone carrier systems, the hydrocarbons will have migrated updip to fill any traps encountered 'en route'. The migration pathway into the Cyrus Field is not known for certain. However, it is likely that the field was filled from the south or southeast.

Hydrocarbons

A total of four wells have been drilled on the structure, three of which were tested. The reservoir fluid is an under-saturated oil with a density of 860 kg m^{-3} and a viscosity of 1.0 cP. The gas–oil ratio is 239 SCF/STB and no gas-cap is present in the reservoir.

Reserves

The aquifer providing pressure support to the Cyrus reservoir is believed to be regional. It is not anticipated that water injection will be required to maintain pressure in the reservoir.

The development of the Cyrus reservoir is planned using two horizontal wells drilled from the same sea-bed location. The two wells are to be tied back, via a common riser, to the SWOPS vessel (Single Well Oil Production System). The reservoir will be produced to the SWOPS vessel which, in reaching capacity, will sail to an appropriate port for discharge of the oil. The first of the two development wells, 16/28-11, has been successfully drilled with a horizontal section of around 520 m. The second well, 16/28-13, was drilled in 1989 with a horizontal section within the reservoir of c. 700 m (Fig. 9).

Production from the reservoir is constrained by the fluid handling capacity of the SWOPS vessel (maximum oil flowrate: 15 000 BOPD). The ultimate recovery from Cyrus is estimated to the approximately 12 MMBBL.

Reference

DEEGAN, C. E. & SCULL, B. J. (Compilers). 1977. *A standard lithostratigraphic nomenclature for the Central and Northern North Sea*. Institute of Geological Sciences Report No. 77/25. HMSO, London.

Cyrus Field data summary

Trap
Type	Compaction anticline
Depth to crest	2511 m (8238 ft) SS
Lowest closing contour	2551 m (8369 ft) SS
Hydrocarbons contacts	OWC: 2551 m (8369 ft) SS
Gas column	None
Oil column	40 m (131 ft)

Pay zone
Formation	Andrew Fm
Age	Palaeocene
Gross thickness (average)	250 m (820 ft)
Net/gross ratio (average)	90%
Porosity (average)	20%
Hydrocarbon saturation (average)	65%
Permeability (average)	200 md
Productivity index	20–50 STB/D/psi

Hydrocarbons
Oil Gravity	860 kg/m^3 (35° API)
Oil type	Low sulphur
Bubble point	1300 psi
Gas/oil ratio	239 SCF/BBL
Formation volume factor	1.19

Formation water
Salinity	90 000 ppm
Resistivity	0.028 ohm m

Reservoir conditions
Temperature	112°C
Pressure	3425 psi
Pressure gradient in reservoir	1.05 psi/m (0.32 psi/ft)

Field size
Area	13.3 sq km (3286 acres)
Gross rock volume	122 × 10^6 m^3 (4308 MMCF)
STOIIP	75 MMBBL
Recovery factor: oil	16%
Drive mechanism	Aquifer support
Recoverable hydrocarbons: oil	12 MMBBL

Production
Start-up date	1990
Development scheme	SWOPS
Production rate	15 000 BOPD (max)
Cumulative production to date	None
Number/type of wells	2 horizontal oil producers
Secondary recovery methods	None

The Forties Field, Block 21/10, 22/6a, UK North Sea

J. M. WILLS

BP Exploration, 301 St Vincent Street, Glasgow G2 5DD, UK

Abstract: The Forties Field is located 180 km (112 miles) ENE of Aberdeen, predominantly in UK Licence Block 21/10. It was discovered in 1970, when an exploration well encountered hydrocarbons in Palaeocene sandstone within an anticlinal structure. Four appraisal wells confirmed the existence of a major oilfield, with an area of approximately 90 sq km (35 sq miles) at a depth of approximately 2000 m (6500 ft). The reservoir occurs in thick Late Palaeocene sandstones deposited in two major sand-rich submarine fan sequences.

The field has been producing oil since September 1975. Stock Tank Oil Initially in Place (STOIIP) has been calculated as 4343 MMBBL, and original reserves estimated as 2470 MMBBL, representing an overall recovery of 57%. The field came off plateau production of 500 000 BOPD in 1981, and by mid-1989 production had declined to about 250 000 BOPD. After fourteen years production, the field has produced more than 2 billion STB. Remaining reserves are about 500 MMBBL which includes recovery via artificial lift. Field life has been projected to extend beyond the year 2000.

Water injection into the aquifer commenced in 1976, and continues at the present day at an average rate of 390 000 BWPD. There is also a significant contribution to pressure support from the underlying aquifer.

At present there are 103 available wells in the field, 81 producers and 22 injectors.

The Forties Field is located 180 km (112 miles) ENE of Aberdeen, in Blocks 21/10 and 22/6a of the UK Sector of the Central North Sea. Water depth ranges from 104 m to 128 m (342–420 ft). The field is a broad, four-way dip-closed anticlinal structure, 90 km^2 (35 sq miles) in area with a vertical relief of 190 m (590 ft), at a depth of approximately 2000 m (6500 ft). The reservoir occurs in thick Late Palaeocene sandstones deposited in two major sand-rich submarine fan sequences. The field is named after Sea Area Forties, an area of relative shallows in this part of the North Sea.

History

The original licensees were BP (100%) in Block 21/10, and Shell and Esso (50% each) in Block 22/6. The blocks were awarded in the UK Second Licensing Round in 1965. In 1983 and 1984, BP assigned certain fixed interests in Block 21/10 to 22 minor partners. Subsequent company and interest changes have resulted in there being now 22 interest owners in the field.

The field was discovered in October 1970 by well 21/10-1, drilled on the axis of a large structural nose recognised on seismic at Base Tertiary level (Thomas *et al.* 1974). The well encountered 119 m (390 ft) of oil-bearing Palaeocene sands at a depth of 2131 m (6991 ft) subsea. Following the acquisition of a 1.5 km × 1.5 km seismic grid over the structure, a further four appraisal wells were drilled during 1971–72, which confirmed the existence of a major oil field. Shell/Esso also drilled a successful well, 22/6-1, in 1971, which demonstrated the southeastern extension of the field into Block 22/6.

The development plan evolved during 1972–75. Four fixed steel drilling/production platforms were installed on the Main Field in Block 21/10 in 1974 and 1975, with a total facility for 108 wells, a producing system capable of 500 000 BOPD, and a seawater injection system with a capacity of 600 000 BWPD. Development drilling began in June 1975 and production started in September 1975. Plateau production of 500 000 BOPD was reached in 1978 and lasted until 1981. A fifth platform was added over the southeastern extension (South East Forties) during 1985, with production commencing in March 1987.

Major additional seismic data were acquired over the field in 1980–81, and a regional grid overlapping the periphery of the field in 1986. A full 3-D seismic survey was acquired over the entire field in 1988.

Unit Participation Interests have been finally agreed as 94.78% in Block 21/10 (BP 83.13%, 19 minor partners 11.65%) and 5.22% in Block 22/6a (Shell 2.61%, Esso 2.61%):

Amerada Hess (Forties) Ltd	0.447783
Aran Energy Exploration Ltd	0.25
BP Petroleum Development Ltd	83.132734
Clyde Expro plc	0.75
Clyde Petroleum (Exploration) Ltd	0.25
Elf UK plc	1.50
Elf Oil & Gas Ltd	1.25
Montrose Energy Ltd*	0.25
Enterprise (E & P) Ltd	0.25
Esso Exploration and Production UK Ltd	2.61085
Gas Council (Exploration) Ltd	0.25
Hardy Oil and Gas plc	0.50
Hardy Oil and Gas (UK) Ltd	0.947783
Industrial Scotland Energy Ltd†	0.25
Monument Petroleum Mitre Ltd	1.25
Norsk Hydro Oil and Gas Ltd	0.75
Ranger Oil (UK) Ltd	0.50
Union Jack Oil plc‡	0.50
Repsol Exploration (UK) Ltd	0.25
Shell UK Ltd	2.61085
Sovereign Oil & Gas plc	0.50
Ultramar Exploration Ltd	1.00

* Montrose Energy Ltd is wholly owned by Elf.
† Industrial Scotland Energy Limited is wholly owned by Total.
‡ Union Jack Oil plc is controlled by Ranger.

Field stratigraphy

The field is situated close to the crest of the underlying Forties–Montrose Ridge, at the northern end of the Central Graben (Fig. 1). Major NNW–SSE trending Mesozoic growth faults form the boundaries to the ridge. These faults became inactive by Early Cretaceous times, and the structure is expressed at the top Cretaceous only as a series of monoclinal ramps.

In the Forties area, the oldest rocks penetrated by drilling are Middle Jurassic volcanics and associated sediments. The volcanics attain maximum thickness just to the north of the field and consist of basaltic lavas and tuffs interbedded with argillaceous, calcareous siltstone and tuffaceous sandstone (Howitt *et al.* 1975; Deegan & Scull 1977). The volcanics are overlain by a thin Kimmeridge Clay Formation consisting of silty mudstones and siltstones. The formation probably thins as a result of deposition over the structurally

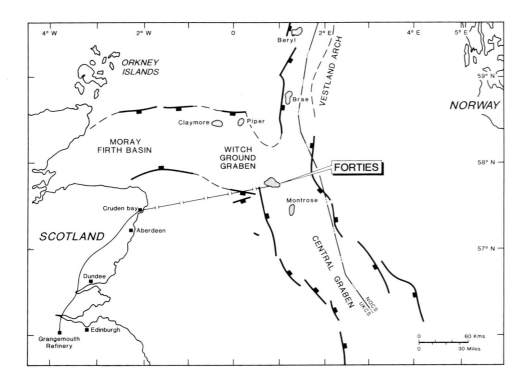

Fig. 1. Location map of the Forties Field.

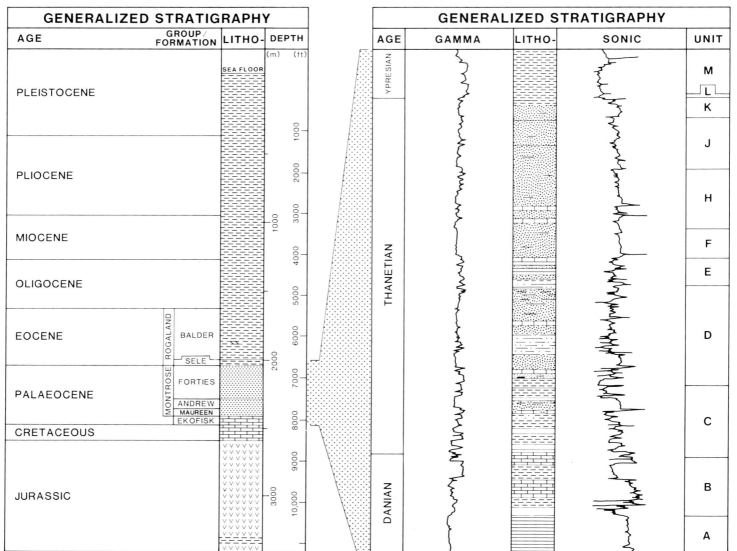

Fig. 2. Generalized stratigraphic column and discovery well 21/10-1 type log.

emerging ridge, together with erosion associated with the regional base Cretaceous unconformity. The unconformity surface was successively overlain by thin or condensed Early Cretaceous mudstones and sandstones containing volcanic rock fragments, and then the massive chalks and marls of the Late Cretaceous and basal Palaeocene (Danian), laid down in the broad downwarping Central North Sea.

The early Tertiary marked the transition from Late Cretaceous and Danian pelagic carbonate sedimentation with low terrigenous input to the accumulation of vast thicknesses of clastic sediment deposited in deltaic, shelf and submarine-fan environments (Stewart 1987). This rapid change is believed to be the result of uplift, erosion and the eastward tilting of the Shetland Platform, located to the north-west, associated with an early phase of the opening of the North Atlantic Ocean (Parker 1975; Ziegler 1978). The pre-Tertiary structural configuration controlled the deposition of the early Tertiary sediments so that early clastic sediments accumulated in the axial regions of the Mesozoic grabens, with no or little sedimentation over the previously emergent highs such as the Forties–Montrose Ridge. By the late Palaeocene, however, clastic sedimentaton had smothered all topographic features so that the regional structure was expressed as a series of monoclinal ramps. This resulted in the development of a thick and extensive sand-rich submarine fan system which both dips and thins in a southeasterly direction along the Central Graben. Sediments were derived from a shelf area established to the west of the field area. The Forties Field is part of this major fan system.

At the end of the Palaeocene, a relative rise in sea-level resulted in the deposition of the regional basinal mudstones of the Sele Formation. These fine-grained sediments form the seal to the reservoir. Further fluctuations between highstand mudstones and lowstand submarine fan clastics occurred throughout the Eocene. In the Forties area, extensive basin-slope erosional channels were developed during the Middle Eocene, some of which are sand-filled and oil-bearing.

From late Eocene to Holocene, continuing mudstone deposition onlapped and overlapped the underlying shelf-delta-fan system in the Forties area and infilled the entire basin with thick, monotonous mudstones.

The generalized stratigraphic column for the field area is shown in Fig. 2. The lithostratigraphy over the Forties area is consistent with the regional nomenclature established by Deegan & Scull (1977).

Geophysics

The field structure was originally recognized from regional seismic data acquired in 1967 (Fig. 3a). The most recent delineation of the field (Fig. 3b) was based on a triangular seismic grid of approximately half to one kilometre spacing shot across the field in the winter of 1980–81. The data were reprocessed in 1985. A later regional survey in 1986 provided better quality data over peripheral areas of the field. Prior to the 1988 3-D survey, total seismic coverage was approximately 750 line kilometres. The 3-D programme consists of 3865 sail line kilometres of dual-streamer data, providing a 50 m subsurface line spacing. Interpretation of the new data will commence once processing is complete in Summer 1989.

All reservoir maps have been generated by isopaching down from the Top Sele horizon, the most reliable seismic event in the area. Eight intra-reservoir seismic horizons are mapped using a combination of well data, seismic sequence and time isopach mapping.

Seismic data have been a fundamental part of a number of successive seismic-stratigraphic reservoir models developed since 1982, culminating with the current reservoir model. The intra-reservoir seismic horizons have been interpreted and tied, using synthetic seismograms and well logs, to a comprehensive micro-palaeontological/palynological biozonation in key cored wells. Recognizable sequence and subsequence boundaries are combined with observed sedimentological sequence breaks in the cored wells and production barriers recognized from reservoir monitoring data, to establish a field-wide stratigraphy. Seismic sequence correlation and log correlation were then used to extend the stratigraphic framework to all wells in the field.

Trap

The field is a broad, four-way dip-closed anticlinal structure with two culminations (Fig. 4). There is an approximate coincidence between the structural spill points and the initial oil–water contact at 2217 m (7274 ft) subsea. Depth to crest is 2030 m (6660 ft) subsea, and vertical relief is about 190 m (590 ft). The structure is very gentle and nowhere in the field are dips in excess of 10°. Structural style is similar at all levels within the reservoir and aquifer (Fig. 5).

The structure has resulted from a combination of sedimentary drape, synsedimentary faulting, and compaction over the buried Forties–Montrose Ridge. Minor faults penetrate the reservoir, particularly on the southern flank where evidence of syndepositional slumping exists in South East Forties. However, the current seismic and well database does not permit resolution; field performance does suggest, however, that they do not significantly influence reservoir continuity or production.

The reservoir seal is provided by the overlying and conformable Eocene mudstones of the Sele Formation.

Reservoir

The field occupies a minor part of a large Palaeocene submarine fan system (Stewart 1987). This fan system consists mostly of stacked sandstone bodies with fine-grained sediments restricted to a narrow distal fringe. It depicts a sand-dominated ramp depositional system as described by Heller & Dickinson (1985).

Sands were deposited in elongate, topographically confined fairways which generally trend NW–SE or N–S. The controlling influences on the thickness and direction of the fairways are the boundary faults of the Forties–Montrose Ridge and differential compaction over previously deposited sequences. Lobes and offshoots from the main fairways occur as a result of overspill from one filled low to an adjacent low; these are still essentially topographically controlled. Depositionally controlled lobate sandbodies formed in the waning flow periods at the ends of depositional cycles or at the distal end of a sand fairway.

The fairways are composed of massive sandstones deposited from high-density turbidity currents. The sandstones are characterized by elongate sheet-like geometry, with high net-to-gross ratios, and good lateral and vertical connectivity. Classical turbidites of interbedded mudstones and sandstones were deposited away from the fairways, often on local topographic highs.

Eight lithofacies have been recognized from studies of available data.

The fine- to medium-grained sandstones of lithofacies B constitute approximately 80% of the field lithology. Using seismic-stratigraphic techniques, as previously described, the reservoir has been subdivided into eleven units, commencing at the top of the Ekofisk Formation (Unit A) and extending upward to the top of the Sele Formation (Unit M) (Fig. 2). Oil occurs in the six topmost reservoir units.

Most production is from the Main Sand (Units D, E, F, H, J and K) in the centre of the field (Fig. 6). The Charlie Sand (Units J and K) in the west is in pressure isolation from the Main Sand. Production in South East Forties is also from Units J and K, which are also almost in pressure isolation at reservoir level.

The petrography of the reservoir units is uniform throughout the field. The sandstones comprise very fine-, fine- and medium-grained sandstones, moderately sorted, with a low detrital clay content. Locally they grade into pebbly sandstones and conglomerates. The sandstones are generally poorly lithified with only minor cementation.

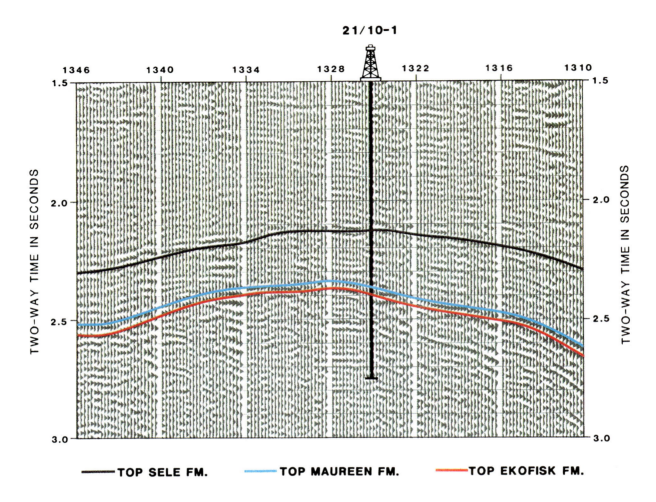

Fig. 3 (**a**). Seismic line (1967) through discovery well, 21/10-1. (**b**) Comparable 1980 seismic line through 21/10-1.

The primary depositional characteristics are the most important parameters which control the poroperm distribution in the reservoir. Essentially, porosity and permeability decrease as grain size decreases and detrital illite/illite–smectite content increases. Diagenetic kaolinite, illite and chlorite are not abundant and have only a minor detrimental effect on reservoir quality. The coarsest-grained sandstones, deposited by high density turbidity currents, tend to have average porosities around 27% and permeabilities generally greater than 300 md, with high permeability zones in excess of 1000 md common. The finer-grained sandstones, deposited by low density turbidity currents, have lower porosities (23–25%) and significantly lower permeabilities (100–200 md).

Mudstone and shale intervals can be laterally extensive, resulting in permeability barriers to vertical fluid flow with significant pressure differentials across them. Several important barriers to lateral fluid movement also exist in the reservoir where sandstones thin and pinchout into areas of low net-to-gross siltstones and mudstones. The most significant lateral barriers are those between the Main Sand and the Charlie Sand, and between the Main Field in Block 21/10 and South East Forties.

After fourteen years of production and predominantly bottom-up water flooding, most of the remaining reserves are now concentrated in the crestal regions of Units H, J and K.

Source

Forties oil was sourced from the Jurassic Kimmeridge Clay Formation. The present depth of the Kimmeridge defines two source kitchens: one to the south of the field which drains into South East Forties, and the other to the northeast draining into the northern area of the field. The oil migrated vertically and then laterally through the Palaeocene sandstones into the trap. The main phase of expulsion was from Middle Eocene to Miocene times (50–10 Ma) in the southern kitchen, and from Late Eocene/Oligocene times to the present day (40–0 Ma) for the northern kitchen.

Hydrocarbons

Forties crude is a low sulphur (0.3%), medium wax (5%), undersaturated light oil of 37° API gravity. Gas chromatogram and GC-MS analyses indicate a moderately mature crude derived from a clastic marine source.

Although analysis of oils from across the field reveal no major compositional differences, there are variations in the oil properties. The occurrence of two major permeability barriers has prevented the complete homogenization of the oils and has resulted in the

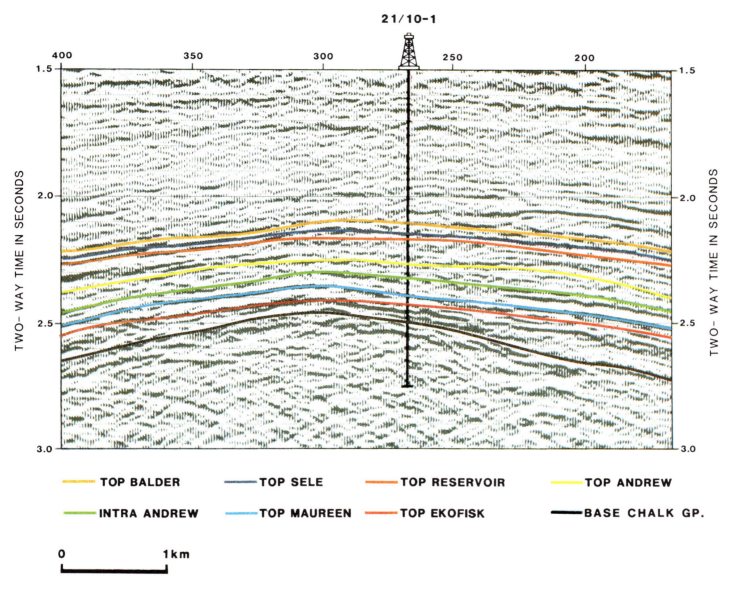

Fig. 3 (b)

establishment of three distinct PVT areas through time, corresponding to the three areal regions of the field: South East Forties, the Main Sand in the centre, and the Charlie Sand in the west. These variations correspond with the source kitchen locations and maturities. The more mature gaseous oil (392 SCF/STB) in South East Forties contrasts with the less mature, less gaseous (300 SCF/STB) oil in the western area of the field. Between these the oil has intermediate but constant properties.

The original datum reservoir pressure was 3210 psig at 2175 m (7136 ft) subsea.

Reserves/resources

Oil in Place (STOIIP) has been calculated as 4343 MMBBL, and original reserves, including those recovered through artificial lift, estimated at 2470 MMBBL, representing an overall recovery of 57%. After fourteen years of production, the field has produced more than 2000 MMBBL (including gas liquids), approximately 80% of the total available, with remaining reserves estimated of the order of 500 MMBBL. Field life has been projected to extend beyond the year 2000.

The reservoir performance has been characterized by excellent vertical sweep through the clean sands by the underlying very mobile aquifer, supported by peripheral water injection. Development of the reservoir was initially planned with approximately 80 production and 20 injection wells from the four main-field platforms. Development of South East Forties with 14 production wells followed at a later date. At present there are no plans to add water injection to South East Forties.

There are sufficient permeability restrictions between South East Forties and the Main Sand, and between the Main Sand and the Charlie Sand, to cause significant pressure differences between the three regions. The management of the reservoir has been based on

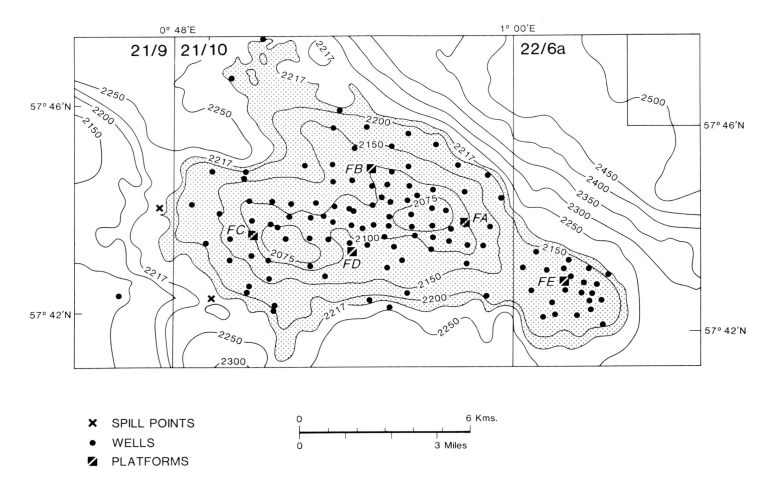

- ✕ SPILL POINTS
- • WELLS
- ▰ PLATFORMS

Fig. 4. Reservoir structure map showing platforms, wells and spill points.

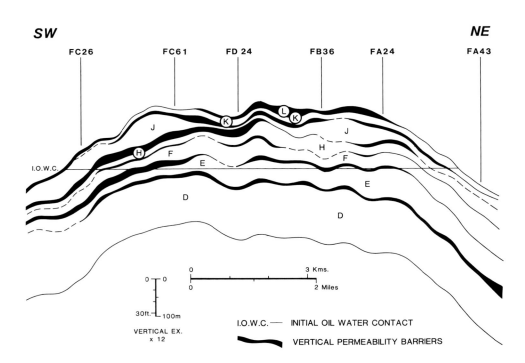

I.O.W.C. — INITIAL OIL WATER CONTACT

▬ VERTICAL PERMEABILITY BARRIERS

Fig. 5. SW–NE cross-section through the Forties Field.

treating the three regions separately. Generally, there is more significant edge water movement in the Charlie Sand and in South East Forties than in the Main Sand.

The reservoir sandstone is friable in most areas of the field, which has resulted in significant sand production causing erosion to down hole tubulars and to surface facilities. Consequently, there is a requirement to periodically wash out the separators. Since the onset of seawater breakthrough in certain producers, there has been an increase in the deposition of barium sulphate scale in downhole tubulars and some surface pipework.

Plateau production of 500 000 BOPD was maintained from 1978 to 1981. Average well rates were 11 000 BOPD, with a range of 3000–24 000 BOPD. Average production in mid-1989 was still about 250 000 BOPD. Injection commenced in 1976, approximately one year after the start of production, and currently averages around 390 000 BWPD, with average injection rates of 25 000 BWPD, and a range between 5000 and 40 000 BWPD.

In the first 5–7 years, pressure declined by 800–1000 psi below original. However, as the aquifer support has become more effective, coupled with the increasing water injection, reservoir pressure has been rising at rates of 5–10 psi/month since 1984.

Forties Field data summary (March 1989)

Trap
Type	Four way dip closed anticline
Depth to crest	2030 m SS (6660 ft SS)
Lowest closing contour	2217 m SS (7274 ft SS)
Initial oil–water contact	2217 m SS (7274 ft SS)
Oil Column	187 m (614 ft)

Pay zone
Formation	Forties Formation/Andrew Formation
Age	Palaeocene
Gross thickness (average/range)	354/199–469 m (1161/653–1539 ft)
Net/gross ratio (average)	0.65
Net/gross cutoff	4.3 md Ka, 10 pu
Porosity (average/range)	0.27/0.10–0.36
Oil saturation	0.85
Permeability (average/range)	700/30–4000 md
Productivity index (average/range)	25/5–70 STB/D/psi

Hydrocarbons
Oil gravity	0.73 g cc^{-1} at 96°C & 1181 psig
Oil type	Low sulphur
Bubble point	1142–1390 psig at 96°C
Dew point	NA
Gas/oil ratio: total (average/range)	303 (297–394) SCF/STB
Condensate yield	4% of total production
Formation volume factor	1.24–1.32 RB/STB at 96°C & 1181 psig

Formation water
Salinity	55 500 mg l^{-1} Cl$^-$
Resistivity	0.034 ohm m at 96°C

Reservoir conditions
Temperature	96°C at 2175 m SS (7135 ft SS)
Initial reservoir pressure	3215 psig at 2175 m SS (7135 ft SS)
Pressure gradient in reservoir	1.08 psi m^{-1}

Field size
Area	23 000 acres (93.4 sq km)
Gross rock volume	6032 MMcu.m (64 382 mmcft)
Hydrocarbon pore volume	5296 RB
Recovery factor: oil	57%
Drive mechanism	Predominantly bottom drive aquifer with peripheral water injection
Recoverable oil	2470 MMBBL

Production
Start-up date	September 1975
Development scheme	103 production wells/5 platforms natural lift, then by ESP/gas lift
Production rate (plateau/current)	500 000 BOPD/250 000 BOPD
Cumulative production to date	1924 MMBBL (end February 1989)

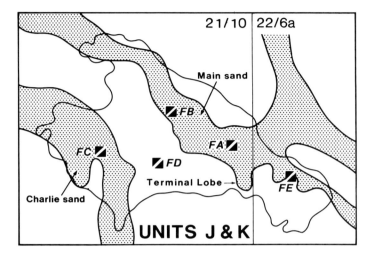
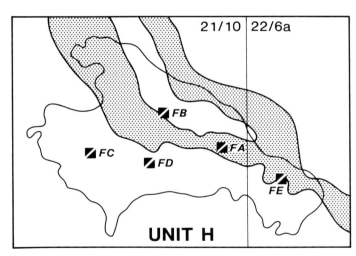
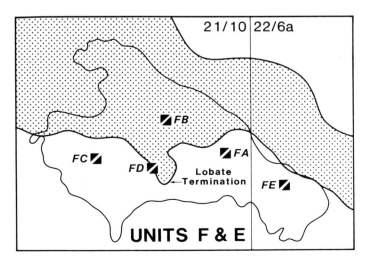

Fig. 6. Distribution of major sand bodies in the Forties Field.

The Fulmar Field, Blocks 30/16 & 30/11b, UK North Sea

C. P. STOCKBRIDGE[1] & D. I. GRAY[2]

[1] *Esso Exploration and Production UK Ltd, 21 Dartmouth Street, London SW1H 9BE, UK*
[2] *Shell UK Exploration and Production Ltd, 1 Altens Farm Road, Nigg, Aberdeen AB9 2HY, UK*

Abstract: The Fulmar Field is located on the southwestern margin of the Central Graben in Blocks 30/16 and 30/11b of the UK sector of the North Sea. The field is a partially eroded anticline with steeply dipping flanks formed by the withdrawal of deeper Zechstein salt. The reservoir consists of thick Upper Jurassic, shallow marine, very bioturbated sandstones of the Fulmar Formation and deep marine turbidites interbedded within the Kimmeridge Clay Formation. Seal to the reservoir is provided by the Kimmeridgian shales in the west and Upper Cretaceous chalks which unconformably overlie the Fulmar Formation in the east.

The reservoir section has been subdivided into seven members and 14 reservoir units. Reservoir quality is generally excellent, although there are lower-energy sandstone facies found in the eastern part of the field. The Fulmar oil is highly undersaturated and a secondary gas cap has been created by gas injection.

Two exploration wells were drilled before the field was declared commercial. Development is from a 36 slot steel platform and a six slot template. Oil evacuation is by a floating storage unit and gas evacuation is via the Fulmar gas pipeline. Total STOIIP is 815 MMBBL and ultimate recovery is 462 MMBBL oil and 264 BCF gas. Production started in 1982 and 319 MMBBL oil and 121 BCF gas have been produced by year-end 1988. A total of 80 BCF gas has been re-injected for conservation purposes.

The Fulmar Field is located on the southwestern edge of the Central Graben in the UK sector of the North Sea approximately 170 miles southeast of Aberdeen (Fig. 1). The average water depth is 270 ft. The field is a steep-flanked anticline with a relatively small areal extent (11.4 sq km), but with a large oil column in excess of 900 ft. The reservoir consists of shallow marine sandstones of the Upper Jurassic Fulmar Formation and overlying deeper marine turbidites interbedded within the Kimmeridge Clay Formation.

The field is situated mostly in block 30/16 (Shell/Esso) but extends onto block 30/11b (Amoco Exploration Co., North Sea Inc., and Amerada Hess Ltd). The field is unitized and is operated by Shell. The STOIIP is estimated to be 815 MMBBL oil with a gravity of 40° API and a GOR of 614 scf/stb. Ultimate recovery is 462 MMBBL oil of which some 319 MMBBL (69%) had been produced by year-end 1988 through a total of 28 development wells from a 36 slot steel platform and a six slot subsea template and wellhead jacket. Detailed descriptions of the field have been given by Johnson *et al.* 1986, Mehenni & Roodenberg (1990) and van der Helm *et al.* (1990).

History

Early 2-D seismic shot over the Auk–Fulmar area in 1970 and 1974 indicated a small closure beneath the Late Cimmerian Unconformity on the downthrown side of the Auk Horst fault (Fig. 2a). The structure was tested in 1975 by the Shell/Esso well 30/16-6 located slightly southwest of the structural crest (Fig. 3). The objective of the well was to test prognosed Upper Jurassic sandstones overlying Zechstein evaporites.

The well encountered an Upper Jurassic sequence consisting of 495 ft of Kimmeridgian shales (with some interbedded sandstones) beneath the Late Cimmerian Unconformity and a 900 ft thick sequence of massive, late Oxfordian to early Kimmeridgian sandstones which were later called the Fulmar Formation. Reservoir quality was excellent. A 668 ft oil column (607 ft net sands) was penetrated above an OWC at 10 830 ftss. The well penetrated a further 1600 ft of Triassic red-brown shales and occasional silty sandstones before reaching a total depth of 12 800 ftss in Zechstein carbonates.

A single appraisal well (Shell/Esso 30/16-7) located 600 m southwest of the discovery well was drilled in 1977. It was designed to test the steeply dipping flank of the structure identified on the seismic data. This well encountered a similar, but water-bearing, section of the Fulmar Formation. The well also proved a second oil-bearing reservoir within the Kimmeridge Clay section which was 139 ft thick and separated from the main Fulmar sands by 94 ft of shale. A 3-D seismic survey was shot in 1977 and this has formed the basis of all subsequent interpretations.

Further appraisal was considered unnecessary. The field was declared commercial and named Fulmar after the sea bird. Annex B approval was obtained from the Secretary of State at the Department of Energy in 1978.

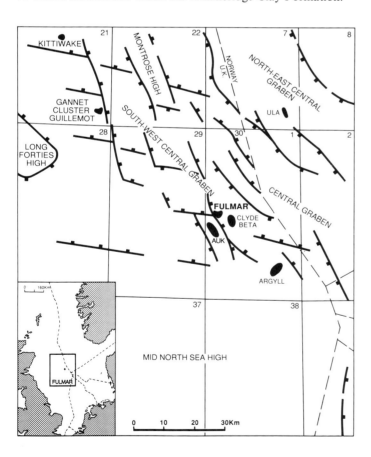

Fig. 1. Tectonic framework of the southwest Central Graben and location of the Fulmar Field.

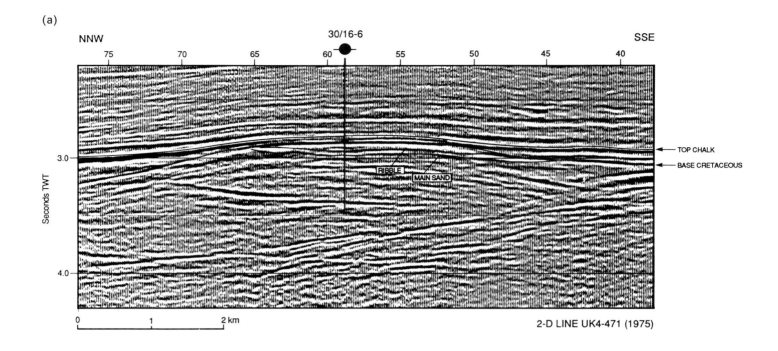

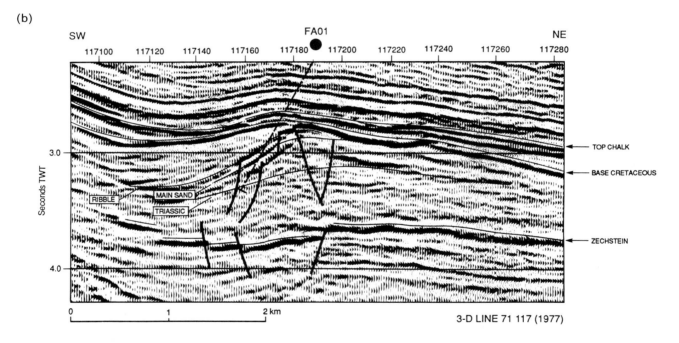

Fig. 2. (a) 2-D seismic line through the Fulmar Field discovery well 30/16-6; 1974 vintage. (b) Annotated representative SW–NE 3-D seismic line across Fulmar, 1979 vintage; indicating the steep western flank of Fulmar, truncation by the Base Cretaceous Unconformity and onlap of the Chalk Group. Note the clear Ribble and top Fulmar reflectors on the west flank. The far southwest is dominated by the Auk Horst boundary fault.

A six slot template was installed in 1978 and four wells were drilled for early production. These wells showed that the field was more heterogeneous than indicated by the two exploration wells. These wells discovered (1) a shallower OWC in the north, (2) truncation of the Fulmar sandstones by the Late Cimmerian Unconformity in the north, and northeast and (3) poorer quality reservoir rocks in the eastern part of the field.

A wellhead jacket was placed over the template in 1979 and the main 36 slot steel platform was installed in 1980. The template wells were tied to the platform for first production in early 1982 with evacuation via the nearby Fulmar Floating Storage Unit (FSU). Prior to the completion of the Fulmar gas pipeline, surplus gas was re-injected into the reservoir creating a secondary gas cap. Plateau oil production of 165 000 BOPD oil was reached in mid 1987 along with gas export of 88 MMSCFD starting in mid 1986.

Field stratigraphy

The general stratigraphic sequence of the Fulmar Field is shown in Fig. 4. Rifting in the Central Graben probably started in the early Permian with normal faulting and regional downwarping. This gave rise to a rapid transgression of the Zechstein sea and subsequent deposition of a thick sequence of carbonates and evaporites. These are the oldest rocks so far penetrated under the Fulmar Field (Shell/Esso well 30/16-6). Movement and removal of Zechstein salt associated with graben faulting is considered to have been the main control on the thickness of facies present in the overlying Triassic and Jurassic rocks.

The Triassic Smith-Bank Formation comprises non-marine, red-brown shales and silty sandstones. Extensive erosion by the Mid Cimmerian Unconformity was followed by a marine transgression and deposition of the Upper Jurassic Fulmar Formation which is the main productive formation on the Fulmar Field. Salt withdrawal, which probably started during the Triassic, resulted in lenticular sand bodies that progressively migrated to the west.

A major transgressive event marks the end of Fulmar Formation deposition and the beginning of deposition of the Kimmeridge Clay Formation. A series of turbiditic sandstones, called the Ribble Member, are located near the base of the Kimmeridge Clay. Extensive erosion during the Late Cimmerian truncated a large part of the reservoir on the eastern and northern flanks of the field. The Fulmar Field was a positive feature during most of the Cretaceous. It was eventually onlapped and covered by Upper Cretaceous chalks. The period from the Tertiary to the present day has been one of continual subsidence and enhancement of the Fulmar structure by differential compaction.

Geophysics

The Fulmar Field was first identified using 2-D seismic surveys shot in 1970 and 1974. A 3-D seismic survey was shot and processed in 1977 covering an area of 62 sq km and consisting of 96 lines with a spacing of 75 m. Interpretation was done initially on paper sections but is now done using an interactive workstation.

Data quality is considered moderate by present day standards (Fig. 2b). The Late Cimmerian Unconformity, top Ribble, top Fulmar Formation and Zechstein reflectors are well defined; but the top Triassic (base Fulmar) is very weak and discontinuous. Fault identification within the reservoir is difficult particularly where faults are truncated by the Late Cimmerian Unconformity to the north and east of the field. Time to depth conversion is achieved by mapped interval velocities of three layers to the top Fulmar and by interval velocity functions to deeper horizons.

Trap

The Fulmar Field is a small (11.3 sq km) anticlinal closure (Fig. 5) with its crest at 9900 ftss. It has an approximately triangular shape as illustrated on the top Fulmar sands structure map (Fig. 3). The west flank of the field dips steeply at 25°, elsewhere dips are less than 12°. The Fulmar Formation sandstones are overlain conformably by Kimmeridge shales which provide the seal in the south and west. To the north and east there is progressive truncation of the reservoir which is sealed by unconformably Upper Cretaceous chalk.

The field is heavily faulted with two main suites identified.

(1) A NNW–SSE trending suite of moderately low angle faults

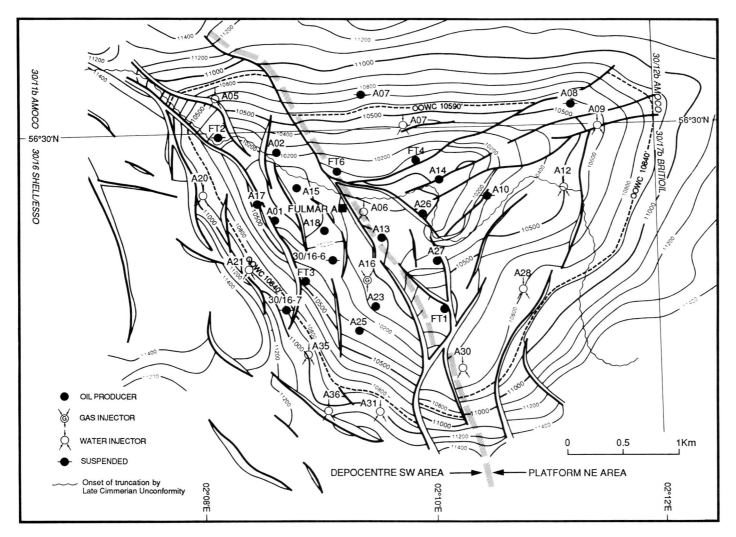

Fig. 3. Top Fulmar Formation structural depth map, with well locations and original oil–water contact. The 'hinge-line' between the depocentre and more stable platform area to the northeast is indicated, along the onset of reservoir subcrop against the base Cretaceous unconformity.

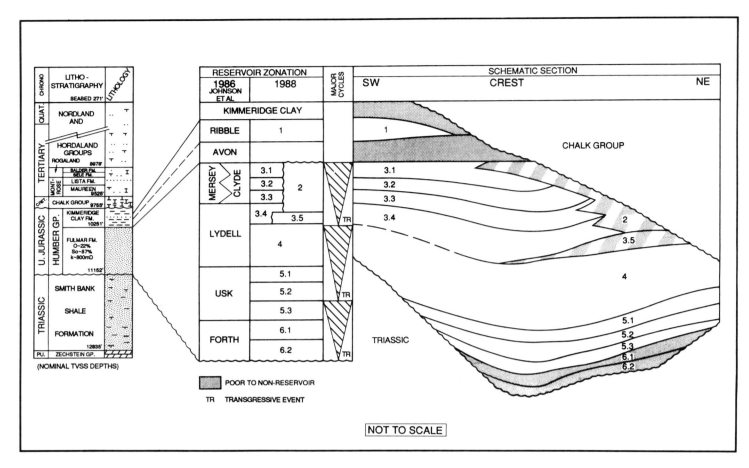

Fig. 4. Sketch of Fulmar Field stratigraphy, reservoir zonation and a schematic cross-section. A degree of historical review to the reservoir subdivision is indicated.

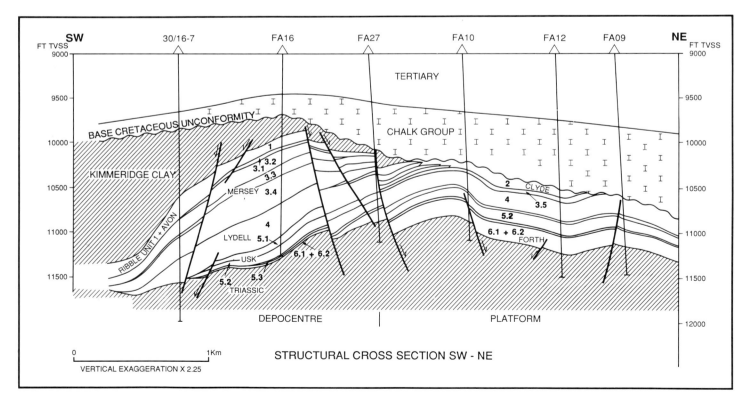

Fig. 5. SW–NE structural cross-section indicating the mapped extent and geometry of the reservoir units.

that generally sole out in the Triassic. These faults are related to the Auk Horst and Central Graben boundary fault. They dip to both east and west with throw of up to 300 ft.

(2) A second series of faults have an ENE–WSW trend and are particularly common in the eastern side of the field where they form a horst feature. The faults are steeper than the NNW–SSE trending faults, cut the Zechstein and may have a component of strike slip motion which further supports transtensional Late Cimmerian tectonics.

The generally lenticular shape of the Fulmar sandstones in cross-section (Fig. 5) is considered to have been caused by extensive mobilization of Zechstein salt which originally underlay the Fulmar area. Salt withdrawal resulted in local downwarping of the Triassic surface and the generation of rim-synclines which moved progressively westwards during the Triassic and Jurassic.

During the deposition of the Fulmar sandstones, salt withdrawal was more rapid in the west of the present field area. This resulted in a NNW–SSE 'hinge line' (Fig. 3) which runs through the centre of the field and separates thicker, homogeneous sands in the west and thinner, more stratified sands in the east. Further salt withdrawal and reactivation of the Auk Horst during the Upper Jurassic formed a 'deeper water' basin where deposition of a thick sequence of Kimmeridgian shales occurred.

Reservoir

The Fulmar Formation is Oxfordian to Kimmeridgian in age and consists predominantly of fine- to medium-grained, shallow marine, arkosic sandstones which were probably derived from Triassic sediments on the Auk Horst. There is a pronounced variation in both thickness and facies across the field related to the 'hinge line' as illustrated by the type logs shown in Fig. 6. To the southwest, where subsidence was greatest, the sands are up to 1200 ft thick, massive and of excellent reservoir quality. To the northeast, where sedimentation was more restricted, the sands are thinner, more interbedded and generally of poorer reservoir quality. The overall stratigraphy represents a major regressive sequence in which three

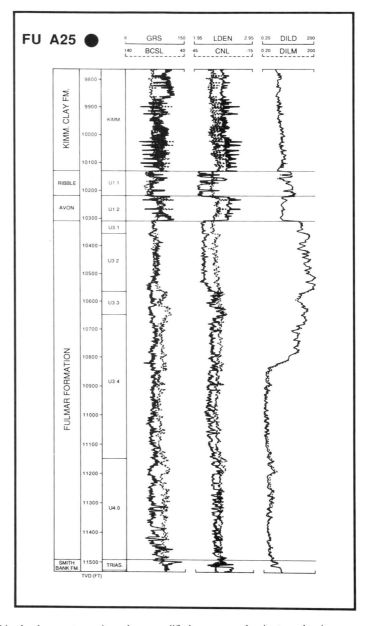

Fig. 6 (a) Reservoir type log. Well FA-25 is typical of the thick succession within the depocentre region where amplified sequences dominate and unit boundaries tend to be subtle. (b) Reservoir type log. Well FA-12 is typical of the thinner succession developed on the north-eastern 'platform' where a more pronounced layer-cake stratigraphy is visible and where the upper 'cycle' of the Fulmar Formation, the finer-grained sands of the Clyde Unit 2., are strongly developed.

transgressive/regressive cycles can be recognized. Bioturbation is common throughout and often destroys the primary sedimentary structures. Marine turbidites form a separate reservoir within the Kimmeridge shales on the western flank.

Diagenesis occurred soon after deposition and resulted from burial within a closed aquifer. Dolomite cement is locally common. It occurs as early nodules and later pore filling rhombs. Authigenic kaolinite is almost absent. Silica cementation, associated with dissolution of sponge spicules, is common in the poorer reservoir rocks on the northeast flank. Fracturing and brecciation deformation, is common; particularly in the area of maximum subsidence in the western part of the field.

The Fulmar reservoir was originally subdivided into six members named after British rivers and based on the FULMAR mnemonic. These are called, from the base: Forth, Usk, Lydell, Mersey, Avon and Ribble (Fig. 4). A seventh, the Clyde member, is found in the east of the field and is a shaley equivalent of the Mersey member. Following the acquisition of more well data, the section has been further subdivided into 14 reservoir units (Fig. 4) which reflect the depositional cycles in the overall succession. A stratigraphic correlation across the field is illustrated in Fig. 7.

Forth–Lower Usk (Units 6.2–5.3)

The lowest regressive sequence, the Forth and the Lower Usk Units, has a maximum thickness of 265 ft in the east, but is absent due to non-deposition in the west. The Forth Member (Units 6.2 and 6.1) consists of bioturbated (*Chondrites* and *Zoophycus*), very fine-grained, argillaceous, glauconitic siltstones and shales. These rocks are all considered to be non-reservoir (average porosity = 14% and K_{air} = 0.3 md). The Lower Usk (Unit 5.3) is a thin sand that can be correlated over a large area. This unit has a distinctive funnel-shaped log character corresponding to a rapid, upward increase in grain-size and an improvement in reservoir quality.

Middle Usk–Lydell (Units 5.2–4)

The Middle Usk (Unit 5.2) was deposited during a transgressive event and comprises interbedded shales and poor to moderate quality sandstones. There is a rapid coarsening upwards through the Upper Usk (Unit 5.1) into the Lydell (Unit 4). The Lydell Member is a thick (maximum 350 ft), fine- to medium-grained, well supported sandstone with excellent reservoir quality (average porosity = 23 percent and K_{air} = 100–4000 md). These sandstones are very bioturbated and are interpreted to be a complex of stacked sand ridges.

Mersey–Clyde (Units 3.5–2)

The impact of the 'hinge line' is very noticeable on the third regressive cycle. Subsidence continued in the southwest with deposition of massive sandstones of the Mersey Member (Units 3.1–3.5) which have similar reservoir properties to the Lydell Member. Northeast of the 'hinge line' was more stable. Distal, lower energy, shoreface sedimentation occurred here. These argillaceous, fine-grained sandstones are called the Clyde Member (Unit 2). Although the Clyde Member has an average porosity of 21%, the sandstones are rich in siliceous sponge spicule debris which have been largely redistributed within the pore space and has resulted in a much reduced permeability (average = 4 md). Figure 8 illustrates the different porosity–permeability relationships of the Mersey and Clyde Members.

Avon and Ribble (Unit 1 reservoir)

The Ribble sands (Unit 1) are separated from the main reservoir by laminated, rarely burrowed, Kimmeridgian age Avon Member shales. The sandstones are up to 180 ft thick. These fine- to

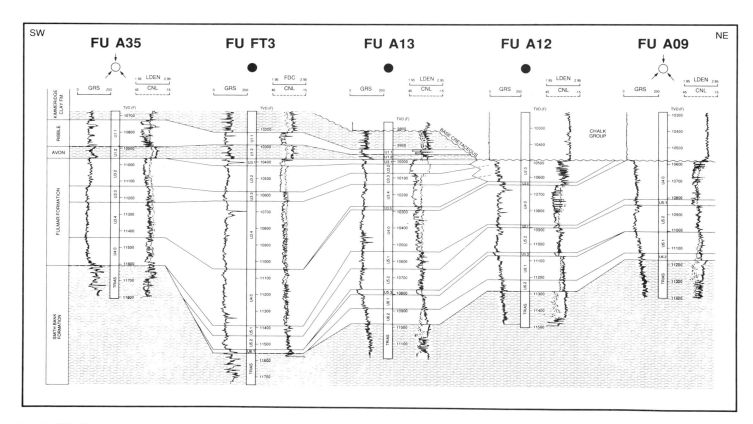

Fig. 7. SW–NE stratigraphic correlation section across the Fulmar Field.

medium-grained subarkosic sandstones commonly contain clay clasts and are interbedded with laminated shales. Ribble reservoir quality is excellent. Average porosity is 24% and K_{air} is 1000–10 000 md. Deposition is interpreted to have been from high energy gravity flows from the Auk horst. Several distinct sandstones can be recognized all of which thin and pinch out toward the crest of the structure.

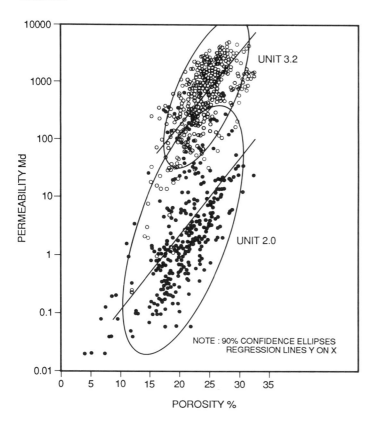

Fig. 8. Porosity–air permeability relationships of the Mersey Unit 3.2 compared with the Clyde Unit 2. sandstones. Unit 3.2 has an average porosity of 24.7% and K_{air} of 674 md, whilst Unit 2. has a comparable porosity of 20.5%, but a K_{air} of 2.8 md.

Source

The mature, organic-rich (average TOC = 5%) shales of the Kimmeridge Clay formation are the source of the oil in the Fulmar Field. Oil migration occurred during the early to middle Tertiary from the deeper parts of the Central Graben with further migration from the overlying Kimmeridge shales.

Hydrocarbons

The Fulmar crude has a gravity of 40° API and is highly undersaturated (P_b = 1800 psi). Original reservoir pressure was 5700 psi and temperature was 285° F at the datum depth of 10 500 ftss. The field production GOR is 614 SCF/STB in the Fulmar sandstones and 515 SCF/STB in the Ribble sandstones (Unit 1). Oil viscosity is 0.4 cP and the original oil formation volume factor is 1.43 rb/stb.

The exploration wells encountered an OWC in good quality Lydell sands at 10 830 ftss which is coincident with the structural spill point. Later wells drilled in the north of the field have encountered an OWC at 10 590 ftss, 240 ft higher. Formation pressures indicate that there is communication through the oil column between these two separate oil pools and this has been confirmed by later production history.

The shallow OWC appears to be restricted to a fault bounded block in the north of the field. During migration of oil the formation water was displaced until it was trapped by the northern chalk seal and fault juxtaposition of the reservoir sands against non-reservoir Forth and Triassic rocks. The resultant 'perched water' model allows communication through the oil column. The OWC in the Ribble Member occurs at 10 875 ftss which is 45 ft deeper than that in the main sands units.

Gas injection started soon after production in 1982. By the time the Fulmar gas pipeline became operational in 1986 a total of 80 BCF gas had been injected which had formed a secondary gas cap with a current GOC at about 10 180 ftss.

Fulmar Field data summary

Trap	
Type	Salt induced eroded anticline
Depth to crest	9900 ftss
Hydrocarbon contacts	Main field area OOWC 10 830 ftss
	Ribble OOWC 10 875 ftss
	Northern area OOWC 10 590 ftss
Oil column	930 ft
Pay zone formations	Kimmeridge Formation (Ribble Member)
	Fulmar Formation
Age	Oxfordian–Kimmeridgian
Gross thickness	Maximum Fulmar Formation 1200 ft
Net/gross ratio	Average 94%
Cut-off for net/gross	1 md air permeability
Porosity	Average 23%
Hydrocarbon saturation	Average 79%
Permeability	Average 50–800 md
Productivity index	Average 80 stb/d/psi
Hydrocarbons	
Oil gravity	40° API
Oil type	Undersaturated
Oil viscosity	0.42 cP
Bubble point	1800 psi
Gas/oil ratio	614 SCF/STB
Formation volume factor	1.43 RB/STB
Formation water	
Salinity	138 000 ppm NaCl equivalent
Resistivity	0.018 at 285°F
Reservoir conditions	
Temperature	285°F at 10 500 ftss
Pressure	5700 psi at 10 500 ftss
Presure gradient in reservoir	Oil: 0.29 psi/ft (0.64 gm/cc)
	Water: 0.45 psi/ft (1.10 gm/cc)
Field Size	
Area	2825 acres
Gross rock volume	877 500 acre-ft
Hydrocarbon pore volume	150 500 acre-ft
Recovery factor	57%
Drive mechanism	Water flood
Production	
Development scheme	Platform and wellhead jacket
	Oil: FSU/SALM; and gas: pipeline
Production rate	Plateau oil 165 000 BOPD
	Plateau gas export 88 MMSCFD
Cumulative production at 31.12.88	Oil: 319 MMBBL
	Gas: 121 BCF
	Water: 20 MMBBL
Cumulative injection at 31.12.88	Water: 360 MMBBL
	Gas: 80 BCF

The effective field aquifer appears to be relatively limited in extent and is estimated to be two to three times the reservoir volume. Formation water salinity is very high at 138 000 ppm NaCl equivalent, which equates to an Rw of 0.018 ohm m at reservoir temperatures. The high salinity allows very effective monitoring of the oil–water contact movement using pulsed neutron capture logs.

Reserves/resources

The total field STOIIP is estimated to be 815 MMBBL, 78% of which is contained in the Mersey and Lydell Members (Units 4–3.1) and a further 9% in the Ribble (Unit 1). The ultimate recovery is estimated to be 463 MMBBL oil and 264 BCF wet gas.

The large oil column, steeply dipping flanks and favourable mobility ratio make Fulmar very suitable for development by flank water injection and crestal oil production. A total of 17 oil producers, 10 water injectors and 1 gas injector have been drilled. Plateau oil production is 165 000 BOPD, associated gas production is 102 MMSCFD and water injection is 205 000 BWPD. At the end of 1988, 319 MMBBL of oil and 121 BCF gas have been produced with 360 MMBBL water and 80 BCF gas injected. Field life is expected to last until the year 2000.

The data and interpretations presented in this paper are the result of many years of study by Shell and Esso staff. The authors wish to thank the managements of Shell UK Exploration and Production, Esso Exploration and Production UK Ltd, Amoco Exploration Co., Amerada Hess Ltd and North Sea Inc. for permission to publish this paper. It should be noted that the interpretations presented are those of Shell and Esso and do not necessarily represent the views of all the field owners.

References

JOHNSON, H. D., MACKAY, T. A. & STEWART, D. J. 1986. The Fulmar Oil Field (Central North Sea): Geological Aspects of its Discovery, Appraisal and Development. *Marine and Petroleum Geology*, 3, 99–124.

MEHENNI, M. & ROODENBERG, W. Y. 1990. Fulmar Field, Central North Sea, UK Sector. *AAPG Atlas of Hydrocarbon Reservoirs* (in press).

VAN DER HELM, A. A., GRAY, D. I., COOK, M. A. & SCHULTE, A.M. 1990. Fulmar: The development of a large North Sea field. *In:* BULLER, A. T. ET AL. (eds) *North Sea Oil and Gas Reservoirs II*. Graham & Trotman, London, 25–45.

The Glamis Field, Block 16/21a, UK North Sea

A. R. FRASER & P. C. TONKIN

Sun International E and P Company Ltd, 80 Hammersmith Road, London, UK

Abstract: The Glamis oilfield lies within UK block 16/21a, 140 miles off the northeast coast of Scotland. The structure has the general form of an east–west oriented tilted fault block. The reservoir consists of clean sandstones of Late Kimmeridgian to Volgian age, overlain by the Kimmeridge Clay Formation but effectively sealed by Cretaceous marls. The sandstones were deposited directly on the eroded Devonian surface as the result of a late Jurassic marine transgression from the southwest.

The reservoir is subdivided into four facies: a basal conglomerate (unit 2b); an overlying fine- to medium-grained sandstone of shallow marine nearshore origin (unit 2a); a succeeding facies of coarsening upwards sandstones of barrier bar origin (unit 1b), and an uppermost localized silty bar-flank transitional facies (unit 1a). Units 2a and 1b constitute most of the reservoir section, having an average porosity of 15% and permeabilities up to 1.5 Darcies.

Glamis contains recoverable reserves of 17.5 MMBBL of 41.5° API undersaturated oil with an initial gas/oil ratio of 1037 SCF/STB. Oil production is from two wells, 16/21a-6 and 16/21a-8, with reservoir pressure maintained by water injection in well 16/21a-17z. Dedicated flowlines exist from each of the producers to a manifold on the Balmoral floating production vessel 4 miles northeast of Glamis.

The Glamis Field lies within UK block 16/21a, 140 miles from the northeast coast of Scotland in a water depth of 480 ft. It is elongated in an east–west direction, measuring 3.4 miles by 0.4 miles. The reservoir consists of marine sandstones of Volgian age, with recoverable reserves estimated to be 17.5 MMBBL. Oil is trapped in a structural high having the general form of a tilted fault block (Fig. 1). Glamis is named after the Scottish castle in Tayside, which was the childhood home of the Queen Mother. Other Sun operated fields in block 16/21 (Balmoral, Blair and Stirling) are also named after Scottish castles.

History

Block 16/21 was awarded to a consortium headed by North Sea Sun Oil Co. Ltd (now Sun Oil Britain Ltd) as part of the Fourth Licensing Round in March 1972. Block 16/21, together with block 211/22 were awarded under licence P.201. Following the mandatory 50% relinquishment of the block in March 1978, North Sea Sun Oil Co. Ltd retained 13 462 acres as block 16/21a. The relinquished portion became block 16/21b. Glamis Field was discovered in October 1982 by the drilling of well 16/21a-6. Four further appraisal or development wells were drilled between 1983 and 1986.

The field lies wholly within block 16/21a, for which the current percentage holdings for each participating company are Sun Oil Britain Ltd (62%), Deminex Balmoral Ltd (15%), Clyde Petroleum (Balmoral) Ltd (5%) Clyde Expro plc (10%) and Lasmo (TNS) Ltd (8%).

Well 16/21a-6 was drilled in August 1982 to test a Jurassic fault block lying 3.5 miles southwest of Balmoral Field. The well successfully tested oil from 66 ft of Upper Jurassic sandstone, with rates up to 8560 BOPD through a 2 inch choke. An oil–water contact was

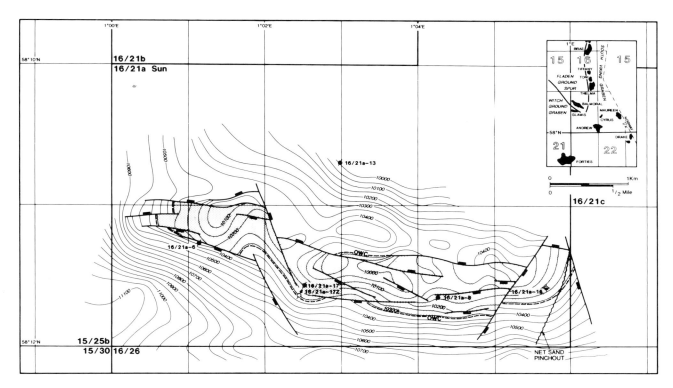

Fig. 1. Glamis Field depth structure map. Top Glamis Sandstone.

From Abbotts, I. L. (ed.), 1991, United Kingdom Oil and Gas Fields, 25 Years Commemorative Volume, Geological Society Memoir No. 14, pp. 317–322

identified at a depth of 10 306 ftss. The well reached a total depth of 11 206 ft in Devonian shales and pebbly sandstones. The Jurassic discovery was subsequently named Glamis Field.

Well 16/21a-8 was drilled in 1983 to appraise the easterly extension of the Glamis structure. It encountered an 81 ft oil column in Upper Jurassic sandstones similar to those in the 16/21a-6 well. No oil–water contact was apparent in this well. The well tested oil from the Jurassic at a maximum rate of 8035 BOPD through a 96/64 inch choke. A short oil column was also tested in Palaeocene sandstones, with oil, water and sand recovered from an interval perforated over 22 feet.

Well 16/21a-13 was drilled in 1984 to test a separate Jurassic fault block north of the Glamis structure. It penetrated water wet Upper Jurassic sandstones of good reservoir quality. The Palaeocene sandstones in this well tested oil at a rate of 3500 BOPD, apparently from the same reservoir as was tested in 16/21a-8. This Palaeocene oil accumulation was subsequently named Blair Field.

Additional seismic data was acquired in 1984 and the entire Glamis area was remapped. Well 16/21a-15 was drilled to evaluate the eastern limit of the field. Only 8 ft of Volgian sandstones were penetrated, and although this section was oil-bearing, the overall reservoir quality was considerably diminished. No drillstem tests were performed and the well was abandoned. Results of the well were used to map the eastern limit of Glamis Field corresponding to the pinchout of the Upper Jurassic sandstone.

Reservoir modelling studies carried out in 1985 showed that satisfactory recovery efficiency could be achieved with a simple development programme comprising two producers and one water injection well. Field development was considered to be economically viable due to the proximity of the Balmoral Floating Production Vessel 4 miles to the northeast. The water injection well 16/21a-17 was drilled late in 1985 at a position midway between wells 16/21a-6 and 16/21a-8. The well came in high to prognosis and was subsequently sidetracked to a point closer to the oil–water contact. Test injection resulted in a final average water injection rate of 15 300 BWPD.

Production from Glamis started in September 1989 and the field is expected to maintain a plateau production rate of 10 200 BOPD for 3 years. A single dedicated flowline will be run from each of the two producers to a Glamis manifold mounted on the Balmoral template. The Glamis oil will be directed on to the Floating Production Vessel (FPV) through flexible risers, where it will be comingled with Balmoral production prior to separation in the Balmoral two-stage train. The oil will be exported from the Balmoral FPV via a 14 inch line to the main Brae–Forties trunkline, and then to Cruden Bay, Scotland. The two production wells, 16/21a-6 and 16/21a-8, are predicted to cut water after approximately 2–3 years of production. The wells will then be sidetracked to crestal locations in order to maximize recovery.

Field stratigraphy

The oldest rocks identified in the Glamis area are continental mudstones, sandstones and conglomerates of Devonian age (Fig. 2). These were uplifted, tilted and eroded as a result of Hercynian movement, and no sediments of Carboniferous, Permian or Triassic age are known in block 16/21.

Cimmerian movements from late Triassic to middle Jurassic times marked the onset of major rifting and the formation of the Witch Ground, Central and Viking Graben systems. The Glamis area remained as a positive feature throughout the Bajocian to Early Kimmeridgian, and deposits within this age range have not been observed, with the exception of 3 ft of Middle Jurassic Rattray Formation volcanics penetrated by the 16/21a-8 well.

During Volgian times, the Glamis area was inundated from the southwest by a marine transgression. The shallow marine sandstone which forms the Glamis reservoir was deposited at this time, directly overlying tilted Devonian strata. The 'Glamis Sandstone', as it has been informally named, is overlain by the Kimmeridge Clay Formation which in this area comprises Volgian to Ryazanian shales. After the deposition of the Kimmeridge Clay Formation and renewed Cimmerian uplift, the Glamis area was again exposed to erosion and as a result, the Kimmeridge Clay Formation section is thin or locally absent, particularly in the crestal part of the field.

The area stayed above sea level until Barremian times when a thin sequence of claystones and marls of the Valhall Formation was deposited at the onset of the Cretaceous marine incursion. Deepening and extension of the Cretaceous sea led to deposition of marls and chalks of the Flounder, Tor and Ekofisk Formations which attain a combined thickness of more that 1000 ft in the Glamis area. In the axial zone of the subsiding basin clays, sands and reworked chalk of the Maureen Formation were deposited.

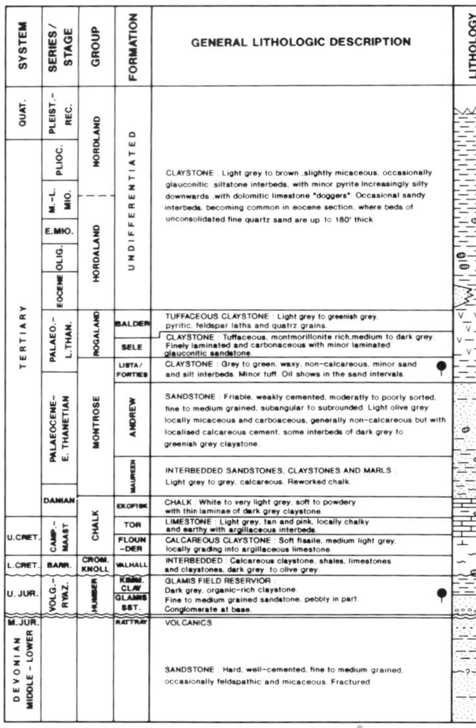

Fig. 2. Glamis Field stratigraphy.

With the cessation of chalk deposition at the end of the early Palaeocene, there followed a rapid and continual influx of clay and sand which, by the end of the Cenozoic, had filled the subsiding graben areas with 7000–10 000 ft of sediment. The sands were sourced from areas to the north and west and locally from intrabasinal highs around salt domes and along fault scarps. The Andrew Formation consists predominantly of sandstones of submarine fan origin sourced from the west. In block 16/21, the thickness increases from about 600 ft in the Balmoral area to more than 1000 ft in the southern Glamis area. Subsequent compaction and drape of the Andrew Formation over a Devonian high produced the broadly anticlinal Balmoral structure 3.5 miles northeast of Glamis.

The Lista Formation claystones and Forties Formation sandstones were deposited only thinly in block 16/21, the latter as submarine fan deposits sourced from the Orkney–Shetland platform. In the axial part of the graben muds of the Sele Formation were deposited in dysaerobic conditions. Deposition of tuff and mud of the Balder Formation marks the end of Palaeocene deposition.

The Hordaland Group, of Eocene to late Miocene age, and the Nordland Group of late Miocene to Pliocene age, were deposited as basinal muds with localized channel sands and turbidity flows. This entire sequence is 5000 to 6000 feet thick and provided sufficient depth of burial of the Kimmeridge Clay Formation for maturation and oil generation during the Oligocene.

Geophysics

Geophysical mapping is based on an interpretation of data from the Glamis Seismic Survey shot in 1984 using a Starjet source. The survey comprises 78 lines of 60 fold data, totalling 723 km in length with dip lines at a spacing of 250 m. Approximately 55% of the lines are in block 16/21; the remainder are in portions of blocks 15/25b, 15/30 and 16/26.

The lowest correlatable event seen on seismic sections is the reflection from the Base Cretaceous Unconformity (Fig. 3). The pick of this event was used in deriving the top reservoir map for Glamis Field. The reflection is a generally strong peak associated with a marked decrease in sonic velocity and density at the top of the Kimmeridge Clay Formation. Its character and amplitude are affected by variations in thickness of the Barremian Cromer Knoll Group above the unconformity, and the Kimmeridge Clay Formation and 'Glamis Sandstone' below the unconformity, all of which are close to tuning thickness.

Where the Kimmeridge Clay Formation is locally thin or absent as at several places on the crest of the Glamis structure, the reflection is of low amplitude because there is no significant acoustic contrast between the Chalk or Cromer Knoll Groups and the 'Glamis Sandstone'. Such areas are limited in extent and represent only a minor interpretational problem. The Kimmeridge Clay/ 'Glamis Sandstone' and 'Glamis Sandstone'/Devonian interfaces are acoustically significant and theoretically resolvable by seismic. However, the units are only thinly developed in the Glamis area, with the result that reflections from the interfaces are masked by interference from the stronger Top Kimmeridge Clay Formation reflection. The structure map of the Top 'Glamis Sandstone' was therefore derived indirectly, by the addition of Kimmeridge Clay Formation thickness to the seismic depth of the Base Cretaceous Unconformity.

Depth conversion of Base Cretaceous reflection times used a linear velocity function of the form:

V (Base Cretaceous) $= V$ (2500) $+ k\ T$, where V (2500) is the velocity at 2.5 seconds two way time, and k the acceleration factor are spatially variable. The method is applicable because of almost uniform velocity over the interval 2.5 seconds to Base Cretaceous observed in Glamis area wells. The interval corresponds approximately to the Flounder Formation.

The thickness of Kimmeridge Clay Formation over the Glamis structure was estimated by a combination of well control, and modelling of the amplitude and frequency characteristics of the Base Cretaceous reflection. Acoustic logs, synthetic seismograms and formation thicknesses in well 16/21a-8 were used as a basis for simple one dimensional modelling, in which thickness of units around the Base Cretaceous Unconformity were varied discretely, and the resulting reflectivity series convolved with a number of wavelets. The aim was to investigate the effect of thickness variations on the Base Cretaceous reflection, in particular the changes in amplitude and frequency resulting from thickness variations of the Kimmeridge Clay Formation. By comparing one dimensional models it was possible to generate estimates of Kimmeridge Clay Formation thickness, which upon integration with well control allowed an isopach map to be contoured over the Glamis structure. Addition of this map to the Base Cretaceous Unconformity depth map resulted in the top reservoir map for Glamis Field (Fig. 1).

Trap

The Glamis oil accumulation occupies an east–west oriented structural trap approximately 3.4 miles long and 0.4 miles wide. The trap has the general form of tilted fault block bounded in the north by normal faults of significant throw, and in the south by a combination of southerly dip and normal faults of relatively small throw (Fig. 4).

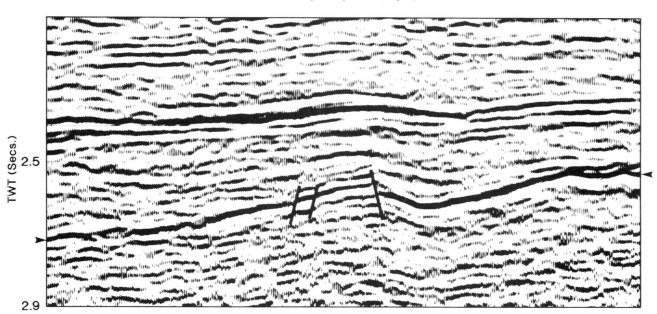

PORTION OF LINE SU 84-118 ACROSS GLAMIS FIELD. REFLECTION PEAK CORRESPONDING TO BASE CRETACEOUS UNCONFORMITY IS INDICATED BY ARROWS.

Fig. 3. Portion of line SU 84-118 across Glamis Field. Reflection peak corresponding to Base Cretaceous unconformity is indicated by arrows.

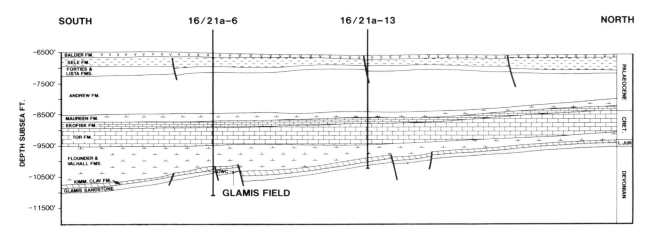

Fig. 4. Geological cross-section through Glamis Field.

The central part of the Glamis structure is displaced southwards by about 1000 ft relative to the western and eastern portions. This is interpreted as due to strike-slip movement along faults trending NNW and NE respectively. Most faulting is WNW trending, parallel to faults controlling the early to middle Jurassic opening of the Witch Ground Graben. It is likely that the Glamis structure formed during the latter stages of this Mesozoic phase of continental extension. Faulting is tentatively placed in the age range late Jurassic to early Cretaceous, following deposition of the Kimmeridge Clay Formation but preceding the deposition of an onlapping sequence of Cretaceous marls and claystones.

Over most of its area, the Glamis structure is overlain by thin shales of the Kimmeridge Clay Formation. These do not form a continuous seal as they are evidently absent at structural crests and along the planes of north hading faultings. Effective seal is thus provided by marls of the Flounder Formation which have a thickness of more than 500 ft in the Glamis area.

The oil–water contact identified in wells 16/21a-6, 17 and 17z is at 10 306 ftss, approximately equivalent to the lowest closing contour on the top reservoir structure map.

Closure is not dependent on fault seal or Devonian bottom seal, and the possibility exists for an oil column within the fractured Devonian rocks above 10 306 ftss. However, no definite Devonian oil column has been encountered in the field and the low porosities and permeabilities measured in Glamis wells suggest that a significant contribution to oil production from the Devonian section is unlikely.

Reservoir

The Glamis Sandstone of late Kimmeridgian to Volgian age, consists of a basal conglomerate overlain by medium-grained sandstones with some pebbly horizons and an uppermost unit of siltstones with minor interbedded fine-grained sandstones (Fig. 5). It is younger than the Piper Formation sandstone to the west, though similar in depositional environment. The sandstones were deposited on the shoreface and inner shelf area of the emergent Fladen Ground Spur. The sequence of facies indicates a marine transgression advancing from the southwest. The entire upper Jurassic sequence pinches out to the northeast of Glamis where it onlaps the Balmoral high.

The Glamis Sandstone has been subdivided into four reservoir units (Fig. 5). From oldest to youngest these are named Unit 2b, Unit 2a, Unit 1b and Unit 1a. Unit 2b is a basal conglomerate that directly overlies the Devonian and is present in wells 16/21a-6, 13 and 17. It has a maximum observed thickness of 12 ft in the 16/21a-6 well and thins eastwards, being absent at the eastern end of the field. It consists of clasts of Devonian sandstone, quartz, schist and mudstone within a pyritic sandstone matrix. The clasts range from pebble to boulder size and are subrounded and poorly sorted. Unit 2b represents deposition at the onset of the late Jurassic marine transgression and is largely derived from reworking of the weathered Devonian surface; coastal cliff erosion probably provided the boulder size material. Where core measurements have been made, porosities and permeabilities are low (7–12% and up to 80 md respectively), so the unit is not believed to contribute significantly to the reservoir.

Unit 2a is a fine- to coarse-grained sandstone, up to 92 ft in thickness, which thins northeastwards. It is present in all wells except 16/21a-15. The base of the unit generally consists of coarse-grained quartz sandstones above which the sequence fines upwards to moderately sorted fine-to medium-grained sandstones. These sandstones are cross-bedded, with minor bioturbation and local concentrations of glauconite and heavy minerals. Peaks on the gamma ray log in several wells correspond to bands with a high concentration of heavy minerals.

The unit is of shallow water nearshore origin and was subject to reworking by wave action, as evidenced by the lack of mudstone beds, and the presence of wave ripple laminations and low-angle cross-bedding. Sponge spicules and pelecypod fragments have been observed in cores. The upper part of unit 2a consists of fine-grained argillaceous sandstone, commonly with carbonaceous laminae. Coarser pebbly horizons in the upper part of the unit are high energy deposits of probable storm surge origin; they appear to be of limited lateral extent. These sandstones of Unit 2a are exceptionally clean having net to gross rates of at least 0.97 in all wells. Porosities range from 13–15% and permeabilities from 40–800 md, the higher permeability values occurring in the coarser-grained facies.

Unit 1b is a clean, well-sorted, massive sandstone, with discrete upward coarsening cycles. The top of the unit is fine-grained and represents a post-depositional phase of winnowing and reworking related to a sea level stillstand. The massive, coarser sandstones are extensively bioturbated, with rare cross-bedding and localized storm lag deposits. The unit varies in thickness from 8–65 ft, being thickest in a zone extending north from the centre of the field towards well 16/21a-13. A north–south barrier bar is suggested by this geometry, also evidenced by the coarsening upward cycles which are interpreted as shoals related to bar formation. The sand was supplied by offshore and storm-generated currents from source areas to the north and west of Glamis. Porosity is highest for unit 1b along the bar axis, up to 18%, decreasing to 14% towards the bar flanks. Permeabilities range from 500–1500 md. Like Unit 2a, net to gross ratios exceed 0.97 and Units 2a and 1b together make up the bulk of the Glamis reservoir.

Unit 1a is only observed in the 16/21a-6 and 16/21a-15 wells, where it is a fine-grained sandstone/siltstone sequence with an overall fining upwards trend. It is argillaceous and micaceous, with

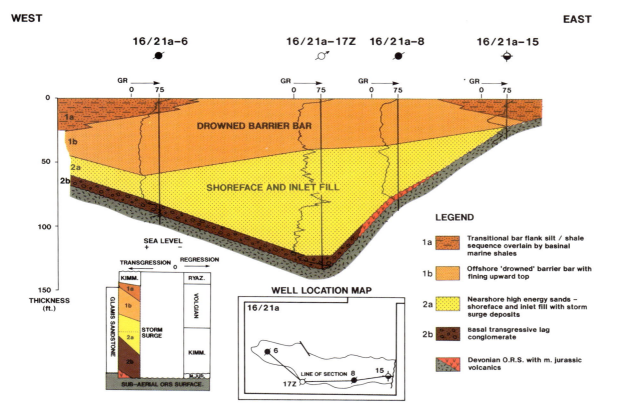

Fig. 5. Schematic subsurface stratigraphic cross-section of Glamis Sandstone.

traces of pyrite and carbonaceous material. Wave ripple laminations and low-angle cross-bedding are present, as are vertical and horizontal burrows. The unit has a sharply defined contact with the overlying Kimmeridge Clay, and is underlain by clean sandstones of unit 1b. From its distribution and lithology the unit appears to be a transitional bar flank facies marking the end of the late Jurassic transgressive event. The continued rise in sea level subsequent to this time must have been rapid to preserve the barrier bar and bar flank facies without extensive reworking by wave action. Porosities and permeabilities are generally less than 9% and 3 md respectively so the bulk of Unit 1a is non-reservoir. Unit 1a is directly overlain by claystones of the Kimmeridge Clay Formation.

Source

Analysis shows that the source rock for Glamis oil is the Kimmeridge Clay Formation, as indicated by the presence of bisnorhopane in oil from the 16/21a-6 well. However, when compared with oils from the Balmoral field, the Ts/Tm ratio and C28/C30 hopane ratio show that the Glamis oil was generated from a more mature interval and from a different facies of the source formation.

The Kimmeridge Clay Formation thickness varies from 23 ft in well 16/21a-17 to 75 ft in well 16/21a-13 just north of Glamis. Analysis of core material from the 16/21a-17 well shows that the claystones are high in oil-prone kerogen, but are only at a stage of early maturity. It is clear that oil has migrated into the reservoir from thicker, more deeply buried intervals of the source rock in the Witch Ground Graben to the southwest. Burial history curves indicate that oil migration occurred during Eocene to Oligocene times, at about the same time that the Balmoral structure was charged.

Hydrocarbons

The Glamis oil is undersaturated with an API gravity of 41.5° (0.818 g/cc) and an original gas/oil ratio of 1037 SCF/STB, at a reservoir pressure of 4562 psig. The average reservoir temperature is 246°F at a datum depth of 10 150 ftss. Formation water salinities are typically 84 500 ppm NaCl equivalent. Hydrogen sulphide levels of the separated gas are low (3.7 ppm in the 16/21a-17z well, and below the limit of detection in the 16/21a-8 well).

Two tests were conducted in the Jurassic reservoir of the 16/21a-6 discovery well in October 1982. DST 4 was conducted over the interval 10 376–10 390 ft MD, and flowed a maximum of 8650 BOPD. DST 5 was performed over the topmost silty reservoir section from 10 320–10 335 ft MD, and recorded no flow to surface. The 16/21a-8 well was tested in September 1983. DST 2 averaged 5320 BOPD over the interval 10 210 to 10 240 ft MD and flowed briefly at a maximum rate of 8035 BOPD on a 96/64 inch choke.

Drillstem tests were carried out in the 16/21-17z well during February/March 1986. DST 1 flowed 1650 BOPD from a short oil column between 10 382–10 386 ft MD. DST 2 was conducted over the interval 10 382–10 522 ft MD to clean out the perforations.

Reserves/resources

Reservoir simulation studies indicate that 17.5 MMBBL of oil will be recovered over a period of 7–8 years, with an overall recovery efficiency of 50%.

Production from Glamis commenced in September 1989 and reached a plateau rate of 10 200 BOPD which will be maintained until 1992. Production rates will decline rapidly as the water cut increases and after 2–3 years wells 16/21a-6 and 16/21a-8 will be sidetracked updip towards crestal locations in the field. Water injection via well 16/21a-17z commenced at the start of production to maintain reservoir pressure.

A maximum of 10.6 MMSCF/D of gas will be produced from Glamis over the period 1990 to 1991, declining over the life of the field. This will be used to sustain gas lift operations and provide fuel gas for Balmoral. The excess gas will be flared.

Glamis Field data summary

Trap
Type	Structural; asymmetric horst
Depth to crest	10 000 ft ss at Top 'Glamis Sandstone'
Lowest closing contour	10 300 ft ss at Top 'Glamis Sandstone'
Oil–water contact	10 306 ft ss
Oil column	81 ft maximum (16/21a-8 well)
Maximum vertical closure	306 ft

Pay zone
Formation	'Glamis Sandstone'
Age	Volgian
Gross thickness	20–140 ft
Lithology	Sandstone
Porosity (average)	15%
Hydrocarbon saturation (average)	88%
Permeability	40–1500 md
Net to gross (average)	0.91

Hydrocarbons
Density	41.5° API (0.818 g/cc)
Gas/oil ratio (initial)	1037 SCF/STB
Formation volume factor	1.635 RB/STB
Bubble point (initial)	3210 psig

Formation water
Salinity (average)	84 500 ppm NaCl equivalent
Resistivity (average)	0.028 ohm m at 246°F

Reservoir conditions
Temperature	246°F at datum depth of 10 150 ft ss.
Pressure (initial)	4562 psig

Field size
Net area	790 acres
Recoverable oil	17.5 MMBBL
Recovery factor	50%

Production
Commencement	September 1989
Development	2 producers, 1 water injection well
Plateau production rate	10 200 BOPD

This article is based on in-house and contractor studies. The authors would like to thank Sun International Exploration and Production Company Ltd and its partners in the Glamis Field, for giving permission to publish this paper.

Acknowledgement is given to J. Dyer and P. Jordan of Sun International E & P Co. Ltd who prepared much of the material for reservoir modelling studies and development planning documents.

The Highlander Field, Block 14/20b, UK North Sea

M. WHITEHEAD[1] & S. J. PINNOCK

Texaco Ltd, 1 Knightsbridge Green, London SW1X 7QJ
[1] *Present address: Enterprise Oil plc, 5 Strand, London WC2N 5HU, UK*

Abstract: Highlander Field, discovered in 1976, is a small oil accumulation located $7\frac{1}{2}$ miles northwest of the Tartan Platform and 114 miles northeast of Aberdeen in UK Block 14/20b. The Field lies on the NW–SE-trending Claymore–Highlander Ridge which forms the southern margin of the Witch Ground Graben. Upper Jurassic sandstones of the shallow marine Piper Formation and deeper marine turbidites (the 'Hot Lens Equivalent') within the Kimmeridge Clay Formation form the principal reservoirs. An additional important reservoir occurs within Lower Cretaceous turbidite sandstone and a small crestal accumulation occurs in Carboniferous deltaic sandstone. The structure is a tilted NW–SE-trending fault block downthrown to the northeast. The sandstone reservoirs all dip to the south and southwest and become thin due to onlap or truncation to the north. The Field has a combined structural–stratigraphic trap configuration. Seal is provided by Upper Jurassic siltstone and Lower Cretaceous calcareous claystone. The accumulations have been sourced from the Kimmeridge Clay Formation in adjacent basins. Eight wells delineate the structure and production is currently 30 000 BOPD. Ultimate recoverable reserves are 70 million barrels of crude oil. Development has been achieved utilizing an innovative remote subsea system, connected to the Tartan Platform $7\frac{1}{2}$ miles to the southeast.

The Highlander Field is located in UK Block 14/20b, 114 miles northeast of Aberdeen and $7\frac{1}{2}$ miles northwest of the Tartan Platform in 430 feet of water (Fig. 1). It is situated on a prominent structural feature, the Claymore–Highlander Ridge, on the southern flank of the Witch Ground Graben and is separated from the Halibut Horst, 8 miles distant to the southwest, by the Scapa–Highlander Sub-Basin (Fig. 1). The Field consists of a tilted fault block with three separate black oil accumulations. The principal accumulation is contained within sandstone reservoirs of late Jurassic age in a combination structural–stratigraphic trap. A second accumulation is contained within sandstone of early Cretaceous age, trapped stratigraphically by onlap onto the Highlander structure. A third, small, crestal accumulation is trapped structurally within sandstone of Carboniferous age. The ultimate recoverable reserves of the field are currently estimated to be 70 MMBBL of crude oil.

History

Texaco North Sea UK Ltd's 100% interest in Block 14/20 was obtained through licence P237 acquired in the Fourth Round of Licensing in March 1972. In Feburary 1979, a small portion of Block 14/20 (designated 14/20a) which formed part of the Tartan Field was carved out of licence P237. Subsequently, the remaining acreage within Block 14/20 (designated 14/20b) was granted a new licence, P324.

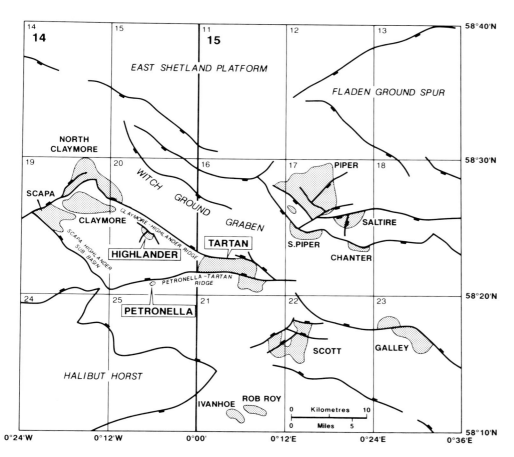

Fig. 1. Simplified tectonic framework of the Outer Moray Firth area, North Sea, showing the location of Highlander Field and surrounding discoveries.

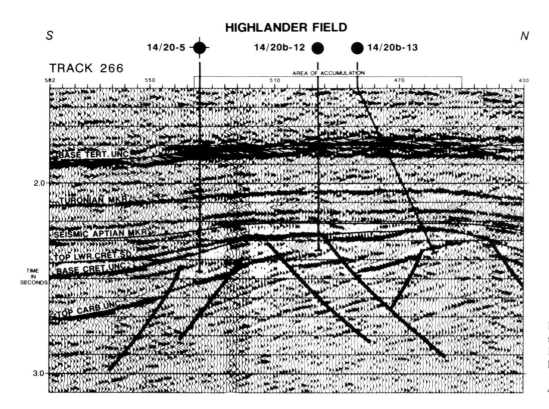

Fig. 2. Extract of north–south 3-D seismic line through the Highlander Field showing the principal seismic horizons and the locations of wells 14/20-5, 14/20b-12 & 13. (See Fig. 4 for location of line.)

Between 1974 and 1979 an approximate 1 km by 1 km 2-D seismic grid was recorded over Block 14/20. The first well on the Highlander Field was 14/20-5, which was drilled in April 1976 on a seismically defined fault closure on the southern flank of the Claymore–Highlander Ridge (Fig. 2). It encountered 441 gross feet of water-wet Upper Jurassic sandstone and 258 gross feet of Lower Cretaceous sandstone and established an oil–water contact for the Lower Cretaceous sandstone at −9546 ft TVDSS. The Lower Cretaceous sandstone tested 3296 BOPD of 29° API oil on a 36/64 inch choke from 47 ft of perforations. The well was subsequently plugged and abandoned as an oil discovery.

Well 14/20-8 was drilled in an attempt to delineate the accumulation first tested by 14/20-5 and assess the updip potential of the thick Upper Jurassic sandstone sequence. The well was drilled in August 1976 on the downthrown side of what is now recognized as the main bounding fault of the Highlander Field (Fig. 3). It encountered 294 gross feet of Lower Cretaceous sandstone with only poor oil shows and 52 gross feet of water-wet Upper Jurassic sandstone.

Following the re-processing of the 2-D seismic grid over Block 14/20 in 1979, in order to enhance the resolution within the Jurassic envelope, the Upper Jurassic discovery well, 14/20b-12, was drilled in July 1980 updip of 14/20-5 (Fig. 4). It encountered 254 gross feet of oil-bearing Upper Jurassic sandstone with a further 12 feet of sandstone beneath the oil–water contact. When tested, the well flowed at a rate of 6136 BOPD of 33.7° API oil on a $\frac{3}{4}$ inch choke and the well was suspended. Appraisal wells 14/20b-13, 14 and 15 were directionally drilled from the same surface location as 14/20b-12. Wells 14/20b-13 and 14 were drilled between September 1980 and August 1981 at updip locations to the western and eastern margins of the field respectively (Fig. 4). Both wells encountered oil-bearing Upper Jurassic sandstone. Well 14/20b-15 was drilled in April 1983 as a water injection well and established an oil–water contact for the Upper Jurassic reservoir at −9486 feet TVDSS. Wells 14/20b-12, 13, 14 and 15 were suspended on completion. Well 14/20b-H1 was the first to be directionally drilled from the template. It was located updip and to the west of 14/20-5 and enabled the first oil to be produced from the Lower Cretaceous Sandstone in September 1987.

The development of the field was undertaken in two phases. Initially, wells 14/20b-12, and 14 were connected back to the Tartan Platform via two 8 inch service lines and the first oil arrived at Tartan in February 1985. In the second phase these wells along with 14/20b-13 and 15 were tied back to the drilling/production template which was installed in July 1987.

Field stratigraphy

The earliest sediments penetrated within the Highlander Field are deltaic/paralic deposits of the Lower Namurian to Visean Forth Formation (Fig. 5). Deposition of this formation ceased with the onset of Hercynian tectonism in the late Carboniferous which consequently inverted the basin and led to erosion of the earlier sediments. Therefore, a major Hercynian unconformity separates the Carboniferous sediments from the overlying Zechstein carbonates of the Halibut Bank Formation which were deposited following a phase of late Permian subsidence and marine incursion.

Following the withdrawal of the Zechstein sea, a thick sequence of continental red siltstone and shale of the Triassic Smith Bank Formation was deposited prior to and partly synchronous with the onset of early Cimmerian tectonic activity. Late Triassic to mid-Jurassic time was a period of regional tectonic disturbance and is recorded by a major unconformity at the base of the Upper Jurassic sequence. There are no Lower or Middle Jurassic sediments within the Highlander Field and the Triassic sequence is overlain directly by Kimmeridgian high energy marine sandstone of the Piper Formation. Following this period of shallow marine sand deposition during the Kimmeridgian, continued rifting, possibly at the onset of the late Cimmerian, caused local uplift and partial erosion of the Piper Formation in the Highlander Field area. It is also probable that the block and basin structural configuration of the Halibut Horst, Scapa–Highlander Sub-Basin and Claymore–Highlander Ridge was well established by this time.

Subsequent deepening of the basin and further widespread transgression led to the onset of deposition of black, organic-rich clay and silt of the Kimmeridge Clay Formation in restricted basinal conditions during early Volgian times. These sediments unconfor-

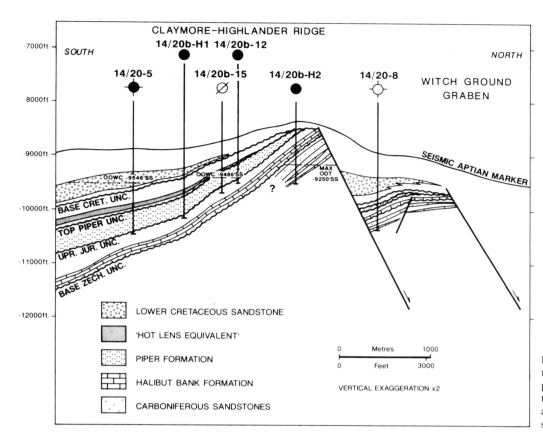

Fig. 3. Generalized cross-section through the Highlander Field with well locations projected onto it showing the structure and trapping styles of the hydrocarbon accumulations. (See Fig. 4 for line section.)

mably overlie the shallow marine Piper Formation. Periodically, coarser-grained clastic influxes resulted in the deposition of siltstone and fine- to medium-grained sandstone during early to mid-Volgian times by gravitational processes in the Scapa–Highlander Sub-Basin. This led to a thick sequence of sands being deposited in the Claymore Field, and a thin equivalent sand unit onlapped onto the uplifted Highlander structure. This thin turbidite unit close to the base of the Kimmeridge Clay Formation is informally known as the 'Hot Lens Equivalent' in the Highlander Field. Following a reduction in the supply of sand, a sequence of thin to very thin bedded, very fine-grained sandstone, siltstone and claystone was deposited during the remainder of mid-Volgian times. After a brief hiatus from latest mid-Volgian to earliest late Volgian times, quiet anaerobic basinal conditions prevailed and pelagic to hemi-pelagic, organic rich claystone accumulated over the Highlander Field. This high gamma ray claystone, known informally as the 'Hot Shale', was deposited from latest Volgian times to the Ryazanian.

Continued late Cimmerian tectonic activity caused the final collapse of the Witch Ground Graben from Valanginian to Barremian times. This intense tectonic activity coupled with inversion of the water column allowed the deposition of calcareous claystone in aerobic bottom waters. The calcareous claystone is unconformably overlain by coarse, gravity driven sediments in the Scapa–Highlander Sub-Basin. However, calcareous claystone deposition continued over basin margin highs. These clastic sediments were probably sourced from the Halibut Horst. Deposition continued from the Valanginian through to the Hauterivian, with minor influxes of sand and limestone deposition continuing into the Barremian.

Following this final period of tectonic disturbance, the remainder of Cretaceous time was marked by basin subsidence. A thick sequence of calcareous claystones of the Sola and Rodby Formations accumulated over the Highlander Field during the rest of the early Cretaceous. A thick sequence of late Cretaceous limestone of the Chalk Group marked the drowning and cut-off of clastic supply into the basin. Renewed clastic input into the basin led to the accumulation of over 6000 ft of Tertiary sediments in the Highlander Field area. Quaternary sediments outcrop on the seafloor.

Geophysics

A 1 km by 1 km 2-D seismic grid was shot over Block 14/20 in various stages between 1974 and 1979. This data allowed the identification of several structural prospects within Block 14/20, and was used in the early mapping and location of exploration and appraisal wells on the Highlander Field. A 5600 km 3-D survey covering the entire block was shot in the summer of 1981. This detailed survey provided good seismic control and maps of nine principal seismic horizons were produced in the main covering the entire block. The horizons mapped were as follows: Base Tertiary, Turonian Marker (Top Plenus Marl), Base Upper Cretaceous, Seismic Aptian Marker (seismic marker near top Barremian), Top Lower Cretaceous Sand, Base Cretaceous, Upper Jurassic Unconformity, Top Zechstein and Top Carboniferous.

The top of the Piper Formation is not discernible on reflection seismic records in this area. Therefore, an appropriate Kimmeridge Clay Formation isochore has to be hung beneath the Base Cretaceous seismic horizon in order to generate this important surface.

Trap

The trap for the Highlander Field oil accumulations exhibits both structural and stratigraphic elements. The Highlander Field is situated on the upthrown side of a major northwest to southeast trending fault which downthrows to the north (Figs 3 & 4). The first evidence of movement along this fault occurs in early Volgian times although the majority of the throw occurred during the early Cretaceous when up to 1200 ft of vertical displacement took place. The associated tilted fault block dips to the south and southwest and is transected by a series of NE–SW-trending faults, the throws on which rarely exceed the thickness of the Upper Jurassic reservoir sequence (Piper Formation and 'Hot Lens Equivalent').

The Carboniferous accumulation is trapped by upthrown fault closure with the seal being provided by calcareous claystone of the Valhall Formation. Interbedded claystone and siltstone may also contribute to the seal of this reservoir. No oil–water contact has

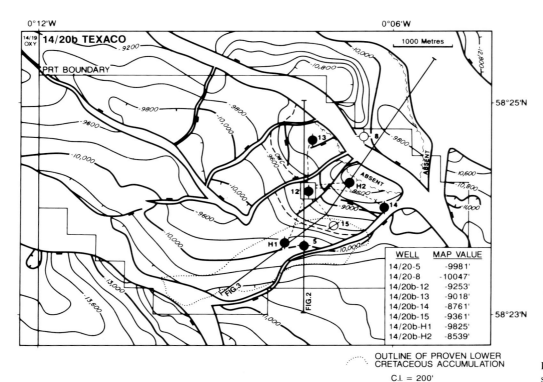

Fig. 4. Top Piper Formation depth structure map of the Highlander Field.

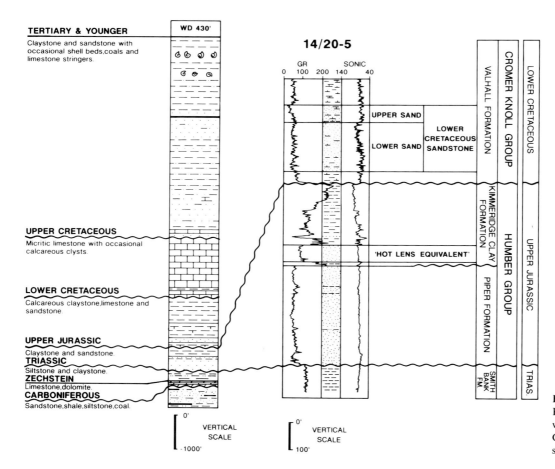

Fig. 5. Generalized stratigraphy of the Highlander Field (based on composite well data) with detail of 14/20-5 Lower Cretaceous and Upper Jurassic reservoir sections.

been proven for the Carboniferous accumulation, however the accumulation may have an oil column up to 450 ft thick (Fig. 3).

The Upper Jurassic accumulation is trapped structurally within the fault block with erosion of the reservoirs beneath the Base Cretaceous Unconformity occurring along the crest of the structure. However, the onlap of the 'Hot Lens Equivalent' during the early Volgian onto the embryonic structure intoduces stratigraphic elements to the trapping mechanism (Fig. 3). The seal for both reservoirs is provided by siltstone of the Kimmeridge Clay Formation and calcareous claystone of the Valhall Formation. The crest for the Upper Jurassic accumulation is at -8500 ft TVDSS and an oil-water contact has been established at -9486 ft TVDSS. This gives a hydrocarbon column of 986 ft. Although the Upper Jurassic accumulation is contained within two separate reservoirs (Piper Formation and 'Hot Lens Equivalent'), they are in pressure communication and are believed to share the same oil-water contact.

The Lower Cretaceous accumulation is trapped stratigraphically by onlap of the Lower Cretaceous sandstone onto the Claymore–Highlander Ridge, the enclosing calcareous claystone of the Valhall Formation and siltstone of the Kimmeridge Clay Formation providing the seal. The accumulation has an oil-water contact at -9546 ft TVDSS and a maximum oil column of 506 ft.

Reservoirs

Lower Cretaceous sandstone

The Lower Cretaceous sandstone is Valanginian to Hauterivian in age. The sandstone consists predominantly of fine- to medium-grained, massive to parallel-laminated units which are interbedded with, or have tops of, very thinly-bedded to laminated, bioturbated siltstone, claystone and very fine-grained sandstone. Within the sandstone, small clay clasts may be common and, locally, thin matrix and clast-supported pebble-conglomerates occur. The reservoir is subdivided into an Upper Sand unit and a Lower Sand unit. The Upper Sand is predominantly of poor reservoir quality due to the extensive development of calcium carbonate cements which occlude most of the porosity. However, in thin zones where the Upper Sand is not cemented by carbonate, porosities of 20–30% and good permeabilities have been recorded. The boundary between the Upper and Lower Sand units is marked by an increase in porosity coincident with the incoming of good quality sandstone. The Lower Sand is a good quality reservoir with high porosity and high permeability and only occasional nodular carbonate cements.

Lateral continuity of individual beds cannot clearly be demonstrated. However, the gross internal stratigraphy of the sandstone can be correlated between wells. The Lower Cretaceous Sandstone of the Highlander Field can be correlated, at least in part, with the Scapa Sandstone Member of the Valhall Formation as defined for the Scapa Field to the northwest (Boote & Gustav 1987; Harker *et al.* 1987; O'Driscoll *et al.* 1990).

The sandstone is interpreted to have been deposited in an aggrading, submarine fan environment, possibly within channels and sand lobes in a tectonically active, narrow basin. This may have been an extension of the Scapa system to the northwest, with the sands being predominantly transported axially along the Scapa–Highlander Sub-Basin. As a result, the sandstone thins mainly due to onlap onto the Claymore–Highlander Ridge and thickens off-structure into the adjacent low (Fig. 3).

'Hot Lens Equivalent'

The 'Hot Lens Equivalent' is a subarkosic, massive, fine- to medium-grained sandstone. It is separated from the underlying Piper Formation by a thin unit of very thinly interbedded siltstone and claystone at the base of the Kimmeridge Clay Formation. This intervening unit is thin to absent on the crest of the Highlander structure, and thickens off-structure into the adjacent Scapa–Highlander Sub-Basin.

The 'Hot Lens Equivalent' is believed to have been deposited by mass flow/turbidity currents. It is correlatable with the youngest Jurassic sandstone within the Claymore Field and was probably part of the same turbidite complex which comprises both the Claymore Sandstone in the Scapa–Highlander Sub-Basin and the 'Hot Lens A' of the Tartan Field. Rotation of the Highlander fault block had already begun at the time of 'Hot Lens Equivalent' deposition and the sandstone onlaps onto the structure (Fig. 6). The sandstone is therefore thin or absent along the crest of the structure and thickens off-structure to a maximum penetrated thickness of 80 ft. Where it is well-developed, the 'Hot Lens Equivalent' has porosities of 15–19% and permeabilities in the range of 10–100 md.

Piper Formation

The Piper Formation of the Highlander Field consists entirely of clean, subarkosic sandstone which is predominantly coarse-grained but may vary in grain size (especially towards the base of the sandstone) from fine- or medium-grained to a granule conglomerate. The formation contains a marine microfossil assemblage and, with the exception of common bioturbation towards the base and some rare local cross-bedding, the sandstone is generally structureless. It is believed to have been deposited in a very high energy marginal marine environment.

The coarsening upward cycles which characterize the Piper Formation of the Tartan Field (Coward *et al.* this volume) are not developed over Highlander, making correlation between the fields difficult. However, biostratigraphic data do suggest that the Piper Formation of the Highlander field may be equivalent to the Main and '15/16-6' Sand cycles of the Tartan Field.

The Piper Formation sandstone forms a thick clastic wedge which thickens off-structure with both the top and base of the formation being unconformities (Fig. 6). The sandstone is believed to be absent along the crest of the structure due to erosion below the Base Cretaceous Unconformity. It has fair to excellent porosity and permeability with no laterally continuous low permeability zones within the main body of the reservoir. Lateral and vertical continuity within the reservoir is therefore excellent. The one exception is found in sandstone in the highest structural positions which had the highest original permeability. In these sandstones extensive sulphate (mainly barite) mineralization has taken place. The sulphates appear to nucleate and grow poikilotopically, eventually coalescing and completely occluding the porosity. However, high permeability channels have been maintained between such patches, so that even when the overall porosity of the sandstone is reduced to 4 or 5%, permeabilities can still exceed 1000 md. Wells 14/20b-14 and 14/20b-H2 have both been extensively cemented in this way yet it can be demonstrated that up to 90% of the flow within the well is contributed by the more heavily cemented portions of the sandstone.

Carboniferous sandstone

The Carboniferous sandstone is generally medium- to coarse-grained, arkosic and poorly sorted. It commonly fines upwards and is interbedded with a typical assemblage of siltstone, micaceous siltstone, shale (both marine and non-marine) and coal with occasional rootlet beds. The sandstone is interpreted as being deposited within a deltaic environment. Individual units appear to be of limited lateral extent and as such, may have a ribbon geometry forming discrete sandbodies. Porosities are moderate to low and permeability is greatly reduced by kaolinite cements (up to 15% bulk rock volume) formed by breakdown of the feldspathic detrital components.

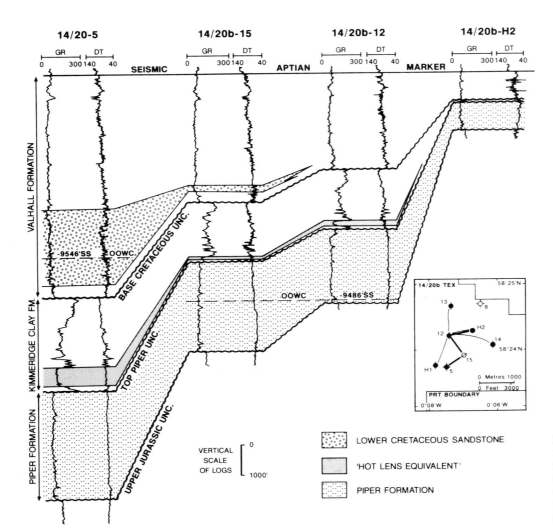

Fig. 6. Well correlation through 14/20-5, 14/20b-15, 12 & H2 hung beneath Seismic Aptian Marker showing gamma ray and sonic log response. Major unconformities are illustrated along with thickness variations within the Lower Cretaceous and Upper Jurassic sandstone reservoirs.

Source

The source rock for the hydrocarbons within the Highland Field consists of organic-rich siltstone and claystone of the Upper Jurassic Kimmeridge Clay Formation. This hydrogen-rich, oil-prone source rock has an average TOC of over 2% by weight and is believed to attain optimum maturity for oil generation and release below a depth of 10 000 ft. Burial to this depth has been reached in the adjacent Witch Ground Graben and parts of the Scapa–Highlander Sub-Basin. Time/depth reconstructions suggest that expulsion of oil from the Kimmeridge Clay Formation may have started during Eocene times and is still proceeding today (Bissada 1983).

Hydrocarbons

There are three separate oil accumulations in the Highlander Field. These are in the Upper Jurassic (Piper Formation and 'Hot Lens Equivalent'), Lower Cretaceous and Carboniferous reservoirs.

The Upper Jurassic and Lower Cretaceous reservoirs contain low sulphur black oil accumulations of similar gravity, gas/oil ratios and low bubble points. Formation water salinities for the two reservoirs are slightly different, with the Upper Jurassic having an NaCl equivalent concentration of 87 000 ppm as compared to 50 000 ppm in the Lower Cretaceous.

Detailed information has been obtained from five drillstem tests within the Highlander Field. Of these tests, two were in the Cretaceous reservoir and three were in the Upper Jurassic reservoir. An additional test within the Carboniferous reservoir flowed at subeconomic rates.

Under natural flow the Cretaceous tests produced between 3296 and 5200 BOPD of dry oil and the Upper Jurassic tests produced between 4000 and 6620 BOPD with no water.

Reserves

The Upper Jurassic reservoirs have an essentially infinite aquifer to the south and southwest and, as such, pressure support via water injection has not been required to date. However, if and when required, water injection can be initiated immediately in existing downdip wells. The productivity of the Lower Cretaceous reservoir has not proven to be as high as in the Upper Jurassic. Therefore gas lift was installed at an early stage, raising production from 7000 to over 10 000 barrels per day. With declining reservoir pressure and increasing water cut in off-structure wells in the Upper Jurassic, gas lift has been installed in all of the producing wells. This has maintained well capacities at about 10 000 barrels per day. Test data, poor reservoir quality and low recoveries make production from the Carboniferous uneconomic at the present time. Recoverable reserves for the Upper Jurassic and Lower Cretaceous reservoirs of the Highlander Field are estimated to be 63 and 7 million barrels of crude oil respectively.

The authors wish to thank Texaco Ltd for the co-operation received and permission to publish this paper. In particular, special thanks to our colleagues in the Development Geology Group for their support and encouragement.

Highlander Field data summary

	Upper Jurassic	Lower Cretaceous
Trap		
Type	Structural–stratigraphic	Stratigraphic
Depth to crest	−8500 ft	−9040 ft
Lowest closing contour	−8700 ft	−9546 ft
Oil–water contact	−9486 ft	−9546 ft
Oil column	986 ft	506 ft
Pay zone		
Formation	Piper Formation	
Age	Upper Jurassic	Lower Cretaceous
Gross thickness (average/range)	141 ft/91–247 ft	109/24–154 ft
Net/gross ratio	1	0.55
Cut off for N/G	N/A	10 md
Porosity (average/range)	15%/7–25%	24%/3–30%
Hydrocarbon saturation	86%	78%
Permeability (average/range)	1350/13–8200 md	180/0–1200 md
Productivity index	100	12
Hydrocarbons		
Oil gravity	36	35
Oil type	Black	Black
Gas gravity	1.4	1.5
Bubble point (psia)	800	650
Dew point	N/A	N/A
Gas/oil ratio	85	70
Condensate yield	N/A	N/A
Formation volume factor	1.2	1.1
Formation water		
Salinity (ppm NaCl eq.)	87 000	50 000
Resistivity	0.087 at 68°F	0.134 at 60°F
Reservoir conditions		
Temperature (°F)	200	200
Pressure (psia at OWC)	4250	4250
Pressure gradient in reservoir (psi/ft)	0.33	0.33
Field size		
Area (acres)	600	500
Recovery factor	56%	25%
Drive mechanism	Combination (gaslift/water drive)	
Recoverable hydrocarbons		
oil (MMBBL)	63	7
gas (MMSCF)	5255	490
Production		
Start-up date	February 1985	September 1987
Development scheme	Subsea satellite to Tartan	
Production rate (BOPD)	27 000	3000
Cumulative production to end March 1989		
Crude (MMBBL)	31.9	2.0
NGL	0	0
Liquids (MMBBL)	31.9	2.0

References

BISSADA, K. K. 1983. Petroleum generation in Mesozoic sediments of the Moray Firth Basin, North Sea area. *In*: BJOROY, M. (ed.) *Advances in geochemistry*. John Wiley, London, 7–15.

BOOTE, D. R. & GUSTAV, S. H. 1987. Evolving depositional systems within an active rift, Witch Ground Graben, North Sea. *In*: BROOKS, J. & GLENNIE, K. W. (eds) *Petroleum Geology of North West Europe*. Graham & Trotman, London, 819–833.

COWARD, R. N., CLARK, N. M. & PINNOCK, S. J. The Tartan Field, block 15/16, UK North Sea. *This Volume*.

HARKER, S. D., GUSTAV, S. H. & RILEY, L. A. 1987. Triassic to Cenomanian stratigraphy of the Witch Ground Graben. *In*: BROOKS, J. & GLENNIE, K. W. (eds) *Petroleum Geology of North West Europe*. Graham & Trotman, London, 809–818.

O'DRISCOLL, D., HINDLE, A. D. & LONG, C. D. 1990. The structural controls on Upper Jurassic and Lower Cretaceous reservoir sandstones in the Witch Ground Graben, UK North Sea. *In*: HARDMAN, R. F. P. & BROOKS, J. (eds) *Tectonic Events Responsible for Britain's Oil and Gas Reserves*. Geological Society, London, Special Publication, **55**, 299–323.

The Ivanhoe and Rob Roy Fields, Blocks 15/21a-b, UK North Sea

R. H. PARKER

Amerada Hess Ltd, 2 Stephen Street, London W1P 1PL, UK

Abstract: The Ivanhoe and Rob Roy Fields are located in the Outer Moray Firth Basin, seventy nautical miles off the northeast coast of Scotland. The Ivanhoe Field was discovered in 1975, and the Rob Roy Field in 1984. The reserves in both fields occur in tilted fault block traps of Upper Jurassic, Piper Sandstone Formation. Estimated total recoverable reserves amount to 100 MMBBL and 62 BCF. The fields are separated by a water corridor approximately 1 km wide. Both fields contain two reservoir sandstone units, an upper and lower, locally termed the Supra Piper Sandstone and Main Piper Sandstone respectively. The reservoirs in both fields exhibit excellent rock properties with porosities up to 28% and permeabilities of several Darcies.

Each field is developed via a subsea manifold surrounded by a cluster of production and injection wells, of which two were pre-drilled on Ivanhoe and six pre-drilled on Rob Roy. This allowed rapid achievement of the 60 000 BOPD plateau oil production rate soon after commissioning of facilities in July 1989. The two subsea manifolds are tied into a single subsea production manifold which connects with a Floating Production Facility. Crude oil is exported to the Claymore A Platform and gas to the Tartan A Platform.

The Ivanhoe and Rob Roy Fields are located 70 nautical miles off the northeast coast of Scotland in the eastern part of the Moray Firth Basin. The Ivanhoe Field lies completely within block 15/21a, whilst the Rob Roy Field has a small extension into block 15/21b (Fig 1). Water depths average 460 ft across the two fields. The two adjacent fields have a NW–SE orientation and are separated by a water corridor approximately 1 km wide.

The fields comprise two southwesterly tilted fault blocks trapping oil in Upper Jurassic Piper Sandstones which contain recoverable reserves in the order of 100 MMBBL and 62 BCF. The field names are taken from the novels of Sir Walter Scott. Rob Roy was published in 1817 and set in the 1715 Jacobite Rising, whilst Ivanhoe was published in 1819 and set in the rule of Richard the First (1189–1199).

History

Exploration rights to Block 15/21 were originally awarded in the 4th Round of UK offshore licensing. The initial six year period commenced on 15 March 1972, with licence P218 being awarded to a group operated by Monsanto Oil Company of the UK, Inc.

The Ivanhoe discovery well, 15/21-3, was drilled in 1975 to test a structure identified at Base Cretaceous. The well penetrated an Upper Jurassic interval consisting in part, of two sandstone reservoirs separated by a shale some 80 ft thick. The upper or Supra Piper Sandstone reservoir was 90 ft thick and oil-bearing with 87 ft net pay, whilst the lower or Main Piper Sandstone reservoir was 177 ft thick and water-bearing. The Supra Piper Sandstone tested oil (30° API) at rates up to 3730 BOPD. The next well, 15/21-4 was

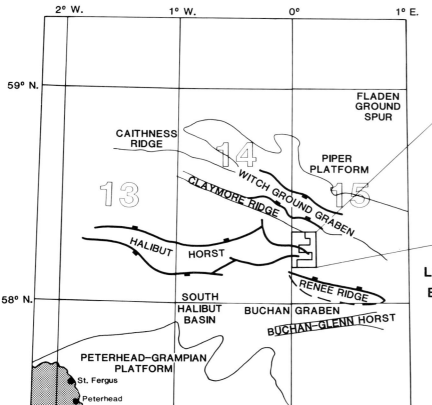

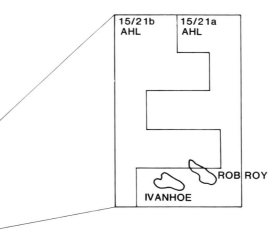

Location Map of Ivanhoe & Rob Roy Fields.
Blocks 15/21a & 15/21b U.K. North Sea.

Partnership:

Amerada Hess Limited (Operator)	42.084%
Deminex U.K. Oil & Gas Limited	43.333%
Kerr McGee Oil (U.K.) PLC	10.833%
Pict Petroleum plc	3.750%

Major Structural Elements Influencing Ivanhoe & Rob Roy Fields.

Fig. 1. Location map.

drilled in 1977 and tested a structure to the northeast of Ivanhoe. The Main Piper Sandstone was water-bearing, while the Supra Piper Sandstone, although containing oil shows, was of poor reservoir quality and not tested. In 1978 50% of the licence area was relinquished with the 15/21a part block being retained by the group.

Improved seismic data, acquired between 1977 and 1984, combined with a greater understanding of the Upper Jurassic stratigraphy, led to a better geological model and more accurate mapping. Key seismic events below the Base Cretaceous could now be mapped for the first time and the Top Jurassic Volcanics structure map showed a series of low relief tilted fault blocks overlain by Upper Jurassic sediments.

The Ivanhoe Field was confirmed by well 15/21a-8 drilled in 1983, testing oil from both the Supra and Main Piper Sandstones. Two further appraisal wells delineated the Ivanhoe Field (15/21a-9, 10) and gave a better understanding of the tectonic history and the influence of faulting on Piper Sandstone depositional environments. The discovery well for the Rob Roy Field is considered to be the 15/21a-11 well drilled in 1984 updip of the earlier 15/21-4 well. The 15/21a-11 well tested at prolific rates in both the Supra Piper Sandstone (up to 5275 BOPD) and in the Main Piper Sandstone (up to 8510 BOPD), enhancing the commercial viability of both fields.

The structural cross-section (Fig. 2) and seismic line (Fig. 3) illustrates the structural relationship between the two oil accumulations which are separated by the water corridor defined by well 15/21a-13. The Rob Roy Field was delineated by the 15/21a-12A, 13 and 14 appraisal wells completed during 1984. The results of the drilling programme confirmed that commercially viable reserves were present and an Annex B was submitted in June 1985.

In December 1985, Amerada Hess Limited purchased Monsanto Oil Company of the UK, Inc and assumed operatorship of the P218 Licence. In January 1986, government approval for the project was obtained and subsequently the relinquished part-block 15/21b was regained in the Tenth Offshore Licensing Round in May 1987 (Licence P588). The current composition of the licence in both blocks is Amerada Hess Limited 42.08%, Deminex UK Oil and Gas Limited (43.33%), Kerr McGee Oil (UK) plc (10.83%), Pict Petroleum plc (3.75%).

The development plan involves a semi-submersible vessel which has been converted into a Floating Production Facility (FPF) which is moored centrally between the two fields. Production from both fields is via production manifolds located over each field. Fluids collected at the production manifolds are routed through flexible flowlines to a riser base manifold and then onto the FPF. Processed products are exported from the FPF via the riser base manifold by pipelines. Crude oil is pumped to the Claymore 'A' platform and then onshore to the Flotta Terminal. Gas is exported to the Tartan 'A' platform and ultimately to shore at the St Fergus Terminal. Prior to commencement of production, six development wells were drilled on Rob Roy and two on Ivanhoe. This enabled a plateau production rate of 60 000 BOPD to be reached soon after commencement of production on 7 July 1989.

Field stratigraphy

The Ivanhoe and Rob Roy Fields are located within the Outer Moray Firth Basin. Figure 1 depicts the major structural features, in the general area of block 15/21, which may have influenced sediment deposition during the Jurassic. The fields lie within a shallow embayment between the southern part of the Halibut Horst and the Renee Ridge.

The stratigraphy of the fields is summarized in the stratigraphic log of Fig. 4. Three major episodes of basin formation are recorded within the sedimentary succession of the Moray Firth Basin. Thick,

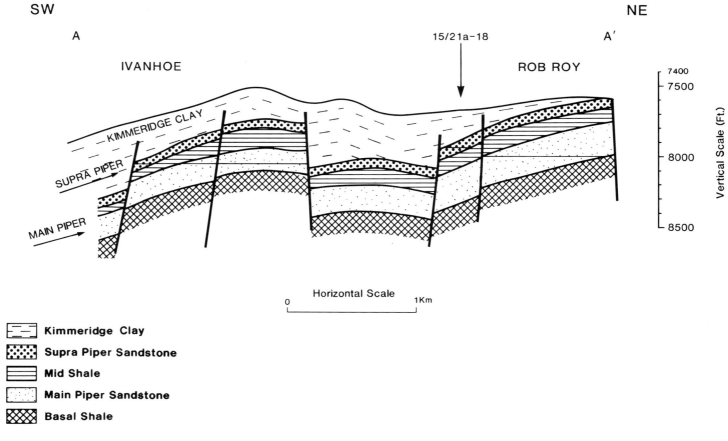

Fig. 2. Structural-cross section A–A' across Ivanhoe and Rob Roy Fields, Block 15/21. See Fig. 5.

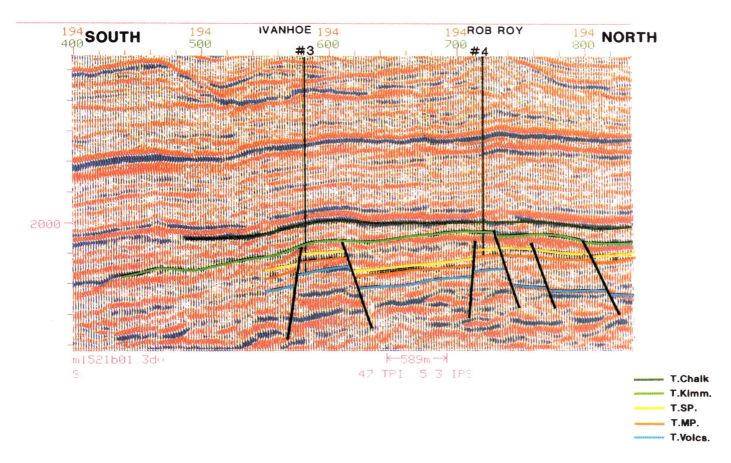

Fig. 3. Interpretation on seismic line MNS 3D-194.

Devonian and Upper Carboniferous non-marine sequences were deposited in a transtensional, later extensional tectonic setting, with the Hercynian orogeny terminating this early sequence. Continental red bed sequences were then deposited when sedimentation recommenced in the early Permian and continued throughout the Triassic. A major Early Jurassic marine transgression brought shallow marine shelf sea deposition to the Moray Firth area.

The late Lower to early Middle Jurassic involved thermal updoming of the Outer Moray Firth resulting in widespread erosion. This Mid-Cimmerian uplift terminated the second basin-fill sequence and culminated in the formation of the Central North Sea Dome. A major phase of extrusive volcanism commenced during the Bajocian, accompanying the collapse of the Central North Sea Dome. The centre of activity was located in the area in the south of Quadrant 15 and north of Quadrant 21. This Bajocian volcanic activity generated the volcanic and sedimentary sequences of the Fladen Group which is represented in the Ivanhoe and Rob Roy areas by the Rattray Formation. This Formation has greater than 50% volcanic lithologies and is a significant seismic marker in this area, and a key reflector in mapping the fields.

A final transgression during the Oxfordian led to the deposition of the Basal Shale Member of the Piper Formation, which is overlain by the Main Piper Sandstone reservoir.

The Main Piper Sandstone Unit is of late Oxfordian age and accumulated in a marine, inner shelf environment. A significant unconformity separates the Piper Formation from the overlying Kimmeridge Clay Formation. Deposition of organic, black shales continued through the Kimmeridgian into the early Volgian.

Major extensional tectonic activity occurred in the early Volgian giving rise to the tilted fault block and basin terrain which developed in the Outer Moray Firth and are typified by the Ivanhoe and Rob Roy Field structures. Thick sequences of syn-tectonic clastics accumulated as a result of rapid, fault-controlled subsidence during Kimmeridge Clay times. The source of these sandstones is thought to have been the intra-basinal fault blocks, such as the Halibut Horst. These Claymore Sandstones form important hydrocarbon reservoirs to the north, but are absent in the Rob Roy and Ivanhoe Fields. Late Cimmerian tectonism continued until mid Cretaceous times with Kimmeridge Clay facies continuing into the Lower Cretaceous (early Ryazanian). By late Lower Cretaceous times, tectonic stability returned with the deposition of shallow marine sediments. Sedimentation continued across the Albian–Cenomanian boundary with a major transgression during the Cenomanian causing the deposition of the Chalk facies. Chalk sedimentation continued into the early Palaeocene, further blanketing the Cimmerian fault block terrain.

The Palaeocene of the field areas is divided into two groups: the Montrose Group consisting of numerous submarine fan sequences, and the overlying Moray Group, consisting of a shelf deltaic sequence. These groups follow the regional Palaeocene stratigraphy outlined by Deegan & Scull (1977). The Palaeocene delta was drowned by a marine transgression in the late Palaeocene/early Eocene. Post-Palaeocene, Tertiary sedimentation was dominated by slow subsidence and sediment progradation from the west. Renewed rapid subsidence during the late Pliocene and Pleistocene, and rising sea levels have resulted in the present-day extent of the Basin.

Geophysics

A number of seismic surveys were acquired over the fields from 1972 to 1984. The 3-D survey of 1984 is the definitive study used to delineate the structures of the two fields. This survey provides full 3-D coverage over the field areas with a grid of 75m × 25 m, and 2-D tie lines to surrounding wells. Depth conversion is by a two layer velocity model, using a single layer average velocity model to Top Kimmeridge Clay Formation, with an interval velocity from Top Kimmeridge Clay Formation to Top Rattray Volcanics.

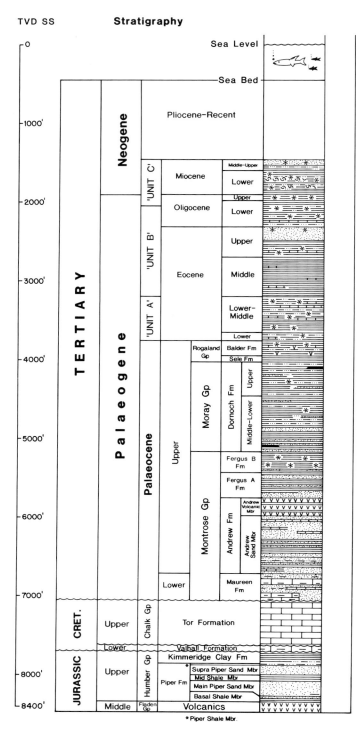

Fig. 4. Ivanhoe and Rob Roy Fields general stratigraphic log.

The early (1985) geophysical interpretation of the 3-D was concerned with determining the structural configuration at the Top Rattray Volcanics, and isopaching upwards through the Piper Sandstones to produce a map on Top Main Piper Sandstone. Other reflectors mapped were Top Kimmeridge Clay, Top Chalk and Top Palaeocene. A more detailed interpretation took place in 1988, whereby, with the benefit of greater well control, maps on top of both the Supra and Main Piper reservoirs were produced. The Top Kimmeridge Clay event is variable in quality, being easily recognized on the flanks but less well defined on the crests due to interference from the overlying Top Chalk event. With the exception of the Rob Roy boundary fault, none of the older faulting offsets the Top Kimmeridge Clay within the two fields.

The Top Rattray Volcanics is the most important event in defining the fields. It is very variable in character and continuity, changing from a strong, continuous event as at 15/21a-9, to a weaker one overlying stronger events as at 15/21-3 & 4. This is due to the lateral variation in the uppermost volcanic rocks from hard basalts to soft weathered tuffs. The event representing the Top Volcanics horizon is not phase consistent throughout the fields, but can generally be correlated to a trough overlying a series of lower frequency events within the volcanics.

The Supra and Main Piper events are both of low to medium frequency, with the Main Piper easier to map than the Supra Piper. The seismic character varies from good, on the flanks, to fair to poor on the crests of the structures. In areas of bad data quality the events can, with caution, be 'ghosted' from the top volcanics pick to which they sometimes have a subparallel relationship. These events clearly exhibit the influence of several tectonic episodes which do not penetrate the overlying Kimmeridge Clay event.

Trap

The Ivanhoe and Rob Roy Fields lie in fault bounded blocks which are tilted to the southwest. The northern boundaries of the fields are defined by NW–SE-trending faults, parallel to the major late Cimmerian Witch Ground Graben. Subordinate NE–SW-trending faults cut across the fields, these are parallel to major Caledonian regional fault trends. The Rob Roy Field is an elongate, rectangular structure with approximate dimensions of 3 km by 0.8 km, whilst the Ivanhoe Field measures approximately 2.2 km by 0.8 km. Both fields are elongated NW–SE, as seen on the top reservoir map (Fig. 5).

The Ivanhoe and Rob Roy Fields both contain two reservoir intervals, the Supra Piper and Main Piper Members of the Piper Formation. In Ivanhoe, faulting has caused juxtaposition of Supra to Main Piper across non-sealing faults, and consequently a common oil–water contact of 8052 ft TVSS. However, within the Rob Roy Field, the Main Piper has an oil–water contact at 7994 ft TVSS whereas the Supra Piper has water up to 7937 ft TVSS.

These oil–water contacts form the westerly limits to both fields, with neither structure being filled to spill point. The overlying Kimmeridge Clay Formation forms the seal over most of the field area, with the exception of a limited area of Rob Roy where Lower Cretaceous marl overlies the Supra Piper Sandstone, as in well 15/21a-31.

Reservoir

As discussed, the Piper Sandstones of Oxfordian to Kimmeridge age form the reservoir intervals in both Ivanhoe and Rob Roy. In the field area the Piper Formation can readily be subdivided into five units that have locally been given Member status:

Piper Shale Member
Supra Piper Sandstone Member
Mid Shale Member
Main Piper Sandstone Member
Basal Shale Member.

The Basal Shale is informally subdivided (in ascending order) into Coaly Series, Paralic Shales and Marine Shales which show an overlapping relationship.

The Coaly Series and Paralic Shales are the lateral equivalent of the Sgiath Formation (Harker et al. 1986) and represent marginal

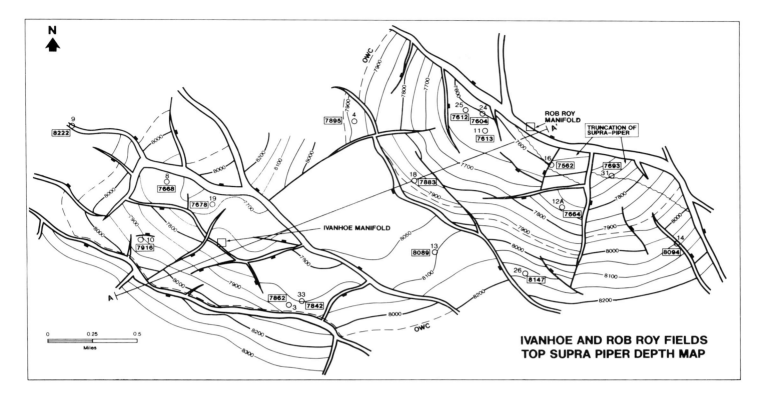

Fig. 5. Map on top reservoir.

marine sedimentation which preceded the development of the fully marine conditions of the Piper Formation. The coal and carbonaceous shales of the Sgiath Formation are restricted to the northern part of the Rob Roy Field, with only the overlying Marine Shales being present over the rest of the field area.

The Marine Shales consist of silty, micaceous shales containing some parallel laminated graded beds, often destratified by bioturbation. Deposition occurred in localized lagoons and ponds on the weathered, uneven surface of the Rattray Volcanics.

In the reservoir sandstones eight lithofacies have been identified and these are summarized in Table 1. The lithofacies generally represent deposition in a marginal marine, coastal environment affected by a major regression and punctuated by minor transgressive events. A composite well log summary is presented in Fig. 6. The Main Piper Sandstone Member commences with a coarsening upward, bioturbated silty sandstone unit with occasional wave-generated ripple forms. This is interpreted as a lower shoreface sequence. Above this lie medium- to coarse-grained sandstones with planar, cross-stratified foresets and parallel stratification. These upper shoreface, or barrier bar sandstones are the most common lithofacies in the Main Piper Sandstone, and form the best reservoir layers as in layers V, VIb and VIII. The coarsest sequences seen in the Main Piper Sandstone fall into two categories. First, the foreshore to beach sequences laid down under very high energy conditions, consisting of quartz and volcanic pebbles up to 3 cm in diameter. These are observed in the upper part of layer VIII in well 15/21a-11. The second type of coarse sequences are developed in fining upwards units typical of layer VII. The high permeability conglomerates within layer VII are restricted to the Rob Roy Field and are fine- to very coarse-grained, cross-laminated with pebbly/gravel beds. Large coalified wood fragments and high pollen and plant cuticle content suggest deposition in a distributary channel. Permeabilities of greater than 10 Darcies have been recorded in this horizon.

However, within layer VII of wells 15/21a-31 and 15/21a-12A a tightly silica cemented interval was encountered, up to 20 ft thick. The zone was cored in 15/21a-12A and petrographic studies indicate precipitation of these cements in the vadose zone forming a silcrete horizon. The areal extent of this silcrete appears confined to the eastern part of the Rob Roy Field where it may act as a vertical barrier to flow.

Table 1. *Ivanhoe and Rob Roy lithofacies*

Lithofacies	Composition	Interpretation
i	Fine–medium, well sorted massive to cross laminated sandstone	Upper shoreface
ii	Fine, bioturbated, micaceous sandstone with glauconite	Lower shoreface
iii	Mixture of i and ii, moderately sorted	Transitional i and ii
iv	Fine to very coarse poorly sorted	? Washover sandstone
v	Fine to coarse poorly sorted argillaceous sandstone	Reworked i
vi	Fine to coarse grained, pebbly, gravel beds with scoured bases	Submerged distributary
vii	Micaceous pyritic/calcareous, bioturbated shale/siltstone	Marine and lagoonal
viii	Glauconitic siltstone/mudstone poorly sorted, burrowed	Submarine non-depositional and winnowing

Fining upwards sequences up to 2 ft thick, commencing with a scoured, erosive base and passing upwards into argillaceous, heavily bioturbated sandstones, are also recognized. These units occur in stacked sequences up to 15 ft thick and are interpreted as either back barrier deposits possibly washover fan complexes, or alternatively minor fluvial or tidal channel reworking. Deposition of the Main Piper Sandstone Member was terminated by a major marine transgression, and the uppermost part of the sand body is interpreted as a reworked barrier complex (layer IV).

The Main Piper Member reaches a maximum thickness of 258 ft (15/21a-25) generally thickening to the northwest in Rob Roy and thinning to the northwest in Ivanhoe. A NNE–SSW trend is seen for the Main Piper gross thickness within the Ivanhoe and Rob Roy

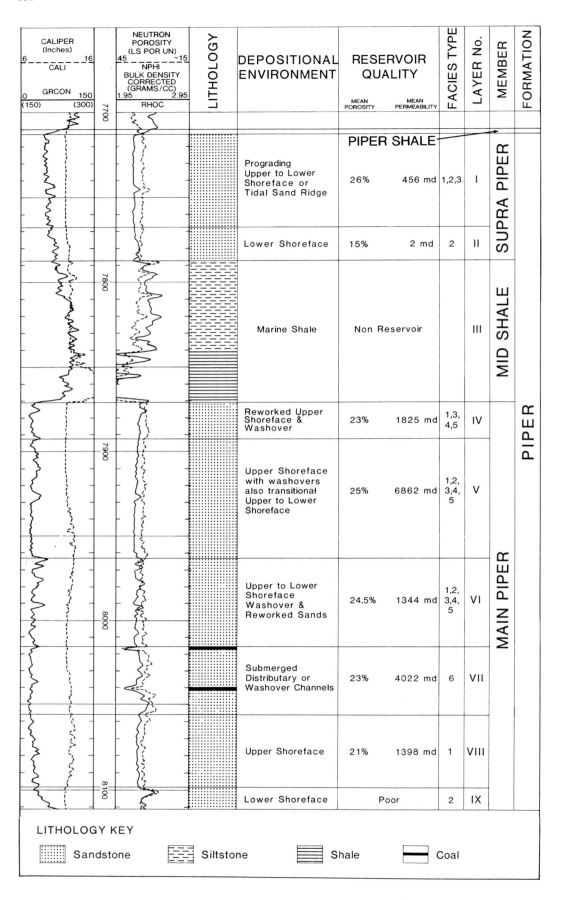

Fig. 6. Composite log of well 15/21a-11 over reservoir.

Field Area with a depocentre around the crestal part of the Rob Roy Field.

The Mid Shale Member marks a major marine transgression across the area. Deposition recommenced in a fully marine environment with anoxic bottom conditions as indicated by abundant pyrite nodules and high organic content of the shales. Gradually renewed progradation occurred with the deposition of intensely bioturbated silty clays. These progressively coarsen upwards into argillaceous sandstones that remain heavily bioturbated and show evidence of reworked storm deposits. Finally, the Mid Shale grades upwards into a lower shoreface sequence at the base of the Supra Piper Sandstone.

The Supra Piper Sandstone Member consists of a coarsening-upward sequence up to 90 ft thick and reflects either progradation into deeper water (water depth up to 200 ft) or significant syn-depositional downwarp. The lithofacies seen are similar to the Main Piper, but the lower shoreface sequence is much thicker. The Supra Piper Member thickness varies between 15 ft (15/21a-31) and 96 ft (15/21a-26). It generally thins and shales out towards the northwest and the elongation of the contours suggests NE–SW-trending sand bodies.

The uppermost member of the Piper Formation is the Piper Shale. This is typically some 15 ft thick and consists of a fining upwards sequence of fine sandstone to mudstone. The unit is poorly sorted, glauconitic, reworked, and intensely burrowed, suggesting a period of submarine non-deposition and winnowing, probably associated with erosion. It is believed this facies marks a major transgressive surface which may represent a significant stratigraphic break.

Sedimentological evaluation of the overall sequence suggests a depositional environment with a barrier bar coastline, forming part of a wave dominated delta with progradation from south to north. The interpretation is consistent with other published reports of the Piper Sandstones (Maher 1980; Boote & Gustav 1986). The reservoirs are subdivided into a number of layers based on sedimentological units. This generates a relatively detailed subdivision for a reservoir which has fairly uniform porosity and permeability distribution.

There has been little modification of the Piper Sandstones by diagenesis; quartz overgrowths had minor effects in the reduction of pore space, whilst carbonate cementation can be severe in localized nodules or doggers. Kaolinite is the dominant clay mineral, with minor illite–smectites which can significantly influence porosity and permeability within the reservoir.

Source

The Upper Jurassic Kimmeridge Clay Formation is the principal source rock for the Ivanhoe and Rob Roy Fields. It reaches a thickness of 200 ft over the field areas, and thickens to 500 ft in the North Renee and Witch Ground Grabens. Thermal maturation modelling has been carried out indicating that Kimmeridge Clay as a whole is a mid to late mature. Oil generation began some 30 million years b.p. with onset of the main generating phase some 10 million years b.p.

Geochemical analyses of oil samples from wells 15/21a-8 and 15/21a-10 in the Ivanhoe Field and wells 15/21a-11 and 15/21a-12A in the Rob Roy Field indicate that the oils in the two fields are different. The Ivanhoe oils are mixtures of a biodegraded oil and a live oil similar to the live oils in the Rob Roy Field. The Rob Roy oils do not have the biodegraded component. While both oils are typical of oils sourced by the Kimmeridge Clay Formation, the source of the Rob Roy oils is less anoxic and more mature than the live oils from the Ivanhoe Field.

Hydrocarbons

In the Ivanhoe Field the oil properties are the same in both the Main and Supra Piper reservoirs as listed in the data summary. However, within the Rob Roy Field the two reservoir intervals display marked difference in oil properties, with a GOR of 1391 SCF/STB in the Supra, and 631 SCF/STB in the Main Piper.

The Rob Roy oils have significantly higher API gravities than the Ivanhoe oils. In both fields, the Supra Piper oils are slightly lighter than their counterparts in the Main Piper. These differences are attributable to fractionation during migration.

Nine drill stem tests were performed in the oil zones of the appraisal wells. Oil rates of up to 12 965 BOPD for Main Piper and 9451 BOPD for Supra Piper were obtained from 15/21a-12A. All reservoirs are undersaturated except the Rob Roy Supra Piper which is 100 psi above its saturation pressure. The reservoir fluid properties are taken from single stage flash studies conducted on 15/21a-16 and 19.

Reserves/resources

The Ivanhoe Field contains 95 MMBBL and the Rob Roy Field 104 MMBBL of stock tank oil initially in place. The breakdown of stock tank oil, and recoverable reserves by reservoir is presented in the data summary. Total recoverable reserves are expected to be 100 MMBBL and 62 BCF.

The Ivanhoe and Rob Roy Fields are both expected to have good aquifer support in the Main Piper reservoir. Water injection does however assist in maintaining reservoir pressure above 3000 psia, enabling wells to flow naturally until artificial lift is required to handle increasing water cut. Water injection into each reservoir was initially delayed to allow evaluation of aquifer support. It is expected that the Rob Roy Field should produce adequately, even with significant watercuts, due to the extremely high productivity of the reservoir. However, Ivanhoe well production will be limited without artificial lift after water breakthrough. Gas lift is included for artificial lift as it requires less maintenance and wellbore access than other alternatives. This is a major consideration due to the difficulty of re-entering subsea wells. The Rob Roy Supra Piper completions will not require gas lift; the high GOR of this reservoir will provide the kick off gas for the other reservoirs following a field shut down.

Full field simulation studies have been carried out to optimize development well locations. On commencement of oil production, such simulation studies will be refined to incorporate reservoir decline and interference data and this information will be used in locating the future wells to be drilled to improve reservoir sweep efficiency and provide extra well capacity.

The author wishes to thank S. R. Boldy and R. Warren of Amerada Hess Ltd who provided valuable input to this paper. The author also wishes to thank Amerada Hess Limited, Deminex UK Oil and Gas Limited, Kerr-McGee Oil (UK) plc and Pict Petroleum plc for their permission to publish this paper.

References

BOOTE, D. R. D. & GUSTAV, S. H. 1986. Evolving Depositional Systems Within an Active Rift, Witch Ground Graben, North Sea. *In*: BROOKS, J. & GLENNIE, K. W. (eds) *Petroleum Geology of North West Europe*. Graham & Trotman, London, 819–833.

DEEGAN, C. E. & SCULL, B. J. 1977. *A standard Lithostratigraphic Nomenclature for the Central and Northern North Sea*. Report of the Institute of Geological Sciences, No. 77/25.

HARKER, S. D., GUSTAV, S. H. & RILEY, L. A. 1986. Triassic to Cenomanian Stratigraphy of the Witch Ground Graben. *In*: BROOKS, J. & GLENNIE, K. W. (eds) *Petroleum Geology of North West Europe*. Graham & Trotman, London, 809–818.

MAHER, C. E. 1980. Piper Oil Field. *In*: HALBOUTY, M. T. (ed.) *Giant Oil and Gas Fields of the Decade: 1968–1978*. American Association of Petroleum Geology, Memoir 30, 131–172.

Ivanhoe and Rob Roy Fields data summary

	IVANHOE		ROB ROY	
	Supra	Main	Supra	Main
Trap				
Type	structural		structural	
Depth to crest (ft TVSS)	7670	7730	7560	7725
Oil–water contact ft TVSS	8052	8052	7937 WUT*	7994
Oil column (ft)	382	322	377	269
*water up to				
Pay zone				
Formation	Piper			
Age	Upper Jurassic			
Gross thickness (average) ft	60	165	81	193
Net/gross ratio % (average)	68.6	99.7	80.9	92.5
Cut off for N/G %	18.0	16.0	18.0	16.0
Porosity % (average)	24.6	23.6	24.3	23.4
Hydrocarbon saturation %	91.0	87.3	94.0	91.5
Permeability (average) md	928	1759	508	2408
Productivity index (ST/D/PSI)	12	74	48	189
Hydrocarbons				
Oil gravity (°API)	31	29	41	39
Oil type	low sulphur crude			
Bubble point (psia)	1800	1800	3460	1900
Gas/oil ratio SCF/STB	360	360	1391	613
Formation volume factor (at initial conditions) RB/STB	1.218	1.218	1.802	1.374
Formation water				
Salinity	90 990 ppm			
Resistivity	0.102 ohm m at 60°F			
Reservoir conditions				
Temperature (°F)	175	175	175	175
Pressure (psia)	3510	3510	3510	3510
Pressure gradient in reservoir (psi/ft)	0.345	0.345	0.262	0.307
Field size				
Area (acres)	693	461	707	537
Recovery factor: oil	0.30	0.54	0.41	0.62
Drive mechanism	aquifer & water injection			
Recoverable oil MMBBL	11.23	31.22	14.20	43.29
Recoverable gas BCF	4.04	11.24	19.75	26.53
Production				
Start-up date	July 1989		July 1989	
Development scheme	Floating production facility (FPF)			
Production rate	60 000 BOPD			
Number/type of wells	2P 1I	1P 1I	2P 2I	3P 2I
Secondary recovery methods	waterflood			

The Kittiwake Field, Block 21/18, UK North Sea

K. W. GLENNIE[1] & L. A. ARMSTRONG[2]

[1] Consultant, 4 Morven Way, Ballater, Grampian, UK
[2] Shell International Petroleum Co. Shell Centre, London, UK

Abstract: Kittiwake was discovered by well 21/18-2 within a 7th Round block, part of Production Licence P351. Highly undersaturated oil is present in the Fulmar Formation and Skagerrak Formation reservoir sequences; 70 MMBBL of reserves is in Fulmar sandstones whereas oil in the Skagerrak is mostly immovable. The field will be developed from a single 16-slot platform with initially 5 producing and 5 water-injection wells. Solution gas is removed via the Fulmar Field pipeline to St Fergus and, as from September 1990, the oil is loaded onto tankers from a single-buoy mooring.

The Kittiwake Field is located 100 miles ENE of Aberdeen, 20 miles SSW of the Forties Field (Fig. 1); here, the North Sea is about 285 ft deep. The reservoir is at a depth of some 10 000 ft TVSS, where it forms an east–west-trending elongate dome-shaped feature involving elements of both structural and stratigraphic trapping (Fig. 2A & B). Virtually all the recoverable oil is contained in the Upper Jurassic Fulmar Formation sandstone reservoir; immovable oil is also present in the underlying Triassic Skagerrak Formation. The field came into production in September 1990, and is estimated to contain recoverable reserves of 70 MMBBL oil and 25 BCF of solution gas. Kittiwake follows the Shell/Esso tradition for North Sea oil fields in being named after a sea bird.

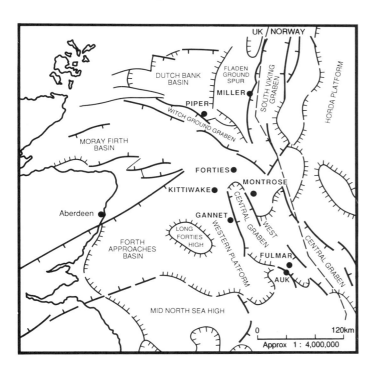

Fig. 1. Location map. Kittiwake Field in its megatectonic setting.

History

Kittiwake lies entirely within 7th Round Block 21/18, which forms part of Licence P351, awarded to the Shell/Esso partnership in December 1980. The discovery well was sited on the crest of a structural high created by diapiric movement of Zechstein salt (Fig. 2). The slope of the sea bed caused drilling difficulties with 21/18-2, which had to be abandoned. 21/18-2A was relocated 120 ft away and became the discovery well. Drilled in 1981, it found the Upper Jurassic Fulmar sandstones and the Triassic Skagerrak Formation oil-bearing. On test, this well produced oil at a rate of 10 000 BOPD from a 50 ft interval in the middle of the 207 ft thick Fulmar Formation, but at only 35 BOPD from a 90 ft interval of the 260 ft thick Skagerrak Formation.

Appraisal wells

In 1982, the appraisal well 21/18-3 was drilled threequarters of a mile NW of the discovery well (Fig. 3). It confirmed the high productivity of the Fulmar Formation with a combined flow rate from two horizons of 11 484 BOPD. There was an oil-down-to of 10 170 ft TVSS at the base of the productive part of the Fulmar reservoir but no flow of oil from the Skagerrak Formation.

To establish the height of the oil column, a second appraisal well, 21/18-4, was located about 1 mile east of 21/18-2. It found the Fulmar reservoir thinner (187 ft gross) and water-bearing below 10 434 ft TVSS, which was within the uppermost (low permeability) part of the Fulmar. An updip sidetrack (21/18-4A), however, established an oil-down-to at 10 430 ft TVSS, which is now accepted as a field-wide oil–water contact. A small culmination was mapped to the west of the then proven field. This was tested in 1985 with well 21/18-6 (Fig. 3A). The well showed that the culmination is an extension to the main field, connected to it by a saddle above a common water table. The thickness of good reservoir in this well is barely one third of that in the discovery well (Unit 2: Fig. 3C).

Development programme

Kittiwake was declared commercial in 1987, and came on stream in September 1990. Production is through a slimline 6000 tonne steel jacket supporting a fully integrated drilling and production deck of some 7000 tonnes. Both jacket and drilling/production facilities were pre-fabricated onshore, installation being planned for the 1990 summer 'weather window'. The drilling of two crestal oil producers began in the late Summer of 1990, and will be followed by alternating water injectors and producers. All wells will be drilled through the Skagerrak reservoir. The platform will have 6 extra well slots to allow for reservoir uncertainties. With the drilling of the second production well, an average plateau production rate of 29 000 BOPD should be achieved during 1991 and will be maintained for some 4 years, after which there will be a rapid decline caused by increasing water production. Field depletion at an average rate of 15% per year will give an expected economic life of 15 years. Provisions have been made to allow the sub-sea development of nearby small accumulations, which could extend the field's life.

Oil export is achieved by using dynamically positioned tankers loading from a single-buoy mooring sited over one mile to the east of the production platform. There are no permanent oil-storage

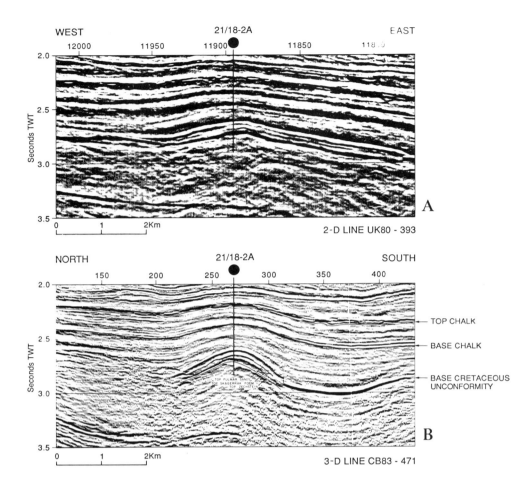

Fig. 2. (A) 1980 2-D line at the exploration phase through the Kittiwake discovery well 21/18-2A. (B) 1983 3-D seismic dip line for comparison.

facilities, so production is shut down during the intervals between tanker loading and ullage shortfall; delays caused by exceptionally severe sea-states can be partially compensated for because the reservoir is inferred to have a substantial excess production capacity. Solution gas and associated NGL is exported to St Fergus via a 4 inch line connected to an existing T-junction in the Fulmar Gas Pipeline south of the platform (Fig. 3C).

Field stratigraphy

The Kittiwake Field is located on the platform area west of the Central Graben. The Fulmar and Skagerrak Formations occur at a depth of around 10 000 ft TVSS. The latter formation is underlain by the thick Upper Permian Zechstein halite (Fig. 4), diapiric movement of which can be seen to have controlled deposition of the Skagerrak Sandstone, preservation of the Fulmar Formation, and creation of the structural aspects of the Kittiwake trap.

The Late Triassic age of the unfossiliferous continental Skagerrak Formation is based on its lithology and facies. Deposition of the Skagerrak Formation was probably entirely pre-Rhaetian. The Skagerrak is overlain directly by shallow-marine sandstones of the Upper Jurassic Fulmar Formation.

The Fulmar Formation grades up into the Kimmeridge Clay Formation, which, is partly of source-rock quality (Figs 4 & 5). The Late Cimmerian Unconformity separates the Kimmeridge Clay from the 9500 ft of overlying strata summarized in Fig. 4.

Geological history

In the Permian, Rotliegend desert conditions were followed by the Zechstein transgression and precipitation of halite up to 6000 ft thick. Triassic and later reactivation of faults near the platform edge probably triggered diapiric movement of Zechstein halite, which played a major role in controlling sedimentation and preservation of the reservoir sequence and in eventual trap formation.

Early Triassic sedimentation over much of the Western Platform comprised lacustrine mudstones of the Smith Bank Formation. These mudstones are not present beneath the Kittiwake Field, where Late Triassic sandstones lie directly above Zechstein halite. Seismic evidence indicates that over much of this area, the Smith Bank mudstones sank differentially into halokinetic rim synclines flanking a system of diapiric swells and ridges of Zechstein salt. A similar process is interpreted to have taken placed in the vicinity of the Fulmar Field (Johnson et al. 1986). Diapirism was probably triggered by reactivation of faults in the pre-Zechstein basement in response to an early phase of extension across the Central Graben, which was largely coeval with Smith Bank sedimentation. It is suspected that sub-aerial and sub-surface solution of salt at the crests of linear diapirs helped to limit the distribution of flood waters carrying the Skagerrak sands to relatively narrow areas between those of earlier Smith Bank deposition.

The central North Sea area was subjected to mid Jurassic regional uplift and erosion. Late Jurassic subsidence brought a return to marine conditions of deposition and a sequence of prograding shoreline and shallow-marine sands, the Fulmar Formation (Johnson et al. 1986), which forms the only producible reservoir in the Kittiwake Field. Like the Skagerrak sandstones, those of the Fulmar Formation were also preserved in the vicinity of the Kittiwake Field. Sandstones of the Fulmar Formation grade up into gas-prone claystones of the Heather Formation overlain by the oil-prone Kimmeridge Clay.

Seismic evidence indicates that another phase of rim-syncline subsidence began during the Early Cretaceous. Gentle diapiric uplift of the Kittiwake structure, and mild subsidence of adjacent areas, continued until the Mid Eocene, leading to the Kittiwake

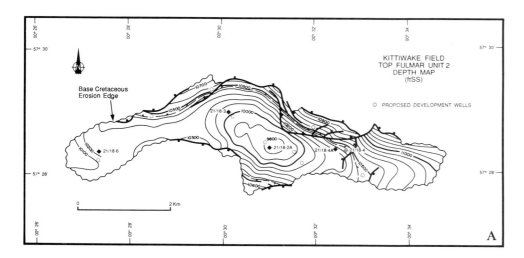

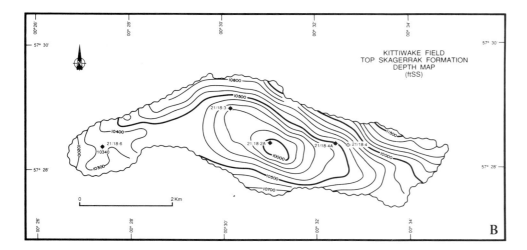

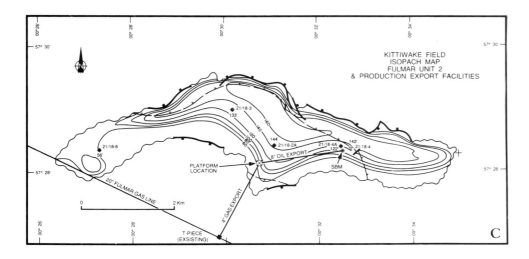

Fig. 3. Depth maps of (A) top Fulmar and (B) top Skagerrak Formations, and (C) isochore map of Fulmar Unit 2, with projected hydrocarbon exporting facilities superimposed.

structural configuration as we now know it. Regional subsidence brought the Kimmeridge Clay source rock to a depth and temperature sufficient for oil generation by the later Tertiary.

Geophysics

The Kittiwake Field was discovered using conventional 2-D seismic data shot between 1974 and 1979. This data is of relatively poor quality and does not cross the crest of the Jurassic structure. In 1983, however, 600 miles of dense 3-D seismic was shot for Shell/Esso by SSL. During processing, the original 75 m line spacing was reduced to 25 m by interpolation. Figure 2B is a 3-D dip line through the discovery well, taken from that survey.

For interpretation of the 1983 3-D survey, the following horizons were picked and correlated: (1) Top Chalk; (2) Base Chalk; (3) Late Cimmerian Unconformity; (4) Top Fulmar Unit 2; (5) Top Skagerrak Formation; (6) Top Zechstein anhydrite.

The Fulmar Formation sandstones are characterized by a distinctive continuous black loop beneath the Late Cimmerian Unconformity event. The lateral extent of the Fulmar Formation is sometimes difficult to assess where it thins beyond seismic resolution.

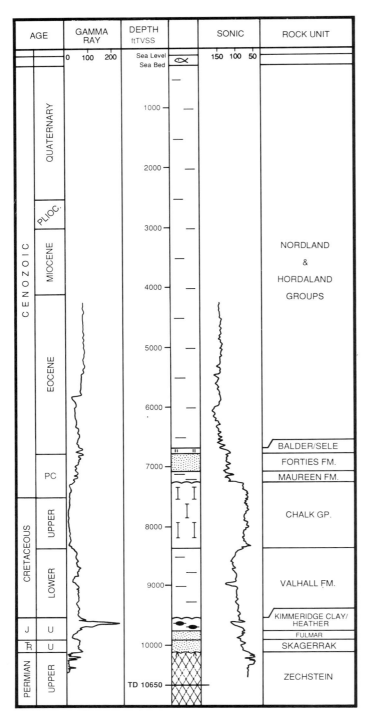

Fig. 4. Simplified stratigraphic column of the Kittiwake area.

The Skagerrak Formation extends laterally beyond the marine Fulmar sandstones until it also wedges out. Neither Fulmar nor Skagerrak sandstones can be recognized down flank (both north and south), where the Kimmeridge Clay shales seem directly to overlie the top Zechstein anhydrite (Fig. 4).

A layer-cake model was used to determine interval velocities between marker horizons from the sea bed to top Zechstein anhydrite; these were corrected to match correlated depths at well locations.

Trap

The Kittiwake trap is an elongate, E-W-trending domal structure some four miles long and generally less than one mile wide (Figs 2 & 3). The crest of the structure is 657 ft above the oil-water contact (10 430 ft TVSS), which is common to both the Fulmar and Skagerrak reservoirs; a spill point seems to be present within the upper Skagerrak sandstones in the western part of the field. A core area of non-reservoir Zechstein evaporites extends about 300 ft above the level of the oil-water contact.

The reservoir rocks were deposited in a solution zone overlying the exposed crest of a Zechstein salt ridge. Extended diapiric uplift throughout the Cretaceous and Early Tertiary followed the line of the original fault-induced movements (see Fig. 2) and deformed the reservoir rocks into a structural trap. Thinning of the reservoir sandstones towards the edge of the original area of deposition has resulted in a degree of stratigraphic trapping in the western and southern parts of the field.

The bottom and top seals comprise Zechstein evaporites and the Upper Jurassic claystone sequence (Heather and Kimmeridge Clay Formations) respectively. Laterally, the reservoir sandstones pinch out between shales of the Smith Bank and Heather Formations.

Reservoirs

The trap contains two oil-bearing reservoir sandstones of the Skagerrak and Fulmar Formations, but only the middle sandstones of the Fulmar Formation are capable of giving up their oil with any ease. Their log character, main reservoir parameters and the energy conditions controlling sedimentation, porosity and permeability are summarized in Fig. 5 and Table 1.

Skagerrak Formation Sandstones: mostly red, fine-grained sandstones and shales that can be divided into three facies associations: alluvial sheet-flood sandstones with mudstone intraclasts; overbank red sandstones, siltstones and claystones, and clean, well-sorted, cross-bedded fluvial-channel deposits. In Kittiwake, the sheet-flood and overbank deposits form the dominant facies in the upper part of the Skagerrak Formation. Skagerrak sandstones also contain many cemented layers, giving the formation negligible reservoir potential.

The Fulmar Formation comprises: shoreline to shallow-marine sandstones. The principal depositional mechanism was probably periodic storm-induced currents rather than persistent longshore or tidal currents, creating submerged offshore ridges or sheet sands on a shallow marine shelf. The overall rate of deposition was slow enough to permit complete destratification by bioturbation. The sandstones of the Fulmar Formation can be divided into three reservoir units (Table 1).

Units 1 and 3 have very poor reservoir properties; attention focuses on *Unit 2* with Darcy permeabilities. Unit 2 has five sub-divisions, with *Unit 2.1* having only moderate reservoir quality reflecting its very fine grain size, the abundance of clay-lined and clay-filled burrows and the presence of silica and dolomite nodules and layers; the permeability profile over this unit is typically irregular. *Units 2.2 and 2.3* have the best and most homogeneous reservoir properties. Differences in reservoir properties between these two sub-units reflect variations in sorting, grain size, argillaceous content and carbonate cementation. *Units 2.4 and 3.5* have poor to moderate reservoir properties in response to the intensity and degree of dolomite cementation.

The thickness of Unit 2 decreases from the central area towards both east and west, while Unit 3 decreases from east to west. The overall reservoir properties of Unit 2 improve from the eastern to the central area; this reflects the thicker development of the good reservoir Units 2.2 and 2.3 in the central area and the relatively thick, poor reservoir Units 2.4 and 2.5 in the east.

Source rock

The Kittiwake oil is fairly light (38° API) and is very undersaturated in gas. Kimmeridge Clay is the only source rock for this oil.

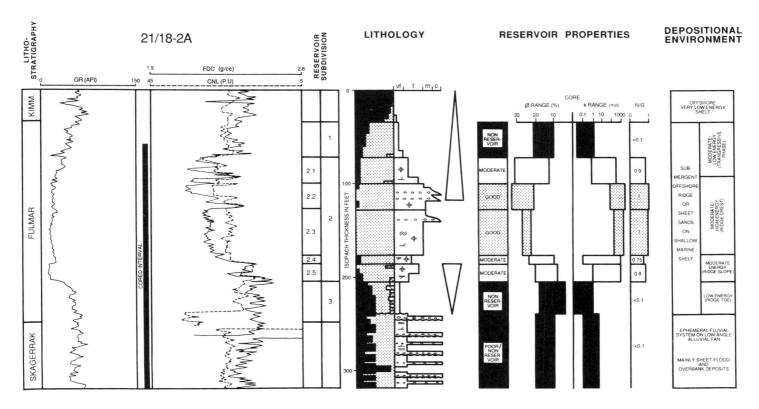

Fig. 5. Wireline log character (21/18-2A) and generalized factors controlling the Kittiwake reservoir quality.

Table 1. *Reservoir parameters of the Kittiwake Oilfield*

Reservoir Units	Porosity %	Permeability md	Net-to-Gross	Sedimentary features
Fulmar				
1 & 3	<18	<1	<0.2	Low-energy sands
2.1	12–30	1–700	0.8–0.9	Fine-grained sands of high-energy ridge crests; clay-lined and filled burrows
2.2	18–32	<3200	1.0	High-energy ridge-crest sands. Dominant in middle; thins to west
2.3	20–26	<2000	1.0	Highest-energy most homogeneous ridge-crest sands. Dominant in middle; thins to west
2.4 & 2.5	<20	<1000	0.75	Moderate-energy ridge slope. Unit 2.5 thicker in east
Skagerrak	6–20	0.05–10	0.1–0.4	Sheet-flood, overbank and fluvial-channel sands, the first two dominating

Reservoir Unit 2 is probably in direct contact with the source rock across the northern faulted margin. At a depth just in excess of 10 000 ft, the Kimmeridge Clay will be mature for the generation of oil but not gas (cf. Cornford 1990). The lack of gas and the highly overpressured state of the reservoir indicates a local, closed source–reservoir system with efficient seals provided by claystones of the Kimmeridge and Smith Bank Formations; this excludes any relatively long-distance migration from a Kimmeridge Clay kitchen within the Central Graben.

Hydrocarbons

The 38° API oil in the Fulmar Formation reservoir has a specific gravity of 0.837, contains a 0.64% by weight of sulphur, 0.14% asphaltene and 6.97% of paraffin wax; the wax has a congealing point of 58.5%C. The oil has a pour point of $-12°C$ to $-32°C$, and a viscosity at 25°C of 4.71cP, 5.74cS. An H_2S concentration of up to 40 ppm was found in separator gas during production testing.

Kittiwake Field data summary

	Fulmar Formation	Skagerrak Formation
Trap		
Type	Combined structural/stratigraphic	
Depth to crest	9773 ft to top Fulmar Unit 2	9942 ft TVSS
Lowest closing contour	10 000 ft TVSS	10 400 ft TVSS
Hydrocarbon contacts	Fault-bounded to north	Uncertain
	Thins and pinches out to south	
Gas column	Nil	Nil
Oil column	650 ft TVSS	Not producible
Oil–water contact	10 430 ft TVSS	10 430 ft TVSS (and higher)
Pay zone		
Formation	Fulmar Formation Unit 2	Skagerrak Formation (not proven producible)
Age	Late Jurassic	Late Triassic
Gross thickness (average/range)	120 ft/<60–213 ft	Up to 367 ft?
Net/gross ratio (average/range)	0.54/0.39–0.66	0.2/0.1–0.4
Cut off for Net/Gross	17%	16%
Porosity (range)	16–30%	6–20%
Hydrocarbon saturation	30–89%	28%
Permeability (range)	<0.1–3000 md	0.05–10 md
Productivity index	80 BOPD/psi with removal of skin factor	
Hydrocarbons		
Oil gravity	38° API	
Oil type	Highly undersaturated	
Gas gravity	Average 0.962 gm/l	
Bubble point	1150–1480 psi at 224–250°F	
Dew point	None	
Gas/oil ratio	300–380 CF/BBL	
NGL yield	18.0 tonnes/MMSCF	
Formation volume factor	1.20	
Formation water		
Salinity	Regional value 130 000 NaCl equiv	
Resistivity	0.02 ohm m	
Reservoir conditions		
Temperature	244°F	
Pressure	6520–6540 psi at 10 100 ft	
Pressure gradient in reservoir	0.30 psi/ft oil gradient	
	0.45 psi/ft water gradient	
Field size		
STOOIP	173 MMBBL	
SGIIP	62 BCF	
Recovery factor: oil	40%	
Recovery factor: gas	40%	
Drive mechanism maintenance	Water drive pressure	
Recoverable hydrocarbons:		
Oil	70 MMBBL	Nil
Gas	25 BCF	
Production		
Start-up	On stream September 1990	
Development scheme	5 producers	
	5 water injectors	
Production rate	Plateau of 29 000 BOPD	
Cumulative production to date	Not available	

Reserves/resources

Of the calculated expectation of 173 MMBBL of oil in place, the recoverable reserves will be about 70 MMBBL, giving a recovery efficiency of 40%. The recovery factors are based on the result of an 11-layer full-field simulation study carried out to determine the oil recovery and to investigate sensitivities on the proposed development plan for the field. Because of their low permeabilities (<1 md), no reserves have been included for Units 1 and 3.

The limited extent of the Fulmar sandstones, together with the overpressured nature of the accumulation, indicates that only limited aquifer support to oil production is likely. It was assumed, therefore, that water-injection wells will be necessary.

To enable producing wells to flow to high water-cuts for maximum oil recovery, a pressure maintenance level of 5000 psia will be adopted as an average reservoir pressure. This is sufficient to give flow rates of oil per well in excess of 4000 BOPD towards the end of field life even at high water-cuts. The initial overpressure of about 2000 psi gives scope for high rates of production prior to establishing water injection.

The writers thank Shell UK Exploration and Production and Esso Exploration and Production UK Limited for permission to publish this paper. They wish to make special tribute to A. H. ten Have for his interpretation of the sedimentology, and also to acknowledge those other contributions made to the present understanding of the field; the explorers who discovered it and the many geologists, geophysicists and engineers who prepared the way for it to come on stream in 1990 in the wake of the 1986 oil-price collapse.

References

CORNFORD, C. 1990. Source rocks and hydrocarbons of the North Sea. *In*: GLENNIE, K. W. (ed.) *Introduction to the Petroleum Geology of the North Sea*. Blackwell Scientific Publishers, Oxford, 294–361.

JOHNSON, H. D., MACKAY, T. A. & STEWART, D. 1986. The Fulmar Oilfield (Central North Sea); geological aspects of its discovery, appraisal and development. *Marine and Petroleum Geology*, **3**, 99–125.

The Maureen Field, Block 16/29a, UK North Sea

P. L. CUTTS

Phillips Petroleum Company (UK) Ltd, Phillips Quadrant, 35 Guildford Road, Woking, Surrey GU22 7QT, UK

Abstract: The Maureen Oilfield is located on a fault-bounded terrace in Block 16/29a of the UK Sector of the North Sea, at the intersection of the South Viking Graben and the eastern Witch Ground Graben. The field was discovered in late 1972 by the 16/29-1 well, and was confirmed by three further appraisal wells. The reservoir consists of submarine fan sandstones of the Palaeocene Maureen Formation, deposited by sediment gravity flows sourced from the East Shetland Platform. The Palaeocene sandstones, ranging from 140 to 400 ft in thickness, have good reservoir properties, with porosities ranging from 18–25% and permeabilities ranging from 30–3000 md. Hydrocarbons are trapped in a simple domal anticline, elongated NW–SE, which was formed at the Palaeocene level by Eocene/Oligocene-aged movement of underlying Permian salt. The reservoir sequence is sealed by Lista Formation claystones. Geochemical analysis suggests Upper Jurassic Kimmeridge Clay shales have been the source of Maureen hydrocarbons. Estimated recoverable reserves are 210 MMBBL. Twelve production wells have been drilled on the Maureen Field. A further seven water injection wells have been drilled to maintain reservoir pressure.

The Maureen Field is located in Block 16/29a of the UK Sector of the North Sea, approximately 163 miles northeast of Aberdeen, in 325 ft of water. Minor portions of the field extend into Blocks 16/29b and 16/24b (Fig. 1). Oil is trapped in Lower Palaeocene submarine fan sandstones of the Maureen Formation, laid down immediately following Cretaceous chalk deposition. Recoverable reserves are estimated to be 210 MMBBL.

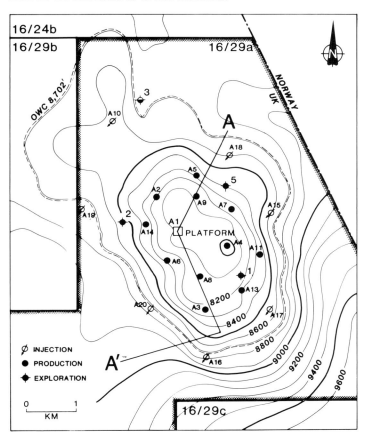

Fig. 1. Structure on top Maureen Formation with well locations.

History

In June 1970, Block 16/29, part of Licence P.110, was awarded in the Third Round of licensing to the Phillips group, consisting of the Phillips Petroleum Company UK (33.78% and Operator), Fina Exploration UK Ltd (23.96%), Agip (UK) Ltd (17.26%), Century Power and Light Ltd (9%), Ultramar Exploration (6%) and British Electric Traction (5%). In 1983 BET sold its interest to Century (now Gas Council Exploration) and Ultramar; each increased its existing interest by 2.5%. In June 1976, 50% of Block 16/29 was relinquished; Block 16/29a remained as the licence area containing the Maureen prospect.

Interpretation of seismic data shot over Block 16/29 in late 1970 confirmed a closed domal structure in the Lower Tertiary to Mesozoic interval, in the northeastern part of the block. In November 1972 the first exploration well 16/29-1 was spudded near the structural crest, with Palaeocene and Eocene sandstones as primary objectives. The well was successful, encountering 95 ft of hydrocarbon-bearing Palaeocene sandstone which flowed 3588 BOPD on testing.

The 16/29-2 appraisal well, located 1.6 miles WNW of the discovery well, was drilled to evaluate the western flank of the accumulation, and to establish an oil–water contact. Completed in May 1974, the well encountered 317 ft of gross sandstone with an oil–water contact at 8700 ft TVSS. A maximum flow rate of 10 832 BOPD from a single perforated zone was achieved.

Appraisal continued in June 1974 with the 16/29-3 step-out well, located 2.6 miles NNE of the discovery well. Completed in August 1974, the well encountered the Palaeocene reservoir in a structurally low position. From a 200 ft gross Palaeocene section, 17 ft of pay at the top of the reservoir proved to be tight, and tested only a minor quantity of oil.

The 16/29a-5 well, drilled in early 1978, completed the Maureen appraisal programme, the 16/29a-4 well having tested a separate structure to the southwest. 16/29a-5 encountered a full oil column of 172 ft within the Palaeocene reservoir section, which flowed at a maximum oil rate of 5328 BOPD on testing.

In December 1978 a development plan was submitted to the Department of Energy. The programme involved the installation of a 24-slot subsea template, pre-drilling of the production wells from a semi-submersible rig, positioning of a gravity tripod structure with integral oil storage over the template and the laying of a pipeline to a remote tanker mooring and loading system. Facilities for reservoir water injection and artificial gas lift were also included in the design.

Development drilling commenced in June 1979 with 12 production and 7 water injection wells completed by February 1983 (16/29a-A1 to A11, and A13 to A20; Fig. 1). Well 16/29a-A12 was abandoned after encountering mechanical problems. By June 1983 the Maureen platform was positioned over the template and the field was put on production on 14 September 1983.

A successful application was made in 1987 in the Tenth Round of licensing for Block 16/24b and 16/29b. Both blocks contain part of the northwest extension of the field.

From Abbotts, I. L. (ed.), 1991, *United Kingdom Oil and Gas Fields,*
25 Years Commemorative Volume, Geological Society Memoir No. 14, pp. 347–352

Field stratigraphy

The Maureen Field is located on the Maureen Shelf (Fig. 2), a fault-bounded terrace bordered by the Vestland Arch to the east and limited to the northwest, west and southwest by the basinal areas of the South Viking Graben and the southeast Witch Ground Graben. A generalized stratigraphic column is shown in Fig. 3.

The oldest strata penetrated during drilling in the Maureen Field are of Zechstein age and comprise a fractured dolomite/anhydrite section, overlying thick halite. A tilted fault block is thought to have developed in the Maureen area during the Triassic as a result of early Cimmerian tectonic movements along north–south trending faults. Triassic continental shales and siltstones of the Smith Bank Formation, which are thinly developed or absent (16/29-1 well) over the crest and eastern flank of the Maureen structure, are seen to thicken significantly to the north and west, where they are overlain by up to 700 ft (16/29-3 well) of sandstone, thought to be of alluvial origin.

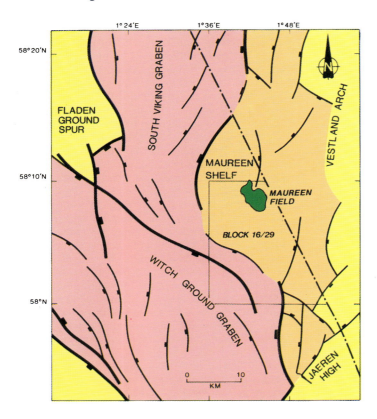

Fig. 2. Regional tectonic elements.

Early to mid-Jurassic regional uplift resulted in the erosion or non-deposition of Lower Jurassic strata over the Maureen area; a widespread thin development (50–120 ft) of Middle Jurassic Pentland Formation delta top shales, coals and sandstones rest unconformably on the Triassic.

With the onset of late Cimmerian extensional rifting, the north–south Viking Graben fault trend was cut by west-northwesterly faults associated with the evolution of the Witch Ground Graben. Marginal marine sandstones of the Hugin Formation, encountered in the 16/29-3 well, mark the end of the Middle Jurassic regressive phase and the start of the Upper Jurassic marine transgression. The continued and accelerated subsidence of the region combined with an eustatic sea level rise in late Jurassic is marked by the transition from grey/brown Heather Formation marine silts and shales to the grey/black shales of the Kimmeridge Clay Formation, reflecting an increase in water depth and the onset of restricted anaerobic bottom conditions.

Lower Cretaceous sediments were deposited in more open marine conditions, with up to 300 ft (16/29a-A3 well) of grey and red–brown shales and marls of the Cromer Knoll Group developed over the Maureen Field. By the close of the Early Cretaceous, extensional tectonics were replaced by post-rift subsidence, with chalks of the Hod, Tor and Ekofisk Formations deposited through the Late Cretaceous and Early Tertiary. Biostratigraphic evidence suggests that the Ekofisk/Tor Formation boundary is unconformable and that the lowermost part of the Palaeocene is absent.

Palaeocene uplift of the Scottish Highlands/Orkney Shetland Platform to the west of Maureen resulted in a massive supply of coarse clastics that was transported eastwards by fluvio-deltaic processes onto a shallow marine shelf before being reworked. The shelf break coincided with the western margin of the present South Viking Graben, a fault-controlled Mesozoic half-graben system. The reworked clastic rocks were periodically shed into the deeper graben area as sediment gravity flows, forming arcuate offlapping fans, thinning to the east up the easterly dip slope of the former half graben. There was a tendency within the graben to develop axial flows downslope towards the regional basin centre to the south. Most of the Maureen reservoir sandstones reached the Maureen shelf area by this depositional pathway, with reworked units of Ekofisk and Tor Formation chalk incorporated in the lower sandstone sections. The Maureen Formation containing the reservoir sandstones attains a maximum drilled thickness of 400 ft (16/29a-A20 well) in the western part of the field, and thins to 140 ft (16/29a-A15) in the east.

Subsequent sediment deposition over the Maureen Field area consisted predominantly of claystones associated with North Sea basin development through the remainder of the Tertiary; Lista Formation claystones overlie and seal the Maureen Formation reservoir sandstones; claystones of the Sele Formation and tuffaceous claystones of the Balder Formation complete the Upper Palaeocene sequence. The overlying Tertiary sediments consist of monotonous sections of clays and silts with stringers of sandstone and limestone. Thick sandstones equivalent to the Utsira Formation are present over Maureen, marking the base of the Miocene. The present day configuration of the Maureen structure can be seen in the NE–SW cross-section in Fig. 4. The thinning of all of the major units over Maureen on this section testifies to the long term, periodic nature of salt movement.

The Maureen Formation is named by Deegan & Scull (1977) after the Maureen Field. However, the type Maureen Formation sequence defined by Deegan & Scull in the 21/10-1 well, consisting of an irregularly distributed mixture of reworked chalks, claystones and sandstones, is not lithologically representative of the Maureen Field reservoir sandstone section. In several Maureen Field wells, sandstone occurs in the upper chalk sections of the Ekofisk Formation, and this carbonate/clastic mixture is lithologically comparable with the Maureen Formation in the 20/10-1 type well. In this document, the name 'Maureen' is reserved for the largely Danian age submarine fan sandstone package found in the Maureen Field vicinity; the representative type well is 16/29-2 (Fig. 5), where the Maureen Formation sequence is complete and largely cored.

Geophysics

The first seismic coverage over Block 16/29 was acquired in 1970 and consisted of five north–south and five east–west lines, totalling 135 km. Subsequent to the drilling of the 16/29-1 discovery well, 132 km of additional seismic were shot in 1973, infilling the 1970 survey, to give a 1.5 × 1.5 km north–south grid. A further 107 km of seismic data were acquired between 1973 and 1978 and from this database, time and depth structure maps were produced for the top Palaeocene sandstone, top Danian Chalk and top Triassic horizons. At the Palaeocene sandstone level, the Maureen structure was interpreted as a simple unfaulted dome, elongated NW–SE.

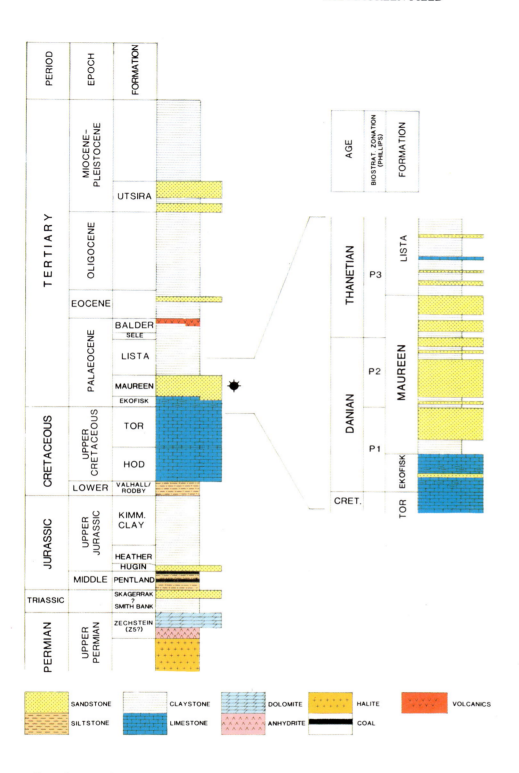

Fig. 3. Generalized stratigraphic column.

In order to enhance the Palaeocene seismic interpretation over Maureen, the P.8102 survey was acquired in 1981, consisting of a grid of 45 lines covering 370 km. Predominant grid orientation was ENE–WSW, with 250 m spacing in the dip direction. The data were processed by Phillips and involved a detailed velocity analysis, finite difference migration and wavelet processing designed to broaden the bandwidth of the basic seismic wavelet, while minimizing the noise content of the data.

Prior to construction of a 3-D reservoir simulation model, a seismic interpretation revision was undertaken in September 1983. To facilitate the creation of reservoir maps for simulation purposes, the reservoir was subdivided into units that could largely be seismically defined. Based primarily on biostratigraphic age dating, a three-layer subdivision was employed (denoted P1, P2, and P3 with P1 being the oldest). The top P1 and P2 interfaces are chronostratigraphic surfaces associated with local unconformities. Four horizons corresponding to the top of the three reservoir layers and the top Chalk were interpreted in time and synthetic seismograms generated for 23 Maureen Field wells were used to establish ties and seismic polarity.

The top Chalk was the first horizon to be depth converted with the three reservoir layers subsequently isopached and subtracted to arrive at successive depth maps. A compaction (or acceleration) factor was determined at each well based on seismic times and well depths. A map of the compaction factor combined with the top Chalk time map produced the top Chalk depth values. The three reservoir layers were determined from well control to have a constant interval velocity regardless of depth or location.

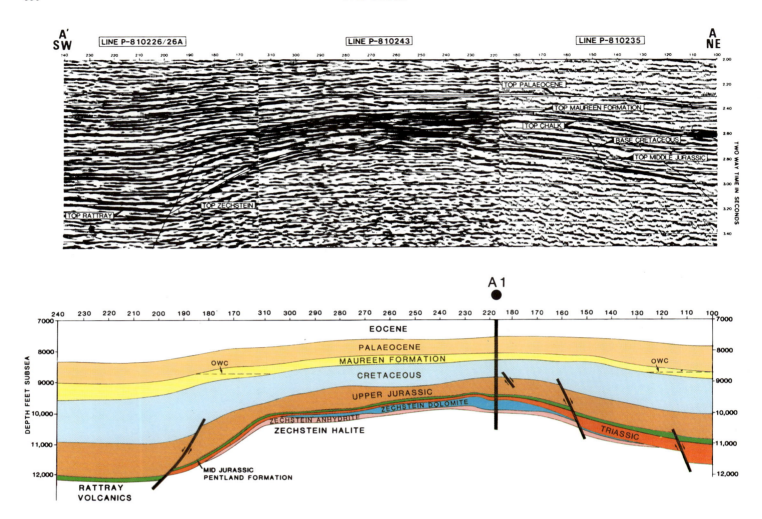

Fig. 4. Composite seismic section with interpretation. Line of section is shown on Fig. 1.

Trap

The Palaeocene reservoir is contained in a simple domal anticline, with nearly 1000 ft of unfaulted structural relief at top Palaeocene level. A fieldwide oil–water contact is taken as 8702 ft TVSS. This results in an oil accumulation which is 4.4 miles long in a NNW direction, 2.4 miles wide in a WSW direction, has an areal extent of 4650 acres and a maximum net oil column of 650 ft. The structure, which is effectively filled to spill point, is sealed by overlying Upper Palaeocene Lista Formation claystones.

The reservoir sandstones were deposited over an evolving structure originally formed by early Cimmerian tectonic events. Zechstein salt movement, probably initiated during the Triassic, continued intermittently through the Mesozoic, maintaining the structure as a local high. A final post-Palaeocene episode of salt movement, ending in the Miocene, resulted in the simple anticlinal dome seen at Palaeocene level.

Reservoir

Depositional model

The Palaeocene age Maureen Formation reservoir sandstones were deposited by sediment gravity flows within a submarine fan environment. Sedimentological analyses of over 1200 ft of core, were conducted, but the three basic rock types, namely massive sandstones, interbedded sandstones and claystones, and silty claystones remained as the most useful lithofacies subdivisions.

(1) Massive sandstones are moderately sorted, predominantly fine- to medium-grained, with subangular to rounded grains. The sandstones occur in units 3–50 ft thick. The sandstone/shale ratio is generally high, exceeding 0.9, although there are thin, laterally discontinuous claystone layers separating the sandstone units. There is no evidence of fauna or bioturbation in the sandstone and sole markings are rare. Pebbles and granules of claystone are occasionally concentrated at unit bases, but are commonly dispersed throughout the sandstone. Dewatering (dish) structures are abundant. This lithofacies was deposited from fluidized flows of low transport efficiency, resulting in thick amalgamated clastic lobes. Rapid dewatering caused the lobes to 'freeze' as sandstone rich deposits. This lithofacies is the dominant type developed at the Maureen Oilfield.

(2) Interbedded sandstones and claystones occur as units from less than 1 in to 2 ft in thickness. Grain size in the sandstones is, in general, slightly finer as compared with the massive sandstones although the degree of sorting is similar. Rare poorly developed sole markings are present and beds are often distorted by compaction. This lithofacies was deposited from high to low density turbidite flows associated with the lower energy tails of bypassing gravity flows, and is less well developed in the Maureen Oilfield than the massive sandstone lithofacies.

(3) Calcareous, bioturbated claystones with thin interbeds of siltstone, sandstone and limestone are interbedded with the main sandstones and are also well developed in the Lista Formation which overlies and seals the Maureen Formation reservoir. The facies was deposited as a hemipelagic 'rain' of clay size detritus.

In the context of a fan model (Ricci-Lucchi 1975), the massive

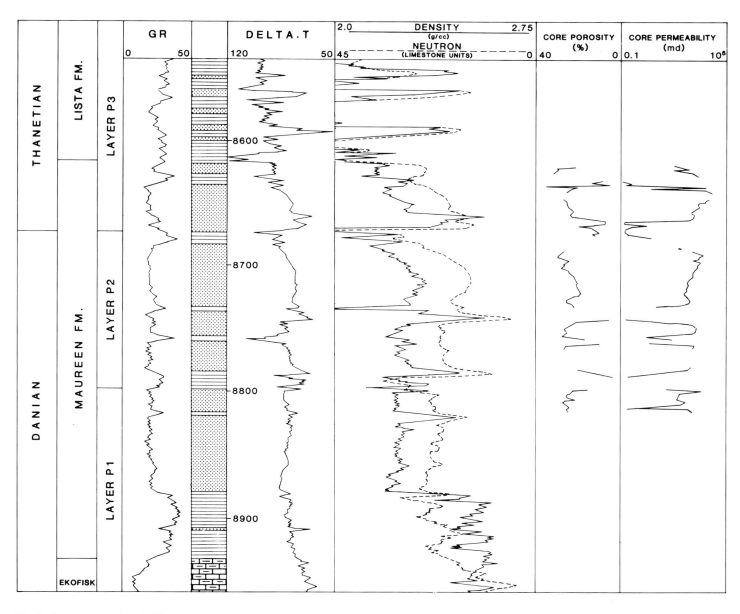

Fig. 5. Reservoir type log: 16/29-2.

sandstone lithofacies was deposited in a channlized setting, probably on the inner or middle fan. The sandstone/shale lithofacies formed in interchannel areas as overbank and levee deposits and the claystone lithofacies, which was also well developed in interchannel areas, blankets all areas of the fan.

Mineralogy and diagenesis

The detrital mineralogy is dominated by simple strained quartz grains. Feldspar (8–12%), rock fragments (2–3%) and detrital clay matrix (1–2%) constitute the remainder. Quartz grain contacts are occasionally straight or concave–convex, but are usually tangential. Feldspars include microcline, slightly sericitized orthoclase, plagioclase and rare perthite. Rock fragments consist mainly of altered or partially dissolved volcanics rocks with rare granite, metaquartzite, chert and reworked sandstone grains with authigenic quartz overgrowths. Chalk and claystone grains are widespread throughout the massive sandstones, ranging up to a few millimetres in length. Heavy minerals include zircon, tourmaline and locally garnet.

X-ray diffraction analysis of the detrital clay matrix indicates the presence of illite/mica mixtures, mixed layer clays, chlorite and kaolinite. They occur as grain rims and occasionally as pore space fillings, but never in sufficient quantity to have a major effect on porosity or permeability; some of the clay may be authigenic.

The diagenetic history of the Maureen reservoir includes the following events (starting with the earliest):

early calcite cementation;
minor quartz overgrowths;
minor kaolinite cementation;
chlorite cementation;
late calcite cementation;
minor grain dissolution/welding (throughout
diagenetic history).

Calcite cementation within massive sandstone units at sandstone/shale contacts is shown clearly in Fig. 6 by the 16/29-2 well log responses at 8678 ft and 8784 ft. Chlorite cementation has been the principal diagenetic event, and although occurring in minor quantities, it has a significant detrimental effect on permeability by occluding small pore throats. Overall, diagenesis has been mild resulting in very good preserved porosities (18–25%) through most of the reservoir. Permeability values can vary from 30–3000 md, but range, more typically, from 100–300 md.

Correlation

The Maureen sandstones have been correlated on a framework of biostratigraphic zonation. The sandstones themselves are devoid of diagnostic foraminiferal assemblages, but abundant fauna in the interbedded claystones have allowed indirect age determination of the sandstones. In conjunction with seismic interpretation the Maureen Palaeocene reservoir interval has been subdivided into the three layers previously mentioned. P1 and P2 represent the Early and Late Danian, respectively, and P3 the Thanetian. These are shown petrophysically on Fig. 5 and in cross-section on Fig. 6.

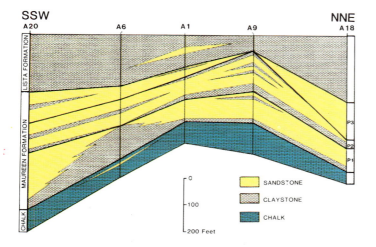

Fig. 6. Correlation of reservoir layers. Locations shown on Fig. 1.

The gross P1 sequence consists of two units, a basal claystone overlain by sandstone. Two P1 clastic lobes have been identified over the Maureen Field, both with a northwest axial orientation. A thin P1 interlobe area is situated over the present structural crest, suggesting that the palaeostructure has influenced sediment deposition. Varying between 45 and 200 ft thick, the P1 sandstones contain about 25% of Maureen Oilfield reserves.

P2 sandstone deposition followed without a major break in sedimentation. This layer is interpreted as a broad easterly trending clastic wedge, ranging in thickness from 200 ft in the northwest part of the field to 40 ft in the southeast and it accounts for around 40% of the reserves.

The P3 layer comprises two main clastic lobes, one to the northeast with a northwest axial trend similar to the P1 trends and one to the southeast, which is interpreted as a broadly east–west trending wedge, similar to the P2 layer. The lobes are separated by an area of basin plain mudstones. Thickness varies from 144 ft to less than 12 ft; around 35% of Maureen Oilfield reserves are contained in this layer.

The reservoir geometry shown in Fig. 6 illustrates rapid sandstone pinchouts and apparently laterally extensive claystones within the field area. However, production history over five years strongly suggests that these non-reservoir units do not form vertical or lateral barriers to fluid flow; the reservoir has behaved to date as a single unit.

Source rocks

Geochemical analyses of the Maureen oil have shown a good correlation with hydrocarbon extracts from Upper Jurassic source rocks of the region, primarily the Kimmeridge Clay Formation shales. Upper Jurassic rocks with source potential lie in the deep southeast sector of the Witch Ground Graben, some 10 miles to the southeast of the Maureen Field. In this source kitchen, early maturity was reached in the Mid- to Late Cretaceous, with peak oil generation commencing during the Early Tertiary.

Hydrocarbons migrated vertically via basement rooted graben bounding faults into Palaeocene sandstones. Having suffered relatively little deformation and representing laterally continuous, normally pressured aquifer systems, the Palaeocene sandstones have allowed lateral oil migration of up to 15 miles (Cayley 1987) from the source kitchen to the Maureen Field. Hydrocarbons have accumulated in the Maureen Field since the Oligocene, when a suitable trap at the Palaeocene level had evolved.

Hydrocarbons

The best test of the Palaeocene reservoir was from the 16/29-2 appraisal well. Using a multiple choke configuration, a flow rate of 10 832 BOPD was achieved from a single, 75 ft thick, perforated interval. The PVT description in the present reservoir simulation model is based on subsurface DST samples from the 16/29a-5 well. In this test, the Palaeocene interval flowed at 4285 BOPD through a 7/8 inch choke. The following fluid properties taken at 8300 ft TVSS datum, are considered representative of the Maureen Oilfield:

Oil gravity	35.8° API
Gas gravity	0.782
Initial solution GOR	392.7 SCFPB
Initial reservoir pressure	3792 psia
Reservoir temperature	243°F

Reserves

The current estimate for Maureen Field original oil-in-place is 393 MMBBL. Estimated primary and secondary recoverable reserves of 210 MMBBL amount to a recovery of 53.9%.

Production, which began in September 1983, peaked at 29.3 MMBBL per year in 1984.

A peripheral water injection system, from seven wells, supports the natural water drive. Water injection was started in February 1984 and has produced a sweep from the west, where most water is injected. In December 1988 the water injection rate averaged 58 250 BWPD. Artificial gas lift is employed to aid in lifting water producing wells and also to increase production from selected wells. Maximum gas lift capacity is 10 million SCFD.

The author thanks Phillips Petroleum Company UK Limited, the Phillips Petroleum Company, and Licence P.110 coventurers (Fina, Agip, Gas Council and Ultramar) for permission to publish this paper.

References

CAYLEY, G. T. 1987. Hydrocarbon migration in the Central North Sea. *In*: BROOKS, J. & GLENNIE, K. W. (eds) *Petroleum Geology of North West Europe*. Graham & Trotman, 549–555.

DEEGAN, C. E. & SCULL, B. J. 1977. *A standard lithostratigraphic nomenclature for the Central and Northern North Sea*. Institute of Geological Sciences, Report 77/25.

PHILLIPS PETROLEUM COMPANY (UK) LTD 1985. *Maureen Field, UK North Sea, Annex B (Revised) Plan of Development*.

RICCI-LUCCHI, F. 1975. Depositional Cycles in Two Turbidite Formations of the Northern Apennines (Italy). *Journal of Sedimentary Petrology*, **45**, 3–43.

The Petronella Field, Block 14/20b, UK North Sea

P. WADDAMS & N. M. CLARK

Texaco Ltd., 1 Knightsbridge Green, London SW1X 7QJ, UK

Abstract: Petronella Field is a small oil and gas accumulation located 110 miles northeast of Aberdeen in UK Block 14/20b. The field lies on the highest part of the east–west-trending Petronella Ridge approximately 6 miles southwest of the Witch Ground Graben axis. The reservoir is Upper Jurassic in age and lies some 7500 ft below sea level. It comprises shallow marine sandstone of the Piper Formation ('Principal Reservoir Sequence') overlain by deeper marine turbidites ('Hot Lens Equivalent') of the Kimmeridge Clay Formation. The structure is a tilted fault block which is bounded to the north by a major fault system, downthrown to the north. Sandstone units dip to the south and thin or are truncated to the north as a result of erosion of the crest of the structure. Seal is effected by Upper Jurassic siltstone and Lower Cretaceous calcareous claystone. The accumulation has been sourced from maturation of the Kimmeridge Clay Formation below approximately 10 000 ft in adjacent basins. The Field was discovered in February 1975 and is delineated by six wells. Current production of 13 000 BOPD comes from one well and uses an innovative remote subsea system controlled from, and with pipelines to, the Tartan Platform 6.4 miles to the east. Ultimate recoverable reserves from the main portion of the Field are 17 MMBBL of crude oil.

Petronella is a small oil and gas field located 110 miles northeast of Aberdeen near the southern boundary of UK Block 14/20b in 430 ft of water (Fig. 1). The field lies on the Petronella Ridge 5 miles northeast of the Halibut Horst and 6 miles southwest of the Witch Ground Graben axis. The accumulation is approximately 1 mile long in an east–west direction and covers an area of 240 acres. The structure comprises a tilted fault block bounded to the north by a major fault which downthrows to the north. Within the fault block beds dip approximately 15° to the south. The reservoir consists of Upper Jurassic sandstone deposited in a high energy shallow marine environment (the 'Principal Reservoir Sequence') overlain by deeper marine turbidites (the 'Hot Lens Equivalent').

'Petronella' is the name of a Highland dance and maintains the Scottish theme traditionally adopted by Texaco and other operators for commercial discoveries in this area.

History

Block 14/20 (P.237) was awarded to Texaco North Sea UK Limited (100%) during the Fourth Round of licensing in March 1972. In February 1979 a small portion of the block (14/20a) became included in the Tartan PRT boundary (see Coward *et al.*, this volume). Block 14/20b was subsequently awarded a new licence (P.324) and it is within this partial block that the Petronella Field lies.

In February 1975 well 14/20-1, drilled near the crest of the structure, discovered a thick Jurassic sandstone comprising 768 gross ft of which 275 net ft were hydrocarbon bearing, and established the OWC at −7648 ft TVSS. The well tested at a stabilized rate of 4537 BOPD (40.2° API) plus 4.7 MMSCFD. Wells 14/20-2 and 14/20-3A, to the west and east respectively, both

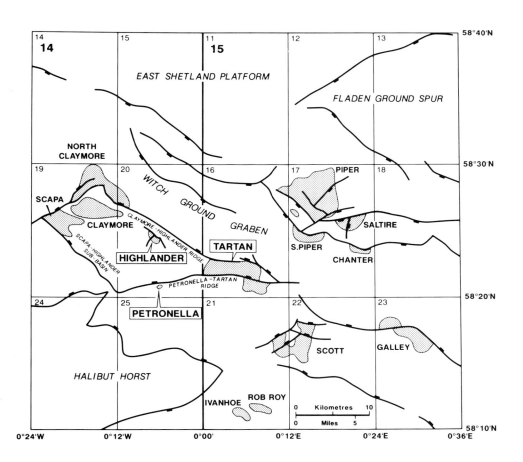

Fig. 1. Simplified tectonic framework of the Outer Moray Firth area, North Sea, showing the location of Texaco's Petronella, Tartan and Highlander Fields and surrounding discoveries.

From Abbotts, I. L. (ed.), 1991, *United Kingdom Oil and Gas Fields, 25 Years Commemorative Volume*, Geological Society Memoir No. 14, pp. 353–360

found thick Upper Jurassic sandstone sections but were dry. Well 14/20b-11, down-dip to the southwest, confirmed the OWC established by 14/20b-1. Well 14/20b-16, drilled approximately 1000 ft to the north of 14/20-1, contributed to a greater understanding of the reservoir behaviour of the Petronella area and established a GOC at −7316 ft TVSS.

2-D seismic surveys were recorded between 1974 and 1979 and reprocessed in 1979. A 3-D seismic survey was shot in 1981 over the entire block and forms the basis for the current structural interpretation.

Annex B approval for Petronella was given on 25 April 1986 with production start-up in November 1986. The field has been developed as a subsea satellite to Tartan. Production comes from a single well 14/20b-16. Fluids are transported to the Tartan Platform via a trenched seven mile long 8 inch diameter pipeline where they are metered, separated and processed.

Field stratigraphy

The oldest sediments penetrated by the Petronella wells comprise Carboniferous sandstone, mudstone and coal of the Lower Namurian to Upper Visean Forth Formation (860 ft penetrated in 14/20-1). During the late Westphalian terminal phase of the Hercynian orogeny, the Moray Firth Basin became inverted. Post-Lower Namurian strata were eroded and are absent from the Petronella area (Fig. 2).

With the relaxation of the late Hercynian stress systems, the area began to subside during the early Permian. In 14/20-2, 343 ft of anhydrite with minor claystone record the southerly transgression of the Zechstein sea from the Norwegian–Greenland Sea rift. During the latest Permian, the Zechstein seas withdrew from the North Sea area (Ziegler 1988).

By early Triassic time, rifting activity accelerated creating a

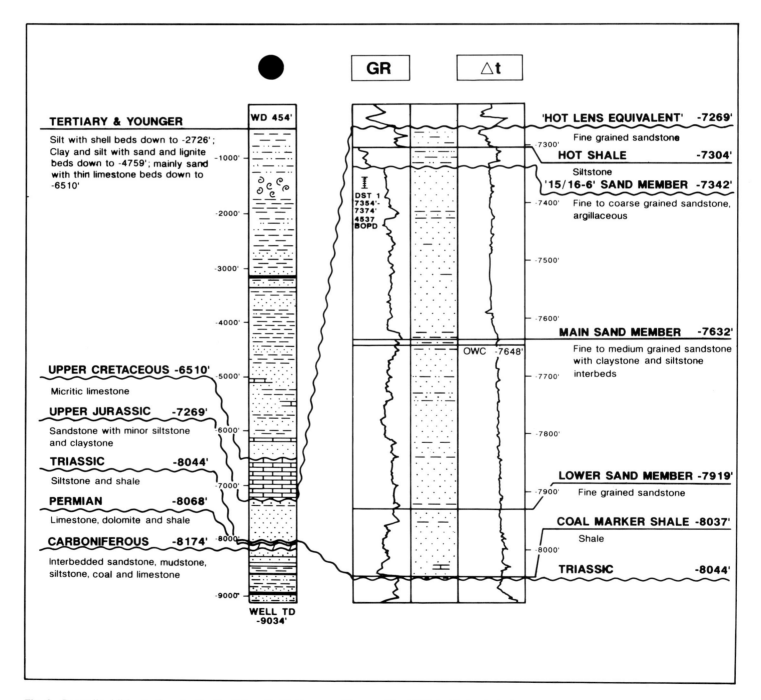

Fig. 2. Generalized lithostratigraphy for the Petronella Field area as illustrated by 14/20-1. All depths in feet sub-sea.

complex series of variably-orientated grabens. The NW–SE-trending Witch Ground Graben may have developed at this stage as a result of the reactivation of Hercynian features (Harker *et al.* 1987). All Petronella wells have penetrated variable thicknesses of red or grey mudstone, shale and siltstone of the Triassic Smith Bank Formation.

Lower and Middle Jurassic rocks are not present in the Petronella Field area owing either to non-deposition or erosion. However, 14/20-3A down-dip to the east of Petronella, penetrated 20 ft of Middle Jurassic volcaniclastic siltstone (Fig. 3). This facies thickens progressively eastwards and 15/16-7, 4 miles to the east, reached total depth in Bathonian tuffaceous sediments (86 ft penetrated). Subsidence during Upper Jurassic times led to the deposition of the widespread shallow marine sands of the Piper Formation (Late Oxfordian to Kimmeridgian in age). Uplift and erosion of the Halibut Horst and Fladen Ground Spur provided a plentiful supply of sand. By early Volgian times, deeper water conditions were established. Organic-rich silt and clay were deposited under anoxic conditions and these now form the principal source rock for the Petronella Field accumulation. Locally, some areas such as the Halibut Horst continued to supply sand to the Witch Ground Graben. Turbidites which form the 'Hot Lens Equivalent' of the Petronella Field were deposited at this time.

At the transition from the Jurassic to the Cretaceous Period, a relative drop in sea level was accompanied by an acceleration of rifting activity. Most of the displacement across Fault 'A' (Fig. 4), which defines the northern boundary of the Petronella Ridge, occurred at this time. This tectonism resulted locally in the erosion of the crests of some fault scarps and the accumulation of submarine sand fans now referred to as the Scapa Sand Member (Boote & Gustav 1987; Harker *et al.* 1987). Lower Cretaceous marl, shale and minor sandstone of the Cromer Knoll Group thicken away from the crest of the Petronella structure and usually onlap on to the flanks of the Witch Ground Graben. A more detailed account of the Upper Jurassic and Lower Cretaceous evolution of the Witch Ground Graben is given in O'Driscoll *et al.* (1991). The basal part of the Upper Cretaceous Chalk Group comprises argillaceous limestone overlain by the laterally extensive Turonian Marker

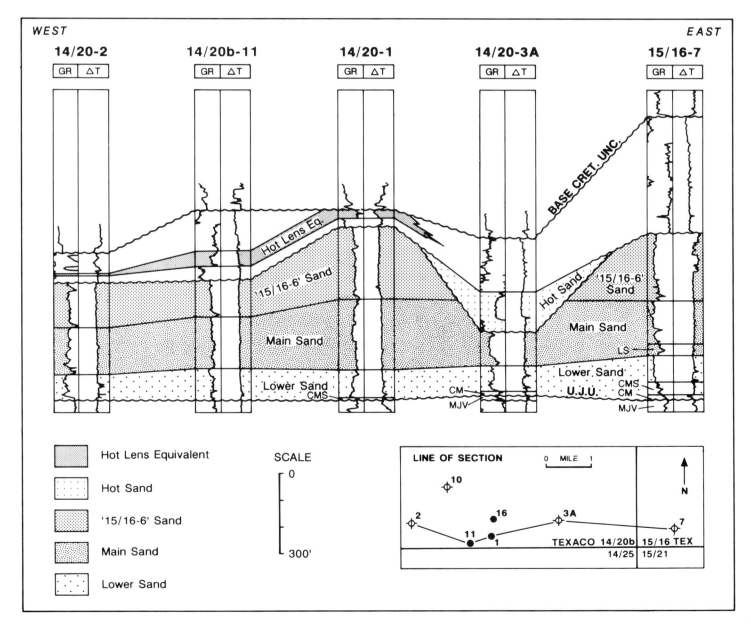

Fig. 3. Simplified west–east log correlation of the Upper Jurassic section in wells 14/20-2, 14/20b-11, 14/20-1, 14/20-3A and 15/16-7. UJU, Upper Jurassic Unconformity; LS, Lower Shale; CMS, Coal Marker Shale; CM, Coal Marker; MJV, Middle Jurassic volcanic rocks.

(Plenus Marl). The latter consists of shale or claystone deposited during a basin-wide anoxic event. Subsequent late Cretaceous biogenic micritic limestone deposition accompanied gentle post-rift subsidence (Fig. 2).

During the Palaeocene and Eocene, the Petronella field occupied an area dominated by a thick wedge of coastal clastic sediments comprising mainly unconsolidated sand, silt, clay and lignite. The Upper Palaeocene sediments are occasionally tuffaceous, recording Thulean volcanic activity. The position of Petronella close to the northern flank of the stable Halibut Horst ensured relatively thin accumulations of post-Eocene sediment.

Geophysics

A total of 740 km of 2-D seismic data were acquired between 1974 and 1979 giving a 1 km × 1 km grid over the prospective areas of Block 14/20. The 3-D survey, shot in 1981, covers the entire block and comprises 5600 km with a line spacing of 50 m. Eight seismic horizons have been mapped: Base Tertiary, Turonian Marker, Base Chalk, Seismic Aptian Marker, Base Cretaceous, Upper Jurassic Unconformity (UJU), Top Zechstein and Top Carboniferous. With the exception of the Seismic Aptian Marker, Base Cretaceous and UJU horizons, these events are generally of good quality. However, over the northern part of the Petronella Field where the Upper Jurassic reservoir section has been eroded, all the pre-Cretaceous events become difficult to identify (Figs 5 and 6).

The top of the Piper Formation reservoir sequence does not provide a consistent mappable seismic event (Fig. 5). The horizon was mapped by adding to the Base Chalk map (the nearest reliable seismic event) the isochores representing the Lower Cretaceous section and the Upper Jurassic Kimmeridge Clay Formation. Similarly, the structure map for the top of the 'Hot Lens Equivalent' reservoir was constructed from the Top Piper Formation depth map by subtracting isochores of the 'Hot Lens Equivalent' and the shale beneath it.

Trap

The Petronella Field occupies the highest structural point along the east–west-trending, southerly-dipping, tilted fault block called the Petronella Ridge (Fig. 4). The major fault zone which defines the Petronella Ridge (Fault 'A') downthrows to the north and developed mainly during Lower Cretaceous times. This is the only fault in the Field area which cuts the Base Cretaceous. Other normal faults which define the Petronella structure are those designated 'B', 'C', 'D' and 'E' which downthrow to the northwest, southwest, southeast and northeast respectively.

Uplift provided by faults 'A' and 'E' exposed the northern area of the Petronella Ridge to deep erosion. Closure to the north and northeast is therefore effected by truncation of the Principal Reservoir Sequence beneath the Base Cretaceous Unconformity. Fault 'D' has significant throw at the Top Piper Formation level and

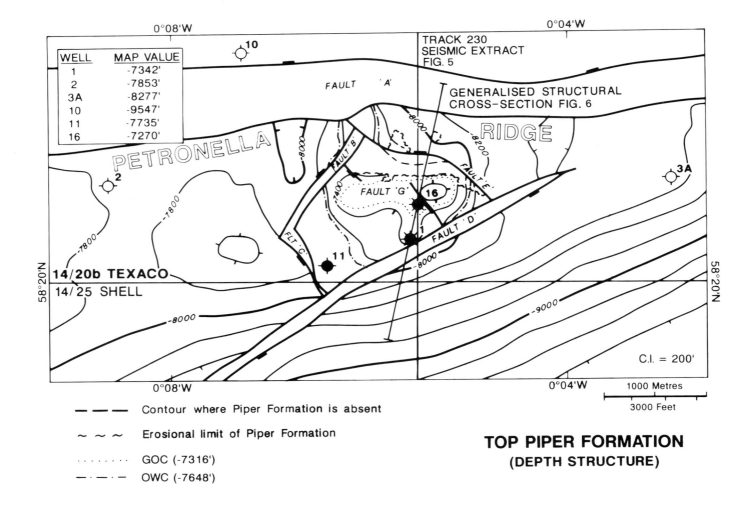

Fig. 4. Top Piper Formation depth structure map showing well location, fault designation and position of cross-sections shown in Figs 5 and 6. Contour interval 200 feet. All depths in feet sub-sea.

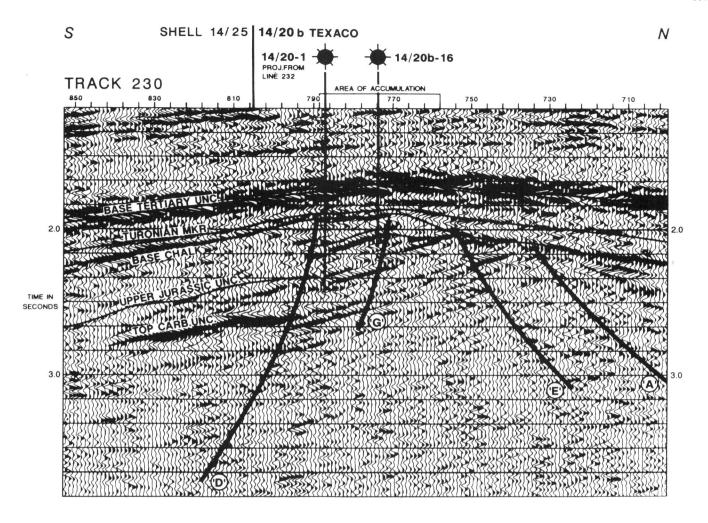

Fig. 5. A portion of the 3-D north–south seismic line (Track 230) through well 14/20b-16 to show the nature of the principal seismic events (well 14/20-1 projected 165 feet from the west). See Fig. 4 for location of seismic line and fault designation.

provides upthrown fault closure for the accumulation to the southeast. Closure to the south and west is effected by dip.

The reservoir is sealed above by siltstone of the Kimmeridge Clay Formation on the southern flank and by thin Lower Cretaceous calcareous siltstone over the crest and to the north (Fig. 6). Basal seal is provided by thick Triassic claystone. The OWC occurs at −7648 ft TVSS in 14/20-1 and 14/20b-11. In 14/20b-16 the GOC occurs at −7316 ft TVSS and is considered to occur at the same depth in 14/20-1 within siltstone of the Kimmeridge Clay Formation.

Reservoir

The Upper Jurassic Humber Group sediments of the Petronella Field area consist of a lower, sand-dominated sequence of Upper Oxfordian to Kimmeridgian age (the Sgiath and Piper Formations or Principal Reservoir Sequence) and an upper, shale-dominated sequence of Volgian to Ryazanian age (the Kimmeridge Clay Formation or Caprock Sequence). Thin sandstone beds occur within the Caprock Sequence and are here termed the 'Hot Lens Equivalent'.

The sub-division of the Humber Group is an extension of the scheme developed for the nearby Tartan Field (see Coward *et al.*, this volume), thereby allowing consistent correlations to be made. A lithostratigraphic sub-division was adopted since, to date, the biostratigraphy is too insensitive to allow the construction of a reliable, alternative scheme. It is probable that the individual members described below are diachronous, but until a finer biostratigraphic zonation scheme is erected, this cannot be confirmed.

Sgiath and Piper Formations (Principal Reservoir Sequence)

The Principal Reservoir Sequence sandstones are essentially subarkoses. The most abundant detrital minerals are quartz and feldspars, with subordinate quantities of mica, rock fragments, opaque and heavy minerals and plant debris. Detrital clay is a minor constituent of all but the finer grained sandstones. Within the area of accumulation, the diagenetic changes that have taken place have not been seriously detrimental to reservoir quality. Within the relatively clean Piper Formation sandstone, quartz overgrowths constitute the main authigenic phase. Feldspar overgrowths and patchy authigenic calcite have also been observed.

Four phases of deposition can be identified: Coal Marker to Lower Sand Members (bottom); Lower Shale and Main Sand Members; '15/16-6' Sand Member; and Hot Sand Member (top). The first three phases comprise marginal to shallow-marine coarsening-upwards mega-cycles, well illustrated by the gamma response in 15/16-7 (Fig. 3). The last phase records deeper water sedimentation. Each phase is described briefly below.

Coal Marker to Lower Sand Members. Deposition of the Principal Reservoir Sequence in the Witch Ground Graben area began with

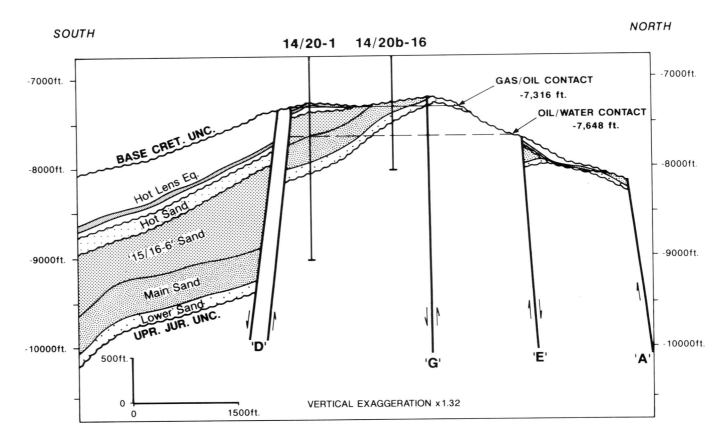

Fig. 6. Generalized north–south seismic line (Track 230) through well 14/20b-16 and 14/20-1 to show the truncation of the Principal Reservoir Sequence on the crest of the Petronella Ridge. See Fig. 4 for the location of section.

the fresh water Coal Marker and continued with the brackish water Coal Marker Shale. The Coal Marker and Coal Marker Shale both equate to the Sgiath Formation, defined by Harker *et al.* (1987) and dated by them as Oxfordian. No Sgiath Formation sandstone has been identified on Block 14/20. The Coal Marker has not been penetrated by any of the Petronella Field wells due to the structurally high position of this area at the time of deposition. However, it is represented in wells 14/20-3A and 15/16-7 by 11 ft and 18 ft respectively of coal and siltstone (designated CM in Fig. 3). The Coal Marker Shale (designated CMS in Fig. 3) is recorded as an upward-decreasing gamma peak at the base of the Lower Sand Member. It comprises dark siltstone, four feet thick in 14/20b-16. It is absent in 14/20b-11 but thickens eastwards to 15/16-7 where it consists of 53 ft of siltstone and fine-grained sandstone (Fig. 3).

Overlying the Coal Marker Shale, the Lower Sand is present throughout the area and was deposited under nearshore shallow marine conditions in a tidal flat or possibly beach environment. Porosity of the Lower Sand ranges from 24.8% in 14/20b-16 on the crest of the structure, to approximately 16% on the northern and southern flanks.

Within the Lower Sand in 14/20b-16 an anomalously large separation between the neutron and bulk density curves was recorded between -7415 ft and -7423 ft TVSS. Such a separation is usually considered to be indicative of gas. However, DST and RFT results confirmed this zone to be oil-bearing. Particularly high core porosities (22–31%) and permeabilities (527–13 000 md) occur in this interval. In such zones a protective mud-cake builds up very rapidly on the borehole wall during drilling, resulting in extremely low invasion of the formation by mud filtrate. In zones which have in addition good vertical permeabilities, segregation of the mud filtrate and virgin formation fluid will occur after approximately 24 hours. The denser mud filtrate will be displaced downwards by the less dense virgin formation fluid (the virgin formation fluid is oil in the Lower Sand in 14/20b-16). The effect will be enhanced if the formation is steeply-dipping (in Petronella the Piper Formation dips approximately 15° to the south). The response of the neutron and density tool (which have shallow depths of investigation) will therefore be strongly hydrocarbon-affected. Tool response can be simulated if the formation adjacent to the borehole is filled with 67% oil and 27% mud filtrate, with a water saturation of 6% (K. F. Wichtl, pers. comm.).

Lower Shale to Main Sand Member. The Lower Shale Member comprises about 60 ft of grey brown silty shale in the Tartan Field 6.4 miles to the northeast (see Coward *et al.*, this volume). In the Petronella area it is poorly developed and is probably represented by the silty micaceous sandstone (producing the gamma peak labelled LS in Fig. 3) at the base of the Main Sand Member.

The Lower Shale to Main Sand interval is characterized by an upward increase in grain size. The sediments pass upwards from siltstone and fine-grained sandstone of the Lower Shale Member, through flaser-bedded sandstone, and into clean locally cross-bedded sandstone. The sequence records a gradually shallowing environment passing from offshore and lower shoreface to upper shoreface and possibly foreshore environments. The Main Sand has been completely eroded from the extreme northern crest of the Petronella structure and partially eroded from the area around 14/20b-16. Porosity ranges from 19.3% in 14/20-10 to 23.3% in 14/20b-11.

'15/16-6' Sand Member. The '15/16-6' Sand Member comprises a coarsening-upwards facies reflecting progressively shallower water. The basal part consists of offshore and lower shoreface deposits, accumulated during both storm and fair-weather conditions. These pass upwards into upper shoreface and foreshore deposits with a strong component of tidal channel reworking. The '15/16-6' Sand

Member is absent through erosion from the 14/20b-16 area and has been partially eroded from the area around 14/20-1 (Fig. 6). It thickens to the east, west and south away from the Petronella structure. Porosity averages 25% on the Petronella Ridge but decreases to 14.1% in the downthrown well 14/20-10.

Hot Sand Member. The Hot Sand Member comprises fine-grained, argillaceous, heavily bioturbated sandstone which accumulated in shallow marine, lower shoreface and offshore environments as the sea level rose and the shoreline retreated. It passes upwards through progressively finer-grained sediments into the Caprock Sequence above. The Hot Sand Member only occurs in 14/20-3A where the porosity is 16.3%. Elsewhere in the Petronella Field it is absent through erosion (Fig. 3).

Kimmeridge Clay Formation (Caprock Sequence)

During Volgian to Ryazanian times, black or dark brown, organic-rich silt and clay accumulated throughout the outer Moray Firth area. In the Petronella area, massive, sharply based, turbidite sandstone beds were deposited locally and are considered from log response and sediment character to be similar to the 'Hot Lenses' of the Tartan Field. Consequently this sandstone has been termed the 'Hot Lens Equivalent'. In 14/20b-11 it is 68 ft thick (28 ft net oil), but has been reduced to 35 ft in 14/20-1 as a result of erosion. Porosities range from 21% in 14/20b-11 to approximately 15% in 14/20-1 and 14/20-10. The 'Hot Lens Equivalent' appears to have been sourced from a southerly direction.

The complicated faulting on the flanks of the Petronella structure, together with the thick interval of Kimmeridge Clay Formation separating the 'Hot Lens Equivalent' from the 'Principal Reservoir Sequence', indicates the likelihood that both sandstone reservoirs may be in pressure communication. It is therefore anticipated that the majority of the Hot-Lens Equivalent will be drained by production from well 14/20b-16.

Source

Most oils in the Outer Moray Firth area have been sourced from hydrogen-rich, oil-prone kerogens present in the Kimmeridge Clay Formation. Total organic carbon contents for this formation range from 1.0% to 9.0% by weight. Thermal maturation data suggest that the Jurassic attains optimum maturity for effective oil generation and release between 9500 and 11 500 ft in the Witch Ground Graben. Expulsion of oil from the Kimmeridge Clay Formation started 40–50 million years ago and probably is proceeding vigorously at the present time (Bissada 1983).

Hydrocarbons

DSTs conducted on 14/20-1 and 14/20b-16 gave an early indication of hydrocarbon properties, reservoir conditions and production performance.

The rapid build-up during DST 1 on well 14/20-1 indicated a high-permeability reservoir. The test produced 4537 BOPD (40.2° API) with no water from a perforated interval of 20 ft. A maximum flow rate of 7605 BOPD was achieved at the end of the test.

In 14/20b-16 DST 1A was intended to resolve a major log anomaly that indicated an additional gas accumulation within the oil column. The test produced black oil with a normal GOR thus proving the log response to be anomalous. DST 2 was the main production test and added perforations in the oil column to those of DST 1A. During the multi-rate test (at 3000 BOPD, 6400 BOPD and 11 000 BOPD) the GOR remained stable and oil had a gravity of 33.5° API. Analysis showed the formation to be highly productive. During DST 3, a short interval was perforated in the oil column just below the GOC. The well was flowed at increasing rates and the rising GOR behaviour was recorded.

These data were used to create a radial coning simulation model.

Reserves

Recoverable reserves within the Principal Reservoir Sequence are estimated to be 17 MMBBL of crude oil. This excludes the reserve potential of the Hot Lens Equivalent or the potential of a hitherto

Petronella Field data summary

Trap	
Type	Combined structural/stratigraphic
Depth to crest	−7200 ft
Lowest closing contour	−7800 ft
Gas–oil contact	−7316 ft
Oil–water contact	−7648 ft
Gas column	116 ft
Oil column	332 ft
Pay zone	
Formation	Piper Formation
Age	Upper Jurassic
Gross thickness (average/range)	341 ft/188–306 ft
Sand/shale ratio	N/A
Net/gross ratio (average/range)	1
Cut-off for N/G	N/A
Porosity (average/range)	21%/16–25%
Hydrocarbon saturation	81%
Permeability (average/range)	212/10–629 md
Productivity index	140
Hydrocarbons	
Oil gravity	39
Oil type	Black
Gas gravity	1.0
Bubble point (psia)	3350
Dew point	N/A
Gas/oil ratio	1250
Condensate yield	N/A
Formation volume factor	1.7
Formation water	
Salinity (ppm NaCl equivalent)	76 000
Resistivity (Rw at 160°F)	0.045
Reservoir conditions	
Temperature (°F)	160
Pressure (psia at OWC)	3450
Pressure gradient in reservoir (psi/ft)	0.28
Field size	
Area (acres)	240
Recovery factor: oil	51%
Drive mechanism	Aquifer support
Recoverable hydrocarbons: crude oil	17 MMBBL
Production	
Start-up date	November 1986
Development scheme	Subsea satellite to Tartan
Production rate	13 000 BOPD
Cumulative production to end March 1989	
Crude oil	6.6 MMBBL
NGL	0.6 MMBNGL
Liquids	7.2 MMBBL hydrocarbons
Number/type of wells	1 producer

untested small extension of the field. Petronella production began in November 1986 at a rate of 5300 BOPD and is currently producing around 13 000 BOPD. If economically feasible, any gas remaining in the reservoir can subsequently be produced through the Tartan Platform.

As a contingency, a water injection line was laid between Petronella and the Tartan Platform. However, the Field continues to experience strong acquifer support and water injection is not anticipated to be required in the near future.

Texaco Inc gave permission to publish this paper which is based extensively on the work of a large number of Texaco employees who have been involved in the detailed evaluation of Block 14/20 since 1972.

References

BISSADA, K. K. 1983. Petroleum generation in Mesozoic sediments of the Moray Firth Basin, North Sea area. *In*: BJORØY, M. (ed.) *Advances in Geochemistry*. John Wiley, London, 7–15.

BOOTE, D. R. D. & GUSTAV, S. H. 1987. Evolving depositional systems within an active rift, Witch Ground Graben, North Sea. *In*: BROOKS, J. & GLENNIE, K. W. (eds) *Petroleum Geology of North West Europe*. Graham & Trotman, London, 819–833.

COWARD, R. N., CLARK, N. M. & PINNOCK, S. J. 1991. The Tartan Field, Block 15/16, U.K. North Sea. *This volume*.

HARKER, S. D., GUSTAV, S. H. & RILEY, L. A. 1987. Triassic to Cenomanian stratigraphy of the Witch Ground Graben. *In*: BROOKS, J. & GLENNIE, K. W. (eds) *Petroleum Geology of North West Europe*. Graham & Trotman, London, 809–818.

O'DRISCOLL, D., HINDLE, A. D. & LONG, D. C. 1991. The structural controls on Upper Jurassic and Lower Cretaceous reservoir sandstones in the Witch Ground Graben, UK North Sea. *In*: HARDMAN, R. F. P. & BROOKS, J. (eds) *Tectonic Events Responsible for Britain's Oil and Gas Reserves*. Geological Society, London, Special Publication, **55**, 299–323.

ZIEGLER, P. A. 1988. *Evolution of the Arctic-North Atlantic and the Western Tethys*. American Association of Petroleum Geologists, Memoir 43.

The Piper Field, Block 15/17, UK North Sea

H. R. SCHMITT[1] & A. F. GORDON[2]

[1] *Occidental International Exploration and Production Company, 1200 Discovery Drive, Bakersfield, CA 93389, USA*
[2] *Occidental Petroleum (Caledonia) Ltd, 1 Claymore Drive, Bridge of Don, Aberdeen AB2 8GB, UK*

Abstract: The Piper Field is located in Block 15/17 and was discovered in 1972 by the 15/17-1A well, which tested 37° API oil from the Late Jurassic Piper Formation. The field is situated on a shelf on the northern margin of the Witch Ground Graben and comprises four tilted fault blocks dipping gently to the northeast away from the graben. The trap is formed by a combination of dip and fault closure. The reservoir comprises sandstones of the paralic Sgiath Formation and the shallow-marine wave-dominated deltaic Piper Formation. The original productive area of the field was 7350 acres with a maximum oil column of 1210 ft, containing approximately 1400 MMBBL STOIIP. Following cumulative production of 834 MMBBL, a productive area of 2500 acres remains. Ultimate recovery of approximately 70% of STOIIP is forecast, with estimated remaining reserves of 118 MMBBL. Hitherto, production has been through a fixed 36-slot platform which at the time of the platform explosion had 27 producing wells and 8 injectors. Production averaged over 250 000 BOPD for the first three years of the field's life and was reduced to approximately 200 000 BOPD in 1980 to improve reservoir management and ultimate recovery. Production had declined to about 120 000 BOPD by July 1988, water-injection rates having reached an average of 260 000 BPD.

The Piper Field is in UK North Sea Licence Block 15/17, 110 miles northeast of Aberdeen and 125 miles ESE of the Flotta Oil terminal in the Orkneys (Fig. 1). The field lies in 474 ft of water and is situated on a shelf on the northern margin of the Witch Ground Graben (WGG) in the Outer Moray Firth. The WGG is a north-westerly trending graben branching off from the intersection of the N–S trending Viking and Central Grabens, 50 miles southeast of Piper Field.

The field contained an estimated STOIIP of approximately 1400 MMBBL. The reservoir comprises sandstones of the shallow-marine Piper Formation and the paralic Sgiath Formation, sealed by Upper Jurassic claystones and Cretaceous marls (Fig. 2). Oil is trapped in four northeasterly tilted fault blocks. These have a common oil–water contact of 8510 ft TVSS except for the small anticlinal accumulation in the southwestern block, which has a deeper oil–water contact at 9200 ft TVSS (Figs 3 & 4).

The name Piper refers to a player of the bagpipes, the Scottish musical instrument.

History

Occidental began their regional seismic and geological studies of the petroleum potential of the onshore and offshore United Kingdom in 1969 in anticipation of a forthcoming licensing round. In 1971, Occidental, as operator, formed a consortium with Getty Oil International (England) Ltd, Allied Chemical (Great Britain) Ltd and Thomson Scottish Associates. Production Licence P220 was awarded to the Occidental Group in March 1972. The Piper Field was discovered in December 1972 by the 15/17-1A well. The discovery well location was based on mapping of 38 miles of 2-D data. The well encountered a 200 ft oil-bearing section and tested 5266 BOPD from 143 ft of perforations.

Following the discovery well, an appraisal drilling programme commenced. Well 15/17-2, drilled 1.6 miles northwest of 15/17-1A, encountered a slightly thicker reservoir section and tested oil at rates of 15 257 and 16 873 BOPD from two intervals. The 15/17-3 well was drilled on the downthrown side of a normal fault 1.4 miles

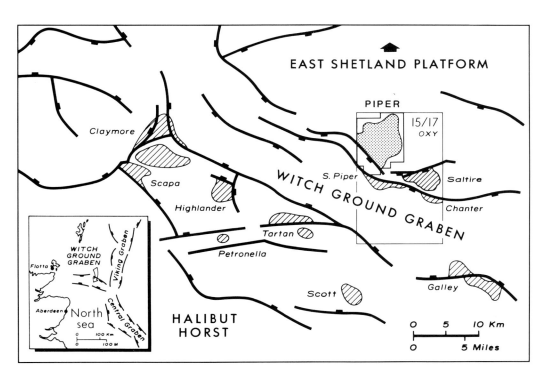

Fig. 1. Structural elements, Witch Ground Graben.

From Abbotts, I. L. (ed.), 1991, *United Kingdom Oil and Gas Fields, 25 Years Commemorative Volume*, Geological Society Memoir No. 14, pp. 361–368

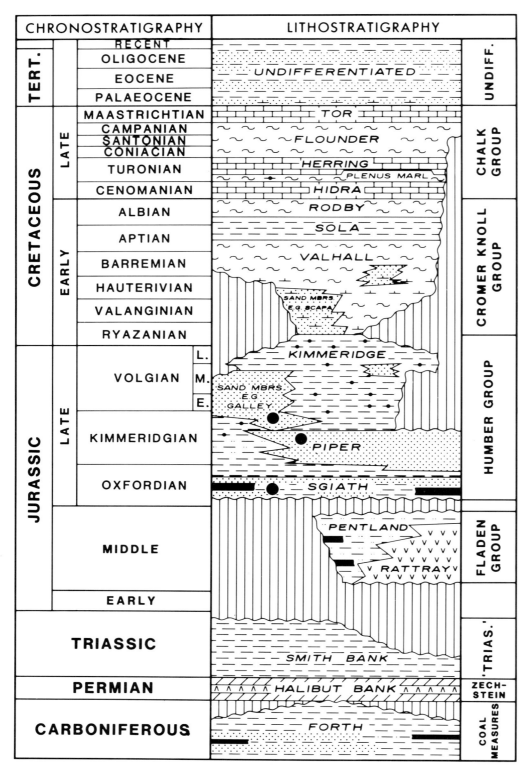

Fig. 2. Piper stratigraphy.

NOTE: VERTICAL SEQUENCE NOT TO SCALE.   ● OIL RESERVOIRS

south of 15/17-1A. In this well, the upper part of the Piper sandstone was missing because of Late Jurassic to Early Cretaceous erosion.

At this time, a bottom-hole contribution was made to the Burmah 15/12-1 well drilled just north of the 15/17 Block boundary (Fig. 3). It encountered thick reservoir sandstones but they were wet. The 15/17-4 well was then drilled updip and established an oil–water contact at 8510 ft TVSS.

Wells 15/17-5 and 15/17-6 confirmed that the Piper Field could be fully developed from one centrally located platform. Both encountered the same oil–water contact as 15/17-4 (8510 ft TVSS). The last appraisal well, 15/7-7, was drilled to the southwest and across a major fault (the 'A' fault) in late 1973. The Jurassic sandstones were present but significantly deeper with an oil–water contact at 9200 ft TVSS.

The discovery and appraisal programme had established the size and extent of the field, and during initial development geology studies, the field was interpreted to be a layered reservoir. A 36-slot platform was centrally located over the field in June 1975 and development drilling commenced in October 1976.

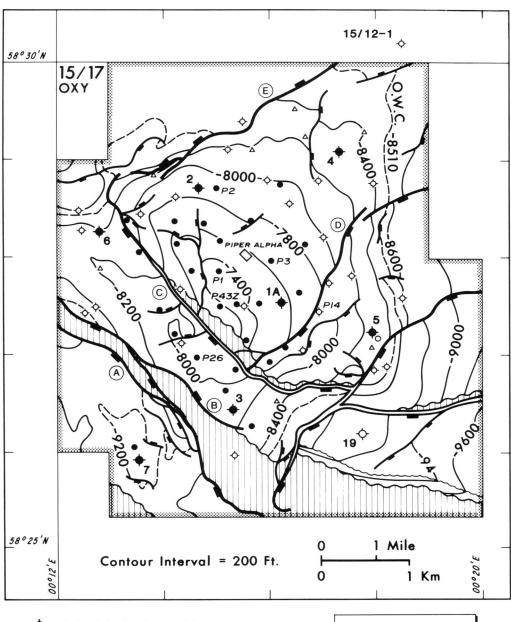

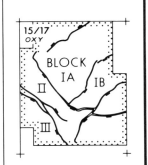

Fig. 3. Top Piper Sandstone depth structure map.

The P1 production well was spudded on 10th October 1976 and established commercial production on 7th December 1976 at more than 30 000 BOPD (Fig. 4). After the field rate was reduced in 1980 to enable more efficient reservoir management, wells were generally produced at rates of less than 25 000 BOPD.

Field stratigraphy

The Piper Field reservoir comprises sandstones of the Piper and Sgiath Formations within the Late Jurassic (Oxfordian–Lower Kimmeridgian) Humber Group (Fig. 2). A geological history of the area is included in the Chanter paper in this volume.

Geophysics

The seismic coverage in Block 15/17 consists of both 2-D and 3-D data. The 2-D seismic lines were acquired prior to 1978. The 2-D data were superseded by the 3-D survey shot in late 1982 and processed in 1983. This N–S orientated survey covered the entire block and showed a marked improvement over previous 2-D data.

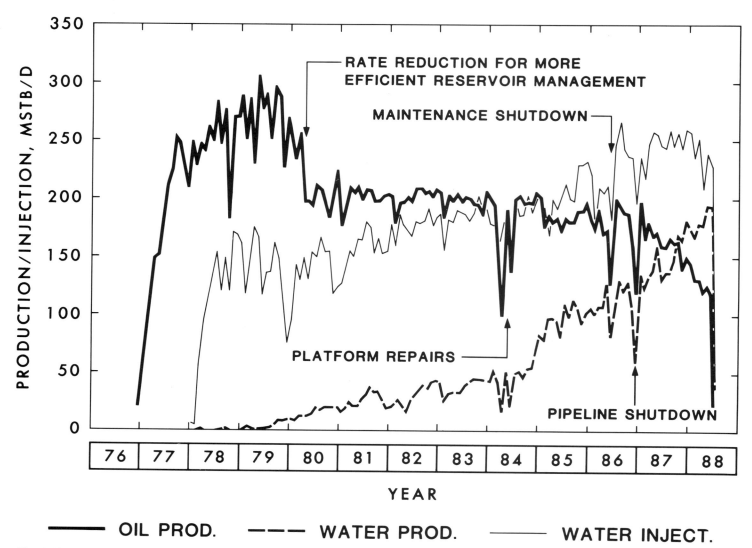

Fig. 4. Piper Field performance history.

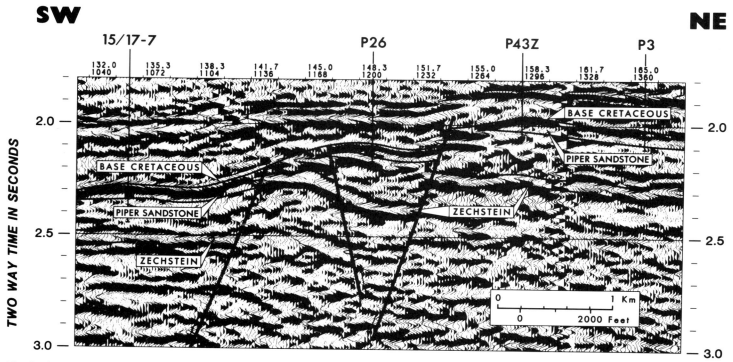

Fig. 5. Piper Field SW–NE interpreted seismic line.

The geophysical definition of the Piper reservoir has been good since the acquisition of the 3-D survey (Fig. 5). The major faults within and around the field are easily discernible at Zechstein level. The main problem with the seismic interpretation is that any top Piper reflection is generally masked by the strong reflection from the overlying Kimmeridge shale, which thickens downflank, and is actually absent over the crest of Block 1A. This problem was previously circumvented by picking top Kimmeridge shale and adding a two-way time value dependent on Kimmeridge shale thickness. However, the interactive seismic interpretation system acquired in 1986 has enabled random tracks linking all of the wells to be generated, and the top of the Piper reservoir to be mapped directly.

Trap

The Piper Field comprises four tilted, folded, fault blocks (Fig. 6). To the southwest, closure is provided by the 'A' fault, a major NW–SE-trending fault zone. To the northeast, dip closure results from the gentle tilting of the shelf away from the WGG. Northwest and southeast closure is provided by gentle folding about a NE–SW axis together with drape associated with NE–SW Caledonian trend faults. The southern block contains a small anticlinal closure with a deeper oil–water contact.

The field extended across 7350 acres with a gross reservoir column of 1210 ft from the crest of the field at 7300 ft TVSS to the original oil–water contact in the main field at 8510 ft TVSS. The small four-way dip closure to the southwest had a deeper oil–water contact at 9200 ft TVSS.

The top and lateral seals are principally the shales of the Kimmeridge Clay Formation. Along the 'C' fault, where these shales are eroded, onlapping Campanian marls form the seal. Although structural closure of the blocks occurred at the end of the Jurassic, the trap for the main part of the field was not formed until the Campanian marls onlapped the Piper sandstones along the 'C' fault. Most of the fault movement occurred after the deposition of the Piper sandstone in response to Late Cimmerian movements, but some was syndepositional. Within the Piper Field, faults seal only when total offset of the permeable section of the reservoir occurs.

The original oil–water contact of 8510 ft TVSS for 'Blocks I and II' equalized across all of the faults, and the trap was full to the spill point, which is in the northwest of the field. The small anticlinal accumulation in 'Block III' was also filled to spill point, which is along its northeastern side.

Reservoir

The reservoir comprises sandstones of the Piper and Sgiath Formations within the Late Jurassic Humber Group (Fig. 2; Harker *et al.* 1987). Reservoir properties are given in the data summary.

Sgiath Formation

Throughout the field, the Sgiath Formation lies unconformably on the Middle Jurassic Pentland Formation, and the upper boundary is a conformable contact with the Piper Formation at the base of the 'I shale'. The Sgiath is composed of five divisions: from base to top, the 'M sandstone', 'M shale', 'L sandstone', 'K shale' and 'J sandstone' (Fig. 7).

Sgiath lithology consists of coals, silty bioturbated shales and planar and trough cross-bedded, rippled, medium- to coarse-grained, poorly to well-sorted sandstones with occasional bioturbated units. It is interpreted as a lower delta plain to interdistributary bay deposit. Deposits occurred over a surface of very low relief, where local topography was gradually infilled by a series of distributary channels and mouth bars building out into a restricted marine environment. These sand bodies were periodically abandoned during minor brackish-water transgressions. The 'M sandstone' was deposited as a basal transgressive sand following the initial transgression over the Middle Jurassic continental sequence.

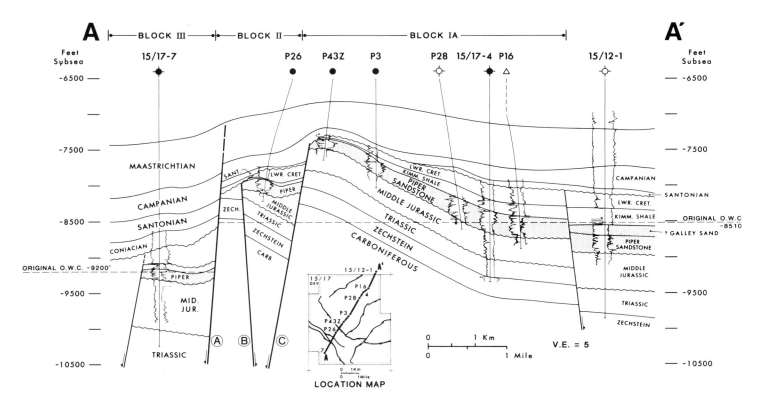

Fig. 6. Piper Field SW–NE structural cross-section.

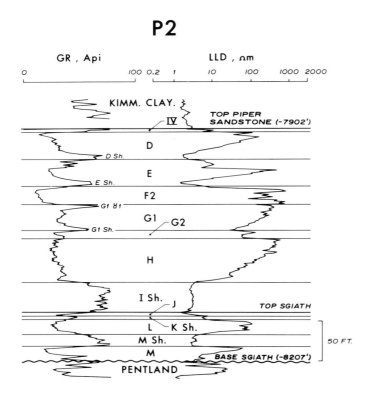
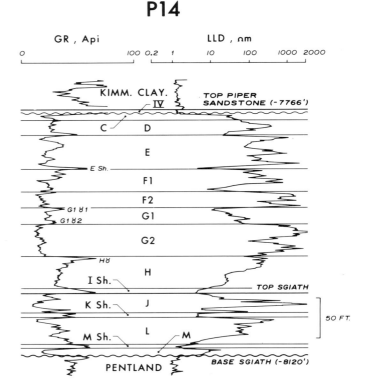

Fig. 7. Piper Field type wells.

Sgiath correlations are relatively straightforward, although lateral continuity of sandstone units is not as good as in the overlying Piper Formation. Sandstone units range in thickness from 5 to 60 ft with the thickest sandstone developed in the 'J' and 'L' mouth-bar facies.

Piper Formation

The Piper Formation extends from the field-wide 'I shale' to the basal transgressive sandstone ('Unit IV') below the Kimmeridge Clay Formation. Twelve major sub-divisions have been recognized within the Piper Formation (Fig. 7). The uppermost 'A and B sandstones' are not present in wells P2 and P14 as they are only developed down-flank along the eastern margin of the field (Fig. 8). They are now considered to be Galley sand equivalents rather than units of the Piper Formation.

The Piper Formation consists of a series of stacked sandstones, each generally 20–80 ft thick, interrupted by occasional shale horizons typically 5–10 ft thick. Sandstone units are normally organized in overall coarsening-upward cycles, but fining-up cycles have been identified locally within the 'F and E sandstones'. Grain size ranges from fine to pebbly, and sorting from poor to excellent. Bioturbation is pervasive except in the coarsest units, and has destroyed many of the primary sedimentary structures. Calcite doggers occur sporadically but are most common in the 'E sandstone'. The sandstones are friable to poorly cemented with calcite and silica cement.

The Piper Formation is interpreted as a high-energy wave-dominated delta system (Boote & Gustav 1987). As the sands prograded seaward, individual cycles were interrupted by marine transgressions from the west, represented by the intervening shales. The direction of the marine incursions is inferred from the thinning of the 'I shale' to the east, and facies analysis of the sand bodies, which suggests a more landward influence to the east.

The 'I shale' is interpreted as a prodelta shale. The overlying sandstones generally show a coarsening/cleaning-upward profile, reflecting the transition from lower to higher energies. These sandstones are interpreted as lower to upper-shoreface sediments. Upper-shoreface to foreshore sandstones are developed locally within the 'F sandstone'. Possible distributary channels have also been identified in the 'F and E sandstones'. 'Unit IV' is interpreted as a basal transgressive sand, which was deposited at the onset of the major transgression which resulted in the deposition of the Kimmeridge shales over the Piper Field. The facies interpretation of the 'A and B sandstones', which occur only in the extreme eastern flank of the field, is uncertain. The most likely interpretation is that they represent deep-water marine turbidites, which onlapped the main Piper Field from the east.

The Piper sandstones constitute a well organized reservoir with good lateral and vertical continuity. The Piper section exhibits most variability in the east–west orientated correlations, the sequence becoming more marine westwards. The shales between the sandstone units provide a framework for the correlations. These shales are good correlatable units as they represent minor marine transgressions. Correlation of the Piper Formation is also aided by the presence of three radioactive horizons in the 'G and H sanstones', which represent lag deposits produced by winnowing currents (Maher 1981).

The porosity of both the Piper and Sgiath sandstones is almost entirely intergranular. The most significant effect of burial diagenesis, other than compaction, is porosity reduction by quartz overgrowths. Detailed studies have indicated textural characteristics typical of secondary porosity (Burley 1986) but other studies have suggested that secondary porosity resulting from dissolution of cement is of minor significance.

Source

The organic-rich shales of the Kimmeridge Clay Formation are the source of the Piper oil. These shales contain 4–16% TOC comprising predominantly amorphous marine kerogen. The most likely migration path is from the south out of the WGG and updip into the Piper reservoir.

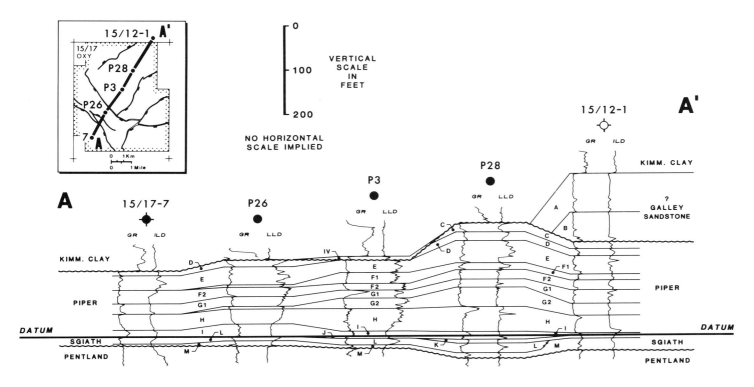

Fig. 8. Piper Field stratigraphic correlation section.

The migration could not have taken place before the Campanian marls onlapped the Piper sandstones and formed the seal along the 'C' fault during Late Cretaceous times. Most evidence indicates a Late Cretaceous to Early Tertiary maturation and migration phase, coinciding with the period of most rapid burial of the Kimmeridge shales in the source kitchen area to the south.

Hydrocarbons

Pertinent test results were briefly discussed in the preceding History section. Both the Piper and Sgiath sandstones have exceptionally good flow characteristics, Piper Field wells having typical productivity indices of 150–450 BPD/psi. Reservoir conditions and fluid characteristics are given in the data summary.

Very low irreducible-water saturations (Swi) are present in the field, due to high permeability, a thick oil column and a minor percentage of small pores within the pore network. In the very permeable sands, the Swi is less than 10% within 20 ft of the oil–water contact. On the crest of the structure, these same sands have Swi as low as 1%.

Reserves/resources

A series of contributory geological factors, combined with efficient reservoir management, will result in approximately 70% of the Piper Field STOIIP being recovered. Early in the production history of the field, it was apparent that the field had a strong aquifer drive, with water influx calculated at approximately 250 000 BPD. By mid-1977, it became clear that water injection would be necessary to maintain reservoir pressure at projected production rates of 250 000–300 000 BOPD. A peripheral injection pattern was selected and the first injectors were drilled with injection commencing in early 1978. These and subsequent injectors halted the decline in reservoir pressure, even at production rates above 300 000 BOPD.

In 1980, the field rate was reduced to allow for selective completion and injection and more efficient reservoir management (Fig. 4). This enabled the natural influx and injected water to be balanced against offtake from each reservoir layer. Further improvements in recovery were also made following the installation of a gas lift system in 1977 and high-volume submersible pumps in 1982.

Initial ultimate recovery for the Piper Field was put at 618 MMBBL, but subsequent reservoir performance has resulted in a steady increase to 952 MMBBL. At the time of the accident in July, 1988, 834 MMBBL had been produced, remaining reserves thus being 118 MMBBL.

The authors wish to thank Occidental Management and their Piper operating group partners for permission to publish this paper. The interpretations expressed in this paper are entirely the responsibility of the authors. A special vote of thanks is extended to C. Maher and S. Green, whose work on a previous and, as yet, unpublished paper in 1988 proved of great assistance in writing this paper. We also wish to thank the Occidental drafting department for their invaluable aid in completing this paper.

References

BOOTE, D. R. D. & GUSTAV, S. H. 1987. Evolving Depositional Systems within an Active Rift, Witch Ground Graben, North Sea. *In*: BROOKS, J. & GLENNIE, K. (eds) *Petroleum Geology of North West Europe*. Graham & Trotman, 819–833.

BURLEY, S. D. 1986. The Development and Destruction of Porosity within Upper Jurassic Reservoir Sandstones of the Piper and Tartan Fields, Outer Moray Firth, North Sea. *Clay Minerals*, 21, 649–694.

HARKER, S. D., GUSTAV, S. H. & RILEY, L. A. 1987, Triassic to Cenomanian Stratigraphy of the Witch Ground Graben. *In*: BROOKS, J. & GLENNIE, K. (eds) *Petroleum Geology of North West Europe*. Graham & Trotman, 809–818.

MAHER, C. E. 1981. The Piper Oil Field. *In*: ILLING, L. V. & HOBSON, G. D. (eds) *Petroleum Geology of the Continental Shelf of North West Europe*. 2nd Conference Proceedings of the Institute of Petroleum, 358–370.

Piper Field data summary

	Formation	
Reservoir Parameters	*Sgiath*	*Piper*
Max. gross thickness (ft)	140	320
Net/gross	0–0.7	0.7–0.9
Porosity (%)	20–28, average 24	18–30, average 24
Permeability (d)	0.2–4, average 1.2	0.5–10, average 4
Average water saturation (%)	8–25, average 12	7–15, average 10
Reservoir conditions		
Datum (ft TVSS)	8100	
Temperature (°F)	175	
Pressure (psia)	3700	
Oil–water contact (ft TVSS)	8510 (9200 in Block III)	
Hydrocarbons		
Crude type	Naphtheno-paraffinic	
Oil gravity (°API)	37	
GOR (SCF/STB)	430	
Bubble point (psia)	1600	
Formation volume factor	1.263 (1.3 in Block III)	
Viscosity (cp)	0.47	
Sulphur content (%)	1	
Pour point (°F)	21.2	
Formation water		
Salinity (Cl ppm)	44 000	
Resistivity (ohm m)	0.045 at 175°F	

The Scapa Field, Block 14/19, UK North Sea

G. J. McGANN[1], S. C. H. GREEN[2], S. D. HARKER[2] & R. S. ROMANI[2]

[1] *Occidental Petroleum (Caledonia) Ltd, 16 Palace Street, London, UK*
[2] *Occidental Petroleum (Caledonia) Ltd, 1 Claymore Drive, Aberdeen, UK*

Abstract: The Scapa Field is located in UK North Sea Block 14/19 in the Witch Ground Graben, 112 miles northeast of Aberdeen. The field was discovered in 1975 by the 14/19-9 well which tested 32° API crude from the Scapa Sandstone Member of the Early Cretaceous Valhall Formation. The field is a combination structural/stratigraphic trap situated in a NW–SE trending syncline. Updip limit to the NE is by onlap termination of the reservoir sands onto the Claymore tilt block, and to the southwest by fault closure and/or sand pinch-out into tight conglomerates associated with the Halibut Shelf boundary fault. Two thinly bedded, fine- to medium-grained turbidite sand units, in partial pressure communication, form the oil-bearing zone within the Scapa Sandstone Member.

Original oil in place was 206 MMBBL. In 1984, prior to development, a long-term production test was conducted via a deviated well drilled from the Claymore platform. Subsequent wells were thus drilled in a dynamic reservoir-pressure environment. Field development utilizes an integrated production/injection subsea template system tied back to the Claymore platform. Template production commenced in 1986 from currently estimated proved ultimate recoverable reserves of 63 MMBBL and averaged 28 000 BOPD in June 1988 from four production wells supported by four injection wells.

The Scapa Field is located in UK North Sea Block 14/19, 112 miles northeast of Aberdeen, Scotland, in a water depth of 385 ft (Fig. 1). Field STOIIP is estimated at 206 MMBBL, with proved originally recoverable reserves of 63 MMBBL contained within turbidite sandstones of the Early Cretaceous Scapa Sandstone Member of the Valhall Formation. The sandstones are located in a SE–NW-trending syncline located between the Halibut Horst to the south and Claymore tilted fault block to the north (Figs 1 & 2). Trap geometry consists of onlap termination of the reservoir sands onto the Claymore tilt block to the northeast, fault and pinch-out termination to the south and dip closure to the northwest and southeast, the top seal being marls of the Valhall Formation.

The Scapa Field is named after the location of the Flotta oil terminal in Scapa Flow in the Orkney Islands, where Scapa oil is landed. 'Scapa' is an Orcadian blend of two Norse words meaning sailing ship and isthmus.

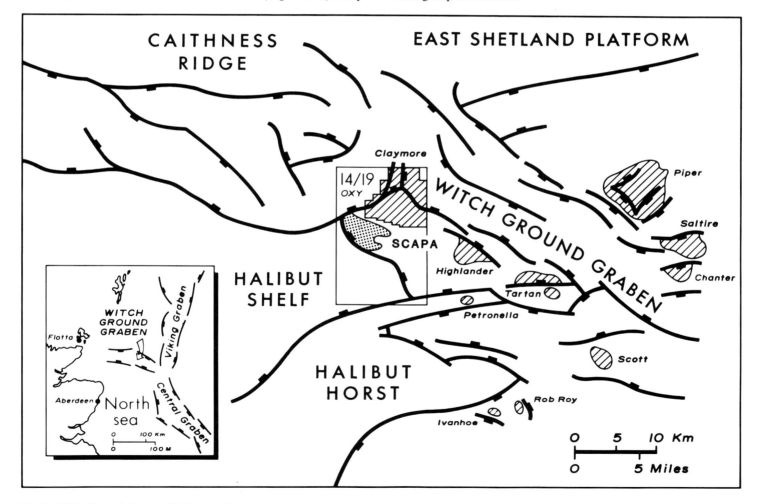

Fig. 1. Witch Ground Graben oilfields and fault trends map.

From Abbotts, I. L. (ed.), 1991, *United Kingdom Oil and Gas Fields,*
25 Years Commemorative Volume, Geological Society Memoir No. 14, pp. 369–376

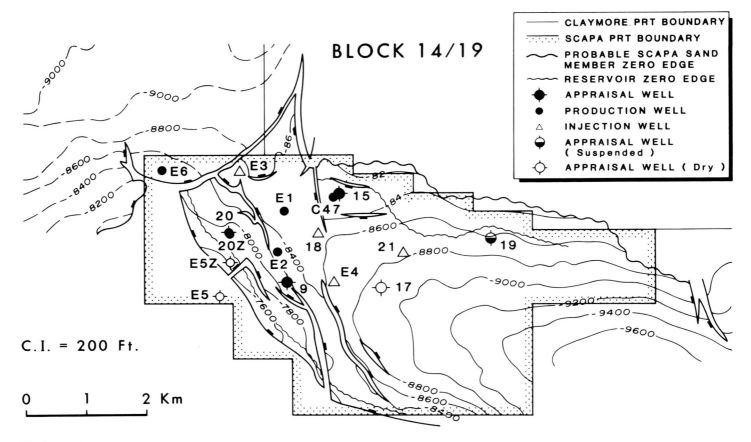

Fig. 2. Top Scapa Sandstone Member depth structure map.

History

Exploration and appraisal

Block 14/19 was awarded to Occidental and Partners (Thomson North Sea, Getty, Allied Chemical (Great Britain) Ltd) in 1972 in the UK Fourth Licensing Round.

Current licensees and percentage holders in the Scapa Field are as follows: Occidental Petroleum (Caledonia) Ltd, 36.5%; Lasmo North Sea plc, 20%; Texaco Britain Ltd, 23.5% and Union Texas Petroleum Ltd, 20%. Texaco Britain Ltd attained their present share via the acquisition of Getty Oil Co., in 1984, and Lasmo theirs by the acquisition of Thomson North Sea in 1989. Union Texas Petroleum was formerly known as Allied Chemical (Great Britain) Ltd.

The Scapa Field was discovered by the 14/19-9 Claymore appraisal in July 1975, which encountered 414 ft of Early Cretaceous sands and conglomerates. This interval contained approximately 40 ft of net oil sand, which tested 1260 BOPD of 31.2° API gravity oil from perforations between 8399 and 8464 ft below Kelly Bushing (KB). No oil–water contact was observed.

The full implications of the synclinal nature and northern extension of the Scapa Field became established with the 14/19-15 well drilled in early 1981 on the northern flank of the Scapa syncline. This well encountered approximately 86 ft of net oil sand and tested 32.5–33.5° API gravity oil at rates between 277 and 2021 BOPD from three intervals between 8368 and 8518 ft below KB. As with 14/19-9, no oil–water contact was encountered.

Prior to further appraisal, field status separate from the Claymore field was applied for, this being granted in January 1982. The next appraisal well, 14/19-17, was completed in May 1982 in the plunging synclinal axis, and encountered water-bearing Scapa sandstone. In order to evaluate more accurately the production capability of the Scapa reservoir, a long-reach deviated well, C47, was drilled from the Claymore platform. This well was completed at a location close to that of the 14/19-15 appraisal well (Fig. 2). Extended production testing of C47 commenced in April 1984 and continued until a total of 750 000 STBBL/oil had been produced. Material-balance calculations based on this extended test gave a clearer indication of the size of the Scapa accumulation than had previously been available. As a result of production and pressure depletion from C47, all subsequent appraisal and development wells were drilled in a dynamic reservoir-pressure environment.

The appraisal programme continued in 1984 with the 14/19-18 well located between 14/19-15 and 14/19-17, which encountered an oil–water contact at 8812 ft subsea. Well 14/19-19, completed in January 1985, appraised the northeast flank of the syncline, finding mostly tight, water-bearing sandstones. Some 10 ft of net oil sand was present, however, below the water-bearing interval, at a level deeper than the oil–water contact seen in the 14/19-18 well.

14/19-20 was a deviated appraisal well drilled in 1985 from a surface location near 14/19-18; it was designed to test the potential of a structurally high block in the northwestern portion of the Scapa Field. This well encountered a separate lower sandstone within the Scapa Sandstone Member, not seen in previous appraisal wells. The upper sandstone tested at a maximum rate of 2183 BOPD of 32.3° API gravity oil, while the lower sandstone tested at a maximum rate of 4648 BOPD of 31.6° API gravity oil.

Development

Development plans for Scapa Field were announced on 11th September 1985; they involved the construction of a 360 tonne, 8-slot subsea production template to be installed during the summer of 1986 and tied back to the Claymore platform. Oil would then be exported to the Flotta terminal via the Claymore installation. These plans included the addition of new topside facilities on the Claymore platform to provide pipeline end stations, crude-oil heating, gas testing facilities, separation and metering equipment, injection water, gas lift and chemical lines.

Production commenced from the C47 well at approximately 10 000 BOPD immediately following Department of Energy approval of the development plans on September 11th 1985. The subsea template was installed in July 1986, subsea installation being completed in October 1986. Production from the suspended 14/19-18 appraisal well commenced on 14 November 1986 with the tie-in of the off-template wells. Production in June 1988 averaged 28 000 BOPD from four production wells, with pressure maintenance from four injection wells. Cumulative production to 6 July 1988 was 17.1 MMBBL of oil.

Field stratigraphy

The general stratigraphy encountered in the Scapa Field is illustrated in Fig. 3 and in an account presented in Harker et al. (1987). The Early Cretaceous sediments, including the Scapa Sandstone Member, were deposited in a half-graben initiated in the latest Jurassic by the southerly rotation of the fault block now containing the Claymore Field (Boote & Gustav 1987). Tilting was accommodated by major down-to-the-north normal faults that form the margin of the Halibut Shelf to the south. Less significant fault movement occurred concurrently along a NE–SW trend of reactivated Caledonian faults. The developing half-grabenal relief was infilled with a thick conglomerate wedge, which prograded north and northeast from the faults bounding the Halibut Shelf, starting in the latest Ryazanian. The Scapa Sandstone Member, consisting of better-sorted clastic material, represents Valanginian–Hauterivian phases of turbidite sand deposition within the Scapa half-graben, in part, at least, derived from a shallow marine environment on the Halibut Shelf to the south. Conglomerates continued to be deposited close to the Halibut Shelf fault zone, although occasional debris flows are encountered within the Scapa Sandstone Member in the axis of the depocentre and towards the northern flanks of the field.

By the Barremian, only gentle relief appears to have been present. This period was one of relative tectonic quiescence and marine transgression onto the adjoining Halibut Shelf and across the Claymore tilt block. Limestones and marls were deposited across the Scapa and surrounding area, and provide caprock integrity for the Scapa reservoir. A high gamma response near the top of the Valhall forms a correlatable event throughout the Scapa–Claymore area and is termed informally the 'Vipers Tongue' (Fig. 4).

With increased loading during the Late Cretaceous and Tertiary, the Early Cretaceous conglomerates deposited close to the Halibut Platform compacted more slowly than the laterally equivalent marls and sands closer to the Claymore tilt block. This resulted in differential compaction, giving the current synclinal geometry, which probably started developing in the Hauterivian (Fig. 5).

Geophysics

The Scapa Field is covered by a grid of 2-D seismic originally shot to cover the adjacent Claymore Field (Harker et al. This Volume). The original 1974 1 × 2 km spacing was infilled in stages eventually giving a 500 × 500 m coverage. Current Scapa geophysical mapping utilizes data from a 3-D survey shot in 1982 on a 75 m line spacing orientated north–south. This survey covers most of Block 14/19 and was shot primarily for use on Claymore Field. The structural configuration at Scapa, however, means that the N–S orientation of the 3-D survey is not ideal. The use of a 3-D seismic workstation since late 1986 to manipulate the data has considerably enhanced the structural and stratigraphic interpretation. The ability to generate seismic displays in any orientation has proven essential for accurate fault interpretation. Most of the Scapa appraisal wells and all production wells were located on the basis of the 3-D seismic interpretation.

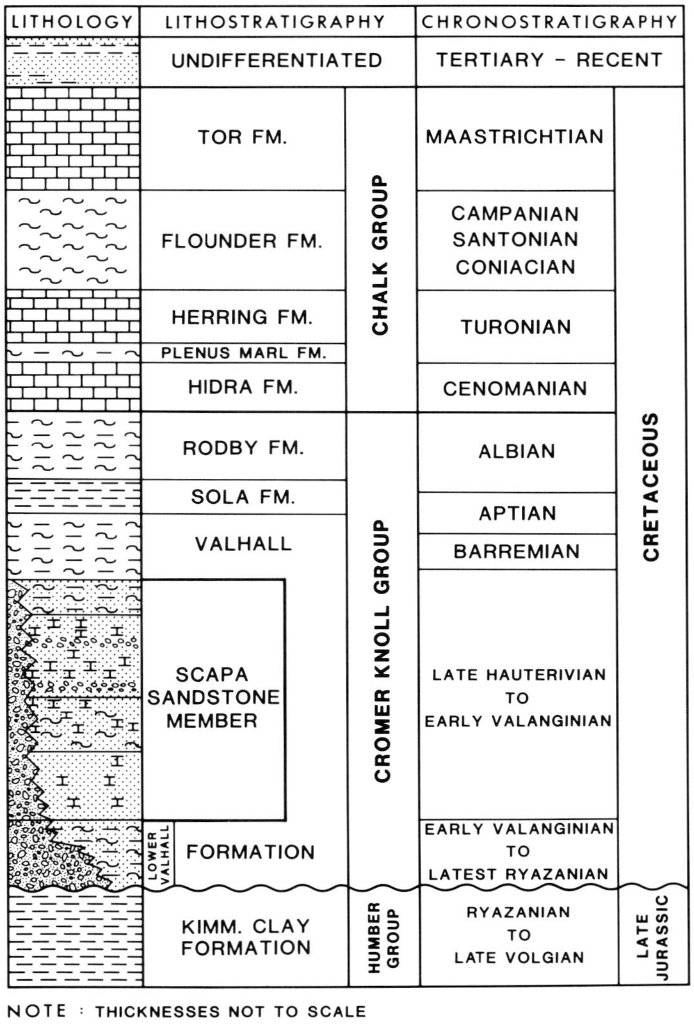

LITHOLOGY	LITHOSTRATIGRAPHY		CHRONOSTRATIGRAPHY	
	UNDIFFERENTIATED		TERTIARY – RECENT	
	TOR FM.	CHALK GROUP	MAASTRICHTIAN	CRETACEOUS
	FLOUNDER FM.		CAMPANIAN SANTONIAN CONIACIAN	
	HERRING FM.		TURONIAN	
	PLENUS MARL FM.			
	HIDRA FM.		CENOMANIAN	
	RODBY FM.	CROMER KNOLL GROUP	ALBIAN	
	SOLA FM.		APTIAN	
	VALHALL		BARREMIAN	
	SCAPA SANDSTONE MEMBER		LATE HAUTERIVIAN TO EARLY VALANGINIAN	
	LOWER VALHALL FORMATION		EARLY VALANGINIAN TO LATEST RYAZANIAN	
	KIMM. CLAY FORMATION	HUMBER GROUP	RYAZANIAN TO LATE VOLGIAN	LATE JURASSIC

NOTE : THICKNESSES NOT TO SCALE

Fig. 3. Scapa stratigraphy.

Despite the moderate data quality, the top of the Scapa Sandstone Member can be mapped with reasonable confidence over most of the field. The updip stratigraphic pinch-out of the Scapa Sandstone Member is seismically well defined (Fig. 6). This seismic section also shows the dramatic thinning of the Lower Valhall section away from the Halibut Shelf and the distinctive synclinal shape of the Scapa Field. In the northwestern part of the field the two oil-bearing reservoir sands (SA and SD) are seismically distinct events.

Multiple reflections present in the 1982 3-D data present a serious problem in certain areas of the field, and some trial reprocessing was undertaken in early 1989. The successful suppression of multiples and enhanced resolution of the reservoir zone was considered sufficient encouragement to reprocess the entire Block 14/19 3-D survey. The seismic line shown in Fig. 6 is from the reprocessed data set.

Trap

The structure of the Scapa Field at reservoir level is a relatively simple southeast-plunging syncline (Figs 2 & 5). Very little faulting is interpreted within the field itself at Scapa Sandstone level, except for the major faults that delineate the southern field limits. The field is terminated to the north by onlap onto the dip slope of the

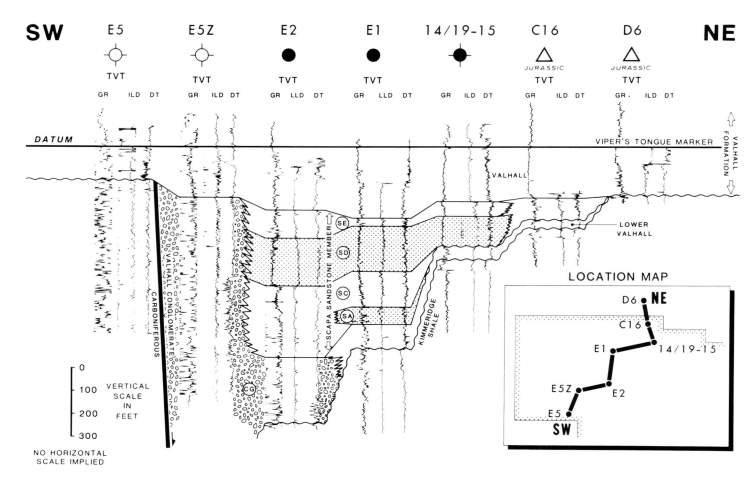

Fig. 4. Scapa Field stratigraphic correlation section.

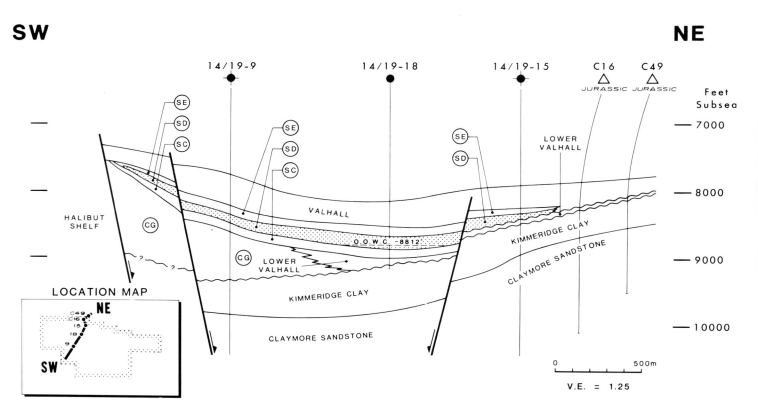

Fig. 5. Scapa Field schematic structural cross-section.

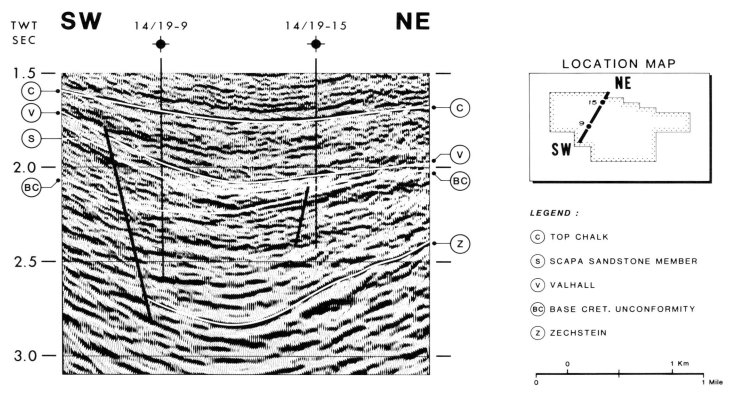

Fig. 6. Scapa Field SW–NE seismic line.

Claymore tilt block and by facies change of the sandstones to non-reservoir marls. The actual location of the reservoir edge is constrained by excellent well control on the southern flank of the Claymore Field. Closure to the south is provided by a combination of major down-to-the-north faults and a facies change of the sandstones to non-reservoir conglomerates. Closure to the southeast is provided by the dip of the plunging syncline. The full extent of the Scapa Field to the northwest has yet to be determined. Recent development drilling has shown the original fault closure mapping to be incorrect. Trapping in this direction probably results from a combination of dip closure and stratigraphic pinch-out of the reservoir sandstones, although further appraisal drilling is required to confirm this.

The field oil–water contact is defined at 8812 ft subsea in the 14/19-18 well, and the Scapa structure is not full to spill-point.

Reservoir

The Scapa Sandstone Member is of Early Valanginian to Late Hauterivian age, and has been subdivided informally into 6 alphabetical units, SA through SF (Figs 4 & 7). Sands of reservoir quality are developed in the SA, SD and SF units, but only the SA and SD sands are productive in Scapa, the SF sand being at virgin pressure and water-bearing. Figure 8 illustrates the type log for the SA and SD sands. As shown in Fig. 5, the type well for the Scapa Sandstone Member, 14/19-15 (Harker et al. 1987), contains only the SD unit. The SD unit is present field-wide but the SA unit is confined to the west of the field (Fig. 7).

The Scapa reservoir sandstones are typically fine- to medium-grained turbidites, cemented by calcite and silica, with a low clay content. Occasional shale matrix-supported coarse conglomerates occur within the sequence, together with thin mudstone and marl intercalations. The SD unit contains abundant laterally discontinuous calcareous concretions which result in an average bed thickness of between 6 inches and 12 inches in well penetrations. Bed-for-bed log correlation between wells is not possible, and recent palynological studies indicate that the SD sandstones are diachronous, becoming younger towards the west (Fig. 7). The sandstones are thickest in the axis of the Scapa syncline.

Deposition by gravity-flow processes is indicated for the Scapa reservoir sands by the presence of common massive sandstones, often fining upwards, containing floating shale clasts and disrupted by various water-escape features. The vertical sequences developed, however, are not typical of those described in the literature from a variety of turbidite settings. In particular, the common occurrence of traction features suggests that the sands were probably subjected to further reworking by bottom currents (E. Mutti, pers. comm.).

Average log porosity for the SD unit is 18.2% with a range of 16.6–19.3%. Average core permeability is 111 md with a range between 0.02 and 1447 md. The SA unit is of somewhat better quality with an average log porosity of 22.6%, and a range of 21.6 to 24.4%. Average permeability is 390 md with a range between 0.02 and 2338 md.

The SB unit is in a calcite-cemented non-reservoir facies; it is found at the base of the SD unit in the eastern part of the field, which appears, in part, to be time-equivalent to the top of the SA unit (Fig. 7). The SC unit, recognized in the western part of the field, consists of the non-reservoir marls, limestone, conglomerates and tightly cemented sandstones that separate the SA and SD units. The SE unit consists of tightly cemented sandstones and interbedded sandy marls of non-reservoir quality at the top of the Scapa Sandstone Member in the central part of the field, and separates the SD from the SF units in the eastern part of the field (Fig. 7).

Despite their apparent vertical and areal separation within the field, pressure data have shown that the SD and SA sandstones are in partial pressure communication. Studies are continuing in order to address the problem of where and how this communication occurs.

Two major problems complicating the development of the Scapa field concern reservoir heterogeneity and determination of the initial oil in place. As described in the reservoir section, the Scapa reservoir, particularly the SD sand, consists of thinly (6–12 inch) intercalated, tight, calcite cemented sandstone and porous sand-

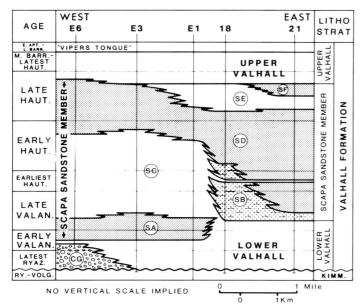

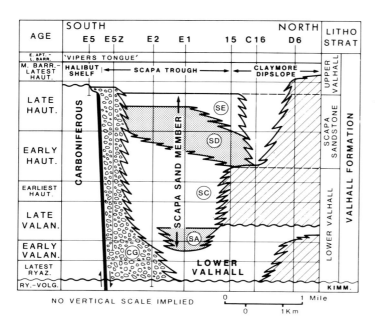

Fig. 7. Schematic Valhall stratigraphy: Scapa dip section (top); Scapa strike section (bottom).

stone. The vertical resolution of standard logging tools does not allow accurate measurement of the true porosity of such beds. This results in log-derived porosities and net-pay figures that are inconsistent with those derived from core, and in water saturations that are too high over the reservoir section. A number of methods have been used to address these inconsistencies (McGann et al. 1988) including enhancement of the density and neutron resolution by using the near-detector signal in uncompensated mode, reducing logging speeds and increasing sample frequency. This has resulted in a better core-to-log comparison than was previously possible.

The data from Scapa appraisal wells proved insufficient to determine the initial oil-in-place accurately enough for a development decision to be made. This problem was overcome by the extended production test of C47 (Chen 1988).

Source

Geochemical analyses on a number of wells in the Scapa area indicate that the source of the oil is the Kimmeridge Clay Formation. Migration paths are envisaged as being directly updip from the mature source rock into the overlying Early Cretaceous clastic sequence. Migration is thought to have occurred during the early Tertiary.

Hydrocarbons

Test results

Tables 1 and 2 summarize the test results from the 14/19-20 well, which encountered both SA and SD sands.

Table 1. *14/19-20 DST 1: Summary of the test results Scapa A (SA)*

Interval tested	10 260–10 400 ft RKB
Reservoir pressure	3638 psia at 8436 ft ss
Reservoir pressure at datum	3650 psia at 8500 ft ss
Permeability–thickness	42 600 md ft
Permeability	367 md
Total skin factor	−0.5
Skin factor due to well deviation	−0.75
Damage skin factor	0.25
Wellbore storage coefficient	0.08 BBL/psi
Productivity index	24.7 BBL/day/psi
Ideal productivity index	25 BBL/day/psi
Stabilized flow rate	4079 BOPD
Choke size	44/46 inch
Cumulative oil production at main shut-in	4380 STBBL
Flowing bottomhole pressure	3463 psia
Flowing wellhead pressure	458 psia
Separator pressure	209 psia
Separator temperature	58°F
Gas:oil ratio	129 S ft^3/STBBL
Oil gravity	31.9° API
Separator gas gravity	0.801
Shrinkage factor (separator-tank)	0.961

Table 2. *14/19-20 DST 2: Summary of test results Scapa D (SD)*

Interval tested	9780–9920 ft RKB
Reservoir pressure	3346 psia at 7971 ft ss
Reservoir pressure at datum	3529 psia at 8500 ft ss
Permeability–thickness	16580 md ft
Permeability	169 md
Total skin factor	−0.1
Skin factor due to well deviation	−0.6
Damage skin factor	0.5
Wellbore storage coefficient	0.02 BBL/psi
Productivity index	10.7 BBL/day/psi
Ideal productivity index	11.5 BBL/day/psi
Stabilized flow rate	2150 BOPD
Choke size	9/16 inch
Cumulative oil production at main shut-in	1466 STBBL
Flowing bottomhole pressure	3145 psia at 7971 ft ss
Flowing wellhead pressure	344 psia
Separator pressure	190 psia
Separator temperature	77°F
Gas:oil ratio	160 S ft^3/STBBL
Oil gravity	32.3° API
Separator gas gravity	0.89
Shrinkage factor (separator-tank)	0.987

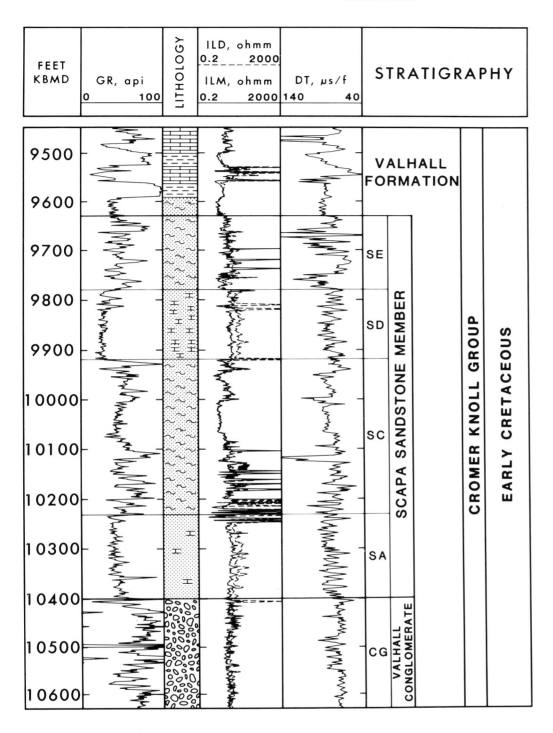

Fig. 8. 14/19-20 Scapa type well.

Fluid types

Table 3 summarizes the data from the PVT analyses carried out on samples from Scapa wells 14/19-9, 14/19-15, 14/19-18, 14/19-19, 14/19-20 and C47. Reservoir temperature and pressure are related to a datum of 8500 ft subsea. Table 4 summarizes the water analysis from the 14/19-18 well.

Reserves/resources

The Scapa Field contained an estimated 206 MMBBL oil with estimated ultimate proved recovery of 63 MMBBL oil. A maximum vertical oil column of 761 ft has been proved. Recovery factors are 26.4% for the Scapa A and 31.8% for the Scapa D units.

A limited natural water influx from the southeast is supplemented by pressure support from four water injection wells.

Table 3. *Scapa PVT Analysis*

API gravity	= 32.5° API
Bubble point	= 1104 psia at 185°F
GOR (flash)	= 228 S ft^3/STBBL at separator conditions of 80°F and 120 psig
Formation volume factor	= 1.164 Res BBL/STBBL at 3800 psia
Oil viscosity	= 1.35 cp at 185°F and 3800 psia

Table 4. *Formation water composition and resistivity Well 14/19-18 Scapa Field*

Type of water	Produced water
Well	Scapa 14/19-18
Test date	12.09.84
H_2O volume recovered	Total = 117 BBL
Density	1.052 at 60°F
pH	7.24 at 60°F
Total dissolved solids (mg/l)	77 010
Anions (mg/l)	
Chloride	45 790
Sulphate	190
Carbonate	nil
Bicarbonate	1040
Sulphide	nil
Hydroxyl	nil
Cations (mg/l)	
Sodium	28 130
Potassium	310
Calcium	1000
Magnesium	315
Iron (total/dissolved)	24/0.1
Barium	8.8
Strontium	225
Additional elements	
Boron	58
Aluminium	0.1
Silica	20
Phosphorus	0.1
Resistivity Measurements	
Rw	0.127 at 60°F
Rw converted to 160°F	0.0508

Scapa Field data summary

Reservoir parameters		
Reservoir	SD	SA
Average porosity (%)	18.2	22.6
Average permeability (md)	111	390
Maximum net pay (ft)	162	157
Average water saturation (%)	23	22
Reservoir conditions (SD/SA at datum)		
Datum (ft TVSS)	8500	
Temperature (°F)	185	
Pressure (psia)	3384	
Fluid contact (ft TVSS)	8812	
Volumetrics		
Oil originally in place	206 MMBBL	
Recovery factor	26.4% SA, 31.8% SD	
Original recoverable reserves	63 MMBBL	

The writers wish to thank Occidental Management and their Scapa operating group partners for permission to publish this paper. The interpretations expressed in this paper are entirely the responsibility of the authors. In addition we wish to thank A. Downton and his drafting group for their aid in completing this paper.

References

BOOTE, D. R. D. & GUSTAV, S. H. 1987. Evolving depositional systems within an active rift, Witch Ground Graben, North Sea: *In*: BROOKS, J. & GLENNIE, K. (eds) *Petroleum Geology of Northwest Europe*. Graham & Trotman, London, 819–833.

CHEN, H. K. 1988. Field Development of the Scapa Field: A Marginal North Sea Field SPE 18347. *SPE European Petroleum Conference*, London Oct 16–19 1988. 553–562.

HARKER, S. D., GREEN, S. C. H. & ROMANI, R. S. 1991. The Claymore Field, Block 14/19, UK North Sea. *This Volume*.

——, GUSTAV, S. H. & RILEY, L. A., 1987. Triassic to Cenomanian stratigraphy of the Witch Ground Graben: *In*: BROOKS, J. & GLENNIE, K. (eds) *Petroleum Geology of Northwest Europe*. Graham & Trotman, London, 809–818.

McGANN, G. J., RICHES, H. A. & RENOULT, D. C. 1988. Formation evaluation in a thinly bedded reservoir. A case history: Scapa Field, North Sea *Transactions SPWLA 29th Annual Logging Symposium*, San Antonio, Texas.

The Tartan Field, Block 15/16, UK North Sea

R. N. COWARD, N. M. CLARK & S. J. PINNOCK

Texaco Ltd, 1 Knightsbridge Green, London SW1X 7QJ, UK

Abstract: Tartan Field lies 118 miles northeast of Aberdeen in UK Block 15/16 on the southern flank of the Witch Ground Graben. The Field consists of two major southerly-dipping rotated fault blocks, the 'Upthrown Block' to the south and the 'Downthrown Block' to the north. The primary reservoir comprises Oxfordian–Kimmeridgian shallow marine sandstone of the Piper Formation deposited during a regional marine transgression. The secondary reservoir consists of Volgian turbidites (the 'Hot Lenses') within the Kimmeridge Clay Formation. The accumulations have been sourced from maturation of the Kimmeridge Clay Formation below approximately 10 000 ft in adjacent basins. The Piper Formation exhibits markedly different petrophysical properties within each block. A relatively homogeneous intergranular porosity system is present throughout the oil zone of the 'Upthrown Block'. Porosities are lower and more variable in the 'Downthrown Block' as a result of cementation and the presence of an intense compaction fabric. The trapping mechanism is a combination of structural and stratigraphic elements. The Field was discovered in December 1974 by the 15/16-1 well which penetrated 263 ft of Piper Formation sandstone full to base with oil. A further ten straight holes and three sidetracked holes delineated the structure. Tartan Field came onstream in January 1981 and is currently producing 30 000 BOPD through eight platform producers, assisted by six subsea water injection wells. Ultimate recoverable reserves are currently estimated at 116 MMBBL of crude oil.

The Tartan Field is located in UK Block 15/16 in the Outer Moray Firth area of the North Sea, 118 miles to the northeast of Aberdeen in 465 ft of water (Fig. 1). The single conventional steel drilling and production platform is situated some 9 miles to the southwest of the Piper Field and 14.5 miles to the southeast of the Claymore Field. Two satellite developments are tied back to the Tartan Platform, the Highlander Field 7.5 miles to the northwest and the Petronella Field 6.4 miles to the west.

The black-oil accumulations which together comprise the Tartan Field are contained within sandstone reservoirs of Upper Jurassic age, situated within two separate fault blocks. The older shallow marine sandstone reservoir, the Piper Formation, is Oxfordian–Kimmeridgian in age. The younger sandstone reservoir comprises Volgian age turbidites within the Kimmeridge Clay Formation. The accumulation in the southerly 'Upthrown Block' has an area of approximately 985 acres, whilst the more northerly 'Downthrown Block' has a productive area of 1996 acres. The recoverable reserves of the field are estimated to be 116 MMBBL of crude oil.

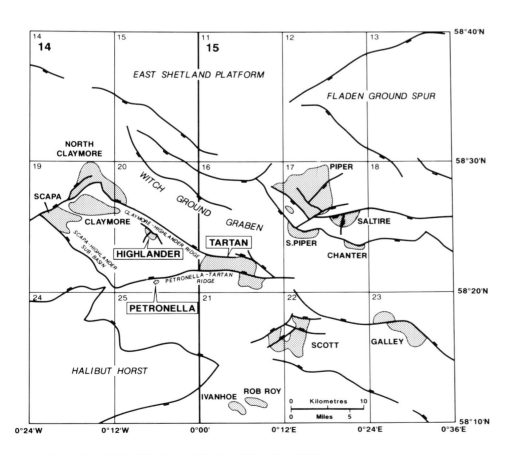

Fig. 1. Simplified tectonic framework of the Outer Moray Firth area, North Sea, showing the location of the Tartan Field and surrounding discoveries.

History

Texaco North Sea UK Ltd gained a 100% interest in Block 15/16 through the Fourth Round of licensing in March 1972. The original licence was P237 which included Blocks 13/22, 14/20, 15/7, 15/16, 15/23, and 20/5. Under a 1970 Illustrative Agreement the licensee's beneficial interest was conveyed to Texaco North Sea UK Company which was appointed operator. Of the total acreage held under licence P237 some 50% was relinquished in March 1978. When the Tartan Field Annex B was approved on 15 February 1979 the field retained licence P237 and all the remaining acreage was granted a new licence P294.

The exploration and appraisal phase of drilling on the Tartan Field structure spanned the period 1974 to 1977. Hydrocarbons were first discovered within a sandstone reservoir of Upper Jurassic age within the 'Upthrown Block'. Once that accumulation had been defined by drilling then activity was transferred to the 'Downthrown Block' where a separate accumulation was discovered and subsequenlty delineated. All of these early wells were plugged and abandoned.

The Tartan Field discovery well, 15/16-1, was spudded in September 1974 at a location close to the crest of the structure, defined seismically at the Base Cretaceous level. The well encountered sandstones of Upper Jurassic age at a depth of −9815 ft TVSS. The sandstone sequence had a gross thickness of 263 ft and was oil-bearing to base. Following this discovery the first step out well to the south was dry (15/16-2) but was sidetracked northwards as 15/16-2Z and established an oil–water contact at a depth of −10 305 ft TVSS. The westerly extent of the accumulation was defined by the results of wells 15/16-3 and 3Z and the easterly truncation of the reservoir was demonstrated by well 15/16-4A. Another prospect within the 'Upthrown Block' was tested by well 15/16-7 but without success.

The 'Downthrown Block' accumulation discovery well, 15/16-5, was spudded in August 1975 at the structurally highest eastern end of the prospect. Oil bearing Upper Jurassic Sandstone (Piper Formation) was encountered at a depth of −11 528 ft TVSS. The sandstone proved to be oil-bearing to base with a gross thickness of 378 ft. A long step out to the west, well 15/16-6, showed that the reservoir thickened in that direction (gross thickness 754 ft). The next appraisal well 15/16-8, was drilled midway between wells 5 and 6. At this location the reservoir was encountered structurally lower than anticipated and was water-wet. The possibility of a major extension to the proven accumulation westwards into adjacent Block 14/20 (100% Texaco) was tested by well 14/20-7. The Upper Jurassic reservoir sands were penetrated at a structurally high elevation (−10 980 ft TVSS) but proved to be entirely water-bearing. This result left a question mark over the nature and position of the westerly closure of the 'Downthrown Block' accumulation. The uncertainty was reduced by the results of the final appraisal well 15/16-9. This well was spudded in June 1977 at a location in the western part of the 'Downthrown Block' and demonstrated the continuity of the proven accumulation.

The Tartan platform was installed in summer 1980 prior to the start of development drilling. The platform comprises a four-legged manned production and drilling platform of conventional design fabricated in steel. All crude oil and natural gas liquids from the platform are exported as a high vapour pressure crude to the Flotta terminal by means of the Occidental (Claymore) pipeline. All associated gas in excess of the platform operating requirements is exported to St Fergus via the gas collection point MCP-01 and the Frigg trunkline.

Development drilling in the Tartan Field commenced in 1980 and to date 13 platform wells and 6 subsea wells have been drilled. Development of the 'Upthrown' and 'Downthrown Blocks' did not occur sequentially but for ease of discussion the development of the two blocks will be dealt with separately.

The first development well in the 'Upthrown Block' was the 15/16-11 well spudded in March 1981. This well was drilled as a twin to the discovery well 15/16-1 and penetrated 241 ft of Piper Formation full to base. The well was completed as a subsea producer and tied back to the platform via a 6 in flowline. This was followed between December 1981 and October 1982 by two deviated subsea injection wells, 15/16-13 and 16, which were drilled to the south from a surface location close to the 15/16-11 and 1 wells. The 15/16-16 well encountered a well defined oil–water contact in the Piper Formation at −10 328 ft TVSS. This compares to −10 305 ft TVSS established by the 2Z well. There is no additional evidence to suggest that this is a real difference and is assumed to reflect the greater accuracy of the more recent generation of well directional survey tools. The figure of −10 328 ft TVSS has therefore been accepted as the oil–water contact in the 'Upthrown area' of accumulation. The 15/16-16 well penetrated a large fault at the base of the Piper Formation which juxtaposed Triassic shales against a Carboniferous sandstone–shale–coal sequence. These sandstones were found to be hydrocarbon-bearing and tested at a rate of 760 BOPD through a three-quarter inch choke. After the test a bridge plug was set and the Piper Formation perforated for water injection. To date three platform wells have been drilled into the 'Upthrown Block' in the period November 1982 to December 1985; these are the 15/16-T8, T10A, and T10Y wells.

In May 1980 the first development well 15/16-T1 was drilled to the 'Downthrown Block' from the Platform. The well encountered 700 ft of Kimmeridge Shale directly overlying Middle Jurassic Volcanics. The absence of the targeted Piper Formation, interpreted to be due to faulting in the well, caused the well to be plugged and sidetracked northwards. The sidetrack 15/16-T1Z well reached a bottom hole location just north of exploration well 15/16-6 and was completed as the first production well on the Tartan Field. Between November 1980 and February 1986 a further eleven development wells have been drilled to the 'Downthrown Block', eight of which were drilled from the platform. The first of these the 15/16-10 well, confirmed the oil–water contact at −12 169 ft TVSS. In 1983 the 15/16-T6 well which was initially completed in the Piper Formation was recompleted in the Hot Lens 'A' thereby providing the first production from this reservoir.

Field stratigraphy

The oldest sediments penetrated in the Tartan Field are the deltaic sandstones, shales, and coals of the Carboniferous Forth Formation. Deposition of the Forth Formation ceased with the onset of Hercynian tectonism in the late Carboniferous which led to erosion of the earlier sediments. Following the late Permian marine incursion Zechstein anhydrite, clays, silts and sands were deposited in the Tartan Field area.

In the Moray Firth Basin, and in the incipient Witch Ground Graben, fluvial and alluvial sediments accumulated during the Triassic. In the Tartan Field these sediments are represented by brick red and light grey claystones with occasional thin sandstone beds.

Following a regional transgression, marginal marine sediments were deposited in the Lower Jurassic. However, these sediments are absent over the Outer Moray Firth area.

The Middle Jurassic was a period of regional uplift or doming with volcanic centres located at the triple junction of the Witch Ground, Viking and Central Grabens. As a result of the contemporaneous regression, fluvio-deltaic and alluvial sediments with abundant plant debris accumulated throughout the Witch Ground Graben. In the Tartan Field the resulting Fladen Group comprises interbedded sandstone, shale, coal and volcaniclastic sediments derived from the volcanic centre to the southeast.

During Upper Jurassic times, subsidence and active graben development combined with a regional transgression to impart marginal marine and eventually basinal conditions to the area. During Callovian to early Oxfordian times the transgression initially produced swamp conditions across much of quadrants 14 and 15 including the area in which the Tartan Field now lies. During the Oxfordian with more widespread transgression, shallow marine environments developed. Uplift of the Halibut Horst and Fladen Ground Spur at this time exposed areas to erosion and provided a plentiful supply of sand. Following a period dominated by sand deposition, continued subsidence and the widespread late Cimmerian transgression resulted in the deposition of finer-grained clastic deposits, indicative of more restricted or deeper water conditions. The resulting sediments comprise the Kimmeridge Clay Formation (Caprock Sequence of Tartan). The late Jurassic to early Cretaceous interval was a period of intense tectonic activity. Locally this resulted, during Lower Cretaceous times, in the erosion of the emergent fault scarps of the rotated blocks and the accumulation of submarine fans. During Upper Cretaceous times large-scale fault movement ended and the whole area subsided slowly. The Chalk Group and Tertiary sediments accumulated across the area with only minor easterly regional tilting.

The stratigraphy and the reservoir zonation of the Upper Jurassic Humber Group sediments in the Tartan Field is illustrated by the log of the 15/16-9 well (Fig. 2). Lithostratigraphically the strata are divisible into a lower sand dominated sequence, the Principal Reservoir Sequence (PRS), and an upper shale dominated sequence, the Caprock Sequence. These comprise the Piper Formation and the Kimmeridge Clay Formation respectively. Biostratigraphically the PRS is Callovian to Lower Kimmeridgian in age whilst the Caprock Sequence is Middle Kimmeridgian to Volgian in age. The Upper Jurassic is overlain by Lower Cretaceous calcareous claystone throughout the Tartan area. The Piper Formation contains several members which are lithologically distinct. These are the Coal Marker, Coal Marker Shale, Lower Sand, Lower Shale, Main Sand, '15/16-6' Sand, and the Hot Sand Members. In addition the Caprock Sequence contains two sandstone members termed Hot Lens 'A' and 'B'.

The Sgiath Formation, as defined in Harker *et al.* 1987, has not been separated from the Piper Formation although its existence is recognized in the Tartan Field. In the 'Upthrown Block' palaeontological studies indicate that the Sgiath Formation would include all the Upper Jurassic members beneath the base of the Lower Shale Member. In the 'Downthrown Block' the Sgiath Formation would include only the Coal Marker Member and the Coal Marker Shale Member.

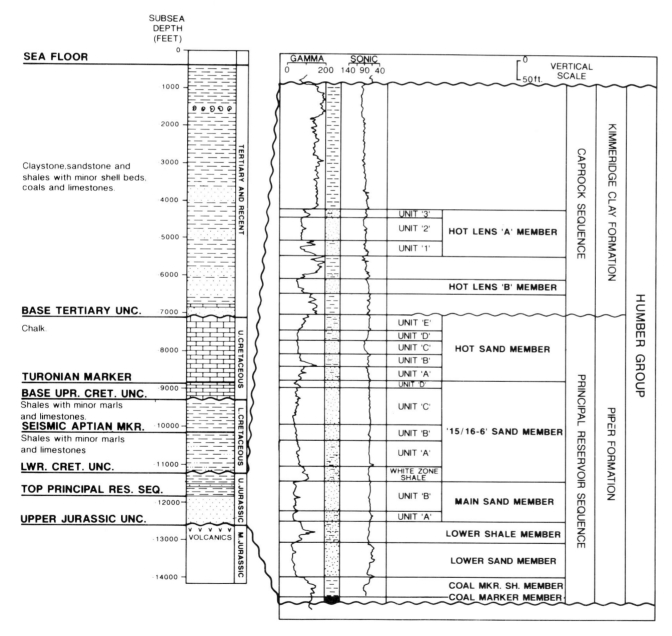

Fig. 2. Stratigraphy and reservoir zonation of the Tartan Field illustrated by the 15/16-9 well.

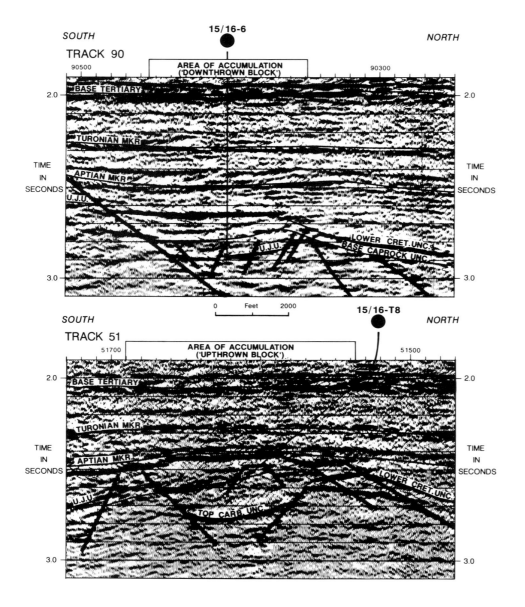

Fig. 3. Extracts from north–south 3-D seismic lines through the 'Upthrown' and 'Downthrown Blocks' of Tartan Field. See Fig. 4 for the location of the lines of section.

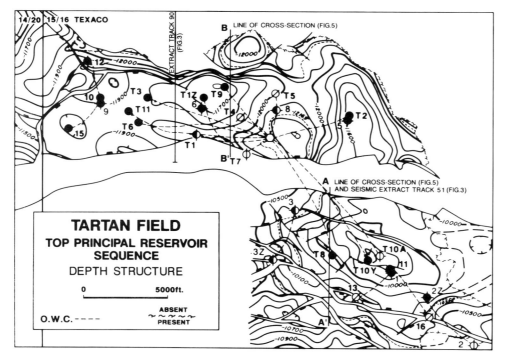

Fig. 4. Top Piper Formation Depth structure map of the Tartan Field.

Geophysics

The discovery well and the following four appraisal wells were positioned based on the interpretation of a grid of 2-D seismic data. This grid comprised speculative data purchased prior to the Fourth Round together with infill lines shot in 1973, giving an overall line spacing of 2.5 km. All of the remaining appraisal drilling was based on a 1 km grid of 2-D seismic data which was acquired in 1975.

A full field area 3-D seismic survey was acquired in 1978, with a line spacing of 75 m and a total line length of 1200 km. Several phases of reprocessing of these data have been undertaken, the last in 1982. Finally, in 1984 the data was merged with the overlapping Block 14/20 3-D survey, to provide better data in the western part of the 'Downthrown Block'.

Seven seismic events are mapped over the Tartan Field area. These are the Base Tertiary, Turonian Marker, Aptian Marker, Base Cretaceous Unconformity, Top Piper Formation, Upper Jurassic Unconformity and the Top Carboniferous reflectors. These events are illustrated in Fig. 3 which shows extracts of north–south lines from the 3-D seismic survey. The Top Piper Formation reflector over the 'Upthrown Block' is generally of poor quality and can only be used as a guideline to geological modelling.

Trap

The Tartan structure which traps the Piper Formation oil accumulation consists of two predominantly south-southwest dipping rotated fault blocks (Figs 4 and 5). The southerly 'Upthrown Block' rises to −9700 ft TVSS and the northern 'Downthrown Block' to −11 600 ft TVSS. Dips range from approximately 2–15° at the Top Piper level. The blocks are separated and constrained to the north by two major graben-bounding fault zones. The zone which separates the blocks has an average throw of approximately 2000 ft and extends to the west to run along the northern edge of Texaco's Petronella Field (see Waddams & Clark, this volume). To the east the fault zone dies out in the block 15/17. The fault zone to the north of the 'Downthrown Block' is the eastern end of the Claymore ridge fault system which borders Texaco's Highlander Field to the west (see Whitehead & Pinnock, this volume). The fault bounding the southern margin of the 'Upthrown Block' accumulation is smaller with throws ranging from 100–400 feet. The rotation of the fault blocks was accompanied by internal deformation within the blocks and this provides the structural dip necessary to constrain the accumulation in the 'Upthrown Block' to the west. Sealing faults act in combination with structural dip to confine the downthrown accumulation to the west and south. As well as structural elements to the trapping mechanism there are also stratigraphic elements. These were provided by the total erosion of the Piper formation to the northeast of the field and by fault scarp erosion along the crests of the blocks during the late Volgian and early Cretaceous. The seal for the trap in the Piper Formation is provided by Kimmeridge Clay. However, Lower Cretaceous calcareous claystone may cap the reservoir along the crest of the 'Upthrown Block'.

The trap for the Hot Lenses in Tartan is stratigraphic. The Hot Lens sandstones are totally enclosed by impermeable Kimmeridge Clay. However, the rotated block structures responsible for the Piper accumulation have influenced the Hot Lens sandstones in that they provided the topographic low in the Volgian which controlled the sand deposition.

The oil–water contacts for the Piper Formation in the 'Upthrown' and 'Downthrown Blocks' are defined by the 15/16-16 and 15/16-10 wells at −10 328 and −12 169 ft TVSS respectively. No OWC has been penetrated to date in the Hot Lenses in the 'Downthrown Block' The 'Upthrown Block' is not full to structural closure. The spill point being to the west of the 15/16-3 and 3Z wells at a depth of −10 500 ft TVSS. No spill point has been mapped in the 'Downthrown Block'.

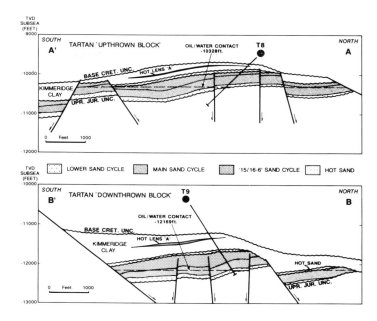

Fig. 5. Schematic structural cross-sections through the 'Upthrown' and 'Downthrown' Blocks of Tartan Field. See Fig. 4 for the location of the lines of section.

Reservoir

The Oxfordian–Lower Kimmeridgian age sandstone of the Piper Formation forms the primary reservoir within the Tartan Field. Volgian sandstones within the Kimmeridge Clay Formation represent a secondary reservoir. The sedimentary evolution of the Upper Jurassic strata in the Tartan Field is recognized to involve two major episodes: deposition of the high energy shallow marine Piper Formation; and deposition of the lower energy 'basinal marine' Kimmeridge Clay Formation. Within the Piper Formation four phases or cycles of deposition are recognized recording the inundation of the Middle Jurassic land surface. These are: the Coal Marker to Lower Sand cycle; Lower Shale to Main Sand cycle; '15/16-6' Sand; and the Hot Sand cycle. The first three cycles mentioned above terminated with an abrupt rise in sea level, whereas the last, the Hot Sand cycle, ended with a progressive increase in water depth. The Caprock Sequence accumulated in deeper water with occasional pulses of clastic input supplying the sands of the Hot Lenses.

Deposition of the Piper Formation began with Coal Marker to Lower Sand cycle of deposition. The Coal Marker comprises drift and rooted coal and shale deposited in the marginal freshwater swamp. Gradual coastal subsidence led to the establishment of brackish conditions in the swamp during the deposition of the carbonaceous clays and muds of the Coal Marker Shale. Eventually the paralic environments were transgressed by fully marine conditions leadng to the deposition of the uniformly fine- and fine- to medium-grained Lower Sand. With continued subsidence and sea level rise the very fine sand, silt and mud which now comprise the lower part of the Lower Shale were deposited, with water depths in excess of mean fairweather wavebase. The first regressive coarsening upwards cycle is recorded by the upward transition from the upper part of the Lower Shale to high energy, coarse-grained shoreface sands at the top of the Main Sand.

Rapid subsidence and a return to more offshore conditions is recorded by the White Zone Shale at the base of the '15/16-6' Sand. The remainder of the '15/16-6' Sand generally becomes cleaner and coarser upwards and is interpreted to represent the second regressive event. It is unclear if the top of the '15/16-6' Sand ever became emergent. However, locally it reached a sufficient elevation to be subject to considerable reworking under high energy conditions producing an unconformable surface overlain by a coarse to very

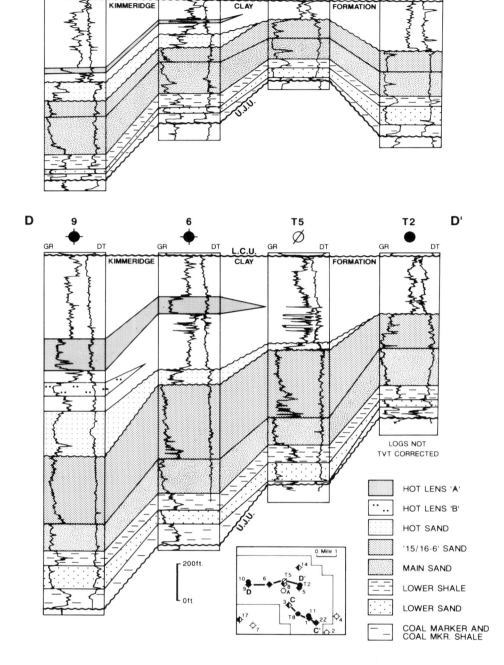

Fig. 6. West–east correlation sections through the 'Upthrown' and 'Downthrown Blocks' of the Tartan Field.

coarse sand (Unit 'D'). After this regressive phase the rate of subsidence and sea level rise exceeded the rate of sediment supply resulting in the deposition of the fine-grained lower shoreface to offshore sediments of the Hot Sand.

With further subsidence the clastic input was progressively reduced and pelagic sedimentation dominated during deposition of the Kimmeridge shale. The Hot Lenses occur within the Kimmeridge Clay Formation in Tartan. The very fine- to fine-grained Hot Lens sandstones are the product of turbidite flows which were channelled along narrow, elongate, pre-existing, tectonically controlled, topographic lows.

Using log character combined with palaeontological studies, core, and engineering data, it is possible to correlate the Piper Formation Members and the Hot Lenses across the Tartan Field. A small panel showing representative correlations in the 'Upthrown' and 'Downthrown Blocks' is shown in Fig. 6. Detailed correlations between the two blocks are more problematical. In particular, recent palaeontological studies indicate that the Lower Sand in the 15/16-T8 well in the 'Upthrown Block' is older than the Lower Sand Member in the 'Downthrown Block'. The Piper Formation Members and the Hot Lenses within Tartan are further sub-divided into reservoir units to enable more realistic reservoir modelling. In the 'Upthrown Block' the division of the members is limited to a simple breakdown of the Main Sand into units I, II, and IIA. In the 'Downthrown Block' there are many more sub-divisions of the Piper Formation and these are displayed in Fig. 2. Hot Lens 'A' has been divided into three units 1, 2, and 3 which represent muddy channel, stacked channel, and distal channel facies respectively.

The Piper Formation exhibits markedly different petrophysical properties in each block. A relatively homogeneous intergranular porosity system is present throughout the oil zone in the 'Upthrown Block'. Porosities are lower and more variable in the 'Downthrown

Block' as a result of cementation and an intense compaction fabric (Burley 1989). Within the Piper Formation of the 'Upthrown Block' gross porosities generally range from 13–19% with permeabilities usually in the 10 md–1 d range. In the 'Downthrown Block' gross porosities range from 9–14% with gross permeabilities usually less than 10 md with exceptions being the Six Sand unit 'D' and Hot sand unit 'C' where permeabilities reach up to 300 md. The Hot Lens sandstones of Tartan have an average gross porosity of approximately 13% and permeabilities in the range 10–150 md.

Source

Geochemical studies by Bissada (1983) for shale samples from wells in quadrants 14 and 15 indicate that the oil was probably sourced by maturation of the Kimmeridge Shale, at depths greater than approximately 10 000 ft in the adjacent Witch Ground Graben. K-Ar determinations on authigenic illite by Burley (1989) suggest that the oil accumulation in Downthrown Tartan was in place at a maximum of 40 million years. Older ages are indicated for the oil accumulation in the 'Upthrown Block'.

Hydrocarbons

Oil from the Piper Formation in the Tartan Field has a gravity of approximately 38° API. Hydrogen sulphide concentrations in the crude are typically in the range 100–500 ppm. A representative formation volume factor for the 'Upthrown Block' is 1.85. In the 'Downthrown Block' the Main Sand, '15/16-6' Sand and Hot Sand Members have formation volume factors of 1.65, 1.57, and 1.39 respectively. Oil from the Hot Lens 'A' reservoir has a gravity of approximately 37° API and a high hydrogen sulphide content (10 000 ppm). The original formation volume factor for the Hot Lens 'A' Member is 1.45.

Formation water samples have been taken from the Piper Formation in the 15/16-16 well in the 'Upthrown Block' and the 15/16-T2 well in the 'Downthrown Block'. The 15/16-16 well sample indicates a relatively fresh connate water (44 000 ppm NaCl); however, it is thought that the sample could have been contaminated by mud filtrate. The 15/16-T2 sample with a salinity of 95 000 ppm NaCl is considered more representative and has been adopted as the Tartan Field average.

Tartan Field data summary

	'Upthrown Block'	'Downthrown Block'
Trap		
Type	Structural/stratigraphic	Structural/stratigraphic
Depth to crest	−9700	−11 600
Oil–water contact	−10 328	−12 169
Oil column	628	569
Pay zone		
Formation	Piper Formation	Piper Formation
Age	Upper Jurassic	Upper Jurassic
Porosity (average/range)	15%/13–19%	12%/9–14%
Hydrocarbon saturation	88%	61%
Permeability (average/range)	150 md/10–1000 md	17 md/0.1–300 md
Productivity index	50	7
Hydrocarbons		
Oil gravity (°API)	38	38
Oil type	Black	Black
Gas gravity	1.1	1.2
Bubble point (psia)	3650	2850
Gas/oil ratio	1450	1100
Formation volume factor	1.85	1.39–1.65
Formation water		
Salinity (NaCl equiv.)	95 000	95 000
Resistivity (ohm m at 225°F)	0.027	0.027
Reservoir conditions		
Temperature (°F)	215	240
Pressure (psi at OWC)	4650	5800
Pressure gradient (psi/ft)	0.26	0.30
Field size		
Area (acres)	985	1996
Drive mechanism	waterflood	waterflood
Recoverable hydrocarbons	68	48
Production		
Start-up date	January 1981	
Development scheme	Fixed steel platform	
Production rate	30 000 BOPD	
Cumulative production to end March 1989:		
Crude	61.0 MMBBL	
NGL	5.9 MMBBL	
Liquids	66.9 MMBBL	

Reserves/resources

Ultimate recoverable reserves within the Tartan Field are currently estimated at 116 MMBBL of crude oil.

Waterflooding is used to provide pressure support for the 'Upthrown Block' Piper Formation and the 'Downthrown Block' Hot Lenses. Originally in the 'Downthrown Block' Piper Formation it was anticipated that the pressure could also be maintained via water injection. However, due to more complicated geology than expected it was not possible to inject sufficient volumes of water to maintain the reservoir pressure. Consequently, the Piper Formation in the 'Downthrown Block' was produced for a time under depletion drive with limited water injection. Now that the reservoir pressure is low, an expanded waterflood is being achieved by the conversion of existing wells to injection and the drilling of new wells.

We wish to thank our co-workers in Texaco Ltd, particularly J. H. Christiansen and L. C. Hart who critically reviewed the paper.

References

BURLEY, S. D. 1989. Timing diagenesis in the Tartan Reservoir (UK North Sea): constraints from combined cathodoluminescence microscopy and fluid inclusion studies. *Marine and Petroleum Geology*, **6**, 97–120.

BISSADA, K. K. 1983. Petroleum generation in Mesozoic sediments of the Moray Firth Basin, North Sea area. *In*: BJOROY, M. (ed.) *Advances in Geochemistry*. John Wiley, London, 7–15.

HARKER, S. D., GUSTAV, S. H. & RILEY, L. A. 1987. Triassic to Cenomanian stratigraphy of the Witch Ground Graben. *In*: BROOKS, J. & GLENNIE, K. W. (eds) *Petroleum Geology of North West Europe*. Graham & Trotman, London, **2**, 809–818.

WADDAMS, P. & CLARK, N. M. 1991. The Petronella Field, Block 14/20b, UK North Sea. *This volume*.

WHITEHEAD, M. W. & PINNOCK, S. J. 1991. The Highlander Field, Block 14/20b, UK North Sea. *This volume*.

The Southern Gas Basin

The Amethyst Field, Blocks 47/8a, 47/9a, 47/13a, 47/14a, 47/15a, UK North Sea

C. R. GARLAND

BP Exploration, Farburn Industrial Estate, Dyce, Aberdeen, AB2 0PB, UK

Abstract: The Amethyst gas field was discovered in 1970 by well 47/13-1. Subsequently it was appraised and delineated by 17 wells. It consists of at least five accumulations with modest vertical relief, the reservoir being thin aeolian and fluviatile sandstones of the Lower Leman Sandstone Formation. Reservoir quality varies from poor to good, high production rates being attained from the aeolian sandstones. Seismic interpretation has involved, in addition to conventional methods, the mapping of several seismic parameters, and a geological model for the velocity distribution in overlying strata.

Gas in place is currently estimated at 1100 BCF, with recoverable reserves of 844 BCF. The phased development plan envisages 20 development wells drilled from four platforms, and first gas from the 'A' platforms was delivered in October 1990. A unitization agreement is in force between the nine partners, with a technical redetermination of equity scheduled to commence in 1991.

The Amethyst Gas Field comprises a group of accumulations on the western flank of the Southern Gas Basin, with an area of 24 000 acres over UK blocks 47/8a, 9a, 13a, 14a and 15a (Fig. 1). Gas is trapped in a gentle NW–SE-trending anticline in the Rotliegendes aeolian sandstones. The field is located 25 miles east of Easington in a water depth of 90 feet. The field was named after the precious stone, reflecting the value placed on it by Burmah Oil Co. in 1972.

History

The field is centred on block 47/14a, which was part of the nine-block licence P005 awarded to a Burmah group in the first round of licensing in 1964. The adjacent block 47/13 was also awarded in the first round as P050 to Trinidad Canadian Oils (100)%.

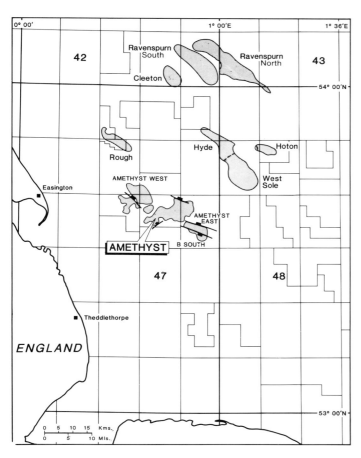

Fig. 1. Amethyst area location map.

The first discovery was made by the Conoco–NCB group who farmed-in to block 47/13a (Conoco gaining operatorship) and drilled well 47/13-1, which flowed 15.8 MMSCFD from the Rotliegendes sandstone. This was followed in 1972 by another discovery, well 47/14a-1, which flowed 22.6 MMSCFD and encouraged Burmah Oil to name the field. A further discovery was made by Amoco in 1973 by well 47/15a-2, which tested 19.7 MMSCFD. The accumulations discovered by these three wells have been termed C, A and B sites respectively, the C site comprising Amethyst West, and the A and B sites comprising Amethyst East (Fig. 2).

Blocks 47/8a and 9a to the north were originally awarded to Gulf Oil (100%) as licence P030 in the first round, but the company sold its interest in the licence (which included Rough Field) to the Amoco group in 1973. Well 47/9a-7, drilled in 1985, tested 25 MMSCFD and proved a northward extension of the field.

Considerable business activity involving these licences has ensured that the partnerships that were awarded the licences are not the same partnerships that currently own the field. However, the situation has been simplified by a unit operating agreement between the now nine partners, and cost-sharing interests, subject to technical redetermination, have been agreed as follows:

	%
Amerada Hess	8.31
Amoco UK Exploration Co.	11.08
Arco British Ltd	12.83
BP Exploration	18.41
Enterprise Oil plc	20.47
Fina Petroleum Development Ltd	1.87
Gas Council (Exploration)	21.43
Murphy Petroleum Ltd	2.80
Ocean Exploration Co. Ltd	2.80

Whilst the discovery wells were all drilled on the crests of time-closures interpreted from seismic data, the early appraisal wells failed to fulfill their promise, for both geological and geophysical reasons.

The first appraisal well, 47/14a-2, encountered a thin reservoir lacking the aeolian facies so productive in the discovery well. A DST in this well flowed 4.6 MMSCFD, probably from a fluvial channel sand. In 1975 another stepout well, 47/14a-3, was drilled to the NE of the suspended discovery well, and tested a flow of 11.3 MMSCFD.

Two further poor appraisal wells, and acquisition of a further seismic survey in 1981 indicated that prediction of reservoir quality was to be as important in the ongoing field appraisal as the accurate mapping of depth to top reservoir. In pursuit of this idea, well 47/14a-6 was drilled in 1984 on a high-amplitude seismic anomaly, and this well not only tested 40 MMSCFD, but also logged a GWC. Seismic amplitudes were also used successfully to locate well

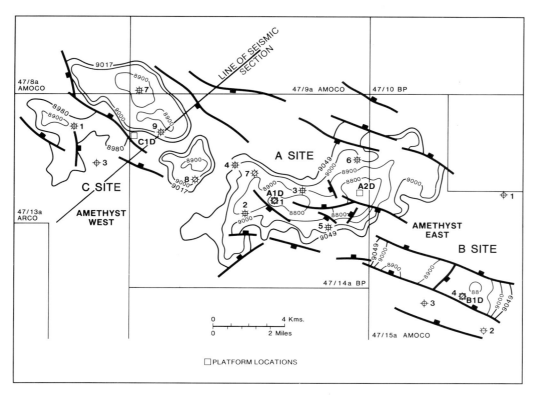

Fig. 2. Amethyst Field depth structure.

47/14a-7 in 1984, which was drilled as a deviated well from a possible platform location, and tested 20 MMSCFD.

The next two wells to be drilled in 1985, 47/9a-7 and 47/14a-8, both tested significant flows of gas (25 and 31 MMSCFD respectively) from the western part of the accumulation at locations showing high-amplitude seismic reflectors: however, well 47/14a-9 was a disappointment and a reminder of the continuing uncertainty associated with the reservoir.

A decision to develop was taken on the basis of the results of the 17 wells drilled in the three blocks, and a phased development plan was devised involving four unmanned platforms and a pipeline to Easington. Approval for development was granted in June 1988; the first development well was spudded in September 1989 from the A2 platform on the 'A' site (Amethyst East). Eleven development wells have been drilled from the two platforms and first gas was delivered in October 1990. Phase II of the development will commence at Platform B1D in May 1991.

Field stratigraphy

The stratigraphy of the Amethyst Field (Fig 3) is a thinner equivalent of the well-documented succession described from the Sole Pit Basin to the east. In particular, the Rotliegendes thins in the south of the field to less than 40 ft, the Zechstein becomes dominated by anhydrite shelf deposits and there is no Muschelkalk Halite in the Triassic. The field overlies an anticline in the Carboniferous which forms a structural high in line with an offshore extension of the Market Weighton Axis.

This anticlinal core, eroded prior to Rotliegendes deposition, contains Namurian sediments comprising a progradational sequence of shales and thin sandy turbidites, passing vertically into more proximal shallow marine clastics. On the flanks of the field, paralic sediments, including coal, of Westphalian 'B' age are preserved. Carboniferous sands are commonly interpreted from electric logs to be gas-bearing, but on test have failed to flow due to very low permeability. The uppermost 100–200 ft of the Carboniferous sediments are reddened, due to oxidation at the exposed bedrock surface prior to Permian deposition in this area.

The Carboniferous is overlain with slight unconformity by the Rotliegendes reservoir which is described in more detail below. This reservoir is overlain and sealed by the late Permian Zechstein Supergroup: above the 2–8 ft thick veneer of Kupferschiefer, the basal Z1 & 2 'Lower Magnesian Limestone' varies from 600 ft of Werranhydrit Formation overlain by 35 ft of Hauptdolomit Formation in well 47/13-2, to a single unit 190 ft thick of Hauptdolomit in well 47/14a-4. The top of this unit is an important seismostratigraphic marker in the area, and the 'Lower Magnesian Limestone' terminology, derived from the Yorkshire and Durham outcrop, is here used informally for simplicity.

These two facies are considered to be 'shelf' and 'basin' equivalents respectively, and the short 1 km (0.6 mile) transition between them has been termed the 'Anhydrite Wall'. The overlying Stassfurt Halite also undergoes a facies change in parallel with that of the Lower Magnesian Limestone: to the southwest the halite is relatively thin with interbedded anhydrite (up to 170 ft of Deckanhydrit) and polyhalite, whilst to the northeast the halite thickens, with the development of potassium salts (Kieserite, Carnallite, Sylvite) but anhydrite is thin or absent (Fig. 4). These variations cause an increase in seismic interval velocity to the southwest which 'raises' the structure seismically in time.

The Triassic is approximately 3000 ft thick, consisting of a well-developed Bacton Group (which lacks any gas shows) overlain by a thin Haisborough Group. The Winterton Formation is represented by 20–40 ft of Rhaetic sandstone.

Lower and Middle Jurassic shales with thin limestone stringers are overlain by the Kimmeridge Clay Formation, which is truncated at the base Cretaceous unconformity towards the northeast. This results from Late Cimmerian uplift and erosion at the margins of the Dowsing Fault Zone, to the east of which the Sole Pit Basin has subsided in Jurassic time. The period of emergence lasted through the Early Cretaceous: the resumption of quiet marine deposition of the Chalk is confirmed by the parallelism of seismic reflectors in this unit with no onlap or internal structure to suggest contemporary topography (Fig. 5).

The similarity between Base Chalk and Top Rotliegendes structure suggests that the main Amethyst closure, was formed in the Tertiary, probably in the Alpine (Eocene–Oligocene) tectonic phase.

Faults in the field (Fig. 6) are confined to the Rotliegendes and Zechstein; they mainly trend WNW–ESE, and are vertical with throws in the order of 100 ft, associated with small secondary

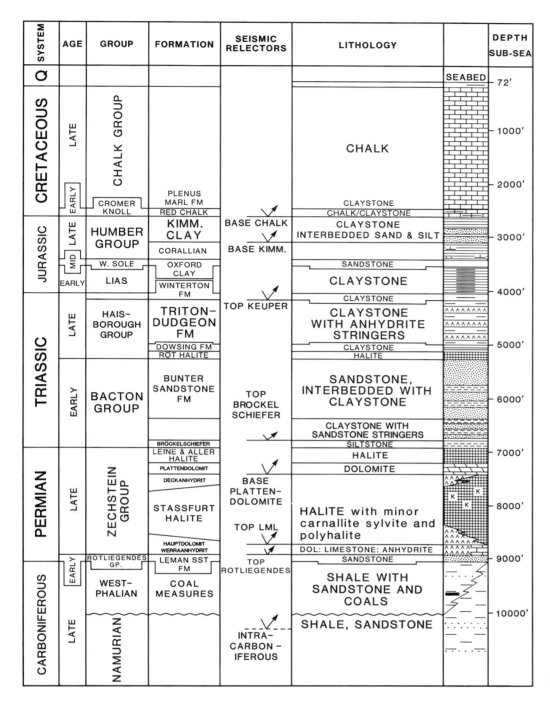

Fig. 3. Amethyst Field stratigraphy

NNW- and NW-trending faults. In the southern part of the field, a suite of faults trending north east may be older, since it appears to be cut and offset dextrally by the main swarm. The absence of a deep basin near the field explains the lack of gravity-driven listric normal faults with block rotation, and implies that all the observed faults have been transmitted through the basement.

The small throw of faults up to two miles in length suggest that they were set up as lines of weakness prior to Rotliegendes deposition and have broken through the Rotliegendes to the Zechstein during later periods of slight reactivation. Thrusting along some of these faults and evidence of a pull-apart zone in the northern part of the field suggests a strike-slip origin.

Geophysics

The watergun acquisition system has proved ideal for this area, and the current interpretation is based primarily on 1983, 1984 and 1985 watergun data sets in a 1.0 km grid over most of the 'A' site and the eastern part of the 'C' site, and the 1981 and 1983 data over the 'B' site. Synthetic seismograms were used to aid identification of each of the picked events. In 1988 a 3000 km high-resolution seismic survey was acquired by SSL, and this data has proved itself to be essential during field development.

Most seismic lines interpreted across the field show the Rotliegendes reflector to have a slightly interrupted monoclinal dip. The time map (Fig. 6) shows limited closure.

The depth conversion was undertaken using a seven-layer model. The Base Cretaceous to Permian intervals are relatively uniform and were depth-converted using simple functions related to time thickness. However, the facies variations within the Zechstein, causing faster velocities towards the south and west, were incorporated into a velocity map. The resultant depth conversion reveals the larger areas of structural closure identified as A, B and C sites (Fig. 2).

The top of the Rotliegendes is picked at a strong trough caused

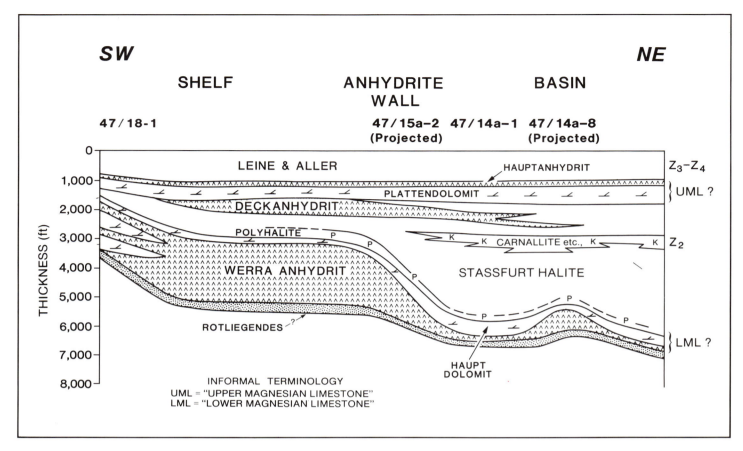

Fig. 4. Amethyst area Zechstein facies variations.

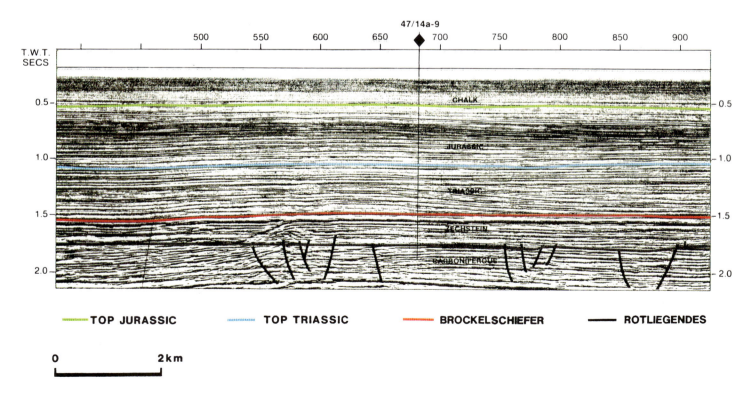

Fig. 5. Amethyst Field seismic line.

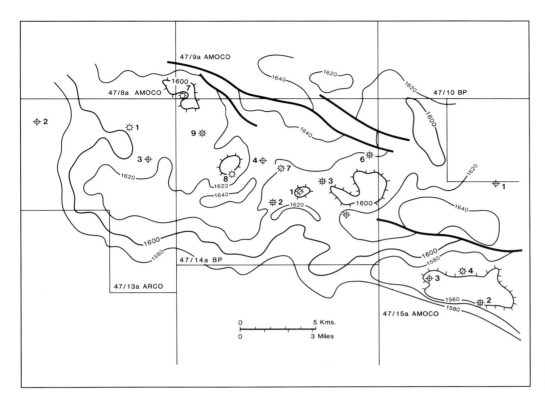

Fig. 6. Amethyst Field two-way time structure.

by the decrease in seismic velocity at the interface between porous aeolian sands and high-velocity anhydrite or dolomite of the 'Lower Magnesian Limestone'. There is a contrasting peak of variable amplitude at the base of the Rotliegendes, but the thin Rotliegendes in this area, coupled with insufficient bandwidth in the older seismic data, led to tuning effects and a non-linear relationship with reservoir thickness. It was found that, in general, higher amplitudes at Top Rotliegendes related to thicker and generally better quality reservoir areas. Only with the 1988 survey has it been possible to map both top and base reservoir consistently across the field, and development wells have been successfully targeted at buildups of aeolian sand.

Trap

The structural dip of all the strata in this area is very slight, so that vertical relief is only 250 ft in a structure with dimensions of 20 × 6 miles (32 × 10 km).

For this reason the locations of the mapped spill-points are very sensitive to the seismic velocities used for depth conversion, and incorporation of a shelf-edge model for the Zechstein lithology has, for the first time, shown that dip closure could be the dominant trapping mechanism.

The current map shows five pools or accumulations (see Fig. 2). In the west, the small accumulation around well 47/13-1 has dip closure to south, west and north, and is separated by a fault from the larger dip-closed accumulation tested by wells 47/14a-9 and 47/9a-7. A common GWC, and common pressure regime, suggest an historical connection with the small accumulation around well 47/14a-8, and these three pools make up the 'C' site or Amethyst West.

The 'A' site is the core of the field but nevertheless epitomizes Amethyst mapping problems. It has been noted above that the area of time closure is small: the Rotliegendes reflector generally rises to the south of the field, where, prior to the current mapping, two alternative trapping geometries were proposed. Initially, the Rotliegendes was believed to onlap the Carboniferous to the south, the seal being provided by impermeable siltstones of Westphalian B age. Alternatively, faulting along the southern flank of the field, accompanied by pervasive anhydrite cementation similar to that encountered in well 47/14a-5, could have provided the seal. At present the southern field boundary is mapped as being faulted along part of its length, showing that at least some fault seal must be present.

Reservoir

The reservoir is 45–120 ft thick, formed of the Permian Leman Sandstone Formation of the Upper Rotliegendes Group. The sands are light grey to pink or brown, generally fine- to medium-grained, moderately well sorted, with subrounded to subangular quartz and 10% feldspar grains in an argillaceous, dolomitic matrix. The sands are commonly crossbedded showing strong sedimentary dip; they are interbedded with argillaceous laminae, thicker shaly bands, and rare zones of fine wavy laminations. There are local zones of poorly-sorted conglomerate and intraformational breccia. The sands are barren of fossils or carbonaceous material, although there is a small proportion of heavy minerals. This association of lithologies is characteristic of a continental desert environment with a water table that was occasionally close to the surface. Four main depositional facies have been identified: aeolian dune sands with good reservoir quality (porosity 25% and average permeability up to 100 md); wadi deposits with poor reservoir quality (porosity 10–12% and permeability 0.5–10 md); sheet sand and interdune deposits with intermediate quality; and reworked deposits. A threefold zonation separates an upper reworked and cemented zone 1 from the aeolian zone 2, with the basal zone 3 comprising wadi and sheet sand deposits. The cross-section (Fig. 7) shows the relationship between the three zones and reservoir facies.

Diagenesis

The Rotliegendes reservoir was significantly affected by diagenesis in several different ways. From the study of cores from sixteen Amethyst wells it is clear that diagenesis can be described in four main categories, and that each of these has a distinct relationship with the reservoir that can be incorporated into the geological model. *Early calcium carbonate* cement is a pervasive cement in the

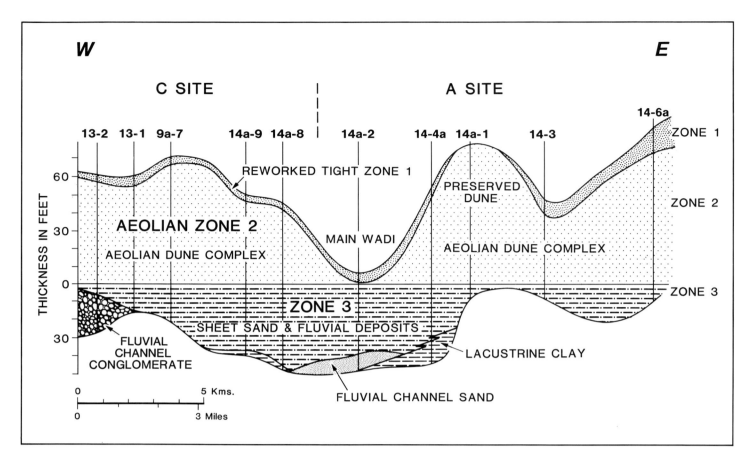

Fig. 7. Amethyst Field reservoir cross-section.

basal reservoir zone 3, and could have been precipitated as calcrete at an early stage; *intermediate dolomitization* affects the uppermost 1–15 m of the reservoir, severely reducing porosity in the aeolian sandstones; *deep burial diagenesis* generated up to 2% of additional porosity by dissolution of feldspar, whilst creating conditions for minor authigenic clay creation; *late anhydrite cementation* is locally severe near to fault zones, and can affect the entire reservoir. Other diagenetic phases (e.g. chlorite, barite and authigenic feldspar) are of minor importance.

Source

Source for the gas in the Amethyst field is undoubtedly Coal Measures, deeply buried and mature in the centre of the Sole Pit basin to the NE, and possibly also down-dip to the south.

Variation in the condensate ratio between Amethyst East and Amethyst West suggested two sources might be present, the more condensate-rich source possibly related to residual oils noted in thin section.

Hydrocarbons

Separator gas samples taken from appraisal wells on test showed at an early stage that the field was divided into at least two accumulations, the eastern with a low condensate ratio and the western with a relatively high condensate ratio. Of greater concern for the development plan is a relatively high content of carbon dioxide, which is corrosive at reservoir temperature and pressure and required that production tubing be manufactured from high-chrome steel. The gas composition and reservoir conditions at 4100 psia and 190°F give a gas expansion factor to standard temperature and pressure of approximately 235. The data summary table details the gas composition.

Reserves/resources

It has been assumed that Amethyst, in common with other North Sea gas fields, will exhibit a pure gas expansion drive when placed on production. The likelihood of lateral aquifer movement is being investigated.

A single-phase reservoir model, incorporating the geological interpretation outlined above and tied to permeability–thickness values calculated from the results of drill stem tests, has been used to determine gas recovery from a total of twenty production wells. A recovery factor of 77% is anticipated with reserves of 844 BCF recoverable over the field's nominal 20-year life.

Despite early development success, there is still some risk attached to reservoir quality prediction. To cover this and other risks, the platforms have been designed to allow extra wells to be drilled if necessary.

The appraisal of Amethyst has been a team effort from the start, and considerable technical work by partners is acknowledged. Conoco, operator until 1987 of 47/13a, and Amoco, operator of 47/15a, shared their interpretations with the operator of 47/14a, and were prime movers in forging a unified partnership. ICI Petroleum were greatly involved in the early studies on seismic amplitudes, and Tricentrol plc suggested a number of the innovations used in the seismic depth conversion. This paper is published by kind permission of the current Amethyst partners. Within BP, the 1988 seismic survey was designed by R. L. Smith, and interpreted by S. J. Marshall and R. P. Steele.

Amethyst Field data summary

Trap
Type	Faulted anticline
Depth to crest	8800 ft ss
Lowest closing contour	9050 ft ss
Gas–water contact	−9049 ft in Amethyst East
	−9017 ft in Amethyst West

Pay zone
Formation	Leman Sandstone Formation
Age	Early Late Permian
Gross thickness	40–120 ft, average 90 ft
Net/gross ratio	0.4–0.98
Cut off for N/G	0.1 md (8% porosity)
Porosity (average range)	11–25%
Gas saturation	50%
Permeability (average range)	1–1000 md

Hydrocarbons
Gas composition	
Methane	91.95 Mol %
Ethane	3.58 Mol %
Propane	0.05 Mol %
Nitrogen	2.22 Mol %
Carbon dioxide	0.64 Mol %
Other hydrocarbons	0.36 Mol %
Condensate yield	7–20 BBL/MMSCF
Formation volume factor	235

Formation water
Salinity	145 000 ppm Cl

Reservoir conditions
Temperature	190°F
Pressure	4100 psia
Pressure gradient in reservoir	0.1 psi/ft

Field size
Area	97 km^2 (24 000 acres)
Recovery factor	77%
Drive mechanism	Gas depletion
Recoverable reserves	844 BCF

Production
Start-up date	First gas October 1990
Development scheme	2 phases, 4 unmanned platforms
Number/type wells	20 gas production wells planned
Production rate	Build up year at 150 MMSCFD (DCQ)*
	6 year plateau at 180 MMSCFD (DCQ)*
Cumulative production to date	40 BCF (end of Feb 91)

* DCQ, Daily Contract Quantity

The Barque Field, Blocks 48/13a, 48/14, UK North Sea

R. T. FARMER[1] & A. P. HILLIER[2]

[1] Esso Expro UK Ltd, 21 Dartmouth Street, London SW1H 9BE, UK
[2] Shell UK, Expro, P.O. Box 4, Lothing Depot, North Quay, Lowestoft, Suffolk NR32 2TH, UK

Abstract: The Barque Field is associated with some of the earliest gas discoveries in the southern North Sea. In the Sole Pit area the reservoir, the Rotliegend Group Leman Sandstone Formation of Lower Permian age, occurs between Carboniferous Coal Measures, which source the gas, and Zechstein evaporites which form an excellent seal. Primarily aeolian, the sandstone has very low permeability resulting from deep burial of the Sole Pit Trough. The deepest burial and hence the maximum diagenetic damage to the reservoir was achieved in the early Late Cretaceous prior to the two-fold main phases of inversion in Late Cretaceous and Mid-Tertiary.

Compaction and diagenesis reduced reservoir permeability to such an extent that parts of the field would be non-productive were it not for the presence of effective natural fracture zones and well stimulation by hydraulic fracturing techniques. Though appraisal and evaluation have been relatively extensive, remaining uncertainties dictate a conservative development in conjunction with the adjacent Clipper Field. A selected initial area was developed for first gas in October 1990. Good reservoir performance may lead to later development of the whole field.

The Barque Field, some 45 miles off the Norfolk coast, lies within the Sole Pit area of the southern North Sea. It is located 60 miles east of Grimsby and a similar distance north of Great Yarmouth.

An elongate feature with a northwest trend, the field has a length of 15 miles and maximum width of 2 miles, and covers an area of more than 9000 acres. Water depth varies between 15 and 25 fathoms (90–150 ft) and the majority of the gas-bearing Rotliegend reservoir is found between 7500 and 8500 ft TVSS.

The Rotliegend is 700–800 ft thick over the field and consists mainly of aeolian quartzose sandstones, which are entirely gas-bearing in the crestal sector. The gas is very dry, consisting of 95% methane with about 1.5% inert gas and a small proportion only of condensate, typical of this area of the North Sea.

The Barque Field is now under development in conjunction with the adjacent Clipper Field. Together with other joint venture southern North Sea gas fields, Galleon and Frigate, these field names commemorate vessels from maritime history.

History

Licence P.008 covering Blocks 48/13, 14, 19 and 20, was awarded to the joint-venture partners Shell UK Expro and Esso Expro UK in the first licensing round, September 1964, with Shell appointed as operator.

The acreage was reduced by 50% in September 1970 and licence expiry is September 2010 for Blocks 48/13a and 48/14, within which the field lies.

Industry's first North Sea discovery, BP's West Sole Gas Field in Block 48/6, was drilled and tested in late 1965 (Butler 1975). Shell/Esso drilling in Sole Pit began in mid-1966 with the discovery of gas in well 48/13-1, located to test a feature on the trend between Leman Field (Block 49/26 and 27) and West Sole Field. This well came in deep, encountering a tight, gas-bearing Rotliegend section in what was later recognized as an independent closure immediately north of the prime target, the Barque Field. The well credited with the Barque Field discovery, 48/13a-2A, was drilled in 1971 in a near-crestal location about 3.5 miles north west of 48/13-1 and though reservoir quality appeared poor the well tested up to 9.7 MMSCFD after stimulation by acid and hydraulic fracturing (Moore 1989).

Appraisal was delayed until 1983 when techniques had advanced and commercial development of a reservoir with such low permeability matrix appeared feasible. That year saw the drilling of a crestal mid-field well 48/13a-4 which tested up to 50 MMSCFD after acidization and a southeastern appraisal, 48/14-1, which tested up to 25 MMSCFD (two zones flowing comingled) after acid and hydraulic fracture stimulation. Appraisal continued in 1984 with the drilling of another mid-field well, 48/13a-5, which tested up to 39 MMSCFD after acid/hydraulic fracture stimulation, followed by 48/13a-6 in the northwest extremity of the field which yielded less than 1 MMSCFD after stimulation. Later that year the sixth Barque well, 48/13a-7A, encountered a reduced section of Leman Sandstone Fm. as a result of seismic uncertainty in mapping the boundary fault. Drilled northward from the location of 48/13a-4, as a highly deviated well, it passed from mid-Rotliegend through the major northern boundary fault to terminate in downthrown Zechstein evaporites. Limited testing in early 1985 failed to establish flow. All appraisal wells were plugged and abandoned after testing.

Conventional core and well analyses confirmed the poor quality of the reservoir with in-situ matrix permeability commonly averaging less than 1 md. It was also apparent, however, that reservoir quality varied considerably and that zones of natural fractures, occurring at intervals throughout the field, increased productivity markedly if conneced to the well bore. Several development plans were considered, the final choice being phased development of the most favourable area of Barque together with a similar favourable area in the adjacent, low-permeability Clipper Field (see Farmer & Hillier, this volume). In the case of Barque, the initial development area is located in the central/northwest part of the field (see Fig. 1) to maximize accessible gas-in-place in an area where fracture zones are likely to occur.

The unmanned 18-slot satellite platform, installed early in 1990, was intended to have an 11 500 ft drainage radius for wells deviated up to 60°. However, the successful application of horizontal drilling techniques may allow this drainage reach to be extended in the future. It will be linked by a 16 inch pipeline to the main production complex on Clipper Field, 12 miles to the southeast. Production from the two fields will be evacuated through a dedicated 24 inch trunkline to Bacton, 46 miles to the south.

Development drilling commenced in January 1990 with a horizontal well predrilled from the platform position. Achieving a test flow rate of 48 MMSCFD, this well confirmed the potential benefits of this technique in tight matrix and, as a consequence, development plans for Barque Field now incorporate extensive horizontal drilling. Platform drilling on Barque commenced in mid-1990 with first gas production in October that year. Initial production from Barque will be low but in the second year, after completing the planned 12 wells, the field will be producing slightly more than 100 MMSCFD.

Field stratigraphy

A typical stratigraphic section from the initial development area of Barque Field is shown in Fig. 2. In addition to group and formation

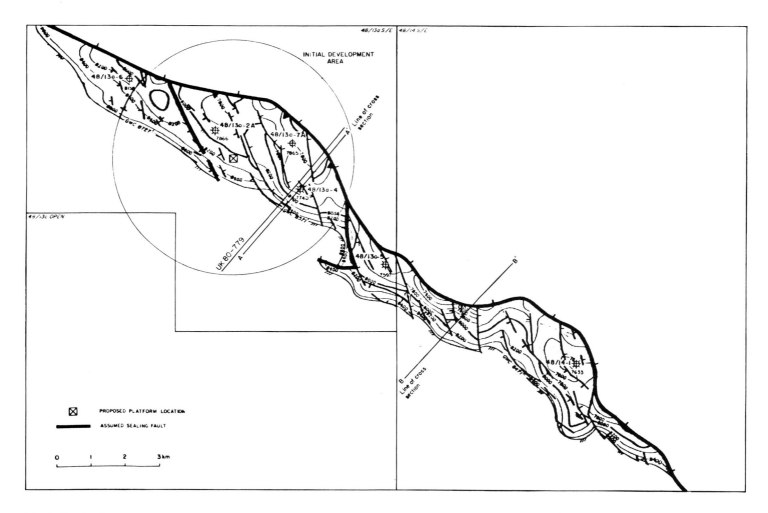

Fig. 1. Top Rotliegend structure map based on 2-D seismic data showing position of initial development area, lines of structural cross-sections and seismic line position.

names, the subdivision of members best suited to seismic and wellsite use is also given.

The Lower Permian Rotliegend Gp in Sole Pit lies unconformably on Carboniferous Coal Measures and is overlain by a thick Zechstein section consisting of cyclic evaporites and carbonates which vary between 1500 and 2500 ft in thickness. Overlying the Zechstein a thin (less than 100 ft) Brockelschiefer siltstone forms the basal member of the Lower Triassic Bacton Group, followed successively by three units, the Bunter Shale Formation, its upper Rogenstein Member and the Bunter Sandstone Formation with a total thickness of 2000 ft in the Barque area. The Bacton Gp is overlain by sediments of the Middle and Upper Triassic Haisborough Group comprising three evaporite cycles (Rot, Muschelkalk and Keuper) and intervening fine-grained clastics. These cycles total some 2500 ft and are generally followed by thin Rhaetic sandstones and shales of the Winterton Formation. The succeeding undifferentiated Lower Jurassic sequence is incomplete, though up to 1000 ft thick, and consists of fine-grained clastics. Where the Rhaetic is absent, the Lias is unconformable upon late Triassic which may itself be eroded down into the Keuper section. Middle and Upper Jurassic and Cretaceous sediments are absent from the Sole Pit structural high. A thin Tertiary–Recent section continues above the Lias to the sea bed.

This stratigraphic sequence represents the late Palaeozoic to Tertiary filling of the southern North Sea Basin interrupted by significant episodes of uplift. The basin formed after early Hercynian uplift and erosion of the Carboniferous and at Barque the Rotliegend is unconformable upon Carboniferous Coal Measures. Subsidence allowed accumulation of continental, predominantly aeolian sediments followed eventually by a marine transgression and the markedly different cycles of the Late Permian Zechstein sediments (Taylor 1984). Basin fill led to further continental sedimentation in the early Triassic and a later return to marginal and fully marine (evaporitic) cycles (Fisher 1984). Marine clastic sedimentation continued through the Jurassic into the Early Cretaceous with accelerated subsidence in sub-basins such as Sole Pit (Lutz et al. 1975).

By the Late Cretaceous the Rotliegend Group had reached its greatest depth of burial, and some local tectonic inversion began (Glennie & Boegner 1981) in response to re-activation of older deep-seated fault systems (van Hoorn 1987). In the Barque area little remains of the post-Triassic sequence as inversion of the Sole Pit high removed most of the Jurassic–Cretaceous sequence.

Compaction studies of the Bunter Shale indicate some 5000 ft or more of post-Liassic sediments were removed during this inversion (Marie 1975; Glennie et al. 1978). Limited Tertiary sedimentation ended with further uplift associated with the Alpine orogeny.

Geophysics

Good seismic control is available over the Barque Field. Surveys shot in 1978, 1979 and 1980 provide a regional 500 m diamond grid of 2-D seismic data. This is complemented by two 3-D seismic surveys, the first being a pilot survey acquired over 15 square miles of southeastern Barque in 1984 and the second in 1987 over the initial development area of the field. Both 3-D surveys have cross-line and inline spacing of 25 m; their location and coverage is shown

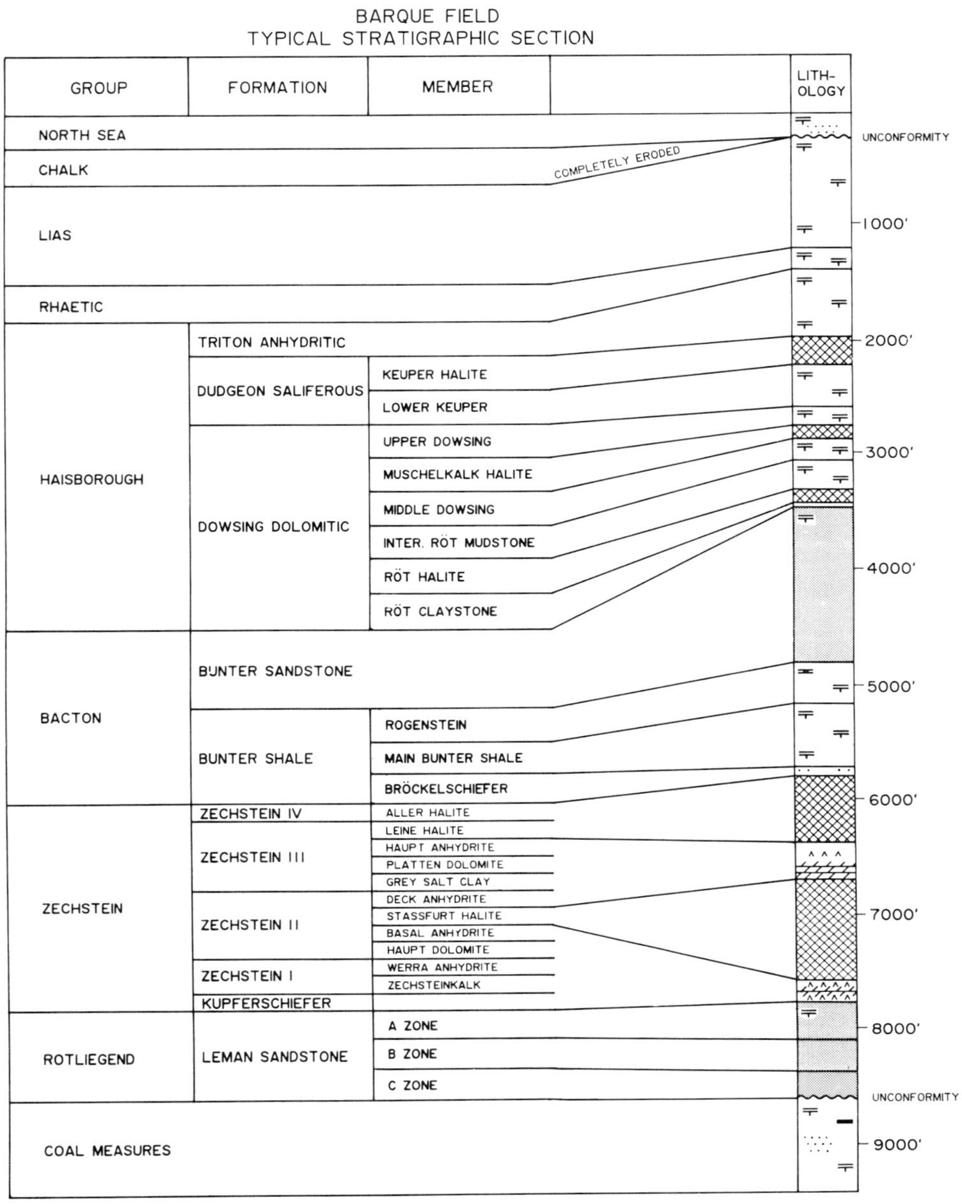

Fig. 2. Stratigraphic sequence of the field showing overburden geology.

in Fig. 3. Earlier seismic suffered both in quality and from uncertain depth conversion, a problem noted by Hornabrook (1975). Processing of the 1987 3-D data was completed in mid-1989 and results were incorporated in revised mapping prior to the beginning of development drilling.

Interpretation difficulties in the Sole Pit area are intensified by post-Triassic faulting in the overburden, nevertheless data quality allow adequate identification of five horizons (Fig. 4). The Top Triassic, often involved in listric fault zones, may be discontinuous and thereby difficult to correlate. The next two, the Rot Halite and Brockelschiefer, are normally continuous and reliable markers. The Platten Dolomite yields a strong reflector and the Basal Anhydrite, near the base of the Zechstein, though less continuous is the deepest mapped horizon. Sporadic reflections about Top Carboniferous level may not be reliably mapped from 2-D but can, with care, be mapped from 3-D.

Time to depth conversion was achieved with a multi-layer velocity model using well velocity data from all wells in the area. The Top Rotliegend structure map was then derived from the Basal Anhydrite depth map with the addition of an isochore for the Anhydrite–Rotliegend interval.

Most faults at reservoir level are interpreted from seismic to be near vertical with the exception of the major northern boundary fault. A segment of this fault was cored in the 48/13a-7A appraisal well, establishing the dip of the fault plane.

Trap

The structural map (Fig. 1) shows that the Barque Field is formed mainly by dip-closure against a sinuous, northwest-trending major fault which seismic interpretation indicates to be a reverse fault in places.

In broader segments of the field the crest shows anticlinal rollover with gentle dips; the extensive flank, however, dips moderately to steeply to the southwest. Within the field, several NNW-trending faults are mapped and some of these are believed to form barriers to fluid flow. The major boundary fault is clearly recognized as sealing where the Rotliegend reservoir is juxtaposed against the Zechstein (see structural cross-sections, Fig. 5).

The top seal is formed by the Zechstein evaporites and tight carbonates immediately above the Rotliegend. Fault compartments within the field, where the throw does not offset the sandstone completely, are believed to result from cataclasis and mineralization

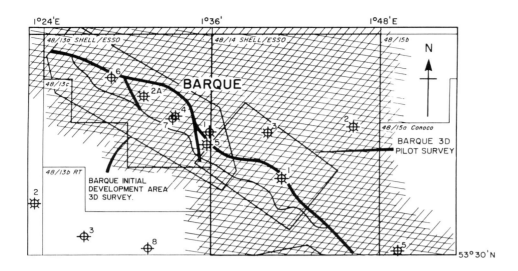

Fig. 3. Map of seismic coverage of the field showing individual 2-D lines and outlines of the 3-D surveys.

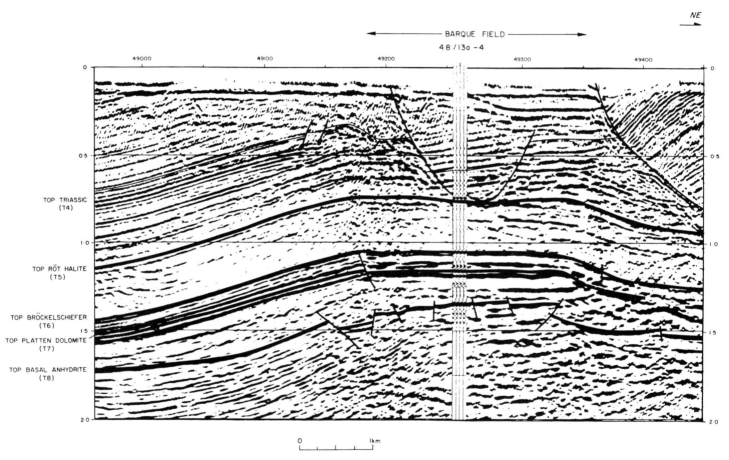

Fig. 4. Seismic line UK80-779 with synthetic seismogram well 48/13a-4 showing overburden faulting and well tie.

along fault zones. Appraisal well results indicate at least three compartments with free water levels about 8540, 8640 and 8770 ft TVSS separated by such sealing faults. These compartments are shown on the Top Rotliegend structure map (Fig. 1). The field is considered to be filled nearly to spillpoint.

Reservoir

Hydrocarbons in the Barque Field are reservoired within the Leman Sandstone Fm which here represents the entire Lower Permian Rotliegend Gp and are known as the Rotliegend after historic precedent. These sandstones are described as moderately mature quartz arenites (Nagtegaal 1979) deposited in a continental desert environment (Glennie et al. 1978). In the Sole Pit area the Leman Sandstone Fm. can be subdivided into three zones ('A', 'B' and 'C' from the top down) for reservoir description purposes. These subdivisions are based upon the proportions of major lithofacies which influence reservoir quality.

The major facies in the Barque Field are dune and interdune sandstones which display the best reservoir properties, fluvial conglomerates, sandstones and siltstones with intermediate properties, and lacustrine and sabkha deposits which generally form the poorest reservoir. The 'A' zone is commonly a mixture of these facies whereas the 'B' zone is predominantly dune sandstone. The basal 'C' zone contains little dune sandstone, being dominated by fluvial and sabkha facies.

Reservoir quality is not dependent upon facies alone. Deep burial of the Rotliegend prior to Cimmerian uplift resulted in compaction and advanced diagenesis, whereby both porosity and permeability were reduced to such an extent that matrix productivity was originally considered uneconomic. Reservoir quality and diagenesis have been described by Glennie *et al.* (1978), Nagtegaal (1979), Rossel (1982) and Seeman (1982), the prime cause of permeability reduction being the formation of diagenetic illite as pore-lining and pore-fill. Late-stage diagenesis involving authigenic quartz of Zechstein-derived cements, has a less-marked but significant effect.

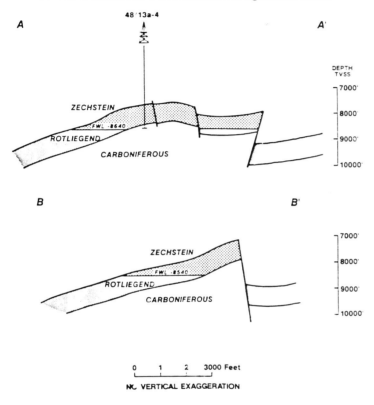

Fig. 5. Structural cross sections through reservoir showing zonation and change in character of the boundary fault.

Reservoir porosity commonly varies between 7 and 15% with occasional intervals below 6% (in cemented sabkha facies) or above 16% (in thick medium-grained dune sandstones). Air permeability varies greatly; dune sandstones may range from 0.1 md to over 100 md whereas wadi facies seldom exceed 5 md and sabkha facies rarely more than 2 md.

Well productivity would be low and probably non-commercial but for the presence of natural fracture zones at intervals within the field. Conventional core was cut in every well and detailed core studies by the joint-venture partners independently established likely models for these fracture systems, their occurrence and behaviour upon stimulation. Comparison with core material from the West Sole Field, 15 miles to the northwest, showed that the Barque Field is less extensively fractured. The reservoir is not considered to be pervasively fractured, hence some wells do not access effective fracture networks. When accessed either directly by the well or from the well bore via a massive hydraulic fracture, they allow high production rates to be achieved. Studies indicate that the extent of fracture development is adequate for commercial production.

Source

In the Sole Pit area, the underlying Carboniferous Coal Measures are the acknowledged source for the gas (Lutz *et al.* 1975) their deep burial leading to maturity and gas generation starting in the Jurassic and continuing at least until the Late Cretaceous uplift. The migration path was directly upward into the overlying Rotliegend. Late Tertiary uplift may have modified fill slightly through re-migration or elevation changes.

Hydrocarbons

Well-test data were summarized in an earlier section. Further details of test analyses for wells 48/13a-2 and 48/13a-5 are given by Moore (1989).

Extensive sampling of reservoir fluids in the Sole Pit area has established the gas composition and expected condensate production. A typical gas analysis from Barque appraisal well 48/13a-5 is given in Table 1. Gas gravity is 0.59 (air = 1) and the gross calorific value is 1019 BTU/SCF. Condensate production will be about 0.5 BBL/MMCF.

Formation water samples are highly saline (180 000–200 000 ppm equivalent) typical of the Rotliegend in this area. The reservoir is at hydrostatic pressure and no encroachment of the water leg is expected throughout field life. In the absence of aquifer support, water production will be limited except in situations where extensive fractures penetrate the water leg. Reservoir pressure is 3850 psi at the datum depth of 8200 ft TVSS and reservoir temperature is 175°F.

Table 1. *Gas composition*

Component	%
Methane	94.59
Ethane	2.73
Propane	0.49
Iso-butane	0.09
N-butane	0.11
Iso-pentane	0.04
N-pentane	0.04
Hexanes	0.05
Hetanes+	0.10
Nitrogen	1.36
Carbon dioxide	0.35

Reserves

Gas recovery will be by pressure depletion. A simulation model constructed for estimates of gas in place was based upon the top Rotliegend structure map, zone isochores, net to gross and porosity maps. Hydrocarbon saturation was derived from height saturation curves for representative porosity groups. Uncertainties associated with these parameters and also the distribution of free water levels led to a range of estimates based on probabilities. Recoverable reserves were derived by modelling the reservoir as a dual porosity/dual permeability system where wells access matrix only, matrix adjacent to fractures or directly to fractures.

The most likely gas in place estimate for the Barque Field initial development area is 674 BCF with 50% in the 'A' zone and 40% in the 'B' zone. Modelling gives an expectation of 316 BCF of recoverable reserves in this area with little recovery from the 'C' zone. Information from development wells and early production history will allow revision of reservoir parameters and more certain estimates of reserves both within the initial development area and over the remainder of the field.

Barque Field data summary

Trap
Type	Dip closure with anticlinal rollover against fault
Depth to crest	7500 ft TVSS
Lowest closing contour	8800 ft TVSS
Free water levels	8770, 8640, 8540 ft TVSS
Gas column	up to 1200 ft

Pay zone
Formation	Leman Sandstone (Rotliegend)
Age	Lower Permian
Gross thickness	700 to 800 ft
Net/gross ratio	0.76
Net sand cut-offs	porosity 6% (approx)
	gamma ray 60°API units (approx)
Porosity	11.1% average all zones
Gas saturation	51% average all zones
Matrix permeability	0.02–100 md
	average less than 1 md all zones

Hydrocarbons
Gas gravity	0.59
Gas type	sweet dry gas
Condensate yield	0.5 BBL/MMSCF

Formation water
Salinity	160 000 ppm chloride
	200 000 ppm sodium chloride equivalent
Resistivity	0.02 ohm m

Reservoir conditions
Temperature	175°F
Pressure	3850 psi at datum 8200 ft TVSS
Pressure gradient	0.08 psi/ft (gas leg)
	0.45 psi/ft (water leg)

Field size
Area	9000 acres (total field)
	4000 acres (initial development area)
Gas expansion factor	229 (SCF/RCF)
Gas-in-place	674 BCF (initial development area)
Drive mechanism	depletion
Recovery factor	47% (all zones)
Recoverable reserves	316 BCF (initial development area)

Production
First gas	October 1990 in conjunction with Clipper Field
Development scheme	Single wellhead jacket, installed April 1990
	two wells commissioned for first gas
	evacuation via Clipper to Bacton.
	12 wells planned for initial development.

This paper has been compiled from the work of many individuals within the operating company, Shell UK Expro, and partner Esso Expro UK and their respective research affiliates. Particular mention should be made of facies and fracture studies, special core analyses, well test analyses and simulation modelling carried out within both companies over a period of several years. The authors gratefully acknowledge the permission of Shell UK Expro Ltd and Esso Expro UK Ltd for publication of this material.

References

BUTLER, J. B. 1975. The West Sole Gasfield *In*: WOODLAND, A. W. (ed.) *Petroleum and the Continental Shelf of North West Europe, Volume 1, Geology*. Applied Science Publishers, London, 213–219.

FISHER, M. J. 1984. Triassic *In*: GLENNIE, K. W. (ed.) *Introduction to the Petroleum Geology of the North Sea*, Blackwell Scientific Publications, London, 85–101.

GLENNIE, K. W. 1984. Early Permian–Rotliegend *In*: GLENNIE, K. W. (ed.) *Introduction to the Petroleum Geology of the North Sea*. Blackwell Scientific Publications, London, 41–60.

—— & BOEGNER, P. L. E. 1981. Sole Pit Inversion Tectonics, *In*: ILLING, L. V. & HOBSON, D. G. (eds) *Petroleum Geology of the Continental Shelf of North West Europe*. Institute of Petroleum, London, 110–120.

——, MUDD, G. C. & NAGTEGAAL, P. J. C. 1978. Depositional Environment and Diagenesis of Permian Sandstones in Leman Bank and Sole Pit Areas of the UK Southern North Sea. *Journal of the Geological Society, London*, **135**, 25–34.

HORNABROOK, J. T. 1975. Seismic Interpretation of the West Sole Gas Field. Proceedings of the Bergen North Sea Conference, Norges Geologiske Undersøkelse, 121–135.

HOORN, B. VAN 1987. Structural Evolution, Timing and Tectonic Style of the Sole Pit Inversion. *Tectonophysics*, **137**, 239–284.

LUTZ, M., KAASCHIETER, J. P. H. & WIJHE, D. H. VAN 1975. Geological factors controlling Rotliegend gas accumulations in the Mid-European Basin. *Proceedings of the 9th World Petroleum Congress*, **2**, 93–103.

MARIE, J. P. P. 1975. Rotliegendes stratigraphy and diagenesis. *In*: WOODLAND, A. W. (ed.) *Petroleum and the Continential Shelf of North-West Europe, Volume 1, Geology*, Applied Science Publishers, London, 205–210.

MOORE, P. J. R. MCD. 1989. Barque and Clipper—Well Test Analysis in Low Permeability Fractured Gas Reservoirs. *Proceedings of the 1989 SPE Joint Rocky Mountain Region/Low Permeability Reservoir Symposium*, Paper No. SPE 18966.

NAGTEGAAL, P. J. C. 1979. Relationship of facies and reservoir quality in Rotliegendes Desert Sandstones, Southern North Sea Region. *Journal of Petroleum Geology*, **2**, 145–158.

ROSSEL, N. C. 1982. Clay Mineral Diagenesis in Rotliegend Aeolian Sandstones of the Southern North Sea. *Clay Minerals*, **17**, 69–77.

SEEMAN, U. 1982. Depositional Facies, Diagenetic Clay Minerals and Reservoir Quality of Rotliegend Sediments in the Southern Permian Basin (North Sea): a Review. *Clay Minerals*, **17**, 55–67.

TAYLOR, J. C. M. 1984. Late Permian–Zechstein *In*: GLENNIE, K. W. (ed.) *Introduction to the Petroleum Geology of the North Sea*. Blackwell Scientific Publications, London, 61–83.

The Camelot Fields, Blocks 53/1a, 53/2, UK North Sea

A. J. HOLMES

Mobil North Sea Ltd. Mobil Court, 3 Clements Inn, London WC2A 2EB, UK

Abstract: The Camelot Gas Fields (Camelot North, Northeast and Central-South) lie in Blocks 53/1a and 53/2 in the Southern North Sea, some 30 miles (48 km) east of Great Yarmouth. Initial sub commercial discovery wells were drilled in 1967, 1969 and 1972. Further exploration and appraisal drilling was carried out in 1987 and 1988. This paper covers the Field history up to the 53/1a-10 appraisal well in June 1988. The Lower Permian, Leman Sandstone Formation is the reservoir, with the gas accumulations trapped in tilted fault terraces. The Leman Sandstone Fm. in the Camelot area is 800 ft thick with a gas column up to 200 ft. Development of the fields will be in two phases. Phase I will consist of 5 wells deviated from the Camelot 'CA' platform to produce reserves from the Camelot North and Central-South Fields. Production commenced in October 1989. Phase II scheduled for 1991/92 will tie-in the Camelot Northeast Field. Gas is exported from the unmanned 'CA' gathering platform via pipeline to Amoco's Leman 'A' complex, from where the gas is transported to shore via the existing Amoco pipeline from Leman to Bacton. Total recoverable reserves for the Camelot Fields are estimated at 215 BCF.

The Camelot Field is the general name for a number of separate gas accumulations located in UK Blocks 53/1a and 53/2 in the Southern North Sea. The accumulations lie entirely within Mobil operated Blocks 53/1a and 53/2, some 30 miles Northeast of Great Yarmouth and approximately 9 miles SSW of the Leman gas field (Fig. 1). Water depth in the Camelot area is highly variable, ranging from 16 ft on the crest of 'Smiths Knoll' sand bank to 154 ft at its base. The reservoir of the Camelot Fields is the Leman Sandstone Formation of the Lower Permian, Rotliegendes Group (Rhys 1974). The various accumulations are structurally simple dip-fault closed horst, and tilted fault terraces (Fig. 2) with structural crests at approximately 6100 ft TVSS. The fields are named after Camelot, the fabled castle of King Arthur and his Knights of the Round Table. Total recoverable reserves in the Camelot Fields are 215 BCF.

History

Blocks 53/1 and 53/2 were awarded to Mobil North Sea Ltd in September 1964 as part of First Round Licence P.025. Mobil is the sole licensee with 100% interest. Block 53/1b was relinquished in 1970.

Exploration drilling commenced in Block 53/1 in October 1967. Well 53/1-1 encountered a 34 ft. gas column in the Leman Sandstone Fm. at a presumed crestal location on what had been mapped as a large NW–SE-trending anticline. This accumulation is now known as Camelot North. Wells 53/2-1 and 53/2-2 were completed in 1968 and proved the extension of the Leman Field into Blocks 53/1 and 53/2. Well 53/2-3, completed in 1969, was drilled on the mapped crest of a large NW–SE-trending faulted anticline and was presumed to be a further extension of the Leman Field. The well encountered 68 ft of gas in the Leman Sandstone Fm. but with a gas–water contact distinct from Leman. This accumulation is known as Camelot Northeast. Well 53/1a-4, drilled in 1972 on a northward dipping tilted fault terrace, encountered 33 ft. of gas in the Leman Sandstone Fm. This structure is now part of the Camelot Central-South Field.

The acquisition of higher quality seismic data in 1982, 1984 and 1987, together with revised depth conversion techniques, indicated untested structural relief updip from the early wells and additional prospects. Well 53/1a-5, drilled in 1987, encountered a 180 ft gas column on an anticlinal horst structure which is a separate culmination of the Camelot Central-South Field. The results from well 53/1a-5 precipitated an extensive cost effective, slim-hole appraisal programme of the earlier discoveries to confirm reserve estimates and to optimize development plans.

Five appraisal wells were drilled between September 1987 and March 1988. Well 53/1a-6 was drilled 6000 ft west, and updip, of 53/1a-4, encountering 181 ft of gas. Well 53/1a-7 encountered 30 ft of gas whilst appraising the western limb of the horst block tested by 53/1a-5. Well 53/1a-8, drilled as an appraisal of Camelot North, encountered 136 ft of gas. Current interpretation indicates that the well has encountered a separate closure to Camelot North. This accumulation is informally designated Camelot Northwest and is not included in the Camelot Development area at this time.

Wells 53/2-6 and 53/2-7 were drilled on the Camelot Northeast structure. Due to faulting, all horizons encountered in well 53/2-6 were water-wet, whereas well 53/2-7, drilled updip of 53/2-3, encountered 110 ft of pay.

In June 1988 well 53/1a-10 was drilled as a crestal appraisal of Camelot North. The well was drilled as a deviated well from the Camelot 'A' platform location and penetrated a 90 ft (TVT) gas column. This well was suspended for future production use.

Following the decision to develop the fields, the outlined development scheme was submitted to the UK Department of Energy in July 1988 and consent obtained in September 1988. Development will be in two separate phases. Core development (Phase I) comprises Camelot North and Central-South structures. This involves a single, unmanned 5-well gas gathering platform, Camelot 'A'. The platform is sited on the Smith's Knoll sand ridge in 43 ft (13 m) of water. The platform is equipped with free water knockout, metering and pigging facilities. Gas and condensate is exported via a 12 in diameter pipeline to Amoco's Leman 27A platform. All 5 production wells were drilled over the pre-installed jacket and suspended, for completion when production facilities became available. The platform will normally be unmanned with remote control telemetry links to Amoco's Bacton and Leman 27A facilities. This development came onstream in October 1989.

Phase II development will consist of the completion of suspended well 53/2-7 in Camelot Northeast. This will entail the installation of Camelot 'B', either a simple single well structure (e.g. a monopod with outrigger piles) or a subsea wellhead. Depending on technical and economic considerations in force at the time, Camelot 'B' will either tie directly into the export line to Leman or will tie-in to Camelot 'A' via a flowline and riser. This phase is scheduled to come onstream in October 1991 or 1992.

Field stratigraphy

The general stratigraphy of the Camelot area (Fig. 3) is similar to the previously described succession in the Southern North Sea Basin (e.g. van Veen 1975, Glennie, 1984a, b).

From Abbotts, I. L. (ed.), 1991, *United Kingdom Oil and Gas Fields,
25 Years Commemorative Volume*, Geological Society Memoir No. 14, pp. 401–408

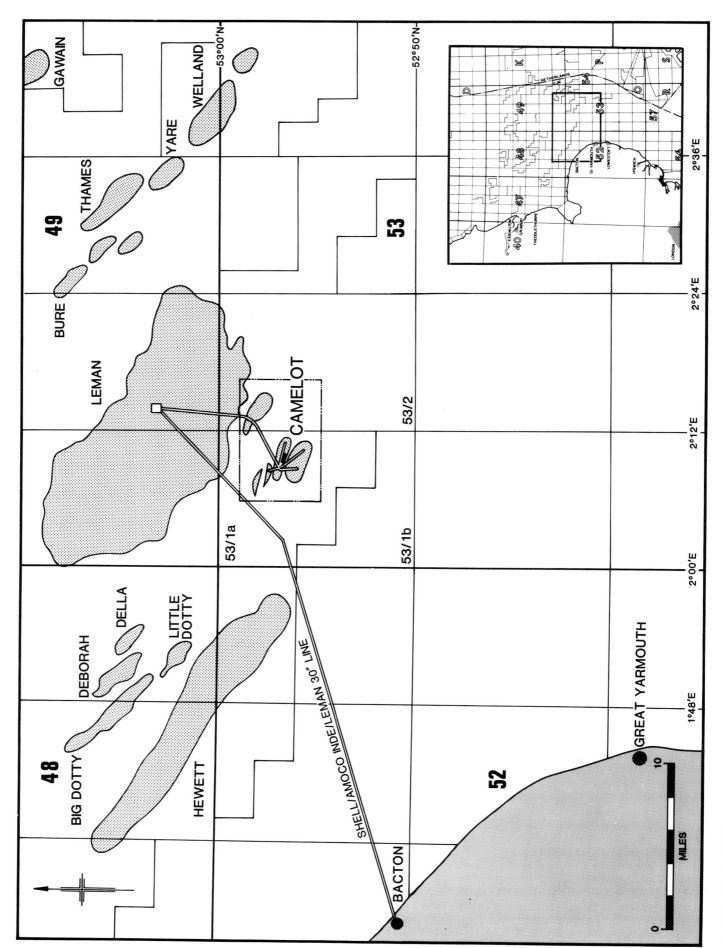

Fig. 1. Camelot Gas Fields: location map and development plan.

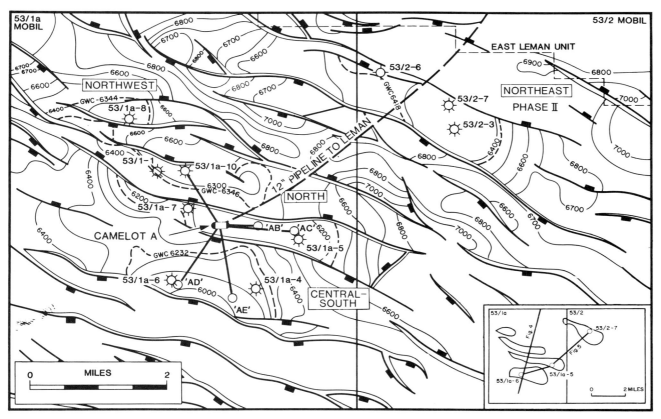

Fig. 2. Top reservoir depth map of the Camelot area.

Fig. 3. Generalized stratigraphy of the Camelot area.

The oldest rocks penetrated in the Camelot area are Late Carboniferous, Westphalian 'B' sediments, representing a fluvio-deltaic sequence of sandstones, siltstones, shales and coals. Extensional Variscan tectonic stresses and resultant NW–SE-trending right-lateral wrench faults, led to the development of a block faulted Permian basin.

In the Camelot area, the Lower Permian Rotliegendes Gp is represented by the Leman Sandstone Fm., which lies unconformably on the peneplained Carboniferous surface and is approximately 800 ft thick over Blocks 53/1a and 53/2.

During the Late Permian, the Camelot area lay towards the southwestern margin of the Zechstein Basin (Taylor 1984), where shelf and shelf edge facies are present and halite development is minor. The total Zechstein Supergroup is 700 to 1000 ft thick with cycles I to III and possibly IV present. The Zechstein Gp is a highly efficient seal to the Leman Sandstone Fm. reservoir.

The Bacton Group sediments of the Triassic are 2000 ft thick and relatively uniform across the Camelot area, consisting of 800 ft of mudstones of the Bunter Shale Fm. (including the siltstones of the Bröckelschiefer Member), overlain by 1200 ft of fluvial Bunter Sandstone Fm. The succeeding Haisborough Group thickens to the NE towards the Sole Pit Basin Axis and is 2200 ft thick in well 53/2-7. The latest Triassic sediments, the Rhaetian Winterton Fm., represent a widespread marine transgression, conformably overlain by Lower Jurassic, Liassic, marine claystones and marls. Preserved thickness of the Liassic sediments increases from 200 ft in the northeast to 700 ft in the southwest. This reflects the influence of Late Jurassic–Early Cretaceous uplift. Late Cimmerian tectonism during this time was taken up by rotation and pull-apart along and between the Dowsing and South Hewett Faults. Accompanying this extension was a regional uplift, which is considered to be a thermally induced event, centred on an axis northeast of the South Hewett Fault (Badley et al. 1989). The resulting Base Cretaceous unconformity erodes progressively deeper to the northeast across the Camelot area (Fig. 4). Only thin Upper Cretaceous (Maastrichtian) Chalk is preserved over Camelot with intra-Chalk reflectors onlapping onto the Base Cretaceous unconformity. Sole Pit inversion to the northeast took place from Late Cretaceous time through

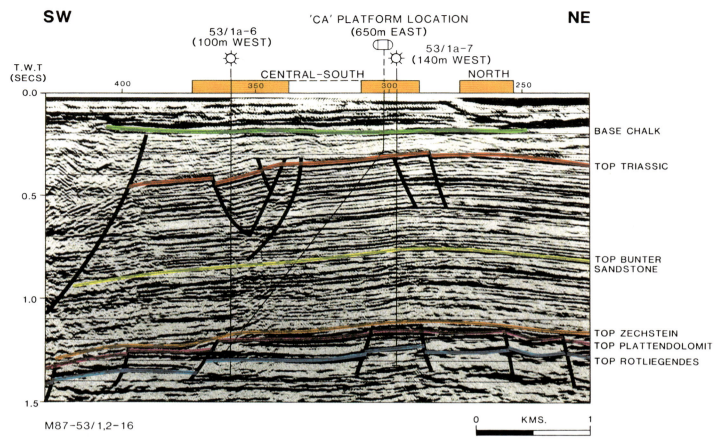

Fig. 4. Seismic line M87-53/1,2-16 (see Fig. 2 for location).

Table 1. *Seismic database*

Year	Acquisition	Fold	km	Quality	Comments
1981	SSL	24	180	Fair	Tails of Leman lines; Amoco processed
1982	Western	48	52	Good	CGG processed; Mobil re-migrated
1982	GECO	48	570	Good	
1984	Mobil	48	230	Good	
1987	Mobil	24	184	Excellent	Western processed
1988	Western	30	18	Excellent	

to the mid-Tertiary in two main pulses (van Hoorn 1987) resulting in low angle onlap of the Upper Cretaceous Chalk by a thin (400–600 ft) Tertiary/Quaternary sequence.

Geophysics

Seismic data covering Blocks 53/1a and 53/2 was acquired from the late 1960s to 1988. Post-1980 data were utilized for the interpretations resulting in the 53/1a-5 wildcat and subsequent appraisal drilling. The seismic database is summarized in Table 1.

The data are largely orientated NE–SW perpendicular to the dominant structural trend. A number of strike lines were acquired and several lines from the 1987 and 1988 survey cut across these orientations to deal with particular fault interpretation problems.

The seismic data form a grid of approximately 0.75 × 0.75 km over the main accumulations and 1 × 2 km elsewhere. Data quality of the 1987 and 1988 surveys is excellent. The major problem with the pre-1987 data is tuning within the Zechstein Gp limiting the Rotliegendes resolution by destructive interference. The 1987 data has spike deconvolution applied before stack, together with predictive deconvolution after stack to decrease the wavelet length and optimize resolution.

Throughout the data set five reflectors were mapped, representing major acoustic interfaces, to aid multi-layer depth conversion. These reflectors were: Base Chalk; Top Triassic; Top Bunter Sandstone; Top Zechstein; Top Plattendolomit; Top Rotliegendes (Fig. 4). A layer-cake depth conversion method was used to account for lateral velocity variation in the post-Rotliegendes sequence. This variation is critical in computing the depth configuration of the low relief Camelot structures. Thickness variations in the low velocity Jurassic interval and high velocity Zechstein intervals, especially across faults, have a significant impact on the time-depth conversion. A seven layer velocity model was used with interval velocities derived from wells only.

Interpretation was carried out on migrated time sections with selected well tie lines depth migrated. Increasing thickness of the low velocity Jurassic sequence modifies the dip in the time of lower horizons. Consequently time migration does not reposition reflectors in their current spatial location. The depth migration technique has been incorporated in the positioning of key faults on the Top Rotliegendes Fm. structure map (Fig. 2).

SW   NE

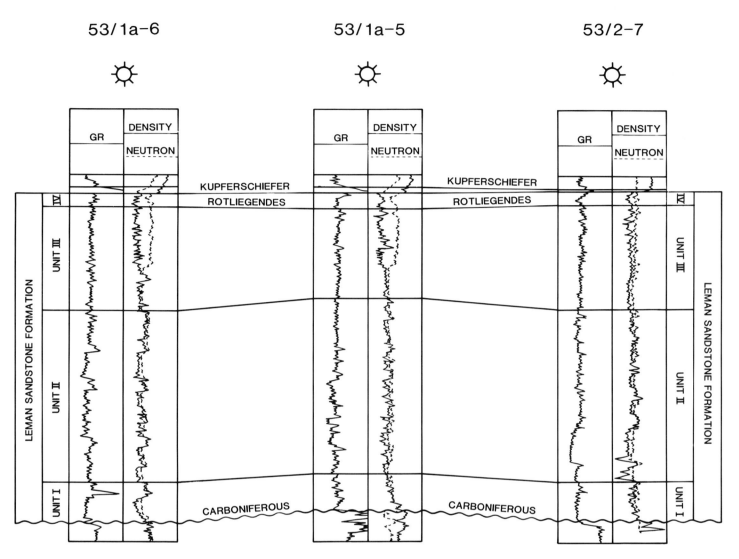

Fig. 5. Wireline correlation section showing principal reservoir units (see Fig. 2 for location).

Accurate picking of the Top Rotliegendes event is not straightforward as it can be an interference event. A 5 ms miss-pick at Top Rotliegendes represents 50 ft in depth and is a significant error given the relatively low relief of the structures. Synthetic seismograms and VSPs has greatly assisted in picking the Top Rotliegendes event.

Trap

The Camelot Gas Fields occur in dip and fault closed horst blocks and fault terraces elongated along a WNW–ESE trend. The dimensions range from 2.2–2.6 miles long (3.5–4.2 km) and 0.3–0.9 miles wide (0.5–1.5 km).

The accumulations are sealed laterally across the bounding faults against downthrown Zechstein Gp evaporites and dolomites. Although predominantly planar normal faults, there is evidence of a wrench component. In Camelot North, the field bounding fault to the north reverses its sense of throw along its length from southeast to northwest, throwing down to the north at its eastern limit and to the south at its western limit. To the east, the accumulation is dip closed. In the zone of inflection, i.e. where there is sandstone-to-sandstone contact in the gas column, hydrocarbons leak across the fault zone and the accumulation is then dip closed to the north.

Initiation of growth on the structures probably occurred in the Upper Permian along pre-existing Carboniferous trends with intermittent movement on these faults continuing throughout the Zechstein and Triassic. It is difficult to quantify the extent of movement on these faults as few of them have any expression in the post-Zechstein succession. Final structural configuration was not achieved until after the inversion of the Sole Pit Axis to the northeast (Late Cretaceous–mid-Tertiary).

The accumulations appear to be full to mapped spill points. Well defined gas–water contacts from well logs and pressure data are apparent in each separate accumulation. These indicate a stepwise shallowing of the gas–water contacts from north to south from 6418 ft TVSS in Camelot Northeast to 6232 ft TVSS in Camelot Central-South.

Reservoir

The Leman Sandstone Formation of the Lower Permian Rotliegendes Group (Rhys 1974), is the hydrocarbon-bearing reservoir in the Camelot Fields. The gross thickness of the Leman Sandstone Fm. is approximately 800 ft, with a net-to-gross ratio of >0.9. Hydrocarbons are confined to the upper 200 ft of the formation where the net-to-gross ratio is effectively 1.

The Leman Sandstone Fm. predominantly comprises a series of stacked aeolian dune sandstones, with a marine influenced upper unit.

Four lithofacies have been described from core data in the upper 300 ft of the Leman Sandstone Fm. in the Camelot area. They are: massive sandstones; dune top sandstones; dune base sandstones; interdune sandstones. A fifth lithofacies, fluvial/wadi sandstones, is inferred from wireline log character (Fig. 5) and regional correlations.

The uppermost lithofacies is typically a light grey sandstone, massive or faintly cross-bedded. These sandstones are identical petrographically to the underlying aeolian dune top sandstones excepting the absence of haematite stained grain rimming clays. They are interpreted as a preserved dune sequence with the faint cross bedding being consistent with dune foresets. The majority of the primary sedimentary structures have been destroyed by in situ degassing/fluidization following the Zechstein marine transgression (Glennie & Buller 1983). There is no evidence to suggest these sandstones have been subject to any marine re-working.

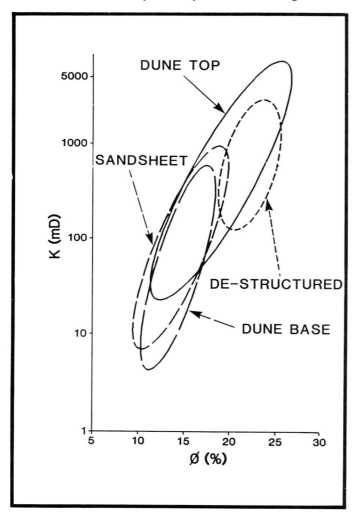

Fig. 6. Core porosity–permeability relationship in individual lithofacies.

The principal hydrocarbon-bearing lithofacies is the dune-top sandstone of the stacked aeolian dune sequence. These are medium–coarse-grained, well-sorted sandstones with high angle (>15°) planar cross bedding. The upper contacts of the sandstones are sharp, reflecting truncation by interdune sandsheets. Lower contacts grade to a dune base sandstone lithofacies. The dune base sandstones are very fine–fine grained, moderate–poorly-sorted with low angle (<15°) planar cross bedding. Upper and lower contacts of these sandstones are gradational to the dune top sandstones and the interdune sandsheets respectively.

Interdune sandsheets, the fourth lithofacies identified from core, consist of red-brown, fine-medium-grained sandstones with low angle (<8°) planar bedding, generally bimodally laminated with thin lenses of coarser sandstone. This facies represents deposition on a low relief interdune surface. Lack of adhesion ripples suggests deposition onto a dry surface with a relatvely low water table.

The Leman Sandstone Fm. in the Camelot Fields has been subdivided into four reservoir units (modified after van Veen 1975). The lowermost Unit I is interpreted as dominantly fluvial/wadi sandstones on log character and regional correlation.

Unit II is an association of the sandsheet, dune top and dune base sandstone lithofacies described above. There is a lower proportion of stacked aeolian dunes than Unit III above. The upper boundary of this Unit is a prominent sandsheet which is mappable field-wide.

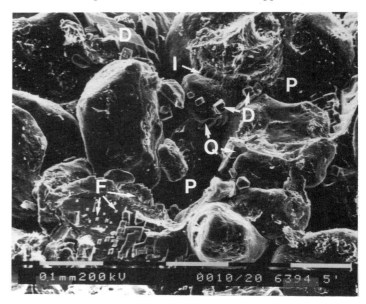

Fig. 7. SEM dune-top sandstone lithofacies. (I, fibrous illite; Q, authigenic quartz; D, ferroan dolomite; F, authigenic K-feldspar; P, intergranular pore).

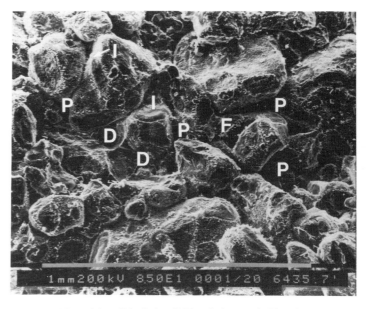

Fig. 8. SEM sandsheet lithofacies. (Abbreviations as in Fig. 7).

It is 50–100 ft thick with associated thin minor dunes. Porosity–permeability characteristics of this sandsheet lithofacies though still good, are relatively lower than the gas-bearing sandstones above. The average porosity of the sandsheet is 16%. In the stacked aeolian dunes above it is 18.7%.

Unit III is the principal hydrocarbon-bearing interval, representing a 300 ft thick stacked aeolian dune sequence of predominantly dune top sandstones with minor dune and sandsheet sandstones.

Unit IV averages 40 ft in thickness and is present in all wells. This is the de-structured unit. Unlike other areas of the southern North Sea, this upper unit is not adversely affected by Zechstein sourced evaporite or carbonate cements.

The core porosity–permeability relationship for the various lithofacies is shown in Fig. 6. There is a high degree of overlap between facies. Overall, the reservoir quality of the Leman Sandstone Fm. in the Camelot Fields is excellent. Dune base and sandsheet sandstones exhibit the lowest porosity and permeability reflecting their fine-grained, laminated nature. Dune top sandstones record the widest variation in porosity and permeability reflecting the marked localized variations in grain size and distribution of grainfall laminae.

The primary control on porosity and permeability is the depositional facies. The principal diagenetic blocky cements (K-feldspars, ferroan dolomites, anhydrite and quartz) have occluded some porosity but not seriously impaired permeability (Fig. 7). Even then, primary depositional control is exerted on the ability of blocky diagenetic phases to reduce porosity e.g. lamination style (grain size, sorting, grain packing) has a direct effect on porosity–permeability relationships. Although grain size and porosity in a dune top sandstone and a sandsheet sandstone may be similar, the

Camelot Field data summary

Trap

Type	Fault terraces, tilted horst blocks		
Depth to crest	North	Central-South	Northeast
	6225 ft TVSS	6050 ft TVSS	6300 ft TVSS
Lowest closing contour	6418 ft TVSS (Northeast)		
Gas–water contacts	6346 ft TVSS	6232 ft TVSS	6418 ft TVSS
Gas column	200 ft		

Pay zone

Formation	Leman Sandstone
Age	Lower Permian
Gross thickness	120–200 ft
Net/gross ratio (average/range)	1.0/0.98–1.0
Cut-off for N/G	8%
Porosity (average/range)	18%/15–21%
Hydrocarbon saturation	Max 80%
Permeability (average/range)	200 md/5–5000 md
Absolute open flow potential	129 MMSCFD

Hydrocarbons

Gas gravity	0.615 relative to air
Condensate yield	1.2 BBL/MMSCF
Gas expansion factor	192 SCF/RCF

Formation water

Salinity	180 000 ppm
Resistivity	0.025 ohm m at 150°F

Reservoir conditions

Temperature	150°F
Pressure	2850 psia at 6200 ft TVSS
Pressure gradient in reservoir	
gas	0.07 psi/ft
water	0.49 psi/ft

Field size	North	Central-South	Northeast
Area	500 acres	1500 acres	740 acres
Gross rock volume	32 200 ac-ft.	191 000 ac-ft.	41 600 ac-ft.
Hydrocarbon pore volume	110×10^6 ft^3	1250×10^6 ft^3	220×10^6 ft^3

Recoverable gas	215 BCF
Recovery factor	70%
Drive mechanism	Water drive

Production

Start-up date	October 1989 (Phase I)
	October 1991/92 (Phase II)
Development scheme	1 platform, 5 wells (Phase I)
Daily contract quantity	70 MMSCFD
Maximum contract deliverability	117 MMSCFD
Cumulative production to date	N/A

permeability map be lower in the sandsheet facies due to smaller pore sizes in the closely spaced laminae (Fig. 8). Blocky cements occlude the smaller pores in these laminae more extensively than the larger pores. Absence of significant illitization of clay minerals has preserved the permeability relationships inherent in the depositional facies.

Source

The Westphalian Coal Measures are the inferred source of the gas in the Camelot Fields. Gas was generated from Late Jurassic until Tertiary times in the adjacent Sole Pit Trough. Migration of gas from the Carboniferous coals into the Leman Sandstone Fm. took place along the conduits provided by the extensional faulting.

Hydrocarbons

Of the post 1970s wells in the Camelot Fields, 53/1a-5 and 53/2-7 were drill stem tested. The 53/1a-5 well was perforated in the top 40 ft. of the gas column and flowed 37 MMSCFD on a 1 in choke. Well 53/2-7 perforated the upper 30 ft of the gas column and flowed 49 MMSCFD on a 80/64 in choke.

The gas in the Camelot Fields is a dry gas with a molecular composition as follows: C_1, 90.7%; C_2, 4.1%; C_3, 1.0%; C_4, 0.4%; N_2, 2.4%; CO_2, 0.1%; C_5+, 1.3%. The gas has a condensate yield of 1.2 BBL/MMSCF. Reservoir temperature is 66°C and the average initial reservoir pressure over the accumulations is 2850 psia at a reference depth of 6200 ft TVSS.

Reserves

Original gas-in-place for the three Camelot accumulations is estimated at 304 BCF. Ultimate recoveries for each field were modelled using a 3-D simulation model. A moderate water drive would represent an ultimate recovery of approximately 70%.

No vertical permeability barriers are apparent in the gas column or the aquifer. However, the increased proportion of laterally extensive lower permeability sandsheets in Unit II may provide partial barriers or 'baffles', to water influx, increasing ultimate recovery.

The author has drawn extensively on the work and thoughts of members of the Producing and Exploration Departments of Mobil North Sea Limited. The author acknowledges K. P. Dean, A. Dodds, T. L. Koonsman, T. M. Levy, A. C. McArthur, A. C. Pierce and J. M. Reilly for their contributions. SEM photographs were provided by Poroperm-Geochem Ltd.

References

BADLEY, M. E., PRICE, J. D., & BACKSHALL, L. C. 1989. Inversion, reactivated faults and related structures: seismic examples from the southern North Sea. *In*: COOPER, M. A. & WILLIAMS, G. D. (eds) *Inversion Tectonics* Geological Society, London, Special Publication, **44**, 201–219.

GLENNIE, K. W. 1984*a*. Early Permian–Rotliegendes. *In*: GLENNIE, K. W. (ed.) *Introduction to the Petroleum Geology of the North Sea*. Blackwell Scientific Publications, Oxford. 41–60.

—— 1984*b*. The Structural Framework and the Pre-Permian History of the North Sea Area. *In*: GLENNIE, K. W. (ed.) *Introduction to the Petroleum Geology of the North Sea*. Blackwell Scientific Publications, Oxford. 17–39.

—— & BUTLER, A. T. 1983. The Permian Weissliegend of NW Europe: The partial deformation of aeolian dune sands caused by the Zechstein transgression. *Sedimentary Geology*, **35**, 43–81.

RHYS, G. H. (Compiler) 1974. *A proposed standard lithostratigraphic nomenclature for the Southern North Sea and an outline structural nomenclature for the whole of the (U.K.) North Sea*. A report of the joint Oil Industry–Institute of Geological Sciences Committee on North Sea Nomenclature. Report of the Institute of Geological Sciences, 74/8.

TAYLOR, J. C. M. 1984. Late Permian–Zechstein. *In*: GLENNIE, K. W. (ed.) *Introduction to the Petroleum Geology of the North Sea*. Blackwell Scientific Publications, Oxford, 61–83.

VAN HOORN. 1987. Structural evolution, timing and tectonic style of the Sole Pit inversion. *Tectonophysics*, **137**, 239–284.

VAN VEEN, F. R. 1975. Geology of the Leman Gas-Field. *In*: WOODLAND, A. W. (ed.) *Petroleum and the Continental Shelf of North West Europe*. Applied Science Publishers, Barking, 223–231.

The Cleeton Field, Block 42/29, UK North Sea

R. D. HEINRICH

BP Exploration, 301 St Vincent Street, Glasgow G2 5DD, UK

Abstract: The Cleeton Gas Field is located in the Sole Pit Basin in the Southern North Sea in UK Block 42/29. The gas is trapped in sandstones of the Lower Permian Lower Leman Sandstone Formation, which was deposited by wind and occasional fluvial action in a desert environment. In contrast to nearby Ravenspurn South, the sands have excellent reservoir properties, particularly in the aeolian sandstones, with porosities around 22% and permeabilities between 10 and 100 md. The trap is a NW–SE-striking faulted anticline, in which the top seal is provided by the early Permian Silverpit shales directly overlying the reservoir. The field has been producing since October 1988 and its use in the Villages Field Project is as a peak shaving producer to the Ravenspurn South Field main output. The initial reserves are 280 BCF and the field life is expected to be 9 years.

The Cleeton Gas Field is located 50 km east of Flamborough Head on the Humberside coast, in the United Kingdom sector of the Southern North Sea (Fig. 1). The field covers an area of 7.4 km² (1829 acres), in Block 42/29. The water depth in this area is about 55 m (180 ft.). Lower Permian, aeolian, sabkha and fluvial sandstones constitute the reservoir. The field is 5 miles southwest of the Ravenspurn South Gas Field: both fields are named after now-drowned villages formerly located on the Humberside coast.

History

Production Licence P001, which included Blocks 42/29 and 42/30, was awarded in 1964; this licence is wholly owned by BP.

Two wells were drilled in the Cleeton field prior to development. The first, well 42/29-2 was drilled in 1983, with the objective of testing the Cleeton structure. The well encountered a 54 m (177 ft) gas column, and on test flowed at a commercial rate of 40

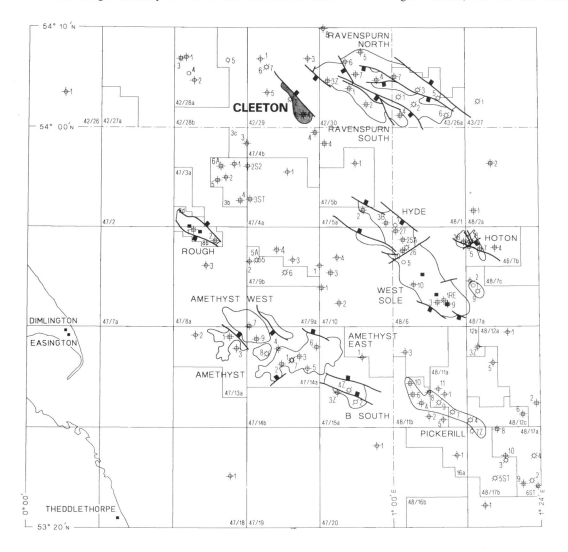

Fig. 1. Location of the Cleeton Gas Field in UK Block 42/29.

MMSCFD without hydraulic fracturing being required. In 1984 well 42/29-4 was drilled to assist in delineating the field extent. The well encountered a gas column of 55 m (180 ft) and again flowed at a commercial rate.

Until 1984 seismic data over the field comprised various 2-D seismic surveys, which combined to give an average line spacing of approximately 0.5 km over the field. In 1984 a 3-D seismic survey was acquired by Western Geophysical, covering the whole field. Since its acquisition this has been the definitive seismic data set for interpretation of the field structure. Much of the survey was reprocessed in 1987.

The Annex B document (covering both Cleeton and Ravenspurn South) was submitted to the Department of Energy in 1984 and approved in 1986. Development drilling commenced in 1987 and was completed in 1988: 5 production wells have been drilled, 24/29-C1, C2Z, C3, C4 and C5 (C2Z is a sidetrack of the unsuccessful 42/29-C2 well). The recovery mechanism is pressure depletion with edge-water in-flow (none of the wells has been hydraulically fractured). First gas flowed in October 1988. Production facilities are one unmanned wellhead tower, bridge-linked to the Process/Quarters platform serving the entire Villages development. The gas is piped ashore to the Dimlington terminal.

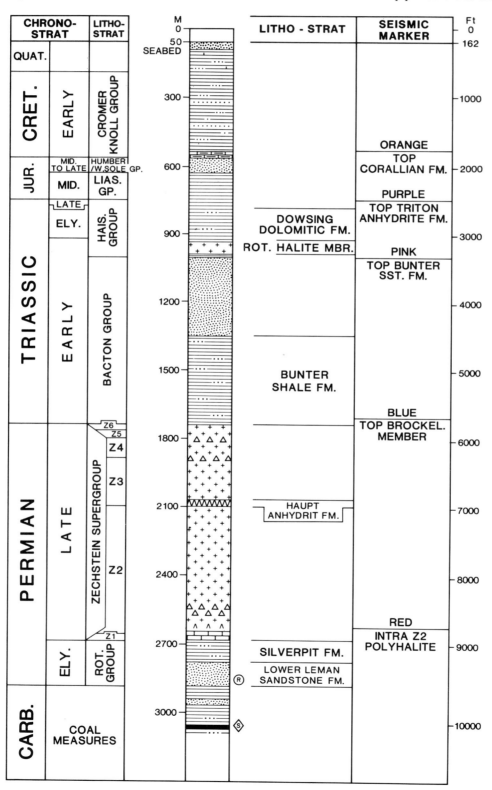

Fig. 2. Generalized stratigraphic column of the Cleeton Gas Field.

Field stratigraphy

Figure 2 shows the lithology of the main stratigraphic units in the Cleeton area.

The basin has been subjected to a number of tectonic events. The Hercynian orogenic events, which occurred in the Late Carboniferous, resulted in the folding, faulting and uplift of Carboniferous strata along the NW–SE structural grain of the future Sole Pit Basin. Uplift resulted in erosion down to Westphalian B levels over Cleeton.

During the later Early Permian E–W tension, associated with oblique-slip fault movements along Hercynian basement faults, led to the initiation of subsidence of the Sole Pit Basin.

In the Cleeton area, Early Permian Rotliegendes sandstones (Lower Leman Sandstone Formation) and mudstones (Silverpit Formation) were deposited upon an irregular Carboniferous unconformity surface. The deposition of these sediments was followed by the continued subsidence of the Sole Pit Basin and the deposition of the Zechstein, Bacton, Haisborough, Lias and West Sole Groups. Mid to Late Jurassic rifting associated with the Mid-Cimmerian tectonic event initiated NW–SE-trending normal faults within the Jurassic and Triassic section.

End Jurassic to Early Cretaceous inversion related to Late Cimmerian tectonism resulted in erosion down to Mid-Jurassic levels over Cleeton. After this phase of uplift, the Cromer Knoll and Chalk Groups were deposited on the resulting unconformity surface during a period of renewed basin subsidence.

The Tertiary was characterized by successive periods of uplift, which began at the end of the Cretaceous and culminated in the main phase of inversion in the Miocene.

Geophysics

Until 1984, mapping of the Cleeton field was based on several vintages of seismic data, constituting approximately a 0.5 km × 0.5 km grid. In 1984 to 1985 a 3-D (three dimensional) seismic survey was shot and processed by Western Geophysical. This survey consists of about 360 SW–NE oriented lines, 37.5 m apart. The Cleeton field has a structurally complex overburden, with large lateral velocity contrasts. This results in poorly imaged seismic reflections at the reservoir level. Extensive seismic reprocessing in conjunction with raytrace modelling was undertaken in 1987, but even with this work it has proved impossible to map the top reservoir structure directly from the seismic data. Instead, a seismic reflector corresponding to the Near Top Z2 Polyhalite has been mapped, and an isochore added to this to yield a top reservoir structure map. An example seismic section across the field is shown in Fig. 4; a contour map of depth to top reservoir is shown in Fig. 5, and a geoseismic interpreted section in Fig. 6.

Trap

The Cleeton field is a structurally closed NW–SE-striking faulted anticline, probably formed as an extensional feature during the

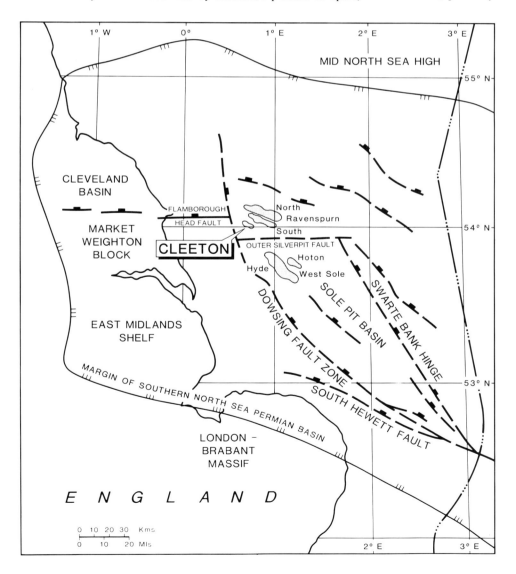

Fig. 3. Structural elements of the Southern North Sea Permian Basin.

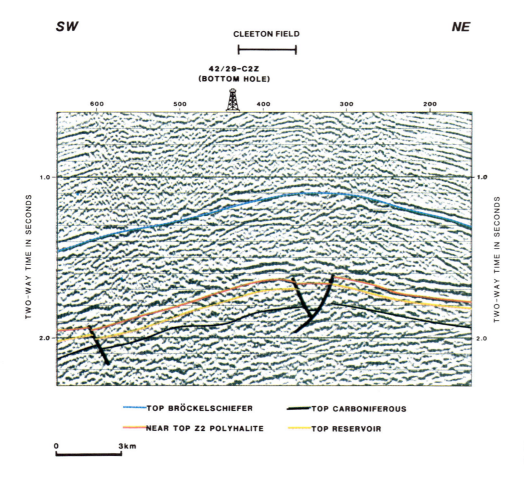

Fig. 4. Seismic section across Cleeton Gas Field; location shown in Fig. 5.

Late Permian to Late Jurassic with subsequent inversion during the Tertiary. The dominant fault trend is NW–SE. The structural relief at top reservoir level (top Lower Leman Sandstone Formation) is approximately 80 m (260 ft) with crest at 9088 ft and the deepest closing contour at 9344 ft. The structure is sealed by the directly overlying Silverpit Shale Formation, but the ultimate regional seal is provided by Zechstein evaporites.

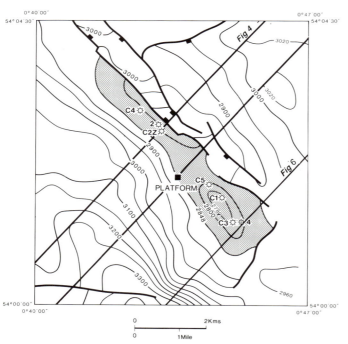

Fig. 5. Contour map of depth to top reservoir (m).

Reservoir

The sandstone reservoir of the Cleeton Field belongs to the Lower Leman Sandstone Formation of Rotliegendes Group. The reservoir thickness varies from approximately 60 to 120 m (200–400 ft) across the field. This is partly due to a progressive shale-out to the NW at the top of the reservoir, and partly to local onlap onto the post-Carboniferous unconformity.

The reservoir comprises a mixture of sabkha (sandy and muddy), fluvial and aeolian facies deposited in an arid to semi-arid environment (see Fig. 7).

Deposition of the Lower Leman sands in the Cleeton area was controlled primarily by changing climatic conditions resulting in 'drier' and 'wetter' periods. Aeolian dunes developed and migrated under prevailing E to NE winds, which resulted in the reworking of mainly fluvial sandstones. During Lower Leman Sandstone times, the southeastern part of the field clipped the NW edge of a major aeolian dune field which extends southeastwards into Block 47/5. Passing laterally north-westwards across the field, the percentage of aeolian sandstone decreases, whilst percentage of fluvial and sabkha sediments increases.

Fluvial deposition was controlled by seasonal rainfall, which resulted in alluvial fans building out from the Dowsing Fault Zone and higher ground to the SW of Cleeton. Sabkha sedimentation developed as a result of water table movement and periodic flooding associated with the gradual southward encroachment of the Silverpit Lake.

The different facies types vary in texture and diagenetic characteristics, which influence the poroperm properties. Diagenetic illite depends largely on the amounts of detrital clay in the original sediment. Mineral cementation (namely by quartz and ferroan dolomite) tends to be less in the aeolian sandstones than in the other facies types.

Aeolian dune sandstones provide the best reservoirs. Porosities are typically around 22%. Horizontal permeabilities range usually

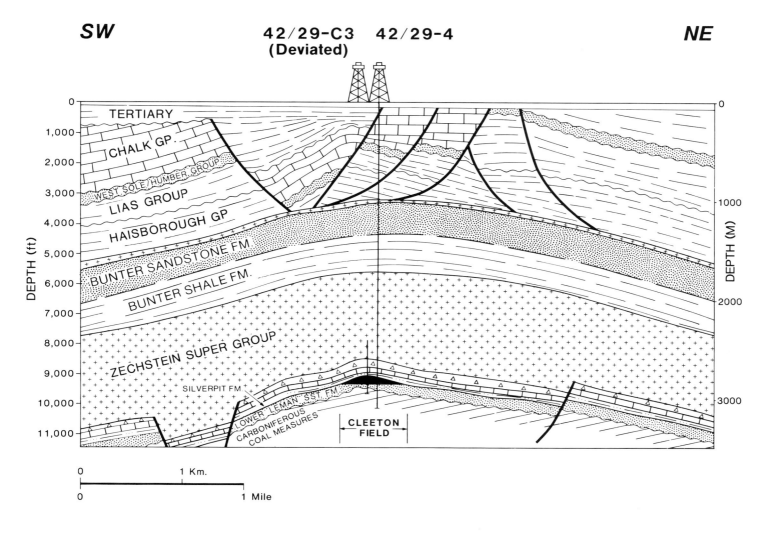

Fig. 6. Interpreted geoseismic section across Cleeton Field.

between 10 and 100 md. Sabkha and fluvial sandstones are more poorly sorted, have finer grain sizes, a higher detrital clay content and a stronger mineral cementation. The typical porosities in these sandstones are between 11 and 15%, the permeabilities between 1 and 10 md.

Reservoir qualities in other 'Villages' gas fields are strongly controlled by diagenetic illite and mineral cements. This is much less the case in Cleeton, mainly for two reasons. (i) Over most of the field kaolinite is the main authigenic clay as opposed to permeability-destroying illite. (ii) Gas emplacement was from the Mid-Jurassic onwards and continued until the Tertiary. This coincides with the critical phases of diagenesis, so that cementation was strongly reduced in the gas leg and poroperm properties were retained.

Source

The source of Cleeton gas was provided by the Late Carboniferous Coal Measures that directly subcrop the Permian reservoir. Maturation and migration took place from the Mid Jurassic onwards and ended with the Tertiary inversion and uplift. Apparently the latter movements did not cause significant changes to the shape or size of the trap as there is no indication of any significant gas-remigration or of changes in the gas–water contact as inferred from clay and mineral-cementation trends.

Hydrocarbons

The Cleeton gas field has a dry gas with the following mol% composition:

Component	Mol%
N_2	1.33
CO_2	0.45
He	0.05
C1	91.55
C2	4.79
C3	0.93
iC4	0.16
nC4	0.21
iC5	0.07
nC5	0.06
C5+–C6	0.08
C6+–C7	0.06
C7+–C8	0.05
C8+–C9	0.04
C9+	traces

The gas viscosity is 0.022 cP at reservoir temperature of 79.4°C and reservoir pressure of 286.93 bar at a depth of 2847.9 m (9343.4 ft). The gas compressibility factor (Z) is 0.947. The approximate

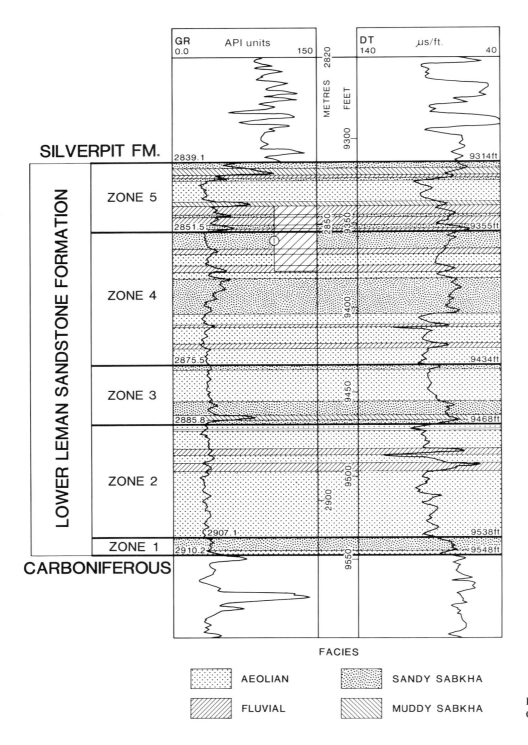

Fig. 7. Zonation and facies of the Cleeton Field reservoir interval.

condensate gas ratio is 0.399 dm³/kmol (3STB/MMSCF). The gravity of the condensate is 45°API.

The accumulation was originally under hydrostatic pressure conditions.

Reserves

The Cleeton and Ravenspurn South gas fields form the Villages project. Cleeton has very high productivity and low reserves. Ravenspurn South contains most of the reserves but has low productivity. Ravenspurn South will be produced preferentially. During peak demand over the winter months, Cleeton will be produced to meet any shortfall from Ravenspurn South. During the first year of production Cleeton was providing all the gas from Villages whilst wells were being drilled on Ravenspurn South and platforms were being commissioned.

The initial gas-in-place of the Cleeton field is 356 BCF, with estimated initial reserves of 280 BCF. Production from the field started in October 1988. Cumulative production to 31 January 1991 was 102 BCF of dry gas and 0.268 MMBBL of condensate. There are five production wells on the Cleeton gas field with productivity being limited by facility constraints. The gas field is underlain by an aquifer.

Cleeton Field data summary

Trap
Type   Dip and fault-closed anticline
Depth to crest (Top Lower
  Leman Sandstone)   2770 mSS (9088 ft)
Gas–water contact   2848 mSS (9344 ft)
Vertical closure   78 m (256 ft)

Pay zone
Formation   Lower Leman Sandstone
Age   Permian
Thickness   60–120 m (average 92 m)
                                                  (200–400 ft, average 300 ft)
Net/gross   89–100% (average 95%)
Porosity   15–20% (average 18%)
Hydrocarbon saturation   81–86% (average 83%)
Permeability   40–100 md (average 95 md)

Hydrocarbons
Gas composition   92% C1, 6% C2
Gas expansion factor   244
Initial condensate/gas ratio   2.6 BBL/MMSCF

Formation water
Water gradient   1.6 psi/m
Salinity   214 000 ppm chlorides
Resistivity   0.0195 ohm m
Other ions   N/A

Reservoir conditions
Temperature   175°F
Initial pressure   4147.8 psig

Field size
Area   7.4 km^2 (1830 acres)
Recovery factor   78.7%
Recoverable gas   280 BCF
Drive mechanism aquifer   Pressure depletion with some support

Production
First gas   1 October 1988
Development scheme   1 central platform
Number/type of wells   5 gas producers
Average production rate   80 MMSCFD (capacity for much
                                                  higher rate during peak demand)
Cumulative production   At end at Jan 1991: 102 BCF dry gas,
                                                  0.268 MMBBL condensate
Secondary recovery methods   None

The Clipper Field, Blocks 48/19a, 48/19c, UK North Sea

R. T. FARMER[1] & A. P. HILLIER[2]

[1] Esso Expro UK Ltd, 21 Dartmouth Street, London SW1H 9BE, UK
[2] Shell UK Expro, PO Box 4, Lothing Depot, North Quay, Lowestoft, Suffolk NR32 2TH, UK

Abstract: The Clipper Gas Field is a moderate-sized faulted anticlinal trap located in Blocks 48/19a and 48/19c within the Sole Pit area of the southern North Sea gas basin. The reservoir is formed by the Lower Permian Leman Sandstone Formation, lying between truncated Westphalian Coal Measures and the Upper Permian evaporitic Zechstein Group which form source and seal respectively. Reservoir permeability is very low, mainly as a result of compaction and diagenesis which accompanied deep burial of the Sole Pit Trough, a sub-basin within the main gas basin. The Leman Sandstone Fm. is on average about 715 ft thick, laterally heterogeneous and zoned vertically with the best reservoir properties about the middle of the formation. Porosity is fair with a field average of 11.1%. Matrix permeability, however, is less than 1 millidarcy on average and is so low that some intervals in the field will not flow gas unless stimulated. Steep dipping zones of natural fractures occur in certain areas of the field; these commonly allow high flow rates to be achieved from large blocks of low-permeability matrix. Expected recoverable reserves from the most favourable part of the field are 558 BCF and Clipper Field is now being developed in conjunction with part of the adjacent Barque Gas Field. Later development of the remainder of Clipper Field will depend upon reservoir performance in the initial development area.

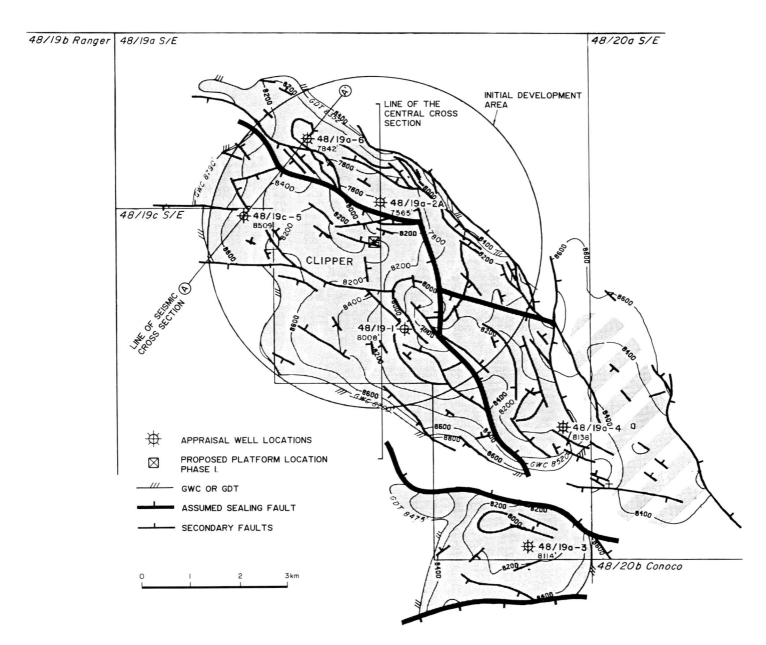

Fig. 1. Top Rotliegend structure map based on 2-D seismic data showing position of initial development area, line of structural section and seismic line.

Located 40 miles off the Norfolk coast in the Sole Pit area of the southern North Sea, the Clipper Gas Field lies in water depths of 70–90 ft. The field is a broad, low-relief, faulted anticline with multiple culminations and differing gas–water contacts, covering some 12 000 acres mainly within Blocks 48/19a and 48/19c with unproven closure in Block 48/20a (see Fig. 1).

The reservoir is the Leman Sandstone Fm. (Rotliegend Gp) in this area and consists mainly of quartzose aeolian sandstones. In the crest of the field the sandstone is fully gas-bearing with a maximum gas column of 900 ft; most gas occurs between 7500 and 8500 ft TVSS. The gas is primarily methane with a minor inert gas content and a low condensate yield. Reserves are estimated at 558 BCF.

Clipper Field is currently being developed in conjunction with the adjacent Barque Field (Farmer & Hillier, this volume). First gas production was in October 1990. In common with several Shell/Esso gas accumulations in the Sole Pit area the field name commemorates vessels from maritime history. The use in this paper of Rotliegend (rather than Rotliegendes as used by Rhys 1975) follows that adopted by Glennie (1984). From historic precedent the term Rotliegend is used in this paper for the Leman Sandstone Fm.

History

Licence P008 covering Blocks 48/19a and 48/20a, was awarded in Round 1, 1964, and will revert to the Crown in September 2010. Block 48/19c was awarded as Licence P465 in May 1983 with a one-third relinquishment being made in May 1989; reversion of the remainder will be in May 2019. The licences are held jointly by Shell UK Exploration and Production and Esso Expro UK with Shell appointed operator in each case.

The Clipper Field discovery well, 48/19-1, was drilled in 1969 on a seismically defined high as part of the ongoing exploration of the southern North Sea Basin. The well established the presence of a thick gas column; however, limited testing showed the Rotliegend reservoir matrix to be fairly tight, flowing about 0.5 MMSCFD after stimulation by a simple hydraulic fracture. The well was abandoned as a non-commercial gas discovery and further appraisal was held in abeyance as fields of this type did not then appear capable of commercial gas production (Glennie & Boegner 1981). By 1983 industry advances suggested development might be feasible and an evaluation programme began that year with the drilling of two wells on the field.

The first of these, 48/19a-2A, located nearly two miles north of the discovery well, confirmed a 750 ft thick gas-bearing Rotliegend section with significant natural fractures. Extensive testing incorporating acid stimulation yielded a maximum flow rate in excess of 50 MMSCFD (Moore 1989). The second well, 48/19a-3, nearly 3.5 miles southeast of the discovery, established a thinner gas column and flowed at rates up to 11 MMSCFD after acid and hydraulic fracture stimulation.

Two further appraisal wells were drilled in 1984, the 48/19a-4 well confirming poor reservoir quality and flowing little more than 7 MMSCFD after acid and hydraulic fracture stimulation. A down-flank well, 48/19c-5, tested less than 1 MMSCFD from a restricted column lacking fractures. The sixth and final appraisal well, 48/19a-6, was drilled in 1985, testing up to 24 MMSCFD comingled flow from two zones, both having been hydraulic fracture stimulated.

Detailed reservoir description based on well, core and test data led to a first phase joint development of the more favourable areas of Clipper and Barque Fields with the option for later development of the remainder of each field. The initial development area of Clipper Field covers the north and central part of the field (see Fig. 1) and includes the crestal block with better matrix quality and several areas assessed as fracture-prone.

Development drilling began in September 1988 with six wells drilled through a subsea template. A platform was installed in late 1989 over the template and the template wells were tied back. Additional wells are now being drilled from the platform whilst the early wells are being produced. Date for first gas was October 1990 with a delivery estimate of about 100 MMSCFD.

A compression platform installed at Clipper Field after two years of plateau production will allow the plateau rate of 200 MMSCFD to be maintained for a further four years. Performance will then decline throughout the remainder of the twenty-year field life. The third (optional) phase of development involves installation of a satellite wellhead platform in the southern area of the field.

Field stratigraphy

The stratigraphic section over Clipper Field typifies the section found over much of the inverted Sole Pit Basin. A thick Permo-Triassic section unconformably overlies eroded Carboniferous sediments and is itself overlain by thin erosional remnants of Jurassic, Cretaceous and Tertiary sediments. The entire sequence represents infilling of a basin whose subsidence began with Late Hercynian extension, was interrupted by Mid- and Late Cimmerian tectonism, and terminated with inversion in Late Cretaceous and Mid-Tertiary times.

A typical sequence encountered in the central part of the field is given in Fig. 2, together with an indication of depth and thickness of the main units recognized as markers in seismic interpretation and drilling. The Carboniferous sediments were deeply eroded following Hercynian uplift and the topmost beds in the Sole Pit area are commonly Westphalian Coal Measures (Glennie & Boegner 1981). The overlying Rotliegend is primarily an aeolian quartz sandstone which accumulated during slow basin subsidence in the later part of the Lower Permian. Marine transgression followed, leading to deposition of the cyclic Zechstein evaporites and carbonates. Basin infill continued in the early Triassic with continental sedimentation of the Bacton Group, followed by a return to marine and marginal marine conditions with cyclic deposition of evaporites of the Haisborough Group, the main halite markers being the Rot, Muschelkalk and Keuper. Little remains of the succeeding marine Jurassic and Lower Cretaceous sequence as a result of Mid- and Late Cimmerian uplift and erosion. However, various estimates have been made as to the thickness of lost section (e.g. Glennie & Boegner 1981) and it is likely that more than 5000 ft were removed in the Sole Pit area.

A partial section of Late Cretaceous Chalk is preserved together with an equally thin section of Late Cenozoic clastics (North Sea Group). Further background to the stratigraphy and structural history of the southern North Sea is given by Siegler (1975) and some details for the Sole Pit area are outlined by Glennie & Boegner (1981) and van Hoorn (1987).

Geophysics

Seismic interpretation in the Sole Pit area was initially hampered by the unexpectedly high velocities encountered in the Triassic section. Depth conversion improved when more extensive coverage of well velocity data became available and the effects of burial and compaction were realized. The Clipper Field discovery well was located by early reconnaissance seismic surveys on a structurally high trend between BP's 1965 West Sole Field discovery (Butler 1975; Walmesley 1977) and Shell/Esso's 1966 Leman Field discovery (van Veen 1975). The well was appropriately located off the crest but found formation tops below the Lias to be deeper than expected due to high interval velocities.

Field appraisal was based on 2-D seismic surveys carried out between 1978 and 1980 using a diamond grid pattern with lines at 500 m spacing. Depth predictions were generally reliable with one notable exception where a flank well came in low due to a higher

CLIPPER FIELD STRATIGRAPHIC SECTION

GROUP	FORMATION	MEMBER		LITH-OLOGY	DEPTH TVSS.
NORTH SEA					
CHALK					
LIAS					1000'
RHAETIC					2000'
HAISBOROUGH	TRITON ANHYDRITIC				
	DUDGEON SALIFEROUS	KEUPER HALITE			
		LOWER KEUPER			
	DOWSING DOLOMITIC	UPPER DOWSING			3000'
		MUSCHELKALK HALITE			
		MIDDLE DOWSING			
		INTER. ROT MUDSTONE			
		ROT HALITE			4000'
		ROT CLAYSTONE			
BACTON	BUNTER SANDSTONE				5000'
	BUNTER SHALE	ROGENSTEIN			
		MAIN BUNTER SHALE			
		BRÖCKELSCHEIFER			6000'
ZECHSTEIN	ZECHSTEIN IV	ALLER HALITE			
	ZECHSTEIN III	LEINE HALITE			
		HAUPT ANHYDRITE			
		PLATTEN DOLOMITE			
		GREY SALT CLAY			
	ZECHSTEIN II	DECK ANHYDRITE			7000'
		STASSFURT HALITE			
		BASAL ANHYDRITE			
		HAUPT DOLOMITE			
	ZECHSTEIN I	WERRA ANHYDRITE			
		ZECHSTEINKALK			
	KUPFERSCHIEFER				
ROTLIEGEND	LEMAN SANDSTONE	A ZONE			8000'
		B ZONE			
		C ZONE			
COAL MEASURES					9000'

Fig. 2. Stratigraphic sequence of field showing relative thickness of overburden geology.

interval velocity than anticipated. The southern extent of the field was delineated by further 2-D lines shot in 1983. The current reserves assessment is also based on the 2-D surveys.

In 1986 a 3-D seismic survey was acquired over the initial development area of the field and this has been used to plan development drilling. The coverage of seismic data over Clipper Field is shown in Fig. 3 with 2-D lines shown individually.

Seismic interpretation is based on five horizons with well ties provided by synthetic seismograms using sonic, density and well velocity logs. The usual seismic character over the field is shown in Fig. 4, which reproduces part of the 2-D line UK9-746 through wells 48/19c-5 and 48/19a-6. The location of this section is shown on Fig. 1. The five mapping horizons indicated are:

- Top Triassic, often affected (as shown) by listric faults in the overburden;
- Top Rot Halite, generally continuous;
- Top Brockelschiefer, generally continuous;
- Top Platten Dolomite, strong seismic expression;
- Top Basal Anhydrite, strong but sometimes obscured by the Platten Dolomite reflector.

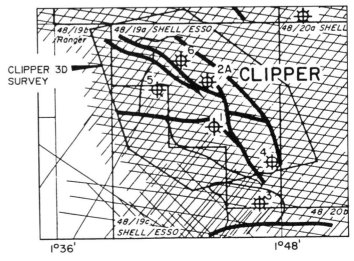

Fig. 3. Map of seismic coverage of field showing individual 2-D lines and outline of the 3-D survey.

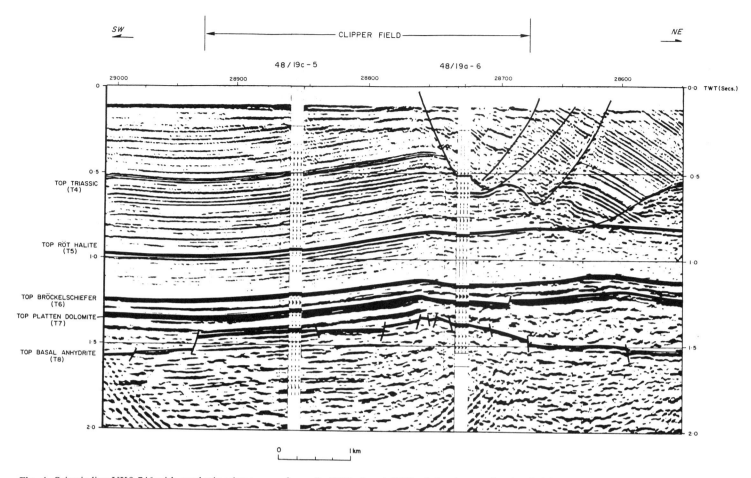

Fig. 4. Seismic line UK9-746 with synthetic seismograms for wells 48/19c-5 and 48/19a-6 showing overburden faulting and well ties to seismic.

The top Rotliegend is commonly mapped by adding an isochore to the Basal Anhydrite depth map. Within the Rotliegend seismic character shows little detail and faults of limited throw are not readily identified. The base Rotliegend is sometimes discernible and a Top Carboniferous map can be derived though reliability suffers. At reservoir level, faults are generally steeply-dipping normal faults reflecting the extensional tectonic environment.

Trap

The Clipper Field is one of several northwest-trending structural highs in the Sole Pit area. Essentially a broad, low-relief, faulted anticline, the trap has both dip and fault closure. Internal faults may have throws of several hundred feet effectively subdividing the field into compartments with differing culminations and gas–water contacts. This complex anticline is believed to have formed at Rotliegend level as an upward diverging flower structure above dextral transcurrent faults in the basement (van Hoorn 1987). The exact timing for structural growth is difficult to determine due to the mobility of the overlying Zechstein interval but most likely involved Late Cimmerian wrench movements followed by two phases of inversion, firstly at the end of the Cretaceous and finally in Mid-Tertiary accompanying the Alpine orogeny. Fault re-activation in the Sole Pit area is noted by Glennie & Boegner (1981) and also by van Hoorn (1987) who observed that earlier movements may be reversed or modified; this carries implications for possible fault seal within a reservoir formed primarily of sandstone.

The field extent is 9 miles by 3 miles and there is sufficient vertical relief for the Carboniferous section to enter the gas leg, although no net sandstones are recognized below the Rotliegend reservoir. Only two appraisal wells established gas–water contacts and three found gas-down-to and water-up-to levels within the Rotliegend. One structurally high well found the Rotliegend completely gas-bearing. By combining these data with pressure information and saturation profiles, four different free water levels have been interpreted for the field. The gas–water contacts and gas-down-to levels are shown on the structure map (Fig. 1) and the interpreted free water levels on the structural cross section (Fig. 5).

The Zechstein Group, comprising mainly halites, anhydrites and tight carbonates, provides excellent top seal for the reservoir as well as lateral seal for gas where the Rotliegend is in juxtaposition. The observed difference of several hundred feet between the shallowest and deepest gas–water contacts in the field is believed to be due to the presence of sealing faults. Though the shale content of the Leman Sandstone Fm. is low and unlikely to lead to fault seal by smear, fault re-activation may provide seal through cataclasis and cementation as observed by Glennie et al. (1978). As a consequence of seal on major faults or reactivated fault zones, reservoir compartments exhibit individual gas–water contacts controlled by local closure and leak points.

Reservoir

The Rotliegend reservoir is considered to be Saxonian or late Lower Permian in age although no precise dating has been obtained from the field. Consisting primarily of aeolian sandstone described by Nagtegaal (1979) as moderately mature quartz arenites, the sandstones are commonly fine-grained with rather low porosity and very low permeability. Reservoir thickness varies between about 650–775 ft with a northward thickening trend. Occasional thin (less than 5 ft) shale beds, found mainly in the lower third of the reservoir, are generally discontinuous and detract little from gross

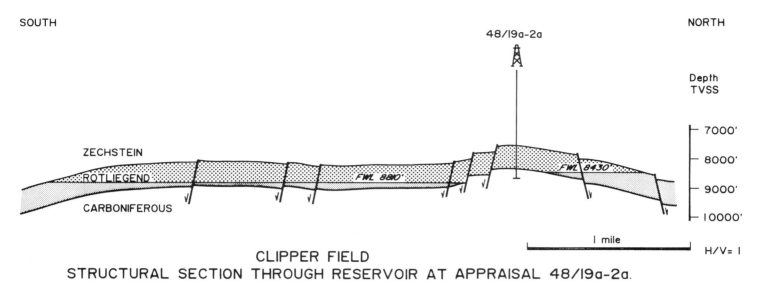

Fig. 5. Structural cross-section through the reservoir showing varying free water levels in different fault blocks.

reservoir volume, seldom comprising more than 3% of the section in Clipper Field wells.

Detailed correlation within the Rotliegend is imprecise due to the limited extent of individual facies units; however, field-wide correlation of groups of facies allows reliable subdivision of the Rotliegend into three zones (see Fig. 2). The uppermost 'A' zone consists mainly of dune, interdune and sabkha sandstone with occasional structureless Weissliegend sandstone (Glennie 1984) at the very top; rare thin lacustrine shales may accompany the sabkha faceis. The middle or 'B' zone consists of cross-bedded dune sandstones of fair to good quality. The lowest or 'C' zone contains predominantly clay-rich wadi sandstone with subordinate very fine-grained dune and interdune sandstone. Reservoir quality in the 'C' zone is poor. The facies subdivisions referred to above have been described by Seeman (1982) and Glennie (1984) amongst others.

The best reservoir quality is found in the 'B' zone with porosities averaging 11–15%; air permeabilities vary from 0.5–100 md. The 'A' zone has slightly lower porosity, averaging 10–12% but permeability is considerably lower, 0.1–1 md with rare thin intervals up to 10 md. The 'C' zone porosities average 7–9.5% with permeability less than 0.5 md. Such widely varying porositiy permeability relationships for facies groups have been described by Glennie et al. (1978) and Seeman (1982) and are ascribed to depositional mode, grainsize, sorting and diagenesis.

Reservoir quality is poor as a result of compaction and diagenesis accompanying deep burial in the Sole Pit Trough prior to Late Cretaceous inversion. Setting aside facies, authigenic cements have probably been responsible for the greatest loss of permeability and these effects have been described in general by Marie (1975) and, more specifically in the Sole Pit area, by Glennie et al. (1978). Further details on clay mineral diagenesis in the southern North Sea are given by Seeman (1982) and Rossel (1982); many of the features reported have been observed in core from Clipper Field. After permeability is corrected for overburden stress, it becomes apparent that matrix permeability in many parts of the reservoir is so low as to preclude gas flow at economic rates and that only the presence of zones of natural fractures allow high production rates to be achieved. The reservoir has been extensively cored and studies have shown that the Rotliegend in Clipper Field is not pervasively fractured. Dilational shear fractures occur in discrete zones measuring tens of feet in width and may be separated from the next fracture zone by several hundred feet of non-fractured matrix. Extension fractures commonly occur in the upper part of the reservoir in the crestal area of the field where fold curvature is greatest. Both groups of fractures can increase productivity greatly if accessed directly by the well or indirectly via hydraulic fracture. Fracture-prone areas have been modelled on a probabilistic basis in Clipper Field, the model being used in simulation studies and in the development drilling programme.

Source

Westphalian Coal Measures directly below the Rotliegend are acknowledged as the source for the gas in the Sole Pit area (Lutz et al. 1975; Glennie 1984). The Coal Measures probably reached maturity during Jurassic time, and continued to generate gas from coals and from organic-rich shales during the Cretaceous. Vitrinite reflectance values greater than 2% are noted in the Sole Pit area and in spite of truncation and erosion of parts of the Westphalian sequence, adequate source volume remains for the known trapped reserves (Cornford 1984). Migration path is considered to be short and direct as suggested by the low nitrogen content of the reservoired gas (see Table 1), Lutz et al. (1975) recognizing high nitrogen content as indicative of a long and tortuous path. Mid-Tertiary structural readjustment may have resulted in some re-migration of gas as noted by Cornford (1984) and may also be responsible for some variation of free water level in the field.

Table 1. *Separator gas composition ex Well 48/19a-6*

Methane	95.76%
Ethane	2.24%
Propane	0.47%
Iso-butane	0.07%
N-butane	0.13%
Iso-pentane	0.04%
N-pentane	0.03%
Hexane	0.03%
Heptane plus	0.04%
Carbon dioxide	0.47%
Nitrogen	0.71%
Helium	0.01%
Hydrogen sulphide	0.00%

Hydrocarbons

A summary of the discovery and appraisal well test results was given earlier when describing the history of the field and analyses of tests on the 48/19a-2A and 48/19a-6 appraisal wells are given by Moore (1989). Development well commissioning and testing is currently in progress (1990) and will not be reported here.

Most appraisal well tests in Clipper Field were carried out in the 'A' zone with the highest flow rates being achieved after acid stimulation of part-cemented fractures in well 48/19a-2A. In less favourable situations, however, even after acid and hydraulic fracture stimulation, 'A' zone intervals have tested less than 1 MMSCFD demonstrating that 'A' zone productivity can be expected to vary greatly in accordance with observed matrix heterogeneity and the presence or absence of natural fractures.

Matrix in the 'B' zone has better permeability and is more homogeneous; test intervals may flow several million cubic feet of gas per day without stimulation. The 'C' zone has not been tested in this field but experience from adjacent fields indicates productivity will generally be low.

Composition of the gas is similar to that in Barque Field (Farmer & Hillier, this volume) and an analysis of separator gas composition from the Clipper Field appraisal well 48/19a-6 is shown in Table 1. Gas gravity is 0.59 and gross calorific value is 1019 BTU/SCF. Condensate recovery is estimated to be 0.5 BBL/MMCF.

Formation water salinity is typically high with a chlorides content of about 160 000 ppm and an equivalent sodium chloride salinity of 200 000 ppm or more. An analysis of produced water is given in Table 2. Formation water resistivity is 0.02 ohm m at 175°F. The reservoir is normally pressured and no water-encroachment of the gas leg is expected during field life.

Table 2. *Produced water analysis ex Well 48/19c-5*

	$Mg\,l^{-1}$
Chloride	165 240
Sulphate	565
Sodium	70 450
Potassium	2350
Magnesium	3500
Calcium	239 000
Barium	4
Iron	215
Strontium	1045
pH	4.78
Density	1.17

Reserves

In the absence of aquifer support, the field drive will be pressure depletion with the majority of gas recovered from the 'A' and 'B' zones during field life. Reserves estimates were derived from a field simulation model built using the Top Rotliegend structure map, zone isochores, net-to-gross and iso-porosity maps. Gas saturations were obtained from height-saturation curves constructed from log-derived and capillary pressure curves. Free water levels for the model were derived from both pressure data and log interpreted gas–water contacts. To allow for uncertainties in parameters leading to construction of the model, a range of probabilities was examined and results were obtained in terms of expectation values.

In the Clipper Field initial development area the most likely value for gas in place is 1331 BCF with a total field gas-in-place of 1678 BCF.

The simulation model incorporates sealing faults within the reservoir as well as fracture zones which may in part be cemented. Recoverable reserves were derived by modelling wells of three types; those accessing matrix only, those located near fracture zones and those in or accessing fracture zones. All wells were considered to be stimulated. Recoverable reserves were based on joint development with Barque Field and a profile with a 4-year plateau and 20-year field life. On this basis reserves in the Clipper Field initial development area are 558 BCF and the recovery factor is 42%.

Clipper Field data summary

Trap
Type	Broad faulted anticline with multiple culminations
Depth to crest	7500 ft
Lowest closing contour	8900 ft (approx)
Free water levels	8430–8810 ft TVSS
Gas column	up to 900 ft

Pay zone
Formation	Leman Sandstone (Rotliegend)
Age	Lower Permian
Gross thickness	650–775 ft
Net/gross ratio	0.81
Net sand cut-offs	Porosity 6% (approx)
	gamma ray 60°API units
Porosity	11.1% average all zones
Gas saturation	49% average all zones
Matrix permeability	0.02–100 md
	less than 1 md average all zones

Hydrocarbons
Gas gravity	0.59
Gas type	Sweet dry gas
Condensate yield	0.5 BBL/MMCF

Formation water
Salinity	160 000 ppm chlorides
	200 000 ppm sodium chloride equivalent
Resistivity	0.02 ohm m

Reservoir conditions
Temperature	175°F
Pressure	3850 psi at datum 8200 ft TVSS
Pressure gradients	0.08 psi/ft (gas leg)
	0.45 psi/ft (water leg)

Field size
Area	12 000 acres (total field)
	8000 acres (initial development area)
Gas-in-place	1331 BCF (initial development area)
Gas expansion factor	229 (SCF/RCF)
Drive mechanism	Pressure depletion
Recovery factor	42%
Recoverable reserves	558 BCF (initial development area)

Production
First gas	October 1990
Development scheme	Single 30-slot wellhead platform bridge-linked to production platform evacuation to Bacton compression platform installed phase II satellite platform optional phase III 10 wells by first gas, 25 for phase I

The authors are grateful to their employers, Esso Expro UK Ltd and Shell UK Expro, for permission to publish this material. In common with most field descriptive papers it represents largely the work of many people and we are grateful to colleagues for their efforts, assistance and advice, especially with reference to studies of facies, fractures, special core analysis, well test analysis and simulation modelling.

References

BUTLER, J. B. 1975. The West Sole Gasfield *In*: WOODLAND, A. W. (ed.) *Petroleum and the Continental Shelf of North West Europe, Volume 1, Geology*. Applied Science Publishers, London, 213–219.

CORNFORD, C. 1984. Source Rocks and Hydrocarbons of the North Sea. *In*: GLENNIE, K. W. (ed.) *Introduction to the Petroleum Geology of the North Sea*. Blackwell Scientific Publications London, 171–204.

GLENNIE, K. W. 1984 Early Permian–Rotliegend *In*: GLENNIE, K. W. (ed.) *Introduction to the Petroleum Geology of the North Sea*, Blackwell Scientific Publications, London, 41–60.

—— & BOEGNER, P. L. E. 1981. Sole Pit Inversion Tectonics. *In*: ILLING, L. V. & HOBSON, D. G. (eds) *Petroleum Geology of the Continental Shelf of North West Europe*, Institute of Petroleum, London, 110–120.

——, MUDD, G. C., & NAGTEGAAL, P. J. C. 1978. Depositional Environment and Diagenesis of Permian Rotliegendes Sandstones in Leman Bank and Sole Pit areas of the UK Southern North Sea. *Journal of the Geological Society, London*, **135**, 25–34.

HOORN, B. VAN, 1987. Structural Evolution, Timing and Tectonic Style of the Sole Pit Inversion. *Tectonophysics*, **137**, 239–284.

LUTZ, M., KAATSCHIETER, J. P. H. & WIJHE, D. H. VAN, 1975. Geological Factors Controlling Rotliegend Gas Accumulations in the Mid-European Basin. *Proceedings of the 9th World Petroleum Congress* **2**, 93–103.

MARIE, J. P. P. 1975. Rotliegendes stratigraphy and diagenesis. *In*: WOODLAND, A. W. (ed.) *Petroleum and the Continental Shelf of North West Europe, Volume 1, Geology*. Applied Science Publishers, London, 205–210.

MOORE, P. J. R. McD. 1989. Barque and Clipper—Well Test Analysis in Low Permeability Fractured Gas Reservoirs. *Proceedings of the 1989 SPE Joint Rocky Mountain Region/Low Permeability Reservoir Symposium*, Paper No. SPE 18966.

NAGTEGAAL, P. J. C. 1979. Relationship of Facies and Reservoir Quality in Rotliegendes Desert Sandstones, Southern North Sea Region. *Journal of Petroleum Geology*, **2**, 145–158.

RHYS, G. H. 1975. A Proposed Standard Lithostratigraphic Nomenclature for the Southern North Sea *In*: WOODLAND, A. W. (ed.) *Petroleum and the Continental Shelf of North West Europe, Volume 1, Geology*. Applied Science Publishers, London, 151–162.

ROSSEL, N. C. 1982. Clay Mineral Diagenesis in Rotliegend Aeolian Sandstones of the Southern North Sea. *Clay Minerals*, **17**, 69–77.

SEEMAN, U. 1982. Depositional Facies, Diagenetic Clay Minerals and Quality of Rotliegend Sediments in the Southern Permian Basin (North Sea): a Review. *Clay Minerals*, **17**, 55–67.

VEEN, F. R. VAN, 1975. Geology of the Leman Gas Field. *In*: WOODLAND, A. W. (ed.) *Petroleum of the Continental Shelf of North West Europe, Volume 1, Geology*. Applied Science Publishers, London, 223–231.

WALMESLEY, P. J. 1977. Ten Years in the North Sea with BP. *Petrole aet Techniques, February 1977*, 7–20.

ZIEGLER, W. H. 1975. Outline of the Geological History of the North Sea. *In*: WOODLAND, A. W. (ed.) *Petroleum and the Continental Shelf of North West Europe, Volume 1*. Applied Science Publishers, London, 165–190.

The Esmond, Forbes and Gordon Fields, Blocks 43/8a, 43/13a, 43/15a, 43/20a, UK North Sea

F. J. KETTER

Hamilton Brothers Oil & Gas Ltd, Devonshire House, Piccadilly, London W1X 6AO, UK

Abstract: The three separate fields, Esmond, Forbes and Gordon form the Esmond Gas Complex in Quadrant 43 in the UK Southern North Sea. The Bunter Sandstone Formation is the reservoir in each of the separate, seismically defined, simple, anticlinal structures. The Bunter Sandstone correlates well across the three structures and is composed of a 400–500 ft thick interval containing individual channel sandstones deposited in an arid environment on an alluvial fan. Porosity and permeability are controlled by original texture and by subsequent diagenesis. Reservoir communication within the Bunter reservoir is good though locally tortuous.

First gas production started in July 1985. Four offshore installations have allowed gas to be produced at full contract (plateau) rates of 200 MMSCFD without major interruption. Ultimate sales gas reserves are approximately 530 BCF.

The Esmond Complex consists of three gas fields designated Esmond, Forbes and Gordon located in Quadrant 43 in the northern portion of the Southern North Sea Basin approximately 75 miles from shore at approximately 54° 35'N latitude and 01° 25'E longitude, (Fig. 1). The Esmond Field is located in Block 43/13a, the Forbes Field is located in Block 43/8a and the Gordon Field is located in Blocks 43/15a and 43/20a. Water depths average approximately 120 feet. These blocks are held under licences P002 and P048.

The fields are each defined by a simple anticlinal structure and produce from the Triassic age Bunter sandstones. The reservoir structures are unfaulted anticlines and communication within each structure is good. Reservoir drive mechanisms are a combination of depletion and natural water drive.

The fields were delineated by a total of eight exploration/appraisal wells. A further nine development wells have been drilled and four appraisal wells have been tied back and completed for production giving a total of 13 wells available for production.

Production started in July 1985 from the Esmond Field; Gordon Field started production in August 1985 and production from Forbes Field started in October 1985. Gas sales at the full contract rate of 200 MMSCFD started on 1st October 1985.

There are four installations in the complex. These are a central production platform, a wellhead tower with twelve slots and two satellite platforms each with nine slots. The central production platform and wellhead tower are located in the Esmond Field. The satellites are located at Forbes and Gordon. All jackets are conventional piled steel jackets. There are no permanent drilling facilities on any of the platforms.

Gas is exported onshore to Bacton by a 127 mile 24 inch subsea pipeline. The pipeline has an estimated maximum capacity of 590 MMSCFD with an inlet pressure of 2000 psig.

The fields are named after three Scottish clans.

History

Exploration

Blocks 8, 13, 15 and 20 in Quadrant 43 were awarded as part of the UK first licensing round in September, 1964. Sub-licences on these Blocks were later granted to the Hamilton Group by the original licensees in 1969. During the pre-discovery years, the Group drilled or participated in three exploration wells in the area and developed an extensive well data base. Extensive seismic surveys, which included both propietary and trade data, were interpreted by Hamilton staff. Geophysical studies indicated closed anticlinal features with amplitude anomalies and flat-spots.

The initial Bunter Sandstone discovery was made by well 43/20a-1, drilled in June 1969.

This well encountered 249 ft of gas-bearing sandstones from a total of 477 ft of Middle Bunter Sandstone at a subsea depth of 5167 ft. A 55 ft interval of the Middle Bunter Sandstone was tested, yielding a maximum test rate of 17.3 MMSCFD. The 43/15a-1 delineation well, drilled on the same structure in April 1970, confirmed the discovery. This well encountered 430 ft of Middle Bunter Sandstone (143 ft of net pay) and established the field gas water–contact at 5404 ft TVSS. Two cores totalling 117 ft (115 ft

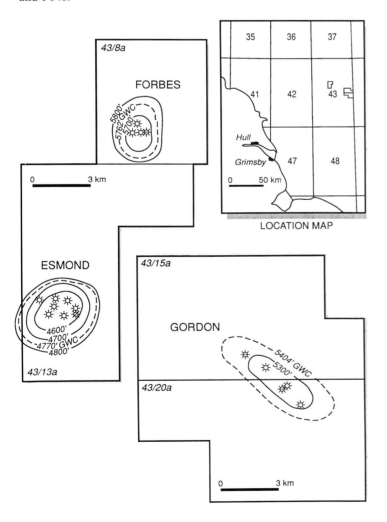

Fig. 1. Esmond Complex, location map.

recovery) were taken in the upper part of the reservoir, but no tests were conducted. This accumulation has been named the Gordon Field.

The 43/8a-1 well, completed in January 1970, discovered gas in a similar structure located approximately 20 miles northwest of the Gordon Field. The well tested 15.05 MMSCFD and 9.57 MMSCFD from two separate intervals in the Middle Bunter Sandstone. A third drill stem test established a gas–water contact at 5762 ft TVSS. This accumulation has been named the Forbes Field.

Between 1971 and 1982, additional seismic surveys were conducted and evaluated to supplement older vintage data and trade data. Regional and local geological studies were also carried out to enhance the knowledge of the Middle Bunter reservoir.

In June 1982, Hamilton Brothers drilled the 43/13a-1 well on a third structure and encountered 302 ft of gas-bearing Middle Bunter Sandstones (302 ft of net pay). This well was cored and tested extensively in the Middle Bunter Sandstones. Two drill stem tests flowed 19.53 and 18.44 MMSCFD from the middle and bottom of the Middle Bunter sandstone interval respectively. The 43/13a-2 deviated hole, drilled from the same surface location, encountered 350 ft (TVD) of gross Middle Bunter Sandstone (190 ft of net pay) and established a gas–water contact at 4770 ft. TVSS by log analysis. This accumulation has been named the Esmond Field.

Appraisal

Three appraisal wells 43/13a-3, 43/8a-3 and 43/15a-2, one in each field, were drilled, cored and tested in 1983. The 43/13a-3 appraisal well confirmed gas reserves on the eastern flank of the Esmond structure.

Appraisal of the Forbes structure by well 43/8a-3 confirmed the reserves on this structure. The well tested gas at a rate of 37.6 MMSCFD. The well was suspended for later completion as a production well.

The 43/15a-2 well was drilled to appraise the southeastern extent of the Gordon structure and tested gas at a rate of 33.3 MMSCFD. The well was suspended for later use as a production well.

Development

Following Annex B approval in May 1984, a wellhead tower and two satellite platforms were fabricated and installed. A central process platform was installed on the Esmond Field adjacent to the wellhead tower and the satellite platforms were installed in Forbes and Gordon Fields. Gas from the three fields is processed on the central process platform and then flows to Bacton via a 24 inch export line. Five development wells, 43/13a-C4, -C5, -C6, -C7 and -C8 were drilled and completed between October 1984 and March 1985 from the CW wellhead tower positioned over the Esmond Field. Two of the existing Esmond Field wells, 43/13a-1 and 43/13a-2, were tied back and completed for production.

Between June and the end of July 1985 three wells were completed for production from the BW wellhead jacket installed in the Gordon Field. Wells 43/15a-B3 and -B4 were drilled and completed and the suspended appraisal well 43/15a-2 was tied back as a producing gas well.

Wells 43/8a-A4 and -A5 were drilled and completed through the Forbes Field wellhead jacket, AW, between August and September 1985. The appraisal well 43/8a-3 was also tied back for use as a production well.

Production commenced on 11 July 1985 from the Esmond Field followed by the Gordon Field (7 August 1985) and finally the Forbes Field (1 October 1985). Production from the fields has continued without problems.

Field stratigraphy

The Middle Bunter Sandstone reservoir is overlain by the Triassic Haisborough Group, Dowsing Dolomite Formation; three members can be recognized within this formation (Fig. 2). The basal Rot Claystone Member forms the seal to the gas reservoir in the Middle Bunter Sandstone of the Esmond Complex. The Rot Claystone Member consists of anhydritic claystones approximately 40–50 ft thick. The Rot Halite Member directly overlies these basal claystones with halites in the lower portion and claystones with minor halite and anhydrite interbeds in the upper section.

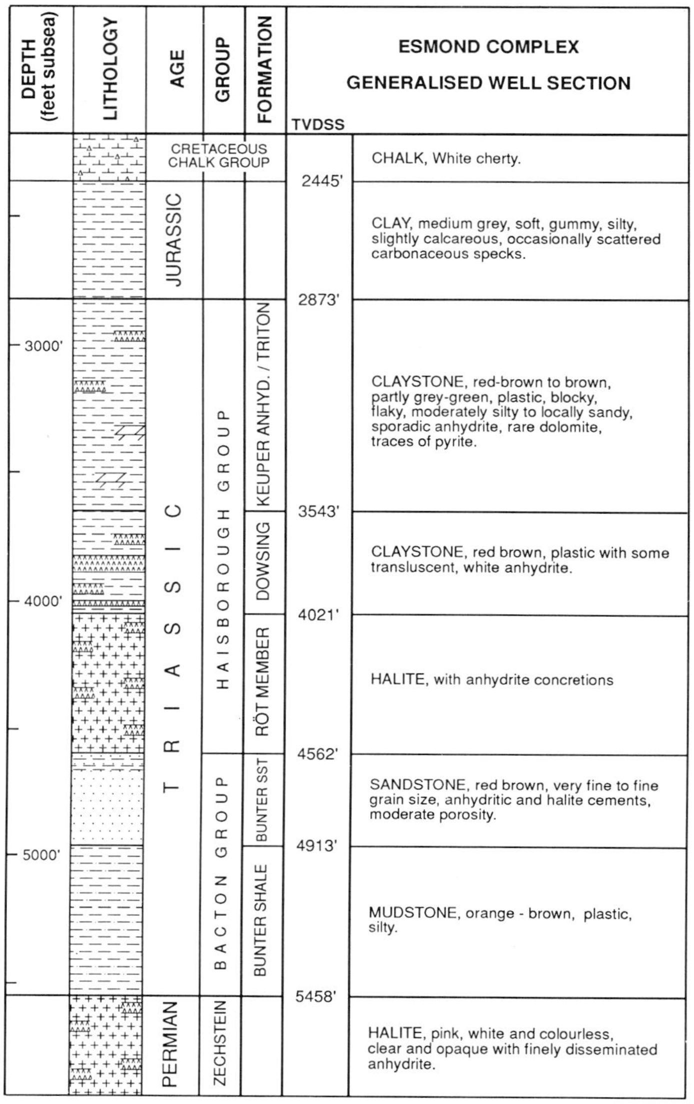

DEPTH (feet subsea)	LITHOLOGY	AGE	GROUP	FORMATION	ESMOND COMPLEX GENERALISED WELL SECTION	
					TVDSS	
		CRETACEOUS	CHALK GROUP		2445'	CHALK, White cherty.
		JURASSIC				CLAY, medium grey, soft, gummy, silty, slightly calcareous, occasionally scattered carbonaceous specks.
3000'					2873'	
		TRIASSIC	HAISBOROUGH GROUP	KEUPER ANHYD. / TRITON		CLAYSTONE, red-brown to brown, partly grey-green, plastic, blocky, flaky, moderately silty to locally sandy, sporadic anhydrite, rare dolomite, traces of pyrite.
					3543'	
4000'				DOWSING		CLAYSTONE, red brown, plastic with some translucent, white anhydrite.
					4021'	
				ROT MEMBER		HALITE, with anhydrite concretions
					4562'	
			BACTON GROUP	BUNTER SST		SANDSTONE, red brown, very fine to fine grain size, anhydritic and halite cements, moderate porosity.
5000'					4913'	
				BUNTER SHALE		MUDSTONE, orange - brown, plastic, silty.
					5458'	
		PERMIAN	ZECHSTEIN			HALITE, pink, white and colourless, clear and opaque with finely disseminated anhydrite.

Fig. 2. Esmond complex, generalized well section.

Above the Rot Halite Member lies the interbedded claystones, dolomites and anhydrite beds of the main Dowsing Dolomite Formation. Within this interval the Muschelkalk Halite Member can be sub-divided, containing variable amounts of halites, anhydrites, dolomite and claystone beds.

The Triassic Dudgeon Formation overlies the Dowsing Formation. It is an interval of claystones and interbedded mudstones up to about 600 ft thick.

The Triton Anhydritic Formation succeeds the Dudgeon Formation and is composed of claystones and interbedded anhydrite beds which can be approximately 300 ft thick in the area. Above this is the Winterton Formation claystones which are about 70 ft thick and form the uppermost formation in the Triassic Haisborough Group.

Lower Jurassic (Lower Lias) shales and thin limestones lie above the Triassic interval and can range from 150 ft to 400 ft thick in the Esmond Complex Area.

About 150 ft of Lower Cretaceous Cromer Knoll Group consisting of chalk grading into claystones lies at the base of the Cretaceous, directly overlying the Jurassic. A monotonous interval of Upper Cretaceous chalk approximately 1500 feet thick overlies the Cromer Knoll Group.

The youngest stratigraphic sequence penetrated by wells in the Esmond Complex consists of up to 800 ft of Tertiary silts and claystones.

Geological history

At the end of the Permian the northward connection to the Zechstein Sea was broken, leaving the Southern North Sea Basin as a featureless peneplane which bacame a site of continental clastic sedimentation. Thick distal clay and mud facies were initially laid down in Triassic time across the area and form the Bunter Shales. Later deposition of the sands and silts of the Middle Bunter Sandstone occurred in an arid to semi-arid continental environment. These sediments were derived from the older Variscan highlands to the south.

Tectonic movements in the mid-Triassic initiated Zechstein halokinesis. Salt movement locally restricted Middle Bunter Sandstone deposition and were the origin of the shallow domal or pillow structures which now form the Esmond, Forbes and Gordon structures, following enhancement by continued tectonic activity in the Tertiary.

The Triassic Haisborough Group represents a period of arid lacustrine/floodplain and low relief continental deposition punctuated by several marine transgressions. The lowest members of the Haisborough Group, the Rot Shale and Rot Salt, form the seal to the Bunter gas accumulations.

In the Rhaetic a marginal marine transgression marked the end of continental deposition. Deposition of shallow marine claystones and limestones blanketed the entire area and continued into the Liassic. The Late Jurassic Cimmerian tectonic event was responsible for uplift, some inversion and removal of parts of the Jurassic section.

The Cretaceous transgression represents renewed rifting with continued inversion on a local scale. Chalk deposition dominated the sedimentary sequence, ending in the Palaeocene. Later uplift or inversion caused erosion of the Mesozoic rocks and resulted in the present-day configuration of the basin.

Geophysics

The following surveys were used in the final mapping for development purposes:

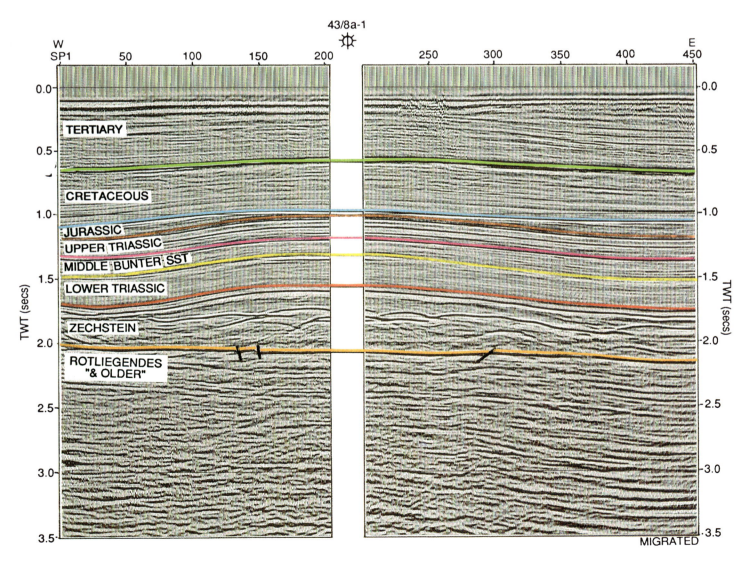

Fig. 3. Forbes Field, seismic line 82/43-27.

Survey	Lines	Contractor	Fold	Source	Navigation	Processing	km
1980	DB-5 to DB-12	Western	48	H.P. Airgun	Pulse 8	Western	77
1982	82-43-17 to 29	SEI	48	Airgun	Pulse 8	SSL	107

Pre-1980 surveys were reviewed but are not incorporated into the mapping due to poor navigation control. The seismic data quality to the base Zechstein is very good on all surveys (Fig. 3).

The 1980 survey has been conventionally processed, incorporating deconvolution before and after stack, time variant filtering and migration. On this data the following horizons have been picked: Top Tertiary; Top Cretaceous; Base Cretaceous Unconformity; Top Rhaetic (43/8a and 43/13a only); Top Keuper (43/15a and 43/20a only); Top Muschelkalk Dolomite; Top Middle Bunter Sandstone; Top Zechstein; Base Zechstein.

The subsequent 1982 survey provided additional control and has been wavelet processed with an extracted signature deconvolution. After stack, a predictive deconvolution, migration, and time variant filter were applied, and the data converted to zero phase. The same horizons have been picked on the 1982 survey as were previously picked for the 1980 survey. On all the lines 'bright spots' (seismic amplitude anomalies) are present over the crest of the structures (see Fig. 3).

On wells 43/13a-1 and 43/20a-1, checkshot surveys were carried out and synthetic seismograms produced.

Constant velocity functions were used for each structure with no interpolation between structures, because: (a) the accumulation areas were small, relative to the distances between the structures; (b) the dips encountered within each accumulation area were relatively flat, being at the crestal position of each anticline.

The computerized depth conversion was performed using the digitized horizons and the interval velocities to produce a top reservoir depth map.

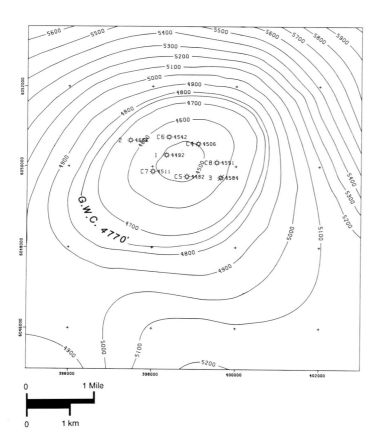

Fig. 4. Esmond Field, Zone 3 depth structure.

Evaluation and structure

The Esmond Complex consists of three, unfaulted, simple, anticlinal structures (see Figs 4 to 6). All three structures have been mapped at the top of the Middle Bunter Sandstone reservoir in time and depth. Each structure shows a prominent 'high' formed by pillowing of the underlying Zechstein evaporites. The Base Cretaceous unconformity progressively eroded the Jurassic section to the northwest. However, the Middle Bunter Sandstone reservoir maintains its full thickness throughout the four blocks.

None of the wells on the three structures penetrate the base Zechstein Group. The Base Zechstein Group depth maps prepared from the seismic data shows no structural closures. It is not possible to pick Rotliegendes Formation or Carboniferous events due to the poor quality of data beneath the base Zechstein in this part of Quadrant 43.

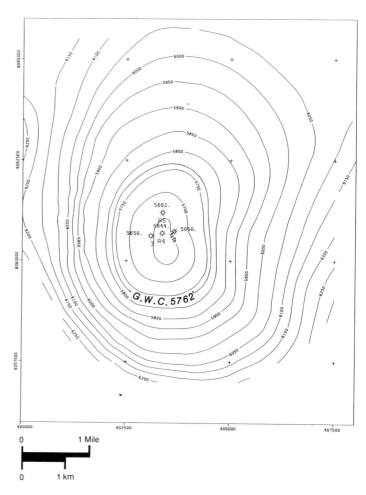

Fig. 5. Forbes Field, Top Zone 3 structure.

Trap

The three fields in the Esmond Complex consist of simple anticlinal structures formed by the pillowing of the underlying Zechstein Group evaporites.

The Esmond and Forbes structures are domal anticlines while the Gordon strucure is elongated in the northwest to southeast direction (Figs 4 to 6).

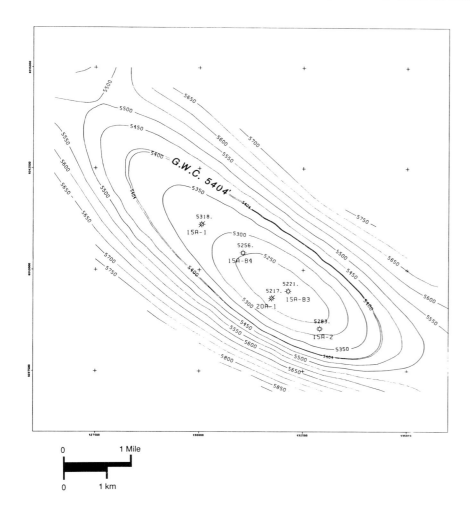

Fig. 6. Gordon Field, Top Zone 3 structure.

The seal to the Middle Bunter Sandstone reservoir is the Rot Claystone Member, below the Rot Halite Member. This claystone is approximately 40 ft thick across the fields and forms an effective seal to the gas below.

The structural traps in the Esmond Complex were charged with gas early in their formation. The traps are not full to spill-point, which may indicate continued movement of the Zechstein halites after gas emplacement.

The following original gas–water contacts have been defined from log and RFT results:

Esmond Field:	4770 ft TVSS
Forbes Field:	5673 ft TVSS (Zone 1)
	5762 ft TVSS (Zones 3–5)
Gordon Field:	5404 ft TVSS

Reservoir

Age

The Middle Bunter Sandstone Formation is of Scythian age. The Middle Bunter Sandstone is a prominent interval which can be correlated across the Southern North Sea.

Depositional facies

The Middle Bunter Sandstone was deposited primarily by fluvial processes. The limited thickness of individual depositional cycles indicates that ephemeral stream channels were quite shallow. These cycles, where complete, are typically composed of:

(1) a basal erosional surface overlain by thin sandy conglomerate beds with red siltstone pebbles and cobbles;

(2) trough cross-stratified, very fine- to fine-grained, sandstone representing channel fill;

(3) tabular and planar cross-stratified sandstones similar to previous deposits interpreted as channel fill;

(4) thin beds of ripple-stratified, very fine-grained, micaceous sandstone or siltstone considered to be overbank levee deposits;

(5) non-channelized sheetflood siltstones and fine- to very fine-grained sandstones which form the background deposits in which the channel sequences are enclosed.

In the Middle Bunter Sandstone of the Esmond Complex the channel cycles are closely superimposed, and the overlying cycle may have removed or re-worked the underlying deposits.

The thin sandy cycles, reflecting shallow channel-dominated processes, suggest that the Middle Bunter Sandstone was deposited in coalesced alluvial fans in the semi-arid to arid climate (Fig. 7).

Individual channels were only a few feet deep, several hundreds of feet wide and relatively straight. The proximal portions of each fan and channel system would take the appearance of a braided system. In the distal parts of this coalesced fan system, as in the Quadrant 43 area, channels would develop into lobes with each channel bifurcating many times cutting previous channels and channel levees. This adds small-scale heterogeneities which disrupt channel connectivity. Lobe abandonment and channel switching would add an additional level of complexity and heterogeneity to the sandstone bodies geometry and confound the reservoir continuity seen in the Middle Bunter Sandstone today.

The Middle Bunter Sandstone is commonly referred to as a sheet sand. It is indeed a generally sandy unit 400–500 ft thick. But it is made up of at least three hierarchies of reservoir heterogeneity,

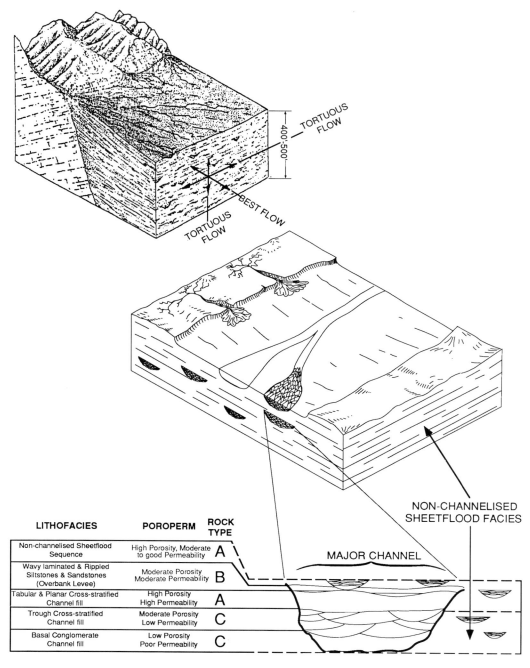

Fig. 7. Esmond Complex, facies and depositional environment.

ranging from sets of cross strata at the smallest scale, individual thin channel units at an intermediate scale, to channel complexes of terminal lobes on the largest scale. This model of sandstone deposition would suggest that fluid flow paths are continuous but tortuous where locally preserved channel-top or overbank silty beds act as baffles. Flow paths are probably more continuous north–south along channel axes than east–west across channels.

Reservoir units and poroperm properties

The Middle Bunter Sandstone reservoir interval has been subdivided into seven zones, for purposes of reservoir analysis (see Fig. 8).

Petrophysical analysis is complicated by the presence of varying quantities of halite cement. Because of their similarity in log response, halite is difficult to distinguish from gas on the majority of wireline logs. This necessitated the use of a multi-mineral multi-log analysis technique, which appears to give consistent and reliable results with regard to identification of halite-cemented layers and levels of calculated porosity.

Formation halite has an adverse effect on the accuracy of core measured porosities since cleaning fluids will either dissolve formation halite leading to an over-estimation of porosity, or not remove halite precipitated from formation water leading to under-estimation.

The high salinity formation water renders shale effects negligible in the calculation of water saturation in the Middle Bunter Sandstones. Archie parameters were obtained from special core analysis.

Net reservoir sand was defined by application of porosity cut-offs of 5% in Esmond Field and 6% in Forbes and Gordon Fields. These values correspond roughly to an air permeability of 1 md in each case, based on the limited core data available.

Zone I. The uppermost sandstone body of the Middle Bunter Sandstone (Zone I) is separated from the main Middle Bunter Sandstone by a silty mudstone unit that ranges from 20 ft to 39 ft thick in wells in the Esmond Complex area. The Zone I sandstone is

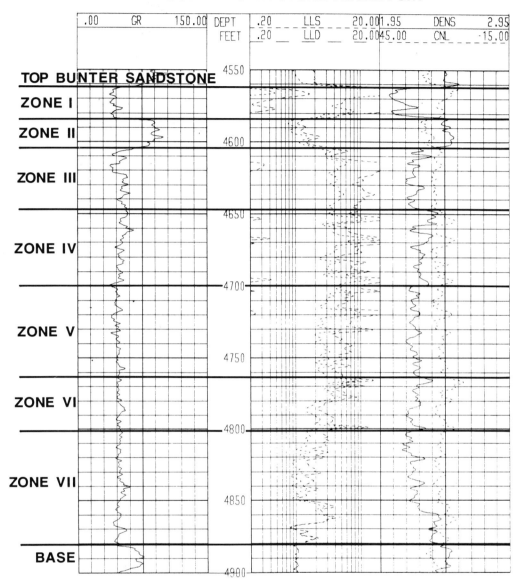

Fig. 8. 43/13a-1, type log in Bunter sandstone reservoir.

generally 12–28 ft thick in the Quadrant 43 area. It is a well-defined unit on gamma ray, density and neutron logs. Porosities in Zone I range from 9% to 23% with an average of approximately 18%. Net to Gross ratios are around 0.90 except where halite cementation occurs.

Zone II. The unit that underlies the uppermost sandstone of Zone I is a deep red mudstone, silty, massive and brittle, containing scattered layers of white anhydrite nodules one or two centimetres in diameter. Zone II provides an effective seal to the gas accumulation in the underlying Middle Bunter Sandstone in the Esmond Complex. It has no reservoir potential.

Zone III. Zone III consistently exhibits the most uniform rock properties of any zone in the Middle Bunter Sandstone reservoir. Thickness ranges from 36 ft to 88 ft with a general thickening to the south-east towards the 43/15a-20a structure. Porosities reflect the improved rock quality and average approximately 20%. Net to Gross ratios are around 0.95.

Zone IV. In contrast to Zone III, lithologies and sedimentary structures in Zone IV are more complex and inconsistent. The depositional environment appears to be of a lower energy regime, and post depositional diagenetic processes are more pronounced. Zone IV varies in thickness from 40 ft to 62 ft, and thins to the north and southeast. Overall, the sandstones are not as well sorted as those in Zone III, contain more authigenic clays and substantially more mica (or illite). Anhydrite and halite cements are moderately to widely abundant and reduce porosity and permeability. Zone IV porosity ranges from 12% to 20% with an average of about 14%. Net to Gross ratios are also lower, in the 0.70 to 0.80 range.

Zones V and VI. Data on lithology and sedimentary structures from core descriptions and log character suggest a return to a higher energy environment, with predominantly complete sequences of 'fining upward' fluvial channels, in Zones V and VI.

Zones V and VI are composed of alternating intervals of high and low porosity and permeability, with broad ranges of values for porosity and Net to Gross ratio similar to Zone III. High values correspond to intervals of very fine- to fine-grained ripple-stratified or flat-laminated sandstones. Layers of larger scale cross-bedding, somewhat coarser-grained, have lower and more erratic porosity and permeabilities due to abundant cementation. It can be inferred that those sediments with initially higher porosity and permeability were eventually selectively altered by emplacement of authigenic cements.

Zone VII. This zone is very similar to the overlying beds; however, the lowermost 20–30 ft have lower porosities and permeabilities due to an increase in silt and anhydrite cements. Porosities average approximately 19%, ranging from 16% to 24%. Net to Gross ratios have a broad range reflecting cementation effects.

Source

The gas found in the Esmond Complex is considered to have been sourced from the coals of the Westphalian. Areas of salt withdrawal, and thinned or absent Zechstein salt, have acted as paths for such gas to migrate into the Triassic Middle Bunter Sandstone. The structures were gas charged relatively early after the initial domal traps were formed by pillowing of the Zechstein evaporites in the mid- to late Triassic. Salt movements continued during the Tertiary enhancing the Esmond, Forbes and Gordon structures. However, there seems to have been no further gas migration, resulting in structures that are not full to spill-point.

Hydrocarbons

The gas in the Esmond Complex is extremely dry with a condensate yield of less than 0.2 BBLS/MMSCF. No hydrogen sulphide has been detected but the gas does contain up to 1 mol% carbon dioxide. The gases from the three fields contain different amounts of nitrogen. Esmond Field has the least with 8% and Gordon Field the most with 14%.

The CO_2 content of the produced gas requires the tubulars to be manufactured with 13% chrome steel to inhibit corrosion. Gas from the three fields is blended onshore and sold to British Gas.

The formation water in all three fields is salt saturated at reservoir conditions with salinities ranging from 130 000 ppm to 205 000 ppm NaCl eq. The specific gravity of the water is 1.21 at 60°F. The water resistivity is estimated as 0.033 ohm m at the approximate bottom hole temperature of 140°F.

Initial reservoir pressures at their appropriate data are:

Esmond: 2280 psig at 4500 ft SS
Forbes: 2797 psig at 5700 ft SS
Gordon: 2617 psig at 5300 ft SS

Reservoir temperatures range from 133 to 153°F. Gas expansion factors vary from 154.8 to 179.8 SCF/NCF.

Reserves

Recovery factors in the three fields of the Esmond Complex range from 60% to 90% of gas initially in place. These values are derived from material balance studies and indicate the drive mechanism to be a combination of depletion with natural aquifer influx.

Since the start of the field production, a reservoir monitoring programme has been conducted. This programme calls for reservoir pressure measurements to be made on an annual basis. A pulsed neutron logging programme has also been conducted on Esmond Complex wells on an annual basis. Time lapse analyses of the logs has been used to monitor the movement of water within the fields. Production logs (PLT) have been run for routine production monitoring.

Water produced in the early life of the fields was water of condensation. Formation water has been produced laterally along higher permeability reservoir layers in the Forbes Field. Remedial workovers are used to control these effects. A gradual vertical rise of the gas–water contact is recognized in Esmond and Gordon Fields.

Compression facilities were commissioned on the Esmond central processing facility during 1987. The installation consists of 4 centrifugal compressors driven by gas turbines. As the reservoir pressure declines the configuration will be modified to increase the compression ratio.

The expected ultimate recoverable reserves in the three fields of the Esmond Complex are just over 533 BCF (15.1 BCM).

The discovery, appraisal and development of the fields in the Esmond Complex was very much a team effort spanning several years and involving numerous disciplines and professionals. The efforts of all Hamilton Brothers Oil and Gas staff involved in the Esmond Complex present or past is recognized and appreciated.

The cooperation of our Partners, Elf Oil and Gas Limited, Ultramar Exploration Ltd and Monument Resources in all aspects of development has ensured the success of the Esmond, Forbes and Gordon Fields projects.

The Hewett Field, Blocks 48/28-29-30, 52/4a-5a, UK North Sea

P. COOKE-YARBOROUGH

Phillips Petroleum Company (UK) Ltd, Phillips Quadrant, 35 Guildford Road, Woking, Surrey GU22 7QT, UK

Abstract: The Hewett Field complex, discovered in 1966, is located in the southwestern part of the Southern North Sea Basin. Gas is currently being produced from four reservoirs: Upper and Lower Bunter sandstones in the Triassic and Rotliegendes sandstones and Zechstein dolomites in the Permian.

The Hewett Unit area straddles five blocks: 48/28, 48/29, 48/30, 52/4a and 52/5a and, at its SW margin, lies about ten miles NE of the Norfolk coast. Figure 1 shows the hydrocarbon distribution of the various Hewett accumulations. Five fields lie within the Hewett Unit area, established in 1969 by the signing of a Unit Agreement between the Phillips and Arpet groups. Interests in the Hewett Unit Area are as follows:

PPCo	18.97%
Fina	16.26%
AGIP	8.13%
Century	3.91324%
Lasmo	4.61784%
Arco British Ltd	4.58%
Arco Oil Production	15.26667%
Canadian Superior UK	1.5266%
Union Rheinische	4.80%
North Sea Sun	10.68667%
Superior UK	9.16%
Plascom	2.30892%.

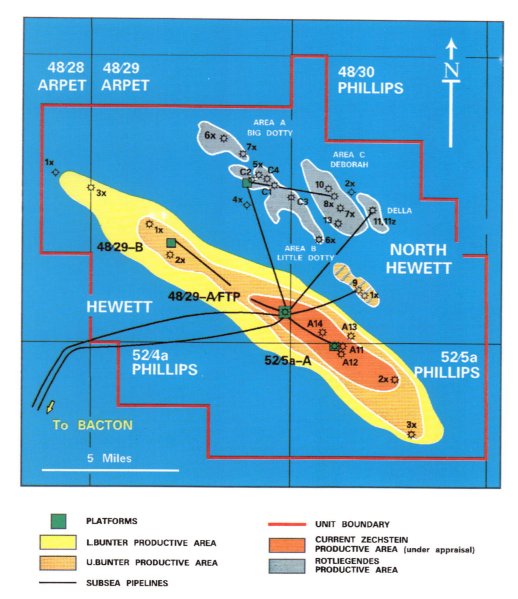

Fig. 1. Hewett Unit area development.

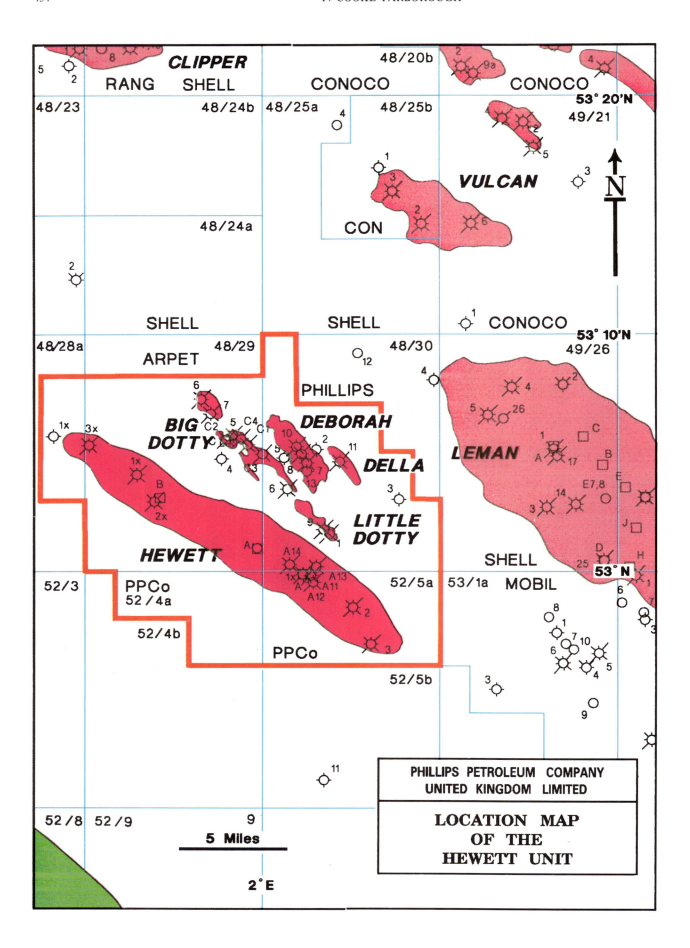

Fig. 2. Location map of the Hewett Unit area.

The main Hewett gas field, the largest in the Unit area, comprises three co-extensive reservoirs, the Upper Bunter sandstone, the Lower Bunter or Hewett sandstone, and the Zechstein carbonate. The field is the shape of a flattened ellipse, aligned NW–SE, has a length of 18 miles, a maximum width of three miles and lies in all five Unit blocks.

The North of Hewett gas fields lie in blocks 48/29 and 48/30 to the NE of the main Hewett gas field. Deborah, Big Dotty and Della are Rotliegendes sandstone reservoirs whilst Little Dotty has both Upper Bunter and Rotliegendes sandstone reservoirs.

The Hewett Field derives its name from a nearby seabed feature known as the Hewett Ledges. This choice of name is unusual for Phillips-operated fields in the UK North Sea which have been named, in general, after geoscientists' wives or daughters. The designated initial letter for this Licensed area is 'D', giving rise to the Deborah, Della and Dotty field names. Hewett is the closest producing field to shore in the United Kingdom North Sea Southern Gas Basin and has a water depth of around 120 ft.

History

Licence P.028, for which Phillips Petroleum Company United Kingdom Limited ('Phillips') is the operator, contains a total of seven blocks designated 47/4a, 47/5a, 48/10a, 48/30, 49/6a, 49/11a and 52/5a comprising 241,268 gross acres.

Licence P.037, for which Arco British Limited is the operator, was also awarded in the first round of licensing (1964) and covers blocks 48/11a, 48/28a, 48/29 and 49/28 comprising 181,128 gross acres.

Licence P.112, which covers block 52/4a and comprises 15,394 gross acres, was subsequently acquired in 1970 on behalf of the Hewett Unit, for which Phillips is the designated operator.

The 48/29-1X Hewett Field discovery well (Fig. 2) was drilled by Arpet in the second half of 1966 and reached T. D. at 7288 ft RKB in the Carboniferous. The Upper Bunter sandstone was encountered at 3034 RKB ft and, on a drill stem test of the uppermost 15 ft, yielded gas at the rate of 18.1 MMSCFD. The Lower Bunter sandstone, encountered at 4178 ft RKB, flowed gas at the rate of 23.4 MMCF/day from a 10 ft perforated interval. A follow-up well, 48/29-2X, confirmed the continuity of these two Triassic gas accumulations. It was the drilling of the 52/5-1X well by Phillips, however, some nine miles southeast of the 48/29-1X well, which gave final confirmation that a field of very significant size had been discovered.

By March 1967, a total of seven wells had been drilled along the long axis of the Hewett field. Subsequently, a further 23 have been drilled from three fixed platforms within the Hewett Field: eight each from the 48/29-A and 48/29-B platforms and seven from the 52/5a-A platform. Beneath each platform, the deviated well bores are some 900 ft distant horizontally from the centre of the cluster at Upper Bunter sandstone level and some 1500 ft distant in the Lower Bunter sandstone.

In 1968, sales contracts were signed with the Gas Council and a 30 inch trunkline linking the Bacton Plant with the Field Terminal Platform was laid. The Lower Bunter (Hewett) sandstone reservoir of the Hewett Field was placed on production on 12 July 1969, less than three years from the spudding of the discovery well.

Desulphurization equipment was installed at Bacton in September 1973 and Upper Bunter sandstone sour gas production from re-completed wells on the 52/5a-A platform commenced shortly thereafter. Onshore compression to maintain the gas supplied to the Gas Council distribution system at 1000 psi pressure has been operational since August 1973.

Exploration of the North of Hewett area followed immediately after the discovery and appraisal of the Hewett Field itself. Three more fields, Big Dotty, Deborah, and Little Dotty were discovered initially and production from them began in 1976, 1978 and 1979, respectively. A fourth field, Della, was discovered in 1987 and began producing in 1988.

Production from the Zechstein carbonates began in 1986 from the 52/5a-A11 well with gas from the first cycle (Z1) Zechsteinkalk. A second well, 52/5a-A12, substantially increased gas production from the Zechstein in 1989.

Long before fixed platform drilling started in December 1967, however, the Arpet and Phillips Groups had been aware that unitization of their respective interests would do much to assure the efficient and economic development and exploitation of the field. With this end in view, negotiations were initiated in April 1967 and culminated in April 1969 with the signing of a Unit Agreement. The Phillips Group equity was fixed at 54.2% and the Arpet Group equity at 45.8% from these discussions, with no provision being made for re-determination

Field stratigraphy

Figure 3 shows a generalized stratigraphic section for Hewett and Fig. 4, four structural cross-sections across the area.

During the Westphalian period of the late Carboniferous, the area occupied a marginal position on an extensive coastal plain that developed to the north of the London–Brabant Massif. With prolonged paralic conditions, very thick coal-bearing sequences were deposited. These sequences are believed to contain the source rock that generated the gas found in the Hewett area.

Rotliegendes sedimentation commenced during the early Permian in a desert environment as the uplifted Carboniferous areas were slowly being eroded. The Rotliegendes increases in thickness towards the NE from 450 ft to over 700 ft. The earliest Rotliegendes sediments consisted of fluvial/wadi sands formed by sheetflood deposits draining off the London–Brabant Massif to the south. These gave way to minor, interbedded intervals of wind-blown sands and eventually dune sands prograded across the entire area. As a result, the upper two-thirds of the Rotliegendes consists of thick dune sequences. Within the Hewett area, the top of the Rotliegendes lies at a depth of between 5000 and 7000 ft and slopes NE towards the centre of the Rotliegendes basin.

Desert conditions were brought abruptly to an end when the Rotliegendes basin was flooded by marine waters of the Zechstein Sea. The base of the Zechstein is well marked by the Kupferschiefer shale and is overlain within the area by four carbonate/evaporite sedimentary cycles, Z1–Z4. Each cycle corresponds to a phase of marine transgression followed by one of regression and, in general, reflects the influence of an increase in salinity. In the ideal case, each cycle would consist of a thin clastic member passing upward, in turn, into limestone, dolomite, anhydrite and halite. This cyclicity is fairly well developed within the Hewett area with the exception that Cycle 4 is present only in a rudimentary form.

The close of the Permian saw the end of widespread marine sedimentation and a return to the dominantly non-marine depositional environments in the Triassic. At the end of the Zechstein, minor uplift of the London–Brabant Massif followed by erosion led to the local development of the Lower Bunter (Hewett) sandstone. This sandstone is extremely limited in extent and rapidly thins from 200 ft thick over the Hewett Field to 20 ft thick in the North of Hewett Fields. The Lower Bunter sandstone is overlain by the Bunter shale, an anhydritic red–brown mudstone with minor, greenish shales deposited in a floodplain environment.

Fine clastic sedimentation was brought to a halt with the abrupt deposition of the Upper Bunter sandstone, probably initiated by further uplift of the London–Brabant Massif. Thick fluvial channel and sheetflood sands were deposited and, with continued subsidence of the basin, increased uniformly in thickness to the NE.

In the Upper Triassic, marine conditions were re-established with the thin Rot clay being deposited within the area as a basal transgressive unit. During the ensuing period of tectonic stability, deposition took place in a floodplain environment, alternating

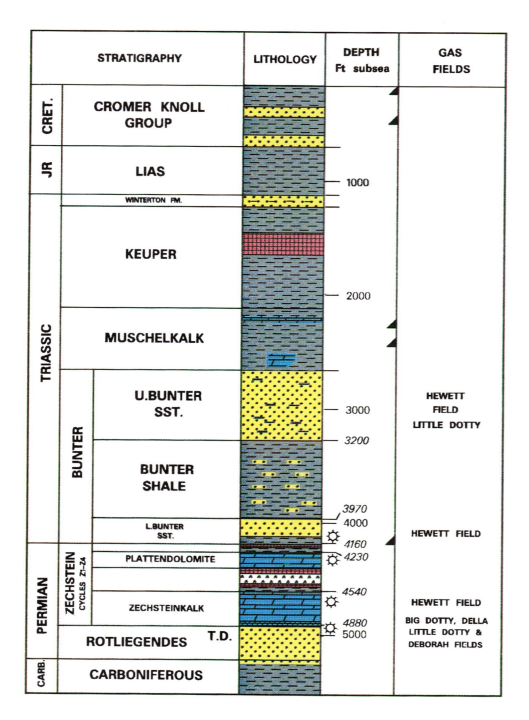

Fig. 3. Hewett Unit area generalized stratigraphic section.

with coastal sabkha or shallow marine conditions. Following the Rhaetic transgression, clays and a sand unit derived from the uplifted London–Brabant Massif were deposited.

Marine conditions persisted into the Jurassic and sedimentation was essentially continuous from the Triassic. Thin sands and limestones were deposited in the Middle Jurassic but more stable marine conditions, accompanied by a deepening sea, resumed in the Upper Jurassic.

Sedimentation continued in the Upper Jurassic/Lower Cretaceous with the deposition of the Spilsby sand, sourced from a recently uplifted area to the northeast of the Sole Pit Axis. With the continuation of passive sedimentation, there commenced a major period of regional subsidence. For the first time the London–Brabant Massif was submerged as the Upper Cretaceous chalk was deposited.

Geophysics

Over the last 25 years, numerous seismic surveys have been acquired within the Hewett area and a number of different geophysical interpretations have been made of the structural configuration. The reason for the differences in interpretation can be seen from the structural complexity of the reservoir horizons and of the sedimentary overburden.

This paper highlights an innovative technique which applies seismic methods better to define reservoir properties.

The Zechsteinkalk dolomite reservoir is one of modest matrix porosity and its successful development hinges on the ability to predict open fracture intensity, orientation and distribution. Thus, well log and core data have been integrated with the structural configuration and fault pattern as mapped from surface seismic

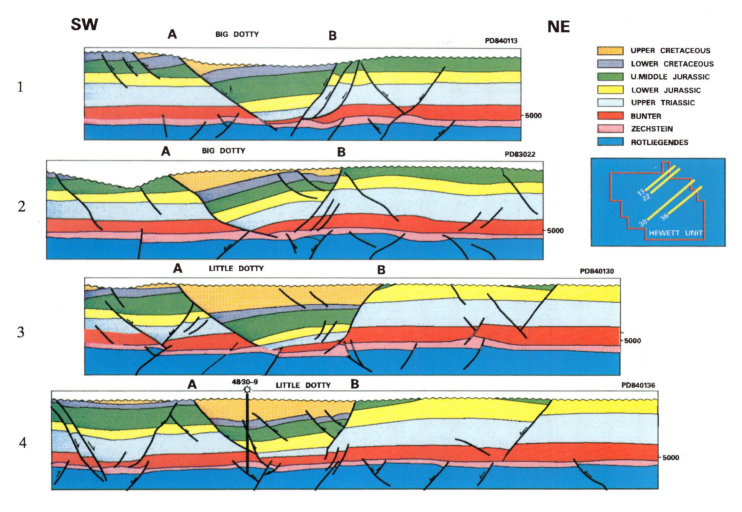

Fig. 4. Schematic cross-sections across the Hewett Unit area.

data to establish a tectonic model which can predict the distribution of the open fracture system. The acoustic and petrophysical relationships through the Zechsteinkalk reservoir interval were evaluated in tandem. Using these relationships, effective porosity can be predicted from acoustic impedance using surface seismic data. In this way, zones of lower porosity with brittle, fractured dolomite which would provide the productivity into the well bores could be identified as well as those areas of higher porosity rock which supply significant gas storage. These zones can be seen on the processed and micromodelled sections in Fig. 5, a dip seismic line through the main Hewett Field.

Trap

The trapping mechanism in the main Hewett Field is entirely structural. Figure 6 shows the top Upper Bunter surface depicting the shape of the Main Hewett and Little Dotty field anticlines. Also shown on this map are outlines of the North of Hewett Rotliegendes accumulations. The three coextensive reservoirs in Hewett are contained in an anticline which is aligned NW–SE and bounded by faults on its NE and SW flanks. The NW-trending alignment of the structure at all three levels is similar to other productive fields in the Southern North Sea. It is probable that the post-Silurian Caledonian movements in the area of Kent and East Anglia were related to those of the mid-European belt all of which produced folds with a NW–SE alignment. It is possible that in these younger features we are seeing later manifestations of such an earlier trend. The main folding associated with the Hewett structure almost certainly occurred contemporaneously with faulting in Cretaceous time.

In the North of Hewett Fields, the key structural events commenced with the development of the NW-trending Dowsing Fault Zone during the Carboniferous with dominant right lateral wrench movement. This was followed by local syndepositional movement along NW-trending faults in the early Permian with subsequent minor NW faulting at the close of the Permian. Initial development of the Big Dotty and Little Dotty structures occurred during the early Triassic in response to movement along the Dowsing Fault Zone. These low relief, fault-bounded anticlines lie at an oblique angle to the wrench zone, as is often the case with wrench-induced folds. Almost certainly the structure was modified during the Cimmerian and Alpine/Sole Pit movements. Major syndepositional faulting in the Jurassic paved the way for the thickness variations seen in the Jurassic across the area. Major folding and faulting occurred in the Hewett area during the Sole Pit inversion period and both the Hewett Graben and the Deborah and Della structures are believed to have been formed at this time.

All of the Hewett accumulations are essentially filled down to, or near, their respective spill points. The Rotliegendes and Zechsteinkalk gas accumulations are sealed by overlying and adjacent Zechstein evaporites. The Lower and Upper Bunter sandstone gas accumulations are sealed, both vertically and laterally, by the overlying Dowsing dolomite and Bunter shale formations.

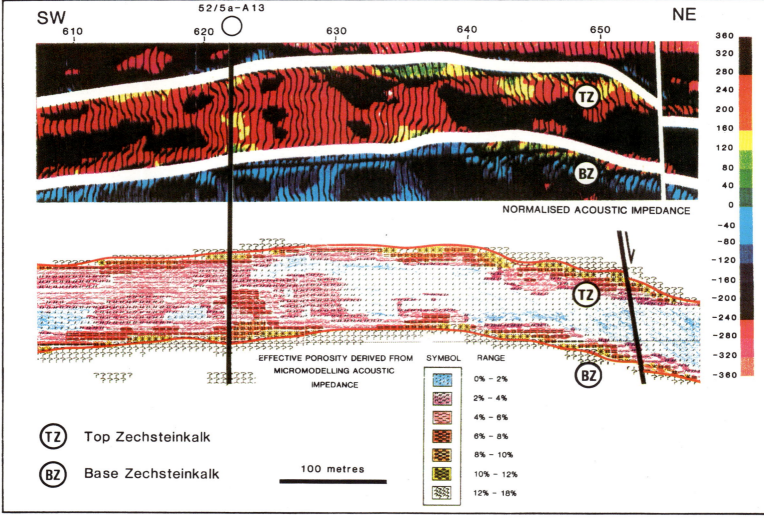

Fig. 5. Zechsteinkalk micromodelling section.

Reservoir

Reservoir properties of the eight Hewett gas accumulations are summarized in Table 1.

Source

Based on geological and geochemical data, the late Carboniferous Westphalian Coal Measures are considered to be the primary source of the gas found in Hewett. This has been corroborated by C-isotope studies.

It is probable that the initial gas migration period was contemporaneous with its generation. The burial history of the basin infers initial generation towards the end of the Jurassic when the gas migrated upwards via faults cutting the Carboniferous. Figure 7 summarizes the timing of the formation of the Hewett Field structures and the generation of the gas which fills them.

The major distinction between the gas of the Upper Bunter sandstone reservoir and that of the Lower Bunter sandstone is the presence in the former of hydrogen sulphide. The hydrogen sulphide is most probably derived from the action of sulphate-reducing bacteria on anhydrite in the presence of hydrocarbons, with carbon dioxide liberated as a by-product of this process. The average carbon dioxide content of the Upper Bunter sandstone gas is four times greater than that in the Lower Bunter sandstone. The likeliest source for the anhydrite which produces hydrogen sulphide is in the Triassic Haisborough Group sediments. Faulting has served to bring these beds into juxtaposition with the Upper Bunter sandstone on the north side of the Hewett Field.

Hydrocarbons

Fields of the Hewett area contain dry gas, with the Upper and Lower Bunter sands of the main Hewett field having liquid gas ratios of 4.8 and 3.2 BBL/MMSCF, respectively. These two reservoirs contain gas of strikingly different compositions; in addition to the differences in the hydrogen sulphide content, the Upper Bunter contains significantly more nitrogen than the Lower Bunter.

The separate fields of Hewett were normally pressured reservoirs prior to production, lying on a water gradient to surface of 0.46 psi/ft.

Reserves

The Hewett Field facilities, shown on Fig. 1, comprise four production platforms: 52/5a-A, 48/29-A and 48/29-B on Hewett and 48/29-C on Big Dotty, all tied back to a central terminal platform 48/29-FTP. Additionally, there are four satellite subsea wells, two of which are tied back to the 48/29-C platform and two to the 48/29-A platform.

Gas flows, via two 30 inch pipelines, to the Phillips-operated Bacton terminal, where onshore compression facilities are located. Offshore compression equipment was installed during 1989 on the

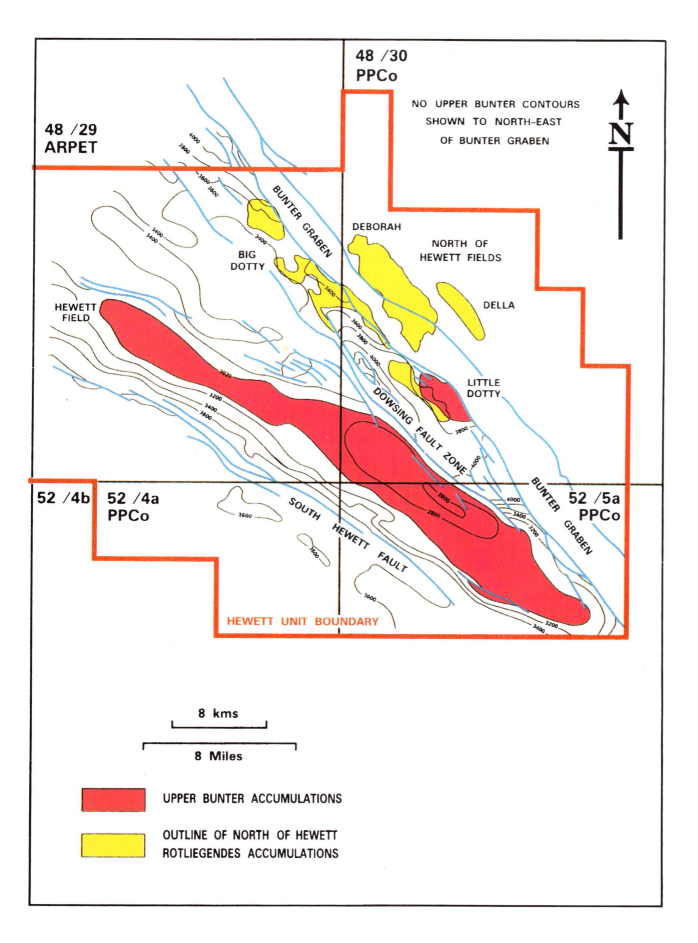

Fig. 6. Hewett Unit: Top Upper Bunter depth map.

Table 1. *Hewett Fields: reservoir properties*

	Main Hewett: Upper Bunter	Main Hewett: Lower Bunter	Main Hewett: Zechsteinkalk	Deborah: Rotliegendes
Formation	Bunter Sandstone	Hewett Sandstone	Zechsteinkalk (Z1)	Leman Sandstone
Age	Lower Triassic	Lower Triassic	Upper Permian	Lower Permian
Depositional environment	Alluvial plain sandstones	Continental alluvial sandstones	Intertidal and upper submarine slope carbonates	Aeolian sandstones
Depth to crest (highest closing contour, ft. TVDSS)	2,600	4,026	4,500	5,500
Net/gross ratio (average)	0.96	0.88	0.66	0.98
Porosity % (average)	21	23	5–8	14.2
Permeability (md) (average)	500	1000	<1	75

Table 1. *Continued*

	Big Dotty: Rotliegendes	Little Dotty: Upper Bunter	Little Dotty: Rotliegendes	Della: Rotliegendes
Formation	Leman Sandstone	Bunter Sandstone	Leman Sandstone	Leman Sandstone
Age	Lower Permian	Lower Triassic	Lower Permian	Lower Permian
Depositional environment	Aeolian sandstones	Alluvial plain sandstones	Aeolian sandstones	Aeolian sandstones
Depth to crest (highest closing contour, ft. TVDSS)	5,600	3,500	5,800	5,800
Net/gross ratio (average)	0.98	0.95	0.98	0.99
Porosity % (average)	18.8	21	18.8	12.7
Permeability (md) (average)	250	350	450	50

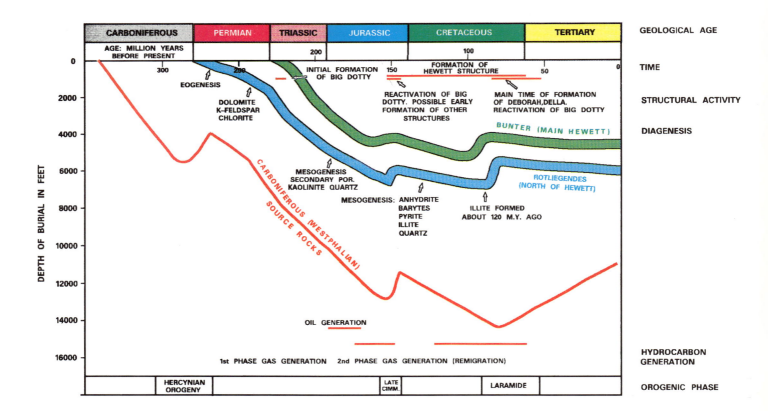

Fig. 7. Hewett Unit: burial history/gas generation.

48/29-A platform. Production from the low pressure wells is also increased by the use of an ejector which uses high pressure gas to lift low pressured wells.

The overall recovery factor from the Hewett Field is very good for a number of reasons: both Triassic reservoirs have good to very good permeabilities in their braided fluvial and sheetflood sandstones; permeabilities are moderate to good in the North of Hewett Rotliegendes aeolian sandstones; there are few laterally extensive permeability barriers within any of the reservoirs; and onshore and offshore compression enable low abandonment pressures to be achieved.

Water influx has threatened to be a production problem, but careful monitoring has minimized its adverse effects. In the Upper Bunter, arrest of the aquifer movement has been effected at an inter-reservoir shale barrier with the result that the life and recovery factor of this reservoir have both recently been increased.

The author thanks Phillips Petroleum Company United Kingdom Limited, Phillips Petroleum Company, Arco British, Fina, Superior, North Sea Sun, Agip, LASMO, Deminex, Acre and Elf for giving permission to publish this paper.

The Indefatigable Field, Blocks 49/18, 49/19, 49/23, 49/24, UK North Sea

J. F. S. PEARSON, R. A. YOUNGS & A. SMITH

Amoco (UK) Exploration Co., Amoco House, West Gate, Ealing, London W5 1XL

Abstract: The Indefatigable Field was discovered in June 1966 by the Amoco Operated Group well 49/18-1. The field is located 55 miles northeast of Great Yarmouth in the central part of the Southern North Sea Basin in water depths of 100 ft. The field lies in Blocks 49/18, 19 and 49/23 and 24. The reservoir section is Lower Permian Rotliegendes sandstones comprising a mixed sequence of aeolian and fluvial sediments at an average depth of 8500 ft subsea. Maximum reservoir thickness is in excess of 420 ft; however, the average pay thickness is of the order of 150 ft. The main reservoir trap is a complexly faulted NW-SE-trending anticlinal horst with a maximum vertical closure of 1300 ft. The field is approximately 17 miles long and 6 miles wide and has an areal extent of some 60 square miles. Recoverable reserves in the Field are estimated to be 4418 BCF of gas. The field is not unitized and is jointly operated by Amoco and Shell.

The Indefatigable Field is located 55 miles northeast of Great Yarmouth and lies in the central part of the Southern North Sea Basin (Fig. 1) within Blocks 49/18, 49/19, 49/23 and 49/24. It is the second largest gas field in the United Kingdom, with an areal extent of 60 square miles. The field is productive from Lower Permian Rotliegendes sandstones trapped in a complexly faulted horst block trending NW-SE. Two smaller structurally separate satellite accumulations tested by wells 49/23-2 and 49/24-2, lie to the southwest and southeast respectively of the main field (Fig. 2).

The field was named after the Indefatigable Banks, a sea-floor shallow sand bank feature in the vicinity.

History

Blocks 49/18 and 49/23 were awarded to the Amoco Operated Group as licence P016 in 1964 as part of the First Round of licensing in the United Kingdom, and the adjacent blocks 49/19 and 49/24 were awarded as licence P007 to Shell/Esso in the same licensing round. The Amoco operated group comprises:

[1] Amoco (UK) Exploration Company 30.77%;
Gas Council (Exploration) 30.77%;
Amerada Hess Ltd 23.08%;
[2] Enterprise Oil plc 15.38%

The field was discovered by the Amoco operated group well 49/18-1 in June 1966, which found 285 ft of gas-bearing Rotliegendes Sandstone (Fig. 3). This well found no gas–water contact and flowed at maximum rates of 18.8 MMSCFD. A subsequent appraisal well 49/19-2A was drilled by Shell/Esso in May 1967 and proved the extension of the structure into the adjacent block.

Sixteen appraisal wells were drilled to delineate the field and development began in 1970. The field came on stream in September 1971 and reached a plateau level of 1132 MMSCFD in the winter of 1975/76. At present a total of nine drilling platform complexes are in operation from which 61 developent wells have been drilled (Fig. 4). All platforms are tied into one field terminal platform from where the gas is sent via a 63 mile 30 inch pipeline to the Bacton terminal.

Field stratigraphy

Figure 5 illustrates the generalized stratigraphic sequence encountered in the Indefatigable Field. The distribution of stratigraphic units and the regional structure were affected by three phases of tectonic movements, the Hercynian in Late Carboniferous times, the Late Cimmerian during the Mesozoic and the Alpine during Late Cretaceous and Early Tertiary times.

The oldest rocks encountered in the Indefatigable Field are Carboniferous Westphalian deltaic sedimentary rocks; these comprise shales, silts, minor coals, and occasional thick sandstones which have moderate reservoir properties. The sequence is rich in organic material. Some sandstones within the Carboniferous are known to be gas-bearing, but their reservoir potential is yet to be fully evaluated.

Unconformably overlying the Carboniferous interval is the Lower Permian Rotliegendes Sandstone, which is the reservoir of the Indefatigable Field. The sandstone was derived from the London–Brabant Massif and the Variscan Mountains to the east and southeast and possibly from as far afield as the Baltic Shield. Abundant accessory minerals especially orthoclase and plagioclase feldspars indicate the source to have been granitic in origin.

The main detrital component is quartz, which is coated with hematite enriched clay in places. In the upper 50–100 ft, the quartz is coated with hematite free kaolinite which resulted from diagenetic changes due to carbonate-rich meteoric waters.

The sandstones were deposited as a series of mixed, massively bedded aeolian and fluvial sands with some good sequences of well-sorted, clean, cross-bedded aeolian dune sandstones.

Overlying these desert sandstones is the Upper Permian Zechstein evaporite sequence. The basal bed of this sequence is the Kupferschiefer, a thin organic shale which forms an excellent marker horizon due to its high gamma ray response; it represents a major marine incursion of the area after which sedimentation occurred in an evaporitic marine environment.

The major part of the Zechstein section comprises a thick evaporite sequence of interbedded halite and anhydrite. These sequences are fairly uniform in thickness although subsequent fault movements and halokinesis have resulted in NW–SE trending salt swells and lows. Minor carbonates are present, the most important in this area being the Platten dolomite which is highly fractured in places. The Zechstein evaporites form the seal for the gas in the Rotliegendes Sandstone.

The Triassic sequence, conformable with the underlying Zechstein was deposited in the period preceding the late Cimmerian tectonism. The earliest Triassic in the Indefatigable area belongs to the Bacton Group which comprises a lower red–brown shale sequence overlain by the Bunter sandstone Formation. The Bunter sandstones have fair reservoir properties (porosity 25%, permeability 450 md) and are productive in the Hewett Field 35 miles to the southwest. Evaporites overlie the Bunter sandstone (the Röt halite) as well as being locally developed in the overlying Dowsing, Dudgeon and Triton Formations.

As a result of the Late Cimmerian erosion, the Upper and Middle Triassic Sections (Winterton, Triton and Dudgeon Formations) are

---

[1] Formerly Amoco (UK) Petroleum Limited
[2] Enterprise acquired Texas Eastern interest on 1 September 1989.

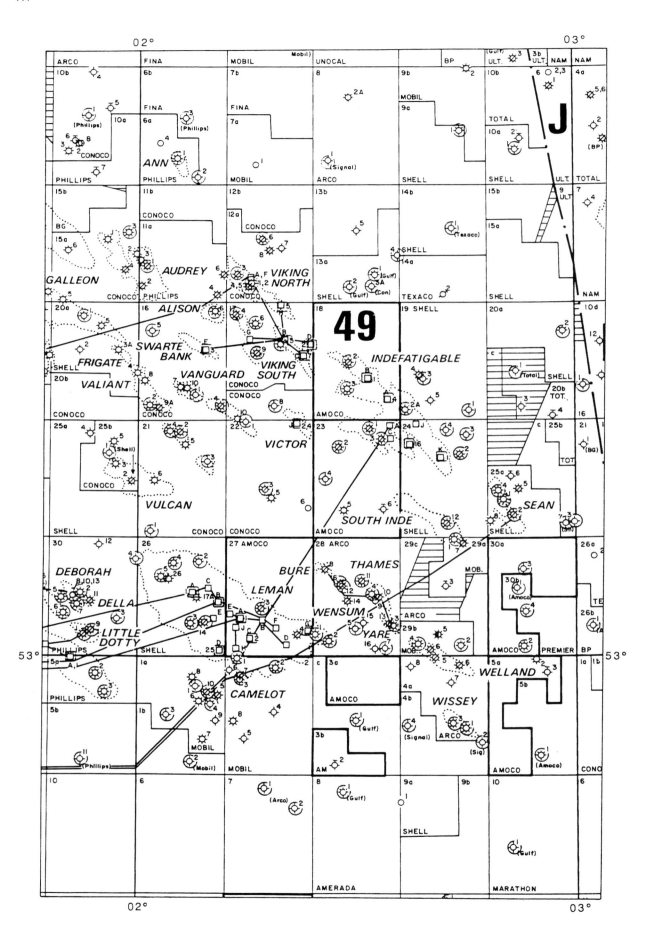

Fig. 1. Regional southern North Sea licence map.

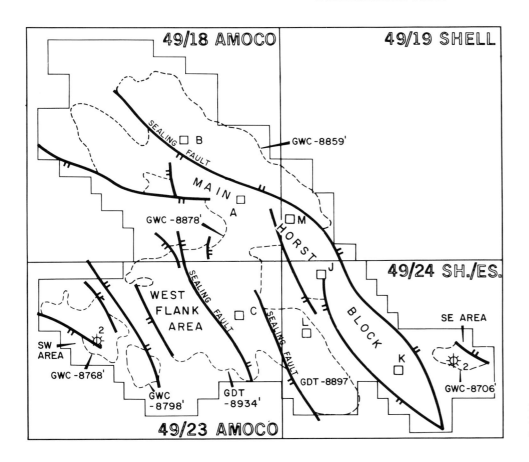

Fig. 2. Structural elements of the Indefatigable Field and gas–water contacts.

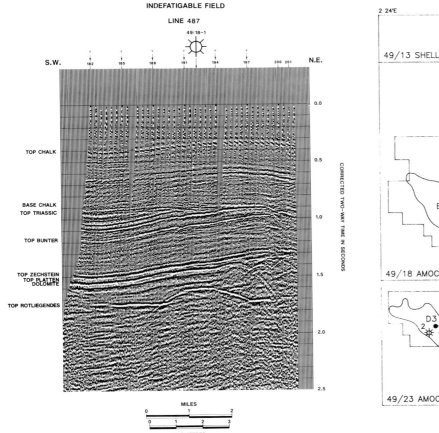

Fig. 3. Seismic section through Discovery Well 49/18-1.

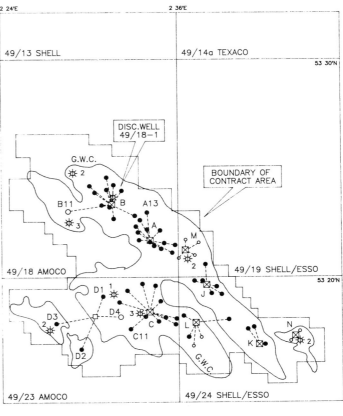

Fig. 4. Development outline of Indefatigable Field.

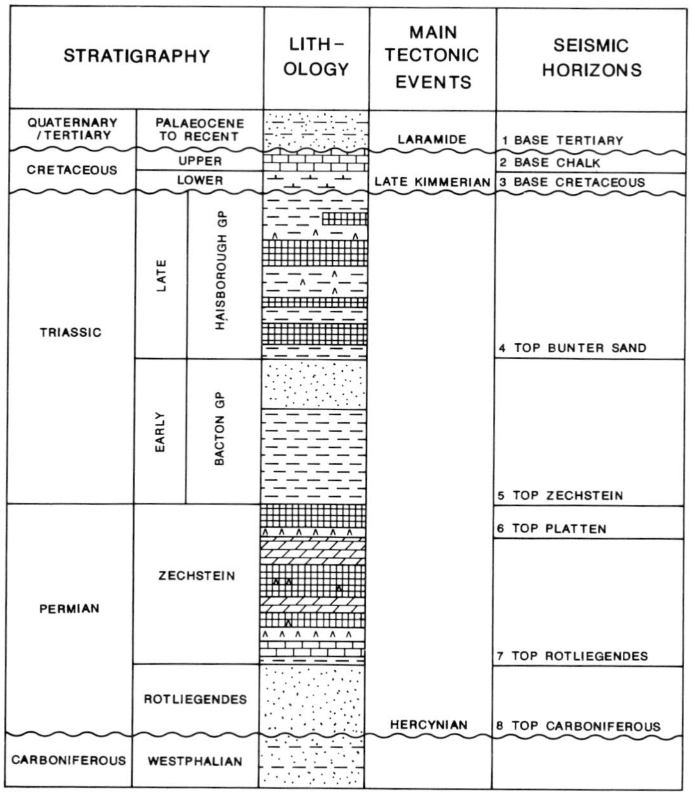

present only in the extreme western part of the field. In the eastern part of the field, erosion has removed the Winterton and Haisborough Sequences so that the Cretaceous interval rests unconformably on Bunter Sandstone.

Jurassic sediments are absent in all wells within the Field as a result of post-Cimmerian erosion. Figure 6 illustrates the effects of post-Cimmerian erosion over the field area.

Cretaceous sediments were deposited across the area as basin-wide subsidence commenced. A thin basal sequence of Lower Cretaceous, Barremian to Aptian, grey shales and glauconitic sandstones were deposited on the Cimmerian erosion surface in shallow water depths. Upper Cretaceous chalks were subsequently deposited uniformly over the area in moderate to deep water as basin-wide subsidence continued. At the same time, minor variations in thickness occurred as a result of tectonic activity and halokinesis.

Alpine movements during Late Cretaceous to Early Tertiary times resulted in uplift and partial erosion of the chalk. The oldest Tertiary rocks in the Indefatigable area are marine shales of Late Palaeocene age, which rest unconformably on the Cretaceous section. The youngest Tertiary sediments are of Pliocene age. Glacial sands and gravels of the Quaternary Period complete the sedimentary sequence.

Geophysics

Seismic data over the Indefatigable Field dates back to the mid-sixties. The surveys used in the latest interpretation are summarized in Table 1, and includes the details of acquisition and processing.

Fig. 5. Stratigraphic column of the Indefatigable Field.

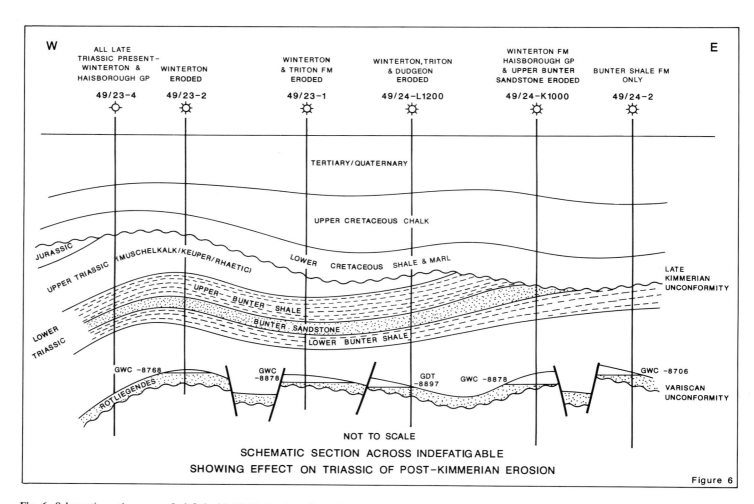

Fig. 6. Schematic section across Indefatigable Field showing effect of post-Cimmerian erosion on Triassic.

Table 1. *Indefatigable Field seismic data summary*

Operator	Survey prefix/year	Acquisition contractor	Acquisition source and navigation	Fold	Survey length in mapped area (km)	Processing contractor	Remarks on processing
Amoco	AUK 73-	SSL	Gasgun N.A.	12	100	SSL	DBS, Migration
Amoco	AUK 78-	S & A	Airgun Trisponder	30	350	S & A	DBS/DAS,FIN DIF MIG. Redisplay (1987) some of data set at 1 : 25 000
Amoco	AUK 80B-	Horizon	Airgun Syledis	36	350	Horizon	DBS/DAS,FIN DIF MIG.
Amoco	AUK 81B-	SEI	Airgun Hydrotrac	48	380	Amoco	DBS/DAS,F/K MIG.
Shell	UK 81-	SSL	Airgun Hydrotrac	48	310	SSL	DBS/DAS,FIN DIF MIG.
Shell	ND 82-	SSL	Airgun Hydrotrac	48	190	SSL	DBS/DAS,FIN DIG MIG.
Amoco	AUK 84B-	SSL	Watergun Pulse 8	44	760	SSL	DBS/DAS,FIN DIF MIG.
					Total: 2440		

Interpretation

Reflector identification. Apart from the top reservoir (Top Rotliegendes), reflectors were chosen that corresponded to major velocity breaks for ease of depth conversion. The following reflectors were therefore chosen:

(1) Base Tertiary
(2) Base Chalk
(3) Top Triassic
(4) Top Bunter Sandstone
(5) Top Zechstein
(6) Top Rotliegendes (Top Reservoir).

Pre-Top Rotliegendes reflectors, including particularly the Base Rotliegendes reflector were also considered for mapping. Interference by multiples, however, does disrupt reflector continuity and reliable character recognition away from the well control.

The reflection character of each horizon was identified using the vertical well synthetic seismograms tied, where possible, to each of the various vintages of data outlined in Table 1.

Interpretation procedure. The AUK84B–watergun dataset formed the framework for the latest interpretation because it is areally extensive, allowing ties with all other datasets, and is also tied to the wells. With the aid of checkshot times and synthetic seismograms, the six horizons were picked and loop-tied on the 1984 dataset. The older datasets were then incorporated and were time shifted as necessary to effect ties. With the exception of the Top Rotliegendes event, the other markers show good reflection continuity and are not significantly masked by multiple interference. The amplitude of the Top Rotliegendes event occasionally falls off making picking difficult. Coupled with the complex faulting at this level, picking was carried out with reference to the nearest dip lines across the field.

Computer-contoured maps of each horizon were generated using an appropriate contour interval. These maps were used to pick out the fault trends and were also used to resolve mistie errors greater than ± 5 m. Faults were then linked up using this information by laying out parallel dip lines to follow fault character. The trends for minor faults that could not be resolved using this approach were drawn following the structural grain. The data was contoured using the linked-up fault-pattern maps. Time structure maps were produced for all horizons.

Depth conversion

The approach adopted for depth conversion over the Indefatigable Field was to use apparent interval velocities for the following six layers:

Mean Sea Level–Base Tertiary
Base Tertiary–Base Chalk
Base Chalk–Top Triassic
Top Triassic–Top Bunter Sandstone
Top Bunter Sandstone–Top Zechstein
Top Zechstein–Top Rotliegendes

Velocities were derived from deviated as well as vertical wells.

Geophysical definition of the trap

The Top Rotliegendes structure is very complex, resulting from continuous structural development through repeated fault movement. Timing these movements is difficult because, with a few exceptions, the faulting is decoupled from the overburden by the Zechstein evaporites. The overall structural grain is one of NW–SE-trending faults superimposed on a mildly folded Rotliegendes surface. A major NW–SE-trending fault defines the eastern boundary of a complex horst that runs the whole length of the field (Fig. 2). In the north (B Platform area) this horst has more of a 'pop-up'-type character with a vertical relief of some 1000 ft, indicative of wrench movement at depth. Towards the southeast, the structure is a classical flat-topped horst. The eastern margin of the field is bounded by faults apart from a down-thrown area northwest of the 49/19-M4 bottom hole location.

Trap

The Indefatigable Field is a structurally complex anticline that resulted from the compounded effects of several periods of tectonic activity. The trapping style is structural. The field is broadly divisible into three structurally separate gas-bearing accumulations (Fig. 1).

(i) the main area
(ii) southwest structure
(iii) southeast structure.

The main area. This comprises a NW–SE-trending horst block approximately 15 miles long, and a flank area which lies to the southwest. The main horst block is anticlinal with flanks dipping towards the boundary faults. The axis plunges gently away from the crest, which is at an approximate depth of 7500 ft. The main horst is divided into several smaller fault blocks by a series of NW–SE sealing faults giving several different gas–water contacts. Maximum vertical relief of the main horst block is 1300 ft. This structure is drained by two platforms of the Amoco Group (A and B) and three Shell/Esso operated platforms (J, K and M).

To the southwest of the main horst block lies a broad flank area of lower relief which is in partial reservoir communication with the main horst block. This has a maximum vertical relief of 500 ft and is partially fault-bounded to the southwest. This feature is drained by the Amoco Group C Platform and Shell/Esso's L platform.

Southwest structure. This feature is separated from the main field area by a small graben, and comprises two separate structures. The larger structure lies to the northwest and was discovered by the 49/23-2 appraisal well. Relief is low, with only 100 ft of vertical closure. This area is drained by the Amoco Operated Group D platform.

Southeast structure. This feature is a small NW–SE-trending horst block, structurally separate from the main area, with a vertical closure of 200 ft. This area is drained by the Shell/Esso N platform.

The Indefatigable Field area as a whole is heavily faulted into several elongated horsts and grabens. The dominant set of faults trend NW–SE with minor faulting in N–S and E–W directions.

The throw on the faults vary, ranging from a maximum of over 1000 ft on the boundary fault of the main horst, to a minimum of less than 50 ft. With the exception of the main boundary faults, there are three sealing faults giving rise to differing gas–water contacts (Fig. 2). As a result, the field area has five defined gas–water contacts and two areas where no contacts have been established, with gas present down to the base of the reservoir (Top Carboniferous).

The main area of the field has a gas–water contact at 8878 ft TVSS, established by wells 49/18-3 and 49/23-C2 (Fig. 2). This part of the field is drained by wells from the A, C (except C11), J, K and M (except M4) platforms. The area lying to the north of a series of sealing faults, and drained by wells from the B platform, 49/18-A2 and 49/19-M4, has a gas–water contact of 8859 ft TVSS. This was established by a small transition zone in 49/18-2, 49/18-B10 and 49/18-B8 and by pressure data from 49/19-M4 (Repeat Formation Tests).

The southwest feature is separated from the main structure by a small graben and comprises two separate structures with two gas–water contacts. The northern structure has a gas–water contact of 8768 ft TVSS, established by 49/23-2 and 49/23-D3, and the southern feature has a gas–water contact of 8798 ft TVSS established by 49/23-D2.

The southeast area, which is also structurally separate from the main field, has a gas–water contact of 8706 ft TVSS, established by 49/24-2.

The area drained by the development wells of platform 49/24-L did not encounter a gas–water contact. Pressure data from the deepest well, however, indicates that there is a gas gradient down to Top Carboniferous at 8897 ft TVSS, which is interpreted as the level of lowest gas in this structure. This structure is separated from the west flank area by a sealing fault.

The part of the west flank area that is drained by wells deviated to the west of platform C and 49/23-D1, has no contact present. The 49/23-C11 well, however, had gas down to the Top Carboniferous at 8934 ft TVSS, and thus the area west of a series of sealing faults is considered to have gas down to this depth. (Fig. 2).

The overlying Kupferschiefer, Zechstein carbonates and evaporites provide an excellent seal for the reservoir and in no part of the field has this seal been breached. The reservoir is full to structural spill point.

Reservoir

The reservoir of the Indefatigable Field is Early Permian Rotliegendes sandstones. These reservoir sandstones tend to be a greyish colour in the upper 50–100 ft and red–brown below. Reservoir quality is good with average porosity for the field of around 15% (ranging 5–22%).

Relatively few shale interbeds are seen in the reservoir sandstone sequence. The unconformable contact with the underlying Carboniferous is sometimes marked by a thin bed of conglomerate or breccia containing clasts of reworked Carboniferous shale. The reservoir is divided into three categories (Van Veen 1975).

(i) An upper section of aeolian sandstones represented by massive cemented sandstones which have been reworked by the transgressive Zechstein Sea.
(ii) A middle section of aeolian sandstones with excellent reservoir characteristics, represented by cross-bedded dune sandstones.
(iii) A lower section of fluvial/wadi-type deposits commonly mixed with interbedded aeolian deposits.

All three zones are not always present and there is considerable variation within each zone, hence correlation of the zones within the field can be difficult. The overall Rotliegendes sandstone, however, is very easily correlated since the top is defined by the regionally present Kupferschiefer, which is represented by a 'spike' on gamma-ray logs, and the base is marked by the high gamma-ray response associated with shales at the top of the Carboniferous sequence.

Following the Variscan orogeny, the folded and uplifted rocks of the Hercynian mountain chain, which lay to the south and southeast, were subjected to erosion. The products of this erosion were deposited in a broad basin bounded by the London–Brabant Massif to the south, and the mid North Sea high to the north (France 1975). The sequence of red aeolian sediments, which form the bulk of the Rotliegendes reservoir, was deposited in a continental environment under very arid climatic conditions.

In the field area, the earliest sediments deposited on the post-Variscan erosional surface were derived from rivers flowing from the southern mountains. These sediments were transported northwards when sporadic heavy rainfall gave rise to flash floods. Intervening arid periods account for the minor aeolian deposits seen locally developed in this part of the section. During the succeeding period of increased aridity, the fluvially derived deposits were reworked into extensive aeolian dune and interdune deposits.

These wind-transported deposits are clean and of good reservoir quality.

At the end of the Early Permian, the Zechstein Sea rapidly transgressed the Rotliegendes, reworking the aeolian deposits. This reworking of the upper aeolian deposits, resulted in the removal of the iron-rich hematite and the sandstones became cemented with early calcite and gypsum (now seen as anhydrite) resulting in the greyish colour (Glennie et al. 1978).

The reservoir section has been extensively cored and for the whole reservoir average core porosity is 15% and geometric average permeability is 30 md. Average porosity for the gas-bearing section tends to be slightly higher at 16%.

Early diagenesis in the area was limited to calcite cementation in the wadi sandstones as the sediments dried, and iron-rich (hematite) clay coating of the grains in the aeolian sandstones where they occurred below the water table. Late diagenesis occurred as a result of deep burial followed by later uplift resulting in recrystallization of clays, authigenic feldspars, dolomite and anhydrite. Diagenesis did not lead to severe porosity reduction of the reservoir in the Indefatigable area.

Source

The source of the gas for the Indefatigable Field is the Carboniferous Westphalian Coal Measures. Unfortunately, the Carboniferous in the field area has only limited penetration (maximum of 380 ft in 49/19-2A). Vitrinite reflectance studies indicate that the Coal Measures are mature in the area, with vitrinite reflectance values ranging from 1.07 to 1.39. Total organic carbon averages 1–2% and the kerogen type is predominantly Vitrinite and Inertinite. Migration of gas from the Carboniferous into the Rotliegendes was via Carboniferous sandstones that subcrop the Rotliegendes in the area, and also along faults that terminate within the Zechstein Section. The time of gas generation is probably Late Jurassic to Early Cretaceous.

Hydrocarbons

The Indefatigable reservoir contains an essentially dry methane gas, consistent with derivation from coal, with a specific gravity of 0.614. The gas is made up of:

methane	92 mol %
ethane	3.4 mol %
nitrogen	2.7 mol %
carbon dioxide	0.5 mol %
hydrogen sulphide	0 mol %

The condensate liquid ratio yield is 2.9661 BBL/MMSCF. Formation water contains 196 200 ppm NaCl equivalent and has a resistivity of 0.0185 ohm m at 195°F and a pH of 6.8.

Initial well potentials were in excess of 50 MMSCFD; test flowrates however were limited to 50 MMSCFD in accordance with safe working practices. In the majority of cases well absolute open-flow potentials were greater than 300 MMSCFD.

Initially, the reservoir was at a pressure of 4122 psia and a temperature 195°F at a datum level of 8500 ft TVSS. Reservoir pressure now varies between 700–3500 psia across the field.

Reserves/resources

The reservoir drive mechanism is primarily by volumetric expansion of the reservoir gas. Rock compressibility and aquifer support has been estimated to provide less than 2% of the reservoir energy.

With the currently installed compression facility of 120 000 horsepower, flowing wellhead pressures of around 250 psia can eventually be attained giving a reservoir abandonment pressure of 450 psia. Recoverable reserves are estimated to be 4418 BCF.

Condensate yield currently averages 1.7 BBL/MMSCF and free water production from most wells is less than 1 BBL/MMSCF, which is easily lifted from the wellbore with the present tubing performance characteristics and therefore does not present a production problem.

Indefatigable Field data summary

Trap
Type	Structural: anticlinal horst
Depth to crest	7500 ft
Lowest closing contour	8950 ft TVSS
Hydrocarbon contacts	Variable; main field at 8878 ft TVSS
Gas column	1360 ft

Pay zone
Formation	Rotliegendes Sandstone
Age	Early Permian
Gross thickness (average/range)	150 ft (115–423 ft)
Sand/shale ratio	1 : 0
Net/gross ratio (average/range)	1 : 0 (above GWC)
Cut-off for N/G	Gas–water contact
Porosity (average/range)	15% (5–22%)
Hydrocarbon saturation	75–80%
Permeability (average/range)	30 md (0.9–1000 md); geometric

Hydrocarbons
Condensate yield	Initial 2.89 BBL/MMSCF; present 1.70 BBL/MMSCF
Formation volume factor	Initial 227.6 scf/cu ft; present 81.3 scf/cu ft

Formation water
Salinity	196 200 ppm NaCl equivalent
Resistivity	0.0185 ohm m at 195°F

Reservoir conditions
Temperature	195°F (at 8500 ft TVSS)
Pressure	Initial 4122 psia (at 8500 ft TVSS); present 700–3500 psia
Pressure gradient in reservoir	Initial 0.074 psi/ft; present 0.026 psi/ft

Field size
Area	60 sq miles
Gross rock volume	205.3×10^9 cu ft
Drive mechanism	Volumetric gas expansion
Recoverable hydrocarbons	
NGL/condensate	8.47 MMBBL
gas	4418 BCF

Production
Start-up-date	September 1971
Development scheme	9 drilling/production complexes with 1 central gathering/compression platform
Production rate	Plateau yearly rate was 678 MMSCFD which corresponds to peak deliverability of 1132 MMSCFD

The authors wish to thank Amoco (UK) Exploration Company and its partners, Gas Council (Exploration), Amerada Hess Ltd, and Texas Eastern North Sea, Inc. and Shell/Esso for their permission to publish this paper. There have been many years of geological, geophysical and engineering investigation in the Indefatigable area. This paper represents only a brief summary of the results of this investigation, and the authors wish to thank all those who have previously worked the area, without whose efforts this paper would not have been possible.

References

FRANCE, D. S. 1975. Geology of the Indefatigable Gas Field. *In*: WOODLAND, A. W. (ed.) *Petroleum and the Continental Shelf of N.W. Europe, Volume 1, Geology*. Applied Science, England, 233–239.

GLENNIE, K. W., MUDD, G. C. & NAGTEGAAL, P. J. C. 1978. Depositional Environment and Diagenesis of Permian Rotliegendes sandstones in Leman Bank and Sole Pit areas of the UK Southern North Sea. *Journal of the Geological Society, London*, **135**, 25–34.

VAN VEEN, F. R. 1975. Geology of the Leman Gas Field. *In*: WOODLAND A. W. (ed.) *Petroleum and the Continental Shelf of N.W. Europe, Volume 1, Geology*. Applied Science, England 223–231.

The Leman Field, Blocks 49/26, 49/27, 49/28, 53/1, 53/2, UK North Sea

A. P. HILLIER[1] and B. P. J. WILLIAMS[2]

[1] Shell UK Expro, PO Box 4, Lothing Depot, North Quay, Lowestoft, Suffolk NR32 2TH, UK
[2] Department of Geology and Petroleum Geology, University of Aberdeen, Aberdeen AB9 1AS, UK

Abstract: Discovered in 1966 and starting production in 1968, Leman was the second gas field to come into production in the UK sector of the North Sea. It is classified as a giant field with an estimated ultimate recovery of 11 500 BCF of gas in the aeolian dune sands of the Rotliegend Group. The field extends over five blocks and is being developed by two groups with Shell and Amoco being the operators. Despite being such an old field development drilling is still ongoing in the field with the less permeable northwest area currently being developed.

The Leman Field is situated in the southern part of the British sector of the North Sea some 30 miles NE of Bacton on the Norfolk Coast. It is classified as a giant gas field with an estimated ultimate recovery of 11 500 BCF gas and an associated 9.7 MMBBL of condensate.

The Leman Field extends into five licence Blocks (UK Blocks 49/26, 49/27, 49/28, 53/1 and 53/2) in which a total of 13 companies have separate interests by virtue of their production licences. These thirteen companies are combined into two production groups, Shell/Esso and the East Leman Unit (ELU), for which Shell and Amoco are the respective operators.

The field comprises a faulted, elongate periclinal structure oriented NW–SE (Fig. 1) and covers an area of some 150 square miles. The field is located at the southern end of the Sole Pit Basin between two major fault zones: the Dowsing/South Hewett faults and the Swarte Bank hinge line (Fig. 3).

The reservoir comprises the Permian Rotliegend Group Leman Sandstone Formation (550–900 ft thick), capped by the overlying Zechstein evaporites (Fig. 2) and sourced from the underlying Westphalian Coal Measures.

History

Interest in exploration in the southern North Sea was first generated by the discovery in 1959 of the giant Groningen Gas Field in the Netherlands. Extrapolation of geological data from this area to the exposures of similar sandstone in Durham, England, led to the belief that the Leman Sandstone would be present beneath the southern North Sea.

Shell UK Exploration and Production, acting as operator for a 50/50 joint venture with Esso Expro UK Ltd in the UK North Sea, took up several first round licences in 1964.

The Leman Field discovery well (49/26-1), the eleventh to be drilled in the UK sector of the North Sea, was spudded on the 17 December 1965 and abandoned on the 18 April 1966 after testing had proved a potentially large gas accumulation.

The play concept tested was simple. A seismic survey on a 5 km grid with fourfold coverage had been acquired over Block 49/26 in 1964. The deepest seismic reflection recorded corresponded to the base of the upper salt (Leine Halite Fm.) in the Zechstein at which level a very large domal structure 13 by 8 miles was mapped; the discovery well was sited on the crest of the dome.

The Permian Rotliegend Leman Sandstone reservoir was found at a depth of 5914 ft TVSS. The reservoir was 921 ft thick with 804 ft of net gas-bearing sand and with a porosity of 14% above a gas–water contact at 6718 ft TVSS (Fig. 1). A production test over the interval 6334–6454 ft TVSS produced good quality gas at a rate of 17 MMSCFD through a 9/16 inch choke.

Following the discovery well fourteen outstep appraisal wells were drilled to delineate the Leman structure (Fig. 1). Extensions into adjacent licence block areas were subsequently proven by wells 49/27-1 in September 1966, 49/28-1 in October 1967 and 53/2-1 in January 1968. Appraisal wells 49/26-2 to -5, -14 and -25 provide

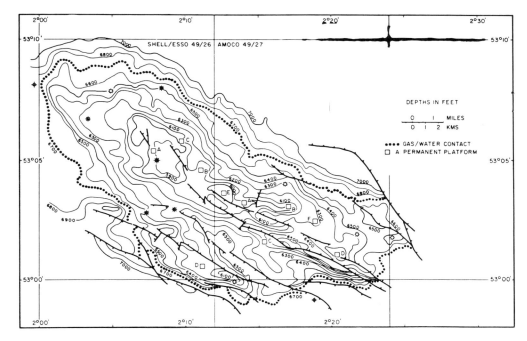

Fig. 1. Structure map of the Leman Field at mid-field development in 1974 showing also the gas–water contact.

From Abbotts, I. L. (ed.), 1991, United Kingdom Oil and Gas Fields, 25 Years Commemorative Volume, Geological Society Memoir No. 14, pp. 451–458

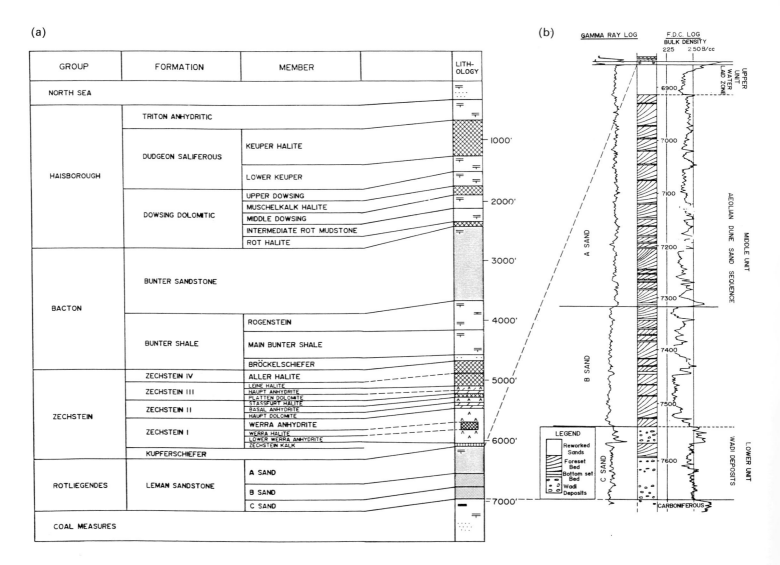

Fig. 2. (a) General lithostratigraphy of the Leman Field. (b) Detailed lithostratigraphy of the Rotliegend sandstone reservoir sequences.

excellent control on the N, NW and SW flanks, whilst the SE flank is defined by Mobil wells 53/2-1 and -2. The northwest flank of the Leman Field was further defined by well 49/26-26 in early 1982.

The Leman Field, following appraisal drilling, now comprises a total of 181 development wells drilled to date from sixteen platforms. Gas delivery onshore is to Bacton via three 30 inch diameter pipelines.

Field stratigraphy

The general stratigraphic sequence of the Leman Field is well known and is illustrated in Fig. 2a. Unconformities are present in the sequence between the Carboniferous and Permian, at the base of the Cretaceous and at the top of the Cretaceous.

Pre-Carboniferous stratigraphy is unknown from either well penetration or seismic interpretation across the field but Old Red Sandstone facies of the Devonian has been encountered in wells to the north and south of the area. The Westphalian Coal Measures sequence comprises mainly purple-grey mudrocks and sandstones with subordinate coals which sourced the gas.

In the UK North Sea the Rotliegend Gp is composed of two formations, the Leman Sandstone Formation which is the reservoir and the Silverpit Formation (Rhys 1975). By common usage, the name Rotliegend has been adopted for the Leman Sandstone Formation.

The Rotliegend sandstone was deposited on the eroded Carboniferous terrain in a hot, arid environment. The Rotliegend is divided into three zones based on depositional environment, waterlain (fluvial), aeolian and a reworked assemblage (Fig. 2b). The source of the fluvial Rotliegend sediment was the London–Brabant Massif, whilst the aeolian dune sands were sourced by denudation of the Variscan Mountains to the east of the field (Glennie et al. 1978).

The Zechstein sediments which cap the Rotliegend are composed of four cycles of evaporites. The Stassfurt Halite in cycle II containing interbedded potassium and magnesium salts exhibits considerable variability in thickness as it exhibits plastic flow in accommodating fault movement in the underlying strata. This unit is important as the major seal for the reservoir.

The Triassic sequence of the Leman field exhibits a continuation of the depositional styles encountered in hot, dry evaporitic settings of hydrologically-closed basins. The basal Triassic unit the Bacton Gp the basal member of which, the Brockelschiefer, is overlain by the lower Bunter Claystone, a red brown claystone, the upper part of which (the Rogenstein Mmbr) is differentiated by the presence of thin oolitic horizons. This in turn is overlain by the Bunter Sandstone Fm., a red brown sandstone with claystone interbeds and lenses, and in which circulation is commonly lost while drilling. The Bunter Sandstone commonly displays good reservoir properties but is not charged with gas in the field.

The upper part of the Triassic, the Haisborough Gp consists of an alternating sequence of red and brown claystones with inter-

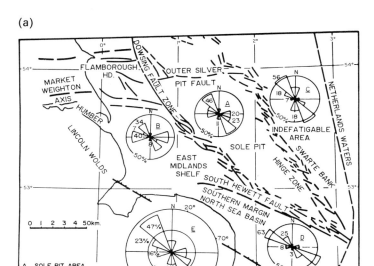

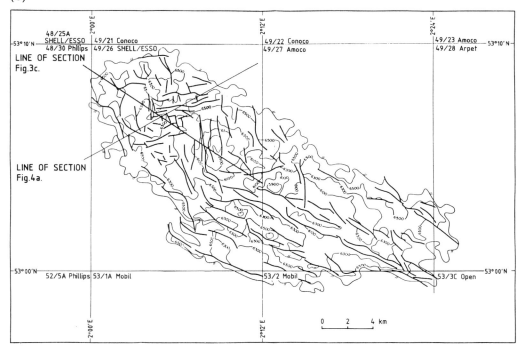

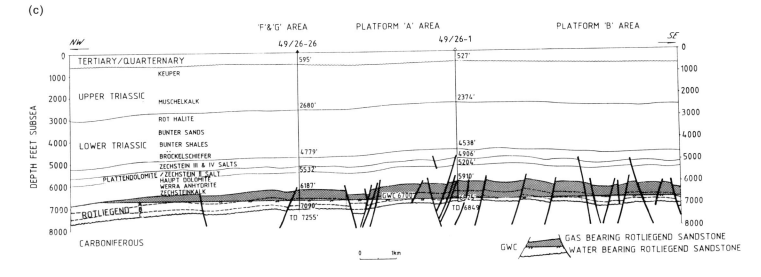

Fig. 3. (a) Tectonic setting of the Leman Field (after Glennie & Boegner 1981). (b) Simplified structural map of the Leman Field based on 1980–81 seismic data interpreted at top Rotliegend sandstone. (c) Structural cross-section of the western half of the Leman Field exhibiting low relief and complexity of fault patterns.

bedded halite and siltstone. A major marine incursion into the area began in the Rhaetic which is followed by the wholly-marine Jurassic claystones and Cretaceous chalk. Although at one time these marine sequences provided extensive cover they have been subsequently eroded over the major part of the Leman field. Undifferentiated Quaternary–Tertiary marine sands and clays occupy the highest stratigraphic position across the field area and comprise some 300 ft or so of sediment directly below the sea bed.

Tectono-stratigraphic history

The southern North Sea Permian sedimentary basin developed over the northern foreland of the Variscan orogenic belt. This mountain belt was created by the northwards movement of Gondwana and its subsequent collision with Laurussia. As a result of these collisional tectonics, the foreland was shortened, uplifted and eroded in the Late Carboniferous.

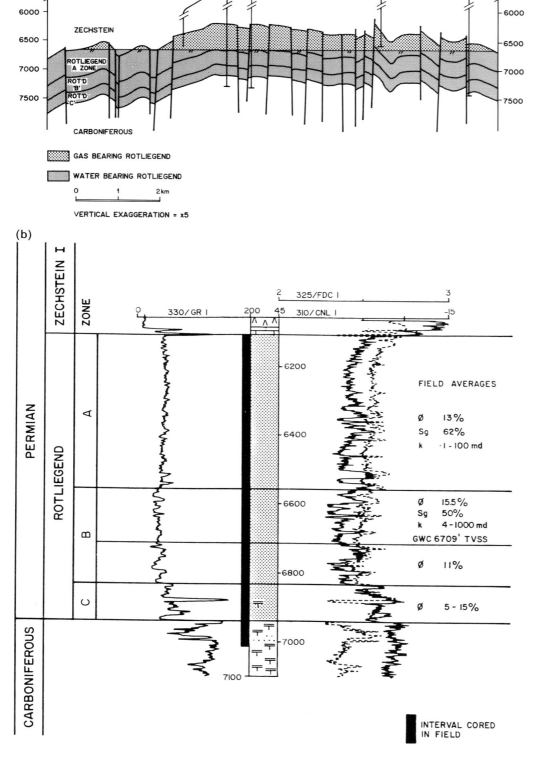

Fig. 4. (a) Structural cross-section of Rotliegend sandstone reservoir, Leman Field showing tripartite reservoir division and gas–water contact. (b) Type log for the Rotliegend sandstone reservoir Leman Field with typical values of poroperm measurements.

The southern North Sea Permian Basin developed on the site of an earlier, Westphalian Basin. By the Late Permian, it became a broad feature, some 900 miles long and 300 miles wide, parallel to and in front of, the east–west Variscan mountain belt.

Sedimentation in the southern North Sea Basin continued into the Late Cretaceous. During the Alpine orogenic phase northwest Europe experienced north–south directed compressional forces which re-activated some of the Variscan structural elements. This gave rise to dextral movement along the existing NW–SE-oriented normal faults. As a result, the Cimmerian–Early Tertiary sedimentary basins were inverted, e.g. Sole Pit and West Netherlands Basins (Ziegler 1982). The inversion resulted in erosion of up to 4000 ft of sediments from the area of the Leman Field. From the Eocene onwards, the compressional stress regime relaxed and the North Sea subsided. A continuous, undisturbed stratigraphic sequence was deposited from Miocene to present.

The Leman Field is located at the southern end of the Sole Pit Basin between two major fault zones; the Dowsing/South Hewett faults and the Swarte Bank Hinge (Fig. 3a). These are part of a NW–SE-trending set of steep normal faults, created during the Variscan orogeny. They were reactivated (with right-lateral strike slip movements) during the Alpine orogeny and were most active during late Cretaceous and early Tertiary (Glennie & Boegner 1981). The Alpine compressive stress regime which was oriented in a NNE–SSW direction, resulted in the right-lateral strike slip movement.

Geophysics

In 1964 seismic data were acquired over the area on a 5 km spacing with only fourfold coverage. This grid delineated structural highs at the deepest visible reflector which was within the Zechstein. From onshore data the Rotliegend was expected to lie more or less conformably below this reflector. Hence the discovery well was located on the crest of the seismic time high.

The refined structural image of the Leman Field was derived from the seismic surveys acquired in 1980 and 1981 (200 m by 400 m grid over the crest of the structure and 400 m by 400 m over the flank areas). These data, which were the first surveys in which the reservoir could be imaged and directly mapped, were first interpreted in 1983. The current map is based on a re-interpretation and remapping study carried out in 1986 and a 1987 study of 217 km of seismic data over the F and G development areas which was reprocessed, interpreted and depth converted.

The following seismic markers were picked and used for depth conversion of the seismic data using a layer-cake model with seismic velocity maps being derived for each layer based on well intersections and well-shoot data. Each depth-converted map was tied in to well data.

Top Röt Evaporite
Top Bröckelschiefer
Top Haupt Anhydrite/Platten Dolomite
Top Rotliegend.

Trap

The Leman Field is a single reservoir, dip-closed anticlinal trap. The maximum closure is 1100 ft with the top and lateral seals being formed by the Zechstein cap rock. The gas–water contact for the field is taken as 6700 ft TVSS, some 200 ft above the deepest dip closure. The exact gas–water contact is difficult to define due to capillary effects in the reservoir which result in long transition zones between gas and water. There are no strategic unconformities within the reservoir with the exception of major aeolian bounding surfaces. The original reservoir pressure of 3200 psi at 6500 ft TVSS was slightly above hydrostatic.

The trap was probably initially formed at the time of the Late Cimmerian inversion of the Sole Pit Basin and has possibly been enhanced by late Cretaceous and Early Tertiary phases on inversion.

Reservoir

In the Leman Field the Rotliegend has been divided into three units based on mode of deposition. From the base upwards these are: wadi, dune and reworked or waterlain sediments (van Veen 1975). These are equivalent to the fluvial, aeolian and 'Weissliegend' of Glennie (1986) (Fig. 2b).

The entire Rotliegend sandstone contains reservoir quality rock. The aeolian dune sands can be divided into two reservoir zones, 'A' and 'B' on the basis of porosity/permeability relationships. A study of the dune sands shows that the upper part, about 400 ft, is more cemented than the lower part (Arthur et al. 1986). Hence for reservoir simulation work a threefold zonation of the reservoir was established with the third zone 'C' being composed of the wadi deposits (Fig. 4a). Table 1 shows average porosity and permeability measurements for these zones. These results were derived from core data from 32 wells in the field which adequately defined the petrophysical parameters (Fig. 4b).

The C zone is of little commercial importance as a reservoir as it is below the gas–water contact over most of the field and has a low porosity due to poor sorting and early cementation exhibited by the sandstones.

The B zone consists of stacked, grey-green or red-brown, fine- to medium-grained aeolian dune sandstones. These sandstones are made up of foreset beds which have millimetre to centimetre laminae dipping at about 25°. Each lamina is itself extremely well sorted; but there is great variation in median grain size from one lamina to the next. This variation causes large changes in permeability (Fig. 5a). Each dune consists of a series of foreset beds separated from the preceding dune by poorly sorted, horizontally stratified bottomsets. The B zone sandstones are in general poorly cemented and comprise the best quality reservoir rocks in the field. The zone varies in thickness between 25 ft and 400 ft.

The A Zone consists of both dune sands and the Weissliegend (Fig. 2b). The A zone is differentiated from the B zone by the increased amount of interstitial cement, particularly grain-coating hematite and illite. The illite, occurring in a fibrous form, tends to

Table 1. *Average core porosity and geometric mean permeability for the zones of the Rotliegend in Leman Field. Note the increase in reservoir properties to the east in the Block 49/27.*

Zone	49/26			49/27		
	Number of samples	Average porosity (%)	Average permeability (md)	Number of samples	Average porosity (%)	Average permeability (md)
A	4389	12.05	1.02	2638	13.70	2.45
B	1675	13.54	6.03	1264	15.06	15.60
C	693	9.30	0.55	340	12.61	4.67

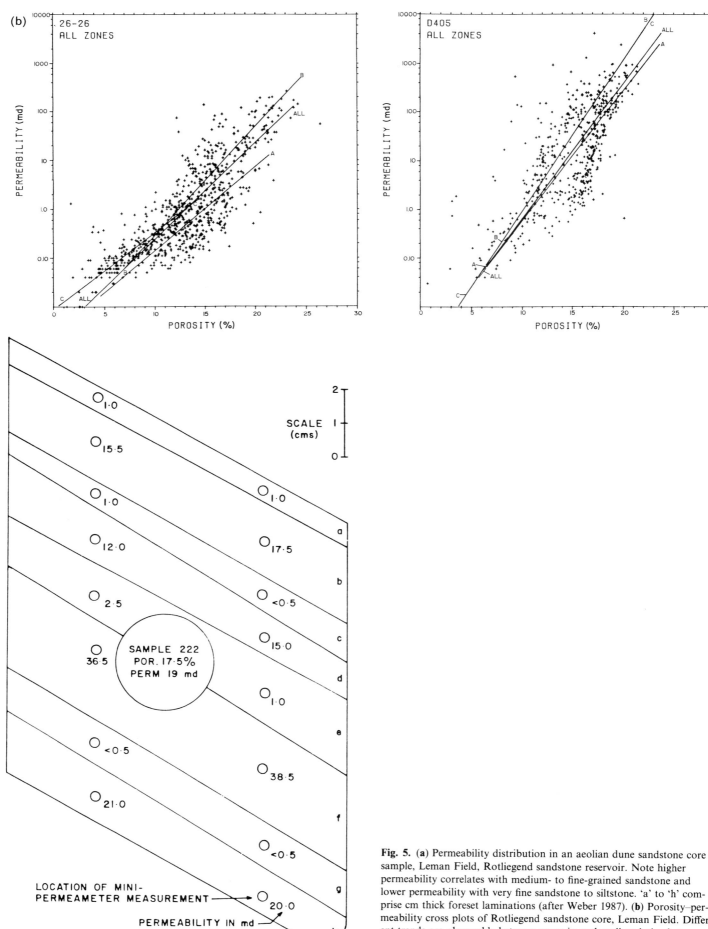

Fig. 5. (a) Permeability distribution in an aeolian dune sandstone core sample, Leman Field, Rotliegend sandstone reservoir. Note higher permeability correlates with medium- to fine-grained sandstone and lower permeability with very fine sandstone to siltstone. 'a' to 'h' comprise cm thick foreset laminations (after Weber 1987). (b) Porosity–permeability cross plots of Rotliegend sandstone core, Leman Field. Different trends are observable between zones in each well and also between wells 49/26-26 in the northwest and D405 in the centre of the field.

have greater effect on the permeability of the rock than would be expected from its effect on the porosity (Stalder 1973). The source of this additional cement is not fully understood but it could be derived from either a large proportion of unstable silicates transported from the hinterland or a greater proportion of detrital clay derived from the sabkha facies which was present in the Sole Pit area to the north. Cementation within the uppermost Rotliegend is further complicated by the influence of the mineral-rich Zechstein pore water (Glennie et al. 1978). The thickness of the A Zone varies between 100 ft and 650 ft. The A Zone, although volumetrically greater than the other two zones, is a poorer reservoir than the B zone due to this increased cementation. It has been shown that the permeability for a given porosity in the B zone is three to seven times higher than that in the A zone.

The porosity in all the Rotliegend zones is intergranular and modified by the diagenetic effects of palaeoburial. As the field was buried unevenly (the northwest some 4000 ft and the southeast some 3000 ft deeper than the present depth) the diagenetic effects are accordingly uneven across the field. Diagenesis consists of quartz overgrowths, dolomite and fibrous illite deposition in the pore throats (Glennie et al. 1978; Arthur et al. 1986). The depositional setting of the field is interpreted to be the main dune field between the eroding London–Brabant Platform and a major desert lake to the north (Fig. 6).

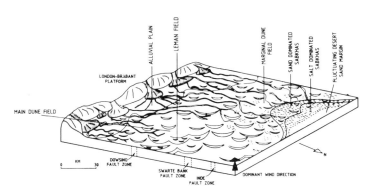

Fig. 6. Palaeographic setting of the Leman Field showing the main deposystems.

Fractures, both open and cemented, are visibly in virtually all the cored wells in the field. Their contribution to the flow of the wells in the majority of the field is insignificant; but in the northwest where the deeper palaeoburial caused more severe diagenetic permeability reduction they have a measurable effect, raising the productivity of wells by up to 50%.

Porosity and permeability varies considerably across the field. Cross plots of porosity and permeability show an areal variation in the relationship: for a given porosity, lower permeability is found in the northwest. Cross plots for two wells are shown in Fig. 5b. The individual samples for a given porosity show a wide variation in permeability. This is due to the depositional environment e.g. well-sorted dune sand lamina with similar porosity. There is, however, a great variation in median grain size and therefore in the permeability. Capillary pressure curves show that there is a very long transition zone above the free water level especially in the A and C zones. The pay zone comprises the entire Rotliegend in the most crestal parts of the field and is up to 275 m thick. However, due to the low general relief of the field some 80% of the gas is contained in the A zone.

The major barriers to flow within the reservoir are related to faulting and depositional style. Faulting, particularly in the southern and eastern parts of the field, created NW–SE trending compartments which have very little pressure communication with each other. On the other hand wind-deposited dune bottomset beds separate the cross-bedded dune foresets. The bottomset beds, which are finely laminated, poorly sorted and carbonate cemented, are relatively poor reservoir rocks. Their permeabilities are one or even two order of magnitude lower than those of the better quality dune avalanche foresets. The dune foreset beds have a strongly anisotropic permeability distribution. The permeability along a dune slip face is much higher than the perpendicular to the slip face. Weber (1987) showed that, by analogy to the Permian de Chelly sandstone in northeastern Arizona, length : width : height ratios of around 200 : 75 : 1 can be expected. Given an average cross-bed height of 15 ft an avalanche set in the Leman Field would be 3000 ft long and 1000 ft wide.

Source

The source for the gas in the Leman Field, and indeed for all the other southern North Sea gas fields (Martin & Evans 1988), is the Westphalian Coal Measures which directly underlie the reservoir. Migration paths are supplied by the sandstones within the Westphalian which have extensive areas of contact with both the coals and carbonaceous shales that are the actual sources. These sandstones form the conduits from the source at generation time to the reservoir. The reservoir unconformably overlies the Westphalian and hence is in very good hydraulic connection.

The time of migration coincided with the time of maximum depth of burial during the Jurassic and Cretaceous. The gas first migrated to the then structurally higher flanks of the Sole Pit Basin. Later, during and after the structural inversion and formation of the trap in the Late Cretaceous, the gas re-migrated back into the field. The fact that the gas remained trapped until the present demonstrates the efficiency of the seal formed by the Zechstein evaporites.

The source has two components, carbonaceous shales (1% TOC) and coals (60%) TOC. The kerogen type is II/III-III. The potential yield for the shale is 0.14 MCF/acre ft and for the coal 7.0 MCF/acre ft (Cornford 1986). The source rock has vitrinite reflectance maturation of 1.6 to 2.1.

Hydrocarbons

The gas is sweet and lean with 95% methane, very little carbon dioxide and a condensate gas ratio of only 1BBL/MMCF. An analysis is given in Table 2.

Reserves/resources

Although no individual fault in the field has sufficient throw to offset the reservoir there is strong evidence that some faults act as permeability barriers within the reservoir. This necessitates a wider spread of drainage points to fully develop the reserves. This aspect was unknown at the time of discovery of the field as the reservoir was not directly mapped from the available seismic data. Given today's seismic techniques, a full three dimensional seismic survey would enable the reservoir to have been imaged in a much more detailed fashion before the development plan was made.

The reduced permeability of the A zone relative to the B zone also allowed large pressure gradients to develop in areas where the B zone was below the gas–water contact, thus effectively sealing off areas of the field from drainage. The original concept of the field behaving like a gas cylinder, and therefore needing only a limited number of centrally located drainage points to develop, has been found to be incorrect. The use of RFT pressure measurements in recent wells has shown that the gas sands are being depleted, but, due to much lower mobility of water relative to gas, the pressures below the gas–water contact are remaining high.

Nevertheless the Leman Field is still regarded as a giant gas accumulation (Carmalt & St John 1986) with an estimated ultimate recovery of 11 500 BCF of gas and an associated 9.7 MMBBL of condensate.

Table 2. *Composition of separator gas, separator oil, (condensate) and reservoir fluid from well 49/26-1 Leman Field.*

Component	Separator gas mol %	Separator oil mol %	Reservoir fluid mol %
Methane	95.05	14.00	94.94
Ethane	2.86	4.41	2.86
Propane	0.49	3.62	0.49
Iso Butane	0.08	1.68	0.08
N-butane	0.09	3.71	0.10
Iso-pentane	0.03	3.02	0.03
N-pentane	0.02	3.70	0.03
Hexanes	0.02	8.54	0.03
Heptanes plus	0.04	57.27	0.12
Helium	0.02	0.00	0.02
Nitrogen	1.26	0.02	1.26
Carbon dioxide	0.04	0.03	0.04
Total	100.00	100.00	100.00

Based on composition of tank gas and tank oil obtained at 77°F (25°C).

Leman Field data summary

Trap
Type — faulted anticline
Depth to crest — 5900 ft TVSS
Lowest closing contour — 6900 ft TVSS
Free water levels — 6717 ft TVSS
Gas column — 800 ft

Pay zone
Formation — Leman Sandstone
Age — Lower Permian
Gross thickness — 800 ft
Net/gross ratio — 1.00
Net sand cut-offs — none used
Porosity — 12.9% average all zones
Gas saturation — 51% average all zones
Matrix permeability — 0.5–15 md

Hydrocarbons
Gas gravity — 0.585
Gas type — sweet dry gas
Condensate yield — 1.0 BBL/MMSCF

Formation water
Salinity — 240 000 ppm sodium chloride equivalent
Resistivity — 0.026 ohm m at 125°F

Reservoir conditions
Temperature — 125°F
Pressure — 3022 psi at datum 6500 ft TVSS
Pressure gradient — 0.08 psi/ft (gas leg)
0.46 psi/ft (water leg)

Field size
Area — 69 456 acres
Gas expansion factor — 229 (SCF/RCF)
Gas-in-place — 13 862 BCF
Drive mechanism — depletion
Recovery factor — 83%
Recoverable reserves — 11 523 BCF

Production
First gas — August 1968
Development scheme — Multiple wellhead platforms on both licence blocks linked to three pipelines to Bacton

Shell UK Exploration and Production, Esso Expro UK Ltd and the East Leman Unit permitted publication of this paper. The advice and assistance of numerous colleagues within Shell and the use of much unpublished data on the field from internal reports has greatly eased the job of the authors. J. Chalmers (University of Aberdeen) is thanked for typing the manuscript under manic conditions imposed upon her by one of the authors!

References

ARTHUR, T. J., PILLING, D., BUSH, D. & MAACHI, L. 1986. The Leman Sandstone Formation in U.K. Block 49/28. Sedimentation, diagenesis and burial history. *In*: BROOKS, J., GOFF, J. & VAN HOORN, B. (eds) *Habitat of Palaeozoic Gas in N.W. Europe*. Geological Society, London, Special Publication, **23**, 251–266.

CARMALT, S. W. & ST JOHN, B. 1986. Giant Oil and Gas Fields. *In*: HALBOUTY, M. T. (ed.) *Future Petroleum Provinces of the World*. American Association of Petroleum Geologists, Memoir, **40**, 11–54.

CORNFORD, C. 1986. Source Rocks and Hydrocarbons of the North Sea. *In*: GLENNIE K. W. (ed.) *Introduction to Petroleum Geology of the North Sea*. Blackwell Scientific Publications, London, 197–236.

GLENNIE, K. W. 1986. Early Permian–Rotliegend. *In*: GLENNIE, K. W. (ed.) *Introduction to Petroleum Geology of the North Sea*. Blackwell Scientific Publications, London, 63–85.

—— & BOEGNER, P. L. E. 1981. Sole Pit Inversion Tectonics. *In*: ILLING, L. V. & HOBSON, G. D. (eds) *Petroleum Geology of the Continental Shelf of North West Europe*. Heyden and Son, London, 110–120.

——, MUDD, G. C. & NAGTEGAAL, P. J. C. 1978. Depositional environment and diagenesis of Permian Rotliegendes Sandstone in Leman Bank and Sole Pit areas of the U.K. Southern North Sea. *Journal of the Geological Society, London*, **135**, 25–34.

MARTIN, J. H. & EVANS, P. F. 1988. Reservoir modelling of marginal aeolian/sabkha sequences, Southern North Sea (U.K. Sector). *Society of Petroleum Engineers*, 473–486.

RHYS, G. H. 1975. A proposed standard lithostratigraphic nomenclature for the Southern North Sea. *In*: WOODLAND, A. W. (ed.) *Petroleum and the Continental Shelf of North West Europe*. Applied Science Publishers, London, **1**, 151–163.

STALDER, P. J. 1973. Influence of crystallographic habit and aggregate structure of authigenic clay minerals on sandstone permeability. *Geologie, en Mijnbouw*, **52**, 217–220.

VEEN, F. R. VAN, 1975. Geology of the Leman Gas Field: *In*: WOODLAND, A. W. (ed.) *Petroleum and the Continental Shelf of North West Europe*. Applied Science Publishers, London, **1**, 223–231.

WEBER, K. J. 1987. Computation of initial well productivities in aeolian sandstone on the basis of a geological model, Leman Gas Field, U.K.: *In*: TILLMAN, R. W. & WEBER, K. J. (eds) *Reservoir Sedimentology*. Society of Economic Paleontologists and Mineralogists, Special Publication, **40**, 333–354.

ZIEGLER, P. A. 1982. *Geological Atlas of Western and Central Europe*. Elsevier, Amsterdam.

The Ravenspurn North Field, Blocks 42/30, 43/26a, UK North Sea

F. J. KETTER

Hamilton Brothers Oil & Gas Ltd, Devonshire House, Piccadilly, London, W1X 6AQ, UK

Abstract: The Ravenspurn North Field is a gas accumulation located in the Southern North Sea, Permian Gas Basin which was discovered in October 1984. It has undergone four years of appraisal well drilling culminating in the approval of the development plan in 1988. Development wells are currently being drilled and three offshore installations are planned; first gas production began in July 1990.

The Ravenspurn North Field is a combined structural and stratigraphic trap. The reservoir is fault closed along a series of anastomosing oblique strike-slip and normal faults. Seals along the faults are provided by the Silverpit Formation mudstones and Zechstein Group evaporites. The reservoir deteriorates to the northwest because of thinning, facies change and increasing authigenic clay content.

The Lower Leman Sandstone Formation of the Rotliegendes Group forms the reservoir. It consists of a sequence of aeolian dune, fluvial sheetflood, fluvial channels and lake margin sabkha deposits. Non-reservoir intervals are formed by playa lake mudstone sequences. Fluvial and sabkha facies dominate in the northwest while aeolian facies dominate in the southeast parts of the Field.

Reservoir quality was initially controlled by lithofacies distribution. Subsequent diagenesis further modified the reservoir rock resulting in variations in the porosity and permeability. Deliverability is a function of variable permeability with two areas identified; the high deliverability area where gaswells have tested sufficient quantities for commercial production without artificial stimulation and a low deliverability area where gaswells require hydraulic fracture stimulation before significant commercial production rates are achieved.

The Ravenspurn North Unit is located 50 miles east of Scarborough in the UK Sector of the Southern Permian Gas Basin, (Fig. 1). It is one of the most northern Rotliegendes gas discoveries to date located at approximately latitude 54° 10′ North 1° 00′ East.

The field covers an area approximately 12 miles by 2 miles straddling the boundary between Quadrants 42 and 43. Ravenspurn North is a dry gas reservoir found in primarily a structural trap. There is an element of stratigraphic trapping caused by deteriorating sandstone quality to the northwest along structural strike.

The Permian Lower Leman Sandstone Formation within the Rotliegendes Group forms the reservoir rock. The overlying claystones of the Silverpit Formation form the seal. The reservoir drive is interpreted to be depletion type with some limited water drive available in certain fault blocks.

The Field was named to commemorate a former seaside village in Humberside which was washed away by the encroaching North Sea.

History

The Ravenspurn North Field encompasses part of two UK Offshore Licences. Block 43/26 constitutes licence P380 and Blocks 42/30, 42/29 and 48/6 constitute licence P001. Blocks 42/30, 42/29 and 48/6 were awarded to BP Petroleum Development Ltd as sole licensee in P001 in 1965.

Block 43/26 was awarded to Hamilton Brothers, Tricentrol, Blackfriars and Trans-European (Kleinwort Benson Energy Ltd) in

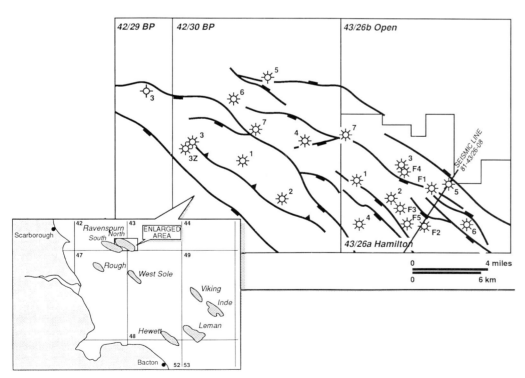

Fig. 1. Ravenspurn North, location map showing appraisal and early development wells.

1981 as part of the UK 7th Round of Offshore Licensing. In 1984 Enterprise Oil plc and Trafalgar House Ltd farmed into the block. The present participants in Licence P380 are:

Hamilton Oil Great Britain plc	15.0
Hamilton Petroleum (UK) Ltd	7.5
Arco British Ltd	25.0
Enterprise Oil plc	20.0
Ultramar Exploration Ltd	15.0
Hardy Oil and Gas UK plc	10.0
Monument Resources	7.5
	100.0

Hamilton Brothers discovered gas in the Rotliegendes Group Lower Leman Sandstone Formation in October 1984 with Well 43/26-1.

Gas was initially discovered in Block 42/30 in the Ravenspurn area in 1976 when BP drilled well 42/30-1 which encountered 267 feet of hydrocarbon-bearing lower Leman Sandstone. Further drilling of wells 42/30-2, -3, -3Z and 42/29-3 extended the field to the northwest and southeast.

Drilling by the Hamilton Group in 1984 confirmed gas in Block 43/26 when well 43/26-1 encountered 255 feet of gas sands. No definitive gas–water contact was identified. Additional drilling of wells 43/26-2, -3, and -5 has extended potential gas production to the north and east. Additional BP wells 42/30-4, -5 and -6 extended the Ravenspurn North Unit to the northwest while Hamilton 43/26-6 extended the gas accumulation further to the southeast and encountered the gas–water contact.

The Secretary of State for Energy designated the Ravenspurn South area and the Ravenspurn North area as a single gas field, and in 1986 a pre-unitization agreement between BP and the Hamilton Group, with Hamilton as operator, established a joint appraisal and development program for the Ravenspurn North Unit. Subsequently BP's Annex B submission for the development of Ravenspurn South and Cleeton was approved in May 1986.

A data exchange area was established for the entire Ravenspurn area and additional appraisal wells were drilled in Blocks 42/30 and 43/26 to define the lateral extent of the lower Leman Sandstones in the Ravenspurn North Unit. Each of these wells has been extensively cored; all have been correlated and zoned through the Lower Leman reservoir. Well 43/26-7 was drilled to investigate the extent of the prime reservoir development, and well 42/30-7 was drilled to test a thick seismically-defined, anomaly at the lower Leman Sandstone level.

The Carboniferous has been drilled and proved to have no reservoir potential.

There are three installations planned for the Ravenspurn North Field. These are as follows:

(1) the Central Complex (installed August 1989) comprising a concrete gravity sub-structure supporting the Central Processing Platform and the Gas Compression Deck, and the bridge-linked Wellhead Tower (WTI), installed in November 1988;
(2) Satellite Tower ST2, installed December 1989;
(3) Satellite Tower ST3, planned installation late-1991.

The Wellhead Tower and the Satellite Towers all have steel framed decks supported on conventional jackets piled in to the seabed.

The Central Complex Wellhead Tower, with 18 slots will be linked to the Central Production Platform by a 50 m bridge.

Satellite Towers ST2 and ST3 with 18 slots each, will be installed approximately 5 km and 13 km respectively to the northwest of the Central Complex.

Development operations began in November of 1988 with the installation of the steel well-tower jacket, WT-1, in the southeast part of the field. Development drilling started in December 1988 and is planned to continue for nearly four years with approximately 45 development wells from three steel well-tower jackets.

Gas from the Central Production Platform will be exported via a 24 inch × 26 km long pipeline to BP's Cleeton Platform, where it will comingle with the gas processed at Cleeton and be transported to shore via BP's 36 in × 61 km trunkline to Dimlington. The gas will then be processed to sales quality at BP's onshore Dimlington terminal, and delivered to the British Gas Distribution Network.

Field stratigraphy

General stratigraphic sequence

The Leman Sandstone Formation is one of two Rotliegendes Group formations characteristic of the Southern Gas Basin; the Silverpit Formation being the other. The Leman Sandstone Formation passes laterally, by interdigitation, into mudstones and siltstones of the Silverpit Formation. A sandstone 'member' is present at the top and/or at the base of the Silverpit Formation. In the Ravenspurn North Unit the lower Leman Sandstone is present and was deposited by both aeolian and fluvial processes, (Fig. 2). The lower Leman Sandstone is between 250 and 300 ft thick and is overlain by approximately 400 ft of Silverpit Formation, siltstones and claystones.

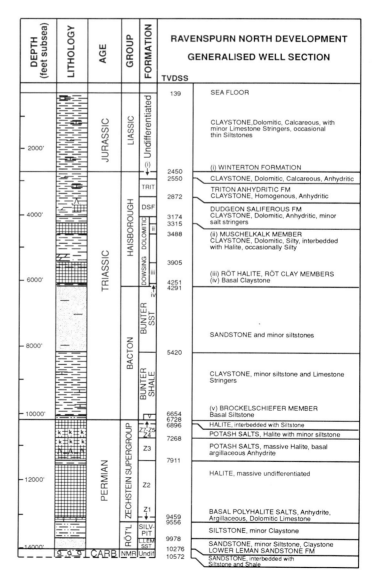

Fig. 2. Ravenspurn North development, generalized well section.

The upper Leman Sandstone is not present at the top of the Silverpit Formation. The base of the Rotliegendes Group is unconformable with the Carboniferous.

Overlying the Silverpit Formation is the Zechstein Supergroup which can be sub-divided into seven Groups. For the purpose of drilling operations it is easiest to divide it into three intervals of varying stability.

The basal interval is composed of interbedded polyhalite salts, anhydrite and dolomitic limestones which can be argillaceous. This division includes the Z1 and the lower part of the Z2 Groups. It does not contain abundant halite sequences and is therefore structurally stable and does not usually cause problems in drilling operations. In the Ravenspurn North area it averages approximately 260 ft thick.

The middle sub-division of the Zechstein Supergroup (upper Z2, Z3, Z4) consists of massive halite, potash salts with minor anhydrites and dolomites. It forms the thickest interval in the Zechstein ranging from 1000 ft to 2500 ft thick in a salt pillow in the southeast part of the field. The middle unit is considered to be a mobile zone and can cause problems during drilling. Intervals of denser anyhydrite and dolomite interbedded in a sequence of otherwise monotonous salt can act as rafts in this matrix of less dense halite. Anhydrite rafts penetrated in the field area are interspersed throughout the mobile interval where they can be overturned and repeated, as well as being hundreds of feet out of place from their original stratigraphic position. The uppermost interval of this middle Zechstein division is a fairly stable sequence of potash salts, interbedded with halites and some siltstones.

The upper division of the Zechstein Supergroup includes the Z5 through to the Z7 Groups, by correlation with marginal sequences found in platform areas surrounding the main Zechstein basin. Halite with interbedded siltstones gives the interval structural stability during drilling phases. It averages approximately 160 ft thick and is uniform across the field area.

Overlying the Zechstein Supergroup is the thick Triassic sequence of the Bunter Shale Formation consisting of claystones with minor siltstones and limestones. The Bunter Shale is generally uniform in thickness averaging approximately 1250 ft thick over the field area.

The Bunter Sandstone Formation overlies the Bunter Shale and consists of sandstones with minor siltstones. Intervals within the Bunter Sandstone can be tightly cemented causing slow drilling rates. This formation averages approximately 1100 ft thick across the entire field area. Though the Bunter Sandstone is known to be gas-bearing in fields to the north, it is not prospective in Ravenspurn North.

The Triassic Haisborough Group overlies the Bunter Sandstone Formation. A lower interval includes the Dowsing Dolomitic Formation consisting of a basal 40 ft thick unit of the Rot Claystone Member, overlain by the Rot Halite Member composed of massive halites approximately 350 ft thick. An interval of dolomitic and silty claystones continues upwards within the Dowsing Dolomitic Formation interrupted by the Muschelkalk Halite Member which varies between 170 ft and 200 ft thick. An interval of non-calcareous claystone about 180 ft thick forms the uppermost part of the Dowsing Dolomitic Formation. In total the Dowsing Dolomitic Formation is approximately 1200 ft thick.

The upper interval of the Haisborough Group consists of three formations which are, from bottom to top, the Dudgeon Saliferous Formation, the Triton Anhydritic Formation and the Winterton Formation. In general, these total approximately 650 ft thick and are composed of claystones with various amounts of calcareous and anhydritic lithologies. The Haisborough Group has no hydrocarbon potential and it presents no problems to drilling operations.

The Lower Jurassic Lias and Middle Jurassic West Sole Groups consist of dolomitic or calcareous claystones with occasional thin limestone stringers which total approximately 2500 ft thick over the central part of the field. These are non-prospective for hydrocarbons and present no problems to drilling.

A thin veneer of Upper Jurassic and Quaternary sediments tens of feet thick overlies the older Jurassic units over central Ravenspurn North. This interval thickens on the flanks up to 500 ft thick. Again no drilling problems are encountered in this interval. Water depths are between 120 ft and 140 ft.

Geological history

Ravenspurn North lies at the northwest end of the geologically complex Sole Pit area, which lies 30–60 miles (50–100 km) east of the Yorkshire, Lincolnshire and Norfolk coasts. The area is bounded by the Pennine uplift to the west, the London–Brabant Massif to the south and the Mid-North Sea High to the north. Ravenspurn North lies immediately north of the Silverpit faults.

The Carboniferous sequence is extensively folded and transected by roughly NW–SE-trending faults generated during the Variscan orogeny. Erosion cut deeply into the Upper Carboniferous sequence. The Variscan NW–SE structural grain controls later development of the southern North Sea Basin. The erosional surface partly controlled early lower Leman Sandstone deposition.

Tectonic trends established during the late Carboniferous and early Permian resulted in subsiding intermontane basins of the Variscan Mountain belt, and in these continental clastics of the Leman Formation sandstones were deposited under increasingly arid conditions. Only minor faulting affected sedimentation. The lower Leman Sandstone deposition terminated with the deposition of the Silverpit Lake siltstones and shales. The Silverpit Formation is considered the reservoir seal in this area.

The Zechstein transgression flooded a basin which shows little contemporaneous fault activity. Late Permian deposition was controlled by widespread regional subsidence in the southern North Sea with possibly some tectonic influence giving increased subsidence in the Sole Pit area.

Triassic sequences show more evidence of tectonic control, with the Sole Pit basin being the main depocentre. Seismic data suggests that the Lower Triassic Bacton Group shows a subtle thickening southwards. Some Middle to Upper Triassic units, such as the minor salt members, were deposited in tectonically-controlled sub-basins. However, there is little evidence of such control in the Ravenspurn North Unit.

Strata of early, mid- and late Jurassic age are present in Ravenspurn North. The sequence from top Triassic Haisborough Formation to top Oxfordian shows pronounced thickness changes on NE-SW seismic lines, with the thinnest sequence preserved in a present structural low. This sequence thickens over the Ravenspurn structure, and the top Corallian limestone marker outcrops on the sea floor over present structural highs. Over most of the southern North Sea Basin there is little information from later Jurassic onwards, and timing of structural uplift is difficult to determine. Late Cimmerian erosion (140 Ma) is extreme in localized areas.

Deposition of Upper Cretaceous Chalk is known to have taken place over parts of the region. Tertiary inversion has resulted in the removal of thick sequences of Cretaceous Chalk and Jurassic sediments across the Ravenspurn North Field.

No Tertiary sediments have been encountered.

Geophysics

Seismic coverage

A dense grid of good quality seismic data covers the Ravenspurn North Area. The seismic coverage available comprises 4045 line km of 3-D seismic data and 1500 line km of 2-D seismic data.

The 2-D seismic surveys were acquired with a higher density of coverage in the northeast–southwest dominant dip direction resulting in a denser than 0.5 km grid across the area. 2-D seismic strike lines were acquired in both northwest–southeast and north–south directions with a line spacing of 1–1.5 km. These data constitute the main seismic control over the southeast flank of the field. The 3-D

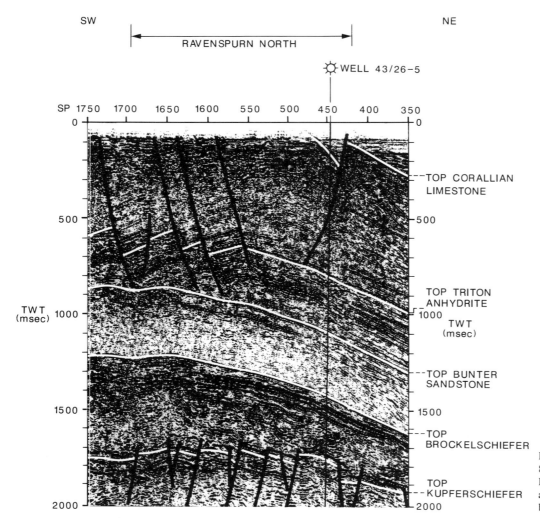

Fig. 3. Representative dip seismic line 81-43/26-08 across SE flank of field. NB: Polarity convention: an increase in acoustic impedance is represented by a black peak.

seismic survey which comprises 301 in-lines and 1104 cross-tracks forms a rectangle 22.5 km by 13.5 km and provides the main seismic coverage over the central and northwest parts of the field.

Data quality is moderate to good on all vintages of seismic data at all levels down to the base of the Zechstein evaporite sequence, but is generally only moderate at deeper levels. Approximately 850 km of 2-D seismic data were reprocessed by GSI in 1985/86 to the same general specification as the latest 1986 2-D acquisition. Good overall improvement in data quality was achieved at all levels down to the reservoir objective.

Synthetic seismograms have been run for all wells in the Ravenspurn North and South areas and include wells 43/26-1 through 7, 42/30-1 through 7 and 42/29-3. All of these wells have been used to confirm the geological identity and continuity of the seismic events throughout the area.

The five principal seismic marker horizons selected for detailed interpretation are as follows: Top Corallian Limestone; Top Triton Anhydrite; Top Bunter Sandstone; Top Brockelschiefer; Top Kupferschiefer. A further four horizons are identified on seismic data at reservoir level in some of the wells, these being: Top Leman Sandstone; Top Reservoir Zone IV; Top Reservoir Zone II; Top Carboniferous.

Seismic interpretation

The five main structural markers interpreted on the seismic data were selected because they represented boundaries between major lithological units associated with marked acoustic impedance contrasts that could be mapped extensively across the field area, (Fig. 3). Time migrated seismic sections were used for the interpretation.

Data quality is moderate to good down to the Top Kupferschiefer seismic horizon for all vintages of seismic data. Data quality is generally only moderate to poor at deeper levels but is of sufficient quality that Top Leman Sandstone and several intra-reservoir picks could be made on the 3-D, and in some cases, on the 2-D seismic data. The main factor controlling overall data quality is the variation in structural complexity of the area. Wherever intense faulting occurs in the shallow post-Triassic section data quality deteriorates at greater depths. This is particularly the case in the vicinity of the circular shaped Zechstein salt 'swell' situated on the southeast flank of the field. This same area is also affected by a higher density of faulting at reservoir level than the northwest of the field. More localized deterioration in data quality is observed in areas where numerous high velocity and high density anhydrite 'rafts' occur. 'Rafting' occurs most commonly across the southeast and central parts of the field and has a greater effect on the quality of the 2-D rather than the 3-D seismic data. Deterioration in data quality at reservoir level also results from destructive interferences between reflections emanating from within and near the top and base of the reservoir as the Leman Sandstone isochore gradually decreases in thickness to the northwest across the field. In addition, data quality deteriorates in response to a reduction in the impedance contrast between the Leman Sandstone and overlying Silverpit Fm. siltstone, also in a northwesterly direction, as the reservoir rocks become progressively more silty.

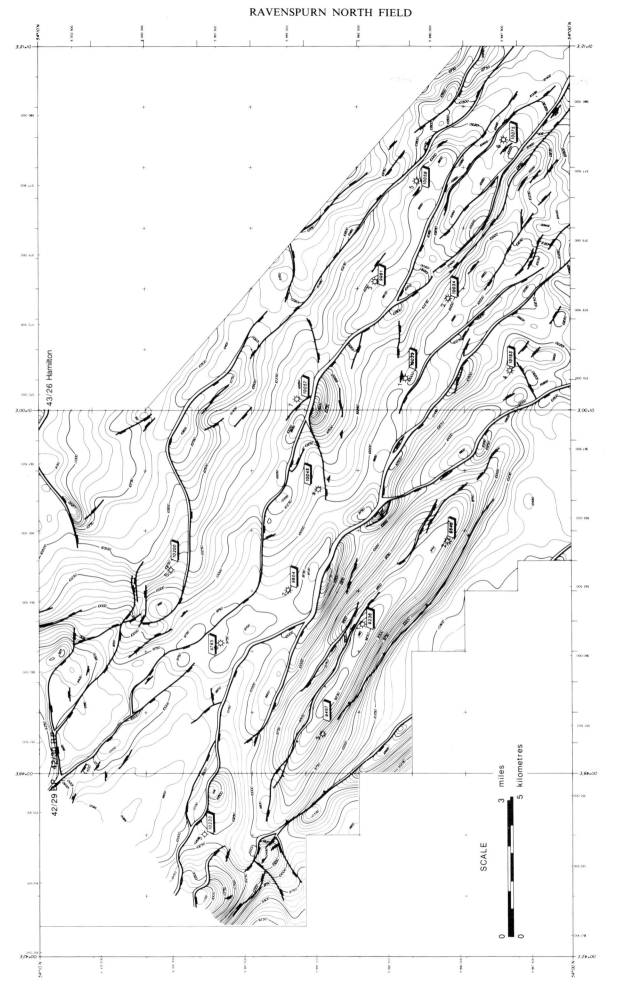

Fig. 4. Top Leman Sandstone depth structure.

The discontinuity and variation in amplitude of the reflectors associated with the reservoir sequence and the relatively thin intervals between them preclude the accurate mapping of these intervals and no precise isochore or structure maps have been prepared from seismic data alone.

A 'layer-cake' method of depth conversion has been applied to the seismic data in the Ravenspurn area, using time grids for the five main seismic markers mapped and their intervening isochrons, as primary input. Interval velocities were calculated between adjacent seismic horizons at each well location, utilizing corrected seismic grid times and TVD corrected subsea well depths.

The depth structure at the Top Kupferschiefer event was used to construct the Top Leman Sandstone depth map, (Fig. 4) by adding an isochore grid of the Silverpit Formation controlled by well data. The fault pattern at Top Leman Sandstone level was interpreted from seismic data where possible or projected down at an angle of 65° from the position of the faults at Top Kupferschiefer level where no pick could be made.

Trap

Ravenspurn North has a combined structural/stratigraphic trap. The field is dip closed to the southeast and fault and dip closed to the northeast and southwest. Fault closure to the southwest is provided by a major anastomosing oblique strike-slip fault system which separates Ravenspurn South from Ravenspurn North. Lateral seal along faults is provided by Silverpit Fm. mudstones and Zechstein Gp evaporites. Reservoir deterioration to the northwest, caused by thinning and pinch-out of aeolian sands and high authigenic illite content, provides a composite stratigraphic/diagenetic seal.

The Ravenspurn North Field consists of two en-echelon NW–SE-trending elongated tilted-fault blocks with their elongated axis dipping to the southeast. The two fault-blocks are separated by a major down-to-the-southwest normal fault exhibiting variations in throw of up to 1100 ft (335 m) and extending almost the length of the field, some 12 miles (20 km).

Crossfaults are normal, or oblique slip, east–west faults, indicative of wrench faulting. As development drilling proceeds the influence of these cross faults will be assessed. They are interpreted to have little consequence on reservoir continuity.

In the area of well 43/26-6 normal faulting has defined a small horst block which has resulted in structural separation from the major fault blocks to the north and south.

Faults are considered sealing unless in juxtaposition with other reservoir quality rock.

Structural synthesis

The present-day structural style of the Ravenspurn area has evolved from a complex tectonic history. Tectonism also has played an important role in influencing reservoir distribution and quality.

Variscan folding, faulting and subsequent erosion controlled the distribution of Carboniferous source rocks and potential hydrocarbon fairways. Post-Variscan structural relief and early Permian wrench movements with associated pull-apart features combined to control early Permian (lower Leman Fm.) reservoir deposition.

During the late Jurassic, north–south extension initiated westerly striking listric faulting within the post-Muschelkalk succession. The presence of low angle normal faults, which sole-out within the Middle Triassic halites, is restricted to the southern part of Block 43/26. Faulting is not observed to affect the Top Bunter Sandstone or Brockelschiefer horizons. Late Jurassic extension locally activated northwesterly striking faults cutting the Leman Fm. reservoir in an oblique left-lateral sense, resulting in enhancement of structures in Ravenspurn North. These late Jurassic structures were charged by early migrating gas which inhibited diagenesis locally and preserved good reservoir quality.

Additionally, high angle, generally east–west-striking reservoir fractures are formed as a result of Jurassic extension. These fractures were subsequently diagenetically sealed reducing reservoir permeability parallel to fracture dip but only marginally along strike. Natural fracturing is insignificant in the field.

Late Cretaceous to Tertiary Alpine inversion resulted in the structural development of the Ravenspurn South anticline as a response to reversal of movement along the previously established northwesterly striking wrench fault system. In Ravenspurn North, however, faults remained in net extension despite undergoing some reverse movements. Gas generation ceased at this time and migration pathways were re-established and reservoirs charged. Reactivation of the intersecting faults in the 43/26-7 area established a localized reservoir fracture network which remained open, possibly due to the presence of gas.

Definition of gas pools

Three separate gas pools are presently defined in Ravenspurn North. Each of these is full to spill point.

The fault block along the northeastern part of the field including the areas defined by appraisal wells 43/26a-3, -5 and -7 is an area of good quality reservoir rock considered to be of high deliverability potential without the need for further fracture stimulation. This area has a gas–water contact of 10 292 ft TVDSS.

The fault block containing appraisal wells 42/30-4, -6 and -7 and wells 43/26-1 and -2 is interpreted to be of lower reservoir quality which will require fracturing in order to obtain commercial rates of gas production. It has a gas–water contact of 10 294 ft TVDSS.

A fault block or fault sliver located between the northeastern and southern pools contains appraisal well 43/26a-6. It is defined as an area of good quality reservoir with a gas–water contact of 10 255 ft TVDSS.

Reservoir

The lower Leman Sandstone Fm. in Ravenspurn North was deposited primarily by fluvial and aeolian processes. While aeolian dune and sandy sabkha prevail in the east (43/26) and fluvial and sabkha facies prevail in the west (42/30) the interval exhibits a complex and varied facies distribution both vertically and laterally. The distribution of sediments according to environments of deposition (facies distribution) affect reservoir quality, zonation and therefore fluid movements across the field.

Facies

Appraisal wells drilled in the Ravenspurn North unit have been extensively cored, described and logged. Variations in lithology, grain size and sedimentary structures have been documented, and the data used for interpretation of a depositional model, and to aid in predicting the vertical and lateral continuity of reservoir units, (Fig. 5).

The following lithofacies have been identified in core and sedimentological analysis: (a) aeolian dune base; (b) aeolian dune; (c) structureless fluvial sheetflood; (d) structured fluvial sheetflood; (e) fluvial channel; (f) lake margin sabkha; (g) playa lake mudstones.

Fluvial and sabkha facies dominate in the northwest while aeolian facies dominate in the southeast.

It should be noted that the term 'sheetflood' is used to denote what are regarded as laterally extensive ephemeral fluvial deposits. Furthermore the term 'channel' is not meant to imply permanent stream flow conditions but rather stacked sequences of coarser-grained sheetfloods which formed under partially confined flow conditions.

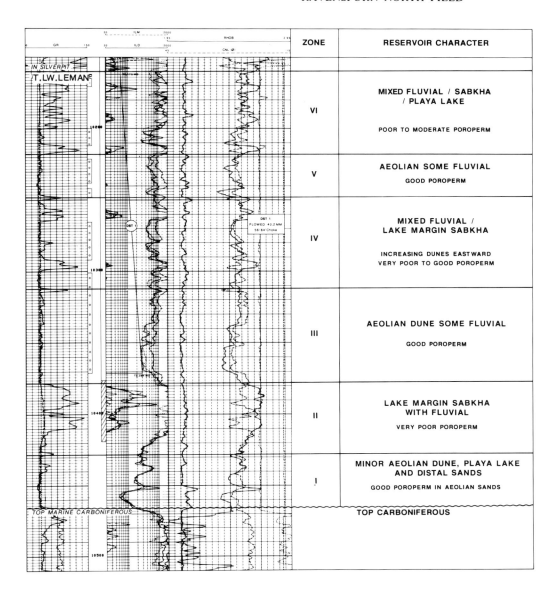

Fig. 5. Ravenspurn North, field type section Block 43/26.

Depositional model

The lower Leman Fm. interval in Ravenspurn North was deposited primarily by aeolian and fluvial processes in marginal proximity to the southern shore of the Silverpit Lake (Fig. 6). Recognition of aeolian dune, fluvial sheetflood, channel, lake margin sabkha and playa lake facies confirm an overall continental desert sedimentary environment.

Early sedimentation was localized and conformed to topographic lows of the deeply eroded Carboniferous landscape (Variscan Unconformity). Aeolian dune migration is generally suspected to be towards the WNW whilst fluvial systems cut normal to dunes and flowed NNE into the Silverpit Lake. Close proximity of the Silverpit Lake, resulting in water table fluctuations, played a major role in controlling sedimentation. Periodic high water in the Silverpit Lake resulted in intervals of sabkha and low water initiated dune development. Laterally dunes predominate in the southeast of the field and sabkha and fluvial in the northwest.

Aeolian sand body geometry is broadly sheet-like and considered to be governed by climatic changes. The fluvial system was ephemeral and probably controlled by seasonal rainfall in the source area. The course of sheetflood and channel deposits, which are recognized in all wells, tend to orient their long axis in a north-northeasterly direction. Fluvial events have reworked some aeolian sands and have been reworked themselves by subsequent aeolian processes.

Sabkha sediments formed by wind blown sand, adhering to a moist ground surface, were deposited at the lake margin or in damp 'interdune' areas.

Across the Ravenspurn Field the lake margin sabkha facies thickens to the northwest and dominates the upper part of the lower Leman Fm. interval as a consequence of the Silverpit Lake encroachment to the south in early Permian times.

The overall thickness of the sedimentary section of the lower Leman Sandstone thins to the northwest. Aeolian sands, which represent the best reservoir quality rock, gradually wedge out towards the northwest across Block 43/26, and thin rapidly in Block 42/30 towards well 42/30-5.

Controls on poroperm characteristics

Reservoir quality of the lower Leman Sandstones was controlled initially by lithofacies distribution. Subsequent diagenesis further modified the reservoir rock resulting in variations in porosity types and authigenic clay types across the field. These variations have affected porosity/permeability relationships particularly between wells in Blocks 43/26 and 42/30.

Concisely, the poorer permeabilities observed in 42/30 wells and 43/26-1 and -2 wells result from a reduced proportion of aeolian sediments and the greater abundance of pore filling 'meshwork or leafy' illite from southeast to northwest.

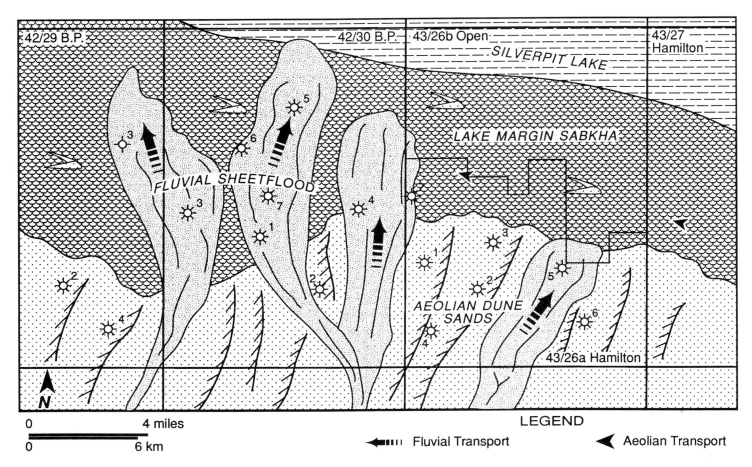

Fig. 6. Schematic diagram to show palaeogeography in Mid-Leman times (Zones III and V).

Porosity types vary from SE to NW. In wells 43/26-3, -5, -6, and -7 secondary-intergranular and dissolution are the main porosity types. The rock fabric still retains an element of primary intergranular porosity. In wells 43/26-1 and -2 and in 42/30-4, -5 and -6 the secondary porosity has become microporous in association with illitic clays.

Diagenetic changes affecting the burial of sandstones included (a) crystallization of authigenic clay, (b) precipitation of mineral cements, and (c) later leaching of rock and mineral components by corrosive or undersaturated pore-fluids to form variations in porosity types.

Reservoir zonation

The lower Leman reservoir has been sub-divided into six reservoir zones based on log and core data. These are summarized in Fig. 5. Zones are defined as a grouping of reservoir (or non-reservoir) quality rock that exhibits similar log responses and fluid flow characteristics. Zonation is a function of reservoir properties that exhibit similar porosity, permeability, net-to-gross ratio, lithofacies, environments of deposition, and continuity.

Zones III and V exhibit the best reservoir quality characteristics. Zone I, where aeolian, exhibits good quality reservoir. Zones II, IV and VI vary in rock quality. Overall rock quality, with respect to porosity and permeability, decreases from southeast to northwest.

All reservoir zones except the lowermost Zone I and II can be correlated across the field area in a layer-cake fashion, (Fig. 7). The basal two zones fill low areas on the Permian land surface. Zone I is of reservoir quality, but Zone II has very little reservoir material.

Source

The source rocks are believed to be the Westphalian Coal Measures with hydrocarbon expulsion initiated during mid- to late Jurassic times with potential source basins lying to the southwest and more speculatively northeast of the Ravenspurn North area. Gas was thought to have been emplaced earlier within localized highs in Block 43/26 preserving the reservoir quality by inhibiting illite cementation. There is no evidence to suggest that this localized structuring occurred elsewhere within the Ravenspurn North Area.

Hydrocarbons

Test results

Three of the appraisal wells in Ravenspurn North were tested at rates ranging between 38.8 and 64.6 MMSCFD. Wells 43/26a-3 and -5 were located in the northern high deliverability area, appraisal well 43/26a-6 is in the southeastern high deliverability area. These test results were sufficient for commercial development without any artificial stimulation and confirmed the higher permeabilities in these areas.

All other appraisal wells 43/26-1, -2 and -7 and 42/30-4, -5, -6 and -7 were tested before and after fracture stimulation. Pre-fracture tests range from 0.1 to 4.2 MMSCFD. Post fracture rates showed significant production increases resulting in rates between 4.8 and 23.2 MMSCFD. No formation water was tested from Ravenspurn North appraisal wells.

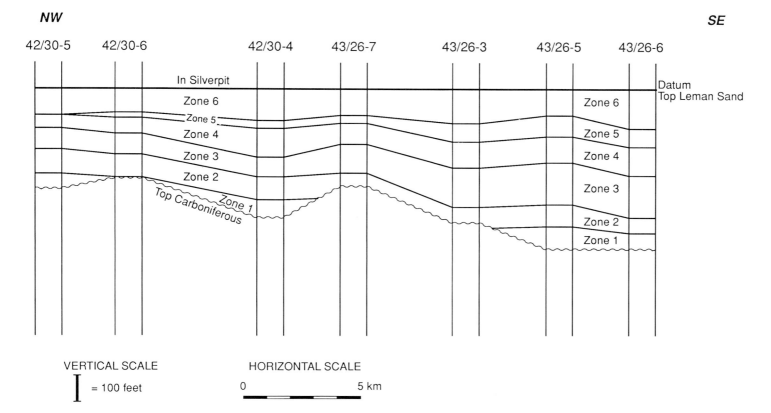

Fig. 7. Ravenspurn North schematic cross-section.

Fracture treatments in the appraisal wells varied in design. The range of specifications of these original fracture jobs are given below:

Pad sizes: 60 000 to 234 000 gallons
Treatment sizes: 51 500 to 328 600 gallons
Proppant grade: 20/40
Proppant concentration: 4 to 12 ppg
Total Proppant: 130 900 to 500 000 pounds

Experience gained in hydraulic fracture these appraisal wells and from other operators performing stimulation on lower Leman Fm. reservoirs will be used to optimize fracturing during the development phase.

Fluid types and specifications

Gas samples produced during tests on appraisal wells in Ravenspurn North show only minor variations in their compositions. The gas contains approximately 2.5 mole % nitrogen and more than one mole % carbon dioxide. All analyses indicate that C7 is greater than C6 due to the presence of benzene in the gas.

All available analyses of the hydrocarbons indicate that the gas will remain single phase throughout the field life consequently the specific gravity will remain constant at 0.60 and the calorific value (wet) will be 995 BTU/SCF. Actual condensate yield at a maximum dew point of 30°F determined from samples taken during tests will be 1.0 BBL/MMSCF. The gas expansion factor is 236 SCF/RCF.

A sample considered to represent formation water was obtained from the nearby Ravenspurn South appraisal well 43/26a-4. This water contained 170 000 ppm NaCl and has a resistivity of 0.060 ohm m at 60°F and 0.018 at reservoir temperature of 219°F. A water gradient seen on RFT pressure data from the 43/26a-6 appraisal well has a clear trend of 0.5 psi/ft. This is the only reliable water gradient established by RFT due to the considerable scatter from super-charging effects typical of relatively low permeability sands in Ravenspurn North.

Reservoir conditions

The initial average reservoir pressure in the lower Leman Sandstone reservoir is 4542 psia at the datum of 10 150 ft SS. A distinct gas gradient can be seen in all wells with RFT data; this gradient is 0.08 psi/ft. As described earlier, there are three separate gas–water contacts.

The average reservoir temperature at datum is 219°F.

Reserves

The Ravenspurn North reservoir is of a depletion drive type. Water influx is not expected in the northwest direction where sandstone properties deteriorate, nor in the northern or southern directions where faults limit the reservoir extent. Within the low deliverability gas areas water drive is not expected because of relatively low permeability to the water phase. The only likely area of water encroachment might be towards the southeast into the high deliverability sandstone near appraisal wells 43/26a-5 and -6.

Based on the results of the stimulation studies the expected field life is 29 years. Owing to the low permeability of the majority of the reservoir the production rate during the last few years of the field life is predicted to be low and to decline slowly. A small change in the assumed abandonment rate or assumed well productivity will result in a change in the expected field life. Such a change would not result in a significant change in ultimate recovery due to the low production rates considered.

Gas-initially-in-place (GIIP) is estimated to range from a proven value of 2.03 TCF to a possible value of 2.25 TCF. The most probable estimate of GIIP is 2.11 TCF. The ultimate recoverable reserves are estimated to be between 1.2 to 1.4 TCF.

The discovery, appraisal and development of the Ravenspurn North Field was a joint effort of all our Partners spanning several years. The efforts of all those involved in the Field, present or past are recognized and appreciated.

We acknowledge the co-operation of our Partners, Arco British Ltd, BP Exploration Ltd, Enterprise Oil plc, Hardy Oil & Gas Ltd, Monument Resources Ltd and Ultramar Exploration Ltd in all aspects of development.

Ravenspurn South Field, Blocks 42/29, 42/30, 43/26, UK North Sea

R. D. HEINRICH

BP Exploration (EPS), Farburn Industrial Estate, Dyce, Aberdeen AB2 0PB, UK

Abstract: The Ravenspurn South Gas Field is located in the Sole Pit Basin of the Southern North Sea in UKCS Block 42/30, extending into Blocks 42/29 and 43/26. The gas is trapped in sandstones of the Permian Lower Leman Sandstone Formation, which was deposited by aeolian and fluvial processes in a desert environment. Reservoir quality is poor, and variations are mostly facies-controlled. The best reservoir quality occurs in aeolian sands with porosities of up to 23% and permeabilities up to 90 md. The trap is a NW–SE-striking faulted anticline: top seal is provided by the Silverpit Shales directly overlying the reservoir, and by Zechstein halites. Field development began early in 1988 and first gas was delivered in October 1989. Production is in tandem with the Cleeton Field, about 5 miles southwest of Ravenspurn South, as the Villages project. Initial reserves are 700 BCF and field life is expected to be 20 years.

The Ravenspurn South Gas Field is located about 40 miles east of Flamborough Head on the Humberside coast, in the United Kingdom sector of the Southern North Sea (Fig. 1). The field covers an area of approximately 36 km² (9000 acres) primarily in Block 42/30, with small extensions into Blocks 42/29 and 43/26. The field abuts the Ravenspurn North Gas Field (operated by Hamilton Brothers plc) which lies within blocks 42/30 and 43/26, and which is the subject of a separate development plan. The water depth in this area is about 150 ft. Aeolian, sabkha and fluvial sandstones of early Permian age constitute the reservoir.

The field is 5 miles northeast of the Cleeton Gas Field: both fields are named after now-drowned villages formerly located on the Humberside coast.

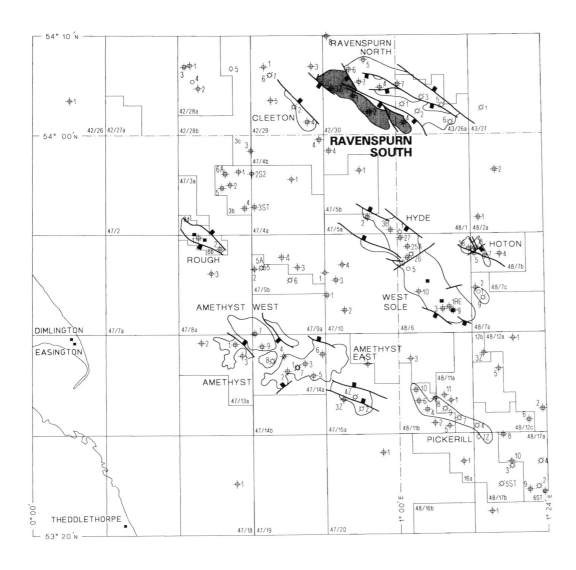

Fig. 1. Location of the Ravenspurn South Gas Field.

From Abbotts, I. L. (ed.), 1991, United Kingdom Oil and Gas Fields, 25 Years Commemorative Volume, Geological Society Memoir No. 14, pp. 469–475

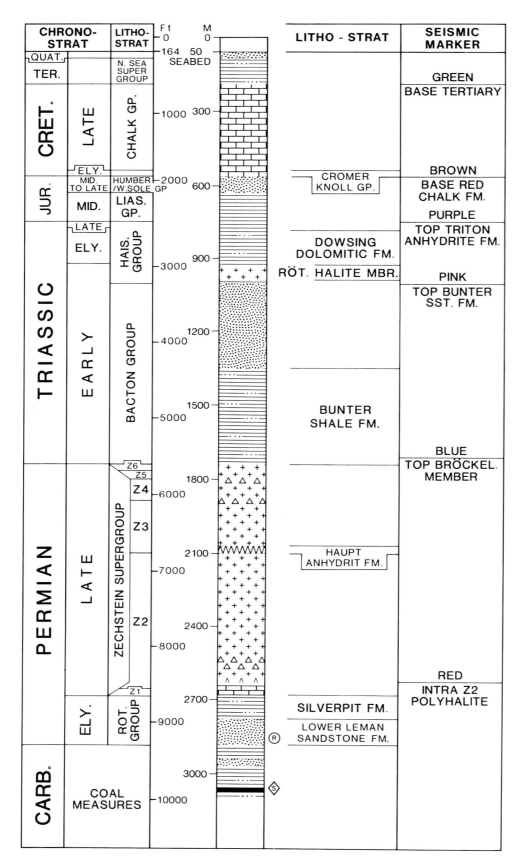

Fig. 2. Generalized stratigraphic column of the Ravenspurn South Gas Field.

History

Production Licence P001, which included Blocks 42/29 and 42/30 was awarded in 1964; this licence is wholly owned by BP. Licence P380 covering Block 43/26 was awarded in 1981; this licence is operated by Hamilton Bros Oil & Gas (22.5%) with partners Arco British (25%), Enterprise Oil (20%), Ultramar (15%), Hardy Oil & Gas UK (10%), and Monument-Oil & Gas (7.5%). A unitization agreement covers both Ravenspurn North and Ravenspurn South fields, which lie within a single P.R.T. area. Both blocks were partially relinquished in 1987.

Six exploration and appraisal wells were drilled in the Ravenspurn South field. The first, 42/30-1, was drilled in 1976, and encountered an 81 m (266 ft) gas column in the Lower Permian Lower Leman Sandstone Formation, but the low flow rates on test caused the field to be considered uncommercial.

However, after a more optimistic interpretation of the test results, well 42/30-2, drilled in 1983, demonstrated that the reservoir could flow at commercial rates after hydraulic fracturing, and in a rapid appraisal programme wells 42/30-3, 42/30-3Z, 42/29-3 and 43/26-4 were drilled to assist in delineating the extent of the field. The first 2 wells encountered gas columns of the order of 90 m (295 ft): well 43/26-4 encountered a gas column of 10 m (33 ft), and well 42/29-3 was a dry hole.

Until 1984 seismic data comprised various 2-D seismic surveys, which combined to give an average line spacing of approximately 0.5 km over the field. In 1984 a 3-D seismic survey was acquired by Western Geophysical, and this has formed the definitive seismic data set.

The Annex B document (covering both Ravenspurn South and Cleeton) was submitted to the Department of Energy in 1984 and approved in 1986. Development drilling commenced in April 1988. A total of 23 development wells have been drilled from 3 platforms. The development plan originally envisaged the drilling of up to 48 wells over a period of up to 6 years. The recovery mechanism will be pressure depletion, with all wells being hydraulically fractured to stimulate flow. First gas was delivered in October 1989. Production facilities are unmanned wellhead towers with sub-sea lines to the Cleeton platform, whence the gas is exported to the Dimlington terminal.

Field stratigraphy

Figure 2 shows the main stratigraphic units and their lithology.

The basin has been subjected to a number of tectonic events. The Hercynian orogenic event which occurred in late Carboniferous times, resulted in the folding, faulting and uplift of a Carboniferous basin along a NW–SE structural grain. Movement was largely strike-slip in response to a N–S stress regime. Uplift resulted in erosion down to the Westphalian over Ravenspurn South. During the late Carboniferous to early Permian times, E–W tension led to oblique slip fault movements along Hercynian basement faults and to the initiation of subsidence of the Sole Pit Basin (Fig. 3). Clastic sediment infill, with coal in the Carboniferous, accompanied the subsidence of the basin.

In the Ravenspurn South area, early Permian Rotliegendes sandstones (Lower Leman Formation) and mudstones (Silverpit Formation) were deposited upon an irregular Carboniferous unconformity surface. The deposition of these sediments was followed by the continued subsidence of the Sole Pit Basin and the deposition of the Zechstein, Bacton, Haisborough, Lias and West Sole Groups. Rifting during Mid- to Late Jurassic times due

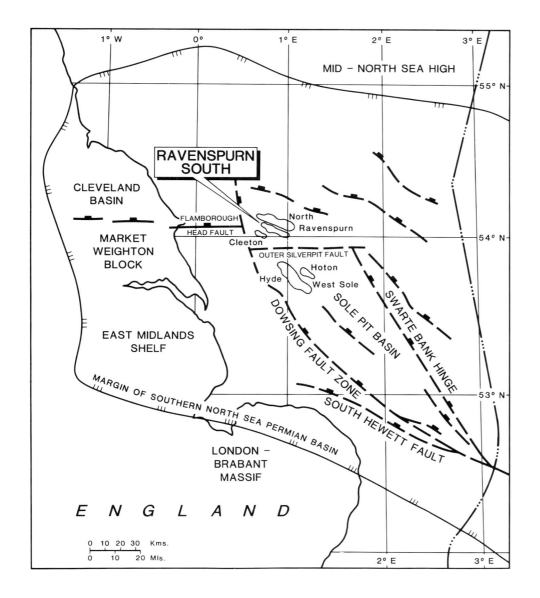

Fig. 3. Structural elements of the Southern North Sea Permian Basin.

to the Cimmerian tectonic event, initiated NW–SE-trending normal faults.

Late Jurassic to early Cretaceous inversion due to the Laramide tectonic event resulted in erosion down to Middle Jurassic levels over Ravenspurn South Field. After this phase of uplift, the Cromer Knoll and Chalk Groups were deposited on the resulting unconformity surface during a period of renewed basin subsidence.

The Tertiary was characterized by successive periods of uplift which began at the end of the Cretaceous and culminated in the main phase of inversion in the Miocene, at which time the Ravenspurn South closure was probably formed. Most of the previously deposited Cretaceous sediments were eroded during this phase.

Geophysics

Until 1984 the mapping of the Ravenspurn South Field was based on several vintages of seismic data, constituting approximately a 0.5 × 0.5 km grid. In 1984 to 1985 a 3-D seismic survey was shot and processed by Western Geophysical. This survey consists of 300 SW–NE oriented lines, 75 m apart. Interpretation of this survey forms the basis for the current development drilling plan. The geological overburden over Ravenspurn South Field is structurally simple and it is possible to map a reflector corresponding to the top of the reservoir interval directly from the seismic data: this is in direct contrast to the situation over the Cleeton Field. However, to date it has not been possible to delineate any structure or lithologies within the reservoir itself. In an attempt to gain more information about the reservoir directly from the seismic data, work is at present under way involving reprocessing and inversion to absolute acoustic impedance of selected data from the 3-D seismic survey. The effectiveness of this technique has still to be fully evaluated.

An example seismic line across the field is shown in Fig. 4; a contour map of depth to the top reservoir is shown in Fig. 5.

Trap

The Ravenspurn South Field is a structurally closed NW–SE-striking faulted anticline (Figs 5 and 6), probably formed as an extensional feature during Late Permian to Late Jurassic with subsequent inversion during the Tertiary. The dominant fault trend is NW–SE. Some degree of strike-slip movement seems possible during either or both phases of faulting. The structural relief at top reservoir level (top Lower Leman Sandstone) is approximately 340 m (1115 ft) with the crest at 2760 m (9055 ft) and the deepest closing contour taken to be at 3100 m (10 170 ft). The structure is sealed by the overlying Silverpit Shale Formation, with ultimate regional seal provided by the Zechstein evaporites that directly

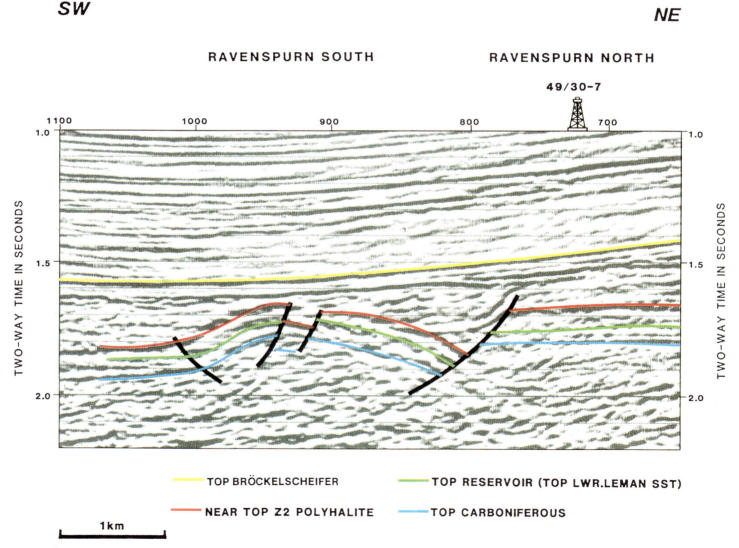

Fig. 4. Seismic line across the Ravenspurn South Gas Field; for location see Fig. 5. Data acquired as part of a 3-D survey shot by Western Geophysical in 1984. This section reprocessed by BP in 1988.

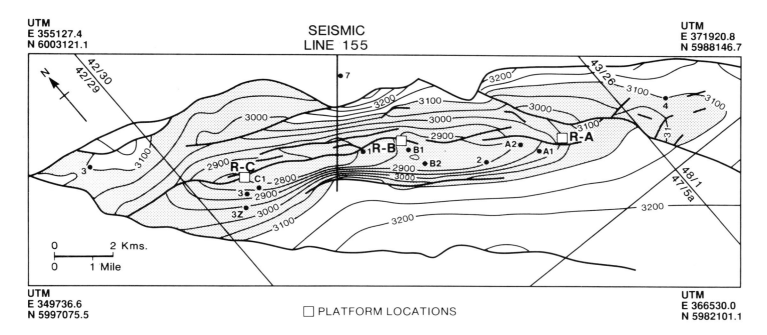

Fig. 5. Top reservoir structure map of the Ravenspurn South Gas Field.

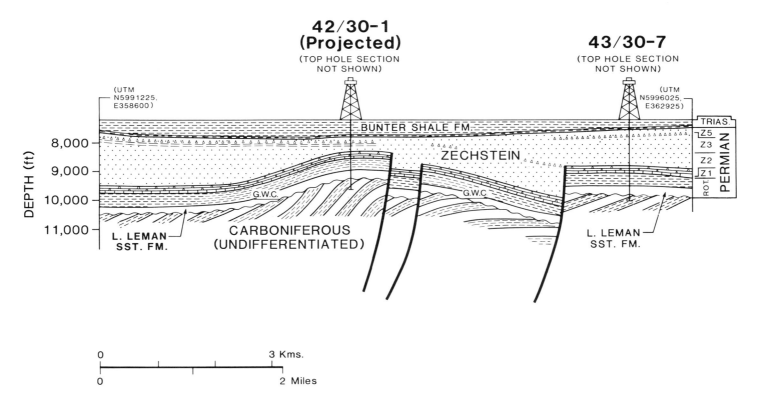

Fig. 6. Geoseismic section through the Ravenspurn South Gas Field.

overlie the Silverpit Shale. Against faults Silverpit mudstones and tight Carboniferous sandstones are regarded as an effective seal; reservoir to reservoir contacts are generally not considered sealing.

Reservoir

The reservoir sandstone of the Ravenspurn South Field belongs to the Lower Leman Sandstone Formation of early Permian age. The reservoir thickness varies from approximately 60–100 m (200–330 ft) across the field. This is partly due to a successive shale-out to the NW at the top of the reservoir, and partly to local topography at the post-Carboniferous unconformity.

The reservoir comprises a mixture of sabkha (sandy and muddy), fluvial and aeolian facies deposited in an arid to semi-arid environment. Figure 7 shows a typical reservoir log.

Lower Leman Sandstone deposition in the Ravenspurn South area was primarily controlled by changing climatic conditions

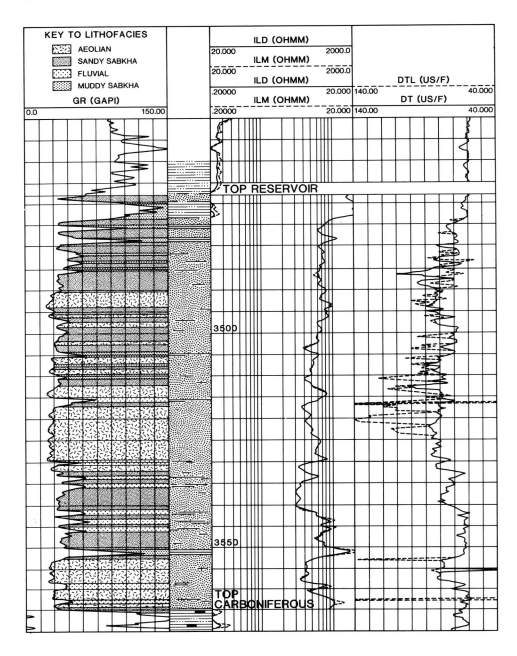

Fig. 7. Typical log through Ravenspurn South reservoir.

resulting in 'drier' and 'wetter' periods. Aeolian dunes developed and migrated under E to NE prevailing winds which resulted in the reworking of mainly fluvial sandstones. During Lower Leman Sandstone times, the southeastern part of the field clipped the NW edge of a major aeolian dune field which extends south-eastwards into Blocks 43/26, 47/5 and 48/1. Passing laterally northwestwards across the field, the amount of aeolian sediments decreases at the expense of fluvial and sabkha sediments.

Fluvial deposition was controlled by seasonal rainfall which resulted in alluvial fans building out from the Dowsing Fault Zone and Hercynian Highlands to the southwest of Cleeton. Sabkha sedimentation developed as a result of water table movements and periodic flooding partly associated with the gradual southward encroachment of the Silverpit Lake.

The different facies types vary in texture and detrital clay content which form the main controls on the poroperm characteristics. Among the diagenetic factors controlling reservoir quality, illite is by far the most important. Its pore-filling and pore-bridging action reduces the permeability significantly in the finer-grained sands, whilst retaining a high proportion of micro-porosity. As most of the illite originates from detrital clay, its presence and abundance is largely controlled by lithofacies. Another diagenetic factor reducing poroperm is mineral cementation, namely by quartz and ferroan dolomite. Two field-wide trends have been observed: vertically, the degree of cementation increases with depth, and, laterally, it increases towards the northwest. The only diagenetic process enhancing poroperm is feldspar dissolution, which in places leads to a significant increase in pore interconnectivity.

Best reservoir qualities are found in coarse-grained aeolian dune sands in which there is little mineral cementation, i.e. in the upper half of the reservoir and in the southeastern half of the field. Porosities in these areas range from 20% to 22% and permeabilities between 10 and 90 md. Poroperms in the non-aeolian fraction, however, are much lower, up to 18% and 1 md respectively.

Source

The source of the Ravenspurn South gas is the late Carboniferous Coal Measures directly underlying and subcropping the Permian reservoir. Maturation and migration took place from the Middle-Jurassic onwards and ended with Tertiary inversion and uplift.

According to structural studies, the inversion resulted in a significant re-shaping of the Ravenspurn North and South structures. The original trap and gas accumulation was in the area of today's Ravenspurn North gas field. The inversion led to a re-migration of the gas and eventually to a separate Ravenspurn South gas accumulation. As a result of this late gas emplacement, the diagenetic effects in Ravenspurn South are much stronger than in Ravenspurn North and Cleeton, where early gas emplacement partly suppressed diagenesis.

Hydrocarbons

The Ravenspurn South gas field has a dry gas which is 92.5% methane. The gas viscosity is 0.02 cP at reservoir temperature of 200°F and reservoir pressure of 4489 psi at 2950 m. The gas compressibility factor (z) is 1.0. The gravity of the condensate is 52° API.

Reserves

Ravenspurn South together with Cleeton forms the Villages project. Ravenspurn South contains most of the reserves but has low productivity. Cleeton has very high productivity and low reserves and therefore acts as a swing producer.

The most likely gas-in-place of the Ravenspurn South field is 1200 BCF, with estimated reserves of 700 BCF. Production will eventually be from 30 wells. Aquifer influence in Ravenspurn South is thought to be negligible.

Ravenspurn South Field data summary

Trap
Type	Dip and fault closed anticline
Depth to crest (Lower Leman Sandstone)	2760 m (9055 ft)
Gas–water contact	Approx. 3111 m (10 206 ft)
Vertical closure	350 m (1148 ft)

Pay zone
Formation	Lower Leman Sandstone Formation
Age	Permian
Thickness	60–100 m (200–330 ft)
Net/gross	39–77% (range for total reservoir)
Porosity	12–14% (range for total reservoir)
Hydrocarbon saturation	51–61% (range for total reservoir)
Permeability	up to 90 md in aeolian sands

Hydrocarbons
Gas composition	N_2 1.78%; CO_2 0.96%; He 0.03% C1 93.15%; C2 3.15%; C3 0.55% C4 0.4%
Gas expansion factor	240
Initial condensate/gas ratio	2 BBL/MCF

Formation water
Water gradient	0.45 psi/m
Salinity	Saturated
Resistivity	No water samples taken
Other ions	No water samples taken

Reservoir conditions
Temperature	200°F
Initial pressure	4490 psig

Field size
Area	36 km² (9000 acres)
Recovery factor	58%
Recoverable gas	700 BCF
Drive mechanism	Pressure depletion

Production
First gas	1 October 1989
Development scheme	3 Platforms
Number/type of wells	23 S-shaped wells, hydraulically fractured
Average production rate	5–30 MMSCFD
Cumulative production	17.7 BCF (October 1990)
Secondary recovery method(s)	None

The Rough Gas Storage Field, Blocks 47/3d, 47/8b, UK North Sea

I. A. STUART

British Gas, 59 Bryanston Street, Marble Arch, London W1A 2AZ, UK

Abstract: The Rough Gas Field is operated by British Gas as the world's first offshore gas storage reservoir. This relatively small field (original reserves 366 BCF) is located on the western margin of the Southern North Sea Basin. The reservoir, the Rotliegendes Gp (Leman Sandstone Fm.), average thickness 95 ft, is developed in a basin-margin facies association in which aeolian and wadi-related fluvial processes have interacted. Small-scale dune and interdune sandstones, with generally excellent permeability, alternate with wadi facies sandstones with variable but mainly poorer permeability. The vertical distribution of these facies permits a three-fold zonation of the reservoir. Flow profiles obtained from Production Logging Tools show a very strong correspondence between sedimentary facies and productivity/injectivity. The Carboniferous subcrop includes several very low-permeability sandstone intervals which contact the Rotliegendes at the angular Base Rotliegendes unconformity. A small amount of gas from these sandstones seeps across the unconformity surface into the Rotliegendes. The Rough Field was discovered by Gulf in 1968, and was operated by Amoco on a conventional depletion basis with 6 development wells from 1975 until the early 1980s. In 1980 British Gas secured 100% ownership of the Field and its facilities, and converted it to storage mode. Excess summer gas supplies are injected into the reservoir, to be produced during the winter to meet the peak demand. Conversion to storage mode has necessitated the installation of substantial new facilities, and the drilling of 23 further wells. First injection of gas was achieved in 1985.

The Rough Gas Field is located on the western margin of the Southern North Sea Basin, about 18 miles ENE of the Humberside coast (see location map, Fig. 1). The name derives from the Rough Bank fishing ground. Water depth is approximately 130 ft. Rough is a moderately sized field in the Permian Rotliegendes Group reservoir (Leman Sandstone Formation), with original reserves of 366 BCF.

Fig. 1. Location of Easington and the Rough Field on the Humberside coast.

In the early eighties Rough was converted from a conventional development to use as a seasonal gas storage facility. This development was the first in the world to use a partially depleted offshore field for summer injection and storage of gas to provide extra supplies at very high rates during periods of peak winter demand. The concept is demonstrated on Fig. 2. An important aspect of this paper will be to present the reservoir geology considerations that are pertinent to the operation of Rough in storage mode.

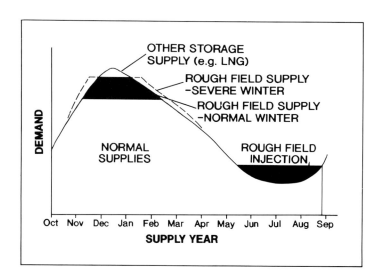

Fig. 2. Position of Rough Field in the gas demand curve. The average seasonal variation in demand is shown. The Rough Field Gas Storage Project was designed to accommodate the peak seasonal increase in demand.

The geology of Rough has been described in Goodchild & Bryant (1986) based on data gathered from the early post-storage-mode development wells, and detailed aspects of the diagenesis in Goodchild & Whitaker (1986). Reservoir engineering aspects were discussed in Hollis (1984), and details of the storage project were given in Anonymous (1985) and Gibbon (1987).

History

The UK Blocks 47/3 and 47/8 were originally awarded to Gulf in 1964, in the first licensing round. The discovery well, 47/8-1, was drilled by Gulf in 1968. It encountered 88 ft (gross vertical thickness) of gas-bearing Rotliegendes Sandstone at a depth of −9141 ft TVSS, and was tested at a rate of 19 MMSCFD on a half inch choke. An appraisal well, 47/8-2, was also drilled in 1968, to the Field's southeastern corner. This well encountered 93 ft (gross) of gas-bearing Rotliegendes at −9210 ft TVSS, and flowed 13 MMSCFD from a restricted perforated interval on a 25/64 inch choke. Both were plugged and abandoned.

In 1973 the licence was transferred to a group operated by Amoco, with Gas Council (Exploration), a subsidiary of British Gas Corporation (BGC), as a partner. Following installation of the AP (processing) and AD (drilling) Platforms, and a 16 in trunk pipeline to the shore terminal at Easington, a 6-well development drilling programme (wells 47/8b-A1 to A6) was carried out between May 1975 and January 1977. Production began in October 1975 and, after the initial build-up period, was on the basis of a Daily Contract Quantity of 104 MMSCFD and a peak winter rate of 146 MMSCFD. This pattern of operation continued until March 1980.

In the late 1970s British Gas recognized the need to expand its gas storage capability to facilitate greater flexibility in meeting the wide seasonal fluctuation in UK gas requirements. The trend towards a higher ratio of domestic to industrial gas consumption was resulting in an increase in the maximum/minimum seasonal demand ratio and a growth in peak winter demand. Conversely an increasing amount of oil-associated gas was being produced from the Northern North Sea at a steady rate throughout the year; therefore during the summer contracted supplies often exceeded demand. In 1979, negotiations began with a view to British Gas securing 100% ownership of the Field for the purpose of converting it to an offshore gas storage facility. Rough was considered the most suitable Southern North Sea gas field for conversion owing to its nearness to shore, its size, and its state of depletion (approximately 43% of reserves produced, and reservoir pressure down to about 2750 psi from a virgin pressure of 4533 psia).

In March 1980 ownership of the Field and its facilities (including the pipeline and onshore terminal) was transferred to British Gas. A new licence was issued to assign the Field area solely to British Gas, which became Field Operator. British Gas gained a licence from the Crown Estate Commissioners to store gas in the Rough Field. Amoco continued to manage the existing facilities on British Gas's behalf on an interim basis, but from 1983 British Gas assumed the management of the facilities. In 1986 British Gas Corporation was privatized as British Gas plc.

Conversion to storage mode has required the installation of substantial new facilities, including 3 additional platforms: BD (drilling and accomodation), BP (process and compression), and CD (wellhead platform). The BD and CD Platforms were designed with 12 drilling slot 500   36 inch trunk pipeline was laid from BP Platform to Easington, also an 18 inch in-field pipeline from AP to BP. The current configuration of the facilities is shown on Fig. 3.

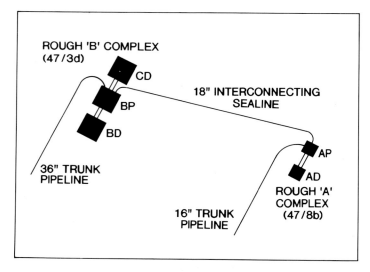

Fig. 3. Rough Field complex and its sub-sea pipelines.

Extensive modifications were required to the original AD and AP Platforms to enable compressed gas to be injected into the AD wells, which were recompleted for gas injection as well as production. All subsequent wells have also been completed in this way. The new development plan called for an expansion in the number of development wells from the original 6 to a total of 24. Between March 1983 and January 1986, 8 CD Platform wells were drilled and completed, and between July 1986 and December 1987 10 BD Platform wells were added. This 24-well development was designed to provide the capacity to supply 1000 MMSCFD for up to 80 days in a severe winter.

Gas injection began in July 1985, and first production from the field after conversion to storage mode was in November 1985. During the severe winter weather of January 1987 the field produced at a peak rate of 639 MMSCFD at a time when 14 of the planned 24 wells had been completed and commissioned.

Six spare slots had been provided for operational contingencies, but had not been used. By the end of 1987 it had been decided that additional capacity could be obtained at relatively little incremental cost by drilling the spare slots. Accordingly, in 1988 2 further wells were drilled from the BD Platform followed by 3 from CD, giving a total of 29 development wells. This provides a notional capacity of 1120 MMSCFD for 80 days, with a peak capacity of 1300 MMSCFD. Given current compressor capacity the maximum injection rate into the field is about 450 MMSCFD.

The Rough Storage Project plays a major role in enhancing the capacity of the British Gas National Transmission System. In gas storage terms the gas content of the Rough reservoir is equivalent to 40 000 large low pressure gas holders (Gibbon 1987).

Field stratigraphy

The stratigraphy of the Rough Field is broadly similar to that of other gas fields in the Southern North Sea (Rhys 1974), and is outlined on Fig. 4. The Plattendolomit is of some interest as it has significant permeability and is overpressured. In a few wells this has given rise to drilling difficulties due to an influx of formation water from the Plattendolomit into the well during drilling.

Geophysics

Several surveys were acquired by the earlier operators between 1967 and 1975. These were largely superseded by a new 0.5 km grid acquired in 1980 by British Gas. A new depth map to Top Rotliegendes (Fig. 5), based on this survey alone, has formed the geophysical basis for all subsequent development drilling and reservoir management of the Field in storage mode.

The Top Rotliegendes event is picked on a peak across the Field. The major bounding faults are seismically well defined. Depth conversion has been achieved using a relatively simple layer model, the shallow events used being Base Chalk, Top Triassic, and Top Zechstein. Depth mapping to Top Plattendolomit has also been carried out, not as a step in the Top Rotliegendes depth conversion, but because of its significance in drilling (see 'Field stratigraphy').

Trap

The Rough structure (Fig. 5) is a WNW–ESE-trending anticlinal horst block at Top Rotliegendes level. The structure has dimensions of about 5.5 by 1.5 miles. The main bounding faults are normal faults with throws of up to 1000 ft. The structural culmination, at about -9000 ft TVSS, is along the upthrown crest of the southern bounding fault. The structure dips away from this crest gently to the north and west. Only in the far northwest is the dip relatively steep (up to 10° west); this is the only part of the trapping mechanism dependent upon dip closure rather than faulting. Internal faulting is minimal, and generally confined to the southeastern corner. The seal is provided by the Zechstein evaporite sequence.

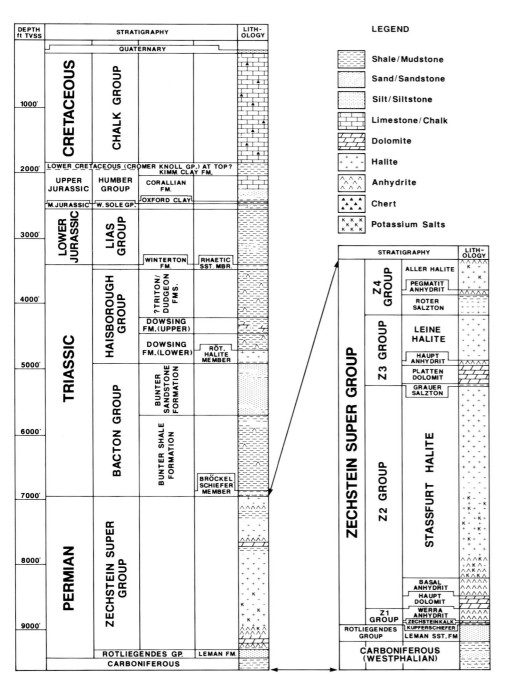

Fig. 4. General stratigraphic column of the Rough Field.

The GWC has never been penetrated within the Rotliegendes. Within the underlying low-permeability Carboniferous sandstones, a GWC has been identified at about −9550 ft TVSS. This depth corresponds to the mapped spill point at Top Rotliegendes level. It is also confirmed by saturation/height considerations.

Reservoir

Management of the Rough Rotliegendes as a storage reservoir, with annual input and output of large volumes of gas, has necessitated a very much more detailed reservoir description than that needed for the original 6-well depletion. The present reservoir description is based on an extensive database of conventional core. The amount of core available from the 6 wells drilled for the initial field development was minimal, since a detailed reservoir description was not a requirement for the simple depletion of this small field. However, upon conversion to storage mode it was realized that the reservoir description would need to be extensively upgraded; therefore the policy was adopted of fully coring 50% of all new development wells. Geological interpretation has been extended to uncored wells by means of the full open-hole and production logging suites run over the reservoir in all wells. Detailed analysis of all these data by reservoir geologists working at British Gas has resulted in a highly reliable predictive model which is outlined below.

The regional setting for the Rough Rotliegendes has been described in Goodchild & Bryant (1986). The key factor is the field's position near the western margin of the Southern North Sea early Permian Basin.

Reservoir deposition can be described in terms of the interaction between aeolian processes associated with the basin as a whole, and complex fluvial processes associated with the basin margin. The Rotliegendes can be subdivided vertically into reservoir units or 'layers', and laterally into regions, characterized by differing degrees of aeolian versus non-aeolian influence. Controls on reservoir properties are discussed in more detail below, but as a good

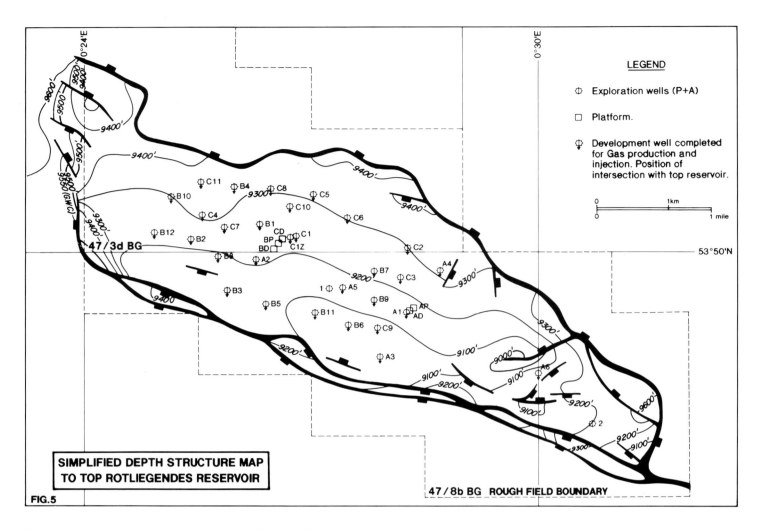

Fig. 5. Simplified depth structure map to Top Rotliegendes Reservoir.

generalization the better permeability is associated with aeolian (and reworked aeolian) sandstones.

Aeolian sandstones are present in varying proportions throughout the reservoir. These consist of cross-stratified intervals of well sorted fine to medium sandstone resulting from the migration of small-scale dunes, and horizontally laminated well sorted fine to medium interdune sheet sandstones. Non-aeolian sands have been carried into the area by fluvial processes and deposited in wadis. These include fine to medium, usually poorly sorted, structureless stacked sandstones representing sheetflood deposits laid down from successive flood events; and very coarse sandy conglomerates organized into intervals, usually fining-upwards, varying from six inches to several feet thick. The latter represent deposition from high-energy proximal flood events, the repeated fining-up being due to the rapidly waning energy associated with each event. Other facies to have been recorded include braided river channel fills, and floodplain mudstones, but these are volumetrically insignificant in terms of the reservoir as a whole.

Lateral facies distribution is controlled by the presence of a major central wadi system debouching northeastwards through the middle of the field (centred on well B5), and fanning out towards the northeastern flank of the field. Further wadi systems have been identified, also running northeastwards, at the southeastern and northwestern ends of the field. Aeolian facies are best developed laterally to these wadi systems (see Fig. 6). These facies contrasts also permit the Rotliegendes to be vertically subdivided into 3 main reservoir units (see Fig. 7) the characteristics of which are as follows.

Unit A

This basal unit, 20–40 ft thick, lies unconformably over the Carboniferous. There is often a thin basal conglomerate, or, in a few wells, a thicker conglomerate interpreted to be a proximal debris flow deposit. The bulk of the unit consists of well developed high permeability dune and interdune sandstones, together with lower permeability fluvial facies comprising sheetflood sandstones and some thin conglomerates at the bases of fluvial intervals. However, within the fluvial-type facies reworked aeolian sandstones occur locally. These originate from sheetflood events and are developed in the intermediate areas along the wadi flanks; they form the highest permeability sandstones in the entire reservoir (up to 1000 md).

Lateral facies distribution in Unit A is controlled by the position of the main wadis. The main central wadi trend extends through wells B3 and B5, then spreads as a fan-like apron through wells C2 and C6, and the lowest permeability is identified in these wells. Wells located on either side display a lesser influence of low permeability wadi-reflected fluvial facies, whilst away from the main wadis aeolian facies are well developed with consequent development of excellent permeability. A further wadi system is identified in the southeast of the Field, though its definition is poor due to the scarcity of wells in this area.

With the general predominance of aeolian facies in Unit A, the porosity and permeability are higher, in most wells, than in Units B or C (see Table 1). However, it is shown in Table 1 that the porosity range is much greater for Unit A than for B or C, and this is directly related to the influence of wadi deposition. In the few wells located

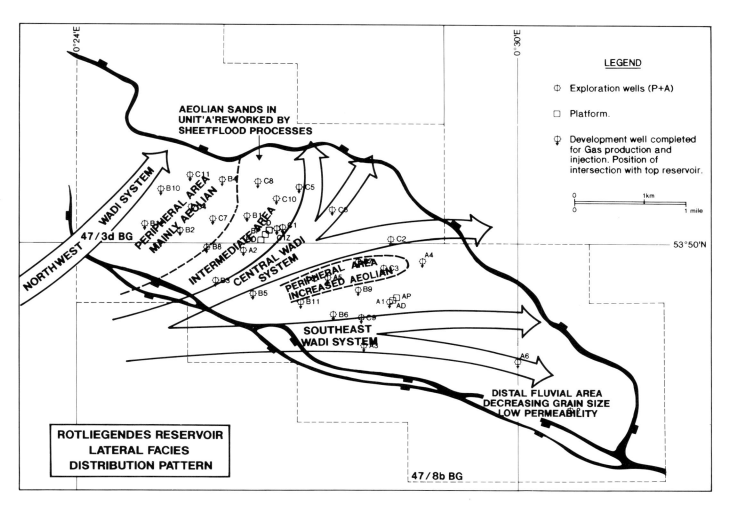

Fig. 6. Rotliegendes Reservoir lateral facies distribution pattern.

Table 1. *Porosity and permeability data (core/log)*

	Porosity (%)		Permeability (md)	
	Average	Range	Average*	Range
Unit A	14.0	9.4–17.3	119	33–291
Unit B	10.8	8.2–12.9	62	21–100
Unit C	13.1	11.5–18.2	41	9–99
Whole reservoir	12.5	10.5–16.1	75	2†–184

* Arithmetic averages.
† Well test data; includes uncored wells.

in the middle of the central wadi trend, porosity and permeability in Unit A can be considerably *lower* than typical Unit B or C values.

Unit B

The characteristic feature of this unit (22–53 ft thick) is the occurrence of very coarse sandy conglomerates and medium to coarse pebbly sandstones. These deposits are organized into crudely fining-upward intervals from 6 in to several feet thick, resulting from proximal high-energy flood events.

The lower two thirds of Unit B comprise a crude composite fining-upward cycle, subunit Bii. The lower part of Bii comprises conglomerates and pebbly sandstones, but the upper part of Bii is usually dominated by sandstone rather than conglomerate for a thickness of about 10 ft, although local, less laterally persistent intervals of conglomerate are still present. These sandstones are mainly sheetflood deposits similar to those of Unit A, though generally coarser and more massive. In a few wells located at maximum distance from the wadi trend, aeolian sandstones occur at this level, their only occurrence in Unit B. These sandstones are overlain by another major conglomerate interval, subunit Bi.

The distribution of facies in Unit B is more uniform than in A or C. The Bii conglomerates form an extensive tabular sheet across the whole field, overlain by the 'upper Bii' sandstone-dominated interval. The Bi conglomerates likewise form a field-wide tabular sheet. The relatively sudden appearance of thick conglomerates at the base of this unit may be related to reactivation of faults in the source area or may be controlled climatically. The repeated occurrence of individual conglomerate beds probably reflects climatically

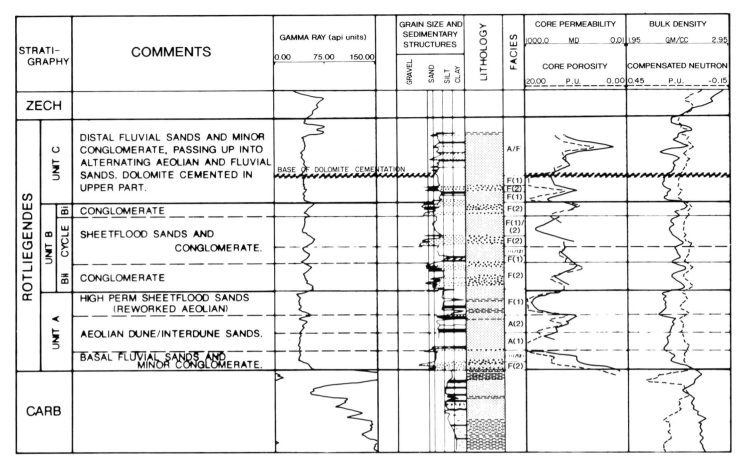

Fig. 7. Rotliegendes Reservoir: vertical facies distribution in a typical well.

controlled processes, but the sudden reintroduction of conglomerates at Bi level suggests control by a second tectonic reactivation and/or major climatic event.

The conglomerate facies invariably display low porosity and permeability, consequently the average reservoir parameters for Unit B are much worse than for Unit A (see Table 1). However, the range of porosity and permeability values is much narrower than for Unit A; this is consistent with the view of a much more uniform facies distribution in Unit B.

Unit C

This unit, 23–45 ft thick, comprises aeolian dune and interdune sandstones interbedded with fine- to medium-grained sheetflood sandstones. Conglomerates occur only rarely and grain size in the fluvial sandstones is generally finer than in Units A or B.

The fluvial influence decreases upwards through the unit, but the rate of decrease varies greatly across the field, so that in wells located furthest from the main wadi trend Unit C is predominantly aeolian throughout. In many wells fluvial influence gives way to wholly aeolian processes in the upper half of the units, whilst in wells located in the middle of the central wadi trend fluvial facies disappear only in the uppermost few feet of the unit. A late-stage wadi system is identified in the northwest of the field.

Unit C is further subdivided into two contrasting diagenetic intervals, the upper of which contains significant amounts of dolomite cement. The boundary between these subunits cuts across, and is independent of, facies boundaries. Porosity and permeability averages and ranges are shown on Table 1. The lower (dolomite free) subunit usually has high porosity, but variable permeability depending on the degree of aeolian influence. Both porosity and permeability are much reduced in the upper subunit due to the cementation. In general the permeability in Unit C is lower, *for a given porosity*, than anywhere else in the reservoir, due to the decreasing grain size in this unit.

Production Logging Data

An extensive body of Production Log data has been obtained from the development wells. PLTs have been run during both production and injection. The flow profiles show an extraordinarily close correspondence between productivity/injectivity and sedimentary facies, with the bulk of flow coming from/going into aeolian and reworked aeolian sandstones. Therefore, in the majority of wells, the greatest flow, more than 50%, is from Unit A. However, in wells located in the central wadi trend, the contribution from Unit A can be negligible. Wells which are poor performers overall are usually those in which the contribution from Unit A is low.

Average contributions from Units B and C are 21% and 23% respectively. Flow from Unit C is reduced to zero in wells where a strong fluvial influence is coupled with a thick dolomite-cemented zone.

Carboniferous

The Rotliegendes reservoir is underlain by rocks of Carboniferous Westphalian A-B age. These comprise alternating intervals of mudstone/siltstone and sandstone, with interbedded thin coals. Little is known about the sandstone intervals in the absence of core data, but they are assumed to have been deposited as delta top distributary channel sandstones. Studies of the closely spaced well control have established that these sandstones can be correlated between wells, and it seems very likely that they are extensive across at least the western half of the Field. Three main correlatable sandstone intervals have been identified.

Although these sandstones occur above the Field GWC and are known to be gas-bearing, they do not form a viable reservoir. Tests on several of the early wells have yielded very low, subeconomic, flow rates.

The interface between the Carboniferous and the Rotliegendes is an angular unconformity, the Carboniferous having been tilted and faulted prior to Rotliegendes deposition. Consequently the Rotliegendes subcrop geometry is complex, with the Carboniferous sandstone intervals directly underlying the Rotliegendes in some areas of the Field. With production from the Rotliegendes a pressure gradient has been established at the sand-to-sand interface. There is evidence to indicate that a slow seepage of gas from the Carboniferous sandstones into the Rotliegendes has occurred, with Repeat Formation Tester data from recent wells showing pressure depletion in Carboniferous sandstones.

Source

The Westphalian coals which underlie the Rough Field are only marginally mature. It is assumed that the Rough native gas was generated from the Westphalian in the Sole Pit Trough, which has been more deeply buried, and migrated up-dip into the Rough structure.

Hydrocarbons

The original 6 wells from the AD Platform were tested at rates between 21 and 48 MMSCFD, the lowest rate being from wells A6, located in the lowest permeability area of the Field. Many of the BD and CD wells have penetrated the higher permeability regions of the reservoir, and most of these have been tested at rates in the range 39–53 MMSCFD.

The main components of the Rough native gas are C_1 (91%), C_2 (3.8%), N_2 (2.4%), C_3 (1.1%), CO_2 (0.6%), and the C_4 to C_{10} components (1.1%). The condensate/gas ratio is 9.2 BBL/MMSCF. On the whole, injected gases have a substantially higher C_1 percentage and a very low proportion of heavier components (C_3 upwards). An important aspect is the degree of mixing between the relatively condensate-rich native gas and the condensate-free injected gas. It is believed that there is very little mixing in the reservoir of injected gas with native gas. Instead injected gas displaced native gas in a piston-like manner. Therefore, some of the new wells produce native gas in their first year of production, but after an injection season a given well produces mainly or entirely injected gas.

No wells have been drilled below the Field GWC and no measurable amounts of free water have been produced. Based on water samples obtained from nearby exploration wells, the salinity of the Rough formation water is believed to be 200 000 ppm total chlorides.

Reserves/resources

Prior to conversion to storage mode, the field had been producing under normal depletion. Had this continued compressors would have been installed on the AP Platform and indeed a compressor was under construction when the field was purchased for conversion to storage mode. Even so, the contract field plateau production rate could have been maintained for only a few more years. Following conversion to storage mode, two compressor trains mounted on the BP Platform are used for injection of gas into the reservoir. Following the end of the storage project, the remaining native gas will be depleted. With the 29-well development it is anticipated that ultimate recovery from the field will be 366 BCF, representing a recovery efficiency of 75%.

Few production problems have been encountered. The most serious problem has been sand production from a number of wells, most probably from the high-permeability reworked aeolian sandstones described above.

Rough Field data sumary

Trap
Type	Anticlinal horst block
Depth to crest	−9000 ft TVSS
Lowest closing contour	−9550 ft TVSS
Gas–water contact	−9550 ft TVSS
Gas column	550 ft

Pay zone
Formation	Leman Sandstone
Age	Permian–Rotliegendes Group
Gross thickness	Average 95 ft (range 78 to 115 ft)
Net : gross ratio	Average .98 (range to .93 to 1.0)
Cut-off for N : G	0.1 md
Porosity:	Average 12.5%
range of well averages	10.5% to 16.1%
Hydrocarbon saturation:	Average 64%
range of well averages	56% to 68%
Permeability:	Average 75 md
range of well averages	2 md to 184 md

Hydrocarbons (native gas)
Condensate gravity	57° API
Gas gravity	0.636 (air = 1) (initial conditions)
Dew point	4360 psia (initial fluid)
Condensate yield	9.2 BBL/MMSCF (initial fluid)
Formation volume factor	255.5 SCF/RCF

Formation water
Salinity	200 000 ppm total chlorides
Resistivity	0.018 ohm m at 197°F

Reservoir conditions
Temperature	197°F
Virgin pressure	4533 psia
Pressure gradient	0.0866 psia/ft (at initial conditions)

Field size
Area	5050 acres
Gross rock volume	25.6×10^9 ft^3
Recovery factor	75%
Drive mechanism	Depletion
Original reserves	366 BCF

Production prior to conversion to storage
Start-up date	October 1975
Development scheme	2 Platforms: AD (drilling), AP (processing); 16 in shoreline
Number of wells	6 gas producers
Production rate	104 MMSCFD
Cumulative production	157 BCF (prior to first injection)
Secondary recovery	None

Storage mode
First injection	July 1985
Development scheme	5 Platforms: AD and AP as above, BD (drilling/accommodation), BP (processing/compression), CD (wellhead); 36 in shoreline
Number of wells	29 completed for gas injection and production
Production rate	1120 MMSCFD (nominal max: 80 days)

The author expresses his gratitude to the management of British Gas for permission to publish this paper. A review paper of this nature inevitably draws upon the work of a great many individual technical specialists, and the author is pleased to be able to acknowledge the following colleagues without whose work it would not have been possible to prepare this contribution. The very important work on the Rotliegendes sedimentology and reservoir description was carried out by D. Ellis and M. Goodchild. J. Ayers-Morgan, D. Ellis, and M. Dearlove conducted work on the geology of the Carboniferous. J. Smith has carried out all seismic interpretation and mapping, and B. Harrison provided the petrophysical analysis. Petroleum engineering information has been provided by A. Hollis, J. Treen, P. Cole, G. Will, and M. Currie. Information on the licence history was provided by J. Vercoe. The work of C. Harwood of G.A.P.S Geological Consultants is also acknowledged. The author also acknowledges staff at British Gas, in particular A. Levison, for providing critical review of the draft paper, N. Davies for typing the text, and finally the staff of the British Gas drawing office for preparing the figures.

References

ANONYMOUS. 1985. Rough Gas Storage Gets Underway. *The Oilman.* August 1985, 35–48.

GIBBON, R. B. 1987. Rough and Ready. *124th Annual General Meeting, The Institution of Gas Engineers*, Communication 1326.

GOODCHILD, M. W. & BRYANT, P. 1986. The Geology of the Rough Gas Field. *In*: BROOKS, J., GOFF, J. C. & VAN HOORN, B. (eds) *Habitat of Palaeozoic Gas in NW Europe.* Geological Society, London, Special Publication, **23**, 223–235.

—— & WHITAKER, J. H. McD. 1986. A petrographic study of the Rotliegendes Sandstone Reservoir (Lower Permian) in the Rough Gas Field. *Clay Minerals,* **21**, 459–477.

HOLLIS, A. P. 1984. Some Petroleum Engineering Considerations in the Changeover of the Rough Gas Field to Storage Mode. *Journal of Petroleum Technology,* **36**, 797–804.

RHYS, G. H. 1974. *A proposed standard lithostratigraphic nomenclature for the southern North Sea and an outline structural nomenclature for the whole of the (UK) North Sea.* Report of the Institute of Geological Sciences, 74/8.

The Sean North and Sean South Fields, Block 49/25a, UK North Sea

G. D. HOBSON[1] & A. P. HILLIER[2]

[1] *7 St John's Avenue, Ewell, Surrey KT17 3BE, UK*
[2] *Shell UK Exploration and Production, Lothing Depot, North Quay, Lowestoft, Suffolk NR32 2TH, UK*

Abstract: Sean North and Sean South are two small prolific gas fields located on the Indefatigable Shelf in the Southern North Sea. They, like most of the other fields in the area, have a Carboniferous source, a Rotliegend aeolian sandstone reservoir and a Zechstein evaporite cap rock. They have been developed to fullfil a peak-shaving role, being produced for only a few days per year in times of high gas demand when they produce at a rate of 600 MMSCFD. Initially thought to be two equally sized accumulations, there is now some evidence from material balance calculations that the Sean South is bigger than North Sean. The contractual recoverable reserves for the two fields are 425 BCF.

The two gas accumulations, Sean North and Sean South, lie in Block 49/25a about 15 km southeast of the major Indefatigable gas field. They are situated some 100 km northeast of the coast of Norfolk, in water approximately 30 m deep, which has allowed drilling by jack-up rigs. The name SEAN is made up from the initial letters of the four partners involved when the discovery wells were drilled: Shell, Esso, Allied Chemical and National Coal Board (ten Have & Hillier 1986).

In many respects Sean North and South are small versions of Indefatigable (France 1975). The reservoir rock is the Permian Rotliegend sandstone, capped by the Zechstein carbonates and evaporites, at about 8400 ft sub-sea. The gas-bearing areas are anticlinal, Sean North being some 4 km long and 1.5 km wide, whereas Sean South is about 3 km long and 2 km wide.

History

The Licence, of which Block 49/25a is a part, was awarded in the Second Round of licensing in 1965 to Allied Chemical and the National Coal Board. Shell and Esso farmed into this by drilling the early wells, and became equal partners with the other two companies. Subsequently Allied Chemical (Great Britain) Ltd changed its name to Union Texas Petroleum Ltd, the National Coal Board's holding became part of BNOC (later being acquired by Britoil), and now Britoil has been absorbed by BP. Shell is the operator.

Well 49/25-1 discovered Sean North in April 1969, and well 49/25-2 found the Sean South gas accumulation in December 1969/January 1970. The discovery wells were sited on seismic time highs. An appraisal well, 49/25-4, was drilled in Sean North in 1982, and an Annex B was submitted to the UK Department of Energy in March 1984. Arising from this well-head platforms were installed on each of the structures, with a process platform linked to the Sean South well-head platform by a bridge. Five development wells were drilled from each well-head platform during 1985, and the fields were brought on-stream in the 1986-87 contract year. The wells are deviated, and each platform has twelve drilling slots, providing for the additional wells which may be needed at some later stage to maintain production rates and to drain the reservoirs adequately.

Sean North is connected to Sean South by a 20 inch diameter inter-field, stainless steel pipeline, 4.7 km long. Dehydration of the combined gas streams takes place on the production platform before export to Bacton by a 107 km long 30 inch diameter pipeline.

Field stratigraphy

The field stratigraphy shown in Fig. 1 is similar to that of Indefatigable Field.

Carboniferous purple to red-brown shales and siltstones with coals and sandstones, of deltaic origin, occur unconformably beneath the Rotliegend, and constitute the oldest rocks penetrated.

Permian Rotliegend consists mainly of a series of aeolian sand dunes, with minor amounts of water-laid sediments, and varies in thickness from 200 ft to 265 ft. The upper part was disturbed by the incoming sea during the Zechstein transgression, giving the Weissliegend. This type of deposit has been described by Glennie (1984) as uncoloured structureless sands, alternating with a grading into highly deformed strata, as well as into beds in which the original aeolian bedding is only faintly preserved. It is considered that these non-depositional features arise either from the escape of air initially trapped below the wetted surface of the aeolian dunes or from the upward escape of water during flooding.

Upper Permian Zechstein Evaporite sequence is up to 1300 ft thick. At the base of the Zechstein I Cycle is the thin ubiquitous Kupferschiefer, a black, hard, brittle, pyritic shale, overlain in succession by a thin Zechsteinkalk unit, followed by the Werra Anhydrite. The Zechsteinkalk is white to grey dolomitic limestone, and the Werra Anhydrite is white and micro-crystalline. Cycle II starts with Haupt Dolomite, grey and argillaceous, followed by the Basal Anhydrite, thin and white. Next the Stassfurt Halite was deposited, consisting of halite interbedded with magnesium and potassium salts, and capped by the thin Deck Anhydrite. The Cycle III sequence is inferred to be incomplete. It starts with the Grey Salty Clay; then the Platten Dolomite follows, being a brown, silty, slightly anhydritic, platy dolomite. Next comes the white to grey Haupt Anhydrite. The thickness of the Zechstein below the top of the Platten Dolomite is comparable with that of the same part of the Zechstein Type Section in well 49/26-4, some 50 km to the south-southwest, yet the overlying Zechstein sequence is much thinner in the Sean wells. This is considered to be a residual or a condensed sequence of the upper part of Z III and/or Z IV. It consists of siltstones and anhydrite beds, with some salt. Taylor (1984) has noted that there can be difficulties in identifying some Zechstein units in going towards the centre of the North Sea Basin, especially when there has been displacement by Z II salt movements. The Upper Zechstein is 330–430 ft thick.

Triassic sedimentation is marked by the appearance of the Brockelschiefer Member of the Bunter Shale formation of the Bacton Group (Fisher 1984), a red-brown slightly anhydritic siltstone, mottled with green. This is followed by the Main Bunter Shale Member of red-brown claystones and capped by the Rogenstein Member with its characteristic ferruginous oolites. Above is the Bunter Sandstone, red, brown, fine and silty in parts.

Following the Late Kimmerian tectonic activity the Jurassic was completely removed, and erosion cut deeply into the Bunter Sandstone and even into the underlying Bunter Shale, leaving only 50–300 ft of Triassic sediments.

Lower Cretaceous, represented by the Cromer Knoll Formation, overlies the unconformity as grey calcareous clay, with traces of

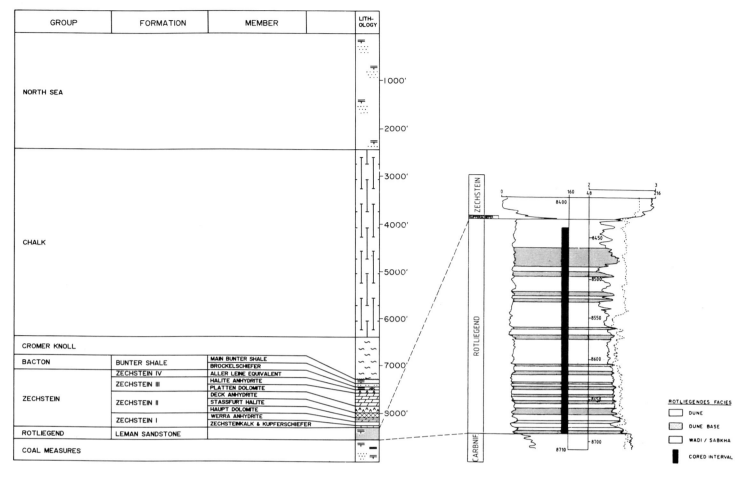

Fig. 1. Stratigraphic section of the Field showing overburden geology and detailed reservoir stratigraphy.

calcareous siltstone, being 600–800 ft thick. The Upper Cretaceous Chalk is 3500–4000 ft thick.

Unconformably above the Cretaceous are some 2000 ft of Tertiary and Quaternary deposits, dominated by marine clays, with some sands.

Geophysics

400 km of seismic lines were shot over Block 49/25a in 1977 in an endeavour to improve on the interpretation of unmigrated data obtained in the previous decade. The resultant 1500 m × 750 m grid made a more precise interpretation possible. An in-fill programme of about 100 km, shot in 1979, covered the gas-bearing structures penetrated by the exploratory wells, 49/25-1 and 49/25-2, together with a prospective trend to the east. This narrowed the grid to 350 m, for the purpose of selecting appraisal locations. A further 200 km of in-fill lines were shot in 1981, giving an overall grid of about 500 m × 200 m, covering the prospective areas, and suitable for selecting production well locations. A comparison between the earliest and latest seismic data is shown in Fig. 2.

Well shoots and synthetic seismograms involving wells 49/25-1, -2, -3 (a dry exploratory well in 49/25b) and -4 facilitated the determination of six main litho-stratigraphic boundaries: (1) Base Tertiary, (2) Base Chalk, (3) Late Kimmerian unconformity, (4) Top Brockelschiefer, (5) Top Platten Dolomite and (6) Top Haupt Dolomite. Iso-velocity maps constructed from the stratigraphic interval velocities calculated at the wells and interpolated between the control points, and used in 'Layer-cake' fashion, yielded a depth contour map for the top of the Haupt Dolomite. The Haupt Dolomite is some 255–400 ft above the top of the Rotliegend in the area covering Sean North and South. A necessarily somewhat speculative isopach map for the Haupt Dolomite/Top Rotliegend interval led to the construction of a Top Rotliegend map, while a further simple isopach map based on three wells lying almost in a straight line, allowed the construction of a tentative depth contour map of the top of the Carboniferous.

Thus, prior to the further drilling undertaken after installing the platforms, it was considered that the important depths might be accurate to 50 ft within a radius of 1 km from each well, and possibly within 100–200 ft some 2–3 km from well control. Inspection suggests that the uncertainty may have been most significant in estimating gas-in-place volumetrically for Sean South. In 1986 a few additional lines were shot over part of Sean South for the better definition of the area.

The two field areas have been remapped to match the additional well data provided by the development wells (Fig. 4). At the top of the Rotliegend the well data embrace an area of 1.75 × 0.7 km for Sean North, and some 2.0 km × 0.5 km for Sean South. Layer-cake depth conversion yielded maps of (1) Top Chalk, (2) Base Chalk, (3) Top Triassic, (4) Top Brockelschiefer, (5) Top Haupt Anhydrite/Platten Dolomite, (6) Top Basal Anhydrite/Haupt Dolomite and (7) Top Rotliegend.

Trap

Sean North and Sean South are in essence anticlinal, with axes striking approximately NNW–SSE, but Sean South has complications on the west side. The two uplifts are separated by faulting (see Figs 3 and 4). Interpretation of the seismic data suggests dips of as much as 21° adjacent to complex faulting in the northwestern sector of Sean South. Sean North is indicated to have fault trapping along most of its eastern boundary, while Sean South appears to be fault-bounded, except in the northwest.

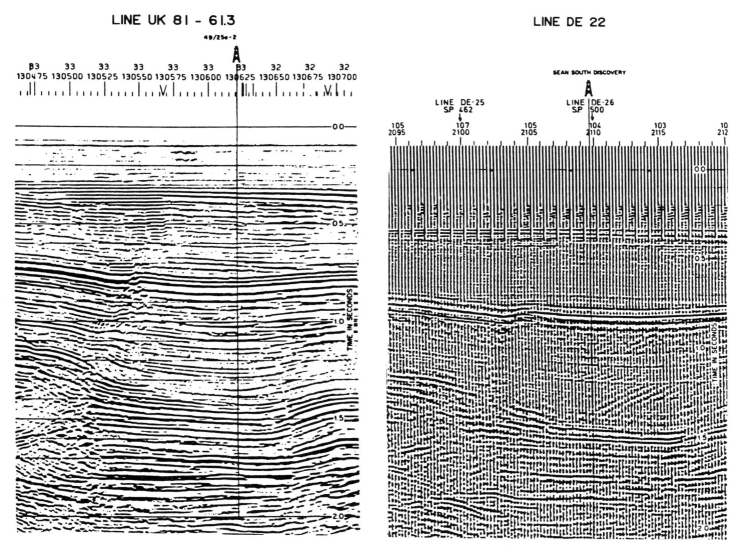

Fig. 2. Seismic sections through Well 49/25-2 showing increase in detail seen from 1968 to 1981, Lines DE 22 (1968) and UK81-61.3 (1981).

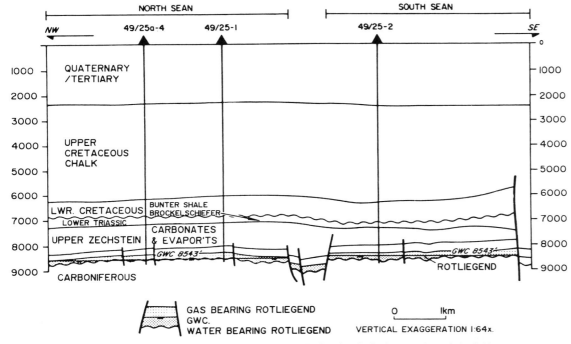

Fig. 3. North–South cross-section through Sean North and South showing faulted separation of the fields.

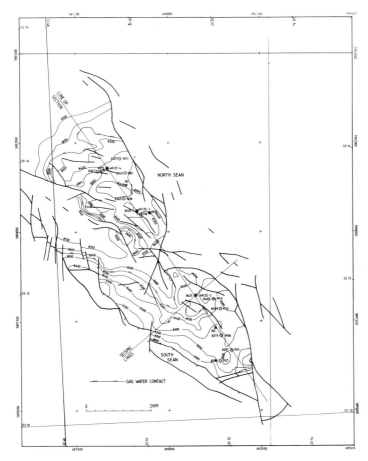

Fig. 4. Structure map of Sean North and South showing relative positions of the fields and lines of cross-section and seismic lines.

Over much of both structures the dips seem to be under 4°. A series of short, almost en echelon, step faults are seen on the west side of Sean North. The bounding faults are downthrown away from the two anticlinal features at reservoir level.

Three main phases of tectonic activity have affected the southeast Indefatigable area in which Sean North and South lie. Hercynian movements gave a dominant northwest-southeast grain to the Pre-Permian formations. In the Late Kimmerian phases Hercynian trends were reactivated and affect Pre-Cretaceous formations. Overburden loading during Cretaceous time caused halokinesis in the Zechstein evaporites, giving overburden faults.

The major northwest-southeast faults bounding the Sean area belong to the Viking-Indefatigable system to the southwest and to the Southeast Indefatigable system to the northeast. The faults are high-angle and sometimes overturned to reverse at Rotliegend level. These major faults extend up through the Zechstein and overlying strata where the sense of throw varies both vertically and laterally, demonstrating the effects of both strike slip movement and salt flow along the fault planes. Sections from Dutch block K-13 published by Roos & Smits (1983) and from the UK well 49/23-4 published by Glennie also show this feature.

There are also subordinate north-south faults which are smaller and die out in the Zechstein.

Intra-reservoir faulting in both Sean North and Sean South is limited to a few normal faults with throws of 30-100 ft. These are unlikely to be barriers to flow, in view of the thickness and general good reservoir properties of the Rotliegend.

Wireline log and pressure data from the wells in the two fields show that there is the same gas-water contact at 8543 ft subsea. The original wells showed a 10 ft difference, but this appears to be a capillary effect, since the latter wells gave a good average value for both fields (Fig. 3).

Reservoir

The reservoir rock of Sean North and South is the Leman Sandstone of the Permian Rotliegend Group (Rhys 1974). As a unit of the Southern Permian Basin of the North Sea it represents deposition under continental desert conditions. The facies recognized are aeolian, wadi, sabkha and lacustrine, as described by Glennie (1984). The aeolian facies in the form of transverse dunes is dominant in the Sean Area. These dunes are perpendicular to the palaeo-wind direction which was generally from the east-northeast. There are cross-bedded fore-set dunes and finely-laminated horizontal bottom-sets. Usually thin wadi-sabkha sediments occur at the base of the sequence, while in the upper part of the sequence inter-dune sabkha, wadi or lacustrine deposits may be found. These intercalations are usually thin and not laterally extensive (Fig. 1). In wells 49/25-1 (Sean North) and 49/25-2 (Sean South), according to data presented by ten Have & Hillier (1986), the breakdown is approximately, in aggregate, 60% dune, 19% dune-base and 21% wadi-sabkha for the former well, and 64% dune, 28% dune-base and 8% wadi-sabkha for the latter well. In general the dune sand units are the thickest.

The reservoir properties are markedly different for the dune, dune-base and wadi-sabkha deposits. The cross-bedded dune sands are fine to medium in grain size, clean and well sorted. Locally calcite cementation impairs the reservoir properties. Porosity averages 21%, and the permeability 650 md. The entry pressure is very low, and the capillary pressure curve has a long plateau. At 50% wetting-phase saturation the pore throat size is about 11 μm.

The dune-base sands (bottom-sets) are commonly less well sorted than the dune sands, with a wider grain-size range. Average porosity is 17% and average permeability 166 md. The capillary pressure curve has a short plateau, and the pore throat size at 50% wetting-phase saturation is 2 μm. Nevertheless, the dune-base sands are good reservoir rocks, although calcite cementation is common, reducing the reservoir quality.

The wadi and sabkha deposits average 11% porosity and 5 md permeability. They are poorly sorted and often argillaceous. The capillary pressure curves show a relatively high entry pressure, lack a plateau, and have a pore throat diameter of 0.3 μm at 50% wetting-phase saturation. The poor quality sections of reservoir rock (wadi, sabkha and lacustrine deposits), which are found at various levels in the Rotliegend, are thin and not considered to be field-wide permeability barriers.

The entire Rotliegend is defined as net sand in all of the production wells, except two in Sean South, in which thin lacustrine shale intervals (11-13 ft) are excluded.

Sand production was feared, and selective perforation and pre-conditioning of the wells was undertaken to prevent excessive sand production. A SANDTEC monitoring system has been installed to detect sand production early, but to date there has been no evidence of excessive sand production.

The average porosity in the water zone of Sean North is 16.3%, compared with 17% average porosity for the gas zone of both Sean North and South and the water zone of Sean South.

Source

The source rock of the gas is believed to be the coal-bearing sequence of the Westphalian in the Sole Pit Trough to the west and/or the Broad Fourteens Basin to the east. In the Viking-Indefatigable area vitrinite reflectance data show the coals to be immature. However, Cornford (1984) has reported vitrinite reflectance values of >2% Ro in both the Sole Pit Trough and in the Broad Fourteens Basin. Glennie & Boegner (1981) generated burial curves based on data derived from sonic logs (Hobson 1960; Marie 1975), other well information and erosion histories for the Sole Pit Trough, which suggest that gas generation would have taken place during the Jurassic and the Cretaceous. Data published by van

Wijhe et al. (1980) indicate that the centre of the Broad Fourteens Basin would have had a main gas-generation phase from about mid-Jurassic to mid-Cretaceous. According to Oele et al. (1981) Late Cretaceous inversion raised that sector of the Basin out of the depth range needed for gas formation, and shifted the depo-centre to the northeast, which, coupled with further sedimentation, led to gas generation taking place beyond that time, in association with a change in the pattern of migration. The very substantial reduction in pressure caused by the uplift during inversion would have released some adsorbed gas from coals and dissolved gas from formation water.

Hydrocarbons

Table 1 shows the composition of the gas in the Sean North and South fields.

Table 1. *Composition of gas in the Sean North and South fields*

Component	Sean North reservoir fluid, mol. %	Sean South well-head fluid, mol. %
Methane	92.06	91.06
Ethane	2.94	3.38
Propane	0.64	0.58
iso-Butane	0.12	0.11
n-Butane	0.13	0.13
iso-Pentane	0.04	0.05
n-Pentane	0.03	0.04
Hexanes	0.03	0.06
Heptanes plus	0.14	0.19
Helium	0.04	
Nitrogen	3.07	3.19
Carbon dioxide	0.76	1.21
	100.00	100.00

The condensate/gas ratio is 1.9 BBL/MMSCF for Sean North and 3 BBL/MMSCF for Sean South. The condensate recovery factor is put at about 10% below that for gas, because of liquid drop-out in the reservoir. However, drop-out is not expected to impair gas flow.

Well tests: 49/25-1 Production test 1–41.8 MMSCFD (max)
2–41.3 MMSCFD (max)
49/25-2 Production test 1–36.1 MMSCFD (max)
In each case the test covered about 20 ft.

The reservoir pressure is hydrostatic, being 3964 psia at a sub-sea datum of 8500 ft tvss.

High pressure gas is found in the dolomites of the Zechstein in vugs and fractures; the volumes are small.

Reserves/resources

Material balance calculations for the Sean North reservoir agreed with the volumetric value for gas-in-place. However, in Sean South the volumetric figure was only 65% of that obtained by material balance. The west side of Sean South has not yet been investigated by drilling, so there is uncertainty about the elevation, while the boundary fault may have been poorly migrated.

The gas contains 0.96 mol. % of carbon dioxide. Batch injection of corrosion inhibitor is carried out to prevent down-hole corrosion. Also stainless steel casing has been installed in all production wells to combat corrosion.

The recovery mechanism has not been conclusively identified. Plots of P/Z against cumulative production show pressure recovery during the summer shut-in. This could arise from water influx, or from gas influx from the less easily accessible parts of the reservoirs.

The recovery factor has been taken as the average of values for natural depletion and from water influx. However, the extensive faulting around both gas accumulations may mean severe limitations on communication with any body of water such as could contribute materially to the recovery mechanism for either of the accumulations.

The combined fields are being used as a 'seasonal' or 'peak-shaving' source of gas supply, capable of producing up to 600 MMSCFD for British Gas Corporation. The customer is entitled to nominate varying quantities from zero up to the maximum daily rate whenever required during the six months of the winter period.

Compression is expected to be needed at some time between 1995 and 2006 depending on the British Gas offtake. This will call for the installation of another platform.

With compression installed Sean North may have an 82% gas recovery factor, but Sean South may give 87% because of the smaller connected aquifer giving rise to lower waterdrive than in the north.

Sean North and Sean South Fields data summary

	Sean North	Sean South
Trap		
Type	Gentle anticlines, with faulting, especially on the west and east	
Depth to crest (ft sub-sea)	8324	8427
Lowest closing contour (ft sub-sea)	8700	8600
Gas–water contact (ft sub-sea)	8543	8543
Pay zone		
Formation	Rotliegend	Rotliegend
Age	Permian	Permian
Gross thickness (ft)	268 (max)	249 (max)
Net/gross ratio (%)	100	100
Porosity (average/range) (%)	17.5 (15–18)	17.1 (16–18)
Hydrocarbon saturation (%)	79	79
Permeability range (md)	130–400	190–420
Hydrocarbons		
Gas gravity (relative to air)	0.618	0.614
Condensate yield (BBL/MMSCF)	1.9	3.0
Formation volume factor (vol./vol.)	218	225
Formation water		
Salinity	225 000 ppm NaCl eq.	
Resistivity	0.017 ohm m	
Temperature	201 F	
Reservoir conditions		
Datum (ft sub-sea)	8500	8500
Temperature (°F at datum)	202	192
Pressure (initial) (psia at datum)	3945	3977
Pressure gradient in reservoir (psi/ft)	0.068	
Field size		
Area (acres)	3379	2207
Gross rock volume (acre-ft)	198 056	326 600
Hydrocarbon pore volume (acre-ft)	34 572	57 765
Recovery factor: gas	82%	87%
Major drive mechanism	Depletion	Depletion
Recoverable hydrocarbons (volumetric estimates)		
NGL/condensate (BBL)	400 000	500 000
Gas (BCF)	192	233
Production		
Start-up date	1986–87 contract year	
Development scheme	Peak-shaving winter only producer	
Production rate	Up to 600 MMSCFD	
Cumulative production to 1/1/89 (BCF)	11.6	22.1

Shell UK Exploration and Production, Esso Exploration and Production UK Ltd, Union Texas Petroleum Ltd and British Petroleum the joint licensees of the field permitted publication of this paper. The paper was prepared by the authors using the work of the many Shell geologists and geophysicists who had worked on the field in the past whose contribution is greatly appreciated.

References

CORNFORD, C. 1984. Source Rocks and hydrocarbons in the North Sea. *In*: GLENNIE, K. W. (ed.) *Introduction to the Petroleum Geology of the North Sea*. Blackwell Scientific Publications, Oxford, 171–204.

FISHER, M. J. 1984. Triassic. *In*: GLENNIE, K. W. (ed.) *Introduction to the Petroleum Geology of the North Sea*. Blackwell Scientific Publications, Oxford, 85–101.

FRANCE, D. S. 1975. The geology of the Indefatigable gas field. *In*: WOODLAND, A. W. (ed.) *Continental Shelf of Northwest Europe Vol. 1, Geology*. Applied Science Publishers Ltd, London, 233–239.

GLENNIE, K. W. 1984. Early Permian–Rotliegend. *In*: GLENNIE, K. W. (ed.) *Introduction to the Petroleum Geology of the North Sea*. Blackwell Scientific Publications, Oxford, 000–000.

—— & BOEGNER, P. L. E. 1981. Sole Pit inversion tectonics. *In*: ILLING, L. V. & HOBSON, G. D. (eds) *Petroleum Geology of the Continental Shelf of Northwest Europe*. Heyden and Son Ltd, London, 110–120.

HOBSON, G. D. 1960. Estimation of relative maximum depths of burial. *Journal of the Institute of Petroleum*, **46**, 89–90.

MARIE, J. P. P. 1975. Rotliegendes stratigraphy and diagenesis. *In*: WOODLAND, A. W. (ed.) *Petroleum and the Continental Shelf of Northwest Europe. Volume 1, Geology*. Applied Science Publishers Ltd, London, 205–214.

OELE, J. A. HOL, A. C. P. J. & TIEMENS, J. 1981. Some Torliegend gas fields in K and L Blocks, Netherlands offshore (1968–1978)—a case history. *In*: ILLING, L. V. & HOBSON, G. D. (eds) *Petroleum Geology of the Continental Shelf of Northwest Europe*. Heyden and Son Ltd, London, 289–300.

RHYS, G. H. *A proposed lithostratigraphic nomenclature for the Southern North Sea and outline stratigraphy for the whole of the (UK) North Sea*. Institute of Geological Sciences, Report No 74/8.

ROOS, B. M. & SMITS, B. J. 1983. Rotliegend and main Buntsandstein gas fields in Block K/13—a case history. *Geologie en Mijnbouw*, **62**, 75–82.

TAYLOR, J. C. M. 1984. Late Permian–Zechstein. *In*: GLENNIE, K. W. (ed.) *Introduction to the Petroleum Geology of the North Sea*. Blackwell Scientific Publications, Oxford, 61–83.

TEN HAVE, A. & HILLIER, A. P. 1986. Reservoir geology of the Sean North and South gas fields, U.K. Southern North Sea. *In*: BROOKS, J. GOFF, J. & VAN HOORN, B. (eds) *Habitat of Palaeozoic gas in N.W. Europe*. Geological Society, London, Special Publication, **23**, 267–273.

WIJHE, D. H. VAN, LUTZ, M. & KAASCHIETER, J. P. H. 1980. The Rotliegend in the Netherlands and its gas accumulations. *Geologie en Mijnbouw*, **59**, 3–24.

The Thames, Yare and Bure Fields, Block 49/28, UK North Sea

OONAGH C. WERNGREN

ARCO British Ltd, London Square, Cross Lanes, (off London Road), Guildford, Surrey GU1 1UE, UK

Abstract: The Thames, Yare and Bure Fields, discovered in 1973, 1969 and 1983 respectively, are located in the UK sector of the Southern Gas Basin. The reservoir is made up predominantly of aeolian dune sandstones of the Rotliegendes Leman Sandstone Formation. The fields lie on the upthrown side of NW–SE-trending fault blocks in the central part of Block 49/28. The Thames complex came on production in October 1986 and is estimated to have a field life of 10 years. Current estimates put gas in place for the three fields at approximately 350 BCF, with recoverable reserves of 240 BCF.

The Thames, Yare and Bure Fields lie in UK Block 49/28 in the southern part of the Gas Basin, approximately 50 miles NE of Bacton and 4 miles NE of Leman Gas Field (Fig. 1). The water depth is 110 ft. The development currently consists of three fields, Thames, Yare and Bure, plus Wensum which is a separate discovery on the same trend. The fields are named after rivers on the east coast of England.

History

Production Licence P.037, which includes Block 49/28, was awarded in 1964 under the terms of the first UK Licensing Round to a group comprising ARPET Petroleum Ltd, British Sun Oil Company Ltd, Superior Oil (UK)Ltd, Canadian Superior Oil (UK) Ltd and North Sea Exploration and Research Company Ltd. Following several mergers, the participating interests in the Thames complex are now:

	% interest
ARCO British Ltd	43.33
Sun Oil Britain Ltd	23.33
Superior Oil (UK) Ltd	20.00
Canadian Superior Oil (UK) Ltd	3.33
Deminex (UK) Oil and Gas Ltd	10.00

Exploration on the block commenced in 1967 with the drilling of the 49/28-1 well, which proved an extension of the Leman Field into 49/28. The 49/28-2 well was drilled on a smaller satellite structure to the east of Leman. The Thames group of structures to the NE of Leman were discovered in 1969 by the 49/28-3 well. The well tested a NW–SE-trending fault block in the eastern part of the block (Fig. 2) and encountered a 75 ft gas column in the Leman Sandstone at a depth of 7921 ft. The well tested gas at a rate of 47 MMSCFD, with a flowing-wellhead pressure of 3174 psig. This accumulation was subsequently named Yare.

The Thames structure was discovered in 1973 with the drilling of 49/28-4, which encountered a gas column of 264 ft at a depth of 7892 ft, and tested at a maximum flow rate of 61 MMSCFD, with a flowing-wellhead pressure of 3653 psig. The 49/28-6 well was the discovery well on a structure known as 'N' which is located to the NW of Thames. The Bure field was discovered in 1983 by the 49/28-8 well, which encountered a gas column of 139 ft at a depth of 8037 ft and flowed at a maximum rate of 31.2 MMSCFD with a flowing-wellhead pressure of 2814 psig. Four appraisal wells were later drilled to provide additional reservoir information across the fields. The 49/28-9 well, drilled in 1984, encountered the Rotliegendes Leman Sandstone on the downthrown side of a fault, 40 ft below the anticipated gas–water contact. The well was consequently sidetracked to a bottom-hole location on the upthrown side of the fault where 40 ft of pay was observed with a gas–water contact identical to that in the Yare discovery well. A directionally drilled

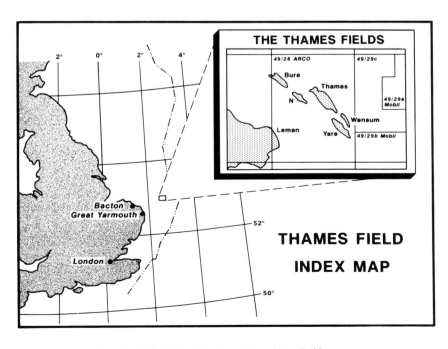

Fig. 1. Thames Field index map.

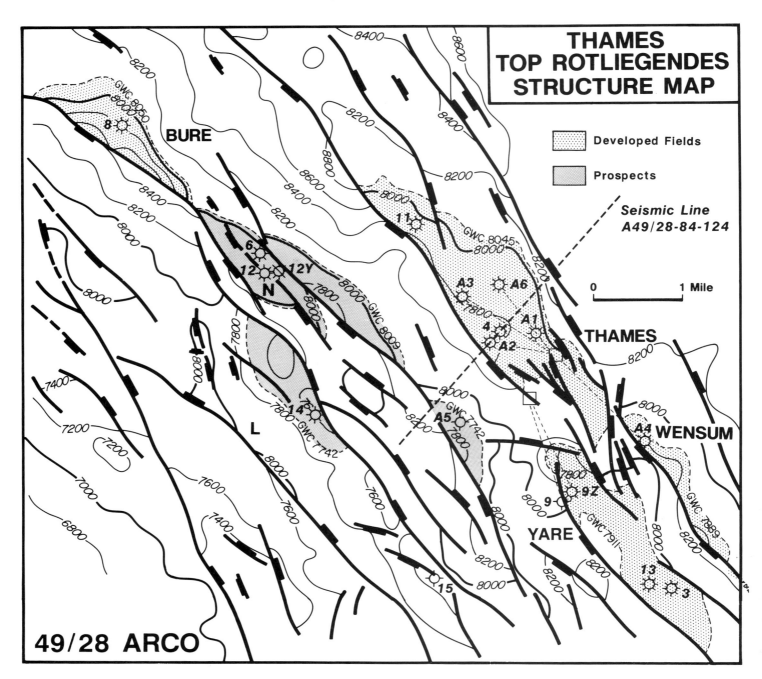

Fig. 2. Thames, Top Rotliegendes structure map.

appraisal well, 49/28-10, was drilled to a southern location in the Thames field to assist mapping of the SE flank. The well encountered a 170 ft gas column and was suspended at the mudline for possible future completion as a producer. The 49/28-11 well, was drilled in 1984 and targeted solely to define the western extent of the Thames field. It was suspended as a gas well. 49/28-12 was drilled to appraise further the 'N' structure. However, this field has not yet been developed. 49/28-13, drilled in 1985 on Yare, was suspended as a gas well to be later completed as subsea producer, having tested 46 MMSCFD of gas with a wellhead flowing pressure of 1934 psig.

The 49/28-A4 development well was a deviated well drilled in 1985 from the Thames template after Annex B approval. The well discovered a small separate Leman Sandstone gas accumulation, now named Wensun, which lies on trend with Thames.

49/28-14 was the last discovery well on the block, drilled in 1987 on the small 'L' structure to the west of Thames. Appraisal of this structure together with the 'N' structure is still underway.

ARCO, as operator for the Thames fields, mobilized a project team to work up the Thames development in 1984, with the Annex B being submitted to the Department of Energy in January 1985. The Thames Complex was subsequently brought on stream in October 1986. The Development (Fig. 3) consists of a wellhead platform (AW) bridge-linked to a process, utilities and quarters platform (AP). The Thames Field is produced via five wells drilled from the AW platform. Four wells are located on the Thames field with one on Wensum. The Yare and Bure fields are produced via single wells completed subsea and tied back to the AW platform via 8 inch flowlines and hydraulic-control umbilicals. All wells were predrilled and the field facilities installed in 1986. To maintain production later in field life, gas-compression modules were installed on the AP platform in the summer of 1989.

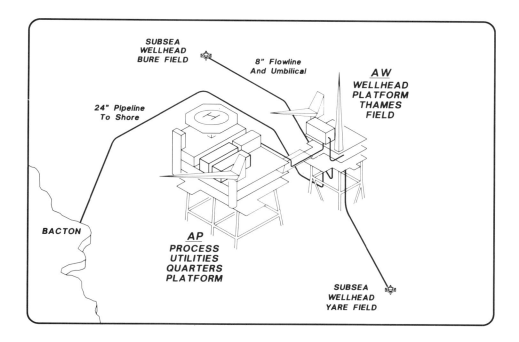

Fig. 3. Thames development complex.

Field stratigraphy

Block 49/28 lies on the southern flank of the Southern North Sea Basin in a narrow zone between the Sole Pit axis and the Cleaver Bank High. The stratigraphy of this area has been well documented by Arthur *et al.* (1986) and conforms to a typical gas-basin sequence as indicated in the stratigraphic column (Fig. 4). The oldest rocks encountered on the block are Carboniferous, Namurian A, found in the 49/28-14 well. However, most wells were drilled to a total depth in the Westphalian just beneath the Base Permian unconformity. The overlying Rotliegendes Leman Sandstone, thickens regionally from NE to SW across the block, reflecting the increasing influence of the Sole Pit axis towards the SW. In the overlying Zechstein sequence, which exhibits the classic Z1–Z4 cycles, facies boundaries trend E–W through the area, indicating a reduction in importance of the Sole Pit axis at this time. While the Zechstein salt generally acts as a seal to the underlying Leman Sandstone reservoir, the halites are observed to have absorbed some of the later fault movements in the area. The Triassic section in Block 49/28 thickens depositionally to the SW and, with the exception of the eastern side of the block, the section is complete.

The overlying Lower Jurassic rocks are only preserved in the centre of the block, in a complex graben zone, with the present-day distribution of the Jurassic being controlled by truncation beneath the Base Cretaceous (Late Cimmerian) unconformity. It is thought that over 1000 ft of Triassic and a significant proportion of Jurassic sediments have been eroded from the eastern part of the block. The major faults that affected Triassic and Jurassic sediments show little movement above the Late Cimmerian unconformity, as indicated on the seismic line A49/28-84-124 (Fig. 5).

During the Base Cretaceous erosional phase, Block 49/28 lay in a narrow graben between the uplifted Sole Pit axis and Cleaver Bank High. Sedimentation resumed in the Early Cretaceous, and a thick sequence of Cretaceous and Tertiary sediments are seen to overlap the Sole–Pit uplift and thicken to the NE. A major fault system is present to the east of the 49/28-4 well, which acted as a hinge zone at this time, and marked thickening of the post-Jurassic section occurs across the fault. Renewed uplift of the Sole–Pit area in early Tertiary times is reflected by structural movement in Block 49/28, and may even be responsible for breaching some of the structures in the southern part of the block.

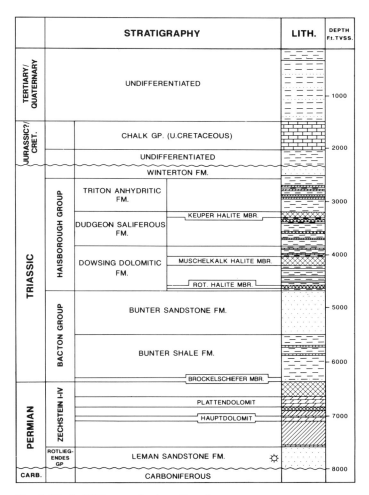

Fig. 4. Block 49/28 generalized stratigraphy.

Geophysics

Approximately 400 km of multifold reflection seismic data were acquired in late 1960s and early 1970s over Block 49/28 in an effort

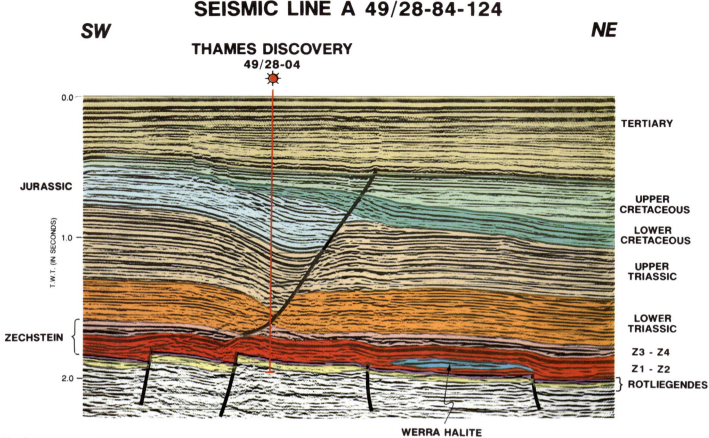

Fig. 5. Seismic line A49/28-84-124.

to discover satellite structures east of the Leman Gas Field. The data initially enabled production of seismic time maps down to Top Plattendolomit level. Depth conversion relied on regional velocity maps. The Top Rotliegendes depth structure was subsequently derived by adding a regional isopach to the Plattendolomit depth interpretation. Subsequent data acquisition in the early 1980s added another 950 km of high-resolution seismic data to the grid. This data, most of which was subsequently reprocessed, enabled direct seismic mapping, of all key horizons in time, down to the Top Rotliegendes reflector. Subsequent surveys in Block 49/28 were acquired in 1985 and 1987 to delineate specific prospects to the north and south of Thames that were not included in the development.

Based on velocity data from wells, seven reflectors were chosen as important velocity interfaces, as follows: Base Tertiary, Base Chalk Group, Base Cretaceous Unconformity, Near Top Bacton, Top Brockelschiefer, Top Plattendolomit and Top Rotliegendes. A layer-cake time-to-depth conversion method was used to allow for effects of lateral velocity variations in each of the layers. These velocity variations influence the depth configuration of the low-relief structural closures at Top Rotliegendes level. The structure at Top Rotliegendes comprises a series of tilted fault blocks progressively downthrown in an eastward direction from the complexly shattered dome that forms the Leman field to the west. The predominant fault trend is NW–SE. The faults are decoupled from the post-Permian overburden by the Zechstein evaporites.

Trap

Detailed mapping of the Thames complex clearly indicates that the fields, all discrete anticlines, are separated by faults and dip slopes. The separation is also confirmed by different gas–water contacts. The Thames field covers an area of 1908 acres. It comprises a northwest-trending eastward-dipping tilted fault block separated from Yare, Bure and 'N' by a major bounding fault on the western flank. Closure to the east and south is by minor faulting, and to the north by gentle dip. Much of the field lies beneath a graben containing a thick Jurassic/Cretaceous sequence with low interval velocities, and this causes some distortion of seismic time data. The Thames field has a gas column of 295 ft with a gas–water contact at −8045 ft TVSS. The Yare field, covering an area of 780 acres to the SE of Thames, forms a narrow elongate anticline, with two low-relief crests separated by a saddle. Closure on the flanks of Yare is by a combination of faults and dip-slope. Yare has a gas column of 161 ft with a gas–water contact at −7911 ft TVSS. Bure, to the NW of Thames, covers an area of 644 acres. The trap is formed by a single elongate anticline, closed to the SW by a major fault. The flanks of the structure are broad and dip gently to the north, but are narrow and steeply dipping to the SE. Bure has a gas column of 200 ft with a gas–water contact at −8050 ft TVSS.

The fields are all sealed both vertically and horizontally by the Zechstein salts. The structures that comprise the Thames field are all Permian tilted fault blocks into which gas migrated during the Cretaceous/Tertiary. The structures are all full to spill point. The timing of fault closure and hydrocarbon migration in this area is complex, as indicated by the presence of a number of dry holes on what appear to be valid structures to the south of Thames.

Reservoir

The reservoir in the Thames field is the Rotliegendes Leman Sandstone Formation. The reservoir comprises some 400–700 ft of aeolian dune sands with a minor fluvial sand component towards the base. Detailed correlation of core and dipmeter data, together with wireline logs, has allowed sub-division of the Leman Sandstone into discrete units characterized by dune height, palaeowind direction

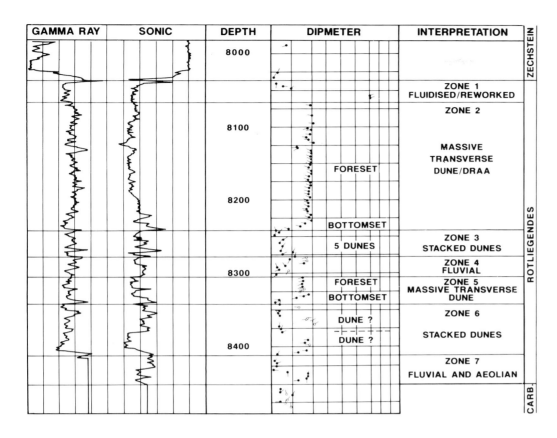

Fig. 6. Typical Leman Sandstone facies and sequence development.

and fluvial sediment content (Fig. 6). The Leman Sandstone Formation in the area comprises very fine- to medium-grained, locally coarse, sublithic and subfeldspathic arenite. Cementation is variable across the field, with the fluvial facies being the most strongly affected. Dolomite, silica, feldspar, anhydrite, iron oxide, illite and kaolinite all contribute to cementation. Porosity and permeability are primarily controlled by depositional facies, with average porosity ranging from 16–22%. Geometric averages from horizontal permeability range from 8.2 and 378 md. The Weissliegendes section, which is present at the top of the Leman Sandstone, contains less illite and, together with the aeolian dune forests, comprises the best quality reservoir rock. In aeolian dune bottom sets, poor sorting and the presence of detrital clays result in the reduction of porosity and permeability. The fluvial facies towards the base of the section, have the worst reservoir characteristics because of a high detrital clay content and moderate to poor sorting, and consequently average porosity decreases with depth. The absence of barriers to vertical flow, and the remarkably uniform nature of the sand across the field, removes the need for complex reservoir models. However, it is evident from the 49/28-9 and 49/28-12 wells that cementation associated with faults and fractures can significantly reduce porosity and permeability. In certain cases the faults may have acted as barriers to fluid migration.

Source

The Thames area is sourced by coals and shales of Westphalian age. Although the sediments near the top of the Carboniferous are not mature for gas in the block itself, migration of hydrocarbons is thought to have originated from the Sole Pit Trough to the NW.

Hydrocarbons

The discovery wells drilled into the Thames, Yare and Bure fields tested gas at rates of up to 61 MMSCFD. The fields contain dry gas of very similar composition, with 92 mol% methane and 0.4 mol% CO_2. Condensate and condensed water are produced equally at about 0.5 BBL/MMSCF of each. The initial reservoir conditions were as follows:

	Thames	Yare	Bure
Mean reservoir datum (ft TVSS)	7925	7850	7975
Initial reservoir pressure (psia)	3707.1	3648.7	3708.0
Initial reservoir temperature (°F)	182	178	177

Reserves/resources

Current field performance reserve estimates indicate smaller gas accumulations on some structures than previously expected based on pre-production mapping. Current estimates put the gas in place for the three fields at approximately 350 BCF. There is apparently a separate aquifer in pressure communication with each field. Initial studies indicated that some degree of water movement could be expected later in field life, but that the depletion would be essentially volumetric. Recovery factors were estimated to be in excess of 80%. Results of TDT logs run on Thames support this assumption. TDT logs run on Wensum revealed significant water influx, and the recovery factor has been reduced accordingly. No completion logs have been run on Yare and Bure as even wireline intervention requires mobilization of a jack-up drilling rig.

The development of the Thames Field is the result of detailed and meticulous work by the exploration and production staff of ARCO British Ltd extending back for some 20 years. Without that work, I would have been unable to write this paper, which summarizes the results of their labours.

Thanks are due to licensees of Block 49/28 (ARCO British Limited, Sun Oil Britain Ltd, Superior Oil (UK) Ltd, Canadian Superior Oil (UK) Ltd and Deminex (UK) Oil and Gas Ltd) for permission to publish this paper. Personal communications and contributions, for which I am grateful, came from A. T. Gregory and A. J. Walker of ARCO British Limited.

Reference

ARTHUR, T. J., PILLING, D., BUSH, D. & MACCHI, L. 1986. The Leman Sandstone Formation in U.K. Block 49/28 Sedimentation, Diagenesis and Burial History. *In*: BROOKS, J., GOFF, J. C. & VAN HOORN, B. (eds) *Habitat of Palaeozoic Gas in NW Europe*. Geological Society, London, Special Publication, **23**, 251–266.

Thames, Yare & Bure Field data summary

	Thames	*Yare*	*Bure*
Trap			
Type	Tilted fault block	Tilted fault block	Tilted fault block
Hydrocarbon contacts	−8045 ft	−7911 ft	−8050 ft
Gas column	295 ft	161 ft	200 ft
Pay zone			
Formation	Leman Sandstone	Leman Sandstone	Leman Sandstone
Age	Permian	Permian	Permian
Gross thickness	270–316 ft	430 ft	400 ft
Net/gross ratio	1.0	1.0	1.0
Porosity	19%	19%	20%
Hydrocarbon saturation	63%	55%	66%
Permeability	135–487 md	38–106 md	78 md
Hydrocarbons			
Gas gravity	0.6	0.6	0.6
Condensation yield (BBL/MMSCF)	0.5	0.5	0.5
Formation water			
Resistivity ohm m at 75°F	0.048	0.048	0.048
Reservoir conditions			
Temperature (°F)	182	178	177
Pressure (psia)	3707.1	3648.7	3708.0
Production			
Start-up date	Oct 1986	Feb 1987	Feb 1987
Number/type of wells	2 exploration 4 development	3 exploration (plus 1 side-track)	1 exploration
Cumulative production to date (31.12.88) BCF	81	17	20
Secondary recovery method	–	–	–

The V-Fields, Blocks 49/16, 49/21, 48/20a, 48/25b, UK North Sea

M. J. PRITCHARD

Conoco (UK) Ltd, 129 Park Street, London

Abstract: The Vulcan, Vanguard, North Valiant and South Valiant gas fields (collectively known as the 4 V-Fields) lie on the eastern flank of the Sole Pit Basin in the southern sector of the UK North Sea. They are contained within Blocks 49/16, 49/21, 48/20a and 48/25b and are operated by Conoco (UK) Limited. The first field to be discovered was South Valiant, in 1970, and exploration drilling continued until 1983 with the discovery of North Valiant, Vanguard and Vulcan Fields. Prominent faults and dip closures define the limits of the fields, and gas is contained within reservoir sands of Early Permian age, which are of desert origin. The gross average reservoir thickness is about 890 feet and porosities vary from 3–23% with permeabilities in producing zones varying from 0.1 md to 1950 md. Initial gas in place is estimated at 2.6 TCF with recoverable reserves of about 1.7 TCF. The fields were brought on-stream in October 1988 and produce, on average, 325 MMSCFD of gas through the LOGGS complex to the Conoco/Britoil terminal at Theddlethorpe, Lincolnshire.

The '4 V-Fields' are four gas accumulations located approximately 75 miles off the Lincolnshire coast, to the southwest of the Conoco/Britoil Viking Field, in the southern sector of the UK North Sea. The fields lie within Blocks 49/16, 49/21, 48/20a and 48/25b and are operated by Conoco (UK) Limited on behalf of Britoil, Arco, Marathon, CanadianOxy, Shell and Esso. The fields vary markedly in areal extent but all lie within about 100 feet depth of water. The total initial gas in place (IGIP) of 2.6 TCF is contained within the Rotliegendes sandstone reservoir below about 7200 ft subsea. Seismic data quality is variable, and depth conversion problematic because of the complexity of overburden structure. The fields are named Vanguard, North Valiant, South Valiant and Vulcan, following the 'V-field' convention established by Conoco for its Southern North Sea Gas Fields.

History

Exploration and appraisal

The South Valiant and North Valiant fields were discovered in 1970 by exploration wells 49/21-2 and 49/16-2 respectively. These wells were drilled to test two of a series of seismically-defined structural closures of Rotliegendes sandstone reservoir to the southwest of the Viking gas field.

Further exploration and appraisal drilling was delayed by uneconomic gas prices until 1982, when well 49/16-7Z was drilled and discovered the Vanguard reservoir. The Vulcan Field was discovered the following year, 1983, by well 49/21-6, and further

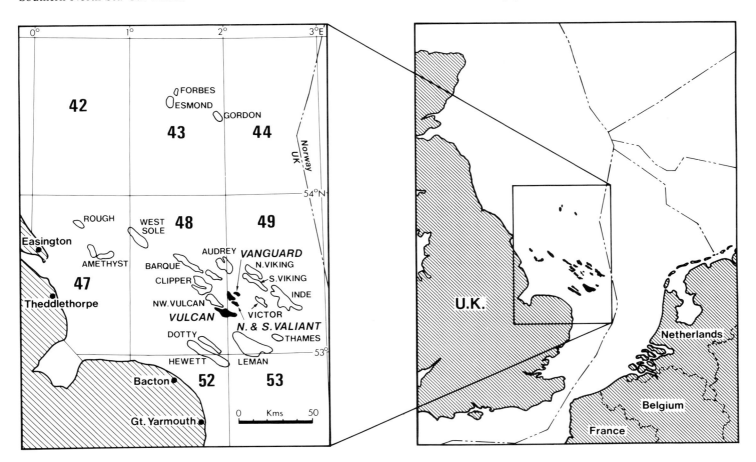

Fig. 1. Location of the V-Fields Gas Complex.

appraisal drilling of the 4 V-Fields continued throughout 1984 and 1985 to confirm the extent of the gas-bearing structures.

Appraisal wells and drill-stem tests indicated favourable reservoir properties for the 4 V-Fields. The deliverability of the wells ranged from 3 MMSCFD to a maximum of 37 MMSCFD, the latter after fracture stimulation.

Development

During 1983, Conoco investigated the options for transporting gas to shore from the unexploited reservoirs of the 4 V-Fields and concluded that the construction of a new system was fully justified on the basis of proven reserves. Final development approval for the 4 V-Fields and Lincolnshire Offshore Gas Gathering System (LOGGS) was granted by the Department of Energy in May 1987. The development scheme comprised six unmanned satellite well platforms linked by subsea flowlines to a manned central gas-gathering station. From the central gas-gathering station, a 75 mile long, 36 inch diameter pipeline runs to shore at the Conoco/Britoil terminal near Theddlethorpe.

The first phase reservoir development philosophy was to create a centrally located group of producers in each field, drilled at locations where structural relief was sufficient to place high quality reservoir zones in the gas column. In order to optimize the locations of the development wells, new seismic surveys were acquired prior to development drilling.

First gas production was contracted for supply to British Gas plc for 1 October 1988. The development plan called for a total of 23 production wells to be drilled through templates at platform jacket locations prior to first gas. These wells would provide a contracted peak gas quantity of 551 MMSCFD during the winter of 1988/89. Drilling a total of 39 production wells was planned throughout the 4 Vs' field life.

Unfortunately, the structural and stratigraphic complexity of the overburden above the Vanguard and North Valiant reservoirs made mapping at reservoir level difficult. Consequently, there were some disappointing well results where structural relief of the reservoir proved to be lower than expected.

In addition to this problem, well tests indicated that reservoir porosity and permeability were more heterogeneous over the area than had been predicted, resulting in less productive reservoir conditions than anticipated in some wells. These disappointments necessitated the drilling of extra wells and fracture stimulation of selected wells to ensure that production requirements were achievable.

The drilling, completion and hook-up of sufficient wells, and platform and pipeline commissioning were completed successfully to enable gas deliveries to commence on 1 October 1988.

Field stratigraphy

Stratigraphic sequence

The geological succession in the 4 Vs area is typical of the UK Southern North Sea. Quaternary and Tertiary cover of variable thickness is underlain by Upper Cretaceous Chalk. Only the Lower

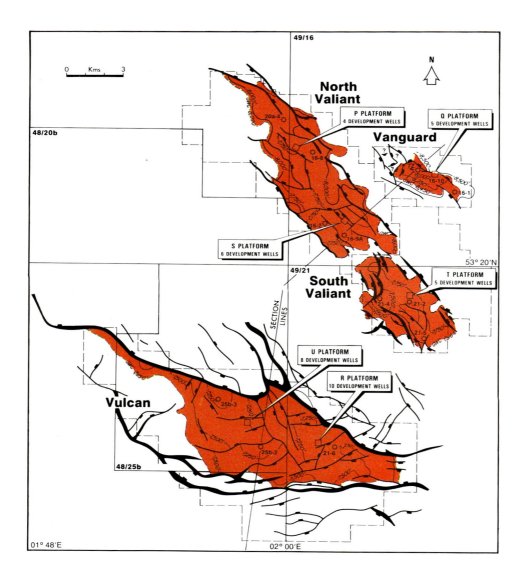

Fig. 2. Top Rotliegendes depth structure map. Contour interval 250 feet.

Jurassic Lias Group siltstones and shales remain of the full Jurassic section, the Middle and Upper Jurassic and Lower Cretaceous sediments having been eroded during the Late Cimmerian. The Lower Jurassic rests conformably upon the Upper Triassic, which consists of evaporites, sands and shales, and this is underlain by Lower Triassic Bunter sandstones and shales. The Permian Zechstein evaporites consist predominantly of thick halites and anhydrites, with variable thicknesses of limestone and dolomite. This evaporite sequence acts as the seal to the underlying Rotliegendes Leman Sandstone which forms the reservoir in the 4 Vs area. The Rotliegendes Group is of Early Permian age and rests unconformably upon Carboniferous Westphalian strata (the source of the gas), which was uplifted and gently folded during the Variscan orogeny.

Geological history

Following minor erosion of the Carboniferous, Rotliegendes deposition commenced in a gently subsiding desert basin, with a NW–SE-trending depositional axis located to the southwest of the Valiant trend.

The Rotliegendes sediments comprise fluvial sands and lacustrine shales overlain by aeolian sands. The source of these sediments was the Carboniferous and Devonian rocks of the London–Brabant Massif, to the SSW of the 4 Vs area, or local structural highs, which had been uplifted during the Variscan orogeny. The fluvial/wadi sediments show a NNE transport pattern away from the Massif, but because of the prevailing Early Permian wind, the aeolian sediments, derived from unconsolidated alluvial fan, wadi and floodplain deposits, show a westward transport direction.

The depositional environment throughout Rotliegendes time was dominated by fluctuations in water height caused by variations in the extent of the Silverpit Lake to the north of the 4 Vs area. Deposition occurred in either aeolian or lacustrine/fluvial-dominated environments depending upon whether the water table was respectively deep or shallow.

Rotliegendes deposition was terminated by the transgression of the Zechstein Sea which rapidly submerged the desert dunes, disrupting sedimentary structures and establishing conditions favourable for cementation of the uppermost reservoir sections.

Differential subsidence ceased during the Zechstein, owing to tectonic stability, and a thick sequence of evaporites was deposited under a series of gradually shallowing cycles. Tectonic stability was maintained during the deposition of a uniformly thick Lower Triassic Bunter Shale sequence, but, thickening of the overlying Bunter Sandstone to the southwest of the 4 Vs area indicates re-establishment of differential subsidence, which continued throughout the Late Triassic. Diapirism and flow of the Zechstein salts was initiated in the Late Triassic, forming NW–SE linear salt swells over the Valiant and Vanguard trends, and variations in sediment thickness and facies within the Upper Triassic. Halokinesis temporarily ceased during the Jurassic, and differential subsidence continued, with conformable deposition of Lower, Middle and Upper Jurassic sediments.

During the late Jurassic and early Cretaceous, the 4 Vs area was uplifted and eroded. Some Lower Cretaceous sediments, together with Jurassic, and in the Vulcan area, uppermost Triassic sediments, were eroded from the 4 Vs area. A major extensional phase was associated with this initial uplift and resulted in listric normal faulting within the post-Permian overburden of the 4 Vs area, forming a 'Collapse Zone'. The listric normal faults of this complex zone exploited weaknesses in sediments overlying the linear salt swells.

A major transgression occurred during the late Upper Cretaceous, resulting in deposition of thick Campanian and Maastrichtian Chalk sequences. In addition, the salt swells beneath the Collapse Zone were partially evacuated, the salt migrating to the northwest of the 4 Vs area. This salt evacuation accentuated Collapse-Zone faulting still further.

Inversion of the Sole Pit anticlinorium occurred in the late Maastrichtian to early Palaeocene, with some associated erosion of Chalk. Further regional uplift caused by Late Alpine compression was followed by uniform subsidence from the Pliocene through to the present day.

Geophysics

Seismic data

A wide variety of seismic data of different vintages was used in the exploration and appraisal mapping of the 4 V-Fields.

This multiple-vintage dataset, with variable acquisition and processing parameters and of variable quality, was considered inadequate for the purposes of detailed field mapping and location of development wells. To provide suitable seismic data for these purposes, two new seismic surveys were acquired, one in 1985 over the Vulcan Field, and the other over North Valiant, South Valiant and Vanguard in 1986. Careful attention was paid to acquisition and processing parameters, and both surveys were shot as detailed 2-D seismic with a dipline spacing of 250 m and a combined 2630 km of coverage. The resulting datasets were of considerably improved quality compared to the original multiple-vintage seismic surveys.

Interpretation methods

Well-velocity logs and Vertical Seismic Profiles were used to define eight seismic horizons for use in mapping of the 4 Vs area. These seismic events were as follows: (a) Top Cretaceous (absent over much of Vulcan and southwest of North Valiant structure); (b) Top Jurassic (absent or very shallow over much of Vulcan structure); (c) Top Triassic; (d) Top Röt Halite; (e) Top Bröckelschiefer; (f) Top Plattendolomit (intermittently absent over parts of accumulations due to Zechstein salt movement); (g) Top Basalanhydrit; (h) Top Rotliegendes.

The Rotliegendes seismic events and faulting are reasonably well defined, except beneath Zechstein salt-swell features and associated complex collapse zone faulting overlying much of Vanguard and North Valiant fields. Mapping of the Top Carboniferous was considered unnecessary since all 4 V-Field gas–water contacts lie within the Rotliegendes reservoir interval.

Depth conversion

Multi-layer depth conversion techniques, of varying degrees of sophistication, were required in order to deal with severe time-to-depth distortions caused by the complexity of overburden structure in the 4 V-Fields. Velocity data from all available wells in the area were incorporated into interval velocity maps, for layers of similar interval velocity, defined by interpreted seismic horizons. These velocity maps were then used in a 'layer-cake' depth conversion method.

Trap

Prominent faults and dip closure define the limits of the fields, and the seal is provided by overlying impermeable Zechstein evaporites. The formation of these structural traps resulted from tectonic movements originating in the late Permian and continuing throughout the Mesozoic. The fields lie on three separate structural trends.

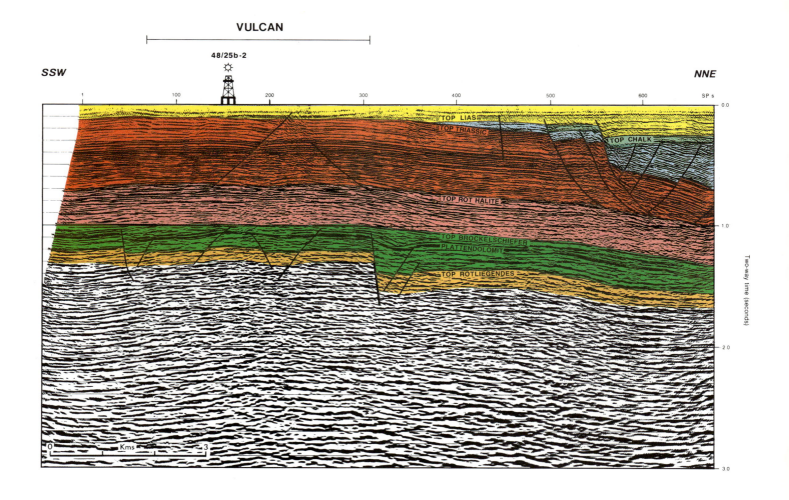
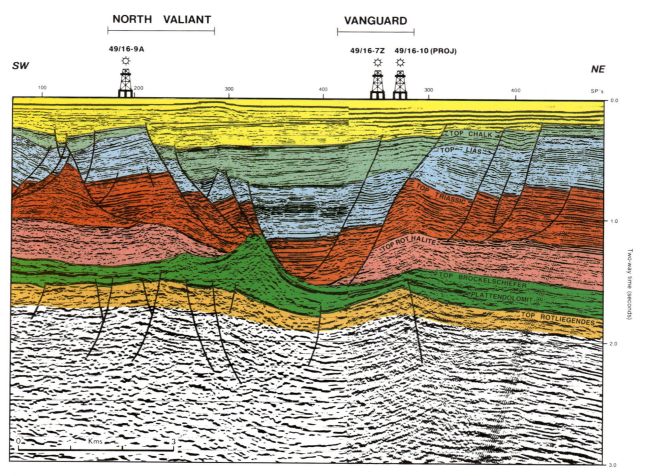

Fig. 3. (a & b) Representative dip-oriented seismic lines across the V-Fields Gas Complex. The location of the lines is shown in Fig. 2.

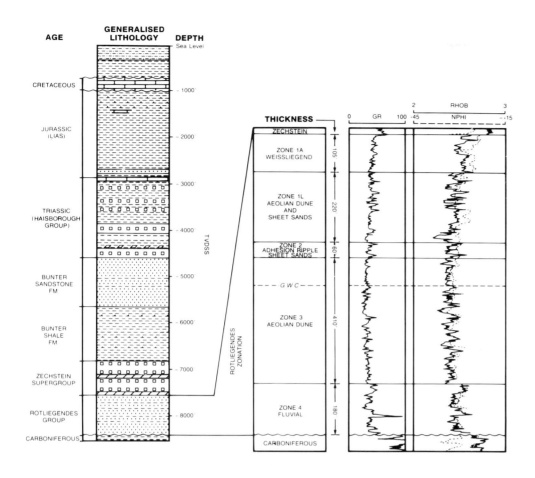

Fig. 4. Generalized lithology and reservoir zonation for the V-Fields Gas Complex.

Valiant trend

From SE to NW, this trend contains several proven structures in addition to the North and South Valiant Fields. The separate culminations are tilted fault blocks dipping to the NE and bounded to the SW by a major fault system; they are separated from each other by NW or SE dip and are probably all filled to spill point. The South Valiant gas–water contact is 7964 ft subsea. The North Valiant Field is divided by a sealing fault into two separate blocks with differing gas–water contacts of 8000 ft subsea for the southerly block and 8090 ft subsea for the northerly block.

Vanguard field

The Vanguard reservoir is contained in a tight NW–SE-trending anticline, bounded on the northeast and southwest by major normal faults which converge to close the structure to the northwest; the reservoir is probably filled to spill point and the gas–water contact is 8450 ft subsea.

Vulcan field

The Vulcan Field consists of a series of NW–SE-trending tilted fault blocks. It is bounded to the northeast by two major en-echelon faults which parallel the Valiant trend. To the south, the field is bounded by a combination of dip closure and an eastward trending, down-to-the-north fault. The gas–water contact is 7659 ft subsea.

Reservoir

The gas-bearing Rotliegendes Leman Sandstone is of Early Permian age and is represented by sediments deposited under desert conditions. The reservoir has been divided into four zones, based on lithology and dominant facies, as a result of log correlation, core description and petrographic studies. To date, more than 20 000 ft of Rotliegendes core have been cut in the 4 V-Fields. Very different relationships between reservoir porosity and permeability exist, which are dependent on facies and the method of cementation. A brief description of each zone, from bottom to top, is given below.

Zone 4

Lies unconformably upon the Carboniferous and was probably deposited when there was a minimum of topographic relief on the Carboniferous surface. It is a unit of remarkably uniform thickness of about 100 ft, and is mainly composed of tight fluvial sandstones. Cementation was caused by dissolved material contained within the transporting waters and is therefore ubiquitous and complete. Porosities average about 6%, and permeabilities are low, ranging between 0 and 10 millidarcies (md). Lacustrine mudstones are common in this zone, related to southward encroachments of the Silverpit Lake. Sedimentary dips are similar to those of the Carboniferous rather than Zone 3, implying a possible disconformity associated with mild tectonism at the end of Zone 4 time.

Zone 3

Overlying Zone 4, this unit is characterized by the widespread build-up of aeolian dune-sand sequences with minor sheet sands and thin adhesion-ripple sediments. In local depocentres over the 4 V area, the lowermost section consists of aeolian sands with frequent fluvial (wadi) sediments indicating the presence of freely available surface water. In equivalent age, topographically higher areas, the lateral equivalent is predominantly aeolian. The upper section comprises blanket aeolian dune sequences representative of a period of fairly constant sediment supply into a gradually subsiding basin. Common sheet sands were produced by redistribution of existing dunes by prevailing winds during temporary interruptions of sediment supply. The occasional occurrence of adhesion ripples indicates rare rises in the water table, but generally it remained low throughout deposition of the unit, resulting in the formation and preservation of excellent reservoir properties. Dune porosities range from 15–20%, and permeabilities are frequently greater than 10 md, often greater than 100 md, and even occasionally greater than 1000 md. The total thickness of the zone is greater than 400 ft in some wells. However, the basal sands of dune sequences have lower porosity as a result of cementation due to the proximity of the desert floor and water table. A regional climatic or tectonic event caused a significant rise of the water table and the end of arid conditions at the close of Zone 3 deposition.

Zone 2

This unit is composed of adhesion-ripple sequences separated by aeolian sheet sands or dunes and is of remarkably uniform thickness of about 60 ft. Deposition of wind transported clastics onto a wet planar surface during a time of sediment starvation, possibly due to an increase in areal extent of the Silverpit Lake, has resulted in reduced reservoir properties, with increased cementation and poor vertical permeabilities. Porosities range from about 12–15% and permeabilities are from 0.1–10 md.

Zone 1

A return to more abundant sediment supply after the impoverished Zone 2 period resulted in the formation of dunes and rare sheet sands. The total zone thickness ranges from about 250–300 ft, but the uppermost 10–200 ft of the zone was variably reworked and locally cemented by waters associated with the Zechstein transgression, resulting in unpredictable decreases to porosity and permeability. The lower dune sands have porosities of the order of 10–12% with permeabilities ranging from 0.1–1 md. The cemented upper sands have porosities of less than 10% and permeabilities of less than about 0.3 md.

Source

The source for the Rotliegendes hydrocarbons is the underlying Carboniferous Westphalian coal beds. The hydrocarbon properties are similar to those in other southern North Sea Fields that undoubtedly have the same source. During the Late Cretaceous to Early Tertiary it is likely that the source rock had reached maturity, the trap had been formed and the gas was able to migrate into the reservoir, probably upwards along fault conduits.

Hydrocarbons

Drill-stem tests in discovery wells flowed gas of 0.6 specific gravity at rates varying from 5.46–34.9 MMSCFD and flowing wellhead pressures from 2080–877 psig. Fracture stimulation of low-productivity wells improved flow rates up to five-fold.

Reserves/resources

The initial recoverable reserves estimate for the 4 V-Fields gas complex is 1654 BCF gas and 3.35 MMBBL Natural Gas liquids and condensate.

The production mechanism is thought to be volumetric depletion.

Daily gas production requirements are currently 324.5 MMSCFD with a swing factor requiring peak production of 551 MMSCFD.

Gas is landed via a 75 mile, 36 inch outside diameter pipeline to the Viking Gas Terminal in Theddlethorpe, where liquids are removed and gas is sold to British Gas. The Natural Gas liquids are pumped from Theddlethorpe to Conoco's Humber refinery.

V-Fields Gas Complex data summary

Trap
Type	Tilted fault blocks
Depth to crest	7200–7900 ft subsea
Lowest closing contour	Fields filled to spill-point
Hydrocarbon contacts	7650–8450 ft subsea
Gas column	450–583 ft

Pay zone
Formation	Rotliegendes Group, Leman Sandstone
Age	Early Permian
Gross thickness (average; range)	890 ft; 790–990 ft
Net/gross ratio (average; range)	0.8; 0.6–1.0
Cut-off for N/G	Variable
Porosity (average; range)	13.5%; 3–23%
Hydrocarbon saturation (average)	60%
Permeability (average; range)	5.4 md; 0.1–1950 md

Hydrocarbons
Gas gravity	0.60
Condensate yield	1.8 STB/MMSCF
Formation volume factor	220 SCF/RCF

Formation water
Salinity	190 000–290 000 ppm
Resistivity	0.017–0.022 ohm m

Reservoir conditions
Temperature	142–177°F
Pressure, initial	3472–3835 psia
Pressure gradient in reservoir	0.07 psi/ft

Field size
Area (total)	49 square miles
Gross rock volume (total)	1.1×10^{12} cubic ft
Initial gas in place	2593 BCF
Recovery factor-gas	70–77%
Recoverable hydrocarbons (total)	
NGL/condensate	3.35 MMSTB
Gas	1654 BCF

Production
Start-up data	Drilling May 1986, Production October 1988
Development scheme	Six unmanned satellite well platforms linked to a manned central offshore gas gathering satation
Production rate (1988–89)	DCQ 324.5 MMSCFD, Peak Prod. 551 MMSCFD
Cumulative production	58.4 BCF; 105 MSTB (28 Feb 1989)

Personal communications and contributions, for which I am grateful, came from N. P. Evans, P. R. Leach, A. R. MacFarlane, and G. R. G. Woodham of Conoco (UK) Ltd. I am indebted to the exploration and production staff of Conoco (UK) Ltd whose ideas and work resulted in the discovery, appraisal and eventual development of the 4 V-Fields and whose papers and documents I referred to continually during the preparation of this paper. Permission to publish this paper was granted by the management of Conoco (UK) Ltd and by Britoil, Arco, Marathon, CanadianOxy, Shell and Esso.

The Victor Field, Blocks 49/17, 49/22, UK North Sea

ROBERT A. LAMBERT

Conoco Inc, 600 North Dairy Ashford, Houston, Texas, 77079, USA

Abstract: The Victor gas field lies in the Southern North Sea Gas Province on the eastern flank of the Sole Pit Basin. The field straddles Blocks 49/17 and 49/22, and is situated approximately 140 km off the Lincolnshire coast. Victor was discovered in April 1972 and is operated by Conoco (UK) Ltd on behalf of BP, Mobil and Statoil. The structure is an elongated tilted fault block, trending NW–SE. The reservoir sands are contained in the Leman Sandstone Formation (Rotliegendes Group) of Early Permian age, and consist mainly of stacked aeolian and fluvial sands with a gross thickness of 400–450 ft across the field. Porosities vary from 16–20%, with permeabilities ranging from 10 md to 1000 md in the producing zones. Initial gas in place is estimated at about 1.1 TCF with recoverable reserves of the order of 900 BCF. The field was brought on-stream in October 1984, and the five producing wells deliver, on average, 200 MMSCFD through the Viking Field 'B Complex' to the Conoco/BP terminal at Theddlethorpe in Lincolnshire.

The Victor gas field is located approximately 140 km due east of Mablethorpe, Lincolnshire in the southern sector of the UK North Sea, and straddles Blocks 49/17 and 49/22 which are operated by Conoco (UK) Ltd on behalf of BP, Mobil and Statoil (Fig. 1). The field is an elongated fault-bounded structure, approximately 11 km long and 3 km wide, and lies in approximately 125 ft of water. Victor is a medium-sized gas field, with approximately 1.1 TCF of gas originally in place, reservoired by Rotliegendes sandstones below 8000 ft sub-sea. Seismic quality is generally good to fair over most of the field, though depth conversion is difficult because of the complexly structured overburden. Two exploration wells and five production wells have been drilled on the field, which came on-stream in October 1984. The field name was chosen following the 'V-Field' convention established by Conoco for its southern North Sea gas fields.

History

The Victor Field was discovered in May 1972 by well 49/22-2. A delineation well, 49/17-8, was drilled in November 1972 and confirmed the northern extent of the reservoir. A deviated appraisal well, 49/22-4a, was drilled in May 1982 from the proposed template surface location adjacent to the discovery well 49/22-2. This well confirmed the extension of the field to the south and was suspended as a future gas production well. Following installation of the 8-slot template during August 1983, a further four development wells were successfully pre-drilled. The production platform was installed in 1984, and the five suspended production wells successfully tied back and completed. Three spare well slots are available should further wells be required. Production commenced in October 1984 with deliveries of gas via a 16 inch pipeline to the Viking B complex

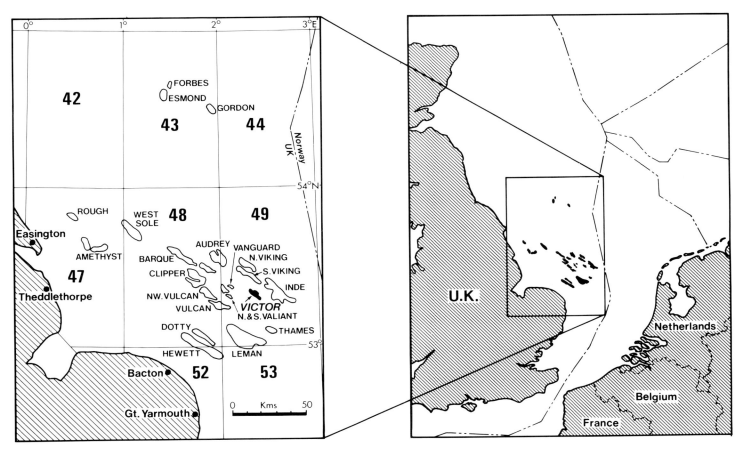

Fig. 1. Location of the Victor Field.

From Abbotts, I. L. (ed.), 1991, *United Kingdom Oil and Gas Fields, 25 Years Commemorative Volume*, Geological Society Memoir No. 14, pp. 503–508

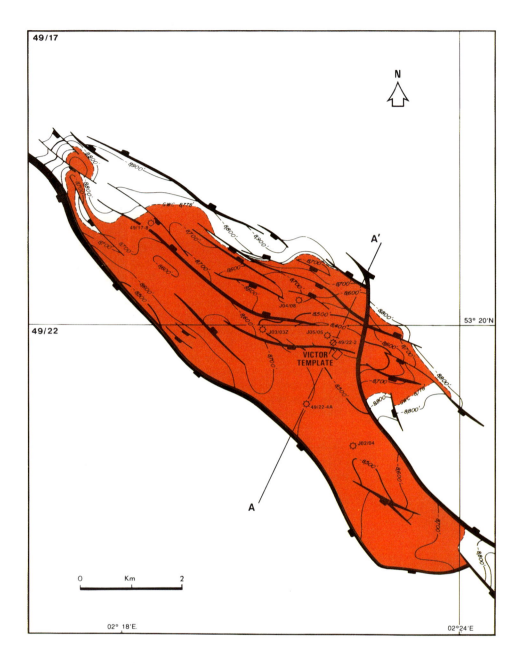

Fig. 2. Top Rotliegendes depth structure map. Contour interval 100 ft.

for metering and initial processing prior to onward transportation to the Conoco/BP gas terminal at Theddlethorpe, Lincolnshire.

Field stratigraphy

Geological succession

The geological succession (Fig. 2) present in Victor is that seen typically throughout the UK southern North Sea. A chalk section of Cretaceous age is underlain by Jurassic siltstones and shales and Triassic evaporites, sands and shales. The Permian Zechstein evaporites consist mainly of thick halites and anhydrites, with variable thicknesses of limestone and dolomite. The Zechstein section acts as the seal to the underlying Leman Sandstone Formation, the reservoir rock in Victor. The Leman Sandstone Formation, Rotliegendes Group, gas reservoir averages some 375 ft in gross thickness and rests unconformably on the Carboniferous Westphalian strata from which the gas was derived.

Stratigraphy

The Leman Sandstone Formation (Rotliegendes Group), of Early Permian age, forms the reservoir over the Victor area and consists mainly of sandstones with occasional silty shales in the basal section. The Rotliegendes rests unconformably on the Carboniferous, which was uplifted, gently folded and eroded during the Variscan Orogeny. Deposition of the Rotliegendes occurred during a period of general quiescence, but minor movements along pre-existing faults have resulted in the formation of a depositional axis through the middle of the Victor Field. Variations, to a greater extent in thickness and a lesser extent in facies, occur away from this line (Figs 3 & 4).

The Rotliegendes sediments of the Southern North Sea Basin were deposited under mainly desert conditions, and consist of dune (aeolian) sands, fluvial (wadi) sands and conglomerates, sabkha shales and sands and lacustrine shales and evaporites. Over the Victor Field area, the Rotliegendes is represented by the Leman Sandstone Formation which consists of thin fluvial sands and lacustrine silty shales, overlain predominantly by dune sands with minor silty sabkha sands and thin fluvial sands.

The various facies present in the Rotliegendes have different characteristics of grain-size, sorting, roundness, cementation and sedimentary structures, which allow them to be distinguished in cores and logs.

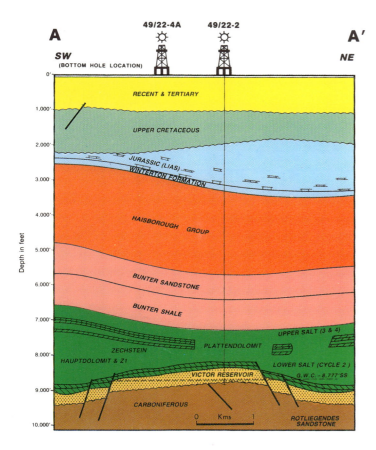

Fig. 3. Representative dip-orientated structural cross section. The location of the cross-section is shown in Fig. 2.

Geological model

Rotliegendes sedimentation followed a period of mild erosion of the Carboniferous. Deposition began in a fluvial environment with sediments sourced from the southwest. Gentle tectonic deformation occurred towards the end of the deposition of these basal sands and, coupled with a regional lowering of the water table, sedimentation became predominantly aeolian, with sediment transport from the northeast.

As the size of the Silver Pit lake to the north increased, the sediment supply was effectively blocked and the regional rise in water table led to the formation of broad wet surfaces characterized by deposition of adhesion-ripple sediments. A gradual return to dry desert conditions followed, and a regional lowering of the water table, coupled with a steady influx of sediment, again from the northeast, allowed the development of aeolian dunes and sheet sands to predominate. Rotliegendes deposition was terminated by the Zechstein transgression, the whole basin being flooded in a relatively short time. Petrographic studies indicate that little erosional disturbance occurred in the Victor Field area, at this time, and original bedding structures remained intact.

Later tectonism, firstly during the Triassic and secondly during the Late Jurassic to Late Cretaceous time span when the major movements occurred, gave rise to the current structural setting of Victor. Localized reductions in porosity have probably been caused by Zechstein mobile evaporitic waters percolating along the fault planes and, under appropriate circumstances, moving into the reservoir and precipitating pore-filling anhydrite and some dolomite.

Geophysics

Seismic data

The Victor reservoir is covered by approximately 200 km of seismic data on a grid of about 500 m × 500 m. The seismic data were acquired at various times between 1969 and 1974. In 1980 most of these data were reprocessed and new filtered and migrated displays were produced. Generally improved data quality over the original set was noted, due primarily to the detailed velocity analyses and improved migration. In 1983, Conoco conducted a regional watergun seismic survey over the area, adding 150 km of seismic data to the Victor Field database. The new data proved to be of superior quality and have been used as the prime dataset for recent field mapping.

Interpretation

The following seismic horizons are prominent throughout the field area and form the basis for geophysical interpretation:

(a) Top Cretaceous Chalk;
(b) Base Cretaceous/Top Jurassic (Liassic);
(c) Intra-Liassic (a seismic velocity change seen in 49/22-2);
(d) Top Rhaetic and Triassic;
(e) Top Lower Triassic Brockelschiefer Member;
(f) Top Zechstein Plattendolomit (Upper Magnesian Limestone);
(g) Top Hauptdolomit (Lower Magnesian Limestone);
(h) Top Rotliegendes Reservoir (Base Zechstein).

All of these seismic events are continuous markers, except the Plattendolomit which is absent over part of the accumulation due to Zechstein salt movements. Fault positions and throws are reasonably well defined at all levels. Mapping of the Base Rotliegendes reservoir (Top Carboniferous) is difficult, and was considered unnecessary in this field since the GWC lies entirely within the reservoir interval, except for a small area around the 49/22-2 well.

Depth conversion

Due to structural complexity of the overburden in the field area, severe time/depth distortion occurs, and a multi-layer depth-conversion technique is necessary. The interpreted seismic horizons are used to divide the geological succession into layers of similar interval velocity. Velocity data from the wells delineating the reservoir, and from surrounding wells in the area, are used to assign interval velocities to these layers for use in a 'layer cake' depth conversion.

Structure

The Victor Field is an elongated and tilted fault-block structure, trending NW–SE. Prominent faults bound the field to the northeast and southwest, with vertical throws of up to 300 ft and 700 ft respectively. Dip closure defines the northwest and southeast limits of the field. The crest of the field is mapped at about −8300 ft TVDSS near the 49/22-2 well, giving a maximum vertical relief

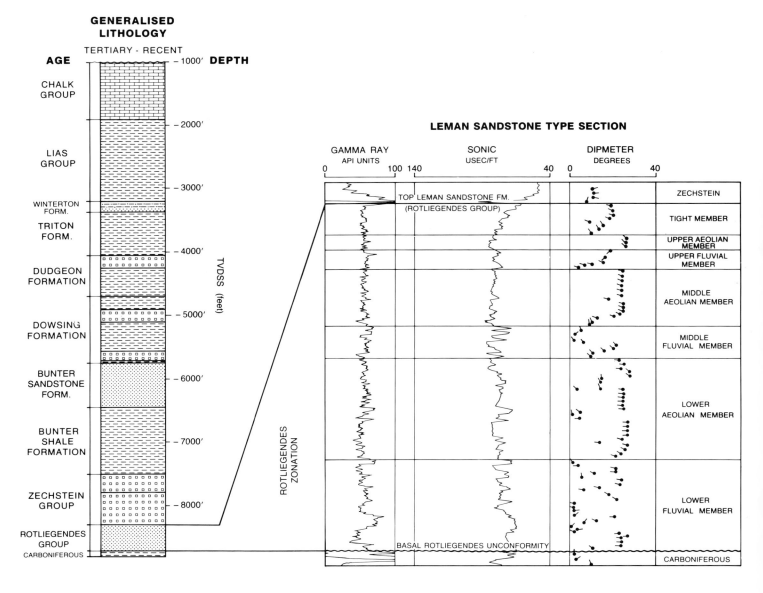

Fig. 4. Generalized lithology and reservoir zonation for the Victor Field.

down to the gas–water contact of about 500 ft. Minor faults within the field generally trend parallel to the boundary faults and have throws ranging up to 150 ft.

Trap

The Victor Gas Field occurs in a structural trap resulting from tectonic movements that probably began in the late Permian and continued throughout the Mesozoic. Evidence exists for reactivation of pre-existing Carboniferous faulting during Zechstein deposition. However, deformation of the area during later regional inversion and related Zechstein halokinesis has overprinted much of the early structural detail. Major tectonic events during the Late Jurassic, Mid Cretaceous and Late Cretaceous have been identified and are apparent as clear unconformities on the seismic data. In particular, erosion following Late Jurassic tectonism, which consisted of block faulting, possible strike slip-movements and inversion, has resulted in the removal of over 2000 ft of Jurassic strata. Using Bunter Shale velocities as a guide, the maximum burial depth of the reservoir can be calculated to be only a few hundred feet more than at present, and probably occurred during the Late Jurassic. The lack of Lower Cretaceous strata reflects significant movement in the Mid-Cretaceous. Little of the Upper Cretaceous chalk has been removed by the Late Cretaceous inversion, which indicates that the Victor Field area was subjected to much less inversion than in the main part of the Sole Pit Basin further to the west.

The trap is a combination of fault seal and dip closure, where the reservoir is juxtaposed against or buried beneath impermeable Zechstein evaporites. A gas–water contact has been identified at -8776 ft TVDDS (Fig. 5) and the structure appears to be filled to spill-point. Critical closure exists at the southeast flank of the field.

Reservoir

The reservoir sandstones of the Victor Field occur within the Leman Sandstone Formation of the Lower Permian Rotliegendes Group. Log correlation, core description and petrographic studies have enabled sub-division of the reservoir into four zones based on lithology and the dominant facies. This layering has been used as a basis for reservoir zonation, petrophysical analysis and volumetric calculations. A brief description of each zone from bottom to top is given below.

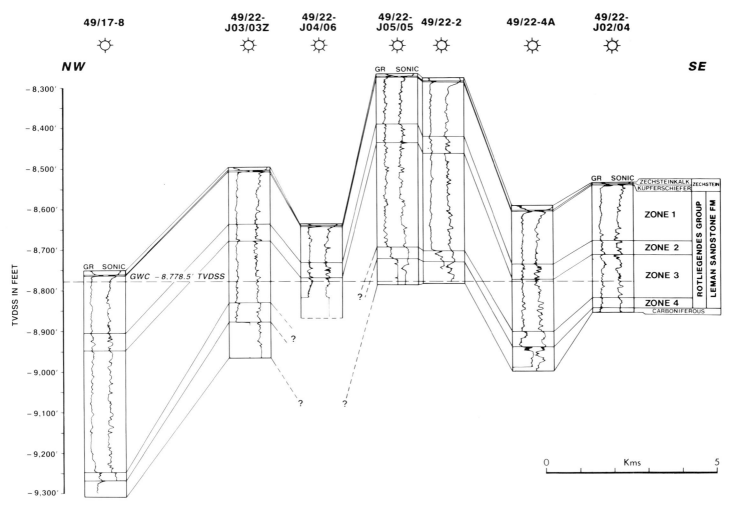

Fig. 5. Reservoir zonation cross-section for the Victor Field.

Zone 4. This zone is a thin extensive unit of remarkably uniform thickness, probably deposited when there was minimum topographic relief on the underlying Carboniferous surface. Over the Victor area it is represented by pale red to purple, fine- to medium-grained, moderately- to well-sorted, tight fluvial sandstones. Source direction was from the south and southwest, with dips more in keeping with the underlying Carboniferous rather than the overlying aeolian sediments.

Zone 3. Overlying Zone 4, this unit represents the first subaerial phase of Rotliegendes sedimentation. It consists of predominantly aeolian dune sand sequences with minor sheet sands and thin adhesion ripple sediments particularly towards the base. Occasional, very localized intra-Rotliegendes fluvial deposits occur, again towards the base, and are representative of a period when surface water was still available. The laterally continuous nature of the dunes suggests abundant and fairly continuous sediment supply; the sheet sands were produced by re-distribution of these dunes on the desert floor by prevailing winds when sediment supply was temporarily interrupted. Throughout the deposition of this unit the water table remained low. Deposition and preservation of aeolian sands in the upper part of the unit have resulted in excellent reservoir properties. A regional event, possibly either climatic or tectonic, caused a rise in the water table at the end of Zone 3 deposition.

Zone 2. This unit is characterized by a series of thin adhesion ripple sediments separated by aeolian sheet or dune sands, and is of remarkably uniform thickness. The unit represents the deposition of wind-transported clastics onto a wet planar surface during a time of sediment starvation, possibly due to an increase in areal extent of the Silver Pit lake to the north.

Zone 1. Indicates a return to more abundant sediment supply after the impoverished period of Zone 2. Dune-top and dune-base sequences predominante with rare, thin sheet sands. The uppermost part of the zone was affected by immediate post-depositional Zechstein transgression. Penetration of the cement-bearing waters was variable. Log response indicates increase in sonic velocity and density towards the top of the zone, but the overall effects upon the reservoir quality have been minimal.

Source

The source of the Rotliegendes hydrocarbons is almost certainly the underlying Carboniferous Westphalian coal beds. The hydrocarbon properties correspond well with those in other southern North Sea fields that undoubtedly have the same source. During the Late Cretaceous to Early Tertiary it is likely that the source rock reached maturity and, with the trap having already been formed, gas migrated into the reservoir, probably upwards along fault conduits.

Hydrocarbons

A drill-stem test of the Victor appraisal well 49/22-4A flowed 18.54 MMSCFD on a 36/64 inch choke. Subsequent tests carried out 2 months after start-up flowed 98.7 MMSCFD through a 60/64 inch choke.

Reserves

The Operator's recoverable reserve estimate for the total field is 920 BCF of which 890 BCF is sales gas. The original gas-in-place was estimated in 1988 by material balance at 1062 BCF.

Victor is a constant-volume depletion reservoir with no evidence of water influx to date. Current Daily Contract Quantity (DCQ) is 150 MMSCFD with a maximum contractual deliverability of 250.5 MMSCFD. Four wells are currently onstream. Cumulative gas production is 248 BCF or 27% of estimated ultimate recovery (as at 28 February 1989).

The current Annual Contract Quantity (ACQ) is 54.8 BCF. This level of sales is the First Plateau Level and is in force until 35% of Economically Recoverable Reserves (ERR) have been produced. The Second Plateau Period follows immediately after the First Plateau Period and is in force until 60% of ERR have been produced. The full Field Life Profile shows an exponentially diminishing production rate over the decline period until 920 BCF have been recovered.

Gas production is from five wells drilled into the Victor reservoir from an unmanned central platform. Gas is transported from the JD platform through a 16 inch pipeline to the Viking 'B' complex 8 miles away. Victor gas is then co-mingled with gas from Viking 'B' and flows through the Viking pipeline system to the Viking Gas Terminal in Theddlethorpe, where liquids are removed and gas is sold to British Gas. Natural gas liquids are pumped from Theddlethorpe to Conoco's Humber refinery.

Personal communications and contributions for which I am grateful came from R. Beham, N. Evans and T. Evans of Conoco (UK) Ltd.

Permission to publish this paper was granted by the management of Conoco (UK) Limited and by BP, Mobil and Statoil.

Victor Field data summary

Trap
Type	Tilted Horst Block
Depth to crest	−8300 ft TVDSS
Lowest closing contour	−8800 ft TVDSS
Hydrocarbon contact	−8776 ft TVDSS
Gas column	480 ft

Reservoir
Formation	Leman Sandstone
Age	Permian
Gross thickness	300–450 ft
Net/gross ratio (average)	0.985
Cut off for N/G	0.5 md
Porosity (average)	16%
Hydrocarbon saturation (average)	73%
Permeability (average)	52 md
Productivity index	17–100 MCF/day/psi

Hydrocarbons
Specific gas gravity	0.612
Calorific value	1019 btu/ft
Condensate yield	1.8 BBL/MMSCF
Specific condensate gravity	0.738 (60 API)

Formation water
Salinity	220 000 ppm
Resistivity	0.017 ohm m

Reservoir conditions
Temperature	192°F
Pressure, initial	4047 psia
Pressure gradient in reservoir	0.075 psi/ft

Field size
Area	5500 acres
Gross rock volume	907 739 acre-feet
Initial gas in place	1062 BCF
Recovery factor	86.63%
Recoverable hydrocarbons:	
gas	920 BCF
condensate	1840 MBBL

Production
Start-up date	October 1984
Current daily contract quantity	150 MMSCFD
Maximum contractual deliverability	250.5 MMSCFD
Cumulative production (28 February 1989)	248 BCF
Number/type of wells	5 gas producers
Secondary recovery method(s)	None

The Viking Complex Field, Blocks 49/12a, 49/16, 49/17, UK North Sea

CLARE P. MORGAN

Conoco (UK) Ltd, 116 Park Street, London W1Y 4NN, UK

Abstract: The Viking complex consists of several separate gas accumulations within Blocks 49/12a, 49/16 and 49/17 of the southern North Sea. The entire field is located on the northeast flank of the Sole Pit Basin, approximately 140 km off the Lincolnshire coast. North Viking was discovered in March 1969; South Viking was discovered in December 1968. Conoco (UK) Ltd operates the Viking complex on behalf of BP. North and South Viking consist of two parallel elongated faulted anticlines trending NW–SE. The gas-producing structures are largely fault controlled. The reservoir comprises Rotliegendes Group sands of Early Permian age. The gas-bearing Rotliegendes consists of stacked aeolian and fluvial sands with a gross thickness range of 430–720 ft (131–222 m, thickest on South Viking). Porosities range from 7–25% and the average permeability range is 0.1 md to over 1000 md in the producing zones. The total Viking gas-in-place is almost 3.2 TCF, with recoverable reserves estimated at 2.83 TCF. North Viking came onstream in October 1972 and is developed by a five-platform complex. South Viking came onstream in August 1973 and is developed by a three-platform complex into which are linked five unmanned satellite platforms. The average daily production in March 1989 was 196 MMSCFD, peaking to 311 MMSCFD with seasonal demand. Gas is piped to the Conoco/BP terminal at Theddlethorpe in Lincolnshire.

The Viking gas complex is located in the Southern North Sea gas province, in 30 m of water, approximately 140 km east of Lincolnshire (Fig. 1). The area covered by the Viking gas field comprises UK block 49/12a and the northern half blocks 49/16 and 49/17. The licences (P033) were held equally by BP and Conoco (UK) Ltd (Operator). The acreage was awarded in September 1964 to the Continental Oil Company as part of the Licensing Round 1. In January 1966 the National Coal Board, (succeeded by the British National Oil Corporation and later Britoil plc), farmed-in 50% to the blocks. The Viking complex contains nine developed accumulations (Reservoirs A, B, C, D, E, F, G, Gn, H; Fig. 2) and is subdivided into North and South Viking. The Viking complex consists of parallel elongate anticlines with NW–SE-trending axes. North Viking field is 16 km × 3 km and consists of an asymmetric anticline hosting a crestal graben. It comprises the producing Viking A, F and H Reservoirs and F south structure. South Viking is structurally complex and includes a wider geographical spread of pools, consisting of several separated traps along the length of the anticline. A NW–SE-trending regional fault divides the anticline. The upthrown block has good structural development and cross-faulting has resulted in independent accumulations, namely Viking B, C and D producing reservoirs. The downthrown anticlinal block is subdued in relief and comprises Viking Gn reservoir. Viking E (6 km × 1.5 km) and G (3 km × 1 km) reservoirs form separate

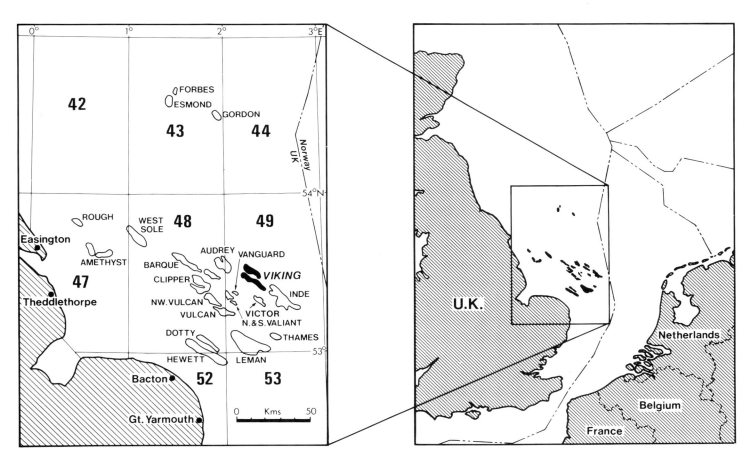

Fig. 1. Location of the Viking Complex Gas Field.

From Abbotts, I. L. (ed.), 1991, *United Kingdom Oil and Gas Fields,
25 Years Commemorative Volume*, Geological Society Memoir No. 14, pp. 509–515

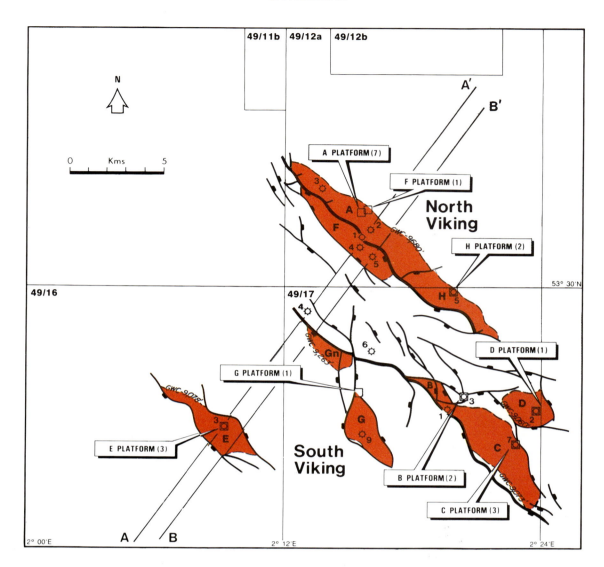

Fig. 2. Schematic outline of the Viking Complex Field showing reservoir accumulations and exploration wells. The number of producing development wells from each platform is shown in brackets. Crestal depths and GWCs for the various reservoirs are addressed in the text.

horsts, also trending NW–SE. The entire Viking complex represents a very large gas accumulation, with approximately 3.2 TCF gas-in-place and supplies on average about 10% of Britain's natural gas requirements. Rotliegendes aeolian and fluvial sandstones, deposited during the Early Permian, form the reservoir. Seismic quality is generally good. The field name reflects the 'V-Fields' convention established by Conoco (UK) Ltd in the early 1970s for its southern North Sea gas fields.

History

The first well drilled on North Viking was 49/12-1 (1969), which did not encounter Rotliegendes but penetrated a large fault of 800 ft (244 m) throw. This well was then sidetracked 500 ft (152 m) to the north. North Viking was thus discovered in March 1969 by well 49/12-2. This well is located on the crestal position of Viking A Reservoir. 49/12-3 (1970) confirmed the extension of the A Reservoir to the northwest. The thickness of Rotliegendes sandstone determined by the early wells on North Viking was a fairly consistent 500 ft (152 m). A gas–water contact (GWC) was not established until 49/17-5 (1969) was drilled into the Viking H Reservoir. Accumulations A and H have the same GWC of 9680 ft (2950 m) sub-sea, but pressure tests indicate separate drainage. In 1973, wells 49/12-4 and 49/12-5 proved the crestal graben of North Viking to be gas-bearing also (F Reservoir), with a separate GWC of 10 140 ft (3091 m). These discoveries were developed by a five-platform complex situated over the A Reservoir. The first production well, AD1, was drilled in January 1971. North Viking came onstream in October 1972.

South Viking was discovered in December 1968 by well 49/17-2. This accumulation is now termed D Reservoir. The earlier 49/17-1 well, completed in December 1965, found 40 ft (12 m) of gas-bearing Rotliegendes sandstone 9050 ft (2758 m) sub-sea on the margin of the present South Viking C pool. 49/17-3 (1968) discovered B Reservoir to the northwest of the C structure with a separate GWC. Subsequent exploration drilling located C, G and Gn accumulations, all with different GWCs. Development of South Viking was complicated by the fact that the gas is located in these separate reservoirs at depths of between 8000 ft (2438 m) and 9700 ft (2957 m) below the seabed. A complex of platforms was needed to tap reserves and South Viking has 3 main platforms and 5 unmanned satellite platforms. The first production well, BD3, was drilled in October 1972. South Viking came onstream in August 1973.

Today, the Viking Field has a total of 13 platforms drawing gas from 20 wells. Gas from the B complex and its computer-controlled satellites is metered and piped 11 km to the A platform through a 24 inch diameter pipeline. At the A complex, all Viking gas comes together and enters a 28 inch pipeline which extends 140 km to

Theddlethorpe. It is then processed, metered and piped to the nearby British Gas Corporation plant where it is metered and odourized before entering the national grid.

Field stratigraphy

General stratigraphic sequence

The geological succession present in Viking is typical of the UK Southern North Sea. The Viking reservoirs comprise continental deposits of the Lower Permian Rotliegendes. The Leman Sandstone Formation dominates the field and passes laterally into the Silverpit Formation of silts and shales to the north. The Rotliegendes sediments lie unconformably on gently folded Carboniferous strata. A large regional Carboniferous anticlinal structure extends NW–SE through blocks 49/16 and 49/17, exposing Westphalian A subcrop along its crest and successive younger Westphalian units on its flanks. Carboniferous Westphalian Coal Measures provide the gas source. Above the Rotliegendes lie the Permian Zechstein evaporites which vary greatly in lithology and thickness, but provide an effective gas seal. Werra, Stassfurt, Leine and Aller Groups, corresponding to Z1, Z2, Z3 and Z4 cycles, are all represented in the Viking area. The bulk of the Zechstein sequence is composed of mobile halites with interbedded anhydrites and carbonates. The Leine Group Plattendolomit Formation occurs extensively over the area, but is discontinuous and frequently rafted out. The Zechstein sequence is overlain by conformable Triassic beds. The Triassic succession falls into two major groups: the Bacton Group represents a phase of clastic deposition of sandstones, shales and mudstones; the Haisborough Group is largely a fine-grained clastic and evaporitic sequence with marked cyclicity. The Triassic is the thickest sequence present and represents a fairly consistent 4000–4500 ft (1219–1372 m) of sediments. The Triassic succession is terminated by the Winterton Formation. Jurassic siltstones and shales of the Liassic Group overlie the Triassic strata. West Sole and Humber Groups are absent.

Isolated occurrences of the Red Chalk Formation of the Lower Cretaceous Cromer Knoll Group are found to the northeast of the North Viking monocline. In contrast, Upper Cretaceous chalk is widespread. It directly overlies the Jurassic sediments along a major unconformity. Tertiary formations are usually present.

Geological history

The Sole Pit Basin, a locally active depocentre within the more regionally extensive Anglo-Dutch Basin, was initiated during the early Permian. The Viking Gas Field is located on the northeast flank of this Sole Pit Basin.

Rotliegendes sediments lie unconformably on the partially eroded surface of a block-faulted Carboniferous which had been gently folded during the Variscan Orogeny (Fig. 3). The Rotliegendes in the Viking area comprises continental aeolian dune and

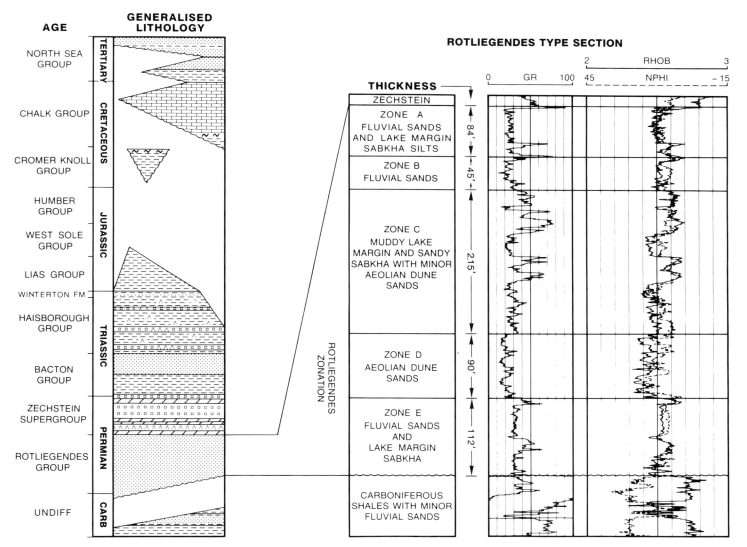

Fig. 3. Generalized lithology and reservoir zonation for the Viking Complex.

fluvial sandstones, associated with major dune belts crossing quadrants 48 and 49. The North Viking Field lies close to the northern limit of this dune belt. Further north and northeast, silts and shales become the dominant Rotliegendes lithology as the margins of the North German Basin are approached. The Variscan NW–SE structural grain, established in the Carboniferous, persisted throughout the Rotliegendes influencing depositional trends and sedimentary environments. These major lineaments also controlled the block faulting of the reservoir during subsequent Late Cimmerian tectonism.

Rotliegendes sedimentation was terminated abruptly by the inflooding of the Zechstein Sea. Low rainfall and limited access to supplies of marine and freshwater resulted in very low sediment influx and deposition lagged behind subsidence. However, high temperatures and evaporation led to chemical precipitation. The Zechstein sequence is the product of at least four deposition cycles.

Continental deposition was re-established at the beginning of the Triassic period when thick sequences of sandstones, shales and evaporites were laid down. The Bunter Shale exhibits remarkably consistent thickness and lithology, suggesting uniform basin subsidence. The Sole Pit basin became the site of more rapid differential subsidence during deposition of the Bunter Sandstone Formation and at various stages during deposition of the succeeding Haisborough Group.

The Upper Triassic Winterton Formation marks the onset of the Jurassic marine transgression. Sedimentation during the Jurassic consisted of marine shales and minor sandstones. A major phase of differential movements and erosion occurred during the late Jurassic to early Cretaceous. This Late Cimmerian tectonic upheaval resulted in relative uplift and erosion along the northeast flank of the Sole Pit Basin, but subsidence continued within the Basin. Later movements caused a reversal of previous subsidence patterns, with the area to the east of Viking becoming an active depocentre from the late Albian to the late Tertiary. Erosion of much of the previously deposited Mesozoic strata is marked by major unconformities which can be traced regionally on seismic data. The Sole Pit Basin was therefore subjected to two phases of inversion in the late Cretaceous and mid Tertiary. Sediments are seen to thin dramatically westwards towards the inversion axis throughout the Upper Cretaceous and Tertiary sequences, with compensatory thickening of units to the east of the North Viking monocline.

Geophysics

Seismic data

The Viking Field has good seismic coverage on a grid of about $1 \text{ km} \times \frac{1}{2} \text{ km}$. The first seismic survey (12 fold, aquapulse) was shot in 1969 Western and was not migrated. Since then several surveys have been acquired. Recent mapping has been based on an extensive 1983 watergun survey and an infill, 1987, airgun dataset. Data quality is very good, in particular the dual-polarity large-scale displays of Conoco's 1987 dataset. Trial reprocessing of selected lines from Conoco's 1983 dataset and older vintages did not improve data quality sufficiently to merit a full reprocessing project.

Synthetic seismograms for all exploration wells were reprocessed in 1987 by SSL and correlate well with seismic character. Most of the wells lack Vertical Seismic Profile (VSP) surveys as they were drilled in the 1970s, but a comprehensive VSP programme was acquired in 1987 on three B-Reservoir development wells.

Interpretation

The early maps on Viking were based on migrated seismic data. Closed structural highs, corresponding to the present North and South Viking Fields, were identified in the late 1960s. Mapping continued through the 1970s as each structure was defined and drilled, with Viking vigorously producing to full contractual demand. As reserves were being depleted and spare capacity was increasing within the present field infrastructure, re-mapping began in earnest in 1986 to: (i) ensure optimum development of the field until the end of the current Viking Contract Agreement and (ii) identify additional structures to replace depleting gas reserves. In early 1987, a regional Top Rotliegendes map was produced based on Conoco's 1983 dataset. The map not only re-defined producing structures and highlighted potential prospects, but also determined the coverage and orientation of the 1987 seismic programme. The 1987 seismic survey initiated a detailed re-mapping phase of each prospect, which is continuing today. Several additional gas reservoirs have already been identified.

The seismic markers picked include the Top and Base Cretaceous Chalk, Top Triassic, Top Rot Halite, Top Brockelschiefer member and Top Rotliegendes (Fig. 4). All these seismic events are prominent throughout the field area. High-frequency dipping reflectors corresponding to intra-Carboniferous formations are also recognized locally, but are unmappable.

Fault positions are best defined on dual-polarity large-scale displays from Conoco's 1987 seismic survey. Fault mapping conforms to the regional NW–SE structural trend, but a secondary system of northeast-trending cross-faults is prevalent.

3-D isometric displays of seismic coloured-amplitude variation have also been used as an interpretation tool and have applications in siting well locations and in detailed reservoir modelling.

The Viking Field is depth-converted in a layer-cake manner, as is typical of fields in the Southern Gas Basin. The interpreted seismic events coincide with major velocity changes and a combination of interval-velocity functions and interval-velocity maps is applied. The good geographical distribution of wells (velocity control points) and a structurally simple overburden across the Vikings support confidence in the proposed velocity model.

Trap

All Viking structures and prospects share the NW–SE structural alignment of the area. Prominent faults control the gas-bearing structures.

North Viking plunges to the northwest at 7° and more gently to the southeast at 1°. North Viking has a crest of 8450 ft (2575 m) on the A Reservoir. The H Reservoir accumulation along strike has a mapped crest of 9400 ft (2865 m). The A and H pools have a common gas–water contact (GWC) of 9680 ft (2950 m). The adjacent F Reservoir is a fault-bounded gas-bearing graben with a crest of 9585 ft (2922 m) and a GWC of 10 140 ft (3091 m). Dip closure defines the northwest and southeast limits of North Viking.

South Viking parallels the elongate northwesterly attitude of North Viking and is dissected by a major regional fault of about 600 ft (183 m) throw. Cross-faults further interrupt the anticline along its axis, creating independent gas traps. B Reservoir, a localized mini-horst, has the shallowest Top Rotliegendes depth of 8050 ft (2454 m), but South Viking crestal values more typically are 8800 ft (2682 m) on C, D, E Reservoirs or 9250 ft (2819 m) on G and Gn structures on the downthrown southwesterly anticline limb. Each trap has a separate GWC. Values range from 9050 ft (2758 m) to 9700 ft (2957 m).

Structural closures are also mapped at Bunter level over the southern limb of the North Viking structure and South Viking C Reservoir, but are unlikely to be gas-bearing due to migration difficulties.

The traps for all Viking gas structures involve tilted fault blocks (Fig. 5). Dip closure defines the limit of the field where faults are minor or absent. In plan view, the structural traps are geometrically cigar-shaped with their long axis trending NW–SE. In cross-section, the largest traps are created where there is good structural development on high relief upthrown fault blocks. The long axis of the North Viking field measures 16 km, the short axis 3 km. The South

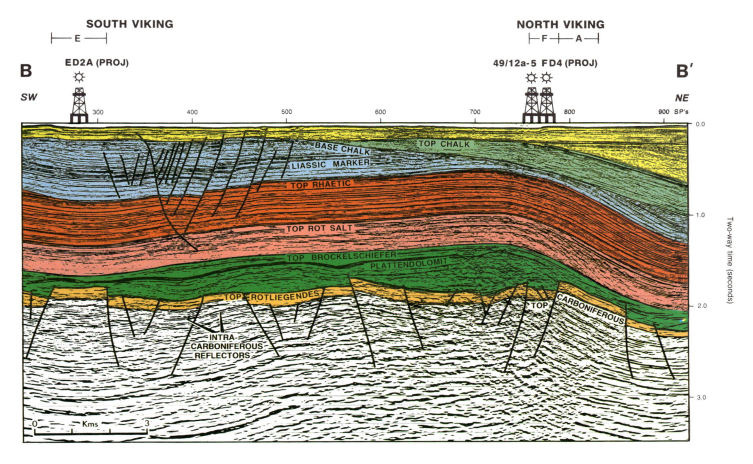

Fig. 4. Representative dip-orientated seismic line across the Viking Complex. The location of the line is shown in Fig. 2.

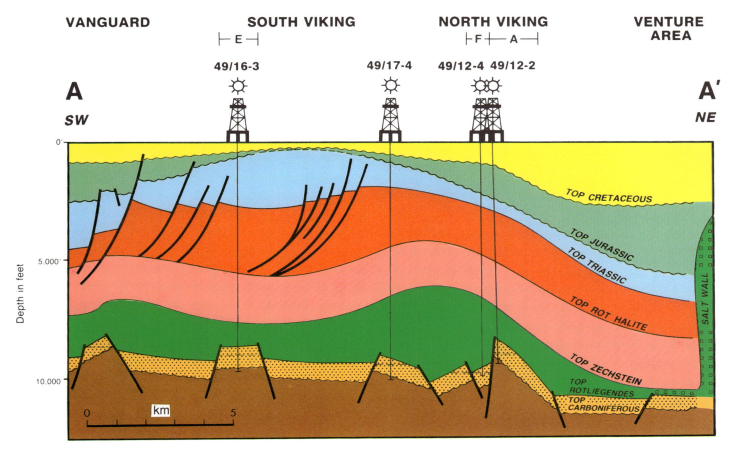

Fig. 5. Representative dip-orientated structural cross-section across the Viking Complex. The location of the line is shown in Fig. 2.

Viking anticline spreads over 20 km in length, hosting several discontinuous accumulations and reaches a maximum width of 2.5 km. Rotliegendes strata were folded, tilted, broken and displaced in the series of tectonic events already outlined in the section on Geological History. Structures must already have been present in the Cretaceous and survived Late Cretaceous and Tertiary tectonics. Vertical and lateral seal is provided by Zechstein evaporites. Exploration history in the Viking area has demonstrated that any closed Rotliegendes structure can be a gas-bearing pool and accumulations appear to be filled to spill-point.

Reservoir

The Leman Sandstone Formation of the Permian Rotliegendes Group forms the reservoir of the Viking Gas Field. The reservoir sediments were deposited in a generally arid, continental environment and comprise two main facies groups: aeolian, providing the optimum reservoir; and locally well-developed fluvial-wadi and lake-margin sabkha facies. The facies' associations were interpreted as transitional between the Silverpit sabkha lake to the north and the major dune fields to the southwest. The Rotliegendes sands thicken regionally to the southwest from 430 ft (131 m) on North Viking to 720 ft (220 m) across the South Viking Field.

Thickening of the Rotliegendes across the major faults and comparison of regional geological sections suggests that faulting exerted some control on sedimentation during the Early Permian.

Log correlation, core description and petrographic studies permit reservoir zonation. The reservoir is divided into five units.

Zone E. The lowermost Rotliegendes consists of fluvial sands with local lake-margin silts and lacustrine shales. The sands are generally poorly sorted and argillaceous, with average porosities in the region of 10–15% and low permeabilities, due to the abundant detrital clays, ranging from <1 md to 100 md.

Zone D. Overlying this unit is a continuous horizon of aeolian dune sediments up to 300 ft (91 m) thick. These sands are generally fine to medium grained, clean and well sorted (typically bimodal) with porosities ranging from 15–25% and permeabilities from 5 md to over 1000 md. In both North and South Viking Zone D is the most productive sandstone unit.

Zone C. A continuous horizon of lake-margin sabkha silts and fluvial sands overlies the aeolian sands and can be correlated across the region. This zone represents a regional climatic change and corresponds to a rise in the groundwater table and a southward expansion of the Silverpit desert lake to the north. Reservoir quality varies widely from moderate to very poor, with porosities averaging 5–15% and permeabilities <0.1 md and 10 md.

Zone B. In the North Viking area, this zone comprises mainly fluvial sands with minor aeolian dune deposits in places. Average porosities vary from 8–12% and permeabilities from 1–100 md. In other areas, particularly in South Viking, this zone comprises predominantly aeolian deposits with correspondingly better reservoir quality.

Zone A. This zone comprises a continuous horizon of reworked fine-grained sandstones (Weissliegendes) underlain by lake margin sandy sabkha sediments. The homogenized and cemented nature of the Weissliegendes results in low porosities and permeabilities. The thickness of this uppermost unit is highly variable and the Weissliegendes appears to be absent in some Viking wells.

Source

The hydrocarbons of both North and South Viking are sourced by underlying Carboniferous Westphalian coals and carbonaceous shales. Migration into Rotliegendes traps probably occurred during the Late Cretaceous to Early Tertiary.

Hydrocarbons

The reservoir fluid characteristics in the Viking Field are relatively constant (see data summary table). The initial drill stem tests in the Rotliegendes sandstone of the 49/12-2 North Viking discovery well flowed 54 MMSCFD on a 58 inch/64 inch choke (FWHP following wellhead pressure 2910 psia). The first North Viking development well 49/12-ADI initially flowed 74 MMSCFD on a 3 inch choke (FWHP 3029 psia). The first South Viking development well deemed commercially viable, 49/17-BD3, flowed at 72 MMSCFD on its initial DST (FWHP 3050 psia).

Viking Field data summary

Trap	
Type	Tilted fault blocks
Depth to crest	Variable 8000–9600 ft
Lowest closing contour	Variable 9000–10 200 ft
Hydrocarbon contacts	Variable 9000–10 200 ft
Gas column	Variable (700 ft max. in Rotliegendes)
Pay zone	
Formation	Leman Sandstone Formation of the Rotliegendes Group
Age	Permian
Gross thickness (range)	400–700 ft
Sand/shale ratio (range)	90:10
Net/gross ratio (range)	50–100%
Cut-off for N/G	Variable
Porosity (range)	7–25%
Hydrocarbon saturation	50–60%
Permeability (range)	0.1–100 md+ highly variable
Productivity index	4–110 MSCF/D/psi
Hydrocarbons	
Specific gas gravity	0.615
Calorific value	1029 btu/SCF
Condensate yield	4–5 BBL/MMSCF
Specific condensate gravity	0.738 (60° API)
Formation water	
Salinity	220 000 ppm
Resistivity	0.017 ohm m
Reservoir conditions	
Temperature	170–200°F
Pressure, initial	4150–4670 psia
Pressure gradient in reservoir	0.075 psi/ft
Field size	
Area	–
Gross rock volume	–
Initial gas in place	3189 BCF
Recovery factor	88.7%
Recoverable hydrocarbons:	
gas	2830 BCF
condensate	12.7 MMBBL (approximately)
Production	
Start-up date	Drilling January 1971
	Production August 1972
Current daily contract quantity	186 MMSCFD
Maximum contractual deliverability	311 MMSCFD
Cumulative production (28/2/89)	2518 BCF

Reserves

The Operator's estimate of recoverable gas reserves for the total Viking Field is 2830 BCF. The original gas-in-place was estimated in 1988 by material balance at 3189 BCF. Production is from the Permian Rotliegendes sandstone at depths between 8000 ft (2438 m) and 9700 ft (2957 m) TVSS.

The Viking Field consists of nine reservoirs (A, B, C, D, E, F, G, Gn and H) with facilities of 13 platforms arranged into two main complexes and outlying satellites. The various reservoirs within the Viking Field show constant-volume depletion with the F, G and Gn reservoirs exhibiting aquifer influx. Daily Contract Quantity (DCQ) peaked at 550 MMSCFD with a maximum contractual deliverability of 918.5 MMSCFD. Current (March 1989) rates are 186 MMSCFD with a maximum contractual deliverability of 310.6 MMSCFD.

Cumulative gas production is 2518 BCF, or 89% of estimated ultimate recovery (as at 28 February 1989).

The current Annual Contract Quantity (ACQ) is 76.2 BCF>. The decline period for the Viking Field began on 11 December 1982 when 60% of the originally agreed recoverable reserves (2960 BCF at that time) had been produced. The remaining 40% recoverable reserves are divided into 16 depletion intervals, the size of which is contractually determined. Viking is presently in the eighth depletion interval.

The current depletion philosophy is to keep the water-drive reservoirs as the base-load reservoirs on production throughout the year. This should maximize the gas recovered from these reservoirs. In addition, where possible, the higher pressure reservoirs are depleted first.

Gas offtake from the Viking Field is by 140 km, 28 inch pipeline, to the Viking Gas Terminal in Theddlethorpe where liquids are removed and the gas is sold to British Gas. Natural gas liquids are pumped from Theddlethrope to Conoco's Humber refinery.

Personal communications and contributions for which I am grateful came from N. P. Evans, T. Evans and R. A. Lambert of Conoco (UK) Limited.

Permission to publish this paper was granted by the management of Conoco (UK)Limited and by BP.

The West Sole Field, Block 48/6, UK North Sea

D. A. WINTER & B. KING

BP Exploration, 301 Vincent Street, Glasgow G2 5DD, UK

Abstract: West Sole is located in the Sole Pit area of the Southern North Sea Permian Basin in UK Block 48/6. The field was discovered in 1965 and was the first commercial discovery in the UK Continental Shelf. Gas Production commenced in 1967. Initial reserves are 1.873 TCF of which 1.335 TCF had been produced by the end of 1989. Gas is trapped in aeolian sandstones of the Permian Lower Leman Sandstone Formation. Three depositional facies are recognized, comprising aeolian dune, fluvial and sabkha. The aeolian dune facies form the principal reservoir sandstones, in units up to 40 m (131 ft) thick. However, permeability is reduced due to pervasive illite cementation, such that it averages 3 md in the dune sandstones. Productivity is enhanced in the southern part of the field by 'open' gas-filled fractures, generated during the Alpine inversion. The trap was also amplified at this stage and comprises a faulted inversion anticline trending NW–SE. The source rock is the Westphalian Coal Measures, lying directly beneath the reservoir.

West Sole is situated approximately 60 km (37 miles) east of Humberside, in the Southern North Sea Permian Basin (Fig. 1). The reservoir is wholly in Block 48/6 where there is an average water depth of 30 m (100 ft). Gas is trapped in a SE-plunging anticline in aeolian sandstones of Permian age. West Sole has been on production since 1967, with gas sold under a long-term contract to British Gas.

History

Block 48/6 is part of licence P001, which was awarded in 1964 during the first round of licensing. The licence, which includes blocks 42/29 and 42/30, is wholly owned by BP.

To date 26 exploration and production wells have been drilled on the field, of which 20 are currently in production.

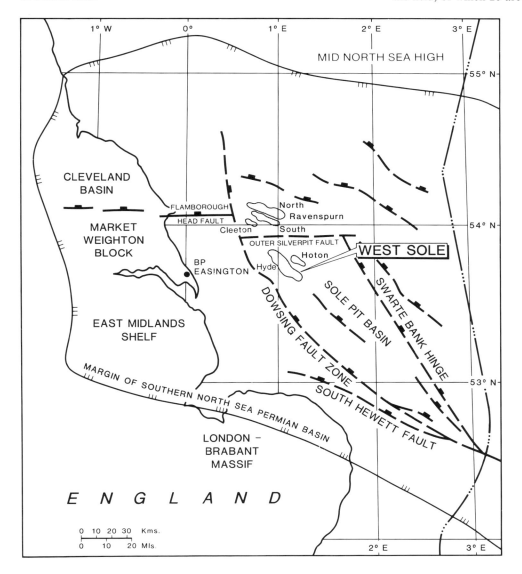

Fig. 1. Location map.

From Abbotts, I. L. (ed.), 1991, *United Kingdom Oil and Gas Fields,*
25 Years Commemorative Volume, Geological Society Memoir No. 14, pp. 517–523

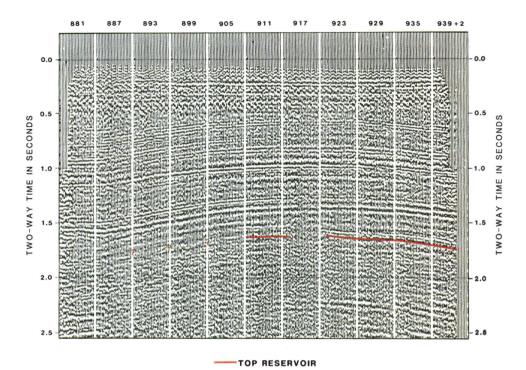

Fig. 2. Early vintage seismic line (486E) across West Sole.

— TOP RESERVOIR

The first well, 48/6-1, was drilled in 1965, with objectives in the Zechstein and the Rotliegendes (Lower Leman Sandstone). The Rotliegendes was gas-bearing, the first commercial discovery in the UKCS. Subsequently twelve production wells were drilled from two platforms in the southern part of the field, which was then termed the Southern Lobe. Structure mapping using well data and poor quality seismic data (see Fig. 2) defined a Northern Lobe. This was tested by well 48/6-16 which confirmed the northerly extension of the field. Four subsequent production wells were drilled from a third platform on the Northern Lobe.

Structural definition of the northern part of the reservoir was originally poor. A production boundary was established representing the limit of the West Sole reservoir, as identified by the nearby NE–SW-trending fault, which inhibits further gas flow from the far north (Hyde) into the production wells.

In 1987 a new seismic survey was acquired. This led to a better understanding of the field and re-evaluation of potential infill sites to optimize production during decline.

Field stratigraphy

West Sole is situated in the Sole Pit area of the Southern North Sea Permian Basin. The basin is bounded to the north by the Mid North Sea High, to the west by the Pennine uplift and to the south by the London–Brabant Massif (Fig. 1). The geological history of the Sole Pit Basin is complex; Fig. 3 is a generalized stratigraphic column.

The Rotliegendes Group was deposited in an arid desert environment within an east–west-trending Permian basin; its fill comprised aeolian dune sands, and the sediments of ephemeral streams and sabkha/playa-lakes. These sediments were deposited onto a deeply eroded Carboniferous sequence, following considerable inversion related to the final phase of Hercynian mountain building. The deepening basin was later inundated by the Zechstein Sea, with the development of marginal marine clastics and cyclic deposition of sabkha carbonates and evaporites of the Zechstein Group.

Rates of subsidence decreased during the Triassic. The basin filled with playa-lake mudstones and fluvial sandstones of the Bacton Group. The development of the Bunter Sandstone Formation indicates that tectonic uplift of the basin margins occurred mainly in the south, the main source area of these sediments. Evaporites in the overlying Haisborough Group mark the onset of a transition to a restricted sea, fully marine conditions being achieved by Rhaetic time.

The onset of the marine conditions at the end of the Triassic (Rhaetic) coincided with the start of tectonic extension of the basin. The E–W extension was related to rifting in the Central and Viking Grabens further north in the North Sea. The NW–SE-trending Dowsing Fault Zone was re-activated and the Sole Pit Basin developed into a half-graben. Active E–W extension continued into the Middle Jurassic with deposition of marine clastics (West Sole Group).

The Cretaceous and Tertiary geological history in the West Sole area is poorly understood because of erosion during Late Cretaceous and Mid-Tertiary inversion. These inversion events resulted in the removal of between 1.5 and 2 km (5000–6500 ft) of section over West Sole. Halokinesis was probably initiated during the Cretaceous (there is no evidence of Jurassic salt movement in the West Sole area), and probably peaked during the inversion events.

Geophysics

The available seismic coverage over West Sole comprises data shot from the mid-1960s up to 1987. Most of the data, however, are of poor quality; an example is shown in Fig. 2. A new seismic survey was acquired in 1987. This comprises some 500 km of 2-D data that cover the whole of the field area on a 0.5 km × 1.5 km grid, and tie into both the Hyde and Hoton 3-D surveys.

The survey is zero-phased, so all example seismic lines included in this paper are zero-phase, and are displayed with the convention that an increase in acoustic impedance is represented as a black peak, and a decrease as a white trough. The data quality of the interpreted horizons is generally very good. There is a deterioration in data quality above the crest of the main salt pillow due to the presence of shallow faulting. Deterioration in the data quality of the deeper reflectors is also observed beneath the salt pillow, partly reflecting the complex shallow structure and partly due to raypath distortion and noise generation within the salt. Example seismic sections are shown in Fig. 4a and b.

The seismic horizons used in the interpretation were chosen

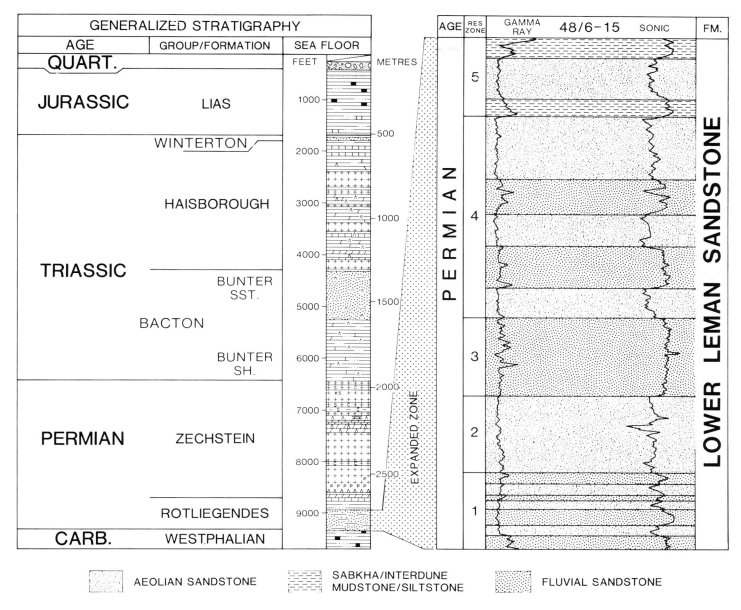

Fig. 3. Generalized stratigraphy in the West Sole area.

because they represent major lithological boundaries and have marked acoustic impedance contrasts. They also have good continuity across most of the West Sole area. The main picked horizons were: Top Winterton (Top Triassic); Top Bröckelschiefer (Near Base Bacton Group); Top Rotliegendes.

The Top Winterton horizon is defined by a trough on the seismic data, and represents the acoustic impedance contrast between the calcareous mudstones of the lower part of the Lias, and the slower velocity, non-calcareous mudstones of the underlying Winterton Formation. This horizon marks the Jurassic/Triassic boundary.

The Top Bröckelschiefer horizon is characterized by a well-defined black peak, and represents an increase in acoustic impedance between the silty Bröckelschiefer Member at the base of the Bunter Shale Formation and the overlying shales. The Bröckelschiefer marks the boundary between the generally conformable Triassic and Jurassic sequences above and, within a few cycles, the chaotic reflectors seen in the Permian Zechstein evaporite sequence below.

The Top Rotliegendes is a clearly defined trough at the base of the Zechstein, the reflection being generated where Z1 carbonates overlie the Upper Leman Sandstone of the Rotliegendes Group with its significantly lower velocity. This horizon marks the top of the West Sole reservoir.

The Zechstein evaporite sequence displays the effects of halokinesis. This is reflected in the time structure of each of the two shallower horizons, which are conformable with the Zechstein. The structure at Top Bröckelschiefer shows, in the east, a NNW–SSE-trending salt pillow, plunging to the SSE within the mapped area. The crest of the structure is faulted, the fault also trending NNW–SSE. The structure at Top Rotliegendes, being detached from the shallower horizons by the Zechstein salt pillow, displays little structural relationship with them.

Trap

The West Sole structure is a SE-plunging inversion anticline trending NW–SE. The anticline is cut by NW–SE-trending reactivated Carboniferous faults. Individual displacements of the larger faults vary from net extension to net compression. The changes in fault geometry are accommodated by a series of NE–SW-trending cross faults, which break the structure into compartments. The seismic

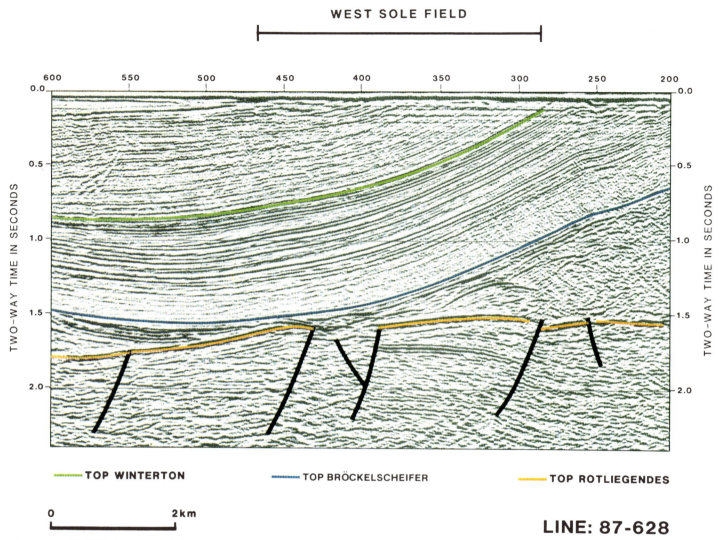

Fig. 4. (a & b). 1987 vintage seismic lines across the West Sole Field.

sections (Fig. 4a and b) together with the top Lower Leman Sandstone structure map (Fig. 5) illustrate the different geometries.

The crest of the structure at Top Lower Leman Sandstone Formation is at 8858 ft. The gas–water contact is at 9678 ft, giving 820 ft of gas column. The structure is full to spill point.

A partial seal to the main reservoir is provided by the playa mudstones and siltstones of the Silverpit Formation. Above this, however, the Upper Leman Sandstone is gas-bearing and, indeed, was productive in one well. The ultimate top seal is provided by the Zechstein evaporites.

Reservoir

The principal reservoir is the Lower Leman Sandstone Formation of the Rotliegendes Group.

The Lower Leman Sandstone can be divided into three depositional facies: (i) aeolian; (ii) fluvial (ephemeral sheet-flood); (iii) sabkha.

Figure 6 is a NW–SE correlation diagram through the field, which shows the facies variations with depth. The figure also shows the five reservoir zones of the field. These zones are based on facies sequences, comprising genetically related packages of a particular facies association. These sequences reflect subtle changes in the depositional environment which were probably climate-controlled.

(i) Aeolian

The principal reservoir-quality sandstones in West Sole are of aeolian dune facies. They occur as units up to 40 m (131 ft) thick in the northern part of the field. Their thickness and the unit-modal dip of cross-sets to the W–NW suggest that transverse dunes were the most commonly developed dune form. The sandstones are pervasively cemented with illite, which drastically reduces the permeability to an average of c. 3 md. The illite does not affect porosity so severely since micro-porosity is preserved; porosity in the aeolian sandstones averages 15%. Thinner aeolian dune sandstones within the fluvial or sabkha facies sequences have generally poor reservoir quality with permeabilities reduced to 1 md. Stacked dune sequences offer the best reservoir quality, these are best developed in Zones 2 and 4c (Fig. 6).

(ii) Fluvial (ephemeral sheet-flood)

Fluvial facies associations consist of argillaceous sandstones, siltstones and rare mudstones which were deposited by ephemeral sheetfloods. These were probably derived from the SW along the margin of the Sole Pit Basin. They are particularly well developed in Zones 1 and 3, which may represent a less arid environment. Reservoir quality is poor with porosities generally less than 10%

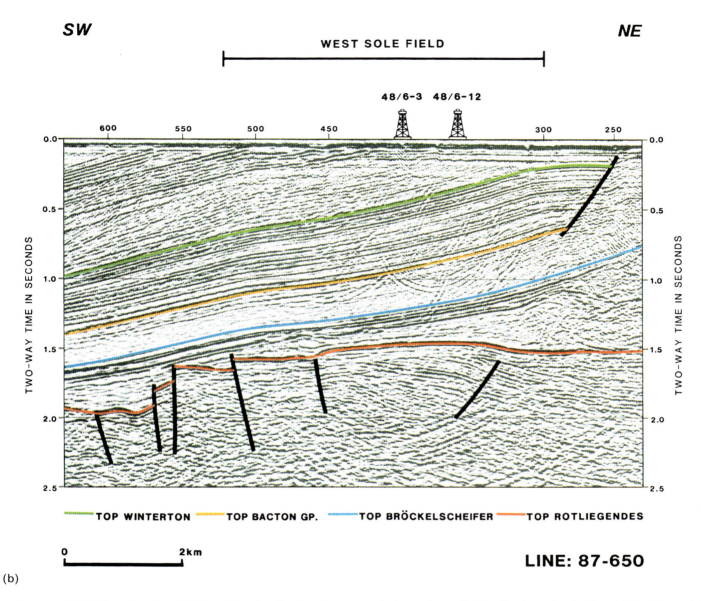

LINE: 87-650

and permeabilities less than 1 md. These thin sheetfloods act as barriers of low permeability within the reservoir.

(iii) Sabkha

The sabkha facies association becomes dominant only in Zone 5, and consists of dune-margin sheet sandstones, interdune sediments and deposits of the playa-lake margin. Their reservoir quality is poor, comparable to sandstones of the fluvial facies. Sediments of the playa-lake margin in Zone 5 reflect the transgression of the Silverpit Lake across the West Sole area.

Productivity is considerably enhanced in the southern part of the field by 'open' gas-filled hair-line fractures. These fractures trend NW–SE parallel to the major faults. They probably formed partly as a release of high fluid pressures immediately after inversion and as a result of bending stresses on the crest of the structure. Fracture permeabilities can be as high as 120 md, but are usually c. 10 md. Shear fractures also occur but are commonly tightly cemented; they reduce productivity by acting as low permeability baffles.

Source

The source of the gas is the Coal Measures of Carboniferous (Westphalian) age. The gas most likely migrated into the structure during or immediately after inversion in the mid-Tertiary as it fills fractures which formed as a result of inversion. It certainly post-dates illite cementation which is thought to be of Late Jurassic age.

Hydrocarbon

West Sole is a dry-gas reservoir. Mole percentage gas composition is shown in Table 1

Table 1. *West Sole gas composition (mol%)*

N_2	CO_2	C1	C2	C3	C4	$C5^+$
1.0	0.8	94.0	3.1	0.5	0.2	0.4

The initial reservoir pressure and temperature were 4265 psia and 185°F at a reference depth of 9000 ft. The gas viscosity is 0.023 cP and the gas deviation factor (Z) is 0.976.

Condensate recovery at the Easington processing terminal is 0.231 tonnes per MMSCF of sales gas (0.00165 moles of condensate to 1 mole of sale gas). The condensate composition in mole percent is shown in Table 2.

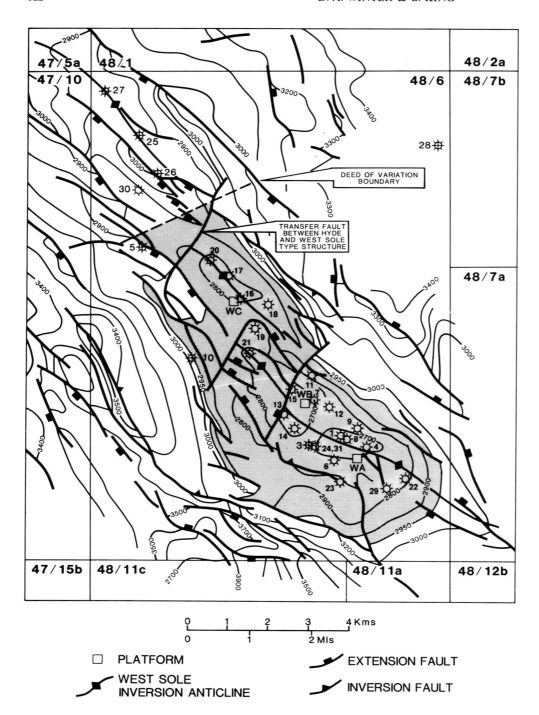

Fig. 5. Structure map, Lower Leman Sandstone, West Sole area.

Table 2. *West Sole condensate composition (mol%)*

C1	C2	C3	C4	C5	C6	C7	C8	C9	C10	C11$^+$
0.4	1.3	1.0	2.4	9.4	3.9	23.0	22.0	13.2	8.3	15.1

The hydrocarbon dew point of the reservoir gas is estimated at 1403 psia at reservoir temperature.

The formation brine has a chloride content of 146 500 mg l^{-1} and has a total salt content of 239 050 mg l^{-1}. The measured specific gravity is 1.17.

Test results are highly varied due to the presence of natural fractures in some wells. Permeability–thickness products (kh) estimated from well test data are in the range 29–36 500 md ft.

Reserves

First production was in 1967 and the plateau production rate of 160 MMSCFD was maintained from 1972 to 1985. The long plateau was maintained by installing onshore compression and an additional pipeline. The current decline rate is 6.5% per year. Initial gas in place is 2.54 TCF. The gas produced to end 1989 was 1.335 (52% initial gas-in-place). Estimated remaining reserves are 0.54 TCF (field life to year 2005) giving a recovery factor of 73.7%. This includes an infill drilling campaign from 1990 to achieve this recovery. There are currently 20 producing wells giving an average drainage area per well of 870 acres. All wells are completed in the Lower Leman Sandstone Formation. Most wells are hydraulically fractured to achieve commercial rates of production, although some wells exhibit natural fractures. The drive mechanism is natural depletion. Current evidence suggests that there is little or no aquifer pressure support.

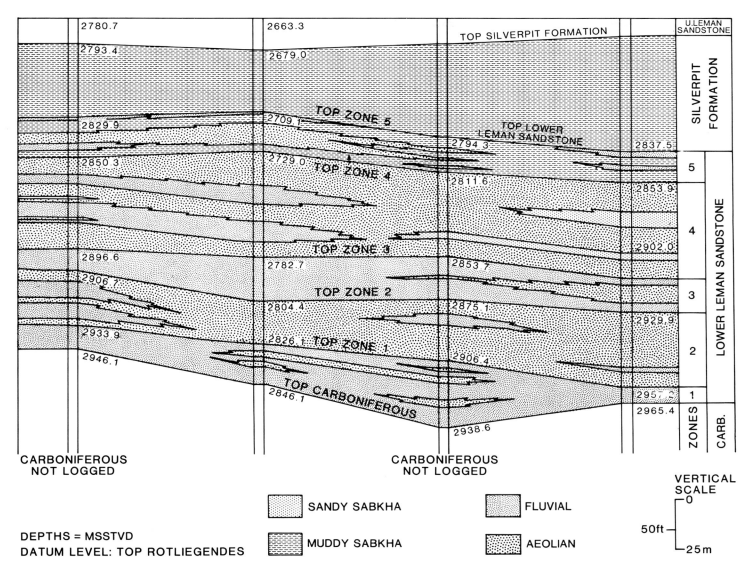

Fig. 6. Reservoir zonation summary.

West Sole Field data summary

Trap
Type — Inversion anticline
Depth to crest — 2700 m (8858 ft) ss
Gas–water contact — 2950 m (9679 ft) ss
Gas column — 250 m (820 ft)

Pay zone
Formation — Lower Leman Sandstone
Age — Permian
Gross thickness — 128.9 m (423 ft)
Sand/shale ratio — c. 0.9
Net/gross ratio — 0.75 average
Cut-off for N/G — 3% porosity, V_{SHALE} 25%, S_w 100%
Porosity — 12.3% average
Hydrocarbon saturation — c. 60%
Permeability thickness — 29–36 500 md ft
Productivity index — Highly variable

Hydrocarbons
Gas gravity — 0.598 SG
Dew point — 1043 psia at 185°F
Condensate yield — 1.5 BBL/MMSCF
Gas expansion factor — 239 SCF/ft^3

Formation water
Salinity — 239 050 mg l^{-1}
Resistivity — 0.0195 ohm m at 185°F

Reservoir conditions
Temperature — 185°F
Pressure — 4265 psia at 2743.2 m (9000 ft TVDSS)
Pressure gradient in reservoir — 0.253 psi/m (0.077 psi/ft)

Field size
Area — 68.8 sq km (17 000 acres)
Recovery factor — 73.7%
Recoverable gas — 1.873 TCF
Drive mechanism — Pressure depletion

Production
Start-up date — 1967
Development scheme — 3 platforms
Number/type of wells — 20 gas producers
Production rate — 160 MMSCFD on plateau to 1985
now declining by 6.5% a^{-1}
Cumulative production — 1.335 TCF (end-1989)
Secondary recovery method — None

The Morecambe Basin

The South Morecambe Field, Blocks 110/2a, 110/3a, 110/8a, UK East Irish Sea

I. A. STUART & G. COWAN

Gas Council (Exploration) Ltd, 59 Bryanston Street, Marble Arch, London W1A 2AZ, UK

Abstract: The South Morecambe Gas Field has been developed as a seasonal supply field to boost supplies to the National Transmission System at times of peak demand. This mode of operation has led to a requirement for exceptionally high reliability in all aspects of the development. This requirement has prompted the generation of an accurate and comprehensive geological model so that reservoir performance can be predicted as reliably as possible, and that wells can be drilled in optimum locations. The exceptional shallowness of the structure (crest at −2400 ft TVSS, GWC −3750 ft TVSS), coupled with the need to drain the reservoir cost-effectively and to minimize the risk of well interference has led to the use of slant drilling techniques for the first time in European waters. The field is located in the East Irish Sea Basin. The Triassic Sherwood Sandstone Gp forms the reservoir, and the Mercia Mudstone Gp provides the seal. The reservoir sands were laid down in a rapidly subsiding basin under continental semi-arid conditions, and comprise a complex interplay of major channel-fill sandstones, secondary channel-fill sandstones associated with non-channelized sheetflood sandstones, and localized, very high permeability (>1000 md) aeolian and reworked aeolian sandstones. A vertical organization of these facies has been observed, with some intervals dominated by channel deposition, others by non-channelized deposits, due to periodic adjustments of the whole basin, and this has permitted the establishment of a reservoir zonation. A complex diagenetic history is recognized, with several phases of dolomite and quartz cementation. Differential compaction is also a major control on the disposition of reservoir properties. The greatest control on permeability (but not porosity) is platy illite which formed beneath a palaeo-GWC at an early stage in the growth of the structure, and which gives rise to a diagenetic layering of the reservoir into a high permeability Illite-Free Layer and a deeper, low permeability Illite-Affected Layer. The data presented herein is based upon the results of development drilling on South Morecambe.

The Morecambe Gas Fields are located in the East Irish Sea Basin, about 26 miles west of Blackpool (Fig. 1). Water depth is about 100 ft, but the tidal range is around 30 ft, resulting in exceptionally strong currents. The reservoir is the Triassic Sherwood Sandstone Gp. The Morecambe structure comprises two fields, South Morecambe and North Morecambe, separated by a narrow but very deep graben. The structures are exceptionally shallow compared with North Sea fields (South Morecambe GWC −3750 ft TVSS). At present only South Morecambe Field has been developed. This paper refers primarily to South Morecambe.

South Morecambe Field was developed as a seasonal supply facility, designed to produce gas at very high rates during periods of peak winter demand. As such the project required extremely high reliability, and the development scheme reflected this requirement. Furthermore, the high reliability requirement has necessitated the development of a comprehensive geological model to permit accurate prediction of reservoir performance.

Various aspects of the geology have been described previously: Colter & Barr (1975), Colter (1978), Colter & Ebbern (1978, 1979), Ebbern (1981), Bushell (1986), Woodward & Curtis (1987) and Levison (1988). Drilling and completion technology have been described by White & Bartle (1987) and Maclean & White (1988), and general aspects of the development by Hughes *et al.* (1985), Brown (1987) and Hughes (1987).

History

The Morecambe Fields lie within the UK blocks 110/2a, 110/3a, and 110/8a. These blocks are now licensed to Gas Council (Exploration) Ltd, a wholly owned subsidiary of British Gas, but until quite recently they were licensed to Hydrocarbons Great Britain Ltd (HGB), another wholly owned subsidiary.

Blocks 110/3 and 110/8 were originally awarded to Gulf in 1966. In 1969 Gulf drilled well 110/8-2 on the flank of what is now recognized as the South Morecambe structure; this was plugged and abandoned. Block 110/2 was licensed to HGB in 1972. Blocks 110/3 and 110/8, after being relinquished by Gulf in 1974, were licensed to HGB in 1976. The discovery well, 110/2-1, was drilled by HGB in 1974 to test a seismically defined high at Top Sherwood Sandstone Gp level. This found 652 ft (gross) of gas-bearing Sherwood Sandstone, and was tested at a rate of 10.25 MMSCFD. Between 1975 and 1978 seven appraisal wells were drilled, including three on the North Morecambe structures. In 1983 two observation wells were drilled, 110/2a-7 on South Morecambe Field and 110/2a-8 on North Morecambe Field.

Development plans for South Morecambe Field were drawn up in the late 1970s and early 1980s. At that time the increase in the domestic: industrial ratio in gas consumption was causing an increasing peak winter demand for gas. The development plan was based on the concept of using the field as a seasonal supply facility, to produce gas at very high rates in the winter. The development was designed to produce at a rate of 1220 MMSCFD, and is capable of producing at even higher rates for short periods.

The first stage of the development was implemented in the mid-1980s with the installation of a Central Processing Platform (CPP1), an Accommodation Platform (AP1), and 3 Slant Drilling Platforms (DP1, DP3 and DP4). The latter are designed for unmanned operation and can be remotely controlled from the CPP1. A 36 inch trunk pipeline was laid from the CPP1 to the shore terminal at Barrow. The remote drilling platforms (DPs) are linked to the CPP1 by 24 inch infield pipelines.

Between August 1984 and August 1987, 19 development wells were drilled and completed from the three DPs. First gas flowed in January 1985. In the severe winter weather of January 1987 the field produced at a peak rate of 970 MMSCFD at a time when 14 wells had been completed.

The first stage of this development has been followed by a second stage involving two further remote DPs, DP6 and DP8, on the northern limb of the South Morecambe structure. Drilling from these commenced in May 1989 and 14 wells are planned in this stage of the development.

Design life for the facilities is about 40 years.

Slant drilling

The South Morecambe reservoir is exceptionally shallow, of very large areal extent, and has a very low permeability lower layer. Using conventional directional drilling from a vertical rig, the

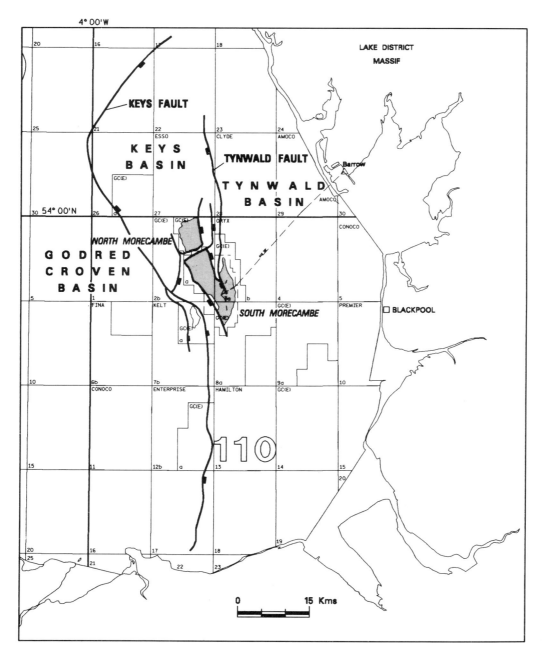

Fig. 1. South Morecambe Field location map.

maximum reach for a typical well penetrating Top Reservoir at −2850 ft TVSS would have been 2300 ft; and less than 2000 ft for wells drilled to the crest of the structure. Conventional development drilling would have resulted in a tight cluster of wells around each platform, giving less than optimum long-term drainage of reserves, particularly from the low permeability Illite-Affected Layer. In addition, interference effects might become significant during extended periods of peak production.

These problems were overcome using 'slant drilling' (Fig. 2), whereby the rig and conductor are slanted at 30° to the vertical. Wells commence at 30° and deviation is quickly built to 60°. By this means, horizontal displacements of 4000 ft or more can be attained for typical wells. Moreover, on wells drilled to the crest of the structure, deviations of up to 72° were achieved, resulting in horizontal displacements of 3300 ft at top reservoir depths of less than −2500 ft TVSS.

Each drilling platform (DP) has four conductor slots in each corner, one vertical and three slanted. Vertical slots can be used for vertical or low angle deviated wells, one of which has been drilled

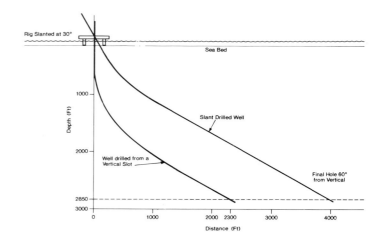

Fig. 2. Slant drilling versus conventional direction drilling.

from each DP. Two slant drilling rigs were built, each consisting of a vertical racking tower and a mast which may be vertical or tilted at 30°. Drilling support is provided by two mobile jack-ups. During operation the jack-up stands alongside the DP. The rig is skidded from the jack-up to the DP and operates on the DP. All services (shakers, pumps, etc.) are sited on the jack-ups.

Field stratigraphy

An outline of the stratigraphy is shown on Fig. 3. The reservoir sandstones belong to the Triassic Sherwood Sandstone Gp. The overlying Mercia Mudstone Gp forms the seal; this comprises mudstones and four correlatable salt horizons which were precipitated during periodic drying of the basin. The Triassic exceeds 8000 ft thickness in the Morecambe area and is devoid of datable fossils. Zonation is therefore purely lithostratigraphic.

Colter & Barr (1975) first described the stratigraphy of the East Irish Sea Permo-Triassic using terminology from the Cheshire Basin. Within the field area, the Sherwood Sandstone Gp was divided into three formations: the St Bees Sandstone, the 'Keuper' Sandstone and the 'Keuper' Waterstones Fms. The 'Keuper' Sandstone was further divided into three members, the Thurstaston, Delamere and Frodsham Mmbrs, in accord with Cheshire Basin stratigraphy. Colter & Barr (1975) used the term 'Keuper' informally. Thompson (1970) suggested that the junction between the St Bees Sandstone and the Keuper Sandstone in the Cheshire Basin was an expression of the Hardegsen Unconformity, but analysis of dipmeter logs from the field shows no evidence of any angular unconformity at this boundary (see below). Warrington et al. (1980) and Jackson et al. (1987) suggested that the 'Keuper' Sandstone and 'Keuper' Waterstones Fms of the East Irish Sea Basin be renamed the Ormskirk Sandstone Fm. However, the terminology proposed by Colter & Barr (1975) has become established for the field and is retained in this paper.

PERIOD	IRISH SEA STRATIGRAPHY	WARRINGTON et al 1980	COLTER & BARR 1975	
TRIASSIC	MERCIA MUDSTONE GROUP	MERCIA MUDSTONE GROUP	MERCIA MUDSTONE GROUP	
	SHERWOOD SANDSTONE GROUP	ORMSKIRK SANDSTONE FORMATION	KEUPER WATERSTONES FM	
			KEUPER SANDSTONE FORMATION	FRODSHAM
				DELAMERE
				THURSTASTON
		ST. BEES SANDSTONE FORMATION	ST. BEES SANDSTONE FORMATION	
PERMIAN	UPPER PERMIAN	ST. BEES EVAPORITES AND EQUIVALENTS	MANCHESTER MARL	
	LOWER PERMIAN	COLLYHURST SANDSTONE AND EQUIVALENTS	COLLYHURST SANDSTONE	
CARBONIFEROUS		WESTPHALIAN	WESTPHALIAN	
		NAMURIAN	NAMURIAN	

Fig. 3. Stratigraphy of Irish Sea basin with reference to the Morecambe Fields.

Geophysics

Seismic data of various vintages from 1969 to 1987 have been acquired. Much of the early data had poor reflection quality and high background noise. Noise problems associated with the geological character of the area resulted in relatively poor seismic character, made worse by the inherent limitations of the air-gun source characteristics due to air bubble oscillations. In the early 1980s a major breakthrough in acquisition techniques in this area was achieved using a water-gun source. The effect of this was to avoid the bubble oscillations associated with air-guns. A first generation of water-gun data was acquired in 1983/4, and in 1986/87 a dense (average 0.5 km) grid of water gun data was acquired on South Morecambe. It was recognized that only true dip lines yield correct two-way time representation as 2-D migration of non-dip

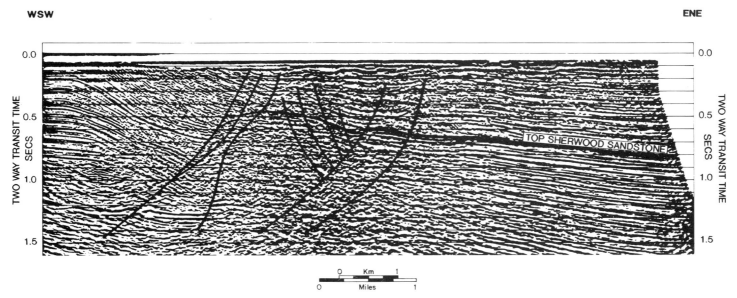

Fig. 4. Typical seismic line through the South Morecambe Field.

lines is an incomplete process. Therefore the 1986/87 grid consisted of dip-orientated lines shot in a radial pattern to follow the shift in dip from ENE in the north to ESE in the south. A typical seismic line through the field is shown on Fig. 4.

Top Sherwood Sandstone is picked on the water-gun data on a clear high frequency peak. The terminations marking the bounding faults are very clear and can be seen to break into a series of step faults along the western boundary of South Morecambe. In-field faults of down to 50 ft throw can be identified.

Trap

The Top Sherwood Sandstone Gp depth structure map is shown on Fig. 5. Regional considerations indicate that the field is situated at the junction of two fundamental structural domains which have controlled the development of the Irish Sea half-graben Basin. The southern domain has been controlled by a dense pattern of north–south-trending faults which dip to the west and produce easterly dipping half grabens. The northern domain is dominated by the

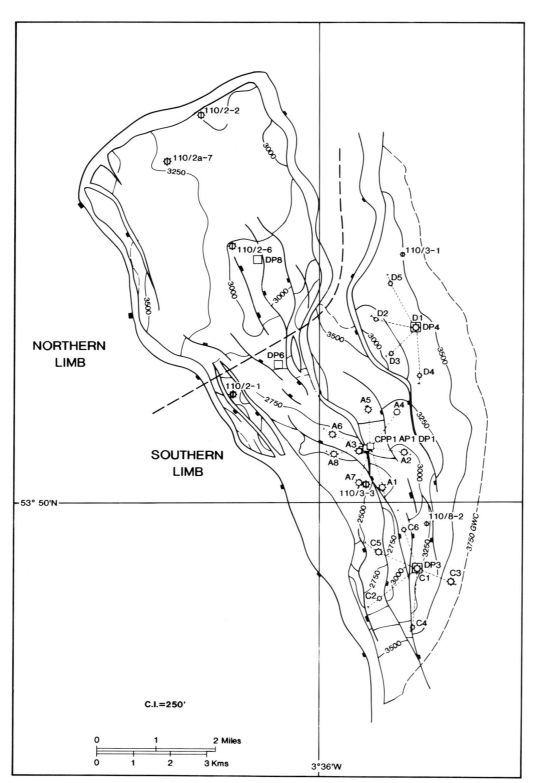

Fig. 5. Simplified depth structure, Top Sherwood Sandstone.

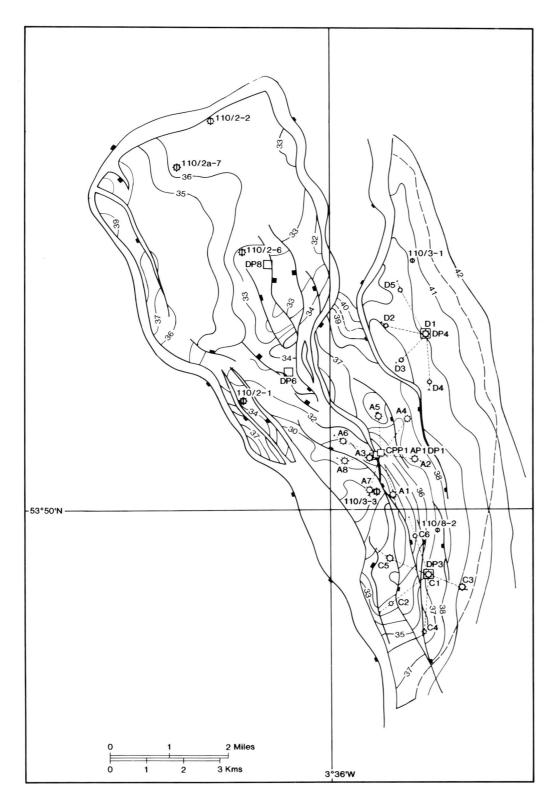

Fig. 6. Top Platy Illite Surface. Faults are shown at their mapped positions at Top Sherwood Sandstone. Only major faults are shown. GWC at Top Sherwood Sandstone. Contour interval is 100 ft.

Keys Fault (Jackson et al. 1987) which lies to the west of the field and has propagated southwards from the northern margin of the East Irish Sea Basin.

Early history of the field is associated with the development of a dome between the Keys Fault and the northward propagating Crosh Vusta Fault. Fracturing of the dome resulted in the creation of faults which bound the east and west of the field.

South Morecambe lies predominantly in the southern domain; it is fault bounded on the western margins by a branch off the Crosh Vusta Fault, with closure on its eastern margin formed by easterly dip. It is believed that the transfer zone between the northern and southern domains passed through what is now the South Morecambe structure. A 'valley' running NE–SW to the north of the DP6 Platform is an early expression of the transfer zone and is thought to represent an incipient graben. The North Morecambe eastern bounding fault propagated southwards so that the northern limb of South Morecambe became fault bounded on its eastern side. Subsequently the transfer zone became established between North and South Morecambe such that a major graben developed between the two fields. The south Morecambe bounding fault propagated northwards so that the northern limb of South Morecambe is now bounded by faults on three sides, and is characterized by a slight westerly dip in contrast to the much steeper westerly dip of North Morecambe and also in contrast to the easterly dip of the rest of South Morecambe.

The major control on reservoir permeability is the presence of platy illite which formed below a palaeo-GWC, now seen as the 'Top Platy Illite Surface'. The structure contemporaneous with platy illite formation comprised certain elements of the present day structure, and the geometry of this palaeo-structure can be assessed by reference to the structural history of the field. By superimposing the effects of 'post-illite' structural events the present configuration of the Top Platy Illite Surface can be predicted and mapped. The 'post-illite' evolution of the field involved continued tilting, associated with extension and late movement on faults, followed by inversion and uplift (see 'Diagenesis and burial history' below). The map of the Top Platy Illite Surface is shown on Fig. 6.

Reservoir

Regional background

The Triassic was a period of active rifting and rapid sedimentation. Thick sequences of continental 'red bed' sediments were deposited in a semi-arid environment. The Sherwood Sandstone Gp is a thick sequence (c. 4800 ft) dominated by sandstones of fluvial origin.

Facies analysis

Five major facies associations can be recognized in cored sections, and their presence inferred from wireline log interpretation in uncored wells. These are summarized graphically on Fig. 7. Over 7500 ft of core has been described and data from 30 wells has now been incorporated into the reservoir model. Individual core runs of over 1000 ft have proved most useful in establishing the reservoir model.

Major channel sandstones (Facies A and A'). Facies A consists of stacked erosively-based intervals of large-scale planar cross-stratified, medium-grained sandstones. Argillaceous intraclasts are often incorporated into foresets, and occasional trough cross-laminated intervals are recorded. These sandstones were deposited as a series of stacked in-channel transverse bars within high energy braided rivers. The Morecambe channel association most closely resembles deposits of the present-day South Platte River in Texas (Miall 1978).

Facies A' exhibits all the characteristics of Facies A sandstones but is composed of a high proportion of coarse, well rounded and well sorted reworked aeolian grains, the presence of which enhances reservoir properties. These sandstones show high sonic transit times, reflecting their higher porosity, and for this reason are easily confused with Facies F in uncored sections.

Facies A and A' sandstones form the main producing horizons of South Morecambe Field.

Secondary channel sandstones (Facies B). Facies B sandstones comprise thinly bedded erosively based intervals, often showing fining-upwards grain-size profiles. Trough cross-lamination is common, and is interpreted as being formed by dune migration within secondary channels. The ephemeral nature of these sediments is evidenced by the fact that they are often overlain by low stage rippled sandstones and desiccated clay abandonment drapes. Facies B channels are often found in association with sheetflood (Facies C) sandstones, and are common in the Delamere Member of the Keuper Sandstone Fm.

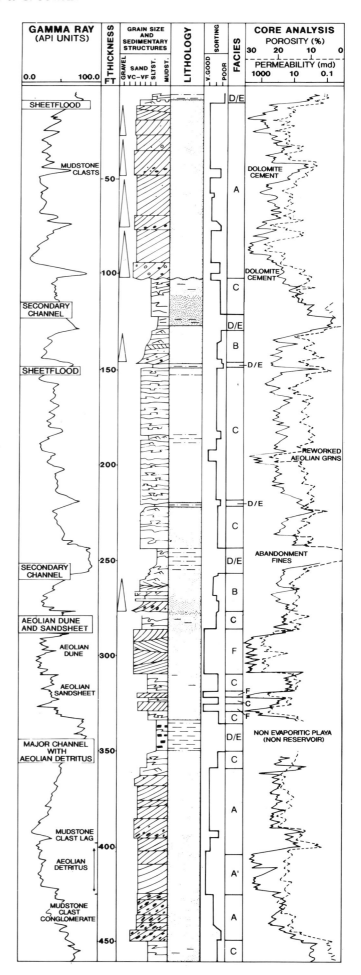

Fig. 7. Typical cored section through major facies associations in South Morecambe Field.

Sheetflood sandstones (Facies C). Facies C consists of tabular intervals of fine- to medium-grained sandstones with interbedded mudstone/siltstone partings. This facies shows distinctive and ubiquitous dewatering structures, and wavy laminae probably caused by the accretion of wind blown sand. Facies C was deposited as non-channelized sheetflood sediment during high flood stage events in a distal fluvial setting with aeolian reworking. Reservoir properties are generally poor but may be locally enhanced by the presence of interbedded aeolian intervals or an abundance of reworked aeolian grains.

Non-reservoir (FaciesD/E). This facies grouping includes abandonment facies mudstones, overbank fines and non-evaporitic playa-lake mudstones. Thin silty mudstone intervals showing desiccation cracks and sandstone injection structures are present throughout the reservoir but are most common in the Keuper Waterstones Fm. and St Bees Sandstone Fm. These were deposited as small non-evaporitic playa deposits and as channel abandonment plugs. They are of limited areal extent, with the exception of a single playa lake interval within the Keuper Waterstones Fm. which was deposited over an area of 200 km^2 in the vicinity of South Morecambe, reaching a thickness of some 30 ft. This interval correlates with marginal deposits of interbedded siltstones, sandstones and mudstones over the rest of the field.

Aeolian facies (Facies F). Thin aeolian deposits are common throughout the field. They are composed of very well rounded, medium- to coarse-grained sandstones exhibiting aeolian style bimodal lamination (Fig. 8), and were deposited as aeolian dune and sandsheet deposits. Facies F sandstones are never more than a few feet thick, but their high permeabilities (up to 11 Darcies) render them an important reservoir facies. The depositional mode of these sandstones is often difficult to interpret in uncored sections where they are easily confused with Facies A'.

Depositional model

Early depositional models for Morecambe reservoir proposed that major braided channel systems were randomly distributed throughout a background of secondary channels and overbank sheetflood fines. However, as data were gathered from the South Morecambe development drilling, an observable layering of facies associations over the field became apparent.

Figure 9 shows the typical sequence encountered in South Morecambe. Layering of the reservoir into units composed dominantly of major channel (Facies A) sandstones, and units composed predominantly of sheetflood (Facies C) sandstones can be seen. These observations indicate that deposition during Sherwood Sandstone Gp times was characterized by periods when major braided channel sandstones were deposited over the entire Morecambe area, and periods when unconfined sheetflood processes were dominant (Fig. 10). This was most probably controlled by the sweeping advance and retreat of the fluvial system, with the sheetflood facies being deposited distally to the channel system. It is believed that the major channel systems were sourced from high ground at the fault-bounded eastern margin of the East Irish Sea Basin and it seems likely that an external control (boundary fault reactivation, climatic change?) was responsible for this migration of the channel/sheetflood system.

Aeolian sandstones and aeolian reworked detritus are common in the Thurstaston Member. This probably reflects reworking of a more extensive aeolian deposit at the beginning of Keuper Sandstone times. Palaeocurrent data from detailed dipmeter interpretation shows that there is a significant shift in palaeocurrent direction between the St Bees and Keuper Sandstone Fms. Palaeocurrents of major channel sandstones within the St Bees Sandstone Fm. show predominantly southward and south-eastward flow directions (this is the reverse of the palaeocurrents derived from regional studies, i.e. Burley 1984). Major channel sandstones in the Keuper Sandstones show a dominantly south westerly palaeoflow. The abundance of aeolian influenced sediments in this part of the section is either a consequence of a post-St Bees hiatus and abandonment of the fluvial system, allowing aeolian reworking of the fluvial sandstones, or the influx of reworked aeolian detritus as a consequence of the change in palaeocurrent and source area.

Palaeocurrents within the Keuper Waterstones reflect a change in sedimentary style prior to the onset of Mercia Mudstone deposition. This is reflected in the increased abundance of argillaceous material within the Keuper Waterstones. The palaeocurrent data shows westward flow directions. This data from the field shows considerable divergence from the measured onshore palaeocurrents for the UK, which show dominantly northwards flow directions, and implies that the East Irish Sea Basin was separated from the Cheshire and Cumbrian Basins.

Diagenetic controls of reservoir properties

A data base of many hundreds of thin sections has been point-counted ($n = 200$) and many hundreds of samples studied by scanning electron microscopy (SEM). Sample spacing is roughly 10 ft for thin sections through the cored intervals. Sidewall cores at between 20 ft and 50 ft intervals were studied in the SEM, mainly to identify the presence of platy illite. Selected samples have been studied using cold cathode cathodoluminescence (CL) and image analysis to examine compaction fabrics. Maturity modelling of the Carboniferous source rocks has been used in an attempt to constrain the diagenetic history, and this has been calibrated by the use of vitrinite reflectance ($R_o\%$) of Carboniferous sediments and apatite fission track analysis of samples from the Sherwood Sandstone Gp.

Diagenesis and burial history

Quartz and early dolomite cements form well over 95% of the authigenic phases present in the reservoir but anhydrite, gypsum, hematite, pyrite, anatase, late stage ankerite, rare calcite, kaolinite and illite are present. The early diagenetic history of the field shows features comparable to 'red-bed' type diagenesis described by Walker *et al.* (1978) and Burley (1984), but the red colouration has been removed by reducing fluids during burial. Bushell (1986) has described the diagenetic history of the field in detail. Early dolomite cements form poikilotopic micronodules which clearly predate most of the quartz cementation. These nodules are ubiquitously overgrown by rhombic ankerite which postdates the quartz cementation. Quartz cements were precipitated during early diagenesis, after the precipitation of carbonate cements as well as after the precipitation of platy illite. Kaolinite is more common in the Illite-Free Layer of the reservoir.

Bushell (1986) and Colter & Barr (1975) estimate post-Triassic inversion to be around 4000–5000 ft, based upon spore colouration and vitrinite reflectance of the underlying Carboniferous. However, recent fission track analysis suggests less post-Cretaceous uplift, and a period of abnormally high temperatures *c.* 60 Ma, probably caused by the early Atlantic rifting. The onshore expression of this event has been described by Green (1986). Independent maturity modelling suggests that around 1500 ft of post-Cretaceous uplift can account for the observed Carboniferous vitrinite reflectance data.

Figure 11 shows a diagenetic flow chart for South Morecambe, and Fig. 12 is a refined version of the burial model presented by Bushell (1986), incorporating the results of this new data.

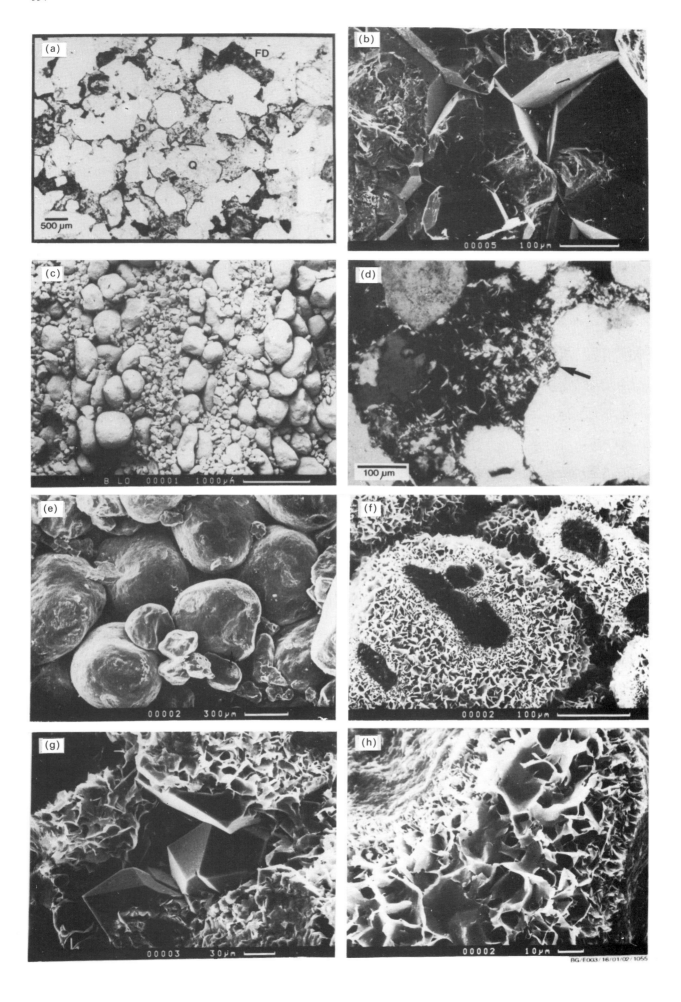

Fig. 8. (a) Thin-section photomicrograph. Plane polarized light. Early dolomite micronodule (D) poikilotopically cementing quartz grains (Q). The dark rim round the edge of the nodule (FD) comprises a later ferroan-dolomite rim. These nodules are often concentrated in the channel bases and were probably seeded by reworked caliche deposits.

(b) Secondary electron image. Fracture surface. Interlocking quartz overgrowths severely reducing porosity in the Illite Free Layer. Much of the quartz precipitated after the platy illite which affects the lower parts of the reservoir.

(c) Backscattered electron image. Fracture surface. Well rounded, high sphericity aeolian grains showing classic bimodal aeolian style lamination. The uncompacted and uncemented nature of these sediments suggest that this porosity is secondary in origin.

(d) Thin-section photomicrograph. Crossed polars. Photomicrograph of Illite-Affected Layer showing large secondary pore infilled with highly birefringent illite. The illite forms a baffle system within the pore which has an obvious effect on permeability. Note also the embayed grain surfaces (arrowed) which suggest that the detrital quartz grains have been etched either by an aggressive cement phase or fluid.

(e) Secondary electron image. Fracture surface. Well-sorted Channel (facies A) sandstones from the Illite-Free Layer. Compare with (b). The well-rounded nature of the detrital grains suggests that the channel detritus is largely composed of re-worked aeolian grains. The clean nature of these sandstones, lack of cementation, excellent sorting and coarse grain size render them excellent reservoirs with permeabilities in the 1–10 darcy range.

(f) Secondary electron image. Fracture surface. Facies A sandstone from the illite-affected zone. Note the coating of grain perpendicular platy illite, the bare patches are pressure dissolution pits. The illite blocks pore throats (see upper centre), and reduces permeability by this mechanism whilst having virtually no effect on porosity. The large surface area offered by these plates also increases water saturations in the Illite-Affected Layer.

(g) Secondary electron image. Fracture surface. Pyramidal quartz overgrowths which have clearly precipitated after the platy illite. Point count data suggest that such cements are much more abundant in the Illite-Free Layer where they significantly reduce porosity. The greater abundance of these cements in the Illite-Free Layer proves that the reservoir lost its hydrocarbon charge after the platy illite was precipitated below a palaeo-GWC.

(h) Secondary electron image. Fracture surface. Close-up of platy illite showing increase in plate size towards the pore centre. Individual plates are over 20 µm in diameter, an exceptional size for this mineral. Transmission electron microscopy (TEM) shows that these coarse illites have a composition close to that of muscovite (Woodward & Curtis 1987).

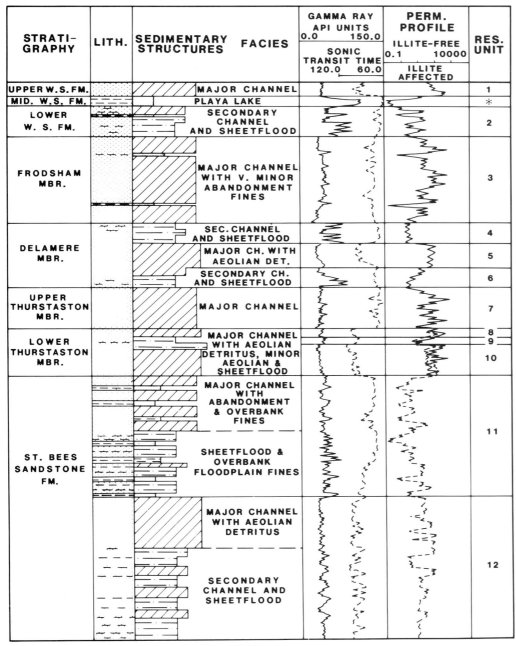

Fig. 9. South Morecambe, summary of sedimentological and petrophysical characteristics.

* LOW PERMEABILITY UNIT – NON RESERVOIR

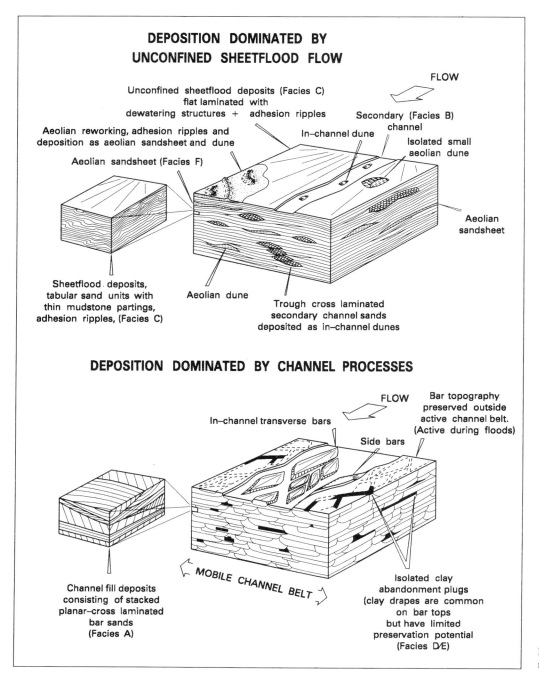

Fig. 10. Block diagram illustrating the main depositional systems.

Platy illite distribution

Much of the diagenetic work on the reservoir has been concerned with the description and delineation of the zone of platy illite, which is believed to have precipitated below a palaeo-GWC (Bushell 1986; Woodward & Curtis 1987). The vertical distribution of platy illite forms the major diagenetic sub-division of the reservoir: the Illite-Free and Illite-Affected Layers, separated by the Top Platy Illite Surface which marks the palaeo-GWC. By a combination of renewed gas generation and post-Cretaceous uplift and gas expansion, the Illite-Affected Layer is now within the gas leg of the reservoir. The Illite-Affected Layer has permeabilities of 2–3 orders of magnitude less than those of the Illite-Free Layer (Fig. 13).

K-Ar dating of this illite indicates that it precipitated c. 180 Ma, after the initial growth of the Morecambe structure. Therefore, this diagenetic layering cuts across the stratigraphic (depositional) layering. In the northern limb of South Morecambe over 1000 ft of Illite-Affected sandstone has been encountered. But in the southern parts of South Morecambe, the Illite-Affected Layer passes downwards after about 450 ft into a layer characterized by poorly developed fibrous illite (or in some wells no illite at all) and consequently higher permeability. This deepest layer is known as the 'Below Platy Illite Layer'.

Cement distribution

Early diagenetic ferroan-dolomite nodules tend to be concentrated near the bases of the Facies A sandstones (Fig. 8a). These probably represent aggradation of channel base lags rich in reworked caliche deposits. These early ferroan-carbonate cements are less abundant in the Illite-Affected than in the Illite-Free Layer (Fig. 14). It is possible that they were partially leached from the Illite-Affected Layer prior to precipitation of platy illite in the palaeo-

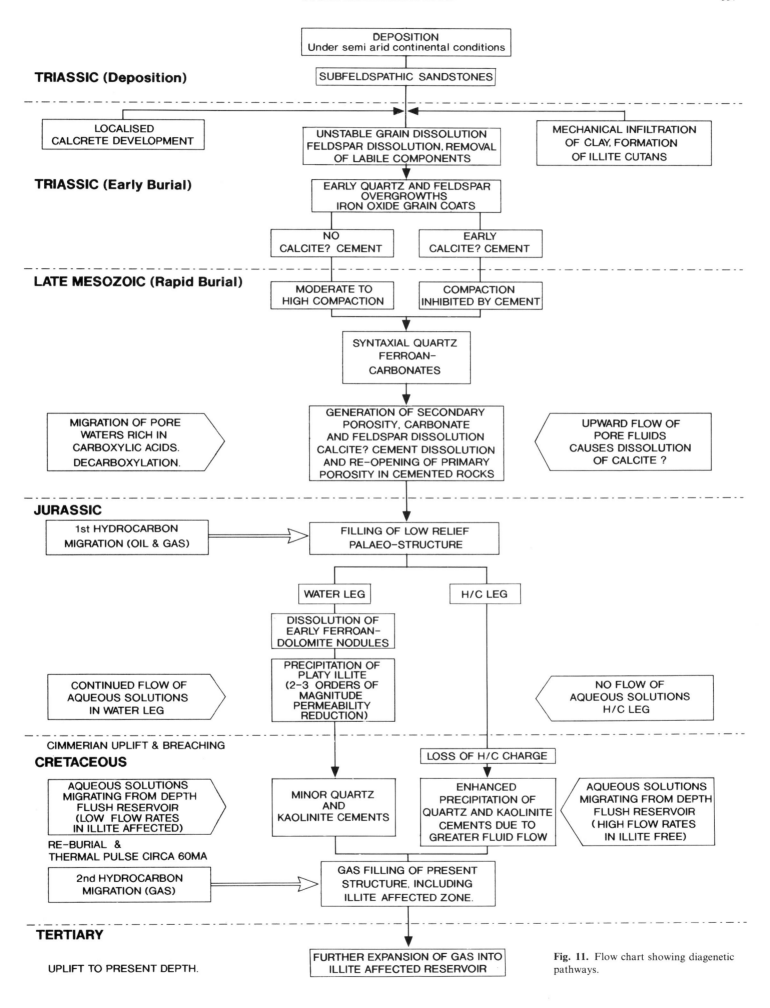

Fig. 11. Flow chart showing diagenetic pathways.

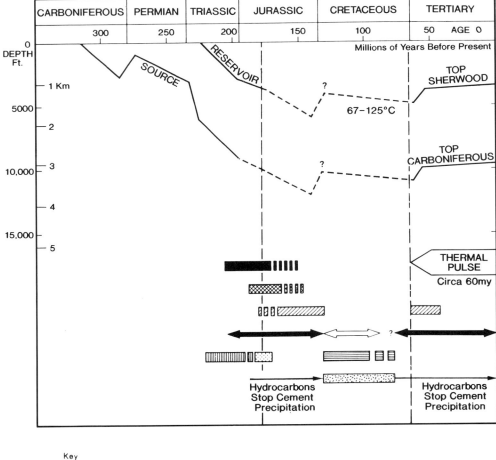

Fig. 12. Burial history curves for the Sherwood Sandstone Reservoir and Carboniferous source rocks showing possible timing of diagenetic events.

water leg. These cements were preserved in the palaeo-gas leg which now forms the Illite-Free Layer.

Evidence that the reservoir was breached is given by comparison of cement-porosity relationships between the Illite-Free and the Illite-Affected Layers, especially by the distribution of post-illite quartz (Fig. 14). Late Stage (post illite) quartz overgrowths are more common in the Illite-Free than in the Illite-Affected Layer. Consequently, the porosity of the Illite-Free (12.5%) is lower than that of the Illite-Affected (14%), although the permeabilities are 2-3 orders of magnitude higher. This decrease in porosity can be accounted for by the preservation of early carbonate cements and intensified precipitation of late quartz cements in the Illite-Free Layer. Clearly, in order to precipitate quartz in the Illite-Free Layer, aqueous solutions must have been free to pass through the reservoir, therefore the first hydrocarbon charge must have been lost.

This enhanced cementation of the Illite-Free Layer can be accounted for by the permeability differential between the two layers, allowing greater fluid flow and therefore greater silica precipitation. Therefore these late quartz cements must have been precipitated during renewed, post-Cimmerian, burial and prior to the second hydrocarbon charge.

Severe quartz cementation occurs locally adjacent to fault planes. These cements were probably sourced by fluids migrating from depth up fault planes. Since the solubility of quartz decreases with decreasing temperature, these fluids would tend to precipitate quartz as they migrate upwards and cool.

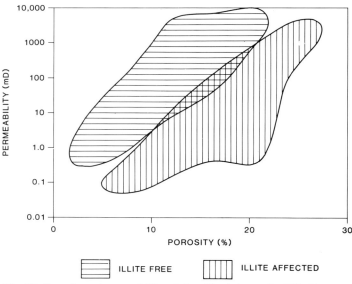

Fig. 13. Porosity and permeability relationships between the Illite-Free and Illite-Affected Reservoir, South Morecambe.

Table 1. *Average total percent cement for typical South Morecambe wells*

	Northern limb		Southern limb	
	110/2-6	110/2a-7	110/3a-A3	110/8a-C1
Quartz	4.5	3.1	4.3	5.9
Carbonate	5.8	3.4	4.0	4.9
Others*	0.3	0.1	0.1	3.7
Visible porosity	3.1	5.0	9.4	6.0
Minus cement porosity	13.7	11.6	17.8	20.5
No. thin sections:	84	56	94	45

* Authigenic feldspar, pyrite, anatase, rutile, hematite

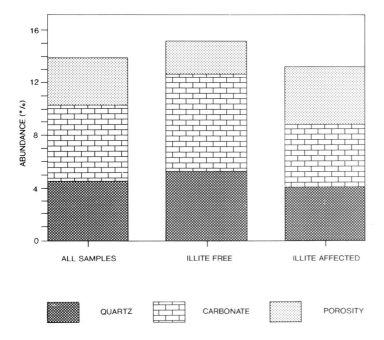

Fig. 14. Cement/porosity distribution in the illite free and illite affected layers of a fully cored well, 110/2-6. Abundance determined from point-counting of closely spaced thin sections.

Aeolian sandstones are characteristically uncemented and uncompacted, being extremely friable in cored sections. Within the Illite-Affected Layer, illite is often absent or poorly developed in aeolian sandstones. It is probable that some early cement, possibly evaporitic, precipitated preferentially in the aeolian facies and prevented further cementation and compaction before being completely removed during deeper burial.

Compaction

Increased compaction in sediments in the northern limb of South Morecambe was observed through analysis of point count data, and confirmed by examination of CL images. Compaction has been identified as the second most important control of reservoir properties after the presence of platy illite and is the major cause of the poorer reservoir quality in the northern limb of the field. Table 1 shows average porosity values for the northern limb and southern areas of South Morecambe, as well as total cement values. Minus-cement porosity values are significantly higher in the southern areas. As has been noted earlier, the South Morecambe structure is situated at the intersection of two discrete structural domains in the East Irish Sea Basin and the differing structural histories of these two domains are probably responsible for the differential compaction across the field.

Secondary porosity

Differential compaction has been observed not only across the field, but also within individual wells. Open, uncompacted 'floating-grain' textures and long grain contact textures can be observed in samples from the same well, and sometimes on the same thin section. This observation implies that cementation has been responsible for maintaining an open framework texture (Burley & Kantorowicz 1986). Dissolution of this cement during burial has restored intergranular porosity which resembles the primary pore system.

Source

The source potential of the Carboniferous has been proved in a number of wells from the East Irish Sea Basin, and geochemical studies have shown that Westphalian coals and shales are the major source of gas in the Morecambe structures. Recent sulphur stable isotope work suggests that gas slightly enriched in traces of H_2S was sourced from the Dinantian Limestone. This gas appears to have migrated up faults in the central part of the field after the main gas charge.

Hydrocarbons

In the exploration/appraisal wells, flow rates were in the range 10–40 MMSCFD, usually from quite restricted perforated intervals. From the development wells, in which the policy has been to perforate all reservoir with permeability above 10 md, flow rates have usually been in the range 50–60 MMSCFD on 80/64 inch choke, or 90–100 MMSCFD on 112/64 inch choke. The restriction on these flow rates is mechanical rather than the reservoir, and on well 110/3a-A2 (DP1 Platform) a rate of 132 MMSCFD was achieved using a 132/64 inch choke; the permeability thickness product (kh) for this well is 277 000 md.ft!

The main components of the South Morecambe gas are C_1 (85%), C_2 (4.5%), C_3 (1.0%), C_4 and above (1.2%), together with N_2 (7.7%) and CO_2 (0.6%). Very minor amounts of H_2S have been recorded, mainly in the range 0–5 ppm.

A review of the (fairly limited) database of water samples in the area of the field indicates a formation water salinity of around 200 000 ppm total chlorides, corresponding to a formation water resistivity of 0.036 ohm m at the reservoir datum temperature of 90°F.

Reserves/resources

South Morecambe Field will be produced under normal depletion, together with compression to be installed when required. Proved reserves for South Morecambe are estimated at 4.57 TCF. So far no undue production problems have been encountered. To April 1989, production has been 112 BCF, 2.5% of the proved reserves.

South Morecambe Field data summary

Trap	
Type	Broad domal horst-structure passing southward to tilted fault blocks.
Depth to crest	−2400 ft TVSS
Lowest closing contour	−4000 ft TVSS
Gas–water contact	−3750 ft TVSS
Gas column	1350 ft
Pay zone	
Formation	Sherwood Sandstone Group
Age	Triassic
Gross thickness	4800 ft
Net:gross ratio	
Illite-Free Layer	Average 0.97 (range 0.91–1.00)
Illite-Affected Layer	Average 0.96 (range 0.89–0.99)
Cut-off for N:G	0.1 md
Porosity	
Illite-Free Layer	Average 12.5% (range 9–15%)
Illite-Affected Layer	Average 14% (range 12–17%)
Hydrocarbon saturation	
Illite-Free Layer	Average 79% (range 65–89%)
Illite-Affected Layer	Average 62% (range 50–80%)
Permeability	
Illite-Free Layer	Average 200 md (range 5–10 000 md)
Illite-Affected Layer	Average 0.8 md (range 0.1–100 md)
Hydrocarbons	
Condensate gravity	72°API
Gas gravity	0.64
Dew point	1650 psia
Condensate yield	3.7 BBL/MMSCF
Formation volume factor	145 SCF/RCF
Formation water	
Salinity	200 000 ppm total chlorides
Resistivity	0.036 ohm.m at 90°F
Reservoir conditions	
Temperature	90°F
Virgin pressure	1860 psia
Pressure gradient	0.05 psi/foot (at initial conditions)
Field size	
Area	20 700 acres
Drive mechanism	Depletion
Original proved reserves	4.57 TCF
Production	
Start-up date	January 1985
Development scheme	1 central processing platform
	1 accommodation platform
	5 wellhead platforms
	19 wells (at April 1989)
	33 wells (planned total)
Production rate	1220 MMSCFD
Cumulative production	112 BCF (to April 1989)

The authors would like to express their gratitude to the management of British Gas for permission to publish this paper. This review has inevitably drawn upon the work of a great many technical specialists who have worked on South Morecambe Field and to all of these the authors express their gratitude. Particular thanks are due to the following colleagues with whom the authors have been closely associated in studies on South Morecambe. Seismic interpretation has been carried out by V. S. Balendran, and petrophysical studies by M. Bowcock. Petroleum engineering information was provided by M. Wannell. S. Cook and R. Sinha of the British Gas London Research Station carried out the SEM work. The work of the staff at Z and S Geologi and of R. Knipe at Leeds University is also gratefully acknowledged. The authors would like to thank A. Levison, D. Ellis, P. Cole and A. Hollis for providing critical review of the draft paper, and the staff of the British Gas drawing office for preparing the figures.

References

BROWN, C. H. 1987. Development of the Morecambe Bay gas field. *124th Annual General Meeting, Institution of Gas Engineers*, Communication 1325.

BURLEY, S. D. 1984. Patterns of Diagenesis of the Sherwood Sandstone Group (Triassic), U.K. *Clay Minerals*, **19**, 403–440.

—— & KANTOROWICZ, J. D. 1986. Thin section and SEM textural criteria for the recognition of cement-dissolution porosity in sandstones. *Sedimentology*, **33**, 587–609.

BUSHELL, T. P. 1986. Reservoir Geology of the Morecambe Field. *In*: BROOKS, J., GOFF, J. C. & VAN HOORN, B. (eds) *Habitat of Palaeozoic Gas in Northwest Europe*. Geological Society, London, Special Publication, **23**, 189–207.

COLTER, V. S. 1978. Exploration for gas in the Irish Sea. *Geologie en Mijnbouw*, **57**, 503–516.

—— & BARR, K. W. 1975. Recent developments in the geology of the Irish Sea and Cheshire Basins. *In*: WOODLAND, A. W. (ed.) *Petroleum and the Continental Shelf of North-West Europe, Vol. 1*. Applied Science Publishers, London, 61–75.

—— & EBBERN, J. 1978. The petrography and reservoir properties of some Triassic sandstones of the Northern Irish Sea Basin. *Journal of the Geological Society, London*, **135**, 57–62.

—— & —— 1979. SEM studies of Triassic reservoir sandstones from the Morecambe Field, Irish Sea, UK. *Scanning Electron Microscopy*, **1**, 531–538.

EBBERN, J. 1981. The geology of the Morecambe Gas Field. *In*: ILLING, L. V. & HOBSON, G. D. (eds) *Petroleum Geology of the Continental Shelf of North-West Europe*, Institute of Petroleum, London, 485–493.

GREEN, P. F. 1986. On the thermotectonic evolution of Northern England: evidence from fission track analysis. *Geological Magazine*, **123**, 493–506.

HUGHES, H. W. D. 1987. Morecambe—discovery and development. *124th Annual General Meeting, The Institution of Gas Engineers*, Communication 1328.

——, MCHUGH, J. & BROWN, C. H. 1987. British Gas presses ahead with Morecambe Field development. *Petroleum Times*, **89**, 22–30.

JACKSON, D. I., MULLHOLLAND, P., JONES, S. M. & WARRINGTON, G. 1987. The geological framework of the East Irish Sea Basin. *In*: BROOKS, J. & GLENNIE, K. (eds) *Petroleum Geology of North West Europe*. Graham & Trotman, 191–203.

LEVISON, A. 1988. The geology of the Morecambe gas field. *Geology Today*, May–June 1988, 95–100.

MACLEAN, M. & WHITE, S. M. 1988. Development drilling, production, and maintenance of slant conductor wells in the Morecambe Gas Field. *17th World Gas Conference, International Gas Union*, A6-88.

MIALL, A. D. 1978. Lithofacies types and vertical profile models in braided river deposits. *In*: MIALL, A. D. (ed.) *Fluvial Sedimentology*, Memoir of the Canadian Society of Petroleum Geologists, **5**, 597–604.

THOMPSON, D. B. 1970. The stratigraphy of the so-called Keuper Sandstone Formation (Scythian–(?)Anisian) in the Permo-Triassic Cheshire Basin. *Quarterly Journal of the Geological Society, London*, **126**, 151–181.

WALKER, T. R., WAUGH, B. & CRONE, A. J. 1978. Diagenesis of first cycle desert alluvium of cenozoic age. Southwestern United States and northwestern Mexico. *Geological Society of America Bulletin*, **89**, 19–32.

WARRINGTON, G., AUDLEY-CHARLES, M. G., ELLIOT, R. E., EVANS, W. B., IVIMEY-COOK, H. C., KNET, P. E., ROBINSON, P. L., SHOTTON, F. W. & TAYLOR, F. M. 1980. *A correlation of Triassic rocks in the British Isles.* Special Report of the Geological Society, London, **13**.

WHITE, S. M. & BARTLE, M. F. 1987. Slant Drilling in the Morecambe Gas Field, Irish Sea, U.K. *Proceedings of the SPE/IADC Drilling Conference (New Orleans)*, 893–902.

WOODWARD, K. & CURTIS, C. D. 1987. Predictive modelling of the distribution of production constraining illites—Morecambe Gas Field, Irish Sea, Offshore UK. *In*: BROOKS, J. & GLENNIE, K. (eds) *Petroleum Geology of North West Europe*. Graham & Trotman, 205–215.

PART SIX

Compilation Tables/Appendix

Table 1. *Triassic/Permian (and older) oil and gasfields*

	Amethyst		Auk	Barque
Trap				
Type	Structural	Structural	Stratigraphic	Structural
Depth to crest ft ss	8800		7337	7500
Lowest closing contour ft ss	9050		7900	8800
GOC or GWC ft ss	9049		—	—
OWC ft ss	—		7606, 7704, 7750	—
Gas column (ft)	—		—	<1200
Oil column (ft)	—		<400	—
Pay zone				
Lithology	Sandstone	Sandstone	Dolomite	Sandstone
Formation	Leman	Rotliegend	Zechstein	Leman
Age	Permian	Permian	Permian	Permian
Gross thickness (ft)	40–120	985	28	700–800
Net/gross ratio (average)	0.40–0.98	0.85	1.00	0.76
Porosity (average, range) (%)	11–25	14–16	12 (2–26)	11
Hydrocarbon saturation (average %)	50	40	60	51
Permeability (average, range) (md)	1–1000	5 (0.2–125)	53 (0.1–10 000)	0.02–100
Productivity index (BOPD/psi)	—	1	50–150	—
Hydrocarbons				
Oil density ° API	—		38	—
Oil type	—		—	—
Gas gravity	—		—	0.59
Bubble point (psig)	—		700	—
Dew point (psig)	—		—	—
Gas–oil ratio	—		190	—
Formation volume factor	235		1.15	229
Formation water				
Salinity (NaCl eq.) (ppm)	145 000		105 000	200 000
Resistivity (ohm m)	—		0.025	0.020
Reservoir conditions				
Temperature (°F)	190		215	175
Pressure (psig)	4100		4067	3850
Pressure gradient (psi/ft)	0.1		0.33	0.08
Field size				
Area (acres)	24 000		16 055	8500
Gross rock volume (acre ft)	—		1 400 000	—
Recovery factor %	77		18	47
Drive mechanism	Pressure Depl.		Water Drive	Pressure Depl.
Recoverable oil (MMBBL)	—		93	—
Recoverable gas (BCF)	844		—	316
Recoverable NGL/condensate (MMBBL)	—		—	—
Production				
Start-up date	1990		1975	1990
Production rate, plateau:				
oil BOPD	—		9500	—
gas MMCFD	180		—	—
Number/type of well	20 producers		—	—

Table 1 *continued*

	Buchan	Camelot N	Camelot C/S	Camelot NE	Cleeton	Clipper
Trap						
Type	Structural		Structural		Structural	Structural
Depth to crest ft ss	9464	6225	6050	6300	9088	7500
Lowest closing contour ft ss	—		6418		—	8900
GOC or GWC ft ss	—	6346	6232	6418	9344	—
OWC ft ss	10 384		—		—	—
Gas column (ft)	—		200		—	<900
Oil column (ft)	1919		—		—	—
Pay zone						
Lithology	Sandstone		Sandstone		Sandstone	Sandstone
Formation	Upper O.R.S.		Leman		Leman	Leman
Age	Dev.–Carb		Permian		Permian	Permian
Gross thickness (ft)	>2182		120–200		200–400	650–775
Net/gross ratio (average)	0.82		1.0		0.95	0.81
Porosity (average, range) (%)	9 (7–10)		18 (15–21)		18 (15–20)	11
Hydrocarbon saturation (average %)	10–68		<80		83	49
Permeability (average, range) (md)	38		200 (5–5000)		95 (40–100)	0.02–100
Productivity index (BOPD/psi)	2–20		—		—	—
Hydrocarbons						
Oil density ° API	33.6		—		—	—
Oil type	Undersat		—		—	—
Gas gravity	—		0.62		—	0.59
Bubble point (psig)	1271		—		—	—
Dew point (psig)	—		—		—	—
Gas–oil ratio	285		—		—	—
Formation volume factor	1.21		192		—	229
Formation water						
Salinity (NaCl eq.) (ppm)	116 528		180 000		214 000	200 000
Resistivity (ohm m)	0.067		0.025		0.020	0.020
Reservoir conditions						
Temperature (°F)	222		150		175	175
Pressure (psig)	7506		2850		4148	3850
Pressure gradient (psi/ft)	0.35		0.07		—	0.08
Field size						
Area (acres)	3527	500	1500	740	1830	12 000
Gross rock volume (acre ft)	—	32 000	191 000	41 600	—	—
Recovery factor %	15–18		70		79	42
Drive mechanism	Pressure Depl. and gas lift		Water Drive		Pressure Depl.	Pressure Depl.
Recoverable oil (MMBBL)	81		—		—	—
Recoverable gas (BCF)	—		215		280	558
Recoverable NGL/condensate (MMBBL)	—		—		—	—
Production						
Start-up date	1981		1989		1988	1990
Production rate, plateau:						
oil BOPD	28 000		—		—	—
gas MMCFD	—		70 (DCQ)		80	—
Number/type of well	9 producers		—		5 producers	—

COMPILATION TABLES

Table 1 *continued*

Indefatigable	Leman	Morecambe S	Ravenspurn S	Sean N	Sean S
Structural	Structural	Structural	Structural	Structural	
7500	5900	2400	9055	8324	8427
8950	6900	4000	—	8700	8600
8878	—	3750	10 206	8543	8543
—	—	—	—	—	—
1360	800	1350	—	—	—
—	—	—	—	—	—
Sandstone	Sandstone	Sandstone	Sandstone	Sandstone	
Leman	Leman	Sherwood	Leman	Leman	
Permian	Permian	Triassic	Permian	Permian	
150 (115–423)	800	4800	200–330	≤268	≤249
1.00	1.00	0.96	0.39–0.77	1.00	1.00
15 (5–22)	12.9	12.5 (9–15)	12–14	17.5 (15–18)	17.1 (16–18)
75–80	51	79 (65–89)	51–61	79	
30 (0.9–1000)	0.5–15	200 (5–10 000)	<90	130–400	190–400
—	—	—	—	—	—
—	—	—	—	—	—
—	0.59	0.64	—	0.62	0.61
—	—	—	—	—	—
—	—	1650	—	—	—
—	—	—	—	—	—
228	229	145	240	218	225
196 200	240 000	—	—	225 000	
0.019	0.026	0.036	—	0.017	
195	125	90	200	202	192
4122	3022	1860	4490	3945	3977
0.07	0.08	0.05	—	0.07	
—	69 456	20 700	9000	3379	2207
—	—	—	—	198 056	326 600
—	83	—	58	82	87
Pressure Depl.	Pressure Depl.	Pressure Depl.	Pressure Depl.	Pressure Depl.	
—	—	—	—	—	—
4418	11 523	4570	700	192	233
8	—	—	—	0.4	0.5
1971	1968	1985	1989	1986–7	
—	—	—	—	—	—
678	—	1200	—	<600	
—	—	—	23 producers	—	

Table 1 *continued*

	Rough	Thames et al.		
		Thames	Yare	Bure
Trap				
Type	Structural		Structural	
Depth to crest ft ss	9000		—	
Lowest closing contour ft ss	9550		—	
GOC or GWC ft ss	9550	8045	7911	8050
OWC ft ss	—		—	
Gas column (ft)	550	295	161	200
Oil column (ft)	—		—	
Pay zone				
Lithology	Sandstone		Sandstone	
Formation	Leman		Leman	
Age	Permian		Permian	
Gross thickness (ft)	95 (78–115)	270–316	430	400
Net/gross ratio (average)	0.98	1.0	1.0	1.0
Porosity (average, range) (%)	12.5 (10.5–16.1)	19	19	20
Hydrocarbon saturation (average %)	64 (56–68)	63	55	66
Permeability (average, range) (md)	75 (2–184)	135–487	38–106	78
Productivity index (BOPD/psi)	—	—	—	—
Hydrocarbons				
Oil density ° API	—		—	
Oil type	—		—	
Gas gravity	0.64		0.60	
Bubble point (psig)	—		—	
Dew point (psig)	4360		—	
Gas–oil ratio	—		—	
Formation volume factor	256		—	
Formation water				
Salinity (NaCl eq.) (ppm)	—		—	
Resistivity (ohm m)	0.018		0.048	
Reservoir conditions				
Temperature (°F)	197	172	170	177
Pressure (psig)	4533	3715	3665	3713
Pressure gradient (psi/ft)	0.09	—	—	—
Field size				
Area (acres)	5050	—	—	—
Gross rock volume (acre ft)	—	—	—	—
Recovery factor %	75	—	—	—
Drive mechanism	Pressure Depl.	—	—	—
Recoverable oil (MMBBL)	—	—	—	—
Recoverable gas (BCF)	366	—	—	—
Recoverable NGL/condensate (MMBBL)	—	—	—	—
Production				
Start-up date	1975	1986	1987	1987
Production rate, plateau:				
oil BOPD	—	—	—	—
gas MMCFD	104	—	—	—
Number/type of well	6 producers	—	—	—

Table 1 continued

V. Gas	Victor	Viking	West Sole
Structural	Structural	Structural	Structural
7200–7900	8300	Variable (8000–9600)	8858
—	8800	Variable (9000–10 200)	—
7650–8450	8776	Variable (9000–10 200)	9679
—	—	—	—
450–583	480	<700	820
—	—	—	—
Sandstone	Sandstone	Sandstone	Sandstone
Leman	Leman	Leman	Leman
Permian	Permian	Permian	Permian
890 (790–990)	300–450	400–700	423
0.8 (0.6–1.0)	0.99	0.50–1.00	0.75
13.5 (3–23)	16	7–25	12
60	73	50–60	60
5 (0.1–1950)	52	0.1–100+	—
—	—	—	—
—	—	—	—
—	—	—	—
0.60	0.61	0.62	0.60
—	—	—	—
—	—	—	1403
—	—	—	—
220	—	—	239
190 000–290 000	220 000	220 000	—
0.017–0.022	0.017	0.017	0.020
142–177	192	170–200	185
3472–3835	4047	4050–4670	4265
0.07	0.08	0.08	0.08
—	5500	—	17 000
—	907 739	—	—
70–77	87	89	74
—	—	—	Pressure Depl.
—	—	—	—
1654	920	2830	1873
3	—	12.7	—
1988	1984	1972	1967
—	—	—	—
324.5	—	—	160
—	5 producers	—	20 producers

Table 2. *Jurassic oil and gasfields*

	Alwyn N.		Beatrice
Trap			
Type	Structural		Structural
Depth to crest ft ss	9875–10 958		5900
Lowest closing contour ft ss	10 367–12 401		—
GOC or GWC ft ss	11 745		—
OWC ft ss	10 315–10 630		6784
Gas column (ft)	787		—
Oil column (ft)	397–440		—
Pay zone			
Lithology	Sandstone	Sandstone	Sandstone
Formation	Statfjord	Brent	Brora, Pentland, Beatrice
Age	JL	JU	JL–JM
Gross thickness (ft)	869	335–361	1100
Net/gross ratio (average)	0.65	0.85–0.90	0.41
Porosity (average, range) (%)	13.5	17	13–22
Hydrocarbon saturation (average %)	80	77–81	78.5
Permeability (average, range) (md)	330	500–800	1–4000
Productivity index (BOPD/psi)	—	—	—
Hydrocarbons			
Oil density ° API	46–48	37–42	38
Oil type	—	Undersat	High Paraffin
Gas gravity	0.34	0.57–0.68	—
Bubble point (psig)	—	—	635
Dew point (psig)	—	—	—
Gas–oil ratio	—	—	126
Formation volume factor	—	—	1.09
Formation water			
Salinity (NaCl eq.) (ppm)	17 000–82 000	20 000–27 000	35 000
Resistivity (ohm m)	0.06–0.20	0.07–0.10	0.062
Reservoir conditions			
Temperature (°F)	225–235	248	176
Pressure (psig)	—	—	2897
Pressure gradient (psi/ft)	—	—	0.33
Field size			
Area (acres)	420–1877	2643	5800
Gross rock volume (acre ft)	—	—	—
Recovery factor %	35 (oil) 70 (gas)	45–56	30
Drive mechanism	Depletion	Water injection	Depletion and Water injection
Recoverable oil (MMBBL)		176	146
Recoverable gas (BCF)		—	—
Recoverable NGL/condensate (MMBBL)		—	—
Production			
Start-up date		1987	1981
Production rate, plateau:			
oil BOPD		90 000	30 000
gas MMCFD		—	—
Number/type of well		—	27 producers 12 injectors

Table 2 continued

	Beryl		Brae N.	Brae C	Brae S
A		B			
Structural/Stratigraphic			Structural/Stratigraphic	Structural/Stratigraphic	Structural/Stratigraphic
9600		10 400	11 830	11 750	11 821
—		—	12 100	12 100	12 100
—		10 900	12 475	—	—
11 300	to	11 550	—	13 426	13 488
—		500	645	—	—
		1950	—	1676	1670
	Sandstone		Sandstone	Sandstone	Sandstone
	(Lewis, Nansen, Linnhe)		Brae	Brae	Brae
	Beryl, (Katrine)				
	(TR, JL) JM (JU)		JU	JU	JU
	500 (200–800)		120 (0–645)	800 (0–1676)	800 (0–1670)
	0.85 (0.4–0.99)		0.64	0.60	0.75
	17 (10–23)		14.5	11.5	11.5
	88		86	80	80
	350 (30–4400)		360 (1–4000)	100 (1–1000)	131 (2–500)
	10–100		—	—	10–40
	37		41–49	33	33–37
	—		Gas Condensate	—	—
	0.71		0.78	—	—
	3200–4900		—	4112	3702
	—		5500–6500	—	—
	900–1500		2500–6500	1415	1343
	1.7		—	1.77	1.73
	—		77 000	79 000	75 000
	—		0.120	0.098	0.120
	207		240	246	253
	4900		6900	7057	7128
	0.28		0.19	0.33	0.30
	12 000		4700	1800	6000
	—		1 571 160	675 000	2 750 300
	—		66 (Cond) 72 (Gas)	40	33
	Sol gas/Gas cap drive		Gas Recycle	Water injection	Water and gas injection
	800 (Beryl Fm 687)		—	64	274
	1600		798	—	—
	—		178	6	38
	1976		1988	1989	1983
	—		81 500	15 500	115 000
	—		—	—	—
	27 producers		16	—	33
	11 injectors				

Table 2 *continued*

	Brent			Chanter	
				Galley	Piper
Trap					
Type		Structural		—	—
Depth to crest ft ss	8240		9000	—	—
Lowest closing contour ft ss	9300		10 700	—	—
GOC or GWC ft ss	8560		9100	—	—
OWC ft ss	9040		9690	12 240	13080
Gas column (ft)	320		100	—	—
Oil column (ft)	480		590	—	—
Pay zone					
Lithology		Sandstone		Sandstone	Sandstone
Formation	Brent		Statfjord	Galley	Piper
Age	JM		JL/TR	Volg	Kimm
Gross thickness (ft)	810		850	—	—
Net/gross ratio (average)	—		—	—	—
Porosity (average, range) (%)	21 (16–28)		23 (16–29)	11.6	13.5
Hydrocarbon saturation (average %)	—		—	87	82
Permeability (average, range) (md)	650 (10–6000)		500 (20–10 000)	250	80
Productivity index (BOPD/psi)	—		—	—	—
Hydrocarbons					
Oil density ° API	—		—	38	52
Oil type	—		—	—	—
Gas gravity	0.74		0.76	—	—
Bubble point (psig)	—		—	3305	5371
Dew point (psig)	—		—	—	—
Gas–oil ratio	—		—	914	—
Formation volume factor	1.80		2.04	1.54	—
Formation water					
Salinity (NaCl eq.) (ppm)	25 000		24 000	44 000	—
Resistivity (ohm m)	0.236		0.270	0.045	—
Reservoir conditions					
Temperature (°F)	204		218	252	252
Pressure (psig)	5785		6020	6196	5860
Pressure gradient (psi/ft)	0.27		0.26	—	—
Field size					
Area (acres)	—		—	—	—
Gross rock volume (acre ft)	—		—	—	—
Recovery factor %	—		—	—	—
Drive mechanism		Water and gas injection		—	—
Recoverable oil (MMBBL)	1815			—	—
Recoverable gas (BCF)	3705			—	—
Recoverable NGL/condensate (MMBBL)	—			—	—
Production					
Start-up date	1976			—	—
Production rate, plateau:					
oil BOPD	334 000			—	—
gas MMCFD	500			—	—
Number/type of well	—			—	—

Table 2 *continued*

	Claymore		Clyde	Cormorant	Crawford	
JU		KL				
	Structural		Structural/ Stratigraphic	Structural/ Stratigraphic	Structural	
	—		12 050	8200	—	
	—		—	10 000	—	
	—		—	—	—	
8655		9400	12 570	10 000	—	
	—		—	—	—	
—		—	520	<1500	—	
	Sandstone		Sandstone	Sandstone	Sandstone	
—		Valhall	Fulmar	Brent	Brent	Skagerrak
JU		KL	Kimm–Volg	Baj	JM	TR
—		—	650	300 (250–550)	—	—
0.50–0.90		0.40–0.95	0.60	0.75	—	—
18–21		20–24	20 (5–30)	13	—	—
—		—	75	70	—	—
0.2–1300		20–4000	1–1000	50–1300	—	—
—		—	—	9–60	—	—
29		34	37–38	34–36	31	25
—		—	Undersat.	Undersat.	—	—
—		—	—	0.838	—	—
680		1550	1722, 2370	1040–2970	—	—
—		—	—	—	—	—
120		408	340, 490	224–770	529	794
1.10		1.23	1.24, 1.34	1.15–1.42	1.32	1.41
49 630		53 400	110 000	—	70000	
0.045		0.039	0.016	0.12–0.17	0.043	
170		190	297	195–225	190	190
3785		4080	6458	4825–5265	3235	3690
—		—	0.9	0.32	—	—
3505		3729	2773	—	—	
—		—	—	—	—	
33		45	38	39	—	
—		—	Depletion and water injection	Water injection	—	
	511		154	572	9	
	—		—	55	—	
	—		—	21	—	
	—		1987	1979	1989	
	—		50 000	122 000	—	
	—		—	—	—	
	—		12 producers 11 injectors	—	—	

Table 2 *continued*

	Cyrus	Deveron	Don	Dunlin
Trap				
Type	Structural	Structural	Structural	Structural
Depth to crest ft ss	8238	8650	10 900	8500
Lowest closing contour ft ss	8369	—	—	9205
GOC or GWC ft ss	—	—	—	—
OWC ft ss	8369	8910	11 409	9205
Gas column (ft)	—	—	—	—
Oil column (ft)	131	260	500	705
Pay zone				
Lithology	Sandstone	Sandstone	Sandstone	Sandstone
Formation	Andrew	Brent	Brent	Brent
Age	Palaeocene	Bath	Bath	Baj–Bath
Gross thickness (ft)	820	450	420	450 (170–650)
Net/gross ratio (average)	0.90	0.12–1.00	0.41–0.55	<1.00
Porosity (average, range) (%)	20	24 (16–30)	16.5	16–30
Hydrocarbon saturation (average %)	65	70	70	35–85
Permeability (average, range) (md)	200	100–3000	2–50	<10 000
Productivity index (BOPD/psi)	20–50	—	4–15	12–110
Hydrocarbons				
Oil density ° API	35	38	39–42	35
Oil type	Undersat.	—	—	Undersat.
Gas gravity	—	—	—	—
Bubble point (psig)	1300	650	1228–2940	960
Dew point (psig)	—	—	—	—
Gas–oil ratio	239	150	335–870	220
Formation volume factor	1.19	1.12	1.25–1.47	1.13
Formation water				
Salinity (NaCl eq.) (ppm)	90 000	23 500	17 000	24 000
Resistivity (ohm m)	0.028	0.264	0.303	0.110
Reservoir conditions				
Temperature (°F)	232	220	265	210
Pressure (psig)	3425	5000	7220	6020
Pressure gradient (psi/ft)	0.32	—	0.36	0.35
Field size				
Area (acres)	3286	512	1390	4446
Gross rock volume (acre ft)	—	49 723	236 000	878 000
Recovery factor %	16	48	22	44
Drive mechanism	Aquifer support	Depletion	Depletion and water injection	Depletion and water injection
Recoverable oil (MMBBL)	12	20	24	363
Recoverable gas (BCF)	—	—	—	—
Recoverable NGL/condensate (MMBBL)	—	—	—	6
Production				
Start-up date	1990	1984	1989	1978
Production rate, plateau:				
oil BOPD	15 000	4800	—	126 000
gas MMCFD	—	—	—	—
Number/type of well	2 horizontal producers	3 producers	—	—

COMPILATION TABLES 555

Table 2 *continued*

Eider	Emerald	Fulmar	Glamis	Highlander	
Structural	Structural	Structural	Structural	Structural/Stratigraphic	
8450	5100	9900	10 000	8500	
9033	5580	—	10 300	9546	
—	5378	—	—	—	
9033	5580	10 830	10 306	9546	
—	—	—	—	—	
—	—	930	<81	<986	
Sandstone	Sandstone	Sandstone	Sandstone	Sandstone	
Brent	Emerald	Fulmar	—	Piper, un-named	
JM	Bath–Call	Oxf–Kimm	Volgian	JU	KL
259	42–64	<1200	20–140	141	109
0.81	1.00	0.94	0.91	1.00	0.55
23	28 (27–33)	23	15	15	24
75	84	79	88	86	78
375	600 (100–1300)	50–800	40–1500	1350	180
40	3–11	80	—	100	12
34	24	40	41.5	36	35
Undersat.	—	Undersat.	—	—	—
—	0.653	—	—	—	—
1050	2425	1800	3210	800	650
—	—	—	—	—	—
205	300	614	1037	85	70
1.15	1.14	1.43	1.64	1.2	1.1
19 000	52 028	138 000	84 500	87 000	50 000
0.11	0.14	0.018	0.028	0.08	0.13
225	140	285	246	200	200
5020	2425	5700	4562	4250	4250
—	0.38	0.29	—	0.33	0.33
865	3650	2825	790	600	500
—	170 000	877 500	—	—	—
42	20	57	50	56	25
Depletion and water injection	Depletion and water injection	Depletion and water injection	Depletion and water injection	Gas lift/Water injection	
85	43	462	17.5	68	7
—	—	264	—	—	—
—	—	—	—	—	—
1988	1990	1982	1989	1985	1987
57 000	40 000	165 000	10 200	27 000	3000
—	—	88	—	—	—
—	7 producers 4 injectors	—	2 producers 1 injector	—	—

Table 2 *continued*

	Heather	Hutton	NW Hutton
Trap			
Type	Structural	Structural	Structural
Depth to crest ft ss	9450	9180	11 000
Lowest closing contour ft ss	—	—	—
GOC or GWC ft ss	—	—	—
OWC ft ss	—	10 090	12 932
Gas column (ft)	—	—	—
Oil column (ft)	>2350	⩽910	1932
Pay zone			
Lithology	Sandstone	Sandstone	Sandstone
Formation	Brent	Brent	Brent
Age	JM	JM	JM
Gross thickness (ft)	210 (125–369)	275 (150–400)	320–550
Net/gross ratio (average)	0.54	0.60	0.45
Porosity (average, range) (%)	10 (0–27)	22 (17–26)	16 (8–24)
Hydrocarbon saturation (average %)	65	67	45
Permeability (average, range) (md)	20 (0.1–2000)	100–4500	0.1–2000
Productivity index (BOPD/psi)	0.2–15	15	—
Hydrocarbons			
Oil density ° API	32–37	34.5	37
Oil type	—	—	—
Gas gravity	0.91	—	—
Bubble point (psig)	1930–3890	715–800	—
Dew point (psig)	—	—	—
Gas–oil ratio	450–1280	140–160	600–800
Formation volume factor	1.3–1.8	1.12–1.125	1.38–1.42
Formation water			
Salinity (NaCl eq.) (ppm)	22 000	—	23 000
Resistivity (ohm m)	0.33	0.36	0.28
Reservoir conditions			
Temperature (°F)	227–242	225	245
Pressure (psig)	4950	6300	7315–7563
Pressure gradient (psi/ft)	0.29–0.31	0.34	0.64–0.66
Field size			
Area (acres)	10 500	6600	13 440
Gross rock volume (acre ft)	2 200 000	848 000	5 376 000
Recovery factor %	23	34	—
Drive mechanism	Water flood	Water injection	Water injection
Recoverable oil (MMBBL)	100	190	145
Recoverable gas (BCF)	61	—	—
Recoverable NGL/condensate (MMBBL)	10	—	—
Production			
Start-up date	1978	1984	1983
Production rate, plateau:			
oil BOPD	38 000	107 000	86 680
gas MMCFD	—	—	—
Number/type of well	29 producers	—	25 producers
	12 injectors	—	9 injectors

Table 2 *continued*

	Ivanhoe/Rob Roy				Kittiwake	
	Structural			Structural	Structural	Stratigraphic
	7670/7730			7560/7725	9773	
	—			—	10 000	
	—			—	—	
	8052			7937/7994	—	
	—			—	—	
	<382			<377	657	
	Sandstone			Sandstone	Sandstone	
	Piper			Piper	Fulmar	Skaggerak
	Supra	JU Main	Supra	JU Main	JU	TR
	60	165	81	193	120 (60–213)	<367
	0.69	1.00	0.81	0.93	0.54	0.2
	25	24	24	23	16–30	6–20
	91	87	94	92	30–90	28
	928	1759	508	2408	0.1–3200	0.05–10
	12	74	48	189	80	—
	31	29	41	39	38	
	—	—	—	—	Undersat.	
	—	—	—	—	—	
	1800	1800	3460	1900	1150–1480	
	—	—	—	—	—	
	360	360	1391	613	300–380	
	1.22	1.22	1.80	1.37	1.2	
		90 990			130 000	
		0.102			0.02	
		175			244	
		3510			6520–6540	
	0.35	0.35	0.26	0.31	0.30	
	693	461	707	537	—	
	—	—	—	—	—	
	30	54	41	62	40	
		Nat. water drive/water injection			Water injection	
	11.2	31.2	14.2	43.3	70	
	4.0	11.2	19.8	26.5	25	
	—	—	—	—	—	
		1989			1991	
		60 000			29 000	
		—			—	
		8 producers			5 producers	
		6 injectors			5 injectors	

Table 2 continued

	Magnus	Miller	Murchison	Ninian
Trap				
Type	Structural/ Stratigraphic	Structural/ Stratigraphic	Structural	Structural
Depth to crest ft ss	9186	13 058	9500	8946
Lowest closing contour ft ss	10 387	—	10 450	—
GOC or GWC ft ss	—	—	—	—
OWC ft ss	10 387	13 418	10 106	10 430
Gas column (ft)	—	—	—	—
Oil column (ft)	$\leqslant$1201	361	<600	167 (average)
Pay zone				
Lithology	Sandstone	Sandstone	Sandstone	Sandstone
Formation	Kimm. Clay	Brae	Brent	Brent
Age	JU	JU	JM	JM
Gross thickness (ft)	0–820	410–820	425 (250–720)	215 (150–370)
Net/gross ratio (average)	0.76	0.80	0.70	0.70
Porosity (average, range) (%)	21 (18–24)	16 (12–23)	22 (16.5–26.5)	20 (16–25)
Hydrocarbon saturation (average %)	83	85	70	60–65
Permeability (average, range) (md)	82 (0–950)	11–1200	500–1000	700 (20–2850)
Productivity index (BOPD/psi)	$\leqslant$200	—	—	—
Hydrocarbons				
Oil density °API	39	38.5	37	36
Oil type	Undersat.	Undersat.	Undersat.	—
Gas gravity	—	—	—	—
Bubble point (psig)	—	4900	1850	1330
Dew point (psig)	—	—	—	—
Gas–oil ratio	750	1813	524	301
Formation volume factor	1.43	1.97	1.31	1.20
Formation water				
Salinity (NaCl eq.) (ppm)	—	70 000	—	—
Resistivity (ohm m)	0.12	0.03	0.31	0.11
Reservoir conditions				
Temperature (°F)	240	250	230	215
Pressure (psig)	6653	7250	6300	—
Pressure gradient (psi/ft)	0.67	0.27	0.31	0.34
Field size				
Area (acres)	8154	11 115	4200	22 000
Gross rock volume (acre ft)	—	—	—	4 760 000
Recovery factor %	40	—	43	35–40
Drive mechanism	Water injection	Water injection	Water injection	Water injection
Recoverable oil (MMBBL)	665	300	340	1045
Recoverable gas (BCF)	—	570	138	150
Recoverable NGL/condensate (MMBBL)	—	—	—	30
Production				
Start-up date	1983	1992	1980	1978
Production rate, plateau:				
oil BOPD	139 000	113 000	127 000	3 150 000
gas MMCFD	—	—	—	—
Number/type of well	—	21 producers 9 injectors	15 producers 11 injectors 1 dual	—

Table 2 *continued*

Osprey	Petronella	Piper		Scapa
Structural	Structural/	Structural		Structural/
	Stratigraphic			Stratigraphic
8389	7200	7300		—
9300	7800	—		—
—	7316	—		—
8918	7648	8510		8812
—	116	—		—
<375	332	<1210		<761
Sandstone	Sandstone	Sandstone		Sandstone
Brent	Piper	Sgiath	Piper	Valhall
JM	JU	JU	JU	KL
375	—	<140	<320	—
0.70	1.00	0–0.7	0.7–0.9	SD SA
25 (22–28)	21 (16–25)	24 (20–28)	24 (18–30)	18 23
83	81	88	90	77 78
300–2100	212 (10–629)	1200 (200–4000)	4000 (500–10 000)	111 390
12–65	140	150–450		11–25
31	39	37		32
—	—	Naphtheno–paraffinic		—
—	1.00	—		—
605	3350	1600		1104
—	—	—		—
89–109	1250	430		208
1.09	1.70	1.26		1.16
22 000	76 000	44 000		—
0.28	0.04	0.045		0.127
214	160	175		185
6000	3450	3700		3384
0.37	0.28	—		—
1112	240	7350		—
—	—	—		SD SA
38	51	70		26 32
Water injection	Nat. Water drive	Water injection		Water injection
60	17	952		63
7	—	—		—
—	—	—		—
1991	1986	1976		1986
25 000	13 000	250 000		28 000
—	—	—		—
—	1 producer	—		4 producers
				4 injectors

Table 2 *continued*

	Tartan Upthrown	Tartan Downthrown	Tern	Thistle
Trap				
Type	Structural	Stratigraphic	Structural	Structural
Depth to crest ft ss	9700	11 600	7800	8500
Lowest closing contour ft ss	—	—	8260	—
GOC or GWC ft ss	—	—	—	—
OWC ft ss	10 328	12 169	8064 (N), 8260 (S)	9322
Gas column (ft)	—	—	—	—
Oil column (ft)	628	569	—	822
Pay zone				
Lithology	Sandstone		Sandstone	Sandstone
Formation	Piper		Brent	Brent
Age	JU		JM	JM
Gross thickness (ft)	—		284	200–550
Net/gross ratio (average)	—		0.62	0.20–1.00
Porosity (average, range) (%)	15 (12–19)	12 (9–14)	22 (20–24)	24 (16–30)
Hydrocarbon saturation (average %)	88	61	71	78
Permeability (average, range) (md)	150 (10–1000)	17 (0.1–300)	350	40–4000
Productivity index (BOPD/psi)	50	7	60	—
Hydrocarbons				
Oil density ° API	38	38	33	38
Oil type	—	—	Undersat.	—
Gas gravity	—	—	—	—
Bubble point (psig)	3650	2850	895	960
Dew point (psig)	—	—	—	—
Gas–oil ratio	1450	1100	177	290
Formation volume factor	1.85	1.39–1.65	1.13	1.18
Formation water				
Salinity (NaCl eq.) (ppm)	95 000	95 000	21 000	23 000
Resistivity (ohm m)	0.027	0.027	0.267	0.253
Reservoir conditions				
Temperature (°F)	215	240	199	220
Pressure (psig)	4650	5800	3580	6060
Pressure gradient (psi/ft)	0.26	0.30	0.34	—
Field size				
Area (acres)	985	1996	2470	3970
Gross rock volume (acre ft)	—	—	—	910 000
Recovery factor %	—	—	39	50
Drive mechanism	Water flood		Gas lift & Water injection	Gas lift & Water injection
Recoverable oil (MMBBL)	68	48	175	396
Recoverable gas (BCF)	—	—	—	—
Recoverable NGL/condensate (MMBBL)	—	—	—	—
Production				
Start-up date	1981		1989	1978
Production rate, plateau:				
oil BOPD	30 000		52 000	124 000
gas MMCFD	—		—	—
Number/type of well	—			34 producers 13 injectors

Table 3. *Tertiary/Cretaceous oil and gasfields*

	Arbroath	Balmoral	Forties
Trap			
Type	Structural	Structural	Structural
Depth to crest ft ss	8030	6905	6660
Lowest closing contour ft ss	8250	7110	7274
GOC or GWC ft ss	—	—	—
OWC ft ss	8250	7150	7274
Gas column (ft)	—	—	—
Oil column (ft)	220	145	614
Pay zone			
Lithology	Sandstone	Sandstone	Sandstone
Formation	Forties	Andrew	Forties & Andrew
Age	Palaeocene	Thanetian	Palaeocene
Gross thickness (ft)	330 (260–440)	550–850	653–1161
Net/gross ratio (average)	0.50	0.87	0.65
Porosity (average, range) (%)	24 (3–30)	25	27 (10–36)
Hydrocarbon saturation (average %)	55	81	85
Permeability (average, range) (md)	80 (1–2000)	20–3300	30–4000
Productivity index (BOPD/psi)	—	—	25
Hydrocarbons			
Oil density ° API	38–42	39.9	37
Oil type	—	—	Undersat.
Gas gravity	—	—	—
Bubble point (psig)	1991	1453	1142–1390
Dew point (psig)	—	—	—
Gas–oil ratio	490	366	303
Formation volume factor	1.33	1.26	1.24–1.32
Formation water			
Salinity (NaCl eq.) (ppm)	135 000	72 000	55 500
Resistivity (ohm m)	0.023	0.039	0.034
Reservoir conditions			
Temperature (°F)	245	207	141
Pressure (psig)	3700	3145	3215
Pressure gradient (psi/ft)	—	—	—
Field size			
Area (acres)	7712	3240	23 000
Gross rock volume (acre ft)	555 000	—	—
Recovery factor %	—	45	57
Drive mechanism	Water drive	Depletion and water injection	Water injection/ Gas lift
Recoverable oil (MMBBL)	102	68	2470
Recoverable gas (BCF)	—	—	—
Recoverable NGL/condensate (MMBBL)	—	—	—
Production			
Start-up date	1990	1986	1975
Production rate, plateau:			
oil BOPD	—	35 000	500 000
gas MMCFD	—	—	—
Number/type of well	6 producers 4 injectors	12 producers 6 injectors	103 producers

Table 3 *continued*

	Frigg	Maureen	Montrose
Trap			
Type	Stratigraphic	Structural	Structural
Depth to crest ft ss	5856	—	8040
Lowest closing contour ft ss	6414	—	8250
GOC or GWC ft ss	—	—	—
OWC ft ss	—	8702	8250
Gas column (ft)	525	—	—
Oil column (ft)	30	<650	210
Pay zone			
Lithology	Sandstone	Sandstone	Sandstone
Formation	Frigg	Maureen	Forties
Age	Lower Eocene	Palaeocene	Palaeocene
Gross thickness (ft)	average 180	140–400	330 (260–440)
Net/gross ratio (average)	0.95	—	0.50
Porosity (average, range) (%)	29 (27–32)	18–25	24 (3–30)
Hydrocarbon saturation (average %)	91	—	55
Permeability (average, range) (md)	1500 (900–4000)	30–3000	80 (1–2000)
Productivity index (BOPD/psi)	—	—	—
Hydrocarbons			
Oil density ° API	23	36	40
Oil type	—	—	—
Gas gravity	0.58	0.78	—
Bubble point (psig)	—	—	2348–2737
Dew point (psig)	—	—	—
Gas–oil ratio	—	393	600–800
Formation volume factor	197	—	1.47–1.56
Formation water			
Salinity (NaCl eq.) (ppm)	63 000	—	111 000
Resistivity (ohm m)	—	—	0.027
Reservoir conditions			
Temperature (°F)	142	243	257
Pressure (psig)	—	3792	3744
Pressure gradient (psi/ft)	—	—	—
Field size			
Area (acres)	24 700	4650	9910
Gross rock volume (acre ft)	—	—	748 000
Recovery factor %	75	54	—
Drive mechanism	Depletion/Nat. water drive	Water drive and water injection	Water drive
Recoverable oil (MMBBL)	—	210	98
Recoverable gas (BCF)	—	—	—
Recoverable NGL/condensate (MMBBL)	—	—	—
Production			
Start-up date	1977	1983	1976
Production rate, plateau:			
oil BOPD	—	—	35 000
gas MMCFD	—	—	—
Number/type of well	—	12 producers 7 injectors	15 producers 6 injectors

Appendix. *A list of common abbreviations used in the articles in this volume*

°	Oil gravity, degrees
API	American Petroleum Institute
BBL	Barrel
BCF	Billion cubic feet
BCPD	Barrels of condensate per day
BMSL	Below mean sea level
BOPD	Barrels of oil per day
BWPD	Barrels of water per day
DST	Drill stem test
FIT	Formation interval test
FLAGGS	Far North Liquids and Associated Gas Gathering System
FT SS	Feet, sub-sea
GIIP	Gas initially in place
GOC	Gas–oil contact
GOR	Gas–oil ratio
GWC	Gas–water contact
MMBBL	Million barrels
MMSCF(D)	Million standard cubic feet (per day)
NGL	Natural gas liquids
OWC	Oil–water contact
PSI	Pounds per square inch
PVT	Pressure, Volume, Temperature
RFT	Repeat formation test
RKB	Below kelly bushing
R_o	Resistivity (oil)
R_w	Resistivity (water)
SCF	Standard cubic feet
SS	Sub-sea
STB	Stock tank barrel
STOIIP	Oil, initially in place
TCF	Trillion cubic feet
TD	Total depth
TOC	Total organic carbon
TVD	True vertical depth
TV(D)SS	True vertical depth, sub-sea
VSP	Vertical seismic profile

Index

Acts of Parliament
  Continental Shelf Act 2
  Petroleum Production Act 1
Alwyn Marl 66
Alwyn North field 12
  data summary 31
  geophysics 26
  history 22–3
  hydrocarbon quality 30
  location 22
  reserves 31
  reservoir character 28–30
  source rock 30
  stratigraphy 24–6
  trap structure 27–8
Amethyst gas field
  data summary 393
  geophysics 389–91
  history 387–8
  hydrocarbon quality 392
  location 387
  reserves 392
  reservoir character 391–2
  source rock 392
  stratigraphy 388–9
  trap structure 391
ammonite zones 12
Amundsen Formation 24–5, 29, 39, 65
Andrew Formation 16
  Arbroath 213
  Balmoral 239, 240
  Cyrus 296, 297–9
  Forties 303–4
  Glamis 318
  Montrose 213
Angus Formation 14
Arbroath field
  data summary 216
  geophysics 213–14
  history 211–12
  hydrocarbon quality 216
  location 211
  reserves 216
  reservoir character 215
  source rock 215
  stratigraphy 212–13
  trap structure 215
Argyll field 3, 9, 10
  geophysics 221
  history 219–20
  hydrocarbon quality 224
  location 219
  reserves 224–5
  reservoir character 222–3
  source rock 223–4
  stratigraphy 220–1
  trap structure 221–2
Argyll Formation 229
Auk field 3, 10
  data summary 235
  geophysics 231
  history 227
  hydrocarbon quality 235
  location 227
  reserves 236
  reservoir character 231–4
  source rock 235
  stratigraphy 227–31
  trap structure 234

Auk Formation 221, 223, 228–9

Bacton Group
  Amethyst 388
  Barque 396
  Camelot 403
  Cleeton 411
  Clipper 418
  Indefatigable 443
  Ravenspurn North and South 461, 471
  Rough 479
  Sean 485
  Viking 511
  West Sole 518
Balder field 3
Balder Tuff Formation
  Alwyn North 26
  Arbroath 213
  Auk 230
  Balmoral 239
  Brent 66
  Crawford 288
  Cyrus 296
  Dunlin 96
  Eider 105
  Frigg 120
  Glamis 318
  Maureen 348
  Montrose 213
Balmoral field 16
  data summary 242
  geophysics 239
  history 237–8
  hydrocarbon quality 242
  location 237
  reserves 242
  reservoir character 240–2
  source rock 242
  stratigraphy 238–9
  trap structure 239
Barque gas field
  data summary 400
  geophysics 396–7
  history 395
  hydrocarbon quality 399
  location 395
  reserves 399
  reservoir character 398–9
  source rock 399
  stratigraphy 395–6
  trap structure 397–8
Beatrice field 14–15
  data summary 251
  geophysics 247–8
  history 245–6
  hydrocarbon quality 251
  location 245
  reserves 251
  reservoir character 248–50
  source rock 250
  stratigraphy 246–7
  trap structure 248
Beatrice Formation 247, 249–50
Beryl Embayment 14
Beryl Embayment Group 39–40
Beryl field 4, 14
  data summary 41
  geophysics 35–6

  history 34
  hydrocarbon quality 40
  location 34
  reserves 40–2
  reservoir character 37–40
  stratigraphy 34–5
  trap structure 36–7
Beryl Formation 14, 39
Big Dotty field 434, 440
Brae complex 4, 9, 15 see also Central, North, South Brae
Brae Sandstone Formation 11, 44–5, 50, 56, 159, 161
Brent field 3, 4, 13
  data summary 71
  geophysics 66–7
  history 63–5
  hydrocarbon quality 71
  location 63
  reserves 71
  reservoir character 69–70
  source rock 70
  stratigraphy 65–6
  trap structure 67–9
Brent Group 9, 11, 12–13
  Alwyn North 25, 26, 28–9
  Brent 65, 67–9
  Cormorant 77–9
  Deveron 84, 86
  Don 90, 91–2
  Dunlin 95–6, 97
  Eider 105
  Heather 129, 132, 133
  Hutton 136–7, 138–42
  Murchison 169, 170
  Ninian 176, 179
  Northwest Hutton 146, 147–8
  Osprey 186
  Tern 193
  Thistle 201, 204–5
Broom Formation 12,13
  Alwyn North 25
  Brent 65
  Cormorant 77–9
  Deveron 84
  Don 90, 92
  Dunlin 97
  Eider 107
  Heather 133
  Hutton 136, 138
  Murchison 169, 170
  Ninian 176, 180
  Northwest Hutton 148–9
  Osprey 186
  Tern 193
Brora Formation 14, 247, 250
bubble point see data summary under named field
Buchan field 9
  data summary 259
  geophysics 255
  history 253–4
  hydrocarbon quality 258
  location 253
  reserves 258
  reservoir character 257–8
  source rock 258
  stratigraphy 254
  trap structure 255–6

Bunter Sandstone Formation 11
  Barque 396
  Camelot 403
  Esmond 426, 427, 429–32
  Forbes 426, 427, 429–32
  Gordon 426, 427, 429–32
  Hewett 435, 440
  Indefatigable 443
  Leman 452
  Ravenspurn North 461
  Sean 485
  V complex 499
  Viking 512
  West Sole 518
Bunter Shale Formation
  Barque 396
  Camelot 403
  Esmond 427
  Forbes 427
  Gordon 427
  Hewett 435
  Ravenspurn North 461
  Sean 485
  V complex 499
  Viking 512
Bure gas field
  data summary 496
  geophysics 493–4
  history 491–2
  hydrocarbon quality 495
  location 491
  reserves 495
  reservoir character 494–5
  source rock 495
  stratigraphy 493
  trap structure 494
Burton Formation 24–5, 29, 65

Caister gas field 6
Camelot gas field 6
  data summary 407
  geophysics 404–5
  history 401
  hydrocarbon quality 408
  location 401
  reserves 408
  reservoir character 405–8
  source rock 408
  stratigraphy 401–4
  trap structure 405
cap rock *see* seal
Carbonates breccia 229
cements *see* diagenesis
Central Brae field 15
  data summary 53
  geophysics 51
  history 49
  hydrocarbon quality 53
  location 49
  reserves 53
  reservoir character 52–3
  source rock 53
  stratigraphy 50
  trap structure 51–2
Central Graben 2–3, 9
Chalk Group
  Buchan 254
  Chanter 265
  Cleeton 411
  Highlander 325
  Ravenspurn South 472
  Rough 479
  Sean 488
  Tartan 379
  Victor 504
Chanter field 15
  data summary 268
  geophysics 265
  history 261–3
  hydrocarbon quality 268
  location 261
  reserves 268
  reservoir character 266–7
  source rock 267–8
  stratigraphy 264–5
  trap structure 265
Clair field 9
clay mineralogy
  Brent Group 28–9, 133
  Leman Sandstone Formation 413, 455, 465–6, 495
  Maureen Formation 351
  Montrose Group 304
  Sherwood Sandstone Group 532, 533–6
Claymore field 4, 9, 15
  data summary 277
  geophysics 272
  history 269–71
  hydrocarbon quality 275
  location 269
  reserves 275, 277
  reservoir character 273–5
  source rock 275
  stratigraphy 271–2
  trap structure 272–3
Claymore Sandstone 11, 15
Cleeton gas field
  data summary 415
  geophysics 411
  history 409–10
  hydrocarbon quality 413–14
  location 409
  reserves 414
  reservoir character 412–13
  source rock 413
  stratigraphy 411
  trap structure 411–21
Clipper gas field 10
  data summary 422
  geophysics 418–20
  history 418
  hydrocarbon quality 421–2
  location 418
  reserves 422
  reservoir character 420–21
  source rock 421
  stratigraphy 418
  trap structure 420
Clyde field 16
  data summary 285
  geophysics 282–3
  history 279–81
  hydrocarbon quality 285
  location 279
  reserves 285
  reservoir character 283–5
  source rock 285
  stratigraphy 281–2
  trap structure 283
Cod field 3
Cod Sands *see* Sele Formation
condensate yield *see* data summary *under named field*

continental shelf exploration 2
Cooke Formation 24–5, 29–30, 65
Cormorant field 4
  data summary 81
  geophysics 75
  history 73–4
  hydrocarbon quality 77, 80
  location 73
  reserves 80–1
  reservoir character 77
  source rock 77
  stratigraphy 74–5
  trap structure 75–6
Cormorant Group
  Alwyn North 26
  Beryl 38
  Brent 65
  Cormorant 74, 77
  Don 90
  Dunlin 95
  Eider 104
  Heather 129
  Murchison 169
  Ninian 176
  Northwest Hutton 146
  Osprey 185
  Tern 192
Cousland field 1
Crawford field 14
  data summary 293
  geophysics 288
  history 287–8
  hydrocarbon quality 293
  location 287
  reserves 293
  reservoir character 289–93
  source rock 293
  stratigraphy 288
  trap structure 288–9
Cromer Knoll Group
  Argyll 221
  Brent 70
  Chanter 264
  Cleeton 411
  Clyde 282
  Cormorant 75
  Duncan 221
  Dunlin 96
  Eider 105
  Esmond 427
  Forbes 427
  Gordon 427
  Heather 130
  Innes 221
  Maureen 348
  Ninian 176
  Osprey 185
  Petronella 356
  Ravenspurn South 472
  Rough 479
  Sean 485
  Tern 193
  Viking 511
Cyrus field 16
  data summary 300
  geophysics 296
  history 295–6
  hydrocarbon quality 299
  location 295
  reserves 299
  reservoir character 297–9
  source rock 299

stratigraphy 296
trap structure 296–7

Davy field 6
Deborah gas field 434, 440
Della gas field 12
Deveron field
  data summary 87
  geophysics 84–5
  history 83–4
  hydrocarbon quality 86–7
  location 83
  reserves 87
  reservoir character 85–6
  source rock 86
  stratigraphy 84
  trap structure 85
dew point *see* data summary *under named field*
diagenesis and cementation
  Leman Sandstone Formation 391–2, 465–6, 495
  Sherwood Sandstone Group 533–9
dinocyst zones 12
Don field 13
  data summary 93
  geophysics 90–1
  history 89–90
  hydrocarbon quality 93
  location 89
  reserves 93
  reservoir character 91–3
  source rock 93
  stratigraphy 90
  trap structure 91
Dowsing Dolomite Formation 426, 443, 461
Drake Formation 24–5, 29–30, 65, 70
drilling history 1
drive mechanism *see* data summary *under named field*
Dudgeon Saliferous Formation 426, 443, 461
Duncan field 16
  geophysics 221
  history 219–20
  hydrocarbon quality 224
  location 219
  reserves 224–5
  reservoir character 223
  source rock 223–4
  stratigraphy 220–1
  trap structure 221
Duncan Sandstone Formation 16, 221, 223
Dunlin field 4
  data summary 102
  geophysics 96
  history 95
  hydrocarbon quality 101
  location 95
  reserves 101–2
  reservoir character 97–100
  source rock 100–1
  stratigraphy 95–6
  trap structure 97
Dunlin Group
  Alwyn North 24–5, 26
  Beryl 39
  Brent 65
  Don 90

  Dunlin 95
  Eider 105
  Heather 129
  Murchison 169
  Ninian 176
  Northwest Hutton 146
  Osprey 185
  Tern 193
  Thistle 201
Dunrobin Bay Formation 247, 248–9
Dunrobin Group 246, 248–9

Eakring field 1
Eider field 13
  data summary 109
  geophysics 105
  history 103–4
  hydrocarbon quality 108
  location 103
  reserves 108–9
  reservoir character 105–8
  source rock 108
  stratigraphy 104–5
  trap structure 105
Eiriksson Formation 38–9, 65, 169
Ekofisk field 3
Ekofisk Formation
  Arbroath 213
  Balmoral 239
  Buchan 254
  Crawford 288
  Forties 303
  Glamis 318
  Maureen 348
  Montrose 213
Emerald field 13–14
  data summary 116
  geophysics 113
  history 111–12
  hydrocarbon quality 114–15
  location 111
  reserves 115
  reservoir character 113–14
  source rock 114
  stratigraphy 112–13
  trap structure 113
Emerald Sandstone 112, 113–14
Erskine field 6
Eskdale gas field 1
Esmond gas field
  geophysics 427–8
  history 425–6
  hydrocarbon quality 431
  location 425
  reserves 431
  reservoir character 429–31
  source rock 431
  stratigraphy 426–7
  trap structure 428–9
Etive Formation 12, 13
  Alwyn North 25
  Brent 65
  Cormorant 77–9
  Deveron 84
  Don 92
  Dunlin 98
  Eider 107
  Heather 133
  Hutton 136–7, 138
  Murchison 169, 172
  Ninian 176, 181

  Northwest Hutton 149
  Osprey 186
  Tern 193, 195
  Thistle 206

Farsund Formation 16
Fladen Group 247, 333, 378
Flounder Formation 239, 254, 318
Forbes gas field
  geophysics 427–8
  history 425–6
  hydrocarbon quality 431
  location 425
  reserves 431
  reservoir character 429–31
  source rock 431
  stratigraphy 426–7
  trap structure 428–9
formation volume factor *see* data summary *under named field*
Formby field 1
Forth field 6
Forth Formation 11
  Claymore 271, 275
  Highlander 234
  Petronella 355
  Tartan 378
Forties field 3, 16
  data summary 307
  geophysics 303
  history 301
  hydrocarbon quality 305
  location 301
  reserves 305–7
  reservoir character 303–4
  source rock 304
  stratigraphy 301–3
  trap structure 303
Forties Formation 16
  Arbroath 213, 215
  Balmoral 239
  Forties 303–4
  Glamis 318
  Montrose 213, 215
Franklin field 6
Frigg field 4, 16
  data summary 126
  geophysics 121–3
  history 117–18
  hydrocarbon quality 125
  location 117
  reserves 125
  reservoir character 123–4
  source rock 124
  stratigraphy 119–21
  trap structure 123
Frigg Formation 11, 16, 120
Frobisher field 6
Fulmar field 16
  data summary 315
  geophysics 311
  history 309–10
  hydrocarbon quality 315–16
  location 309
  reserves 316
  reservoir character 313–15
  source rock 315
  stratigraphy 310–11
  trap structure 311–13
Fulmar Formation 16
  Clyde 282, 283–4

Fulmar 311, 313–15
Kittiwake 340, 342

gas column *see* data summary *under named field*
gas condensate recycling 48
gas exploration history 2
gas–oil contact *see* data summary *under named field*
gas/oil ratio *see* data summary *under named field*
gas storage facilities 477
Glamis field 16
 data summary 322
 geophysics 319
 history 317–18
 hydrocarbon quality 321
 location 317
 reserves 321–2
 reservoir character 320–1
 source rock 321
 stratigraphy 318–19
 trap structure 319–20
Glamis Sandstone Formation 318, 320
Gordon gas field
 geophysics 427–8
 history 425–6
 hydrocarbon quality 431
 location 425
 reserves 431
 reservoir character 429–31
 source rock 431
 stratigraphy 426–7
 trap structure 428–9
gravity values *see* data summary *under named field*
Groningen gas field 2
Gryphon field 6

Haisborough Group
 Amethyst 388
 Barque 396
 Camelot 403
 Cleeton 411
 Clipper 418
 Esmond 426, 427
 Forbes 426, 427
 Gordon 426, 427
 Leman 452
 Ravenspurn North and South 461, 471
 Rough 479
 Viking 511, 512
 West Sole 518
Halibut Bank Formation
 Argyll 221, 222
 Claymore 271, 275
 Duncan 221, 222
 Highlander 324
 Innes 221, 222
Hardstoft field 1
Haugesund Formation 16
Heather field
 data summary 134
 geophysics 130–1
 history 127–9
 hydrocarbon quality 133–4
 location 127
 reserves 134
 reservoir character 133
 source rock 133
 stratigraphy 129–30
 trap structure 132–3
Heather Formation 12, 14
 Alwyn North 25, 26
 Beatrice 247
 Beryl 40
 Brent 65, 70
 Don 90
 Dunlin 96
 Eider 105
 Emerald 112
 Heather 130
 Kittiwake 340
 Magnus 153
 Maureen 348
 Murchison 169
 Ninian 176
 Osprey 185
 Thistle 201
Heimdal Formation 112
Helmsdale Boulder Beds 15
Herring Formation 254
Hewett gas field 2
 data summary 440
 geophysics 436–7
 history 435
 hydrocarbon quality 438
 location 433–5
 reserves 438, 441
 reservoir character 438
 source rock 438
 stratigraphy 435–6
 trap structure 437
Hewett Sandstone 11, 440
Hidra Formation 254
Highlander field 15–16
 data summary 329
 geophysics 325
 history 323–4
 hydrocarbon quality 328
 location 323
 reserves 328
 reservoir character 327
 source rock 328
 stratigraphy 324–5
 trap structure 325–7
Hod Formation 348
Hordaland Group
 Balmoral 239
 Buchan 254
 Cormorant 75
 Cyrus 296
 Eider 105
 Frigg 121
 Glamis 318
 Heather 130
Hot Lens Equivalent 327, 356, 381, 382, 383
Hot Shale 26, 325
Hugin Formation 14, 348
Humber Group 12
 Alwyn North 25
 Beatrice 247
 Beryl 40
 Brent 65
 Chanter 266
 Deveron 85
 Don 90
 Dunlin 96
 Eider 105
 Heather 130
 Osprey 185

Petronella 357
Piper 363
Tartan 379
Tern 193
Thistle 201
Hutton Clay Formation 66, 96
Hutton field 13
 data summary 143
 geophysics 137–8
 history 135–6
 hydrocarbon quality 142
 location 135
 reserves 142
 reservoir character 138–42
 source rock 142
 stratigraphy 136–7
 trap structure 138
Hutton Sand Formation 66, 96

I shale 15
Indefatigable gas field 2
 data summary 449
 geophysics 446–8
 history 443
 hydrocarbon quality 449
 location 443
 reserves 449
 reservoir character 448–9
 source rock 449
 stratigraphy 443–6
 trap structure 448
Innes field 10
 geophysics 221
 history 219–20
 hydrocarbon quality 224
 location 219
 reserves 224–5
 reservoir character 223
 source rock 223–4
 stratigraphy 220–1
 trap structure 221
Ivanhoe field 15
 data summary 338
 geophysics 333–4
 history 331–2
 hydrocarbon quality 337
 location 331
 reserves 337
 reservoir character 334–7
 source rock 337
 stratigraphy 332–3
 trap structure 334

Keuper Sandstone Formation 529, 533
Keuper Waterstones Formation 529, 533
Kimmeridge Clay Formation 7, 15
 Alwyn North 25–6, 30
 Amethyst 388
 Argyll 221
 Beatrice 247
 Beryl 40
 Brae complex 44, 47, 50, 56, 57
 Brent 65, 70
 Chanter 264, 267
 Claymore 272, 273
 Clyde 282
 Crawford 288
 Don 90
 Duncan 221
 Dunlin 96

Eider 105
Emerald 112
Forties 301
Fulmar 311
Glamis 318, 321
Heather 130
Highlander 324–5, 327
Innes 221
Ivanhoe 33
Kittiwake 340
Magnus 153, 155–6
Maureen 348
Miller 159
Osprey 185
Petronella 359
Rob Roy 333
Tartan 379, 381–3
Thistle 201
Kittiwake field 16
 data summary 344
 geophysics 341–2
 history 339–40
 hydrocarbon quality 343
 location 339
 reserves 344–5
 reservoir character 342
 source rock 342–3
 source rock 342–3
 stratigraphy 340–1
 trap structure 342
Kupferschiefer Shale
 Amethyst 388
 Argyll 221, 222
 Auk 229
 Claymore 271
 Duncan 221, 222
 Hewett 435
 Indefatigable 443, 448
 Innes 221, 222
 Rough 479

Lacq field 2
Lady's Walk Shales 249
Leine Group 511
Leman gas field 2
 data summary 458
 geophysics 455
 history 451–2
 hydrocarbon quality 457
 location 451
 reserves 457
 reservoir character 455–7
 source rock 457
 stratigraphy 452–5
 trap structure 455
Leman Sandstone Formation 10, 11
 Amethyst 391
 Barque 398–9
 Bure 493, 494–5
 Camelot 403, 405–8
 Cleeton 411, 412
 Dotty complex 440
 Leman 452, 455–7
 Ravenspurn North and South 460–1, 464–5, 471, 473
 Rough 479
 Sean 488
 Thames 493, 494–5
 V complex 499, 501–2
 Victor 504, 506–7
 Viking 511, 514

West Sole 520–1
Yare 493, 494–5
Lewis Formation 38
licensing rounds 4
 1st round (1964) 2
  Amethyst 387
  Arbroath 211
  Barque 395
  Bure 491
  Camelot 401
  Cleeton 409
  Clipper 418
  Clyde 279
  Esmond 425
  Forbes 425
  Gordon 425
  Hewett 435
  Indefatigable 443
  Leman 451
  Montrose 211
  Ravenspurn North and South 459, 470
  Rough 477
  Thames 491
  Viking 509
  West Sole 517
  Yare 491
 2nd round (1969) 2
  Alwyn North 22
  Argyll 219
  Duncan 219
  Forties 301
  Frigg 117
  Innes 219
  Sean 485
 3rd round (1970) 3
  Auk 227
  Brae complex 43, 49, 56
  Brent 63
  Cyrus 295
  Maureen 347
  Murchison 165
 4th round (1972) 3
  Balmoral 237
  Beatrice 245
  Beryl 34
  Buchan 253
  Chanter 261
  Claymore 269
  Cormorant 73
  Crawford 287
  Deveron 83
  Don 89
  Dunlin 95
  Eider 103
  Emerald 111
  Glamis 317
  Heather 127
  Highlander 323
  Ivanhoe 331
  Magnus 153
  Ninian 175
  Northwest Hutton 146
  Osprey 183
  Petronella 354
  Piper 361
  Rob Roy 331
  Scapa 370
  Tartan 378
  Tern 191
  Thistle 200
 7th round (1980) 6

Balmoral 237
Kittiwake 339
Miller 159
10th round 347
Lias Group
 Cleeton 411
 Ravenspurn South 471
 Rough 479
 V fields 499
 Viking 511
Linnhe Formation 14, 39
Lista Formation
 Balmoral 239, 241
 Brent 66
 Dunlin 96
 Frigg 120
 Glamis 318
 Maureen 348
Little Dotty gas field 434, 440

Magnesian Limestone 388
Magnus field 4, 12, 15
 data summary 157
 geophysics 154
 history 153
 hydrocarbon quality 156
 location 153
 reserves 156–7
 reservoir character 155–6
 source 156
 stratigraphy 153
 trap structure 154
Magnus Sandstone 11
Mandal Formation 16
Marnock field 6
Maureen field
 geophysics 348–9
 history 347
 hydrocarbon quality 352
 location 347
 reserves 352
 reservoir character 350–2
 source rock 352
 stratigraphy 348
 trap structure 350
Maureen Formation 16
 Arbroath 213
 Balmoral 239
 Cyrus 296
 Emerald 112
 Forties 303–4
 Frigg 120
 Glamis 318
 Maureen 348, 350–1
 Montrose 213
Mercia Mudstone Group 529
Miller field 6, 11, 15
 data summary 164
 geophysics 160
 history 159
 hydrocarbon quality 164
 location 159
 reserves 164
 reservoir character 161–2
 source rock 162, 164
 stratigraphy 159
 trap structure 160
mineralogy
 Brent Group 28–9, 133
 Emerald Sandstone 114
 Forties Formation 215

Maureen Formation 351
Mo Clay 121
Montrose field 3
   data summary 216
   geophysics 213–14
   history 211–12
   hydrocarbon quality 216
   location 211
   reserves 216
   reservoir character 215
   source rock 215
   stratigraphy 212–13
   trap structure 215
Montrose Group 11, 16
   Arbroath 213
   Auk 230
   Balmoral 240
   Buchan 254
   Cormorant 75
   Cyrus 296
   Dunlin 96
   Eider 105
   Heather 130
   Ivanhoe 333
   Montrose 213
   Ninian 177
   Rob Roy 333
Moray Firth Basin 9, 15, 16
Moray Group 16, 254, 333
Morecambe Basin *see* South Morecambe
   gas field
Murchison field
   data summary 173
   geophysics 169–70
   history 165–6
   hydrocarbon quality 172–3
   location 165
   reserves 173
   reservoir character 170–2
   source rock 172
   stratigraphy 166–9
   trap structure 170
Murdoch field 6

Nansen Formation 38–9, 65, 129, 169
Nelson field 6
Ness Formation 12, 13
   Alwyn North 25, 28
   Brent 65
   Cormorant 77–9
   Deveron 84
   Don 92
   Dunlin 97, 98–9
   Eider 107
   Frigg 119
   Heather 133
   Hutton 136, 142
   Murchison 169, 172
   Ninian 176, 181
   Northwest Hutton 149–50
   Osprey 188
   Tern 193, 195
   Thistle 206
net : gross ratio *see* data summary *under*
   named field
Nevis Formation 14
New Red Group 75
Ninian field 4
   data summary 182
   geophysics 177–8
   history 175–6
   hydrocarbon quality 181
   location 175
   reserves 181–2
   reservoir character 179–81
   source rock 181
   stratigraphy 176–7
   trap structure 178–9
Nordland Group
   Balmoral 239
   Buchan 254
   Cormorant 75
   Cyrus 296
   Eider 105
   Glamis 318
   Heather 130
North Brae field 12, 15
   data summary 47
   geophysics 45
   history 43–4
   hydrocarbon quality 48
   location 43
   reserves 48
   reservoir character 46–7
   source rock 47
   stratigraphy 44–5
   trap structure 45–6
North Ravenspurn *see* Ravenspurn North
North Valiant gas field
   data summary 502
   geophysics 499
   history 497–8
   hydrocarbon quality 502
   location 497
   reserves 502
   reservoir character 501–2
   source rock 502
   stratigraphy 498–9
   trap structure 501
North Viking *see* Viking
Northwest Hutton field
   data summary 151
   geophysics 147
   history 146
   hydrocarbon quality 150
   location 145
   reserves 150–1
   reservoir character 147–50
   source rock 150
   stratigraphy 146–7
   trap structure 147

oil column *see* data summary *under named*
   field
oil gravity *see* data summary *under named*
   field
oil type *see* data summary *under named*
   field
oil–water contact *see* data summary *under*
   named field
Orcadian Basin 9
Ormskirk Sandstone Formation 529
Orrin Formation 14
Osprey field
   data summary 188
   geophysics 185
   history 183–4
   hydrocarbon quality 188–9
   location 183
   reserves 189
   reservoir character 186–8
   source rock 188
   stratigraphy 185
   trap structure 185–6

palynology 121, 123, 292
Parentis field 2
Pentland Formation 247, 250, 348
permeability *see* data summary *under*
   named field
petrography 28–9
Petroleum Revenue Tax (PRT) 4
Petronella field 16
   data summary 360
   geophysics 354–5
   hydrocarbon quality 359
   location 354
   reserves 359
   reservoir character 357–9
   source rock 359
   stratigraphy 355–6
   trap structure 357
Piper field 4
   data summary 368
   geophysics 363–5
   history 361–3
   hydrocarbon quality 367
   location 361
   reserves 367
   reservoir character 365–6
   source rock 366–7
   stratigraphy 363
   trap structure 365
Piper Formation 11, 15
   Chanter 264, 266–7
   Claymore 272, 273
   Highlander 324, 327
   Ivanhoe 333, 334–7
   Petronella 356, 358–9
   Piper 363, 366
   Rob Roy 333, 334–7
   Tartan 379, 381–3
Plenus Marl 356
porosity *see* data summary *under named*
   field
pressure measurements *see* data summary
   under named field
productivity index *see* data summary *under*
   named field

Rannoch Formation 12, 13
   Alwyn North 25
   Brent 65
   Cormorant 77–9
   Deveron 84
   Don 90, 92
   Dunlin 97–8
   Eider 107
   Heather 133
   Hutton 136, 138
   Murchison 169, 170, 172
   Ninian 176, 180
   Northwest Hutton 149
   Osprey 186
   Tern 193
   Thistle 205
Rattray Formation 15, 318, 333
Raude Formation 38–9, 65
Ravenspurn North gas field 6
   geophysics 461–4
   history 459–60
   hydrocarbon quality 466–7

location 459
reserves 467
reservoir character 464-6
source rock 466
stratigraphy 460-1
trap structure 464
Ravenspurn South gas field
  data summary 475
  geophysics 472
  history 470-1
  hydrocarbon quality 475
  location 469
  reserves 475
  reservoir character 473-4
  source rock 474-5
  stratigraphy 471-2
  trap structure 472-3
recovery factor *see* data summary *under named field*
Red Chalk Formation 511
reservoir rocks
  Carboniferous 3, 9, 271, 327
  Cretaceous 3, 234, 271, 327, 373
  Devonian 3, 9, 223, 257
  Jurassic
    Lower 37, 248
    Middle 28, 37, 69, 77, 85, 91, 97, 105, 113, 133, 138, 147, 170, 179, 186, 204, 289
    Upper 37, 47, 52, 62, 155, 161, 266, 271, 283, 313, 320, 327, 334, 342, 357, 365, 381
  Permian 3, 222, 223, 231, 271, 391, 398, 405, 412, 420, 440, 448, 455, 464, 473, 479, 488, 494, 501, 506, 514, 520
  Precambrian 9
  Tertiary
    Eocene 3, 16-17, 123
    Palaeocene 3, 215, 240, 297, 303, 350
  Triassic 3, 37, 74, 289, 342, 429, 440, 532
resistivity *see* data summary *under named field*
Rob Roy field 15
  data summary 338
  geophysics 333-4
  history 331-2
  hydrocarbon quality 337
  location 331
  reserves 337
  reservoir character 334-7
  source rock 337
  stratigraphy 332-3
  trap structure 334
Rodby Formation 254, 325
Rogaland Group 75, 96, 105, 130, 177
Rosnaes Clay 121
Rotliegendes Group
  Arbroath 212
  Argyll 221, 223
  Auk 228, 232-3
  Barque 396, 398-9
  Bure 493, 494
  Camelot 403, 405-8
  Cleeton 411, 412
  Clipper 418, 420-1
  Clyde 282
  Crawford 288
  Duncan 221, 223
  Hewett 435

Indefatigable 443, 448
Innes 221, 223
Leman 452, 455-7
Montrose 212
Ravenspurn North and South 460-1, 471
Rough 479-82
Sean 485, 488
Thames 493, 494
V complex 499, 501-2
Victor 504, 506-7
Viking 511, 512, 514
West Sole 518, 520-1
Yare 493, 494
Rough gas storage field
  data summary 483
  geophysics 478
  history 477-8
  hydrocarbon quality 483
  location 477
  reserves 483
  reservoir character 479-83
  source rock 483
  stratigraphy 478
  trap structure 478-9

St Bees Sandstone Formation 529
salinity *see* data summary *under named field*
saturation *see* data summary *under named field*
Scapa field
  data summary 376
  geophysics 371
  history 370-1
  hydrocarbon quality 374-5
  location 369
  reserves 375
  reservoir character 373-4
  source rock 374
  stratigraphy 371
  trap structure 371-3
Scapa Sandstone 11, 373
Schoonebeck field 1-2
Scott field 6
seal rocks
  Chalk Group 234, 311
  Cromer Knoll Group 179
  Flounder Formation 320
  Heather Formation 75, 185, 342
  Hordaland Grup 121
  Humber Group 36, 85, 191, 193, 203-4
  Kimmeridge Clay 45, 52, 57, 159, 179, 185, 256, 265, 283, 327, 334, 342, 357, 365, 381
  Lista Formation 239, 348, 350
  Mercia Mudstone Group 529
  Rot Claystone 429
  Sele Formation 215, 297, 303
  Shetland Group 154, 179
  Silverpit Formation 464, 472, 520
  Smith Bank Formation 273
  Valhall Formation 256, 273, 325, 327
  Zechstein evaporites 397, 403, 443, 452, 457, 464, 472, 478, 493, 494, 495, 499, 504, 511, 514, 520
Sean North and South gas fields
  data summary 489
  geophysics 486
  history 485
  hydrocarbon quality 489

location 485
reserves 489
reservoir character 488
source rock 488-9
stratigraphy 485-6
trap structure 486-8
Sele Formation
  Arbroath 213, 215
  Auk 230
  Balmoral 239
  Cyrus 296
  Emerald 112
  Forties 303
  Frigg 120
  Glamis 318
  Maureen 348
  Montrose 213, 215
Sgiath Formation 15
  Chanter 264
  Claymore 272, 273
  Petronella 358-9
  Piper 363, 365-6
  Tartan 379
Sherwood Sandstone Group 527, 529, 532-3
Shetland Clay Formation 66, 96
Shetland Group 75, 96, 105, 130
Shetland Marl Formation 66, 96
Shetland Platform 104-5
Silverpit Formation
  Cleeton 411
  Ravenspurn North and South 460-1, 471
  Viking 511
  Silverpit Lake 465, 466, 499, 505, 514
Skagerrak Formation
  Claymore 272
  Crawford 289-91
  Kittiwake 340, 342
slant drilling 528
Sleipner Formation 291-3
Slochteren field 2
Smith Bank Formation
  Argyll 221
  Auk 229
  Beatrice 246
  Claymore 272
  Duncan 221
  Fulmar 311
  Highlander 324
  Innes 221
  Kittiwake 340
  Maureen 348
  Petronella 335
Sola Formation 325
source rocks
  gas *Coal Measures for all named fields*
  oil
    Brae Formation 61
    Devonian age rocks 9, 250
    Dunlin Group 119, 124
    Dunrobin Bay Formation 250
    Kimmeridge Clay Formation for all other named fields
South Brae field 15
  data summary 62
  geophysics 57
  history 56
  hydrocarbon quality 62
  location 55
  reserves 62
  reservoir character 57-61

source rock 61
stratigraphy 56–7
trap structure 57
South Morecambe gas field
  data summary 540
  geophysics 529–30
  history 527–9
  hydrocarbon quality 539
  location 527
  reserves 540
  reservoir character 531–9
  source rock 539
  stratigraphy 529
  trap structure 530–1
South Ravenspurn *see* Ravenspurn South
South Valiant gas field
  data summary 502
  geophysics 499
  history 497–8
  hydrocarbon quality 502
  location 497
  reserves 502
  reservoir character 501–2
  source rock 502
  stratigraphy 498–9
  trap structure 501
South Viking gas field 2
  see also Viking
Statfjord Group 11
  Alwyn North 24, 26, 29–30
  Beryl 38–9
  Brent 65, 67–9
  Don 90
  Dunlin 95
  Eider 105
  Heather 129
  Murchison 169
  Ninian 176
  Northwest Hutton 146
  Osprey 185
  Tern 193
  Thistle 201
Stinkschiefer 222

Tarbert Formation 12, 13
  Alwyn North 25, 26, 28
  Brent 65
  Cormorant 77–9
  Crawford 291–3
  Deveron 84
  Don 92
  Dunlin 99
  Eider 108
  Emerald 112
  Frigg 119
  Heather 133
  Hutton 136, 142
  Murchison 169, 172
  Ninian 176, 181
  Northwest Hutton 150
  Osprey 185, 188
  Tern 193, 195
  Thistle 206
Tartan field 15
  data summary 383
  geophysics 381
  history 378
  hydrocarbon quality 383
  location 377
  reserves 384
  reservoir character 381–3

source rock 383
stratigraphy 378–80
trap structure 381
tectonostratigraphic framework 9
temperature measurements *see* data summary *under named field*
tension leg platform (TLP) 136
Tern field 13
  data summary 197
  geophysics 193
  history 191–2
  hydrocarbon quality 197
  location 191
  reserves 197
  reservoir character 193–5
  source rock 195
  stratigraphy 192–3
  trap structure 193
Thames gas field
  data summary 496
  geophysics 493–4
  history 491–2
  hydrocarbon quality 495
  location 491
  reserves 495
  reservoir character 494–5
  source rock 495
  stratigraphy 493
  trap structure 494
Thistle field 4, 12
  data summary 207
  geophysics 202–3
  history 200–1
  hydrocarbon quality 206
  location 199
  reserves 206
  reservoir character 204–6
  source rock 206
  stratigraphy 201–2
  trap structure 203–4
Tor Formation 239, 254, 318, 348
trap types
  combined structures 51, 57, 67, 75, 154, 160, 239, 273, 325, 372, 391, 397, 464, 499, 506, 512, 530
  domal anticline 45, 215, 221, 303, 311, 342, 350, 412, 420, 428, 437, 448, 455, 472, 478, 486, 494, 519
  lateral facies change 123
  tilted fault block 27, 36, 65, 91, 97, 105, 113, 132, 138, 147, 170, 178, 185, 193, 203, 221, 222, 234, 248, 255, 265, 283, 288, 296, 319, 334, 357, 365, 381, 405
Triton Anhydrite Formation 426, 443, 461
Turbot Bank Formation 221, 222

Ula Formation 16
Uppat Formation 14
Utsira Formation 348

V gas fields *see* North Valiant; South Valiant; Vanguard; Vulcan
Valhall field 3
Valhall Formation
  Balmoral 238
  Buchan 254
  Claymore 272, 275
  Glamis 318

Highlander 327
Vanguard gas field
  data summary 502
  geophysics 499
  history 497–8
  hydrocarbon quality 502
  location 497
  reserves 502
  reservoir character 501–2
  source rock 502
  stratigraphy 498–9
  trap structure 501
Variscan orogeny 9
Victor gas field
  data summary 508
  geophysics 505–6
  history 503–4
  hydrocarbon quality 508
  location 503
  reserves 508
  reservoir character 506–7
  source rock 507
  stratigraphy 504–5
  trap structure 506
Viking Graben 15
Viking gas field
  data summary 514
  geophysics 512
  history 510–11
  hydrocarbon quality 514
  location 509–10
  reserves 514–15
  reservoir character 514
  source rock 514
  stratigraphy 511–12
  trap structure 512–14
Vipers Tongue 371
Vulcan gas field
  data summary 502
  geophysics 499
  history 497–8
  hydrocarbon quality 502
  location 497
  reserves 502
  reservoir character 501–2
  source rock 502
  stratigraphy 498–9
  trap structure 501

Wensum gas field 491
West Brae 55
West Sole gas field 2
  data summary 523
  geophysics 518–19
  history 517–18
  hydrocarbon quality 521–2
  location 517
  reserves 522
  reservoir character 520–1
  source rock 521
  stratigraphy 518
  trap structure 519–20
West Sole Group
  Cleeton 411
  Ravenspurn North and South 461, 471
  Rough 479
  West Sole 518
Winterton Formation
  Amethyst 388
  Barque 396
  Camelot 403

Esmond 426
Forbes 426
Gordon 426
Indefatigable 443
Ravenspurn North 461
Viking 511, 512

Yare gas field
  data summary 496
  geophysics 493–4
  history 491–2
  hydrocarbon quality 495
  location 491
  reserves 495
  reservoir character 494–5
  source rock 495

stratigraphy 493
trap structure 494

Zechstein Group
  Amethyst 388
  Arbroath 212
  Argyll 222
  Auk 229, 233–4
  Bure 493
  Camelot 403
  Claymore 271, 275
  Cleeton 411
  Clipper 418
  Clyde 282
  Crawford 288
  Duncan 222

  Hewett 435
  Highlander 324
  Indefatigable 443
  Innes 222
  Leman 452
  Montrose 212
  Ravenspurn North and South 461, 471
  Rough 479
  Sean 485
  Thames 493
  V complex 499
  Victor 504
  Viking 511, 512
  West Sole 518
  Yare 493